Chloroplast

Chloroplast

Special Issue Editor

Bartolomé Sabater

MDPI • Basel • Beijing • Wuhan • Barcelona • Belgrade

MDPI

Special Issue Editor
Bartolomé Sabater
University of Alcalá
Spain

Editorial Office
MDPI
St. Alban-Anlage 66
Basel, Switzerland

This is a reprint of articles from the Special Issue published online in the open access journal *International Journal of Molecular Sciences* (ISSN 1422-0067) from 2017 to 2018 (available at: https://www.mdpi.com/journal/ijms/special_issues/chloroplast)

For citation purposes, cite each article independently as indicated on the article page online and as indicated below:

LastName, A.A.; LastName, B.B.; LastName, C.C. Article Title. *Journal Name* **Year**, *Article Number*, Page Range.

ISBN 978-3-03897-336-2 (Pbk)
ISBN 978-3-03897-337-9 (PDF)

Cover image courtesy of Masumitsu Wada.

Contents

About the Special Issue Editor ix

Bartolomé Sabater
Evolution and Function of the Chloroplast. Current Investigations and Perspectives
Reprinted from: *Int. J. Mol. Sci.* **2018**, *19*, 3095, doi: 10.3390/ijms19103095 1

Jianguo Zhou, Xinlian Chen, Yingxian Cui, Wei Sun, Yonghua Li, Yu Wang, Jingyuan Song
and Hui Yao
Molecular Structure and Phylogenetic Analyses of Complete Chloroplast Genomes of Two
Aristolochia Medicinal Species
Reprinted from: *Int. J. Mol. Sci.* **2017**, *18*, 1839, doi: 10.3390/ijms18091839 6

Wenbin Wang, Huan Yu, Jiahui Wang, Wanjun Lei, Jianhua Gao, Xiangpo Qiu and
Jinsheng Wang
The Complete Chloroplast Genome Sequences of the Medicinal Plant *Forsythia suspensa*
(Oleaceae)
Reprinted from: *Int. J. Mol. Sci.* **2017**, *18*, 2288, doi: 10.3390/ijms18112288 21

Wencai Wang, Siyun Chen and Xianzhi Zhang
Whole-Genome Comparison Reveals Heterogeneous Divergence and Mutation Hotspots in
Chloroplast Genome of *Eucommia ulmoides* Oliver
Reprinted from: *Int. J. Mol. Sci.* **2018**, *19*, 1037, doi: 10.3390/ijms19041037 37

Qixiang Lu, Wenqing Ye, Ruisen Lu, Wuqin Xu and Yingxiong Qiu
Phylogenomic and Comparative Analyses of Complete Plastomes of *Croomia* and
Stemona (Stemonaceae)
Reprinted from: *Int. J. Mol. Sci.* **2018**, *19*, 2383, doi: 10.3390/ijms19082383 53

Cuihua Gu, Luke R. Tembrock, Shaoyu Zheng and Zhiqiang Wu
The Complete Chloroplast Genome of *Catha edulis*: A Comparative Analysis of Genome
Features with Related Species
Reprinted from: *Int. J. Mol. Sci.* **2018**, *19*, 525, doi: 10.3390/ijms19020525 69

Ting Ren, Yanci Yang, Tao Zhou and Zhan-Lin Liu
Comparative Plastid Genomes of *Primula* Species: Sequence Divergence and
Phylogenetic Relationships
Reprinted from: *Int. J. Mol. Sci.* **2018**, *19*, 1050, doi: 10.3390/ijms19041050 87

Mei Jiang, Haimei Chen, Shuaibing He, Liqiang Wang, Amanda Juan Chen and Chang Liu
Sequencing, Characterization, and Comparative Analyses of the Plastome of
Caragana rosea var. *rosea*
Reprinted from: *Int. J. Mol. Sci.* **2018**, *19*, 1419, doi: 10.3390/ijms19051419 103

Tao Zhou, Jian Wang, Yun Jia, Wenli Li, Fusheng Xu and Xumei Wang
Comparative Chloroplast Genome Analyses of Species in *Gentiana* section *Cruciata*
(Gentianaceae) and the Development of Authentication Markers
Reprinted from: *Int. J. Mol. Sci.* **2018**, *19*, 1962, doi: 10.3390/ijms19071962 120

Xuan Li, Yongfu Li, Mingyue Zang, Mingzhi Li and Yanming Fang
Complete Chloroplast Genome Sequence and Phylogenetic Analysis of *Quercus acutissima*
Reprinted from: *Int. J. Mol. Sci.* **2018**, *19*, 2443, doi: 10.3390/ijms19082443 135

Zhi-Zhong Li, Josphat K. Saina, Andrew W. Gichira, Cornelius M. Kyalo, Qing-Feng Wang and Jin-Ming Chen
Comparative Genomics of the Balsaminaceae Sister Genera *Hydrocera triflora* and *Impatiens pinfanensis*
Reprinted from: *Int. J. Mol. Sci.* **2018**, *19*, 319, doi: 10.3390/ijms19010319 152

Malte Mader, Birte Pakull, Céline Blanc-Jolivet, Maike Paulini-Drewes, Zoéwindé Henri-Noël Bouda, Bernd Degen, Ian Small and Birgit Kersten
Complete Chloroplast Genome Sequences of Four Meliaceae Species and Comparative Analyses
Reprinted from: *Int. J. Mol. Sci.* **2018**, *19*, 701, doi: 10.3390/ijms19030701 169

Wan-Lin Dong, Ruo-Nan Wang, Na-Yao Zhang, Wei-Bing Fan, Min-Feng Fang and Zhong-Hu Li
Molecular Evolution of Chloroplast Genomes of Orchid Species: Insights into Phylogenetic Relationship and Adaptive Evolution
Reprinted from: *Int. J. Mol. Sci.* **2018**, *19*, 716, doi: 10.3390/ijms19030716 183

Wanzhen Liu, Hanghui Kong, Juan Zhou, Peter W. Fritsch, Gang Hao and Wei Gong
Complete Chloroplast Genome of *Cercis chuniana* (Fabaceae) with Structural and Genetic Comparison to Six Species in Caesalpinioideae
Reprinted from: *Int. J. Mol. Sci.* **2018**, *19*, 1286, doi: 10.3390/ijms19051286 203

Mira Park, Hyun Park, Hyoungseok Lee, Byeong-ha Lee and Jungeun Lee
The Complete Plastome Sequence of an Antarctic Bryophyte *Sanionia uncinata* (Hedw.) Loeske
Reprinted from: *Int. J. Mol. Sci.* **2018**, *19*, 709, doi: 10.3390/ijms19030709 220

Josphat K. Saina, Zhi-Zhong Li, Andrew W. Gichira and Yi-Ying Liao
The Complete Chloroplast Genome Sequence of Tree of Heaven (*Ailanthus altissima* (Mill.) (Sapindales: Simaroubaceae), an Important Pantropical Tree
Reprinted from: *Int. J. Mol. Sci.* **2018**, *19*, 929, doi: 10.3390/ijms19040929 235

Deng-Feng Xie, Yan Yu, Yi-Qi Deng, Juan Li, Hai-Ying Liu, Song-Dong Zhou and Xing-Jin He
Comparative Analysis of the Chloroplast Genomes of the Chinese Endemic Genus *Urophysa* and Their Contribution to Chloroplast Phylogeny and Adaptive Evolution
Reprinted from: *Int. J. Mol. Sci.* **2018**, *19*, 1847, doi: 10.3390/ijms19071847 253

Akira Kawabe, Hiroaki Nukii and Hazuka Y. Furihata
Exploring the History of Chloroplast Capture in *Arabis* Using Whole Chloroplast Genome Sequencing
Reprinted from: *Int. J. Mol. Sci.* **2018**, *19*, 602, doi: 10.3390/ijms19020602 273

Zhitao Niu, Qingyun Xue, Hui Wang, Xuezhu Xie, Shuying Zhu, Wei Liu and Xiaoyu Ding
Mutational Biases and GC-Biased Gene Conversion Affect GC Content in the Plastomes of *Dendrobium* Genus
Reprinted from: *Int. J. Mol. Sci.* **2017**, *18*, 2307, doi: 10.3390/ijms18112307 285

Geng-Ming Lin, Yu-Heng Lai, Gilbert Audira and Chung-Der Hsiao
A Simple Method to Decode the Complete 18-5.8-28S rRNA Repeated Units of Green Algae by Genome Skimming
Reprinted from: *Int. J. Mol. Sci.* **2017**, *18*, 2341, doi: 10.3390/ijms18112341 300

Kangquan Yin, Yue Zhang, Yuejuan Li and Fang K. Du
Different Natural Selection Pressures on the *atpF* Gene in Evergreen Sclerophyllous and Deciduous Oak Species: Evidence from Comparative Analysis of the Complete Chloroplast Genome of *Quercus aquifolioides* with Other Oak Species
Reprinted from: *Int. J. Mol. Sci.* **2018**, *19*, 1042, doi: 10.3390/ijms19041042 309

Huiyu Wu, Narong Shi, Xuyao An, Cong Liu, Hongfei Fu, Li Cao, Yi Feng, Daojie Sun and Lingli Zhang
Candidate Genes for Yellow Leaf Color in Common Wheat (*Triticum aestivum* L.) and Major Related Metabolic Pathways according to Transcriptome Profiling
Reprinted from: *Int. J. Mol. Sci.* **2018**, *19*, 1594, doi: 10.3390/ijms19061594 324

Julia Legen and Christian Schmitz-Linneweber
Stable Membrane-Association of mRNAs in Etiolated, Greening and Mature Plastids
Reprinted from: *Int. J. Mol. Sci.* **2017**, *18*, 1881, doi: 10.3390/ijms18091881 350

Lea Vojta, Andrea Čuletić and Hrvoje Fulgosi
Effects of TROL Presequence Mutagenesis on Its Import and Dual Localization in Chloroplasts
Reprinted from: *Int. J. Mol. Sci.* **2018**, *19*, 569, doi: 10.3390/ijms19020569 365

Ruixin Shao, Huifang Zheng, Shuangjie Jia, Yanping Jiang, Qinghua Yang and Guozhang Kang
Nitric Oxide Enhancing Resistance to PEG-Induced Water Deficiency is Associated with the Primary Photosynthesis Reaction in *Triticum aestivum* L.
Reprinted from: *Int. J. Mol. Sci.* **2018**, *19*, 2819, doi: 10.3390/ijms19092819 378

Qing-Long Wang, Juan-Hua Chen, Ning-Yu He and Fang-Qing Guo
Metabolic Reprogramming in Chloroplasts under Heat Stress in Plants
Reprinted from: *Int. J. Mol. Sci.* **2018**, *19*, 849, doi: 10.3390/ijms19030849 394

Masanori Izumi and Sakuya Nakamura
Chloroplast Protein Turnover: The Influence of Extraplastidic Processes, Including Autophagy
Reprinted from: *Int. J. Mol. Sci.* **2018**, *19*, 828, doi: 10.3390/ijms19030828 416

Yamato Yoshida
Insights into the Mechanisms of Chloroplast Division
Reprinted from: *Int. J. Mol. Sci.* **2018**, *19*, 733, doi: 10.3390/ijms19030733 431

Hiroki Irieda and Daisuke Shiomi
Bacterial Heterologous Expression System for Reconstitution of Chloroplast Inner Division Ring and Evaluation of Its Contributors
Reprinted from: *Int. J. Mol. Sci.* , *19*, 544, doi: 10.3390/ijms19020544 444

Noriyuki Suetsugu and Masamitsu Wada
Two Coiled-Coil Proteins, WEB1 and PMI2, Suppress the Signaling Pathway of Chloroplast Accumulation Response that Is Mediated by Two Phototropin-Interacting Proteins, RPT2 and NCH1, in Seed Plants
Reprinted from: *Int. J. Mol. Sci.* **2017**, *18*, 1469, doi: 10.3390/ijms18071469 457

About the Special Issue Editor

Bartolomé Sabater has served as Chair of Plant Physiology at the University of Alcalá, Spain, for the past 35 years. He is a Member of the American Society of Plant Biologists and the Japanese Society of Plant Physiologists, and has published almost 100 original research papers and more than 20 book chapters and reviews articles, in addition to having authored a textbook with twelve editions.

International Journal of
Molecular Sciences

MDPI

Editorial

Evolution and Function of the Chloroplast. Current Investigations and Perspectives

Bartolomé Sabater

Department of Life Sciences (Ciencias de la Vida), University of Alcalá, Alcalá de Henares, 28805 Madrid, Spain;
bartolome.sabater@uah.es; Tel.: +34-609-227-010

Received: 27 September 2018; Accepted: 8 October 2018; Published: 10 October 2018

Chloroplasts are the place for the major conversion of the sun's radiation energy to chemical energy that is usable by organisms. Accordingly, they account for about 50% of the leaf protein [1], and the enzyme ribulose-1,5-bisphosphate carboxylase of chloroplast is by far the most abundant protein on the Earth [2]. Chloroplasts are not only key in photosynthesis, they are the place of the synthesis of fatty acids in plants and biosynthesis of amino acids, porphyrins, isoprenoids, and secondary metabolites that, sometimes, duplicate parallel biosynthetic pathways in cytosol. Only a small fraction of the involved enzymatic machinery is encoded in the chloroplast DNA. Nuclear DNA encodes most of the chloroplast proteins which, after being synthesised as precursors on cytosol ribosomes, are incorporated into the appropriate chloroplast substructures. Chloroplast evolutionarily derives from a primitive cyanobacteria that was engulfed by non-photosynthetic cells and, progressively, after losing most of its DNA, became the actual chloroplast that retains only a fraction of the original cyanobacterial genes. Most of the original enzyme machinery is now encoded in the cell nuclear DNA that controls the plastid divisions, among others [3].

The structure and dynamics of the chloroplast photosynthesis machinery is a central focus of molecular investigations on photosynthetic productivity and leaf senescence. Functional coordination among chloroplasts, cytosol, nucleus, and other subcellular compartments inspire many active research topics of chloroplast molecular biology. The transition from the engulfed autonomous cyanobacteria to a non-autonomous endosymbiont chloroplast requires new coordination signals, specific for the different plant cells, which have been actively investigated. In fact, the leaf's green chloroplasts are members of the plastid organelles present in all plant cells. All plastids share the same DNA and a few structural features and functions (as the synthesis of fatty acids) and derive from the proplastids present in meristematic cells. The biogenesis of the different plastids, their interconversion and the molecular bases of the functions of some plastids (amiloplasts, chromoplasts, etc.) have also been actively investigated at the molecular level. Obviously, only a few research fields may be reflected in this issue which should be the seed for future issues of a broader scope.

Although most of the original cyanobacterial genes have been lost early in the transition to the endosymbiotic chloroplast, a few of them, like the *ndh* genes [4], have been lost later, in some plant lines, endowing characteristic DNA features to specific plant orders, families, genus, or species.

Therefore, the comparison of the sequences of the chloroplast DNA of different plants, provides valuable information on gene content, reordering in the circular chloroplast DNA and mutational genetic-derive, relevant to the evolution of chloroplast and its relations with ancient environments. Increasing facilities for intense genome sequencing have prompted many laboratories to focus on the chloroplast DNA. Reflecting these efforts, more than half of the articles on this issue deal with the functional or evolutionary investigations, based on sequence analyses of chloroplast DNA. Mainly focussed on phylogenetic comparisons of medicinal plant families, articles on *Aristolochia* [5], *Forsythia* [6], *Eucommia* [7], and the *Stemonaceae* family [8] report and analyse the complete sequences of the corresponding chloroplast DNA. Similar approaches on other economically important plants have been reported in articles on genera *Catha* [9], *Primula* [10], *Caragana* [11], *Genciana* [12], and *Quercus* [13], and on the families of *Balsaminaceae* [14], *Meliaceae* [15], *Orchidaceae* [16], and *Fabaceae* [17]. The chloroplast DNA sequence and the analysis of the species of plant branches, that are still poorly investigated have been reported in articles on the Antarctic Bryophyte *Sanionia uncinate* [18], on *Ailanthus altissima* [19], and on genus *Urophysa* [20].

Adding to the sequences reported elsewhere, the chloroplast DNA sequences described in this issue provide a background for precise bar codes that are useful in plant trade and medicinal applications. In addition, chloroplast DNA sequences are excellent tools for taxonomy and phylogeny. Thus, detailed comparison of the chloroplast DNA sequences of the species of *Arabis* [21] led the authors to propose recent events of the interspecific chloroplast capture and hybridization. Similarly, the comparison of the chloroplast DNA sequences of species of the genus *Dendrobium* led authors to identify the mutational biases affecting the GC (guanine + cytosine) content [22]. Not closely related to the chloroplast title of the issue, but potentially useful for rapid comparison of sequences, is the brief article describing a genome-skimming approach in the chloroplast DNA [23].

Pointing to selective pressures reflected in the chloroplast DNA sequence, the comparison of chloroplast DNAs of *Quercus* and some related species led the authors to propose the intriguing possibility that the specific sequences of the *atpF* gene, encoding a subunit of the thylakoid ATP synthase, show significant positive selection for evergreen sclerophyllous oak species, when compared to the deciduous oak species [24].

Chloroplast gene expression is under a complex post-transcriptional control involving C (cytosine) to U (uracil) editing, intron and polycistron splicing and ribosomal activity [25,26]. Transcriptome analysis of the chloroplast genome is reported by Wu et al. Accordingly, the profile of the chloroplast mRNAs is used to identify genes responsible for wheat yellow leaf colour [27]. Approaching, specifically, the question of the post-transcriptional control of gene expression in chloroplasts, the article by Legen and Schmitz-Linneweber reports that a fraction of the chloroplast mRNAs, associated with ribosomes, are attached to the chloroplast as a direct consequence of translation. Plastid mRNA distribution is stable for different plastid types, enabling rapid chloroplast translation in any plastid type [28].

As pointed above, most chloroplast proteins are encoded in the nucleus. The importation of the nuclear-encoded proteins into chloroplasts is a complex process requiring, among others, the recognition of specific sequences in the amino-ends of the precursor proteins that direct them to the appropriate chloroplast substructure. The amino-end sequence of the amino acids required for chloroplast processing has been investigated for the pre-thylakoid rhodanase-like protein (TROL) and has been reported in this issue. Some researchers have combined the site-directed mutagenesis and *in vitro* reconstructed translation and incorporation systems, in an attempt to clarify the dual localization determinants of this protein [29].

Exposed to the changing environmental conditions, functions in chloroplasts are preserved by adapting their metabolism and molecular structures. The article by Shao et al. [30] reports how nitric oxide (NO) protects the photosynthetic apparatus, against water deficit, by affecting the phosphorylation of diverse proteins, among them some are from the primary reaction centre. In their review article, Wang et al. have revised the actual knowledge on the metabolic adaptation of chloroplasts under heat stress [31]. The turnover of protein is a key to the adaptation of chloroplasts to changing conditions. The influence of the extra-plastidic processes on the turnover of chloroplast proteins is analysed in a review article by Izumi and Nakamura [32].

Present knowledge of the mechanisms of the chloroplast division is updated in two review articles [33,34]. The review by Yoshida dealt with the structure and function of supramolecular machinery involved in the plastid division and its resemblance with the mitochondrial division machinery [33]. The review by Irieda and Shiomi updated the knowledge of the mechanism of chloroplast division, as deduced from the heterologous expression systems [34].

Suetsugu and Wada described, in a brief report, the molecular mechanisms controlling the movement of chloroplasts [35]. They explain how two coiled-coil proteins suppress the chloroplast accumulation response mediated by two phototropin-interacting proteins.

The abundance of chloroplast DNA sequence articles in the issue reflects the predominant research projects in the last few years. Certainly, a refinement of the taxonomy and phylogeny investigations requires high numbers of chloroplast DNA sequences to observe the evolutionary trends. Wide analyses of the DNA sequences reported, surely, mark the starting references for further comparisons which could be useful in phylogeny and trade-vegetable characterisation.

Chloroplast DNA articles do not shadow the variety of chloroplast molecular projects, as shown by the other articles on the issue. New technologies have far surpassed the methods and objectives collected in the pioneering methods book, thirty-five years ago [36].

In the coming months, the *IJMS* (*International Journal of Molecular Science*) will certainly include a number of articles dealing with chloroplasts. Many of the articles in this issue have resulted from collaborations of different laboratories, frequently of different countries, which indicates the universality of the field and the necessity of cooperation among specialised research groups, to attack the complex understanding of chloroplasts, at the molecular level. The key role of chloroplasts in the conversion of radiant energy suggests that basic advances in the field would improve plant productivity and environmental protection.

Conflicts of Interest: The author declares no conflict of interest.

References

1. Peoples, M.B.; Dalling, M.J. The interplay between proteolysis and amino acid metabolism during senescence and nitrogen reallocation. In *Senescence and Aging in Plants*; Noodén, L.D., Leopold, A.C., Eds.; Academic Press: San Diego, CA, USA, 1988; pp. 181–217.
2. Wildman, S.G. Aspects of Fraction I protein evolution. *Arch. Biochem. Biophys.* **1979**, *196*, 580–610. [CrossRef]
3. Martin, W.; Herrmann, R.G. Gene transfer from organelles to nucleus: How much, what happens, and why? *Plant Physiol.* **1998**, *118*, 9–17. [CrossRef] [PubMed]
4. Martín, M.; Sabater, B. Plastid *ndh* genes in plat evolution. *Plant Physiol. Biochem.* **2010**, *48*, 636–645. [CrossRef] [PubMed]
5. Zhou, J.; Chen, X.; Cui, Y.; Sun, W.; Li, Y.; Wang, Y.; Song, J.; Yao, H. Molecular structure and phylogenetic analyses of complete chloroplast genomes of two *Aristolochia* medicinal species. *Int. J. Mol. Sci.* **2017**, *18*, 1839. [CrossRef] [PubMed]
6. Wang, W.; Yu, H.; Wang, J.; Lei, W.; Gao, J.; Qiu, X.; Wang, J. The complete chloroplast genome sequences of the medicinal plant *Forsythia suspensa* (Oleaceae). *Int. J. Mol. Sci.* **2017**, *18*, 2288. [CrossRef] [PubMed]
7. Wang, W.; Chen, S.; Zhang, X. Whole-genome comparison reveals heterogeneous divergence and mutation hotspots in chloroplast genome of *Eucommia ulmoides* Oliver. *Int. J. Mol. Sci.* **2018**, *19*, 1037. [CrossRef] [PubMed]
8. Lu, Q.; Ye, W.; Lu, R.; Xu, W.; Qiu, Y. Phylogenomic and comparative analyses of complete plastomes of *Croomia* and *Stemona* (Stemonaceae). *Int. J. Mol. Sci.* **2018**, *19*, 2383. [CrossRef] [PubMed]
9. Gu, C.; Tembrock, L.R.; Zheng, S.; Wu, Z. The complete chloroplast genome of *Catha edulis*: A comparative analysis of genome features with related species. *Int. J. Mol. Sci.* **2018**, *19*, 525. [CrossRef] [PubMed]
10. Ren, T.; Yang, Y.; Zhou, T.; Liu, Z.-L. Comparative plastid genomes of *Primula* species: Sequence divergence and phylogenetic relationships. *Int. J. Mol. Sci.* **2018**, *19*, 1050. [CrossRef] [PubMed]
11. Jiang, M.; Haimei, M.J.; He, S.; Wang, L.; Chen, A.J.; Liu, C. Sequencing, characterization, and comparative analyses of the plastome of *Caragana rosea* var. rosea. *Int. J. Mol. Sci.* **2018**, *19*, 1419. [CrossRef] [PubMed]
12. Zhou, T.; Wang, J.; Jia, Y.; Li, W.; Xu, F.; Wang, X. Comparative chloroplast genome analyses of species in Gentiana section *Cruciata* (Gentianaceae) and the development of authentication markers. *Int. J. Mol. Sci.* **2018**, *19*, 1962. [CrossRef] [PubMed]
13. Li, X.; Li, Y.; Zang, M.; Li, M.; Fang, Y. Complete chloroplast genome sequence and phylogenetic analysis of *Quercus acutissima*. *Int. J. Mol. Sci.* **2018**, *19*, 2443. [CrossRef] [PubMed]

14. Li, Z.-Z.; Saina, J.K.; Gichira, A.W.; Kyalo, C.M.; Wang, Q.-F.; Chen, J.-M. Comparative genomics of the balsaminaceae sister genera *Hydrocera triflora* and *Impatiens pinfanensis*. *Int. J. Mol. Sci.* **2018**, *19*, 319. [CrossRef] [PubMed]

15. Mader, M.; Pakull, B.; Blanc-Jolivet, C.; Paulini-Drewes, M.; Bouda, Z.H.-N.; Degen, B.; Small, I.; Kersten, B. Complete chloroplast genome sequences of four Meliaceae species and comparative analyses. *Int. J. Mol. Sci.* **2018**, *19*, 701. [CrossRef] [PubMed]

16. Dong, W.-L.; Wang, R.-N.; Zhang, N.-Y.; Fan, W.-B.; Fang, M.-F.; Li, Z.-H. Molecular evolution of chloroplast genomes of orchid species: Insights into phylogenetic relationship and adaptive evolution. *Int. J. Mol. Sci.* **2018**, *19*, 716. [CrossRef] [PubMed]

17. Liu, W.; Kong, H.; Zhou, J.; Fritsch, P.W.; Hao, G.; Gong, W. Complete chloroplast genome of *Cercis chuniana* (Fabaceae) with structural and genetic comparison to six species in Caesalpinioideae. *Int. J. Mol. Sci.* **2018**, *19*, 1286. [CrossRef] [PubMed]

18. Park, M.; Park, H.; Lee, H.; Lee, B.; Lee, J. The complete plastome sequence of an Antarctic Bryophyte *Sanionia uncinata* (Hedw.) Loeske. *Int. J. Mol. Sci.* **2018**, *19*, 709. [CrossRef] [PubMed]

19. Saina, J.; Li, Z.-Z.; Gichira, A.W.; Liao, Y.-Y. The complete chloroplast genome sequence of tree of heaven (*Ailanthus altissima* (Mill.) (Sapindales: Simaroubaceae), an important pantropical tree. *Int. J. Mol. Sci.* **2018**, *19*, 929. [CrossRef] [PubMed]

20. Xie, D.-F.; Yu, Y.; Deng, Y.-Q.; Li, J.; Liu, H.-Y.; Zhou, S.-D.; He, X.-J. Comparative analysis of the chloroplast genomes of the Chinese endemic genus *Urophysa* and their contribution to chloroplast phylogeny and adaptive evolution. *Int. J. Mol. Sci.* **2018**, *19*, 1847. [CrossRef] [PubMed]

21. Kawabe, A.; Nukii, H.; Furiahata, H.Y. Exploring the history of chloroplast capture in *Arabis* using whole chloroplast genome sequencing. *Int. J. Mol. Sci.* **2018**, *19*, 602. [CrossRef] [PubMed]

22. Niu, Z.; Xue, Q.; Wang, H.; Xie, X.; Zhu, S.; Liu, W.; Ding, X. Mutational biases and GC-biased gene conversion affect GC content in the plastomes of *Dendrobium* genus. *Int. J. Mol. Sci.* **2017**, *18*, 2307. [CrossRef]

23. Lin, G.-M.; Lai, Y.-H.; Audira, G.; Hsiao, C.-D. A simple method to decode the complete 18-5.8-28 rRNA repeated units of green algae by genome skimming. *Int. J. Mol. Sci.* **2017**, *18*, 2341. [CrossRef] [PubMed]

24. Yin, K.; Zhang, Y.; Li, Y.; Du, F.K. Different natural selection pressures on the *atpF* gene in evergreen sclerophyllous and deciduous oak species: Evidence from comparative analysis of the complete chloroplast genome of *Quercus aquifolioides* with other oak species. *Int. J. Mol. Sci.* **2018**, *19*, 1042. [CrossRef] [PubMed]

25. Martín, M.; Sabater, B. Translational control of chloroplast protein synthesis during senescence of barley leaves. *Physiol. Plant.* **1989**, *75*, 374–381. [CrossRef]

26. Del Campo, E.M.; Sabater, B.; Martín, M. Transcripts of the ndhH-D operon of barley plastids: Possible role of unedited site III in splicing of the ndhA intron. *Nucleic Acids Res.* **2000**, *28*, 1092–1098. [CrossRef] [PubMed]

27. Wu, H.; Shi, N.; An, X.; Liu, C.; Fu, H.; Cao, L.; Feng, Y.; Sun, D.; Zhang, L. Candidate genes for yellow leaf color in common wheat (*Triticum aestivum* L.) and major related metabolic pathways according to transcriptome profiling. *Int. J. Mol. Sci.* **2018**, *19*, 1594. [CrossRef] [PubMed]

28. Legen, J.; Schmitz-Linneweber, C. Stable membrane-association of mRNAs in etiolated, greening and mature plastids. *Int. J. Mol. Sci.* **2017**, *18*, 1881. [CrossRef] [PubMed]

29. Vojta, L.; Culetic, A.; Fulgosi, H. Effects of TROL presequence mutagenesis on its import and dual localization in chloroplasts. *Int. J. Mol. Sci.* **2018**, *19*, 569. [CrossRef] [PubMed]

30. Shao, R.; Zheng, H.; Jia, S.; Jiang, Y.; Yang, Q.; Kang, G. Nitric oxide enhancing resistance to PEG-induced water deficiency is associated with the primary photosynthesis reaction in *Triticum aestivum* L. *Int. J. Mol. Sci.* **2018**, *19*, 2819. [CrossRef] [PubMed]

31. Wang, Q.-L.; Chen, J.-H.; He, N.-Y.; Guo, F.-Q. Metabolic reprogramming in chloroplasts under heat stress in plants. *Int. J. Mol. Sci.* **2018**, *19*, 849. [CrossRef] [PubMed]

32. Izumi, M.; Nakamura, S. Chloroplast protein turnover: The influence of extraplastidic processes, including autophagy. *Int. J. Mol. Sci.* **2018**, *19*, 828. [CrossRef] [PubMed]

33. Yoshida, Y. Insights into the mechanisms of chloroplast division. *Int. J. Mol. Sci.* **2018**, *19*, 733. [CrossRef] [PubMed]

34. Ireda, H.; Shiomi, D. Bacterial heterologous expression system for reconstitution of chloroplast inner division ring and evaluation of its contributors. *Int. J. Mol. Sci.* **2018**, *19*, 544. [CrossRef] [PubMed]

35. Suetsugu, N.; Wada, M. Two coiled-coil proteins, WEB1 and PMI2, suppress the signalling pathway of chloroplast accumulation response that is mediated by two phototropin-interacting proteins, RPT2 and NCH1, in seed plants. *Int. J. Mol. Sci.* **2017**, *18*, 1469. [CrossRef] [PubMed]
36. Halliwell, B. Methods in chloroplast molecular biology. *FEBS Lett.* **1983**. [CrossRef]

International Journal of
Molecular Sciences

MDPI

Article

Molecular Structure and Phylogenetic Analyses of Complete Chloroplast Genomes of Two *Aristolochia* Medicinal Species

Jianguo Zhou [1], Xinlian Chen [1], Yingxian Cui [1], Wei Sun [2], Yonghua Li [3], Yu Wang [1], Jingyuan Song [1] and Hui Yao [1,*]

[1] Key Lab of Chinese Medicine Resources Conservation, State Administration of Traditional Chinese Medicine of the People's Republic of China, Institute of Medicinal Plant Development, Chinese Academy of Medical Sciences & Peking Union Medical College, Beijing 100193, China; jgzhou1316@163.com (J.Z.); chenxinlian1053@163.com (X.C.); yxcui2017@163.com (Y.C.); ywang@implad.ac.cn (Y.W.); jysong@implad.ac.cn (J.S.)

[2] Institute of Chinese Materia Medica, China Academy of Chinese Medicinal Sciences, Beijing 100700, China; wsun@icmm.ac.cn

[3] Department of Pharmacy, Guangxi Traditional Chinese Medicine University, Nanning 530200, China; liyonghua185@126.com

* Correspondence: scauyaoh@sina.com; Tel.: +86-10-5783-3194

Received: 27 July 2017; Accepted: 20 August 2017; Published: 24 August 2017

Abstract: The family Aristolochiaceae, comprising about 600 species of eight genera, is a unique plant family containing aristolochic acids (AAs). The complete chloroplast genome sequences of *Aristolochia debilis* and *Aristolochia contorta* are reported here. The results show that the complete chloroplast genomes of *A. debilis* and *A. contorta* comprise circular 159,793 and 160,576 bp-long molecules, respectively and have typical quadripartite structures. The GC contents of both species were 38.3% each. A total of 131 genes were identified in each genome including 85 protein-coding genes, 37 tRNA genes, eight rRNA genes and one pseudogene (*ycf1*). The simple-sequence repeat sequences mainly comprise A/T mononucletide repeats. Phylogenetic analyses using maximum parsimony (MP) revealed that *A. debilis* and *A. contorta* had a close phylogenetic relationship with species of the family Piperaceae, as well as Laurales and Magnoliales. The data obtained in this study will be beneficial for further investigations on *A. debilis* and *A. contorta* from the aspect of evolution, and chloroplast genetic engineering.

Keywords: *Aristolochia debilis*; *Aristolochia contorta*; chloroplast genome; molecular structure; phylogenetic analyses

1. Introduction

The traditional Chinese medicine plants, *Aristolochia debilis* and *Aristolochia contorta*, are herbaceous climbers in the family Aristolochiaceae. *Aristolochiae fructus* originates from the mellow fruit of the two species, while *Aristolochiae herba* originates from their dried aerial parts. *Aristolochiae fructus* and *Aristolochiae herba* have been recorded as traditional herbal medicines which can clear lung-heat to stop coughing and activate meridians to stop pain, respectively [1]. Modern pharmacology studies have shown that the primary chemical constituents of the two species are aristolochic acid analogues including aristolochic acids (AAs) and aristolactams (ALs) [2,3]. AAs and ALs have been found among species from the family Aristolochiaceae [4]. Previous researches have revealed that AAs are able to react with DNA to form covalent dA-aristolactam (dA-AL) and dG-aristolactam (dG-AL) adducts [5,6]. With further research, current evidence from studies of AAs has demonstrated that AAs can cause nephrotoxicity, carcinogenicity, and mutagenicity [7–10], especially after prolonged low-dose or shortdated high-dose intake [11,12]. Some nephropathy and malignant tumours including renal interstitial fibrosis, Balkan endemic nephropathy, and upper tract urothelial carcinomas are caused by AAs [13–15]. Currently, there are different degrees of restrictions on the sale and use of AAs-containing herbal preparations in many countries.

Chloroplasts are key and semi-autonomous organelles for photosynthesis and biosynthesis in plant cells [16–18]. The chloroplast genome, one of three major genetic systems (the other two are nuclear and mitochondrial genomes), is a circular molecule with a typical quadripartite structure of 115 to 165 kb in length [19,20]. All chloroplast genomes of land plants, apart from several rare exceptions, are highly conserved in terms of size, structure, gene content, and gene [21–23]. Due to its self-replication mechanism and relatively independent evolution, the genetic information from the chloroplast genome has been used in studies of molecular markers, barcoding identification, plant evolution and phylogenetic [24–26]. In 1976, Bedbrook and Bogorad produced the first chloroplast physical mapping of *Zea mays* by digestion with multiple restriction enzymes [27]. Subsequently, the first complete chloroplast genome sequence of *Nicotiana tabacum* was determined [28]. With the development of sequencing technology and bioinformatics, research into the chloroplast genome has increased rapidly. By now, the number of chloroplast genome sequence recorded in the National Center for Biotechnology Information (NCBI) has reached more than 1,500 plant species [29].

About eight genera and 600 species are classified within Aristolochiaceae, and are primarily distributed in tropical and subtropical regions. Of these plants, there are four genera (one endemic) and 86 species (69 endemic) distributed widely in China. The genus *Aristolochia L.*, comprising about 400 species (45 species in China), is the largest and most representative genus of Aristolochiaceae [30]. However, there are no reports on the chloroplast genomes of the family Aristolochiaceae at present, and this has hindered our understanding and progress in the research of the evolution, phylogeny, species identification, and genetic engineering of Aristolochiaceae.

In this study, we determined the complete chloroplast genome sequences of *A. debilis* and *A. contorta*, which are the first two sequenced members of the family Aristolochiaceae. Furthermore, to reveal the phylogenetic positions of the two species, we conducted a phylogenetic tree using the maximum parsimony (MP) method based on common protein-coding genes from 37 species. Overall, the results provide basic genetic information on the chloroplast of *A. debilis* and *A. contorta*, and the role of the two species in plant systematics.

2. Results and Discussion

2.1. The Chloroplast Genome Structures of A. debilis and A. contorta

Both species displayed a typical quadripartite structure, and the corresponding regions were of similar lengths. The complete chloroplast genome of *A. debilis* is a circular molecule of 159,793 bp in length comprising a large single-copy (LSC) region of 89,609 bp and a small single-copy (SSC) region of 19,834 bp separated by a pair of inverted repeats (IRs), each 25,175 bp in length (Figure 1, Table 1). The complete chloroplast genome of *A. contorta* is 160,576 bp in length, which is divided into one LSC (89,781 bp), one SSC (19,877 bp) and two IRs, each 25,459 bp in length (Figure 2, Table 1).

Table 1. Base composition in the chloroplast genomes of *A. debilis* and *A. contorta*.

Species	Regions	Positions	T(U) (%)	C (%)	A (%)	G (%)	Length (bp)
	LSC	-	32.2	18.7	31.2	17.9	89,609
	SSC	-	34.0	17.4	33.2	15.5	19,834
	IRa	-	28.4	22.4	28.3	21.0	25,175
	IRb	-	28.3	21.0	28.4	22.4	25,175
A. debilis	Total	-	31.2	19.5	30.5	18.8	159,793
	CDS	-	30.9	18.1	30.2	20.8	78,717
	-	1st position	23.5	18.8	30.5	27.2	26,239
	-	2nd position	32.2	20.5	29.2	18.1	26,239
	-	3rd position	36.9	15.1	31.1	17.0	26,239

Table 1. *Cont.*

Species	Regions	Positions	T(U) (%)	C (%)	A (%)	G (%)	Length (bp)
	LSC	-	32.2	18.7	31.2	17.8	89,781
	SSC	-	33.9	17.4	33.3	15.4	19,877
	IRa	-	28.4	22.4	28.2	21.0	25,459
	IRb	-	28.2	21.0	28.4	22.4	25,459
A. contorta	Total	-	31.2	19.5	30.6	18.8	160,576
	CDS	-	30.9	18.1	30.3	20.7	78,765
	-	1st position	23.5	18.8	30.5	27.2	26,255
	-	2nd position	32.2	20.6	29.2	18.1	26,255
	-	3rd position	37.0	15.0	31.1	16.9	26,255

* CDS: protein-coding regions.

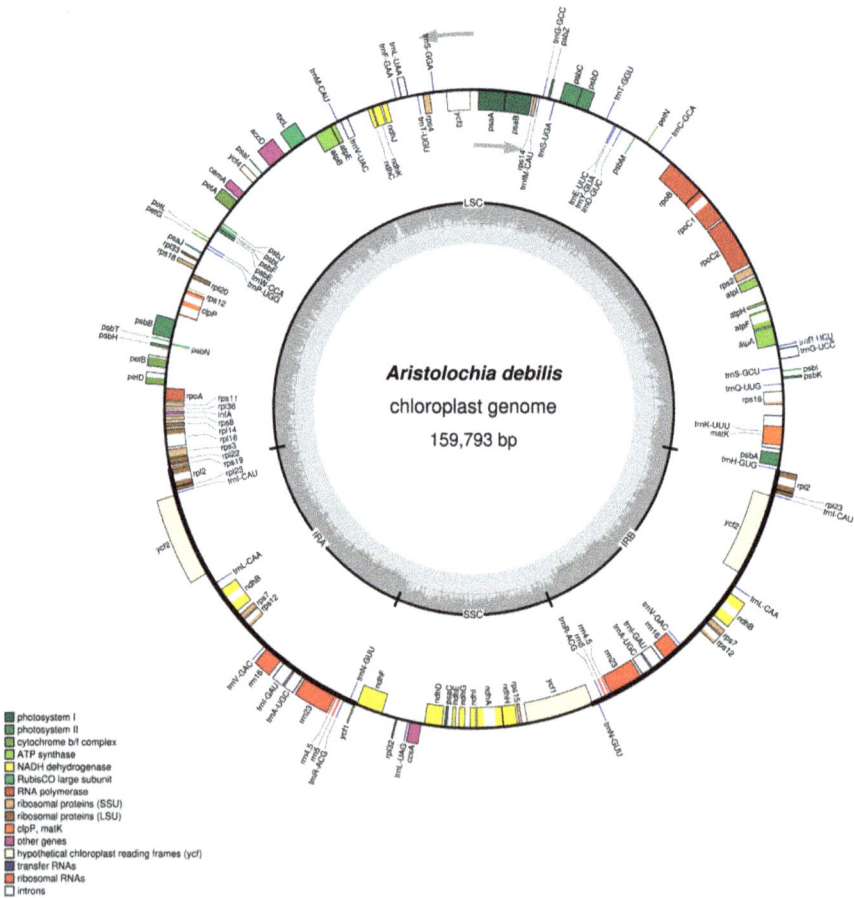

Figure 1. Gene map of the complete chloroplast genome of *A. debilis*. Genes on the inside of the circle are transcribed clockwise, while those outside are transcribed counter clockwise. The darker gray in the inner circle corresponds to GC content, whereas the lighter gray corresponds to AT content.

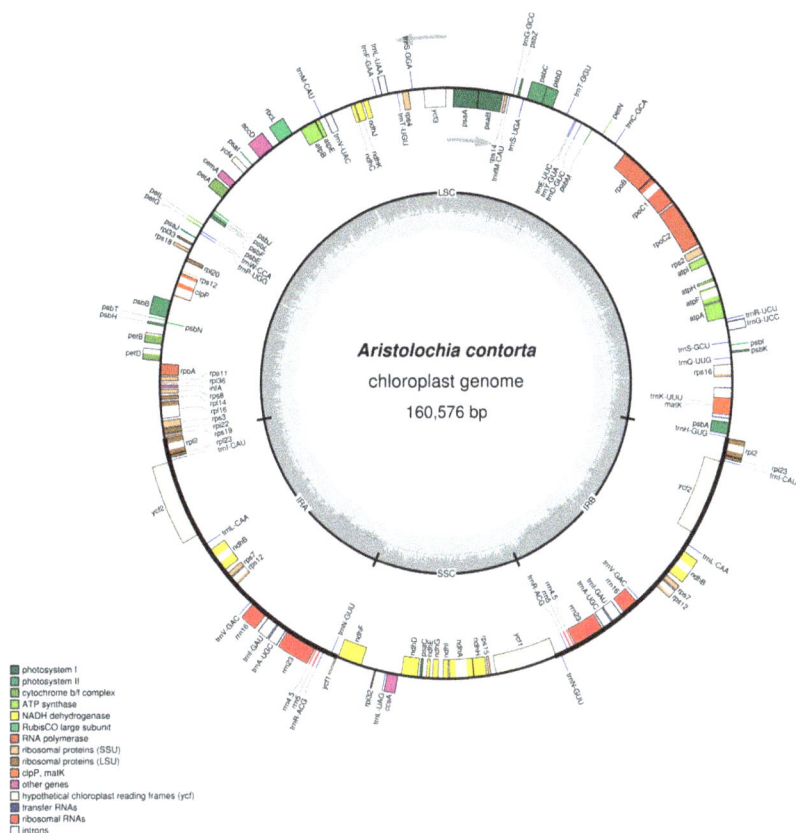

Figure 2. Gene map of the complete chloroplast genome of *A. contorta*. Genes on the inside of the circle are transcribed clockwise, while those outside are transcribed counter clockwise. The darker gray in the inner circle corresponds to GC content, whereas the lighter gray corresponds to AT content.

The analysis results revealed that both species had a GC content of 38.3%. However, this was unevenly distributed across the whole chloroplast genome. In both species, the GC content exhibited the highest values of the IR regions across the complete chloroplast genome, 43.4% both in *A. debilis* and *A. contorta*. The high GC content in the IR regions was the result of four rRNA genes (*rrn16*, *rrn23*, *rrn4.5* and *rrn5*) that occur in this region [31]. In addition, the LSC regions have GC contents of 36.6% and 35.5%, as well as the lowest values of 32.9% and 32.8% are seen in SSC regions in *A. debilis* and *A. contorta*, respectively. Within the protein-coding regions (CDS) of chloroplast genome of *A. debilis*, the percentage of AT content for the first, second and third codon positions were 54%, 61.4% and 68%, respectively (Table 1). A bias towards a higher AT representation at the third codon position has also been observed in other land plant chloroplast genomes [32–34].

A total of 131 genes were identified from each genome including 85 protein-coding genes, 37 tRNAs, eight rRNAs, and one pseudogene (*ycf1*) (Table 2). The functional *ycf1* copy existed encompassing IR-SSC boundary and the other pseudogene *ycf1* copy was on the other IR region. Six protein-coding genes, seven tRNA genes, and all rRNA genes were duplicated in the IR regions. Coding regions including protein-coding genes (CDS), tRNAs, and rRNAs constituted 56.7% and 56.4% in the chloroplast genomes of *A. debilis* and *A. contorta*, respectively; while the non-coding regions including introns, pseudogenes, and intergenic spacers constituted 43.3% and 43.6% of the genome, respectively.

Table 2. Gene contents in the chloroplast genomes of *A. debilis* and *A. contorta*.

No.	Group of Genes	Gene names	Amount
1	Photosystem I	*psaA, psaB, psaC, psaI, psaJ*	5
2	Photosystem II	*psbA, psbB, psbC, psbD, psbE, psbF, psbH, psbI, psbJ, psbK, psbL, psbM, psbN, psbT, psbZ*	15
3	Cytochrome b/f complex	*petA, petB *, petD *, petG, petL, petN*	6
4	ATP synthase	*atpA, atpB, atpE, atpF *, atpH, atpI*	6
5	NADH dehydrogenase	*ndhA *, ndhB *(×2)[1], ndhC, ndhD, ndhE, ndhF, ndhG, ndhH, ndhI, ndhJ, ndhK*	12(1)
6	RubisCO large subunit	*rbcL*	1
7	RNA polymerase	*rpoA, rpoB, rpoC1 *, rpoC2*	4
8	Ribosomal proteins (SSU)	*rps2, rps3, rps4, rps7(×2), rps8, rps11, rps12 **(×2), rps14, rps15, rps16 *, rps18, rps19*	14(2)
9	Ribosomal proteins (LSU)	*rpl2 *(×2), rpl14, rpl16 *, rpl20, rpl22, rpl23(×2), rpl32, rpl33, rpl36*	11(2)
10	Proteins of unknown function	*ycf1, ycf2(×2), ycf3 **, ycf4*	5(1)
11	Transfer RNAs	37 *tRNAs* (6 contain an intron, 7 in the IRs)	37(7)
12	Ribosomal RNAs	*rrn4.5(×2), rrn5(×2), rrn16(×2), rrn23(×2)*	8(4)
13	Other genes	*accD, clpP **, matK, ccsA, cemA, infA*	6

* Gene contains one intron; ** gene contains two introns; (×2) indicates the number of the repeat unit is 2.

Introns play an important role in the regulation of gene expression and can enhance the expression of exogenous genes at specific sites and specific times of the plant [35]. The intron content of genes reserved in the chloroplast genomes of *A. debilis* and *A. contorta* are maintained in other angiosperms [31,36]. Data revealed the presence of 18 genes containing introns in each chloroplast genome, including *atpF, rpoC1, ycf3, rps12, rpl2, rpl16, clpP, petB, petD, rps16, ndhA, ndhB*, and six tRNA genes (Table 3). In addition, the *ycf3* gene and *rps12* gene each contain two introns and three exons. The *ycf3* gene is located in LSC region and the *rps12* gene is a special trans-splicing gene, the 5' exon is located in LSC, while the 3' exon is located in IR, which is similar to that in *Aquilaria sinensis* [25], *Panax ginseng* [36] and *Cistanche deserticola* [37].

Table 3. Genes with introns in the chloroplast genomes of *A. debilis* and *A. contorta* as well as the lengths of the exons and introns.

Species	Gene	Location	Exon I (bp)	Intron I (bp)	Exon II (bp)	Intron II (bp)	Exon III (bp)
A. debilis	*atpF*	LSC	145	805	410	-	-
	clpP	LSC	71	781	292	678	255
	ndhA	SSC	552	1090	540	-	-
	ndhB	IR	777	705	756	-	-
	petB	LSC	6	214	642	-	-
	petD	LSC	6	485	476	-	-
	rpl16	LSC	8	1065	403	-	-
	rpl2	IR	391	657	431	-	-
	rpoC1	LSC	430	776	1622	-	-
	rps12	LSC	114	-	232	536	26
	rps16	LSC	46	853	191	-	-
	trnA-UGC	IR	38	809	35	-	-
	trnG-UCC	LSC	24	761	48	-	-
	trnI-GAU	IR	37	937	35	-	-
	trnK-UUU	LSC	37	2658	35	-	-
	trnL-UAA	LSC	35	521	50	-	-
	trnV-UAC	LSC	39	597	37	-	-
	ycf3	LSC	126	777	228	753	147

Table 3. *Cont.*

Species	Gene	Location	Exon I (bp)	Intron I (bp)	Exon II (bp)	Intron II (bp)	Exon III (bp)
	atpF	LSC	145	771	410	-	-
	clpP	LSC	71	821	292	664	255
	ndhA	SSC	552	1091	540	-	-
	ndhB	IR	777	716	756	-	-
	petB	LSC	6	214	642	-	-
	petD	LSC	7	485	476	-	-
	rpl16	LSC	8	1088	403	-	-
	rpl2	IR	391	657	431	-	-
A. contorta	*rpoC1*	LSC	430	776	1619	-	-
	rps12	LSC	114	-	232	536	26
	rps16	LSC	46	832	221	-	-
	trnA-UGC	IR	38	809	35	-	-
	trnG-UCC	LSC	24	751	48	-	-
	trnI-GAU	IR	37	938	35	-	-
	trnK-UUU	LSC	37	2648	35	-	-
	trnL-UAA	LSC	35	552	50	-	-
	trnV-UAC	LSC	39	605	37	-	-
	ycf3	LSC	126	764	228	760	147

2.2. IR Contraction and Expansion

Although genomic structure and size were highly conserved in Angiosperms chloroplast genomes, the IR/SC boundary regions still varied slightly (Figure 3). The contraction and expansion at the borders of the IR regions are common evolutionary events and represent the main reasons for size variation of the chloroplast genomes [33,38]. From Figure 3, the junctions of the IR and LSC regions of four species including *Arabidopsis thaliana* (accession number: NC_000932), *Nicotiana tabacum* (NC_001879), as well as two *Aristolochia* species were compared. The IRb/SSC border extended into the *ycf1* genes to cause long *ycf1* pseudogenes in all species; however, compared with *A. thaliana* and *N. tabacum*, the length of *ycf1* pseudogene of two *Aristolochia* species were only 171 and 169 bp, respectively. The IRa/SSC border was located in the CDS of the *ycf1* gene and expanded the same length into the 5′ portion of *ycf1* gene as IRb expanded in the four chloroplast genomes. The *trnH* genes were located in the LSC regions in *Nicotiana tabacum*, *Sesamum indicum*, *Arabidopsis thaliana*, and *Salvia miltiorrhiza* [31], while this gene was usually located in the IR region in the monocot chloroplast genomes [39]. Interestingly, the IRa/LSC borders were located in the coding region of *trnH* genes in the two *Aristolochia* species.

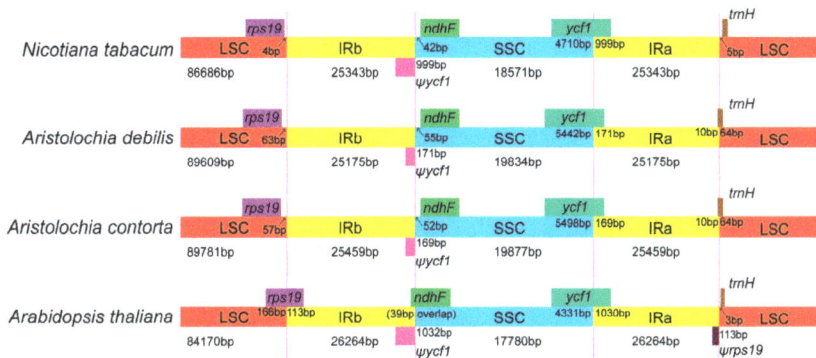

Figure 3. Comparison of the borders of LSC, SSC and IR regions among four chloroplast genomes. Number above the gene features means the distance between the ends of genes and the borders sites. The IRb/SSC border extended intothe *ycf1* genes to create various lengths of *ycf1* pseudogenes among four chloroplast genomes. These features are not to scale.

2.3. Codon Usage and RNA Editing Sites

All the protein-coding genes were composed of 26,239 and 26,255 codons in the chloroplast genomes of *A. debilis* and *A. contorta*, respectively. Among these codons, 2737 encode leucine and 315 encode cysteine, respectively, the most and least universal amino acids in the *A. debilis* chloroplast genome. The codon usages of protein-coding genes in the *A. debilis* and *A. contorta* chloroplast genomes are deduced and summarized in Figure 4 and Table S1. Figure 4 shows that the relative synonymous codon usage (RSCU) value increased with the quantity of codons that code for a specific amino acid. Most of the amino acid codons have preferences except for methionine and tryptophan. The results presented here are similar in codon usage with the chloroplast genomes of species within the genus *Ulmus* [40] and *Aq. sinensis* [25]. In addition, potential RNA editing sites were predicted for 35 genes of the chloroplast genomes of two species. A total of 92 RNA editing sites were identified (Table S2). The amino acid conversion S to L occurred most frequently, while P to S and R to W occurred least. Seventy-six common RNA editing sites were shared in genes of the two species.

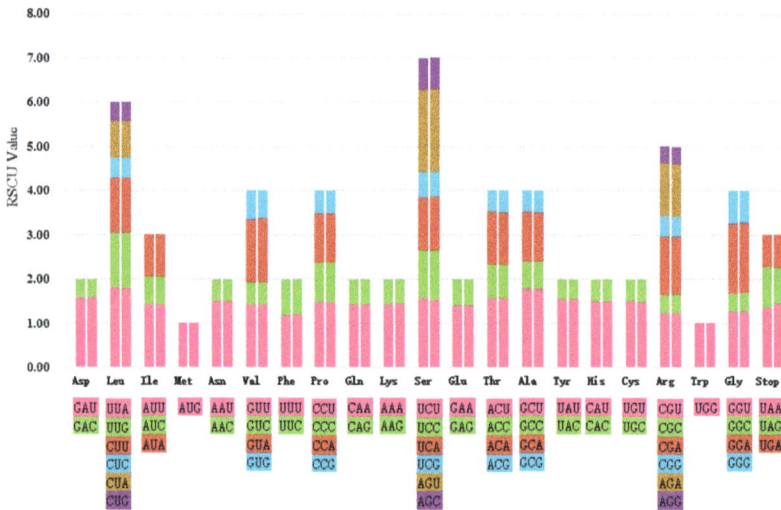

Figure 4. Codon content of 20 amino acid and stop codons in all protein-coding genes of the chloroplast genomes of two *Aristolochia* species. The histogram on the left-hand side of each amino acid shows codon usage within the *A. debilis* chloroplast genome, while the right-hand side illustrates the genome of *A. contorta*.

2.4. Repeat Structure and Simple Sequence Repeats Analyses

The repeats were mostly distributed in the intergenic spacer (IGS) and intron sequences. Figure 5 shows the repeat structure analyses of six species. The results revealed that the repeats of chloroplast genome of *A. contorta* had the greatest number, comprising of 41 forward, 43 palindromic, 29 reverse, and 25 complement repeats. Followed by *A. debilis*, contained 14 forward, 23 palindromic, 23 reverse, and six complement repeats. Simple sequence repeats (SSRs), which are ubiquitous throughout the genomes and are also known as microsatellites, are tandemly repeated DNA sequences that consist of 1–6 nucleotide repeat units [41]. SSRs are widely used for molecular markers in species identification, population genetics, and phylogenetic investigations based on their high level of polymorphism [42–44]. A total of 129 and 156 SSRs were identified using the microsatellite identification tool (MISA) in the chloroplast genomes of *A. debilis* and *A. contorta*, respectively (Table 4; Tables S3 and S4). In these SSRs, mononucletide repeats were largest in number, which were found 81 and 96 times in *A. debilis* and *A. contorta*, respectively. A/T mononucleotide repeats (96.3% and 94.8%, respectively) were the most common, while the majority of dinucleotide repeat sequences comprised of AT/TA repeats (100% and 92.8%, respectively). This result agreed with the previous studies where proportions of polyadenine (polyA) and polythymine (polyT) were higher than polycytosine (polyC) and polyguanine (polyG) within chloroplast SSRs in many plants [24].

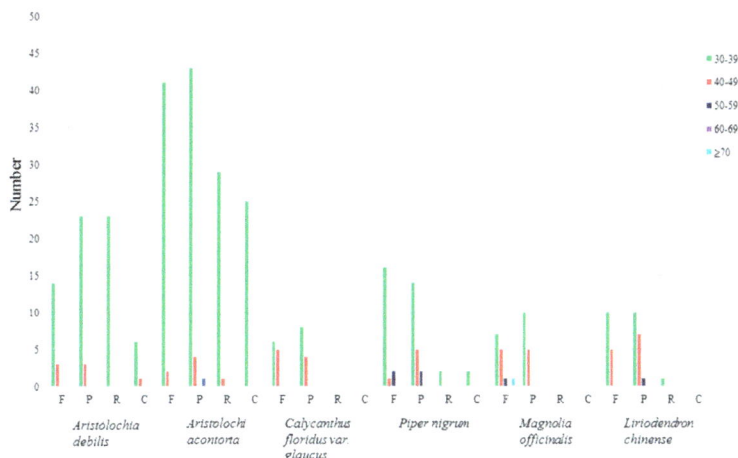

Figure 5. Repeat sequences in six chloroplast genomes. REPuter was used to identify repeat sequences with length ≥30 bp and sequence identified ≥90% in the chloroplast genomes. F, P, R, and C indicate the repeat types F (forward), P (palindrome), R (reverse), and C (complement), respectively. Repeats with different lengths are indicated in different colours.

Table 4. Types and amounts of SSRs in the *A. debilis* and *A. contorta* chloroplast genomes.

SSR Type	Repeat Unit	Amount		Ratio (%)	
		A. debilis	*A. contorta*	*A. debilis*	*A. contorta*
Mono	A/T	78	91	96.3	94.8
	C/G	3	5	3.7	5.2
Di	AC/GT	0	1	0	3.6
	AG/CT	0	1	0	3.6
	AT/TA	19	26	100	92.8
Tri	AAC/GTT	1	1	10	8.3
	AAG/CTT	1	1	10	8.3
	ATC/ATG	1	0	10	0
	AAT/ATT	7	10	70	83.4
Tetra	AAAC/GTTT	2	2	16.7	14.3
	AAAT/ATTT	4	5	33.3	35.7
	AATC/ATTG	1	1	8.3	7.1
	AGAT/ATCT	2	1	16.7	7.1
	AATT/AATT	0	1	0	7.1
	ACAT/ATGT	0	1	0	7.1
	AACT/AGTT	1	1	8.3	7.1
	AATG/ATTC	2	2	16.7	14.3
Penta	AATAT/ATATT	2	2	33.3	50
	AAATT/AATTT	1	0	16.7	0
	AAATC/ATTTG	1	0	16.7	0
	AACAT/ATGTT	0	1	0	25
	AAAAT/ATTTT	2	1	33.3	25
Hexa	AAATAG/ATTTCT	0	1	0	50
	ACATAT/ATATGT	0	1	0	50
	ACTGAT/AGTATC	1	0	100	0

2.5. Comparative Genomic Analysis

The whole chloroplast genome sequences of *A. debilis* and *A. contorta* were compared to those of *Calycanthus floridus* var. *glaucus* (accession number: NC_004993), *Magnolia officinalis* (NC_020316), and *Liriodendron chinense* (NC_030504) using the mVISTA program (Figure 6). The comparison showed that the two IR regions were less divergent than the LSC and SSC regions. The four rRNA genes were the most conserved, while the most divergent coding regions were *ndhF*, *rpl22*, *ycf1*, *rpoC2* and *ccsA*. Additionally, the results revealed that non-coding regions exhibited a higher divergence than coding regions, and the most divergent regions localized in the intergenic spacers among the five chloroplast genomes.

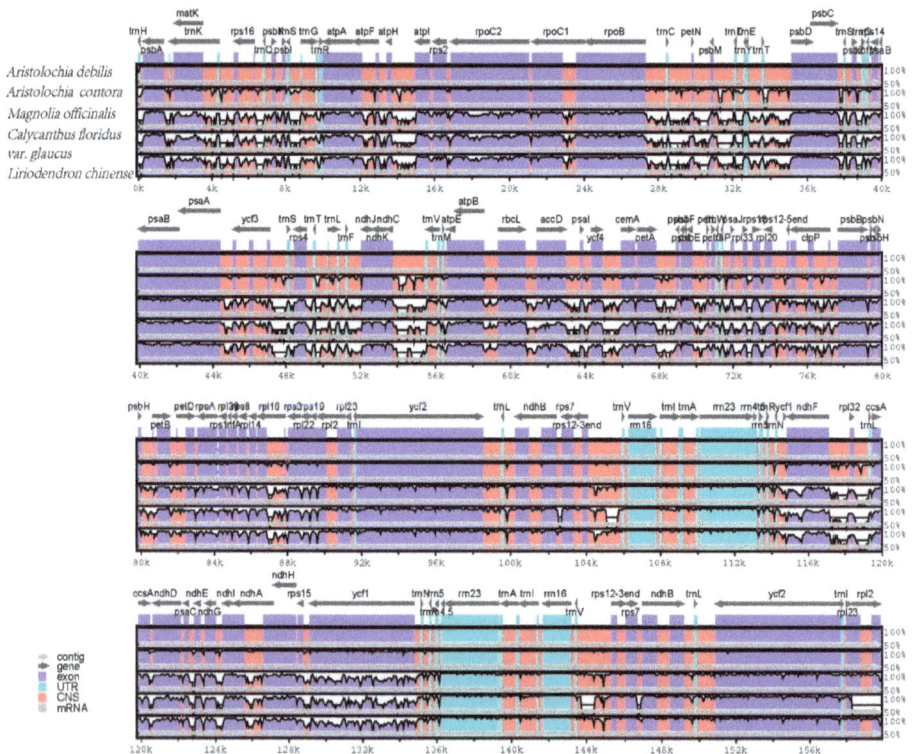

Figure 6. Sequence identity plot comparing the five chloroplast genomes with *A. debilis* as a reference by using mVISTA. Grey arrows and thick black lines above the alignment indicate genes with their orientation and the position of the IRs, respectively. A cut-off of 70% identity was used for the plots, and the Y-scale represents the percent identity ranging from 50% to 100%.

2.6. Phylogenetic Analyses

Chloroplast genomes provide abundant resources, which are significant for evolutionary, taxonomic, and phylogenetic studies [31,45,46]. The whole chloroplast genomes and protein-coding genes have been successfully used to resolve phylogenetic relationships at almost any taxonomic level during the past decade [31,37]. *Aristolochia*, consisting of nearly 400 species, is the largest genus in the family Aristolochiaceae [30]. Phylogenetic analyses employing one or several genes have been performed in previous studies [47–49]; however, these analyses were restricted to the species of Aristolochiaceae and included few species from other families. In this study, to identify the phylogenetic positions of *A. debilis* and *A. contorta* within Angiosperms, 60 protein-coding genes commonly present in 37 species from Piperales, Laurales, Magnoliales, Ranunculales, Fabales, Rosales, Chloranthales, as well as two *Aristolochia* species were used to construct the phylogenetic tree using the Maximum parsimony (MP) method (Figure 7). All the nodes in the MP trees have high bootstrap support values, and 30 out of 34 nodes with 100% bootstrap values were found. The result illustrated that two *Aristolochia* species were sister taxa with respect to four *Piper* species (Piperaceae), and these species were grouped with four species from Laurales and five species from Magnoliales. Additionally, all species are clustered within a lineage distinct from the outgroup. This result (inferred from the chloroplast genome data) obtained high support values, which suggested that the chloroplast genome could effectively resolve the phylogenetic positions and relationships of this family. Nevertheless, to accurately illustrate the evolution of the family Aristolochiaceae, it is necessary to use more species to analyze the phylogeny. This study will also provide a reference for species identification among *Aristolochia* and other genus using the chloroplast genome.

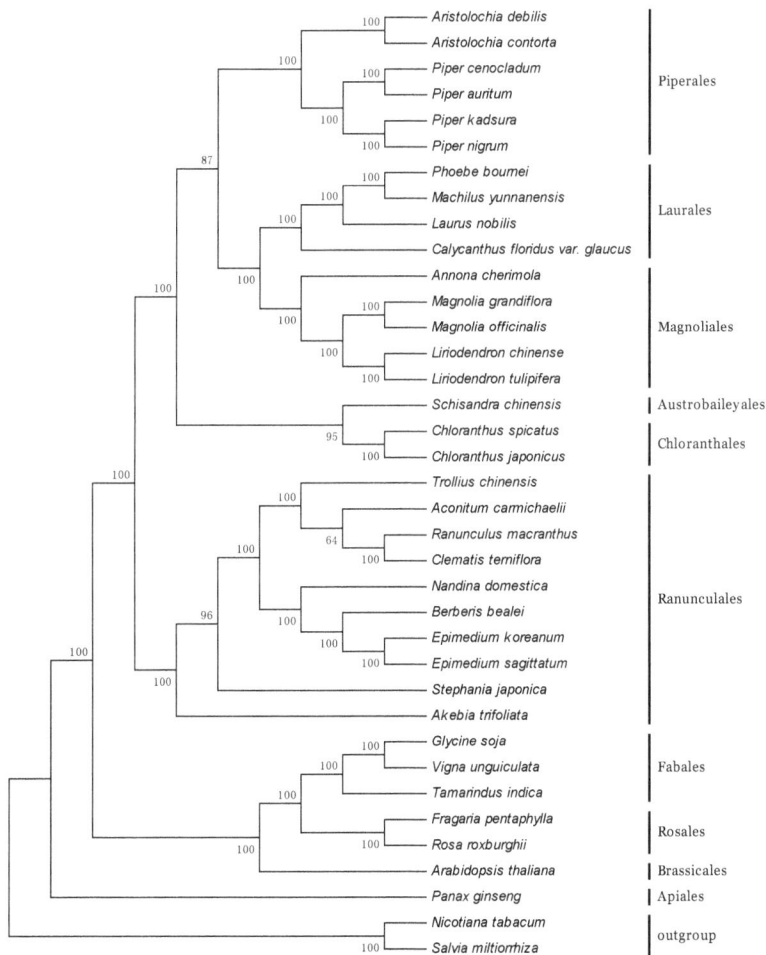

Figure 7. Phylogenetic tree constructed using Maximum parsimony (MP) method based on 60 protein-coding genes from different species. Numbers at nodes are values for bootstrap support.

3. Materials and Methods

3.1. Plant Material, DNA Extraction, and Sequencing

Fresh plants of *A. debilis* and *A. contorta* were collected from Lichuan City in Hubei Province and Tonghua City in Jilin Province, respectively. All samples were identified by Professor Yulin Lin, who is based at the Institute of Medicinal Plant Development (IMPLAD), Chinese Academy of Medical Sciences (CAMS) and Peking Union Medical College (PUMC). The voucher specimens were deposited in the herbarium of the IMPLAD. The leaves were cleansed and preserved in a $-80\ °C$ refrigerator. Total genomic DNA was extracted from approximately 100 mg of samples using DNeasy Plant Mini Kit with a standard protocol (Qiagen Co., Hilden, Germany). Final DNA quality was assessed based on spectrophotometry and their integrity was examined by electrophoresis in 1% (w/v) agarose gel. The DNA was used to construct shotgun libraries with insert sizes of 500 bp and sequenced according to the manufacturer's manual for the Illumina Hiseq X. Approximately 6.3 Gb of raw data from *A. debilis* and 5.8 Gb from *A. contorta* were produced with 150 bp pair-end read lengths.

3.2. Chloroplast Genome Assembly and Annotation

First, we used the software Trimmomatic (v0.36, Max Planck Institute of Molecular Plant Physiology, Potsdam, Germany) [50] to trim the low-quality reads. After quality control, the clean reads were used to assemble the chloroplast genome. All chloroplast genomes of plants recorded in the National Center for Biotechnology Information (NCBI) were used to construct a reference database. Next, the clean reads were mapped to the database on the basis of their coverage and similarity, and the mapped reads extracted. Extracted reads were assembled to contigs using SOAPdenovo (v2, BGI HK Research Institute, Hong Kong, China) [51], and the resulting contigs were combined and extended to obtain a complete chloroplast genome sequence. To verify the accuracy of assembly, four boundaries of single copy (SC) and inverted repeat (IR) regions of the assembled sequences were confirmed by PCR amplification and Sanger sequencing using the primers listed in Table S5.

We used the online program Dual Organellar GenoMe Annotator (DOGMA), (University of Texas at Austin, Austin, TX, USA) [52] and the software Chloroplast Genome Annotation, Visualization, Analysis, and GenBank Submission (CPGAVAS), (Institute of Medicinal Plant Development, Chinese Academy of Medical Sciences and Peking Union Medical College, Beijing, China) [53] coupled with manual corrections to perform the preliminarily gene annotation of chloroplast genomes of two species. The tRNA genes were identified using the software tRNAscan-SE (v2.0, University of California, Santa Cruz, CA, USA) [54] and DOGMA [52]. The gene map was drawn using the Organellar Genome DRAW (OGDRAW) (v1.2, Max Planck Institute of Molecular Plant Physiology, Potsdam, Germany) [55] with default settings and checked manually. The complete and correct chloroplast genome sequences of the two species were deposited in GenBank, accession numbers of *A. debilis* and *A. contorta* are MF539928 and MF539927, respectively.

3.3. Genome Structure Analyses and Genome Comparison

The distribution of codon usage was investigated using the software CodonW (University of Texas, Houston, TX, USA) with the RSCU ratio [56]. Thirty-five protein-coding genes of the chloroplast genomes of two species were used to predict potential RNA editing sites using the online program Predictive RNA Editor for Plants (PREP) suite [57] with a cutoff value of 0.8. GC content was analyzed using Molecular Evolutionary Genetics Analysis (MEGA v6.0, Tokyo Metropolitan University, Tokyo, Japan) [58]. REPuter (University of Bielefeld, Bielefeld, Germany) [59] to identify the size and location of repeat sequences, including forward, palindromic, reverse, and complement repeats in the chloroplast genomes of six species *C. floridus* var. *glaucus*, *M. officinalis*, *L. chinense* and *Piper nigrum* (NC_034692). For all repeat types, the minimal size was 30 bp and the two repeat copies had at least 90% similarity. Simple sequence repeats (SSRs) were detected using MISA software [60] with parameters set the same as Li et al. [61]. The whole-genome alignment for the chloroplast genomes of the five species including *A. debilis*, *A. contorta*, *C. floridus* var. *glaucus*, *M. officinalis*, and *L. chinense* were performed and plotted using the mVISTA program [62].

3.4. Phylogenetic Analyses

A total of 35 complete chloroplast genomes were downloaded from the NCBI Organelle Genome Resources database (Table S6). The 60 protein-coding gene sequences commonly present in 37 species, including the two species in this study, were aligned using the Clustal algorithm [63]. To determine the phylogenetic positions of *A. debilis*, and *A. contorta*, we analyzed the chloroplast genomes of these 60 protein-coding genes. Maximum parsimony (MP) analysis was performed with PAUP*4.0b10 [64], using a heuristic search performed with the MULPARS option, the random stepwise addition with 1000 replications and tree bisection-reconnection (TBR) branch swapping. Bootstrap analysis was also performed with 1,000 replicates with TBR branch swapping.

4. Conclusions

The complete chloroplast genomes of *A. debilis* and *A. contorta*, the first two sequenced members of the family Aristolochiaceae, were determined in this study. The genome structure and gene content were relatively conserved. The phylogenetic analyses illustrated that these two *Aristolochia* species were positioned close to four species from the family Piperaceae and had a close phylogenetic relationship with Laurales and Magnoliales. The results provided the basis for the study of the evolutionary history of *A. debilis* and *A. contorta*. All the data presented in this paper will facilitate the further investigation of these two medicinal plants.

Supplementary Materials: Supplementary materials can be found at www.mdpi.com/1422-0067/18/9/1839/s1.

Acknowledgments: This work was supported by Chinese Academy of Medical Sciences (CAMS) Innovation Fund for Medical Sciences (CIFMS) (No. 2016-I2M-3-016), Major Scientific and Technological Special Project for "Significant New Drugs Creation" (No. 2014ZX09304307001) and The Key Projects in the National Science and Technology Pillar Program (No. 2011BAI07B08).

Author Contributions: Jianguo Zhou, Xinlian Chen, and Yingxian Cui, performed the experiments; Jianguo Zhou, Wei Sun, and Jingyuan Song, assembled sequences and analyzed the data; Jianguo Zhou wrote the manuscript; Yonghua Li, and Yu Wang, collected plant material; Hui Yao conceived the research and revised the manuscript. All authors have read and approved the final manuscript.

Conflicts of Interest: The authors declare no conflict of interest.

Abbreviations

LSC	Large single copy
SSC	Small single copy
IR	Inverted repeat
MP	Maximum parsimony
SSR	Simple sequence repeats
ATP	Adenosine triphosphate
NADH	Nicotinamide adenine dinucleotide

References

1. Chinese Pharmacopoeia Commission. *The Chinese Pharmacopoeia*; Chemical Industry Press: Beijing, China, 2015; pp. 51–52.
2. Chen, C.X. Studies on the chemical constituents from the fruit of *Aristolochia debilis*. *J. Chin. Med. Mater.* **2010**, *33*, 1260–1261.
3. Xu, Y.; Shang, M.; Ge, Y.; Wang, X.; Cai, S. Chemical constituent from fruit of *Aristolochia contorta*. *Chin. J. Chin. Mater. Medica* **2010**, *35*, 2862.
4. Mix, D.B.; Guinaudeau, H.; Shamma, M. The aristolochic acids and aristolactams. *J. Nat. Prod.* **1982**, *45*, 657–666. [CrossRef]

5. Arlt, V.M.; Stiborova, M.; Schmeiser, H.H. Aristolochic acid as a probable human cancer hazard in herbal remedies: A review. *Mutagenesis* **2002**, *17*, 265. [CrossRef] [PubMed]
6. Schmeiser, H.H.; Janssen, J.W.; Lyons, J.; Scherf, H.R.; Pfau, W.; Buchmann, A.; Bartram, C.R.; Wiessler, M. Aristolochic acid activates *RAS* genes in rat tumors at deoxyadenosine residues. *Cancer Res.* **1990**, *50*, 5464–5469. [PubMed]
7. Chen, L.; Mei, N.; Yao, L.; Chen, T. Mutations induced by carcinogenic doses of aristolochic acid in kidney of big blue transgenic rats. *Toxicol. Lett.* **2006**, *165*, 250–256. [CrossRef] [PubMed]
8. Cheng, C.L.; Chen, K.J.; Shih, P.H.; Lu, L.Y.; Hung, C.F.; Lin, W.C.; Yesong, G.J. Chronic renal failure rats are highly sensitive to aristolochic acids, which are nephrotoxic and carcinogenic agents. *Cancer Lett.* **2006**, *232*, 236–242. [CrossRef] [PubMed]
9. Cosyns, J.P.; Goebbels, R.M.; Liberton, V.; Schmeiser, H.H.; Bieler, C.A.; Bernard, A.M. Chinese herbs nephropathy-associated slimming regimen induces tumours in the forestomach but no interstitial nephropathy in rats. *Arch. Toxicol.* **1998**, *72*, 738–743. [CrossRef] [PubMed]
10. Hoang, M.L.; Chen, C.H.; Sidorenko, V.S.; He, J.; Dickman, K.G.; Yun, B.H.; Moriya, M.; Niknafs, N.; Douville, C.; Karchin, R. Mutational signature of aristolochic acid exposure as revealed by whole-exome sequencing. *Sci. Transl. Med.* **2013**, *5*. [CrossRef] [PubMed]
11. Balachandran, P.; Wei, F.; Lin, R.C.; Khan, I.A.; Pasco, D.S. Structure activity relationships of aristolochic acid analogues: Toxicity in cultured renal epithelial cells. *Kidney Int.* **2005**, *67*, 1797. [CrossRef] [PubMed]
12. Tsai, D.M.; Kang, J.J.; Lee, S.S.; Wang, S.Y.; Tsai, I.; Chen, G.Y.; Liao, H.W.; Li, W.C.; Kuo, C.H.; Tseng, Y.J. Metabolomic analysis of complex Chinese remedies: Examples of induced nephrotoxicity in the mouse from a series of remedies containing aristolochic acid. *Evid. Based Compl. Alt.* **2013**, *2013*, 263757. [CrossRef] [PubMed]
13. Grollman, A.P.; Shibutani, S.; Moriya, M.; Miller, F.; Wu, L.; Moll, U.; Suzuki, N.; Fernandes, A.; Rosenquist, T.; Medverec, Z.; et al. Aristolochic acid and the etiology of endemic (Balkan) nephropathy. *Proc. Natl. Acad. Sci. USA* **2007**, *104*, 12129–12134. [CrossRef] [PubMed]
14. Lord, G.M.; Tagore, R.; Cook, T.; Gower, P.; Pusey, C.D. Nephropathy caused by Chinese herbs in the UK. *Lancet* **1999**, *354*, 481–482. [CrossRef]
15. Vanherweghem, J.L.; Tielemans, C.; Abramowicz, D.; Depierreux, M.; Vanhaelen-Fastre, R.; Vanhaelen, M.; Dratwa, M.; Richard, C.; Vandervelde, D.; Verbeelen, D.; et al. Rapidly progressive interstitial renal fibrosis in young women: Association with slimming regimen including Chinese herbs. *Lancet* **1993**, *341*, 387–391. [CrossRef]
16. Dong, W.; Xu, C.; Cheng, T.; Lin, K.; Zhou, S. Sequencing angiosperm plastid genomes made easy: A complete set of universal primers and a case study on the phylogeny of Saxifragales. *Genome Biol. Evol.* **2013**, *5*, 989–997. [CrossRef] [PubMed]
17. Leliaert, F.; Smith, D.R.; Moreau, H.; Herron, M.D.; Verbruggen, H.; Delwiche, C.F.; Clerck, O.D. Phylogeny and molecular evolution of the green algae. *Crit. Rev. Plant Sci.* **2012**, *31*, 1–46. [CrossRef]
18. Raman, G.; Park, S. Analysis of the complete chloroplast genome of a medicinal plant, *Dianthus superbus* var. *longicalyncinus*, from a comparative genomics perspective. *PLoS ONE* **2015**, *10*. [CrossRef]
19. Jansen, R.K.; Raubeson, L.A.; Boore, J.L.; Depamphilis, C.W.; Chumley, T.W.; Haberle, R.C.; Wyman, S.K.; Alverson, A.J.; Peery, R.; Herman, S.J. Methods for obtaining and analyzing whole chloroplast genome sequences. *Methods Enzymol.* **2005**, *395*, 348–384. [PubMed]
20. Wolfe, K.H.; Li, W.H.; Sharp, P.M. Rates of nucleotide substitution vary greatly among plant mitochondrial, chloroplast and nuclear DNA. *Proc. Natl. Acad. Sci. USA* **1988**, *84*, 9054–9058. [CrossRef]
21. Smith, D.R.; Keeling, P.J. Mitochondrial and plastid genome architecture: Reoccurring themes, but significant differences at the extremes. *Proc. Natl. Acad. Sci. USA* **2015**, *112*, 10177–10184. [CrossRef] [PubMed]
22. Tonti-Filippini, J.; Nevill, P.G.; Dixon, K.; Small, I. What can we do with 1,000 plastid genomes? *Plant J.* **2017**, *90*, 808–818. [CrossRef] [PubMed]
23. Wicke, S.; Schneeweiss, G.M.; Müller, K.F.; Quandt, D. The evolution of the plastid chromosome in land plants: Gene content, gene order, gene function. *Plant. Mol. Boil.* **2011**, *76*, 273–297. [CrossRef] [PubMed]
24. Kuang, D.Y.; Wu, H.; Wang, Y.L.; Gao, L.M.; Zhang, S.Z.; Lu, L. Complete chloroplast genome sequence of *Magnolia kwangsiensis* (Magnoliaceae): Implication for DNA barcoding and population genetics. *Genome* **2011**, *54*, 663–673. [CrossRef] [PubMed]

25. Wang, Y.; Zhan, D.F.; Jia, X.; Mei, W.L.; Dai, H.F.; Chen, X.T.; Peng, S.Q. Complete chloroplast genome sequence of *Aquilaria sinensis* (lour.) gilg and evolution analysis within the Malvales order. *Front. Plant Sci.* **2016**, *7*, 280. [CrossRef] [PubMed]

26. Wu, F.H.; Chan, M.T.; Liao, D.C.; Chentran, H.; Yiwei, L.; Daniell, H.; Duvall, M.R.; Lin, C.S. Complete chloroplast genome of *Oncidium* Gower Ramsey and evaluation of molecular markers for identification and breeding in Oncidiinae. *BMC Plant Biol.* **2010**, *10*, 68. [CrossRef] [PubMed]

27. Bedbrook, J.R.; Bogorad, L. Endonuclease recognition sites mapped on *Zea mays* chloroplast DNA. *Proc. Natl. Acad. Sci. USA* **1976**, *73*, 4309–4313. [CrossRef] [PubMed]

28. Shinozaki, K.; Ohme, M.; Tanaka, M.; Wakasugi, T.; Hayshida, N.; Matsubayasha, T.; Zaita, N.; Chunwongse, J.; Obokata, J.; Yamaguchi-Shinozaki, K. The complete nucleotide sequence of the tobacco chloroplast genome. *EMBO J.* **1986**, *4*, 111–148. [CrossRef]

29. NCBI, Genome. Available online: https://www.ncbi.nlm.nih.gov/genome/browse/?report=5 (accessed on 30 June 2017).

30. The Editorial Committee of Flora of China. *Flora of China*; Science Press: Beijing, China; Missouri Botanical Garden Press: St. Louis, MO, USA, 2003; Volume 5, pp. 246–269.

31. Qian, J.; Song, J.; Gao, H.; Zhu, Y.; Xu, J.; Pang, X.; Yao, H.; Sun, C.; Li, X.; Li, C. The complete chloroplast genome sequence of the medicinal plant *Salvia miltiorrhiza*. *PLoS ONE* **2013**, *8*. [CrossRef] [PubMed]

32. Clegg, M.T.; Gaut, B.S.; Learn, G.H., Jr.; Morton, B.R. Rates and patterns of chloroplast DNA evolution. *Proc. Natl. Acad. Sci. USA* **1994**, *91*, 6795–6801. [CrossRef] [PubMed]

33. Yang, M.; Zhang, X.; Liu, G.; Yin, Y.; Chen, K.; Yun, Q.; Zhao, D.; Al-Mssallem, I.S.; Yu, J. The complete chloroplast genome sequence of date palm (*Phoenix dactylifera* L.). *PLoS ONE* **2010**, *5*. [CrossRef] [PubMed]

34. Yi, D.K.; Kim, K.J. Complete chloroplast genome sequences of important oilseed crop *Sesamum indicum* L. *PLoS ONE* **2012**, *7*. [CrossRef] [PubMed]

35. Xu, J.; Feng, D.; Song, G.; Wei, X.; Chen, L.; Wu, X.; Li, X.; Zhu, Z. The first intron of rice EPSP synthase enhances expression of foreign gene. *Sci. China Life Sci.* **2003**, *46*, 561–569. [CrossRef] [PubMed]

36. Kim, K.J.; Lee, H.L. Complete chloroplast genome sequences from Korean ginseng (*Panax schinseng* nees) and comparative analysis of sequence evolution among 17 vascular plants. *DNA Res.* **2004**, *11*, 247. [CrossRef] [PubMed]

37. Li, X.; Zhang, T.C.; Qiao, Q.; Ren, Z.; Zhao, J.; Yonezawa, T.; Hasegawa, M.; Crabbe, M.J.; Li, J.; Zhong, Y. Complete chloroplast genome sequence of holoparasite *Cistanche deserticola* (Orobanchaceae) reveals gene loss and horizontal gene transfer from its host *Haloxylon ammodendron* (Chenopodiaceae). *PLoS ONE* **2013**, *8*. [CrossRef] [PubMed]

38. Raubeson, L.A.; Peery, R.; Chumley, T.W.; Dziubek, C.; Fourcade, H.M.; Boore, J.L.; Jansen, R.K. Comparative chloroplast genomics: Analyses including new sequences from the angiosperms *Nuphar advena* and *Ranunculus macranthus*. *BMC Genom.* **2007**, *8*, 174. [CrossRef] [PubMed]

39. Huotari, T.; Korpelainen, H. Complete chloroplast genome sequence of *Elodea canadensis* and comparative analyses with other monocot plastid genomes. *Gene* **2012**, *508*, 96–105. [CrossRef] [PubMed]

40. Zuo, L.H.; Shang, A.Q.; Zhang, S.; Yu, X.Y.; Ren, Y.C.; Yang, M.S.; Wang, J.M. The first complete chloroplast genome sequences of *Ulmus* species by de novo sequencing: Genome comparative and taxonomic position analysis. *PLoS ONE* **2017**, *12*. [CrossRef] [PubMed]

41. Powell, W.; Morgante, M.; McDevitt, R.; Vendramin, G.G.; Rafalski, J.A. Polymorphic simple sequence repeat regions in chloroplast genomes: Applications to the population genetics of pines. *Proc. Natl. Acad. Sci. USA* **1995**, *92*, 7759–7763. [CrossRef] [PubMed]

42. Jiao, Y.; Jia, H.; Li, X.; Chai, M.; Jia, H.; Chen, Z.; Wang, G.; Chai, C.; Weg, E.V.D.; Gao, Z. Development of simple sequence repeat (SSR) markers from a genome survey of Chinese bayberry (*Myrica rubra*). *BMC Genomics* **2012**, *13*, 201. [CrossRef] [PubMed]

43. Xue, J.; Wang, S.; Zhou, S.L. Polymorphic chloroplast microsatellite loci in *Nelumbo* (Nelumbonaceae). *Am. J. Bot.* **2012**, *99*, 240–244. [CrossRef] [PubMed]

44. Yang, A.H.; Zhang, J.J.; Yao, X.H.; Huang, H.W. Chloroplast microsatellite markers in *Liriodendron tulipifera* (Magnoliaceae) and cross-species amplification in *L. chinense*. *Am. J. Bot.* **2011**, *98*, 123–126. [CrossRef] [PubMed]

45. Jansen, R.K.; Cai, Z.; Raubeson, L.A.; Daniell, H.; Depamphilis, C.W.; Leebens-Mack, J.; Müller, K.F.; Guisinger-Bellian, M.; Haberle, R.C.; Hansen, A.K. Analysis of 81 genes from 64 plastid genomes resolves relationships in angiosperms and identifies genome-scale evolutionary patterns. *Proc. Natl. Acad. Sci. USA* **2007**, *104*, 19369–19374. [CrossRef] [PubMed]

46. Moore, M.J.; Bell, C.D.; Soltis, P.S.; Soltis, D.E. Using plastid genome-scale data to resolve enigmatic relationships among basal angiosperms. *Proc. Natl. Acad. Sci. USA* **2007**, *104*, 19363–19368. [CrossRef] [PubMed]

47. Murata, J.; Ohi, T.; Wu, S.; Darnaedi, D.; Sugawara, T.; Nakanishi, T.; Murata, H. Molecular phylogeny of *Aristolochia* (Aristolochiaceae) inferred from *matK* sequences. *Apg Acta Phyto. Geo.* **2001**, *52*, 75–83.

48. Ohi-Toma, T.; Sugawara, T.; Murata, H.; Wanke, S.; Neinhuis, C.; Jin, M. Molecular phylogeny of *Aristolochia sensu lato* (Aristolochiaceae) based on sequences of *rbcL*, *matK*, and *phyA* genes, with special reference to differentiation of chromosome numbers. *Syst. Bot.* **2006**, *31*, 481–492. [CrossRef]

49. Silva-Brandão, K.L.; Solferini, V.N.; Trigo, J.R. Chemical and phylogenetic relationships among *Aristolochia* L. (Aristolochiaceae) from southeastern Brazil. *Biochem. Syst. Ecol.* **2006**, *34*, 291–302. [CrossRef]

50. Bolger, A.M.; Lohse, M.; Usadel, B. Trimmomatic: A flexible trimmer for Illumina sequence data. *Bioinformatics* **2014**, *30*, 2114–2120. [CrossRef] [PubMed]

51. Luo, R.; Liu, B.; Xie, Y.; Li, Z.; Huang, W.; Yuan, J.; He, G.; Chen, Y.; Pan, Q.; Liu, Y.; et al. SOAPdenovo2: An empirically improved memory-efficient short-read de novo assembler. *GigaScience* **2012**, *1*, 18. [CrossRef] [PubMed]

52. Wyman, S.K.; Jansen, R.K.; Boore, J.L. Automatic annotation of organellar genomes with DOGMA. *Bioinformatics* **2004**, *20*, 3252–3255. [CrossRef] [PubMed]

53. Liu, C.; Shi, L.; Zhu, Y.; Chen, H.; Zhang, J.; Lin, X.; Guan, X. CpGAVAS, an integrated web server for the annotation, visualization, analysis, and GenBank submission of completely sequenced chloroplast genome sequences. *BMC Genom.* **2012**, *13*, 715. [CrossRef] [PubMed]

54. Schattner, P.; Brooks, A.N.; Lowe, T.M. The tRNAscan-SE, snoscan and snoGPS web servers for the detection of tRNAs and snoRNAs. *Nucleic Acids Res.* **2005**, *33*. [CrossRef] [PubMed]

55. Lohse, M.; Drechsel, O.; Bock, R. Organellargenomedraw (OGDRAW): A tool for the easy generation of high-quality custom graphical maps of plastid and mitochondrial genomes. *Curr. Genet.* **2007**, *52*, 267–274. [CrossRef] [PubMed]

56. Sharp, P.M.; Li, W.H. The codon Adaptation Index-a measure of directional synonymous codon usage bias, and its potential applications. *Nucleic Acids Res.* **1987**, *15*, 1281–1295. [CrossRef] [PubMed]

57. Mower, J.P. The PREP suite: Predictive RNA editors for plant mitochondrial genes, chloroplast genes and user-defined alignments. *Nucleic Acids Res.* **2009**, *37*. [CrossRef] [PubMed]

58. Tamura, K.; Stecher, G.; Peterson, D.; Filipski, A.; Kumar, S. MEGA6: Molecular evolutionary genetics analysis version 6.0. *Comput. Mol. Biol. Evol.* **2013**, *30*, 2725–2729. [CrossRef] [PubMed]

59. Kurtz, S.; Choudhuri, J.V.; Ohlebusch, E.; Schleiermacher, C.; Stoye, J.; Giegerich, R. Reputer: The manifold applications of repeat analysis on a genomic scale. *Nucleic Acids Res.* **2001**, *29*, 4633–4642. [CrossRef] [PubMed]

60. Misa-Microsatellite Identification Tool. Available online: http://pgrc.ipk-gatersleben.de/misa/ (accessed on 2 June 2017).

61. Li, X.W.; Gao, H.H.; Wang, Y.T.; Song, J.Y.; Henry, R.; Wu, H.Z.; Hu, Z.G.; Hui, Y.; Luo, H.M.; Luo, K. Complete chloroplast genome sequence of *Magnolia grandiflora* and comparative analysis with related species. *Sci. China Life Sci.* **2013**, *56*, 189–198. [CrossRef] [PubMed]

62. Frazer, K.A.; Pachter, L.; Poliakov, A.; Rubin, E.M.; Dubchak, I. VISTA: Computational tools for comparative genomics. *Nucleic Acids Res.* **2004**, *32*, 273–279. [CrossRef] [PubMed]

63. Thompson, J.D.; Higgins, D.G.; Gibson, T.J. CLUSTAL W: Improving the sensitivity of progressive multiple sequence alignment through sequence weighting, position-specific gap penalties and weight matrix choice. *Nucleic Acids Res.* **1994**, *22*, 4673–4680. [CrossRef] [PubMed]

64. Swofford, D.L. *PAUP*. *Phylogenetic Analysis Using Parsimony (*and Other Methods)*; Version 4.0b10; Sinauer Associates: Sunderland, MA, USA, 2002.

Article

The Complete Chloroplast Genome Sequences of the Medicinal Plant *Forsythia suspensa* (Oleaceae)

Wenbin Wang [1,*], Huan Yu [1], Jiahui Wang [2], Wanjun Lei [1], Jianhua Gao [1], Xiangpo Qiu [1] and Jinsheng Wang [1,*]

[1] College of Life Science, Shanxi Agricultural University, Taigu 030801, China; sxndyh@stu.sxau.edu.cn (H.Y.); leiwanjunleo@sxau.edu.cn (W.L.); gaojh_edu@sxau.edu.cn (J.G.); sxndqiuxiangpo@stu.sxau.edu.cn (X.Q.)

[2] College of Plant Protection, Northwest Agriculture & Forestry University, Yangling 712100, China; wjh1014@nwafu.edu.cn

* Correspondence: sxndwwb@sxau.edu.cn (W.W.); wangjinsheng@sxau.edu.cn (J.W.); Tel.: +86-0354-628-6963 (W.W.); +86-0354-628-6908 (J.W.)

Received: 12 September 2017; Accepted: 25 October 2017; Published: 31 October 2017

Abstract: *Forsythia suspensa* is an important medicinal plant and traditionally applied for the treatment of inflammation, pyrexia, gonorrhea, diabetes, and so on. However, there is limited sequence and genomic information available for *F. suspensa*. Here, we produced the complete chloroplast genomes of *F. suspensa* using Illumina sequencing technology. *F. suspensa* is the first sequenced member within the genus *Forsythia* (Oleaceae). The gene order and organization of the chloroplast genome of *F. suspensa* are similar to other Oleaceae chloroplast genomes. The *F. suspensa* chloroplast genome is 156,404 bp in length, exhibits a conserved quadripartite structure with a large single-copy (LSC; 87,159 bp) region, and a small single-copy (SSC; 17,811 bp) region interspersed between inverted repeat (IRa/b; 25,717 bp) regions. A total of 114 unique genes were annotated, including 80 protein-coding genes, 30 tRNA, and four rRNA. The low GC content (37.8%) and codon usage bias for A- or T-ending codons may largely affect gene codon usage. Sequence analysis identified a total of 26 forward repeats, 23 palindrome repeats with lengths >30 bp (identity >90%), and 54 simple sequence repeats (SSRs) with an average rate of 0.35 SSRs/kb. We predicted 52 RNA editing sites in the chloroplast of *F. suspensa*, all for C-to-U transitions. IR expansion or contraction and the divergent regions were analyzed among several species including the reported *F. suspensa* in this study. Phylogenetic analysis based on whole-plastome revealed that *F. suspensa*, as a member of the Oleaceae family, diverged relatively early from Lamiales. This study will contribute to strengthening medicinal resource conservation, molecular phylogenetic, and genetic engineering research investigations of this species.

Keywords: *Forsythia suspensa*; sequencing; chloroplast genome; comparative genomics; phylogenetic analysis

---

1. Introduction

Forsythia suspensa (Thunb.) Vahl, known as "Lianqiao" in Chinese, is a well-known traditional Asian medicine that is widely distributed in many Asian and European countries [1]. In folk medicine, the extract of the dried fruit has long been used to treat a variety of diseases, such as inflammation, pyrexia, gonorrhea, tonsillitis, and ulcers [2]. In recent years, the dried ripe fruit of *F. suspensa* has often been prescribed for the treatment of diabetes in China [3,4].

Chloroplast (cp) genomes are mostly circular DNA molecules, which have a typical quadripartite structure composed of a large single copy (LSC) region and a small single copy (SSC) region interspersed between two copies of inverted repeats (IRa/b) [5]. The cp genome sequences can provide vast information not only about genes and their encoded proteins, but also on functional implications and evolutionary relationships [6]. Due to high-throughput capabilities and relatively low costs, next-generation sequencing techniques have made it more convenient to obtain a large number of cp genome sequences [7]. After the first complete cp DNA sequences were reported in *Nicotiana tabacum* [8] and *Marchantia polymorpha* [9], complete cp DNA sequences of numerous plant species were determined [6,10–12]. To date, approximately 1300 plant cp genomes are publicly available as part of the National Center for Biotechnology Information (NCBI) database.

Within the Oleaceae family, the complete cp genomes of several plant species have been published [12–15], thereby providing additional evidence for the evolution and conservation of cp genomes. Nevertheless, no cp genome belonging to genus *Forsythia* has been reported. Few data are available with respect to the *F. suspensa* cp genome.

In order to characterize the complete cp genome sequence of the *F. suspensa* and expand our understanding of the diversity of the genus *Forsythia*, details of the cp genome structure and organization are reported in this paper. This is also the first sequenced member of the genus *Forsythia* (Oleaceae). We compare the *F. suspense* cp genome with previously annotated cp genomes of other Lamiales species. Our studies could provide basic data for the medicinal species conservation and molecular phylogenetic research of the genus *Forsythia* and Lamiales.

2. Results and Discussions

2.1. Genome Features

Whole genome sequencing using an Illumina Hiseq 4000 PE150 platform generated 19,241,634 raw reads. Clean reads were obtained by removing adaptors and low-quality read pairs. Then, we collected 662,793 cp-genome-related reads (3.44% of total reads), reaching an average of 636 × coverage over the cp genome. With PCR-based experiments, we closed the gaps and validated the sequence assembly, and ultimately obtained a complete *F. suspensa* cp genome sequence, which was then submitted to GenBank (accession number: MF579702).

Most cp genomes of higher plants have been found to have a typical quadripartite structure composed of an LSC region and an SSC region interspersed between the IRa/b region [5]. The complete cp genome of *F. suspensa* has a total length of 156,404 bp, with a pair of IRs of 25,717 bp that separate an LSC region of 87,159 bp and an SSC region of 17,811 bp (Figure 1). The total GC content was 37.8%, which was similar to the published Oleaceae cp genomes [12–15]. The GC content of the IR regions was 43.2%, which was higher when compared with the GC content in the LSC and SSC regions (35.8% and 31.8%, respectively).

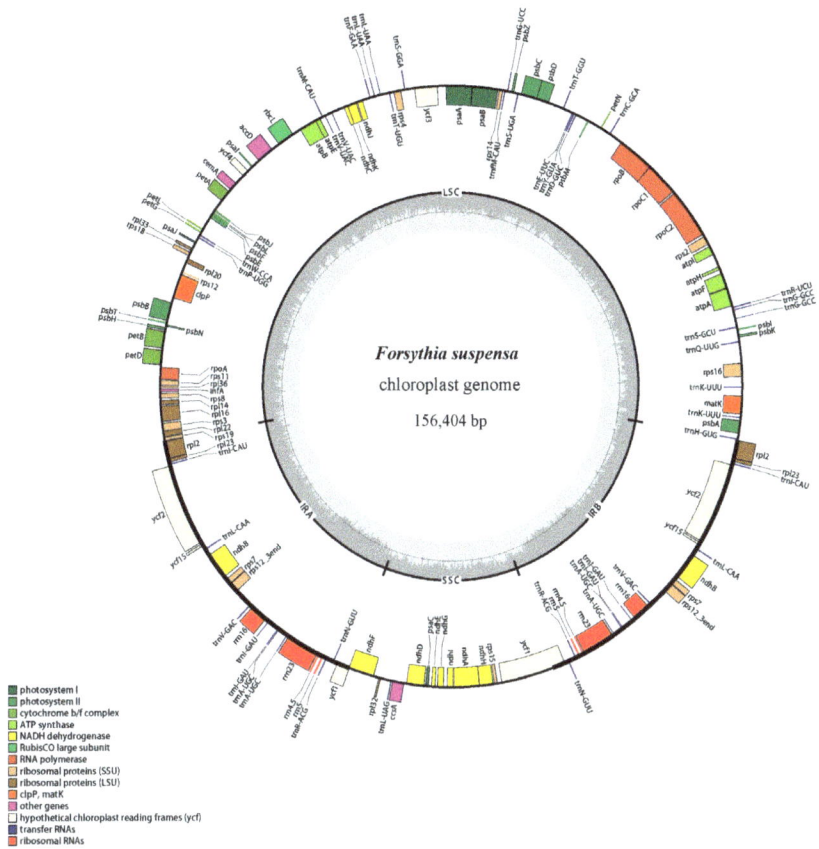

Figure 1. Chloroplast genome map of *Forsythia suspensa*. Genes drawn inside the circle are transcribed clockwise, and those outside are counterclockwise. Genes are color-coded based on their function, which are shown at the left bottom. The inner circle indicates the inverted boundaries and GC content.

The gene content and sequence of the *F. suspensa* cp genome are relatively conserved, with basic characteristics of land plant cp genomes [16]. It encodes a total of 114 unique genes, of which 19 are duplicated in the IR regions. Out of the 114 genes, there are 80 protein-coding genes (70.2%), 30 tRNA (26.3%), and four rRNA genes (*rrn5, rrn4.5, rrn16, rrn23*) (3.5%) (Table 1). Eighteen genes contained introns, fifteen (nine protein-coding and six tRNA genes) of which contained one intron and three of which (*rps12, ycf3,* and *clpP*) contained two introns (Table 2). The *rps12* gene is a trans-spliced gene, three exons of which were located in the LSC region and IR regions, respectively. The complete gene of *matK* was located within the intron of *trnK-UUU*. One pseudogene (non functioning duplications of functional genes), *ycf1*, was identified, located in the boundary regions between IRb/SSC. The partial gene duplication might have caused the lack of protein-coding ability. In general, the junctions between the IR and LSC/SSC regions vary among higher plant cp genomes [17–19]. In the *F. suspensa* cp genome, the *ycf1* gene regions extended into the IR region in the IR/SSC junctions, while the *rpl2* was 51 bp apart from the LSC/IR junction.

Table 1. A list of genes found in the plastid genome of *Forsythia suspensa*.

Category for Genes	Group of Gene	Name of Gene
Photosynthesis related genes	Rubisco	*rbcL*
	Photosystem I	*psaA, psaB, psaC, psaI, psaJ*
	Assembly/stability of photosystem I	*ycf3 *, ycf4*
	Photosystem II	*psbA, psbB, psbC, psbD, psbE, psbF, psbH, psbI, psbJ, psbK, psbL, psbM, psbN, psbT, psbZ*
	ATP synthase	*atpA, atpB, atpE, atpF *, atpH, atpI*
	cytochrome b/f complex	*petA, petB *, petD *, petG, petL, petN*
	cytochrome c synthesis	*ccsA*
	NADPH dehydrogenase	*ndhA *, ndhB *, ndhC, ndhD, ndhE, ndhF, ndhG, ndhH, ndhI, ndhJ*
Transcription and translation related genes	transcription	*rpoA, rpoB, rpoC1 *, rpoC2*
	ribosomal proteins	*rps2, rps3, rps4, rps7, rps8, rps11, rps12 *, rps14, rps15, rps16 *, rps18, rps19, rpl2 *, rpl14, rpl16 *, rpl20, rpl22, rpl23, rpl32, rpl33, rpl36*
	translation initiation factor	*infA*
RNA genes	ribosomal RNA	*rrn5, rrn4.5, rrn16, rrn23*
	transfer RNA	*trnA-UGC *, trnC-GCA, trnD-GUC, trnE-UUC, trnF-GAA, trnG-UCC *, trnG-GCC *, trnH-GUG, trnI-CAU, trnI-GAU *, trnK-UUU *, trnL-CAA, trnL-UAA *, trnL-UAG, trnfM-CAUI, trnM-CAU, trnN-GUU, trnP-UGG, trnQ-UUG, trnR-ACG, trnR-UCU, trnS-GCU, trnS-GGA, trnS-UGA, trnT-GGU, trnT-UGU, trnV-GAC, trnV-UAC *, trnW-CCA, trnY-GUA*
Other genes	RNA processing	*matK*
	carbon metabolism	*cemA*
	fatty acid synthesis	*accD*
	proteolysis	*clpP **
Genes of unknown function	conserved reading frames	*ycf1, ycf2, ycf15, ndhK*

* indicate the intron-containing genes.

Table 2. Genes with introns within the *F. suspensa* chloroplast genome and the length of exons and introns.

Gene	Location	Exon I (bp)	Intron I (bp)	Exon II (bp)	Intron II (bp)	Exon III (bp)
trnA-UGC	IR	38	814	35		
trnG-GCC	LSC	24	676	48		
trnI-GAU	IR	42	942	35		
trnK-UUU	LSC	38	2494	37		
trnL-UAA	LSC	37	473	50		
trnV-UAC	LSC	38	572	37		
rps12 *	LSC	114	-	231	536	27
rps16	LSC	40	864	227		
atpF	LSC	144	705	411		
rpoC1	LSC	445	758	1619		
ycf3	LSC	129	714	228	737	153
clpP	LSC	69	815	291	642	228
petB	LSC	6	707	642		
petD	LSC	8	713	475		
rpl16	LSC	9	865	399		
rpl2	IR	393	664	435		
ndhB	IR	777	679	756		
ndhA	SSC	555	1106	531		

* The *rps12* is a trans-spliced gene with the 5′ end located in the LSC region and the duplicated 3′ end in the IR regions.

2.2. Comparison to Other Lamiales Species

The IR regions are highly conserved and play an important role in stabilizing the cp genome structure [20,21]. For IR and SC boundary regions, their expansion and contraction are commonly considered as the main mechanism behind the length variation of angiosperm cp genomes [22,23]. In this study, we compared the junctions of LSC/IRb/SSC/IRa of the seven Lamiales cp genomes (Figure 2), and also observed the expansions and contractions in IR boundary regions.

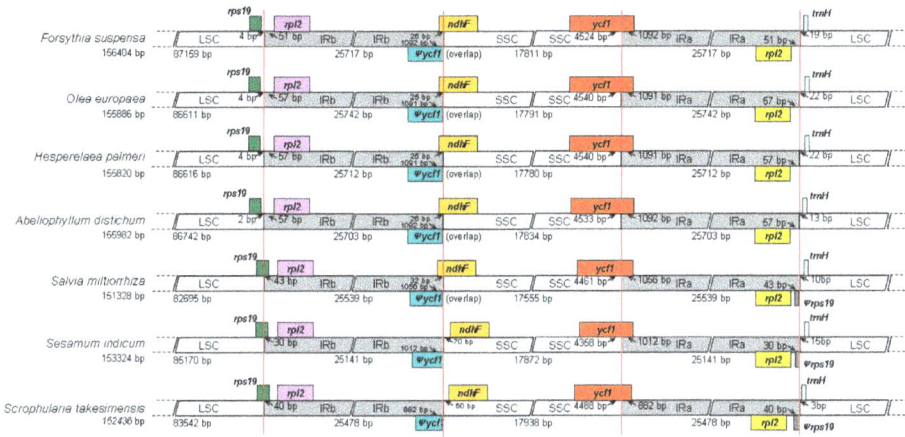

Figure 2. Comparisons of LSC, SSC, and IR region borders among six Lamiales chloroplast genomes. Ψ indicates a pseudogene. Colorcoding mean different genes on both sides of the junctions. Number above the gene features means the distance between the ends of genes and the junction sites. The arrows indicated the location of the distance. This figure is not to scale.

The *rps19* genes of four Oleaceae species were all completely located in the LSC region, and the IR region expanded to the *rps19* gene in the other three genomes, with a short *rps19* pseudogene of 43 bp, 30 bp, and 40 bp created at the IRa/LSC border in *S. miltiorrhiza*, *S. indicum*, and *S. takesimensis*, respectively. The border between the IRb and SSC extended into the *ycf1* genes, with *ycf1* pseudogenes created in all of the seven species. The length of the *ycf1* pseudogene was very similar in four of the Oleaceae species (1091 or 1092 bp), and was longer than that in *S. miltiorrhiza* (1056 bp), *S. indicum* (1012 bp), and *S. takesimensis* (886 bp). Overlaps were detected between the *ycf1* pseudogene and the *ndhF* gene in five cp genomes (except for *S. indicum* and *S. takesimensis*), which also had similar lengths (25 or 26 bp) in four Oleaceae species. The *trnH-GUG* genes were all located in the LSC region, the distance of which from the LSC/IRa boundary was 3–22 bp. Overall, the IR/SC junctions of the Oleaceae species were similar and showed some difference compared to those of Lamiaceae (*S. miltiorrhiza*), Pedaliaceae (*S. indicum*), and Scrophulariaceae (*S. takesimensis*). Our results suggested that the cp genomes of closely related species might be conserved, whereas greater diversity might occur among species belonging to different families, such as one inverted repeat loss in the cp genome of *Astragalus membranaceus* [24] and the large inversions in *Eucommia ulmoides* [25].

2.3. Codon Usage Analysis

The synonymous codons often have different usage frequencies in plant genomes, which was termed codon usage bias. A variety of evolutionary factors which affect gene mutation and selection may lead to the occurrence of codon bias [26,27].

To examine codon usage, the effective number of codons (Nc) of 52 protein-coding genes (PCGs) was calculated. The Nc values for each PCG in *F. suspensa* are shown in Table S2. Our results indicated that the Nc values ranged from 37.83 (*rps14*) to 54.75 (*ycf3*) in all the selected PCGs. Most Nc values were greater than 44, which suggested a weak gene codon bias in the *F. suspensa* cp genome. The *rps14* gene was detected to exist in the most biased codon usage with the lowest mean Nc value of 37.83. Table 3 showed the codon usage and relative synonymous codon usage (RSCU). Due to the RSCU values of >1, thirty codons showed the codon usage bias in the *F. suspensa* cp genes. Interestingly, out of the above 30 codons, twenty-nine were A or T-ending codons. Conversely, the G + C-ending codons exhibited the opposite pattern (RSCU values <1), indicating that they are less common in *F. suspensa* cp genes. Stop codon usage was found to be biased toward TAA. The similar codon usage rules of bias for A- or T-ending were also found in poplar, rice, and other plants [28–30].

Table 3. The relative synonymous codon usage of the *Forsythia suspensa* chloroplast genome.

Amino Acids	Codon	Number	RSCU	AA Frequency	Amino Acids	Codon	Number	RSCU	AA Frequency
Phe	UUU	779	**1.32**	5.59%		UCU	472	**1.76**	
	UUC	405	0.68			UCC	247	0.92	
	UUA	720	**1.93**		Ser	UCA	307	**1.15**	7.59%
	UUG	451	**1.21**			UCG	152	0.57	
Leu	CUU	486	**1.30**	10.56%		AGU	339	**1.26**	
	CUC	129	0.35			AGC	91	0.34	
	CUA	301	0.81			CCU	351	**1.55**	
	CUG	150	0.40			CCC	170	0.75	
	AUU	890	**1.47**		Pro	CCA	269	**1.19**	4.26%
Ile	AUC	377	0.62	8.57%		CCG	113	0.50	
	AUA	548	0.91			ACU	430	**1.63**	
Met	AUG	495	1.00	2.34%	Thr	ACC	201	0.76	4.98%
	GUU	423	**1.48**			ACA	324	**1.23**	
Val	GUC	126	0.44	5.41%		ACG	100	0.38	
	GUA	447	**1.56**			GCU	526	**1.84**	
	GUG	151	0.53		Ala	GCC	177	0.62	5.41%
Tyr	UAU	631	**1.61**	3.70%		GCA	328	**1.14**	
	UAC	152	0.39			GCG	115	0.40	
	UAA	28	**1.62**		Cys	UGU	171	**1.53**	1.05%
TER	UAG	10	0.58	0.25%		UGC	52	0.47	
	UGA	14	0.81			CGU	275	**1.30**	
His	CAU	404	**1.58**	2.42%	Arg	CGC	90	0.42	
	CAC	108	0.42			CGA	284	**1.34**	6.00%
	CAA	595	**1.52**			CGG	97	0.46	
Gln	CAG	186	0.48	3.69%	Arg	AGA	392	**1.85**	
	AAU	796	**1.56**			AGG	133	0.63	
Asn	AAC	224	0.44	4.81%		GGU	493	**1.33**	
Lys	AAA	837	**1.54**	5.15%	Gly	GGC	145	0.39	7.00%
	AAG	253	0.46			GGA	594	**1.60**	
Asp	GAU	690	**1.59**	4.09%		GGG	251	0.68	
	GAC	176	0.41		Glu	GAA	866	**1.54**	5.32%
Trp	UGG	386	1.00	1.82%		GAG	262	0.46	

The value of relative synonymous codon usage (RSCU) > 1 are highlighted in bold.

The factors affecting codon usage may vary in different genes or species. In a relative study, Zhou et al. [30] considered the genomic nucleotide mutation bias as a main cause of codon bias in seed plants such as arabidopsis and poplar. Morton [31] reported that the cp gene codon usage was largely affected by the asymmetric mutation of cp DNA in *Euglena gracilis*. Our result suggested that a low GC content and codon usage bias for A + T-ending may be a major factor in the cp gene codon usage of *F. suspensa*.

The 52 unique PCGs comprised 63,555 bp that encoded 21,185 codons. The amino acid (AA) frequencies of the *F. suspensa* cp genome were further computed. Of these codons, 2237 (10.56%) encode leucine, which was the most frequency used AA in the *F. suspensa* cp genome (Table 3). As the least common one, cysteine was only encoded by 223 (1.05%) codons.

2.4. Repeats and Simple Sequence Repeats Analysis

Repeat sequences in the *F. suspensa* cp genome were analyzed by REPuter and the results showed that there were no complement repeats and reverse repeats. Twenty-six forward repeats and 23 palindrome repeats were detected with lengths ≥30 bp (identity >90%) (Table 4). Out of the 49 repeats, 34 repeats (69.4%) were 30–39 bp long, 11 repeats (22.4%) were 40–49 bp long, four repeats (8.2%) were 50–59 bp long, and the longest repeat was 58 bp. Generally, repeats were mostly distributed in noncoding regions [32,33]; however, 53.1% of the repeats in the *F. suspensa* cp genome were located in coding regions (CDS) (Figure 3A), mainly in *ycf2*; similar to that of *S. dentata* and *S. takesimensis* [34]. Meanwhile, 40.8% of repeats were located in intergenic spacers (IGS) and introns, and 6.1% of repeats were in parts of the IGS and CDS.

Table 4. Repetitive sequences of *Forsythia suspensa* calculated using REPuter.

No.	Size/bp	Type [#]	Repeat 1 Start (Location)	Repeat 2 Start (Location)	Region
1	30	F	10,814 (*trnG-GCC* *)	38,746 (*trnG-UCC*)	LSC
2	30	F	17,447 (*rps2-rpoC2*)	17,448 (*rps2-rpoC*)	LSC
3	30	F	44,547 (*psaA-ycf3*)	44,550 (*psaA-ycf3*)	LSC
4	30	F	45,978 (*ycf3* intron2)	101,338 (*rps12_3end-trnV-GAC*)	LSC, IRa
5	30	F	91,923 (*ycf2*)	91,965 (*ycf2*)	IRa
6	30	F	110,167 (*rrn4.5-rrn5*)	110,198 (*rrn4.5-rrn5*)	IRa
7	30	F	133,335 (*rrn5-rrn4.5*)	133,366 (*rrn5-rrn4.5*)	IRb
8	30	F	149,178 (*ycf2*)	149,214 (*ycf2*)	IRb
9	30	F	149,196 (*ycf2*)	149,214 (*ycf2*)	IRb
10	30	F	151,568 (*ycf2*)	151,610 (*ycf2*)	IRb
11	32	F	9313 (*trnS-GCU* *)	37,781 (*psbC-trnS-UGA* *)	LSC
12	32	F	40,965 (*psaB*)	43,189 (*psaA*)	LSC
13	32	F	53,338 (*ndhC-trnV-UAC*)	53,358 (*ndhC-trnV-UAC*)	LSC
14	32	F	115,350 (*ndhF-rpl32*)	115,378 (*ndhF-rpl32*)	SSC
15	34	F	94,332 (*ycf2*)	94,368 (*ycf2*)	IRa
16	34	F	94,350 (*ycf2*)	94,368 (*ycf2*)	IRa
17	35	F	149,188 (*ycf2*)	149,206 (*ycf2*)	IRb
18	39	F	45,966 (*ycf3* intron2)	101,326 (*rps12_3end-trnV-GAC*)	LSC, IRa
19	39	F	45,966 (*ycf3* intron2)	122,604 (*ndhA* intron1)	LSC, SSC
20	41	F	40,953 (*psaB*)	43,177 (*psaA*)	LSC
21	41	F	101,324 (*rps12_3end-trnV-GAC*)	122,602 (*ndhA* intron)	IRa, SSC
22	42	F	94,320 (*ycf2*)	94,356 (*ycf2*)	IRa
23	42	F	149,165 (*ycf2*)	149,201 (*ycf2*)	IRb
24	44	F	94,340 (*ycf2*)	94,358 (*ycf2*)	IRa
25	58	F	94,332 (*ycf2*)	94,340 (*ycf2*)	IRa
26	58	F	149,165 (*ycf2*)	149,183 (*ycf2*)	IRb
27	30	P	9315 (*trnS-GCU* *)	47,653 (*trnS-GGA*)	LSC
28	30	P	14,359 (*atpF-atpH*)	14,359 (*atpF-atpH*)	LSC
29	30	P	34,338 (*trnT-GGU-psbD*)	34,338 (*trnT-GGU-psbD*)	LSC
30	30	P	37,783 (*psbC-trnS-UGA* *)	47,653 (*trnS-GGA*)	LSC
31	30	P	45,978 (*ycf3* intron2)	142,195 (*trnV-GAC-rps12_3end*)	LSC, IRb
32	30	P	91,923 (*ycf2*)	151,568 (*ycf2*)	IRa, IRb
33	30	P	91,965 (*ycf2*)	151,610 (*ycf2*)	IRa, IRb
34	30	P	110,167 (*rrn4.5-rrn5*)	133,335 (*rrn5-rrn4.5*)	IRa, IRb
35	30	P	110,198 (*rrn4-rrn5*)	133,366 (*rrn5-rrn4.5*)	IRa, IRb
36	30	P	122,764 (*ndhA* intron1)	122,766 (*ndhA* intron1)	SSC
37	34	P	94,332 (*ycf2*)	149,161 (*ycf2*)	IRa, IRb
38	34	P	94,350 (*ycf2*)	149,161 (*ycf2*)	IRa, IRb
39	34	P	94,368 (*ycf2*)	149,179 (*ycf2*)	IRa, IRb
40	34	P	94,368 (*ycf2*)	149,179 (*ycf2*)	IRa, IRb
41	39	P	45,966 (*ycf3* intron2)	45,966 (*ycf3* intron2)	LSC, IRb
42	41	P	122,602 (*ndhA* intron1)	142,198 (*trnV-GAC-rps12_3end*)	SSC, IRb
43	42	P	94,320 (*ycf2*)	149,165 (*ycf2*)	IRa, IRb
44	42	P	94,356 (*ycf2*)	149,201 (*ycf2*)	IRa, IRb
45	44	P	77,475 (*psbT-psbN*)	77,475 (*psbT-psbN*)	LSC
46	44	P	94,340 (*ycf2*)	149,161 (*ycf2*)	IRa, IRb
47	44	P	94,358 (*ycf2*)	149,179 (*ycf2*)	IRa, IRb
48	58	P	94,332 (*ycf2*)	149,165 (*ycf2*)	IRa, IRb
49	58	P	94,340 (*ycf2*)	149,183 (*ycf2*)	IRa, IRb

[#] F: forward; P: palindrome; * part in the gene.

Simple sequence repeats (SSRs) are widely distributed across the entire genome and exert significant influence on genome recombination and rearrangement [35]. As valuable molecular markers, SSRs have been used in polymorphism investigations and population genetics [36,37]. The occurrence, type, and distribution of SSRs were analyzed in the *F. suspensa* cp genome. In total, we detected 54 SSRs in the *F. suspensa* cp genome (Table 5), accounting for 700 bp of the total sequence (0.45%).

The majority of these SSRs consisted of mono- and di-nucleotide repeats, which were found 35 and seven times, respectively. Tri-(1), tetra-(4), and penta-nucleotide repeat sequences (1) were detected with a much lower frequency. Six compound SSRs were also found. Fifty SSRs (92.6%) were composed of A and T nucleotides, while tandem G or C repeats were quite rare, which was in concordance with the other research results [38,39]. Out of these SSRs, 42 (88.9%) and six (11.1%) were located in IGS and introns, respectively (Figure 3B). Only five SSRs were found in the coding genes, including *rpoC2*, *rpoA*, and *ndhD*, and one was located in parts of the IGS and CDS. In addition, we noticed that almost all SSRs were located in LSC, except for (T)19, and no SSRs were detected in the IR region. These SSRs may be developed lineage-specific markers, which might be useful in evolutionary and genetic diversity studies.

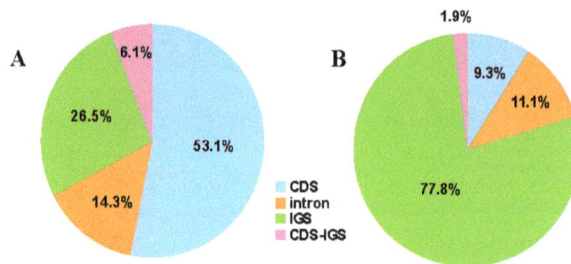

Figure 3. Distribution of repeat sequence and simple sequence repeats (SSRs) within *F. suspensa* chloroplast genomes. (**A**) Distribution of repeats; and (**B**) distribution of SSRs. IGS: intergenic spacer.

Table 5. Distribution of SSR loci in the chloroplast genome of *Forsythia suspensa*.

SSR Type [#]	SSR Sequence	Size	Start	SSR Location	Region
		10	31,855	*psbM-trnD-GUC*	LSC
		10	31,992	*psbM-trnD-GUC*	LSC
	(A)10	10	38,025	*trnS-UGA-psbZ*	LSC
		10	73,886	*clpP* intron1	LSC
		10	85,390	*rpl16* intron	LSC
		10	507	*trnH-GUG-psbA*	LSC
		10	9056	*psbK-psbI*	LSC
	(T)10	10	11,162	*trnR-UCU-atpA*	LSC
		10	59,781	*rbcL-accD*	LSC
		10	66,291	*petA-psbJ*	LSC
		10	69,202	*petL-petG*	LSC
	(C)10	10	5236	*trnK-UUU-rps16*	LSC
		11	19,678	*rpoC2*	LSC
		11	50,871	*trnF-GAA-ndhJ*	LSC
	(T)11	11	61,662	*accD-psaI*	LSC
		11	72,263	*rpl20-clpP*	LSC
p1		11	74,741	*clpP* intron2	LSC
		12	20,216	*rpoC2*	LSC
	(T)12	12	81,254	*rpoA*	LSC
		12	83,666	*rps8-rpl14*	LSC
	(A)13	13	12,741	*atpA-atpF*	LSC
		13	46,877	*ycf3-trnS-GGA*	LSC
		13	14,109	*atpF-atpH*	LSC
	(T)13	13	34,486	*trnT-GGU-psbD*	LSC
		13	37,645	*psbC-trnS-UGA*	LSC
		13	86,860	*rpl22-rps19*	LSC
	(T)14	14	48,630	*rps4-trnT-UGU*	LSC
	(A)15	15	33,163	*trnE-UUC-trnT-GGU*	LSC
	(A)16	16	46,618	*ycf3* intron2	LSC
	(A)19	19	44,559	*psaA-ycf3*	LSC
	(T)19	19	117,928	*ndhD*	SSC
	(A)20	20	29,957	*trnC-GCA-petN*	LSC

Table 5. *Cont.*

SSR Type [#]	SSR Sequence	Size	Start	SSR Location	Region
		10	4646	*trnK-UUU-rps16*	LSC
	(AT)5	10	6558	*rps16-trnQ-UUG*	LSC
		10	21,057	*rpoC2*	LSC
p2	(TA)5	10	69,619	*trnW-CCA-trnP-UGG*	LSC
		12	48,772	*rps4-trnT-UGU*	LSC
	(TA)6	12	49,291	*trnT-UGU-trnL-UAA*	LSC
		12	69,931	*trnP-UGG-psaJ*	LSC
p3	(CCT)4	12	69,371	*petG-trnW-CCA*	LSC
	(AAAG)3	12	73,413	*clpP* intron1	LSC
p4	(TCTT)3	12	31,191	*petN-psbM*	LSC
	(TTTA)3	12	55,102	*trnM-CAU-atpE*	LSC
	(AAAT)4	16	9284	*psbI-trnS-GCU*	LSC
p5	(TCTAT)3	15	9458	*trnS-GCU-trnG-GCC*	LSC
	-	23	17,456	*rps2-rpoC2*	LSC
	-	27	63,589	*ycf4-cemA*	LSC
c	-	33	78,324	*petB* intron	LSC
	-	45	71,570	*rps18-rpl20*	LSC
	-	59	38,501	*psbZ-trnG-UCC*	LSC
	-	90	57,078	*atpB* *	LSC

[#] p1: mono-nucleotide; p2: di-nucleotide; p3: tri-nucleotide; p4: tetra-nucleotide; p5: penta-nucleotide; c: compound;
* part in the gene.

2.5. Predicted RNA Editing Sites in the F. suspensa Chloroplast Genes

In the *F. suspensa* cp genome, we predicted 52 RNA editing sites, which occurred in 21 genes (Table 6). The *ndhB* gene contained the most editing sites (10), and this finding was consistent with other plants such as rice, maize, and tomato [40–42]. Meanwhile, the genes *ndhD* and *rpoB* were predicted to have six editing sites: *matK*, five; ropC2, three; *accD*, *ndhA*, *ndhF*, *ndhG*, and *petB*, two; and one each in *atpA*, *atpF*, *atpI*, *ccsA*, *petG*, *psbE*, *rpl2*, *rpl20*, *rpoA*, *rps2*, and *rps14*. All these editing sites were C-to-U transitions. The editing phenomenon was also commonly found in the chloroplasts and mitochondria of seed plants [43]. The locations of the editing sites in the first, second, and third codons were 14, 38, and 0, respectively. Of the 52 sites, twenty were U_A types, which was similar codon bias to previous studies of RNA editing sites [10,44]. In addition, forty-eight RNA editing events in the *F. suspensa* cp genome led to acid changes for highly hydrophobic residues, such as leucine, isoleucine, valine, tryptophan, and tyrosine. The conversions from serine to leucine were the most frequent transitions. As a form of post-transcriptional regulation of gene expression, the feature has already been revealed by most RNA editing researches [44]. Notably, our results provide additional evidence to support the above conclusion.

Table 6. The predicted RNA editing site in the *Forsythia suspensa* chloroplast genes.

Gene	Codon Position	Amino Acid Position	Codon (Amino Acid) Conversion	Score
accD	794	265	uCg (S) => uUg (L)	0.8
	1403	468	cCu (P) => cUu (L)	1
atpA	914	305	uCa (S) => uUa (L)	1
atpF	92	31	cCa (P) => cUa(L)	0.86
atpI	629	210	uCa (S) => uUa (L)	1
ccsA	71	24	aCu (T) => aUu (I)	1
matK	271	91	Ccu (P) => Ucu (S)	0.86
	460	154	Cac (H) => Uac (Y)	1
	646	216	Cau (H) => Uau (Y)	1
	1180	394	Cgg (R) => Ugg (W)	1
	1249	417	Cau (H) => Uau (Y)	1

Table 6. *Cont.*

Gene	Codon Position	Amino Acid Position	Codon (Amino Acid) Conversion	Score
ndhA	344	115	uCa (S) => uUa (L)	1
	569	190	uCa (S) => uUa (L)	1
ndhB	149	50	uCa (S) => uUa (L)	1
	467	156	cCa (P) => cUa (L)	1
	586	196	Cau (H) => Uau (Y)	1
	611	204	uCa (S) => uUa (L)	0.8
	737	246	cCa (P) => cUa (L)	1
	746	249	uCu (S) => uUu (F)	1
	830	277	uCa (S) => uUa (L)	1
	836	279	uCa (S) => uUa (L)	1
	1292	431	uCc (S) => uUc (F)	1
	1481	494	cCa (P) => cUa (L)	1
ndhD	2	1	aCg (T) => aUg (M)	1
	47	16	uCu (S) => uUu (F)	0.8
	313	105	Cgg (R) => Ugg (W)	0.8
	878	293	uCa (S) => uUa (L)	1
	1298	433	uCa (S) => uUa (L)	0.8
	1310	437	uCa (S) => uUa (L)	0.8
ndhF	290	97	uCa (S) => uUa (L)	1
	671	224	uCa (S) => uUa (L)	1
ndhG	314	105	aCa (T) => aUa (I)	0.8
	385	129	Cca (P) => Uca (S)	0.8
petB	418	140	Cgg (R) => Ugg (W)	1
	611	204	cCa (P) => cUa (L)	1
petG	94	32	Cuu (L) => Uuu (F)	0.86
psbE	214	72	Ccu (P) => Ucu (S)	1
rpl2	596	199	gCg (A) => gUg (V)	0.86
rpl20	308	103	uCa (S) => uUa (L)	0.86
rpoA	830	277	uCa (S) => uUa (L)	1
rpoB	338	113	uCu (S) => uUu (F)	1
	551	184	uCa (S) => uUa (L)	1
	566	189	uCg (S) => uUg (L)	1
	1672	558	Ccc (P) => Ucc (S)	0.86
	2000	667	uCu (S) => uUu (F)	1
	2426	809	uCa (S) => uUa (L)	0.86
rpoC2	1792	598	Cgu (R) => Ugu (C)	0.86
	2305	769	Cgg (R) => Ugg (W)	1
	3746	1249	uCa (S) => uUa (L)	0.86
rps2	248	83	uCa (S) => uUa (L)	1
rps14	80	27	uCa (S) => uUa (L)	1
	149	50	cCa (P) => cUa (L)	1

2.6. Phylogeny Reconstruction of Lamiales Based on Complete Chloroplast Genome Sequences

Complete cp genomes comprise abundant phylogenetic information, which could be applied to phylogenetic studies of angiosperm [11,45,46]. To identify the evolutionary position of *F. suspensa* within Lamiales, an improved resolution of phylogenetic relationships was achieved by using these whole cp genome sequences of 36 Lamiales species. Three species, *C. Arabica*, *I. purpurea*, and *O. nivara* were also chosen as outgroups. The Maximum likelihood (ML) bootstrap values were fairly high, with values ≥98% for 32 of the 36 nodes, and 30 nodes had 100% bootstrap support (Figure 4). *F. suspensa*, whose cp genome was reported in this study, was closely related to *A. distichum*, which then formed a cluster with *H. palmeri*, *J. nudiflorum*, and the Olea species from Oleaceae with 100% bootstrap supports. Notably, Oleaceae diverged relatively early from the Lamiales lineage. In addition, four phylogenetic relationships were only supported by lower ML bootstrap values. This was possibly a result of less samples in these families. The cp genome is also expected to be useful in resolving the deeper branches of the phylogeny, along with the availability of more whole genome sequences.

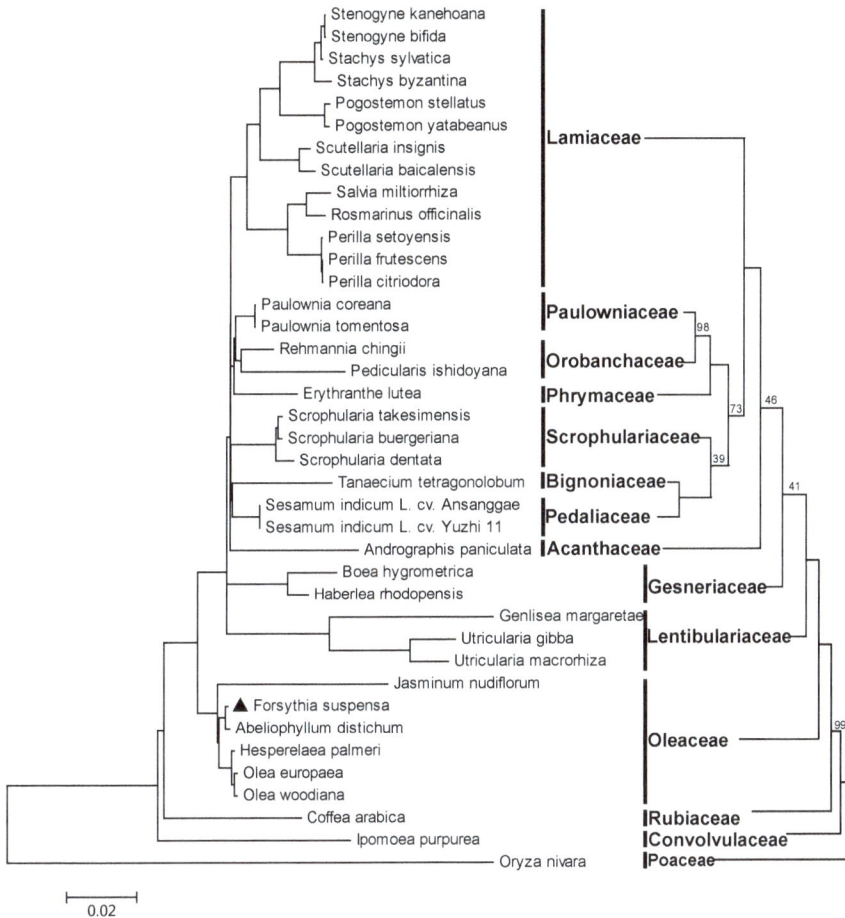

Figure 4. Maximum likelihood phylogeny of the Lamiales species inferred from complete chloroplast genome sequences. Numbers near branches are bootstrap values of 100 pseudo-replicates. The tree on the right panel was constructed manually by reference to the left one, and the distance of branches was meaningless. The branches without numbers indicate 100% bootstrap supports.

3. Materials and Methods

3.1. Plant Materials

Samples of *F. suspensa* were collected in Zezhou County, Shanxi Province, China. The voucher specimens were deposited in the Herbarium of Shanxi Agricultural University, Taigu, China. Additionally, the location of the specimens was not within any protected area.

3.2. DNA Library Preparation, Sequencing, and Genome Assembly

Genomic DNA was extracted from fresh young leaves of the *F. suspensa* plant using the mCTAB method [47]. Genomic DNA was fragmented into 400–600 bp using a Covaris M220 Focused-ultrasonicator (Covaris, Woburn, MA, USA). Library preparation was conducted using NEBNext® Ultra™ DNA Library Prep Kit Illumina (New England, Biolabs, Ipswich, MA, USA). Sample sequencing was carried out on an Illumina Hiseq 4000 PE150 platform.

Next, raw sequence reads were assembled into contigs using SPAdes [48], CLC Genomics Workbench 8 (Available online: http://www.clcbio.com), and SOAPdenovo2 [49], respectively. Chloroplast genome contigs were selected by BLAST (Available online: http://blast.ncbi.nlm.nih. gov/) [50] and were assembled by Sequencher 4.10 (Available online: http://genecodes.com/). All reads were mapped to the cp genome using Geneious 8.1 [51], which verified the selected contigs. The closing of gaps was accomplished by special primer designs, PCR amplification, and Sanger sequencing. Finally, we obtained a high-quality complete *F. suspensa* cp genome, and the result was submitted to NCBI (Accession Number: MF579702).

3.3. Genome Annotation and Comparative Genomics

Chloroplast genome annotation was performed using DOGMA (Dual Organellar GenoMe Annotator) [52] (Available online: http://dogma.ccbb.utexas.edue). Putative protein-coding genes, tRNAs, and rRNAs were identified by BLASTX and BLASTN searches (Available online: http://blast. ncbi.nlm.nih.gov/), respectively. The cp genome was drawn using OrganellarGenomeDRAW [53] (Available online: http://ogdraw.mpimp-golm.mpg.de/index.shtml), with subsequent manual editing. The boundaries between the IR and SC regions of *F. suspensa* and six other Lamiales species were compared and analyzed.

3.4. Repeat Sequence Analyses

The REPuter program [54] (Available online: https://bibiserv.cebitec.uni-bielefeld.de/reputer) was used to identify repeats including forward, reverse, palindrome, and complement sequences. The length and identity of the repeats were limited to $\geq$30 bp and >90%, respectively, with the Hamming distance equal to 3 [55,56]. The cp SSRs were detected using MISA [57] with the minimum repeats of mono-, di-, tri-, tetra-, penta-, and hexanucleotides set to 10, 5, 4, 3, 3, and 3, respectively.

3.5. Codon Usage

To ensure sampling accuracy, only 52 PCGs with a length >300 bp were selected for synonymous codon usage analysis. Two relevant parameters, Nc and RSCU, were calculated using the program CodonW1.4.2 (Available online: http://downloads.fyxm.net/CodonW-76666.html). Nc is often utilized to evaluate the codon bias at the individual gene level, in a range from 20 (extremely biased) to 61 (totally unbiased) [58]. RSCU is the observed frequency of a codon divided by the expected frequency. The values close to 1.0 indicate a lack of bias [59]. AA frequency was also calculated and expressed by the percentage of the codons encoding the same amino acid divided by the total codons.

3.6. Prediction of RNA Editing Sites

Prep-Cp [60] (Available online: http://prep.unl.edu/) and CURE software [61] (Available online: http://bioinfo.au.tsinghua.edu.cn/pure/) were applied to the prediction of RNA editing sites, and the parameter threshold (cutoff value) was set to 0.8 to ensure prediction accuracy.

3.7. Phylogenomic Analyses

ML phylogenetic analyses were performed using the *F. suspensa* complete cp genome and 32 Lamiales plastomes with three species, *Coffea arabica*, *Ipomoea purpurea*, and *Oryza nivara*, as outgroups (Table S1). All of the plastome sequences were aligned using MAFFT program version 7.0 [62] (Available online: http://mafft.cbrc.jp/alignment/server/index.html) and adjusted manually where necessary. These plastome nucleotide alignments were subjected to ML phylogenetic analyses with MEGA7.0 [63] based on the General Time Reversible model. A discrete Gamma distribution was used to model evolutionary rate differences among sites. The branch support was estimated by rapid bootstrap analyses using 100 pseudo-replicates.

4. Conclusions

The cp genome of the medicinal plant *F. suspensa* was reported for the first time in this study and its organization is described and compared with that of other Lamiales species. This genome is 156,404 bp in length, with a similar quadripartite structure and genomic contents common to most land plant genomes. The low GC content of the cp genome might caused the codon usage bias toward A- or T-ending codons. All of the predicted RNA editing sites in the genome were C-to-U transitions. Among several relative species, the genome size and IR expansion or contraction exhibited some differences, and the divergent regions were also analyzed. Repeat sequences and SSRs within *F. suspensa* were analyzed, which may be useful in developing molecular markers for the analyses of infraspecific genetic differentiation within the genus *Forsythia* (Oleaceae). Phylogenetic analysis based on the entire cp genome revealed that *F. suspensa*, as a member of the Oleaceae family, diverged relatively early from Lamiales. Overall, the sequences and annotation of the *F. suspensa* cp genome will facilitate medicinal resource conservation, as well as molecular phylogenetic and genetic engineering research of this species.

Supplementary Materials: Supplementary materials can be found at www.mdpi.com/1422-0067/18/11/2288/s1.

Acknowledgments: This work was supported by the Research Project Supported by the Shanxi Scholarship Council of China (2015-066), the Preferential Research Project Supported by Ministry of Human Resources and Social Security of the People's Republic of China, Doctor Research Grant of Shanxi Agricultural University (XB2009002), the Modernization Base Construction Project of Traditional Chinese Medicine Supported by the Health and Family Planning Commission of Shanxi Province, and Doctor Research Grant of Shanxi Agricultural University (2016ZZ04).

Author Contributions: Wenbin Wang, Huan Yu, Jiahui Wang, Wanjun Lei, Jianhua Gao, and Xiangpo Qiu carried out the experiments; and Wenbin Wang and Jinsheng Wang designed the project and wrote the manuscript.

Conflicts of Interest: The authors declare no conflict of interest.

References

1. Zhao, L.; Yan, X.; Shi, J.; Ren, F.; Liu, L.; Sun, S.; Shan, B. Ethanol extract of *Forsythia suspensa* root induces apoptosis of esophageal carcinoma cells via the mitochondrial apoptotic pathway. *Mol. Med. Rep.* **2015**, *11*, 871–880. [CrossRef] [PubMed]

2. Piao, X.L.; Jang, M.H.; Cui, J.; Piao, X. Lignans from the fruits of *Forsythia suspensa*. *Bioorg. Med. Chem. Lett.* **2008**, *18*, 1980–1984. [CrossRef] [PubMed]

3. Kang, W.; Wang, J.; Zhang, L. α-glucosidase inhibitors from leaves of *Forsythia suspense* in Henan province. *China J. Chin. Mater. Med.* **2010**, *35*, 1156–1159.

4. Bu, Y.; Shi, T.; Meng, M.; Kong, G.; Tian, Y.; Chen, Q.; Yao, X.; Feng, G.; Chen, H.; Lu, Z. A novel screening model for the molecular drug for diabetes and obesity based on tyrosine phosphatase Shp2. *Bioorg. Med. Chem. Lett.* **2011**, *21*, 874–878. [CrossRef] [PubMed]

5. Wicke, S.; Schneeweiss, G.M.; dePamphilis, C.W.; Muller, K.F.; Quandt, D. The evolution of the plastid chromosome in land plants: Gene content, gene order, gene function. *Plant Mol. Biol.* **2011**, *76*, 273–297. [CrossRef] [PubMed]

6. He, Y.; Xiao, H.; Deng, C.; Xiong, L.; Yang, J.; Peng, C. The complete chloroplast genome sequences of the medicinal plant *Pogostemon cablin*. *Int. J. Mol. Sci.* **2016**, *17*, 820. [CrossRef] [PubMed]

7. Shendure, J.; Ji, H. Next-generation DNA sequencing. *Nat. Biotechnol.* **2008**, *26*, 1135–1145. [CrossRef] [PubMed]

8. Shinozaki, K.; Ohme, M.; Tanaka, M.; Wakasugi, T.; Hayashida, N.; Matsubayashi, T.; Zaita, N.; Chunwongse, J.; Obokata, J.; Yamaguchi-Shinozaki, K.; et al. The complete nucleotide sequence of the tobacco chloroplast genome: Its gene organization and expression. *EMBO J.* **1986**, *5*, 2043–2049. [CrossRef] [PubMed]

9. Ohyama, K.; Fukuzawa, H.; Kohchi, T.; Shirai, H.; Sano, T.; Sano, S.; Umesono, K.; Shiki, Y.; Takeuchi, M; Chang, Z.; et al. Chloroplast gene organization deduced from complete sequence of liverwort marchantia polymorpha chloroplast DNA. *Nature* **1986**, 572–574. [CrossRef]

10. Maier, R.M.; Neckermann, K.; Igloi, G.L.; Kossel, H. Complete sequence of the maize chloroplast genome: Gene content, hotspots of divergence and fine tuning of genetic information by transcript editing. *J. Mol. Biol.* **1995**, *251*, 614–628. [CrossRef] [PubMed]

11. Kim, K.; Lee, S.C.; Lee, J.; Yu, Y.; Yang, K.; Choi, B.S.; Koh, H.J.; Waminal, N.E.; Choi, H.I.; Kim, N.H.; et al. Complete chloroplast and ribosomal sequences for 30 accessions elucidate evolution of *Oryza* AA genome species. *Sci. Rep.* **2015**, *5*, 15655. [CrossRef] [PubMed]

12. Besnard, G.; Hernandez, P.; Khadari, B.; Dorado, G.; Savolainen, V. Genomic profiling of plastid DNA variation in the Mediterranean olive tree. *BMC Plant Biol.* **2011**, *11*, 80. [CrossRef] [PubMed]

13. Lee, H.L.; Jansen, R.K.; Chumley, T.W.; Kim, K.J. Gene relocations within chloroplast genomes of *Jasminum* and *Menodora* (Oleaceae) are due to multiple, overlapping inversions. *Mol. Biol. Evol.* **2007**, *24*, 1161–1180. [CrossRef] [PubMed]

14. Zedane, L.; Hong-Wa, C.; Murienne, J.; Jeziorski, C.; Baldwin, B.G.; Besnard, G. Museomics illuminate the history of an extinct, paleoendemic plant lineage (Hesperelaea, Oleaceae) known from an 1875 collection from Guadalupe island, Mexico. *Biol. J. Linn. Soc.* **2015**, *117*, 44–57. [CrossRef]

15. Kim, H.-W.; Lee, H.-L.; Lee, D.-K.; Kim, K.-J. Complete plastid genome sequences of *abeliophyllum distichum nakai* (oleaceae), a Korea endemic genus. *Mitochondrial DNA Part B* **2016**, *1*, 596–598. [CrossRef]

16. Sugiura, M. The chloroplast genome. *Plant Mol. Biol.* **1992**, *19*, 149–168. [CrossRef] [PubMed]

17. Zhang, Y.J.; Ma, P.F.; Li, D.Z. High-throughput sequencing of six bamboo chloroplast genomes: Phylogenetic implications for temperate woody bamboos (poaceae: Bambusoideae). *PLoS ONE* **2011**, *6*, e20596. [CrossRef] [PubMed]

18. Xu, Q.; Xiong, G.; Li, P.; He, F.; Huang, Y.; Wang, K.; Li, Z.; Hua, J. Analysis of complete nucleotide sequences of 12 *Gossypium chloroplast* genomes: Origin and evolution of allotetraploids. *PLoS ONE* **2012**, *7*, e37128. [CrossRef] [PubMed]

19. Wang, S.; Shi, C.; Gao, L.Z. Plastid genome sequence of a wild woody oil species, prinsepia utilis, provides insights into evolutionary and mutational patterns of rosaceae chloroplast genomes. *PLoS ONE* **2013**, *8*, e73946. [CrossRef] [PubMed]

20. Marechal, A.; Brisson, N. Recombination and the maintenance of plant organelle genome stability. *New Phytol.* **2010**, *186*, 299–317. [CrossRef] [PubMed]

21. Fu, J.; Liu, H.; Hu, J.; Liang, Y.; Liang, J.; Wuyun, T.; Tan, X. Five complete chloroplast genome sequences from diospyros: Genome organization and comparative analysis. *PLoS ONE* **2016**, *11*, e0159566. [CrossRef] [PubMed]

22. Chumley, T.W.; Palmer, J.D.; Mower, J.P.; Fourcade, H.M.; Calie, P.J.; Boore, J.L.; Jansen, R.K. The complete chloroplast genome sequence of pelargonium x hortorum: Organization and evolution of the largest and most highly rearranged chloroplast genome of land plants. *Mol. Biol. Evol.* **2006**, *23*, 2175–2190. [CrossRef] [PubMed]

23. Yang, M.; Zhang, X.; Liu, G.; Yin, Y.; Chen, K.; Yun, Q.; Zhao, D.; Al-Mssallem, I.S.; Yu, J. The complete chloroplast genome sequence of date palm (*Phoenix dactylifera* L.). *PLoS ONE* **2010**, *5*, e12762. [CrossRef] [PubMed]

24. Lei, W.; Ni, D.; Wang, Y.; Shao, J.; Wang, X.; Yang, D.; Wang, J.; Chen, H.; Liu, C. Intraspecific and heteroplasmic variations, gene losses and inversions in the chloroplast genome of *Astragalus membranaceus*. *Sci. Rep.* **2016**, *6*, 21669. [CrossRef] [PubMed]

25. Wang, L.; Wuyun, T.N.; Du, H.; Wang, D.; Cao, D. Complete chloroplast genome sequences of *Eucommia ulmoides*: Genome structure and evolution. *Tree Genet. Genomes* **2016**, *12*, 12. [CrossRef]

26. Ermolaeva, M.D. Synonymous codon usage in bacteria. *Curr. Issues Mol. Biol.* **2001**, *3*, 91–97. [PubMed]

27. Wong, G.K.; Wang, J.; Tao, L.; Tan, J.; Zhang, J.; Passey, D.A.; Yu, J. Compositional gradients in gramineae genes. *Genome Res.* **2002**, *12*, 851–856. [CrossRef] [PubMed]

28. Liu, Q.; Xue, Q. Codon usage in the chloroplast genome of rice (*Oryza sativa* L. ssp. japonica). *Acta Agron. Sin.* **2004**, *30*, 1220–1224.

29. Zhou, M.; Long, W.; Li, X. Analysis of synonymous codon usage in chloroplast genome of *Populus alba*. *For. Res.* **2008**, *19*, 293–297. [CrossRef]

30. Zhou, M.; Long, W.; Li, X. Patterns of synonymous codon usage bias in chloroplast genomes of seed plants. *For. Sci. Pract.* **2008**, *10*, 235–242. [CrossRef]

31. Morton, B.R. The role of context-dependent mutations in generating compositional and codon usage bias in grass chloroplast DNA. *J. Mol. Evol.* **2003**, *56*, 616–629. [CrossRef] [PubMed]

32. Nazareno, A.G.; Carlsen, M.; Lohmann, L.G. Complete chloroplast genome of tanaecium tetragonolobum: The first bignoniaceae plastome. *PLoS ONE* **2015**, *10*, e0129930. [CrossRef] [PubMed]

33. Yao, X.; Tang, P.; Li, Z.; Li, D.; Liu, Y.; Huang, H. The first complete chloroplast genome sequences in actinidiaceae: Genome structure and comparative analysis. *PLoS ONE* **2015**, *10*, e0129347. [CrossRef] [PubMed]

34. Ni, L.; Zhao, Z.; Dorje, G.; Ma, M. The complete chloroplast genome of ye-xing-ba (*Scrophularia dentata*; *Scrophulariaceae*), an alpine Tibetan herb. *PLoS ONE* **2016**, *11*, e0158488. [CrossRef] [PubMed]

35. Cavalier-Smith, T. Chloroplast evolution: Secondary symbiogenesis and multiple losses. *Curr. Biol.* **2002**, *12*, R62–R64. [CrossRef]

36. Xue, J.; Wang, S.; Zhou, S.L. Polymorphic chloroplast microsatellite loci in nelumbo (nelumbonaceae). *Am. J. Bot.* **2012**, *99*, e240–e244. [CrossRef] [PubMed]

37. Hu, J.; Gui, S.; Zhu, Z.; Wang, X.; Ke, W.; Ding, Y. Genome-wide identification of SSR and snp markers based on whole-genome re-sequencing of a Thailand wild sacred lotus (nelumbo nucifera). *PLoS ONE* **2015**, *10*, e0143765. [CrossRef] [PubMed]

38. Qian, J.; Song, J.; Gao, H.; Zhu, Y.; Xu, J.; Pang, X.; Yao, H.; Sun, C.; Li, X.; Li, C.; et al. The complete chloroplast genome sequence of the medicinal plant salvia miltiorrhiza. *PLoS ONE* **2013**, *8*, e57607. [CrossRef] [PubMed]

39. Kuang, D.Y.; Wu, H.; Wang, Y.L.; Gao, L.M.; Zhang, S.Z.; Lu, L. Complete chloroplast genome sequence of magnolia kwangsiensis (magnoliaceae): Implication for DNA barcoding and population genetics. *Genome* **2011**, *54*, 663–673. [CrossRef] [PubMed]

40. Freyer, R.; López, C.; Maier, R.M.; Martín, M.; Sabater, B.; Kössel, H. Editing of the chloroplast ndhb encoded transcript shows divergence between closely related members of the grass family (poaceae). *Plant Mol. Biol.* **1995**, *29*, 679–684. [CrossRef] [PubMed]

41. Kahlau, S.; Aspinall, S.; Gray, J.C.; Bock, R. Sequence of the tomato chloroplast DNA and evolutionary comparison of solanaceous plastid genomes. *J. Mol. Evol.* **2006**, *63*, 194–207. [CrossRef] [PubMed]

42. Chateigner Boutin, A.L.; Small, I. A rapid high-throughput method for the detection and quantification of RNA editing based on high-resolution melting of amplicons. *Nucleic Acids Res.* **2007**, *35*, e114. [CrossRef] [PubMed]

43. Bock, R. Sense from nonsense: How the genetic information of chloroplasts is altered by RNA editing. *Biochimie* **2000**, *82*, 549–557. [CrossRef]

44. Jiang, Y.; Yun, H.E.; Fan, S.L.; Jia-Ning, Y.U.; Song, M.Z. The identification and analysis of RNA editing sites of 10 chloroplast protein-coding genes from virescent mutant of *Gossypium Hirsutum*. *Cotton Sci.* **2011**, *23*, 3–9.

45. Jansen, R.K.; Cai, Z.; Raubeson, L.A.; Daniell, H.; Depamphilis, C.W.; Leebens-Mack, J.; Muller, K.F.; Guisinger-Bellian, M.; Haberle, R.C.; Hansen, A.K.; et al. Analysis of 81 genes from 64 plastid genomes resolves relationships in angiosperms and identifies genome-scale evolutionary patterns. *Proc. Natl. Acad. Sci. USA* **2007**, *104*, 19369–19374. [CrossRef] [PubMed]

46. Huang, H.; Shi, C.; Liu, Y.; Mao, S.Y.; Gao, L.Z. Thirteen camellia chloroplast genome sequences determined by high-throughput sequencing: Genome structure and phylogenetic relationships. *BMC Evol. Biol.* **2014**, *14*, 151. [CrossRef] [PubMed]

47. Li, J.; Wang, S.; Yu, J.; Wang, L.; Zhou, S. A modified ctab protocol for plant DNA extraction. *Chin. Bull. Bot.* **2013**, *48*, 72–78.

48. Bankevich, A.; Nurk, S.; Antipov, D.; Gurevich, A.A.; Dvorkin, M.; Kulikov, A.S.; Lesin, V.M.; Nikolenko, S.I.; Pham, S.; Prjibelski, A.D.; et al. Spades: A new genome assembly algorithm and its applications to single-cell sequencing. *J. Comput. Biol.* **2012**, *19*, 455–477. [CrossRef] [PubMed]

49. Luo, R.; Liu, B.; Xie, Y.; Li, Z.; Huang, W.; Yuan, J.; He, G.; Chen, Y.; Pan, Q.; Liu, Y.; et al. SOAPdenovo2: An empirically improved memory-efficient short-read de novo assembler. *GigaScience* **2012**, *1*, 18. [CrossRef] [PubMed]

50. Altschul, S.F.; Madden, T.L.; Schaffer, A.A.; Zhang, J.; Zhang, Z.; Miller, W.; Lipman, D.J. Gapped blast and psi-blast: A new generation of protein database search programs. *Nucleic Acids Res.* **1997**, *25*, 3389–3402. [CrossRef] [PubMed]

51. Kearse, M.; Moir, R.; Wilson, A.; Stones-Havas, S.; Cheung, M.; Sturrock, S.; Buxton, S.; Cooper, A.; Markowitz, S.; Duran, C.; et al. Geneious basic: An integrated and extendable desktop software platform for the organization and analysis of sequence data. *Bioinformatics* **2012**, *28*, 1647–1649. [CrossRef] [PubMed]

52. Wyman, S.K.; Jansen, R.K.; Boore, J.L. Automatic annotation of organellar genomes with DOGMA. *Bioinformatics* **2004**, *20*, 3252–3255. [CrossRef] [PubMed]

53. Lohse, M.; Drechsel, O.; Kahlau, S.; Bock, R. Organellargenomedraw—A suite of tools for generating physical maps of plastid and mitochondrial genomes and visualizing expression data sets. *Nucleic Acids Res.* **2013**, *41*, W575–W581. [CrossRef] [PubMed]

54. Kurtz, S.; Choudhuri, J.V.; Ohlebusch, E.; Schleiermacher, C.; Stoye, J.; Giegerich, R. Reputer: The manifold applications of repeat analysis on a genomic scale. *Nucleic Acids Res.* **2001**, *29*, 4633–4642. [CrossRef] [PubMed]

55. Vieira Ldo, N.; Faoro, H.; Rogalski, M.; Fraga, H.P.; Cardoso, R.L.; de Souza, E.M.; de Oliveira Pedrosa, F.; Nodari, R.O.; Guerra, M.P. The complete chloroplast genome sequence of *Podocarpus Lambertii*: Genome structure, evolutionary aspects, gene content and SSR detection. *PLoS ONE* **2014**, *9*, e90618. [CrossRef] [PubMed]

56. Chen, J.; Hao, Z.; Xu, H.; Yang, L.; Liu, G.; Sheng, Y.; Zheng, C.; Zheng, W.; Cheng, T.; Shi, J. The complete chloroplast genome sequence of the relict woody plant *Metasequoia glyptostroboides* Hu et Cheng. *Front. Plant Sci.* **2015**, *6*, 447. [CrossRef] [PubMed]

57. Thiel, T.; Michalek, W.; Varshney, R.K.; Graner, A. Exploiting EST databases for the development and characterization of gene-derived SSR-markers in barley (*Hordeum vulgare* L.). *TAG. Theor. Appl. Genet.* **2003**, *106*, 411–422. [CrossRef] [PubMed]

58. Wright, F. The effective number of codons used in a gene. *Gene* **1990**, *87*, 23–29. [CrossRef]

59. Sharp, P.M.; Tuohy, T.M.; Mosurski, K.R. Codon usage in yeast: Cluster analysis clearly differentiates highly and lowly expressed genes. *Nucleic Acids Res.* **1986**, *14*, 5125–5143. [CrossRef] [PubMed]

60. Mower, J.P. The prep suite: Predictive RNA editors for plant mitochondrial genes, chloroplast genes and user-defined alignments. *Nucleic Acids Res.* **2009**, *37*, W253–W259. [CrossRef] [PubMed]

61. Du, P.; Jia, L.; Li, Y. CURE-chloroplast: A chloroplast C-to-U RNA editing predictor for seed plants. *BMC Bioinform.* **2009**, *10*, 135. [CrossRef] [PubMed]

62. Katoh, K.; Standley, D. Mafft multiple sequence alignment software version 7: Improvements in performance and usability. *Mol. Biol. Evol.* **2013**, *30*, 772–780. [CrossRef] [PubMed]

63. Kumar, S.; Stecher, G.; Tamura, K. Mega7: Molecular evolutionary genetics analysis version 7.0 for bigger datasets. *Mol. Biol. Evol.* **2016**, *33*, 1870–1874. [CrossRef] [PubMed]

International Journal of
Molecular Sciences

MDPI

Article

Whole-Genome Comparison Reveals Heterogeneous Divergence and Mutation Hotspots in Chloroplast Genome of *Eucommia ulmoides* Oliver

Wencai Wang [1], Siyun Chen [2] and Xianzhi Zhang [3],*

[1] Institute of Clinical Pharmacology, Guangzhou University of Chinese Medicine, Guangzhou 510000, China; wencaiwang@gzucm.edu.cn
[2] Germplasm Bank of Wild Species, Kunming Institute of Botany, Chinese Academy of Sciences, Kunming 650201, China; chensiyun@mail.kib.ac.cn
[3] College of Forestry, Northwest A&F University, Yangling 712100, China
* Correspondence: zhangxianzhi@nwsuaf.edu.cn; Tel.: +86-29-8708-2230

Received: 1 March 2018; Accepted: 25 March 2018; Published: 30 March 2018

Abstract: *Eucommia ulmoides* (*E. ulmoides*), the sole species of Eucommiaceae with high importance of medicinal and industrial values, is a Tertiary relic plant that is endemic to China. However, the population genetics study of *E. ulmoides* lags far behind largely due to the scarcity of genomic data. In this study, one complete chloroplast (cp) genome of *E. ulmoides* was generated via the genome skimming approach and compared to another available *E. ulmoides* cp genome comprehensively at the genome scale. We found that the structure of the cp genome in *E. ulmoides* was highly consistent with genome size variation which might result from DNA repeat variations in the two *E. ulmoides* cp genomes. Heterogeneous sequence divergence patterns were revealed in different regions of the *E. ulmoides* cp genomes, with most (59 out of 75) of the detected SNPs (single nucleotide polymorphisms) located in the gene regions, whereas most (50 out of 80) of the indels (insertions/deletions) were distributed in the intergenic spacers. In addition, we also found that all the 40 putative coding-region-located SNPs were synonymous mutations. A total of 71 polymorphic cpDNA fragments were further identified, among which 20 loci were selected as potential molecular markers for subsequent population genetics studies of *E. ulmoides*. Moreover, eight polymorphic cpSSR loci were also developed. The sister relationship between *E. ulmoides* and *Aucuba japonica* in Garryales was also confirmed based on the cp phylogenomic analyses. Overall, this study will shed new light on the conservation genomics of this endangered plant in the future.

Keywords: *Eucommia ulmoides*; chloroplast genome; heterogeneous divergence; mutation hotspots; whole-genome comparison

---

1. Introduction

There are profuse paleoendemics (e.g., Eucommiaceae) and/or phylogenetically primitive taxa (e.g., Cercidiphyllaceae) in China due to the glaciation refuge role played during the Quaternary period [1,2]. Unfortunately, up to *circa* 5000 flora species are currently endangered in China, some of which have already become extinct [3]. Many plant species with important medicinal values have also been threatened seriously due to the increasing demand for raw materials of medicines, over-harvesting and habitat-loss [4–6]. Conservation of medicinal plants has become one of the most urgent issues faced today in China.

Eucommia ulmoides Oliver, a dioecious woody plant endemic to China, is the sole species in the family Eucommiaceae [7]. *E. ulmoides* has been widely cultivated and used as a herbal drug to reduce blood pressure and strengthen the body in central and southern China for at least 2000 years [8,9]. *E. ulmoides* is also well-known as a "hardy rubber" tree that produces trans-polyisoprene rubber (i.e., gutta or Eu-rubber) in the leaves, bark, and pericarp [10,11]. It has been shown that *Eucommia* fossils occurred widely across the Northern Hemisphere from the Palaeocene onwards [12], which indicates that *E. ulmoides* is a representative model of Tertiary relict species, i.e., living from Tertiary to present. However, *E. ulmoides* may have been extinct in the wild and already listed in the Red List of Endangered Plant Species in China probably due to exhaustive human exploration [13,14]. Therefore, effective strategies are urgently needed to conserve this rare and endangered medicinal plant.

To date, studies on *E. ulmoides* have mainly focused on the morphological variation and the natural products [8,15]. Molecular and population genetics studies of this valuable tree lag behind largely due to limited DNA sequence resources [16,17]. Recently, nuclear microsatellites (nrSSR) were developed to investigate the genetic diversity of *E. ulmoides* [18,19]. Amplified fragment length polymorphism (AFLP) and sequence-related amplified polymorphism (SRAP) have been used to construct genetic maps of *E. ulmoides* [20,21]. The genetic markers of random amplified polymorphic DNA (RAPD), chloroplast microsatellite (cpSSR) and inter-simple sequence repeat (ISSR) have also been uncovered [22–24]. Nevertheless, the variability of these developed fingerprinting markers in *E. ulmoides* is relatively low, with limited population genetics information. A new and promising marker type i.e., Single Nucleotide Polymorphism (SNP) has gained high popularity during the last two decades [25,26]. With the on-going progress of high throughput sequencing techniques, it has become convenient to collect large-scale SNP data for genetic analyses [27,28]. Using SNP markers in conservation genetics studies of endangered plants has attracted much attention; for instance, in *Pinus ponderosa* Douglas ex Lawson [29], and *Sciadopitys verticillata* (Thunb.) Siebold and Zucc [30].

Chloroplast (cp) DNA sequences have been extensively used in the studies of plant population genetics and molecular phylogenetics [31–33]. Typically, cp genomes of land plants have a quadripartite structure with a pair of inverted repeats (IRs) separating a large single-copy (LSC) region and a small single-copy (SSC) region, ranging from 115 to 165 kilobase (kb) [34]. The cp genomes in general are inherited uniparentally, mostly maternally and are essentially recombination-free, leading to a smaller effective population size and a shorter coalescent time than the nuclear genomes [35]. Recently Wang et al. [17] reported a cp genome sequence of *E. ulmoides* with a length of 163,341 bp. Clearly, the availability of additional sequenced cp genomes from *E. ulmoides* would aid our understanding of the cp genome-wide variation at the individual level. Through comparative genomic analysis, polymorphic cpDNA loci with plentiful SNPs and indels i.e., nucleotide insertions and deletions can also be detected, which would be useful for further population genetics studies of *E. ulmoides*.

Genome skimming is currently one of the most economical techniques to obtain plastome sequences [36], through which obtaining complete cp genomes for plant phylogenomics inference becomes convenient [37]. In this study, we generated and characterized one complete cp genome of *E. ulmoides* using the genome skimming approach. By comparing the cp genome generated in this study and the one published previously [17], our main goals were to: (1) test whether the cp genomes in *E. ulmoides* show structural rearrangements; (2) reveal the divergence pattern of the cp genome in *E. ulmoides*; (3) identify highly variable cp genome-wide markers for subsequent population genetics studies of *E. ulmoides*.

2. Results

2.1. Chloroplast Genome Variation in E. ulmoides

About 20 million clean reads (4.72 Gb data) were generated from genome skimming sequencing. Two assembly methods (CLC Genomics Workbench and SPAdes software) both obtained the complete cp genome of *E. ulmoides* with high genome coverage (>180×) and there is no difference between the two assembled sequences, suggesting a high-quality cp genome map was achieved. The final cp genome size was determined to be 163,586 bp, similar to the previously published one (KU204775) (Table 1). The number of protein-coding genes, tRNA genes and rRNA genes, were the same as those in the available *E. ulmoides* cp genome (KU204775). We have deposited the newly sequenced *E. ulmoides* cp genome in GenBank with accession number MF766010. The genome skimming sequencing reads have also been deposited in the Sequence Read Archive (SRA) with the accession number PRJNA399774.

Table 1. Comparison between the newly and previously sequenced chloroplast genomes of *Eucommia ulmoides*.

Item	This Study	KU204775
Chloroplast genome size (bp)	163,586	163,341
LSC [a] length (bp)	86,764	86,592
SSC [b] length (bp)	14,166	14,149
IRa/IRb [c] length (bp)	31,328	31,300
Number of genes (unique genes)	136 (115)	136 (115)
Number of protein-coding genes (unique genes)	89 (80)	89 (80)
Number of tRNA genes (unique genes)	39 (31)	39 (31)
Number of rRNA genes (unique genes)	8 (4)	8 (4)
GC [d] content (%)	38.33%	38.34%
Protein-coding regions (%)	51.91%	51.99%

[a] LSC, large single-copy region; [b] SSC, small single-copy region; [c] IRa/IRb, two identical inverted repeat regions a/b; [d] GC, Guanine and Cytosine.

The whole genome alignments from MAFFT (Figure 1A) and MAUVE (Figure 1B) were consistent. There were no large genome rearrangements in the two cp genomes, which indicated that the cp genome structure in *E. ulmoides* is highly conserved and perfectly syntenic (Figure 1). Interestingly, small-scale nucleotide insertions and deletions were detected in the *E. ulmoides* cp genome. We found 15 insertions with more than ten nucleotides (11–111 bp) in the two cp genomes (Table 2). Five deletions in the range of 16–90 bp were also uncovered. It is worth noting that all of these insertions and deletions were involved in repeat sequence expansions and contractions (Table 2). Furthermore, across the entire cp genome of *E. ulmoides*, the sequence divergences were not uniform but highly heterogeneous (Figure 1).

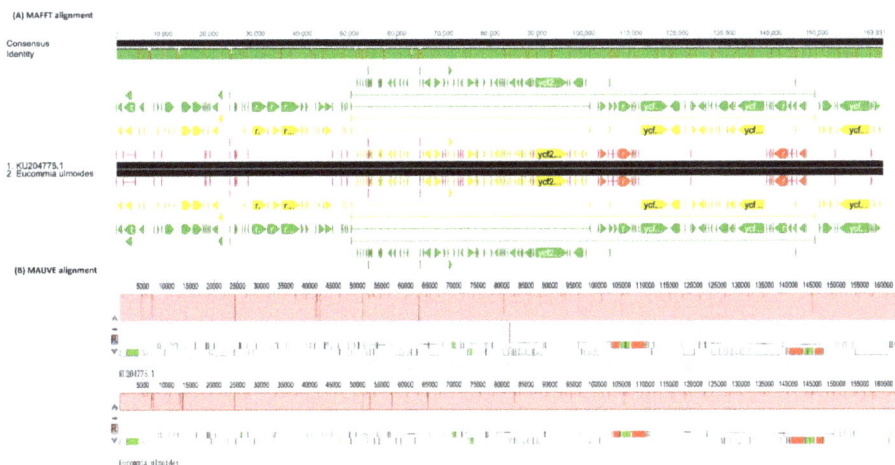

Figure 1. Conserved chloroplast genome structure in *Eucommia ulmoides*. (**A**) Pairwise chloroplast genome alignments derived from Multiple Alignment using Fast Fourier Transform (MAFFT) program. The sequence identity is indicated on the top. Label KU204775.1 represents the *E. ulmoides* chloroplast genome retrieved from GenBank, while label *E. ulmoides* indicates the newly sequenced genome in this study. (**B**) Pairwise chloroplast genome alignments derived from MAUVE software.

Table 2. DNA insertions and deletions with more than 10 nucleotides in the chloroplast genomes of *Eucommia ulmoides*.

No.	Size (bp)	Start Position	Location	Type
1	56	6851	*rps16-trnT(UGU)*	insertion
2	27	7006	*rps16-trnT(UGU)*	insertion
3	45	7196	*rps16-trnT(UGU)*	insertion
4	13	12,693	*ycf3-psaA*	insertion
5	23	12,912	*ycf3-psaA*	insertion
6	111	13,312	*ycf3-psaA*	insertion
7	12	13,471	*ycf3-psaA*	insertion
8	32	24,279	*psbD-trnT(GGU)*	insertion
9	12	26,615	*trnD(GUC)-psbM*	insertion
10	11	51,194	*rps12-rpl20*	insertion
11	17	52,547	*rps18-rpl33*	insertion
12	40	57,075	*psbJ-petA*	insertion
13	18	64,506	*accD-trnM(CAU)*	insertion
14	12	73,717	*trnL(UAA) intron*	insertion
15	14	127,486	*ndhG-ndhI*	insertion
16	16	4673	*trnK(UUU)-rps16*	deletion
17	44	24,700	*trnG(TTU)-trnE(UUC)*	deletion
18	31	41,767	*atpI-atpH*	deletion
19	44	51,109	*rps12-rpl20*	deletion
20	90	62,865	*accD-trnM(CAU)*	deletion

2.2. Molecular Marker Development

A total of 155 mutational events, including 75 nucleotide substitutions (SNPs) and 80 nucleotide indels (insertions and deletions), were detected within 71 loci of the cp genome in *E. ulmoides* (Figure 2). There were 98 mutations (51 SNPs and 47 indels) and 15 mutations (12 SNPs and 3 indels) in the LSC and SSC regions, respectively. In addition, 42 mutations (12 SNPs and 30 indels) were located in the IR region. Distribution patterns of SNPs and indels differed largely in the cp genic and intergenic regions of *E. ulmoides*. Most of the SNPs (59 out of 75) were found in the gene sequences, including 31 protein-coding genes and one tRNA gene. In contrast, indels were mainly (50 out of 80) distributed in the intergenic spacers (Figure 2). Upon further investigation of SNPs and indels in the nine intron-containing protein-coding genes (*atpF*, *ndhA*, *ndhB*, *rpl2*, *rpl16*, *rpoC1*, *rps12*, *rps16*, *ycf3*), we found that all the mutations were located in the intron regions. In all, 40 SNPs and 14 indels occurred in the plastid-coding sequence (CDS) regions.

The proportion of variability in the 71 polymorphic loci ranged from 0.03% to 1.55% with a mean value of 0.37% (Figure 3). The mutation rates in most (53 out of 71) of the loci were between 0.10 and 1.00%. Five of these DNA fragments i.e., *atpF-atpA*, *rps18-rpl33*, *psaJ*, *infA* and *rpl32* had variations exceeding 1.00%. Considering the relatively high percentage of variability and convenience for primer design in PCR (Polymerase Chain Reaction) and sequencing experiments, we chose 20 highly variable loci with length of 200–1500 bp as potential molecular markers for subsequent population genetic studies (Table 3). The percentage of variations in these 20 loci all exceeded 0.25%, among which 16 had a percentage of variable characters (VCs) greater than 0.30% (Table 3).

Through SSR analysis, we found a total of 31 SSR loci in the newly assembled *E. ulmoides* cp genome, among which 27 were shared by the two genomes. Further detection revealed that eight cpSSR loci were polymorphic in *E. ulmoides* (Table 4). All the polymorphic cpSSR loci were mononucleotide repeats, ranging from 10–15 bp in length. Five polymorphic cpSSR loci were located in the LSC region, with another three ones in the IR regions (Table 4).

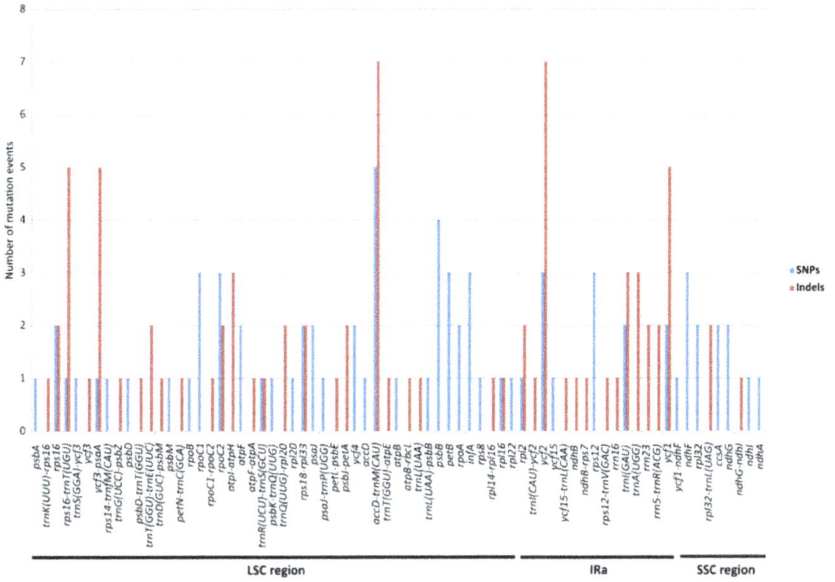

Figure 2. Mutational events (SNPs and indels) detected across the chloroplast genome of *Eucommia ulmoides*. SNPs (single nucleotide polymorphisms) indicate nucleotide substitutions and indels represent nucleotide insertions and deletions. The homologous loci are oriented according to their locations in the chloroplast genome.

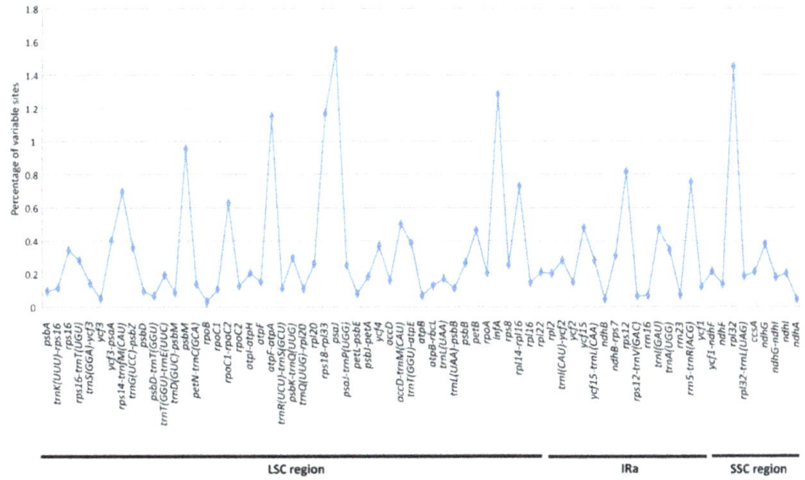

Figure 3. Percentage of variable characters (SNPs and indels) in polymorphic chloroplast loci in *Eucommia ulmoides*. The homologous loci are oriented according to their locations in the chloroplast genome.

Table 3. The 20 chloroplast DNA fragments with relative high genetic divergences identified in *Eucommia ulmoides*.

Region	Aligned Length (bp)	No. VCs [a]	Percentage of VCs (%)
infA	234	3	1.28
rps18-rpl33	343	4	1.17
rps12	369	3	0.81
rrn5-trnR(ACG)	266	2	0.75
ycf15	210	1	0.48
trnI(GAU)	1062	5	0.47
petB	651	3	0.46
ycf3-psaA	1502	6	0.40
trnT(GGU)-atpE	261	1	0.38
ndhG	531	2	0.38
ycf4	546	2	0.37
trnG(UCC)-psbZ	280	1	0.36
rps16	1170	4	0.34
trnA(UGG)	881	3	0.34
ndhB-rps7	325	1	0.31
psbK-trnQ(UUG)	338	1	0.30
trnI(CAU)-ycf2	357	1	0.28
ycf15-trnL(CAA)	359	1	0.28
psbB	1521	4	0.27
rpl20	384	1	0.26

[a] VCs: variable characters, including SNPs and indels.

Table 4. The polymorphic chloroplast SSRs identified in *Eucommia ulmoides*.

No.	SSR Repeat Motif	Length Variation (bp)	Location	Region [a]
1	(G)	10–11	*trnG(UCC)-psbZ*	LSC
2	(A)	12–15	*rpoC2*	LSC
3	(A)	12–13	*ycf1*	IRb
4	(A)	13–14	*rpl32-trnL(UAG)*	IRa
5	(T)	10–14	*psbJ-petA*	LSC
6	(T)	10–11	*trnG(GCC)-trnS(GCU)*	LSC
7	(T)	12–13	*ycf1*	IRa
8	(T)	14–15	*rpl16-rps3*	LSC

[a] LSC, large single-copy region; IRa/IRb, two identical inverted repeat regions a/b.

2.3. SNP Calling and Phylogenomic Inference

The SNPs calling analysis using the previously published *E. ulmoides* cp genome (KU204775) as reference revealed a total of 75 SNPs. This result of SNP occurrence was consistent with the aforementioned molecular marker analysis (75 SNPs, Figure 2), which indicated that the detected SNPs were really present in different individuals of *E. ulmoides*. Further examination of the 40 SNPs in the CDS regions suggested that all these SNPs were synonymous, i.e., no amino acid change at the protein level. There were 34 transitions and six transversions in the protein-coding region SNPs. The average frequency of SNPs occurrence in the *E. ulmoides* cp genome was calculated as 0.46 per kb.

The final length of the supermatrix dataset contained 96,894 unambiguously aligned nucleotide characters. Three methods produced a congruent phylogenetic tree, shown in Figure 4. Eudicots, monocots and magnoliids were all highly supported to be monophyletic (94–100/90–100/1.00). Magnoliids diverged firstly, followed by monocots, then Chloranthales and eudicots, with relatively low (60–72/74–80/0.90–0.99) support values. Two *E. ulmoides* individuals clustered together with high statistical support values (100/100/1.00), which was subsequently sister to *Aucuba japonica* Thunb.

in the Garryales clade with high statistical support values (100/100/1.00). Garryales, Gentianales and Solanales formed as the highly supported lamiids lineage (100/100/1.00), resolved as (Garryales, (Gentianales, Solanales)). Asterales and Dipsacales formed a highly supported clade, campanulids (100/100/1.00), which was sister to lamiids (100/100/1.00) in asterids. Ericales was resolved as the basal most group in the asterids lineage (100/100/1.00). Brassicales, Malvales, Myrtales and Sapindales clustered in the malvids clade (95/98/1.00) with resolution as (((Brassicales, Malvales), Sapindales), Myrtales). Fabales, Malpighiales and Rosales were located in the fabids lineage (98/100/1.00), which was a sister clade to malvids in rosdis.

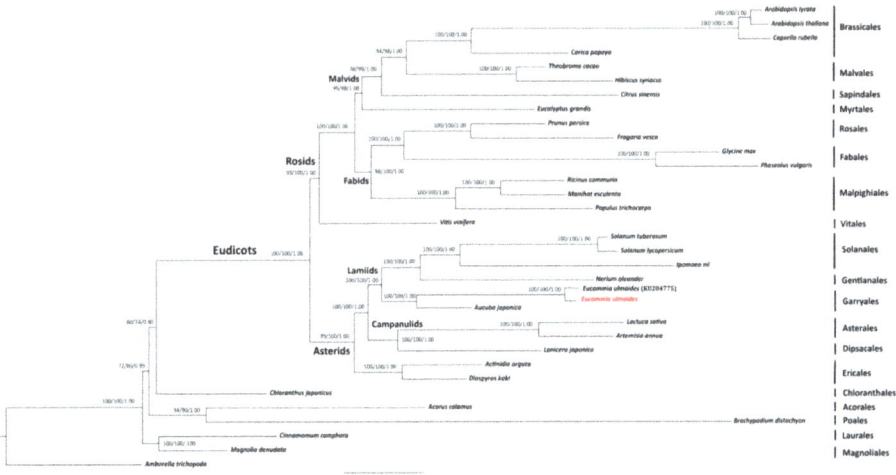

Figure 4. Maximum likelihood (ML) tree for 34 taxa based on 80 unique plastid protein-coding genes of *Eucommia ulmoides*. Values above the branches represent maximum parsimony bootstrap (MPBS)/maximum likelihood bootstrap (MLBS)/Bayesian inference posterior probability (PP). The newly sequenced *Eucommia ulmoides* chloroplast genome is indicated by red color and the previously published *E. ulmoides* chloroplast genome is followed by its GenBank accession number KU204775.

3. Discussion

3.1. Conserved Chloroplast Genome Structure in E. ulmoides

Land plant cp genomes are generally inherited as a haplotype with no recombination, providing useful genetic information to trace relationships between different species [35,38]. Within species cp genome structure was highly conserved [39]. As expected, it is the case in *E. ulmoides* in terms of the contained genes and coding regions in the cp genomes (Table 1). Further whole-genome alignments suggested that the two cp genomes of *E. ulmoides* did not show genome rearrangement having the same linear gene order (Figure 1). As such it is reasonable to use cp genomes for subsequent conservation genomics studies on *E. ulmoides*.

It is noteworthy that the newly sequenced cp genome of *E. ulmoides* (163,586 bp) in this study is 245 bp larger than that of the previously reported one (163,341 bp, [17]) (Table 1). The cp genome size variation within different individuals of the same species has been reported for several other plants, such as in *Camptotheca acuminate* Decne. with its size varied as 157,806 bp [40], 157,877 bp [41] and 162,382 bp [42]. Nuclear genome size variations in plants are mostly caused by the repeats activities (e.g., expansions/contractions) via illegitimate recombination in addition to polyploidy [43–47]. In the two *E. ulmoides* cp genomes, we detected 15 insertions and five deletions, with more than 10 nucleotides for each (Table 2). All these sequences were observed to be part of or the whole DNA repeats. For instance, the repeat sequences in *rps16-trnT(UGU)*, *psbD-trnT(GGU)*, *rps12-rpl20* and *accD-trnM(CAU)* have been detected in the study of Wang et al. [17] as well. Therefore, potentially the illegitimate recombination between repeat regions of *E. ulmoides* cp genome may contribute to the cp genome size variation.

3.2. Heterogeneous Divergence in E. ulmoides Chloroplast Genome

Heterogenous divergence patterns in cp genomes have been reported in several plant groups, such as in Actinidiaceae [37] and in Poaceae [48]. The alignment of the two available cp genomes of *E. ulmoides* revealed highly heterogeneous sequence divergences within this species (Figures 1 and 2). All the identified SNPs and indels from the intron-containing protein-coding genes were located in the intron regions. Due to natural selection CDS regions are in general more conserved than non-coding regions (i.e., intergenic sequences and introns) [49]. In addition, nucleotide substitutions likely have less destructive effect to the integrity of open reading frame (ORF) than indels [50]. We, thus, speculated that this functional constraint may lead to the contrasting occurrences of SNPs and indels in the *E. ulmoides* cp genomes.

The occurrence of synonymous SNPs was more abundant than that of the non-synonymous SNPs in CDS regions because of selection process [51,52]. As expected, all the SNPs detected in protein-coding regions were synonymous. Moreover, since the transitions rather than the transversions usually would generate more synonymous mutations in the CDS the transitions SNPs are more easily retained than the transversion ones [52,53]. In this study we found that the transition SNPs (34) were indeed more frequently detected than the transversion SNPs (6). Previous studies have reported a high level of nuclear genetic diversity at the population level of *E. ulmoides* [18,54]. In this study, we revealed that the frequency of plastid SNPs were 0.46 per kb at the whole cp genome level, lower than the average of 1.02 per kb in the nuclear genes of *E. ulmoides* [54]. The difference of SNP frequency between the cp genome and the nuclear genes could be caused by insufficient sampling in this study and/or different variation rates between the plastid and nuclear sequences.

3.3. Mutation Hotspots in E. ulmoides Chloroplast Genome

In general, protein-coding genes in the cp genome have lower sequence variation than the non-coding loci, for instance in bamboos [55] and mimosoid legume [56]. However, an accelerated variation rate of some plastid protein-coding genes has been reported, such as the *psb* in Poaceae [51], *rps* in Saxifragales [57], and *accD* and *rpl20* in Actinidiaceae [37]. In *E. ulmoides*, we found three protein-coding genes (*infA*, *psaJ* and *rpl32*) that varied the most quickly, all having variations exceeding 1.2% (Figure 3). The gene *infA* encodes translation initiation factor 1 and has been found missing in cp genomes of several plant lineages, e.g., in rosids [58]. The other two genes i.e., *psaJ* and *rpl32* code for photosystem I protein J and ribosomal protein L32, respectively, both of which are short in length with the former having 129 bp and the latter 138 bp (this study and [17]). The relatively high level of variation of these genes in *E. ulmoides* indicates that they are less constrained. Abnormal DNA replication, repair or recombination [59,60] may lead to the elevated divergence of these genes.

The genetic markers of SSR, AFLP and SRAP have been developed and used for the population genetics studies of *E. ulmoides* [18,20,21]. However, the above fingerprinting markers [23,24] may provide insufficient genetic information to resolve the population structure and history of *E. ulmoides*. The relationships among natural and cultivated populations of *E. ulmoides* are elusive at present [18,24]. SNP markers are ample in plant cp genomes, making them useful candidates for population genetics studies. Using cp genome SNP markers has widely received attention during the past few years with the advances of high throughput sequencing techniques [61,62]. Highly variable cpDNA fragments have been mined for phylogenetic and population genetic studies in several species using cp genomes data, such as in kiwifruit [37] and temperate woody bamboos [55]. Given that it would be easy to amplify and sequence DNA fragments with length from *circa* 200 to 1500 bp using Sanger sequencing method [62,63], we thus chose 20 cpDNA loci with relatively high genetic divergences as potential molecular markers (Table 3) for subsequent population genomics studies of *E. ulmoides*. These selected plastid genome-wide loci would genetically be informative for uncovering the genetic relationships among the natural and cultivated *E. ulmoides* populations.

Additionally, 27 cpSSR loci identified by Wang et al. [17] were also confirmed in our SSR analysis, among which eight were further mined as polymorphic cpSSR loci (Table 4). Given that polymorphic cpSSR loci could be applied as useful markers to meet certain study purposes under the circumstances of limited budget [64,65], the newly developed polymorphic cpSSR loci in *E. ulmoides* here would be potential genetic markers to facilitate subsequent population genetics studies in the future.

3.4. Phylogenomic Validation of E. ulmoides

The newly obtained *E. ulmoides* cp genome was further validated via phylogenomic analyses using 34 complete plastomes from 10 major lineages of angiosperms. The resulting phylogenomic tree highly supported the clade of two *E. ulmoides* cp genomes (Figure 4), confirming the validity of the assembled and annotated cp genome of *E. ulmoides* in this study. The sister relationship between *E. ulmoides* and *A. japonica* in the Garryales clade was highly supported, which is consistent with the results derived from five organellar genes [66] and 36 plastid genes [17], supporting the classification of *E. ulmoides* (Eucommiaceae) in the updated APG IV system [67]. *E. ulmoides* and *A. japonica* are both woody and have unisexual flowers in separate individuals, which seem to be morphological synapomorphies for the order Garryales [68]. An average of 92.6% identities between 78 common unique cp protein-coding genes in *E. ulmoides* and *A. japonica* were also detected, suggesting a high similarity between the two species at the molecular level. Garryales was shown to be closely related to the clade of (Gentianales + Solanales) in lamiids, in line with the APG IV system [67].

All 20 sampled orders were highly supported to be monophyletic separately (Figure 4), agreeing with the APG IV system [67]. Within eudicots two large sister clades i.e., asterids and rosids were uncovered, and the relationships among these two lineages were highly resolved as ((campanulids, lamiids), Ericales) and ((fabids, malvids), Vitales), respectively as stated previously [69]. The branching patterns of species within campanulids and lamiids are consistent with recent studies [66,70] as (((Gentianales, Solanales), Garryales), (Asterales, Dipsacales)). Our analyses also resolved the phylogenetic relationships within fabids and malvids as ((((Brassicales, Malvales), Sapindales), Myrtales), ((Fabales, Rosales), Malpighiales)), consistent with the results of previous studies [17,69]. It is noteworthy that the branching orders of magnoliids, monocots, Chloranthales and eudicots only obtained low-level support values here (Figure 4). Further studies with expanded taxon samples are expected to confirm the phylogeny of these lineages. Moreover, plastome is inherited uniparentally in general, which might introduce biases to species phylogeny inference [71,72]. Analyses using orthologous nuclear genes are also needed for studying the evolutionary history of *E. ulmoides* among the flowering plants [54].

4. Materials and Methods

4.1. Plant Materials and DNA Sequencing

Fresh healthy leaves were collected from an adult male individual of *E. ulmoides* growing in the Arboretum of Northwest Agricuture and Forest University in Yangling, Shanxi, China, in April 2015. After collection, the leaves were immediately immersed in liquid nitrogen and then stored at −80 °C until use. The voucher specimen of this tree was deposited at the Trees Herbarium of Northwest A and F University with accession number ZXZ15027.

Total genomic DNA was extracted by the CTAB method [73]. Paired-end (PE) libraries with insert size *circa* 500 bp were constructed from fragmented genomic DNA based on standard Illumina protocols (Illumina Inc., San Diego, CA, USA). Prepared library was then sequenced for PE 100 bp read length on the Illumina HiSeq 2000 platform at the Beijing Genomics Institute (BGI) in Shenzhen, China.

4.2. Genome Assembly and Annotation

Fastq format PE reads were supplied with adaptor sequences removed. Poor quality reads with phred scores lower than 20 for more than 10% of their bases were also removed. Two independent methods were used to assemble the *E. ulmoides* cp genome. (1) The cp genome was de novo assembled using the CLC Genomics Workbench v7.5 software (CLC Bio, Aarhus, Denmark) based on the clean reads. After discarding contigs with length <300 bp and sequences with coverage <50, the remaining contigs were searched against the available cp genome of *E. ulmoides* (GenBank accession number KU204775) that used as the reference by BLAST (http://blast.ncbi.nlm.nih.gov/) with *e*-value <10^{-5}. Aligned contigs with ≥90% similarity and query coverage were determined as cpDNA sequences and ordered according to the reference genome. Small gaps were filled using PE clean reads as conducted in Wang et al. [37]. (2) The clean reads were firstly mapped to the reference cp genome of *E. ulmoides* to determine the proportion of cpDNA using Bowtie v2.3.1 program [74] with a maximum of 3 mismatches. Subsequently, we applied SPAdes v3.9 software [75] with default setting to assemble the cp genome using the determined cpDNA clean reads.

DOGMA software [76] was used for initial cp genome annotation. Start/stop codons and intron/exon boundaries were checked and adjusted manually when necessary by comparing to the reference genome. tRNA genes were confirmed based on tRNAscan-SE 1.21 [77].

4.3. Genome-Wide Comparison and Divergent Hotspot Identification

The previously published cp genome of *E. ulmoides* (accession number: KU204775) was downloaded from GenBank database (https://www.ncbi.nlm.nih.gov/genbank/). This genome was aligned with the *E. ulmoides* cp genome described herein, using MAFFT program [78] and MAUVE software [79], respectively, and manually adjusted where necessary. The obtained pairwise alignment of the cp genomes was visualized in Geneious v9.0 [80]. Moreover, given the genome repeat sequences expansion and contraction may result in genome size variation [81], we examined the DNA insertions and deletions in repeat regions of the two *E. ulmoides* cp genomes.

The two *E. ulmoides* cp genomes were analyzed to identify molecular markers that can be selected in subsequent population genetic studies. We firstly extracted both the genic and intergenic DNA fragments in each cp genome using the "Extract Sequences" option in DOGMA [76]. Then the homologous loci were aligned individually by MUSCLE program (http://www.drive5.com/muscle/) [82] implemented in Geneious v9.0 [80] with default settings. Manual adjustments were made for the alignments where necessary. The proportion of mutational events for each genic and intergenic locus was calculated as follows: the proportion of variation = ((NS + ID)/L) × 100, where NS = the number of nucleotide substitutions (SNPs), ID = the number of indels (insertions and deletions), L = the aligned sequence length.

Polymorphic cpSSR loci were further mined by genome comparison. Firstly, SSRs in the newly sequenced *E. ulmoides* cp genome were detected by MISA perl script (http://pgrc.ipk-gatersleben.de/misa/). The parameter of minimum repeat unit was set as 10 for mono-, 6 for di-, and 5 for tri-, tetra-, penta-, and hexanucleotide SSRs [83]. Then, all the identified SSR loci were compared to the 29 cpSSR loci of Wang et al. [17] to develop polymorphic cpSSR markers in *E. ulmoides*.

4.4. SNPs Validation and Phylogenomic Analyses

To confirm the SNPs identified by the aforementioned cp genome alignment, we here mapped the genome skimming clean reads generated in this study to the previously published *E. ulmoides* cp genome (KU204775) [17] for SNP calling. Picard-tools v1.41 (http://broadinstitute.github.io/picard/) and samtools v0.1.18 [84] were applied to sort and remove duplicated reads and merge the bam alignment results. GATK3 software [85] was further used to perform SNPs identification. Raw vcf files were filtered with GATK standard filter method and other parameters were set as defaults. Moreover, to reveal if the coding region SNPs detected in *E. ulmoides* cp genome caused amino acid substitution on protein level, we firstly translated each protein-coding gene into amino acids in Geneious v9.0 [80]. Then the protein sequences of each gene were aligned, respectively using MUSCLE [82]. The mutational events were checked to uncover the synonymous and nonsynonymous SNPs and the nucleotide transitions and transversions.

Phylogenomic analyses were also conducted to validate the newly assembled and annotated *E. ulmoides* cp genome. 34 plastomes representing 10 major lineages of angiosperms (Table S1) were included for phylogenomic analyses. *Amborella trichopoda* from basal angiosperm lineages was defined as outgroup according to previous studies [66,67]. 80 unique plastid protein-coding genes of *E. ulmoides* (Table S2) were used for the phylogenetic inferences. Each gene was aligned individually by MUSCLE [82] in Geneious v9.0 [80], and then concatenated as a supermatrix. Gaps were not included in the dataset.

Three methods i.e., maximum parsimony (MP), maximum likelihood (ML), and Bayesian inference (BI) were used for phylogenetic reconstruction. We performed parsimony heuristic tree searches in PAUP v4.0b10 [86] with parameters set as 1000 random addition sequence replicates, tree bisection and reconnection (TBR) branch swapping, and MulTrees option in effect. 1000 bootstrap replicates [87] were calculated to evaluate the branch support (MPBS) of the MP tree. RAxML v.8.2.8 [88] and MrBayes 3.2.6 [89] in the CIPRES Science Gateway v3.3.3 [90] were applied for ML and BI analyses, respectively. Supermatrix was partitioned by genes and GTR + G model of nucleotide substitution was used. For the ML tree we conducted 1000 fast bootstrap ML reps to assess the support values (MLBS) of internal nodes. In Bayesian analysis two runs with four chains were carried out up to 50,000,000 generations, sampling one tree every 1000 generations till convergence, i.e., the average standard deviation of split frequencies <0.01. We discarded the first 25% of trees as burn-in, and used the remaining trees to estimate the majority-rule consensus BI tree and posterior probabilities (PP).

5. Conclusions

In summary, in the present study we generated one complete cp genome of *E. ulmoides* using the genome skimming approach. Through comprehensive genome-wide comparative analyses we found that the cp genomes within *E. ulmoides* were highly conserved in terms of structure and content. Nevertheless, obviously heterogeneous sequence divergences were revealed in different regions of the *E. ulmoides* cp genome. A total of 20 polymorphic DNA fragments and eight SSR loci have been identified as potential cpDNA markers for subsequent population genetics studies of this tree species. The phylogenetic placement of *E. ulmoides* in angiosperms was robustly resolved as well based on the cp genomes data, strongly supporting the sister relationship between *E. ulmoides* and *A. japonica* in the asterids lineage. The data presented here will aid further conservation genomic studies and facilitate the development of plastid genetic engineering for *E. ulmoides*.

Supplementary Materials: The following are available online at http://www.mdpi.com/1422-0067/19/4/1037/s1. Table S1: Taxa and GenBank accession numbers included in the phylogenomic analyses, Table S2: List of 80 unique plastid protein-coding genes of *Eucommia ulmoides* included in the phylogenomic analyses.

Acknowledgments: We would like to thank Zhirong Zhang for the help in the experiment performance. This work is supported by the National Natural Science Foundation of China (31600173) and the Basic Science Fund of Northwest A&F University (2452016052).

Author Contributions: Wencai Wang and Xianzhi Zhang conceived and designed the experiments; Wencai Wang performed the experiments; Siyun Chen and Xianzhi Zhang analyzed the data; Xianzhi Zhang contributed reagents/materials/analysis tools; Wencai Wang and Xianzhi Zhang wrote the paper. All authors reviewed and approved the final manuscript.

Conflicts of Interest: The authors declare no conflict of interest.

References

1. Huang, H. Plant diversity and conservation in China: Planning a strategic bioresource for a sustainable future. *Bot. J. Linn. Soc.* **2011**, *166*, 282–300. [CrossRef] [PubMed]
2. Liu, J.; Ouyang, Z.; Pimm, S.L.; Raven, P.H.; Wang, X.; Miao, H.; Han, N. Protecting China's biodiversity. *Science* **2003**, *300*, 1240–1241. [CrossRef] [PubMed]
3. Lópezpujol, J.; Zhang, F.M.; Ge, S. Plant biodiversity in China: Richly varied, endangered, and in need of conservation. *Biodivers. Conserv.* **2006**, *15*, 3983–4026. [CrossRef]
4. Gu, J. Conservation of plant diversity in China: Achievements, prospects and concerns. *Biol. Conserv.* **1998**, *85*, 321–327. [CrossRef]
5. Chen, S.L.; Hua, Y.; Luo, H.M.; Wu, Q.; Li, C.F.; Steinmetz, A. Conservation and sustainable use of medicinal plants: Problems, progress, and prospects. *Chin. Med.* **2016**, *11*, 37. [CrossRef] [PubMed]
6. Huang, L.; Yang, B.; Wang, M.; Fu, G. An approach to some problems on utilization of medicinal plant resource in China. *China J. Chin. Mater. Med.* **1999**, *24*, 70–73.
7. Zhang, Z.Y.; Zhang, H.D.; Turland, N.J. Eucommiaceae. In *Flora of China*; Wu, Z.Y., Raven, P.H., Hong, D.Y., Eds.; Science Press and Missouri Botanical Garden: Beijing, China, 2003; p. 43.
8. Kawasaki, T.; Uezono, K.; Nakazawa, Y. Antihypertensive mechanism of food for specified health use: "*Eucommia* leaf glycoside" and its clinical application. *J. Health Sci.* **2000**, *22*, 29–36.
9. Liu, H.; Hongyan, D.U.; Tana, W. Advances in research on biotechnology breeding of *Eucommia ulmoides*. *Hunan For. Sci. Technol.* **2016**, *43*, 132–136.
10. Suzuki, N.; Uefuji, H.; Nishikawa, T.; Mukai, Y.; Yamashita, A.; Hattori, M.; Ogasawara, N.; Bamba, T.; Fukusaki, E.; Kobayashi, A. Construction and analysis of EST libraries of the trans-polyisoprene producing plant, *Eucommia ulmoides* Oliver. *Planta* **2012**, *236*, 1405–1417. [CrossRef] [PubMed]
11. Du, H.Y.; Hu, W.Z.; Yu, R. *The Report on Development of China's Eucommia Rubber Resources and Industry (2014-2015)*; Social Sciences Academic Press: Beijing, China, 2015.
12. Manchester, S.R.; Chen, Z.D.; An-Ming, L.U.; Uemura, K. Eastern Asian endemic seed plant genera and their paleogeographic history throughout the Northern Hemisphere. *J. Syst. Evol.* **2009**, *47*, 1–42. [CrossRef]
13. Mabberley, D.J. *The Plant Book*; Cambridge University Press: Cambridge, UK, 1989.
14. Fu, L.G.; Jin, J.M. *Red List of Endangered Plants in China*; Science Press: Beijing, China, 1992.
15. Du, H.Y. *China Eucommia Pictorial*; China Forestry Publishing House: Beijing, China, 2014.
16. Wang, L.; Du, H.; Wuyun, T.N. Genome-wide identification of microRNAs and their targets in the leaves and fruits of *Eucommia ulmoides* using high-throughput sequencing. *Front. Plant Sci.* **2016**, *7*, 1632. [CrossRef] [PubMed]
17. Wang, L.; Wuyun, T.N.; Du, H.; Wang, D.; Cao, D. Complete chloroplast genome sequences of *Eucommia ulmoides*: Genome structure and evolution. *Tree Genet. Genomes* **2016**, *12*, 12. [CrossRef]
18. Zhang, J.; Xing, C.; Tian, H.; Yao, X. Microsatellite genetic variation in the Chinese endemic *Eucommia ulmoides* (Eucommiaceae): Implications for conservation. *Bot. J. Linn. Soc.* **2013**, *173*, 775–785. [CrossRef]
19. Zhang, W.R.; Li, Y.; Zhao, J.; Wu, C.H.; Ye, S.; Yuan, W.J. Isolation and characterization of microsatellite markers for *Eucommia ulmoides* (Eucommiaceae), an endangered tree, using next-generation sequencing. *Genet. Mol. Res.* **2016**, *15*. [CrossRef] [PubMed]
20. Li, Y.; Wang, D.; Li, Z.; Wei, J.; Jin, C.; Liu, M. A molecular genetic linkage map of *Eucommia ulmoides* and quantitative trait loci (QTL) analysis for growth traits. *Int. J. Mol. Sci.* **2014**, *15*, 2053–2074. [CrossRef] [PubMed]
21. Wang, D.; Li, Y.; Li, L.; Wei, Y.; Li, Z. The first genetic linkage map of *Eucommia ulmoides*. *J. Genet.* **2014**, *93*, 13–20. [CrossRef] [PubMed]
22. Wang, A.Q.; Huang, L.Q.; Shao, A.J.; Cui, G.H.; Chen, M.; Tong, C.H. Genetic diversity of *Eucommia ulmoides* by RAPD analysis. *China J. Chin. Mater. Med.* **2006**, *31*, 1583–1586.
23. Yao, X.; Deng, J.; Huang, H. Genetic diversity in *Eucommia ulmoides* (Eucommiaceae), an endangered traditional Chinese medicinal plant. *Conserv. Genet.* **2012**, *13*, 1499–1507. [CrossRef]
24. Yu, J.; Wang, Y.; Peng, L.; Ru, M.; Liang, Z.S. Genetic diversity and population structure of *Eucommia ulmoides* Oliver, an endangered medicinal plant in China. *Genet. Mol. Res.* **2015**, *14*, 2471–2483. [CrossRef] [PubMed]
25. Vignal, A.; Milan, D.; Sancristobal, M.; Eggen, A. A review on SNP and other types of molecular markers and their use in animal genetics. *Genet. Sel. Evol.* **2002**, *34*, 275–305. [CrossRef] [PubMed]

26. Bernardi, J.; Mazza, R.; Caruso, P.; Reforgiato, R.G.; Marocco, A.; Licciardello, C. Use of an expressed sequence tag-based method for single nucleotide polymorphism identification and discrimination of *Citrus* species and cultivars. *Mol. Breed.* **2013**, *31*, 705–718. [CrossRef]

27. Elshire, R.J.; Glaubitz, J.C.; Sun, Q.; Poland, J.A.; Kawamoto, K.; Buckler, E.S.; Mitchell, S.E. A robust, simple genotyping-by-sequencing (GBS) approach for high diversity species. *PLoS ONE* **2011**, *6*, e19379. [CrossRef]

28. Kess, T.; Gross, J.; Harper, F.; Boulding, E.G. Low-cost ddRAD method of SNP discovery and genotyping applied to the periwinkle *Littorina saxatilis*. *J. Molluscan Stud.* **2016**, *82*, eyv042.

29. Potter, K.M.; Hipkins, V.D.; Mahalovich, M.F.; Means, R.E. Nuclear genetic variation across the range of ponderosa pine (*Pinus ponderosa*): Phylogeographic, taxonomic and conservation implications. *Tree Genet. Genomes* **2015**, *11*, 38. [CrossRef]

30. Worth, J.R.P.; Yokogawa, M.; Pérez-Figueroa, A.; Tsumura, Y.; Tomaru, N.; Janes, J.K.; Isagi, Y. Conflict in outcomes for conservation based on population genetic diversity and genetic divergence approaches: A case study in the Japanese relictual conifer *Sciadopitys verticillata* (Sciadopityaceae). *Conserv. Genet.* **2014**, *15*, 1243–1257. [CrossRef]

31. Chung, S.M.; Staub, J.E.; Lebeda, A.; Paris, H.S. Consensus chloroplast primer analysis: A molecular tool for evolutionary studies in Cucurbitaceae. In Proceedings of the Progress in Cucurbit Genetics and Breeding Research, Olomouc, Czech Republic, 12–17 July 2004.

32. Ahmed, I.; Matthews, P.J.; Biggs, P.J.; Naeem, M.; Mclenachan, P.A.; Lockhart, P.J. Identification of chloroplast genome loci suitable for high-resolution phylogeographic studies of *Colocasia esculenta* (L.) Schott (Araceae) and closely related taxa. *Mol. Ecol. Resour.* **2013**, *13*, 929–937. [CrossRef] [PubMed]

33. Zhang, Y.; Du, L.; Ao, L.; Chen, J.; Li, W.; Hu, W.; Wei, Z.; Kim, K.; Lee, S.C.; Yang, T.J. The complete chloroplast genome sequences of five *Epimedium* species: Lights into phylogenetic and taxonomic analyses. *Front. Plant Sci.* **2016**, *7*, 696. [CrossRef] [PubMed]

34. Raubeson, L.A.; Jansen, R.K. Chloroplast genomes of plants. In *Plant Diversity and Evolution: Genotypic and Phenotypic Variation in Higher Plants*; Henry, R.J., Ed.; CABI: Cambridge, MA, USA, 2005; pp. 45–68.

35. Birky, C.W., Jr. Uniparental inheritance of mitochondrial and chloroplast genes: Mechanisms and evolution. *Proc. Natl. Acad. Sci. USA* **1996**, *92*, 11331–11338. [CrossRef]

36. Straub, S.C.; Parks, M.; Weitemier, K.; Fishbein, M.; Cronn, R.C.; Liston, A. Navigating the tip of the genomic iceberg: Next-generation sequencing for plant systematics. *Am. J. Bot.* **2012**, *99*, 349–364. [CrossRef] [PubMed]

37. Wang, W.C.; Chen, S.Y.; Zhang, X.Z. Chloroplast genome evolution in Actinidiaceae: *clpP* Loss, heterogenous divergence and phylogenomic practice. *PLoS ONE* **2016**, *11*, e0162324. [CrossRef] [PubMed]

38. Wicke, S.; Schneeweiss, G.M.; Müller, K.F.; Quandt, D. The evolution of the plastid chromosome in land plants: Gene content, gene order, gene function. *Plant Mol. Biol.* **2011**, *76*, 273–297. [CrossRef] [PubMed]

39. Wicke, S.; Schneeweiss, G.M. Next-Generation Organellar Genomics: Potentials and Pitfalls of High-Throughput Technologies for Molecular Evolutionary Studies and Plant Systematics. In *Next Generation Sequencing in Plant Systematics*; International Association for Plant Taxonomy (IAPT): Bratislava, Slovakia, 2015.

40. Chen, S.Y.; Zhang, X.Z. Characterization of the complete chloroplast genome of the relict Chinese false tupelo, *Camptotheca acuminata*. *Conserv. Genet. Resour.* **2017**, 1–4. [CrossRef]

41. Yang, Z.; Ji, Y. Comparative and phylogenetic analyses of the complete chloroplast genomes of three Arcto-Tertiary relicts: *Camptotheca acuminata*, *Davidia involucrata*, and *Nyssa sinensis*. *Front. Plant Sci.* **2017**, *8*, 1536. [CrossRef] [PubMed]

42. Wang, W.; Liu, H.; He, Q.; Yang, W.L.; Chen, Z.; Wang, M.; Su, Y.; Ma, T. Characterization of the complete chloroplast genome of *Camptotheca acuminata*. *Conserv. Genet. Resour.* **2017**, *9*, 241–243. [CrossRef]

43. Puterova, J.; Razumova, O.; Martinek, T.; Alexandrov, O.; Divashuk, M.; Kubat, Z.; Hobza, R.; Karlov, G.; Kejnovsky, E. Satellite DNA and transposable elements in seabuckthorn (*Hippophae rhamnoides*), a dioecious plant with small Y and large X chromosomes. *Genome Biol. Evol.* **2017**, *9*, 197–212. [CrossRef] [PubMed]

44. Rocha, E.P.C. An appraisal of the potential for illegitimate recombination in bacterial genomes and its consequences: From duplications to genome reduction. *Genome Res.* **2003**, *13*, 1123–1132. [CrossRef] [PubMed]

45. Kegel, A.; Martinez, P.; Carter, S.D.; Aström, S.U. Genome wide distribution of illegitimate recombination events in *Kluyveromyces lactis*. *Nucleic Acids Res.* **2006**, *34*, 1633–1645. [CrossRef] [PubMed]

46. Dodsworth, S.; Leitch, A.R.; Leitch, I.J. Genome size diversity in angiosperms and its influence on gene space. *Curr. Opin. Genet. Dev.* **2015**, *35*, 73–78. [CrossRef] [PubMed]

47. Lisch, D. How important are transposons for plant evolution? *Nat. Rev. Genet.* **2013**, *14*, 49–61. [CrossRef] [PubMed]

48. Zhong, B.; Yonezawa, T.; Zhong, Y.; Hasegawa, M. Episodic evolution and adaptation of chloroplast genomes in ancestral grasses. *PLoS ONE* **2009**, *4*, e5297. [CrossRef] [PubMed]

49. Shaw, J.; Lickey, E.B.; Schilling, E.E.; Small, R.L. Comparison of whole chloroplast genome sequences to choose noncoding regions for phylogenetic studies in angiosperms: The tortoise and the hare III. *Am. J. Bot.* **2007**, *94*, 275–288. [CrossRef] [PubMed]

50. Zhang, Z.; Gerstein, M. Patterns of nucleotide substitution, insertion and deletion in the human genome inferred from pseudogenes. *Nucleic Acids Res.* **2003**, *31*, 5338–5348. [CrossRef] [PubMed]

51. Matsuoka, Y.; Yamazaki, Y.; Ogihara, Y.; Tsunewaki, K. Whole chloroplast genome comparison of rice, maize, and wheat: Implications for chloroplast gene diversification and phylogeny of cereals. *Mol. Biol. Evol.* **2002**, *19*, 2084–2091. [CrossRef] [PubMed]

52. Castle, J.C. SNPs occur in regions with less genomic sequence conservation. *PLoS ONE* **2011**, *6*, e20660. [CrossRef] [PubMed]

53. Allegre, M.; Argout, X.; Boccara, M.; Fouet, O.; Roguet, Y.; Bérard, A.; Thévenin, J.M.; Chauveau, A.; Rivallan, R.; Clement, D. Discovery and mapping of a new expressed sequence tag-single nucleotide polymorphism and simple sequence repeat panel for large-scale genetic studies and breeding of *Theobroma cacao* L. *DNA Res.* **2011**, *19*, 23–35. [CrossRef] [PubMed]

54. Wang, W.; Zhang, X. Identification of the sex-biased gene expression and putative sex-associated genes in *Eucommia ulmoides* Oliver using comparative transcriptome analyses. *Molecules* **2017**, *22*, 2255. [CrossRef] [PubMed]

55. Zhang, Y.-J.; Ma, P.-F.; Li, D.-Z. High-throughput sequencing of six bamboo chloroplast genomes: Phylogenetic implications for temperate woody bamboos (Poaceae: Bambusoideae). *PLoS ONE* **2011**, *6*, e20596. [CrossRef] [PubMed]

56. Magee, A.M.; Aspinall, S.; Rice, D.W.; Cusack, B.P.; Sémon, M.; Perry, A.S.; Stefanović, S.; Milbourne, D.; Barth, S.; Palmer, J.D. Localized hypermutation and associated gene losses in legume chloroplast genomes. *Genome Res.* **2010**, *20*, 1700–1710. [CrossRef] [PubMed]

57. Dong, W.; Xu, C.; Cheng, T.; Zhou, S. Complete chloroplast genome of *Sedum sarmentosum* and chloroplast genome evolution in Saxifragales. *PLoS ONE* **2013**, *8*, e77965. [CrossRef] [PubMed]

58. Millen, R.S.; Olmstead, R.G.; Adams, K.L.; Palmer, J.D.; Lao, N.T.; Heggie, L.; Kavanagh, T.A.; Hibberd, J.M.; Gray, J.C.; Morden, C.W. Many parallel losses of *infA* from chloroplast DNA during angiosperm evolution with multiple independent transfers to the nucleus. *Plant Cell* **2001**, *13*, 645–658. [CrossRef] [PubMed]

59. Guisinger, M.M.; Kuehl, J.V.; Boore, J.L.; Jansen, R.K. Genome-wide analyses of Geraniaceae plastid DNA reveal unprecedented patterns of increased nucleotide substitutions. *Proc. Natl. Acad. Sci. USA* **2008**, *105*, 18424–18429. [CrossRef] [PubMed]

60. Dugas, D.V.; Hernandez, D.; Koenen, E.J.; Schwarz, E.; Straub, S.; Hughes, C.E.; Jansen, R.K.; Nageswara-Rao, M.; Staats, M.; Trujillo, J.T. Mimosoid legume plastome evolution: IR expansion, tandem repeat expansions, and accelerated rate of evolution in *clpP*. *Sci. Rep.* **2015**, *5*, 16958. [CrossRef] [PubMed]

61. Bock, D.G.; Kane, N.C.; Ebert, D.P.; Rieseberg, L.H. Genome skimming reveals the origin of the Jerusalem Artichoke tuber crop species: Neither from Jerusalem nor an artichoke. *New Phytol.* **2014**, *201*, 1021–1030. [CrossRef] [PubMed]

62. Downie, S.R.; Jansen, R.K. A comparative analysis of whole plastid genomes from the Apiales: Expansion and contraction of the inverted repeat, mitochondrial to plastid transfer of DNA, and identification of highly divergent noncoding regions. *Syst. Bot.* **2015**, *40*, 336–351. [CrossRef]

63. Shaw, J.; Lickey, E.B.; Beck, J.T.; Farmer, S.B.; Liu, W.; Miller, J.; Siripun, K.C.; Winder, C.T.; Schilling, E.E.; Small, R.L. The tortoise and the hare II: Relative utility of 21 noncoding chloroplast DNA sequences for phylogenetic analysis. *Am. J. Bot.* **2005**, *92*, 142–166. [CrossRef] [PubMed]

64. Huang, J.; Yang, X.; Zhang, C.; Yin, X.; Liu, S.; Li, X. Development of chloroplast microsatellite markers and analysis of chloroplast diversity in Chinese jujube (*Ziziphus jujuba* Mill.) and wild jujube (*Ziziphus acidojujuba* Mill.). *PLoS ONE* **2015**, *10*, e0134519. [CrossRef] [PubMed]

65. Ren, X.; Jiang, H.; Yan, Z.; Chen, Y.; Zhou, X.; Huang, L.; Lei, Y.; Huang, J.; Yan, L.; Qi, Y. Genetic diversity and population structure of the major peanut (*Arachis hypogaea* L.) cultivars grown in China by SSR markers. *PLoS ONE* **2014**, *9*, e88091. [CrossRef] [PubMed]

66. Chen, Z.D.; Yang, T.; Lin, L.; Lu, L.M.; Li, H.L.; Sun, M.; Liu, B.; Chen, M.; Niu, Y.T.; Ye, J.F. Tree of life for the genera of Chinese vascular plants. *J. Syst. Evol.* **2016**, *54*, 277–306. [CrossRef]

67. Byng, J.W.; Chase, M.W.; Christenhusz, M.J.; Fay, M.F.; Judd, W.S.; Mabberley, D.J.; Sennikov, A.N.; Soltis, D.E.; Soltis, P.S.; Stevens, P.F. An update of the Angiosperm Phylogeny Group classification for the orders and families of flowering plants: APG IV. *Bot. J. Linn. Soc.* **2016**, *181*, 105–121.

68. Stevens, P.F. Angiosperm Phylogeny Website. Version 12, July 2012 (and More or Less Continuously Updated Since). Available online: http://www.mobot.org/MOBOT/research/APweb/ (accessed on 1 March 2018).

69. Jansen, R.K.; Cai, Z.; Raubeson, L.A.; Daniell, H.; Depamphilis, C.W.; Leebensmack, J.; Müller, K.F.; Guisingerbellian, M.; Haberle, R.C.; Hansen, A.K. Analysis of 81 genes from 64 plastid genomes resolves relationships in angiosperms and identifies genome-scale evolutionary patterns. *Proc. Natl. Acad. Sci. USA* **2007**, *104*, 19369–19374. [CrossRef] [PubMed]

70. Smith, S.A.; Beaulieu, J.M.; Donoghue, M.J. An uncorrelated relaxed-clock analysis suggests an earlier origin for flowering plants. *Proc. Natl. Acad. Sci. USA* **2010**, *107*, 5897–5902. [CrossRef]

71. Davis, C.C.; Xi, Z.; Mathews, S. Plastid phylogenomics and green plant phylogeny: Almost full circle but not quite there. *BMC Biol.* **2014**, *12*, 11. [CrossRef] [PubMed]

72. Zeng, L.; Zhang, Q.; Sun, R.; Kong, H.; Zhang, N.; Ma, H.; Zeng, L.; Zhang, Q.; Sun, R.; Kong, H. Resolution of deep angiosperm phylogeny using conserved nuclear genes and estimates of early divergence times. *Nat. Commun.* **2014**, *5*, 4956. [CrossRef] [PubMed]

73. Doyle, J.J. A rapid DNA isolation procedure for small quantities of fresh leaf tissue. *Phytochem. Bull.* **1987**, *19*, 11–15.

74. Langmead, B.; Salzberg, S.L. Fast gapped-read alignment with Bowtie 2. *Nat. Methods* **2012**, *9*, 357–359. [CrossRef] [PubMed]

75. Bankevich, A.; Nurk, S.; Antipov, D.; Gurevich, A.A.; Dvorkin, M.; Kulikov, A.S.; Lesin, V.M.; Nikolenko, S.I.; Pham, S.; Prjibelski, A.D. SPAdes: A new genome assembly algorithm and its applications to single-cell sequencing. *J. Comput. Biol.* **2012**, *19*, 455–477. [CrossRef] [PubMed]

76. Wyman, S.K.; Jansen, R.K.; Boore, J.L. Automatic annotation of organellar genomes with DOGMA. *Bioinformatics* **2004**, *20*, 3252–3255. [CrossRef] [PubMed]

77. Schattner, P.; Brooks, A.N.; Lowe, T.M. The tRNAscan-SE, snoscan and snoGPS web servers for the detection of tRNAs and snoRNAs. *Nucleic Acids Res.* **2005**, *33* (Suppl. 2), W686–W689. [CrossRef] [PubMed]

78. Katoh, K.; Standley, D.M. MAFFT multiple sequence alignment software version 7: Improvements in performance and usability. *Mol. Biol. Evol.* **2013**, *30*, 772–780. [CrossRef] [PubMed]

79. Darling, A.C.; Mau, B.; Blattner, F.R.; Perna, N.T. Mauve: Multiple alignment of conserved genomic sequence with rearrangements. *Genome Res.* **2004**, *14*, 1394–1403. [CrossRef] [PubMed]

80. Kearse, M.; Moir, R.; Wilson, A.; Stones-Havas, S.; Cheung, M.; Sturrock, S.; Buxton, S.; Cooper, A.; Markowitz, S.; Duran, C. Geneious Basic: An integrated and extendable desktop software platform for the organization and analysis of sequence data. *Bioinformatics* **2012**, *28*, 1647–1649. [CrossRef] [PubMed]

81. Wang, W.; Ma, L.; Becher, H.; Garcia, S.; Kovarikova, A.; Leitch, I.J.; Leitch, A.R.; Kovarik, A. Astonishing 35S rDNA diversity in the gymnosperm speciesCycas revolutaThunb. *Chromosoma* **2016**, *125*, 683–699. [CrossRef] [PubMed]

82. Edgar, R.C. MUSCLE: Multiple sequence alignment with high accuracy and high throughput. *Nucleic Acids Res.* **2004**, *32*, 1792–1797. [CrossRef] [PubMed]

83. Zhao, H.; Li, Y.; Peng, Z.; Sun, H.; Yue, X.; Lou, Y.; Dong, L.; Wang, L.; Gao, Z. Developing genome-wide microsatellite markers of bamboo and their applications on molecular marker assisted taxonomy for accessions in the genus *Phyllostachys*. *Sci. Rep.* **2015**, *5*, 8018. [CrossRef] [PubMed]

84. Li, H.; Handsaker, B.; Wysoker, A.; Fennell, T.; Ruan, J.; Homer, N.; Marth, G.; Abecasis, G.; Durbin, R. The sequence alignment/map (SAM) format and SAMtools. *Transpl. Proc.* **2009**, *19*, 1653–1654.

85. Van der Auwera, G.A.; Carneiro, M.O.; Hartl, C.; Poplin, R.; Del Angel, G.; Levy-Moonshine, A.; Jordan, T.; Shakir, K.; Roazen, D.; et al. From FastQ Data to High-Confidence Variant Calls: The Genome Analysis Toolkit Best Practices Pipeline. *Curr. Protoc. Bioinform.* **2013**, *43*, 1–33. [CrossRef]

86. Swofford, D.L. *PAUP*: Phylogenetic Analysis Using Parsimony, version 4.0 b10*; Sinauer Associates: Sunderland, MA, USA, 2003.

87. Felsenstein, J. Confidence limits on phylogenies: An approach using the bootstrap. *Evolution* **1985**, 783–791. [CrossRef] [PubMed]

88. Stamatakis, A. RAxML version 8: A tool for phylogenetic analysis and post-analysis of large phylogenies. *Bioinformatics* **2014**, *30*, 1312–1313. [CrossRef] [PubMed]

89. Ronquist, F.; Teslenko, M.; van der Mark, P.; Ayres, D.L.; Darling, A.; Höhna, S.; Larget, B.; Liu, L.; Suchard, M.A.; Huelsenbeck, J.P. MrBayes 3.2: Efficient Bayesian phylogenetic inference and model choice across a large model space. *Syst. Biol.* **2012**, *61*, 539–542. [CrossRef] [PubMed]

90. Miller, M.A.; Pfeiffer, W.; Schwartz, T. Creating the CIPRES Science Gateway for Inference of Large Phylogenetic Trees. In Proceedings of the Gateway Computing Environments Workshop (GCE), New Orleans, LA, USA, 14 November 2010; pp. 1–8.

International Journal of
Molecular Sciences

MDPI

Article

Phylogenomic and Comparative Analyses of Complete Plastomes of *Croomia* and *Stemona* (Stemonaceae)

Qixiang Lu [†], Wenqing Ye [†], Ruisen Lu, Wuqin Xu and Yingxiong Qiu *

Key Laboratory of Conservation Biology for Endangered Wildlife of the Ministry of Education,
and College of Life Sciences, Zhejiang University, Hangzhou 310058, China; 0016616@zju.edu.cn (Q.L.);
yewenqing@zju.edu.cn (W.Y.); reason@zju.edu.cn (R.L.); 21707105@zju.edu.cn (W.X.)
* Correspondence: qyxhero@zju.edu.cn; Tel.: +86-0571-8820-6463
† These authors have contributed equally to this work.

Received: 4 July 2018; Accepted: 3 August 2018; Published: 13 August 2018

Abstract: The monocot genus *Croomia* (Stemonaceae) comprises three herbaceous perennial species that exhibit EA (Eastern Asian)–ENA (Eastern North American) disjunct distribution. However, due to the lack of effective genomic resources, its evolutionary history is still weakly resolved. In the present study, we conducted comparative analysis of the complete chloroplast (cp) genomes of three *Croomia* species and two *Stemona* species. These five cp genomes proved highly similar in overall size (154,407–155,261 bp), structure, gene order and content. All five cp genomes contained the same 114 unique genes consisting of 80 protein-coding genes, 30 tRNA genes and 4 rRNA genes. Gene content, gene order, AT content and IR/SC boundary structures were almost the same among the five Stemonaceae cp genomes, except that the *Stemona* cp genome was found to contain an inversion in *cem*A and *pet*A. The lengths of five genomes varied due to contraction/expansion of the IR/SC borders. A/T mononucleotides were the richest Simple Sequence Repeats (SSRs). A total of 46, 48, 47, 61 and 60 repeats were identified in *C. japonica*, *C. heterosepala*, *C. pauciflora*, *S. japonica* and *S. mairei*, respectively. A comparison of pairwise sequence divergence values across all introns and intergenic spacers revealed that the *ndh*F–*rpl*32, *psb*M–*trn*D and *trn*S–*trn*G regions are the fastest-evolving regions. These regions are therefore likely to be the best choices for molecular evolutionary and systematic studies at low taxonomic levels in Stemonaceae. Phylogenetic analyses of the complete cp genomes and 78 protein-coding genes strongly supported the monophyly of *Croomia*. Two Asian species were identified as sisters that likely diverged in the Early Pleistocene (1.62 Mya, 95% HPD: 1.125–2.251 Mya), whereas the divergence of *C. pauciflora* dated back to the Late Miocene (4.77 Mya, 95% HPD: 3.626–6.162 Mya). The availability of these cp genomes will provide valuable genetic resources for further population genetics and phylogeographic studies on *Croomia*.

Keywords: *Croomia*; *Stemona*; chloroplast genome; comparative genomics; phylogeny; biogeography

1. Introduction

Croomia Torr. ex Torr. et Gray belongs to the monocot family Stemonaceae Engl (Pandanales, Liliidae) and comprises three herbaceous perennial species: *C. pauciflora* (Nutt.) Torr., *C. japonica* Miq. and *C. heterosepala* (Bak.) Oku. Of these three species, *C. japonica* and *C. heterosepala* are endemic to warm-temperate deciduous forests in East Asia, while *C. pauciflora* grows in temperate-deciduous forests in North America [1–3]. There is a considerable difference in morphological traits among this genus. For example, the four tepals of *C. japonica* are homomorphic with a re-curved edge, while those of *C. heterosepala* have a flat edge, and one outside tepal is much larger than the other three [4,5]. Compared to two Asian species, *C. pauciflora* has a smaller flower, shorter petiole, denser underground stem nodes and a more obvious heart-shape leaf base [1]. As the roots of *Croomia* species contain compounds such as pachysamine, didehydrocroomine and croomine groups, they are used as folk medicine to treat cough and injuries [6,7]. *Croomia* can reproduce sexually through seed formation via cross-pollination and asexually through underground rhizomes [1,8]. Due to their limited distribution and small population sizes, the three extant species of *Croomia* are listed as "threatened" or "endangered" in China, Japan and the Americas [9–11]. The other three genera of Stemonaceae are *Pentastemona*, *Stemona* and *Stichoneuron*. The species of *Stichoneuron* are located in India, Thailand and Peninsular Malaysia, while those of *Pentastemona* are only in Sumatra [3,8]. The genus *Stemona* comprises ca. 25 species with the widest distribution from Northeast Asia to Southeast Asia and Australia. The roots of *Stemona* species contain similar medicine compounds as *Croomia* [7]. Although *Croomia* and *Stemona* species have important pharmacological and ecological value, limited molecular markers were available for the utilization, conservation and breeding of these species in the context of population genetics and phylogenetic studies [12].

Croomia exhibits a well-known classic intercontinental disjunct distribution between Eastern Asia (EA) and Eastern North America (ENA) [1,8,13,14]. This continental disjunction pattern was suggested to have resulted from fragmentation of the mid-Tertiary mesophytic forest flora throughout a large part of the Northern Hemisphere, as global temperature cooled down in the late Tertiary and Quaternary [15,16]. For the two East Asian endemics, *C. japonica* is distributed in East China and southern Japan, while *C. heterosepala* is in northern Japan, and they have adjacent ranges in South Japan [17,18]. Therefore, *Croomia* is well suited for testing biogeographic hypotheses about the evolution of both the eastern Asian–eastern North American and eastern Asian–Japanese Archipelago floristic disjunctions. Based on previous molecular phylogenetic analyses using cpDNA sequence variation of the *trn*L-F region, the two Asian species were identified as sister species that likely diverged in the Mid-to-Late Pleistocene (0.84–0.13 million years ago, Mya), whereas the divergence of *C. pauciflora* dates back to the Late Plio-/Pleistocene (<2.6 Mya) [12]. However, the previous cpDNA analysis based on a few parsimony informative sites yielded low bootstrap values for the majority of clades [12]. Thus, it is necessary to develop more highly variable genetic markers for determining the phylogenetic relationships and divergence times for *Croomia*. Nowadays, many phylogenetic relationships that remained unresolved with few loci have been clarified by using whole cp genome sequences [19–21]. Thus, whole cp genome sequences are increasingly being used in phylogeny reconstruction and providing hypervariable genetic markers for population genetic studies, especially in a group of recently-diverged species [22,23].

Here, we sequenced three *Croomia* and two *Stemona* cp genomes using the next-generation Illumina genome analyzer platform. We compared the cp genomes of two Stemonaceae genera to characterize their structural organization and variations and identify the most variable regions. This information on interspecific variability of each region will help guide further systematic and evolutionary studies of Stemonaceae. In addition, we used the whole cp genomes to resolve the phylogenetic relationships of *Croomia* and infer the historical biogeography of the genus.

2. Results and Discussion

2.1. Genome Assembly and Features

Illumina paired-end sequencing yielded 14,163,520–31,094,272-bp clean reads after trimming, and the de novo assembly generated 50,369–123,479 contigs for five Stemonaceae species. With the cp genome of *C. palmata* as a reference, contigs were combined to generate the draft cp genome for each species. The lengths of determined nucleotide sequences were 154,672, 154,407, 155,261, 154,224 and 154,307 bp for *C. japonica*, *C. heterosepala*, *C. pauciflora*, *S. japonica* and *S. mairei*, respectively. (Figure 1, Table S1). All five cp genomes exhibited the typical quadripartite structure of angiosperms, consisting of a pair of IR regions (27,082–27,243 bp) separated by an LSC region (81,844–82,429 bp) and an SSC region (17,889–18,346 bp). The cp genomes of three *Croomia* species and two *Stemona* species were deposited in GenBank (MH177871, MH191379–MH191382).

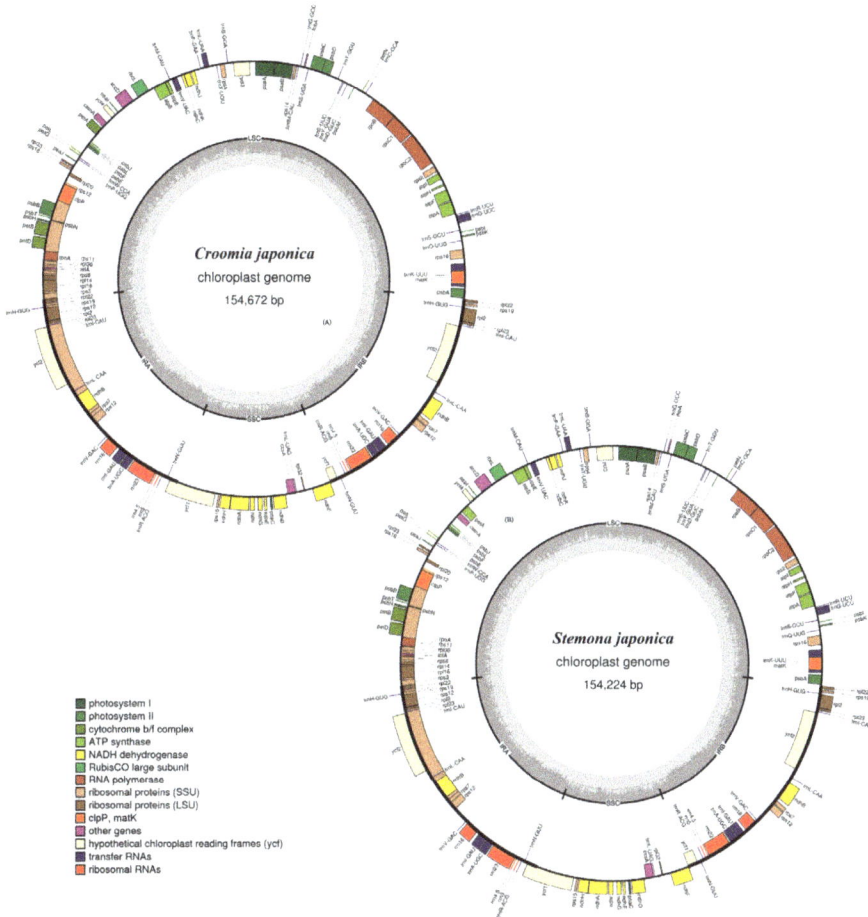

Figure 1. Gene maps of *Croomia* and *Stemona* chloroplast genomes. (**A**) *Croomia japonica*; (**B**) *Stemona japonica*.

These five cp genomes contained 134 genes identically, of which 114 were unique and 20 were duplicated in IR regions (Table S1). Those 134 genes were arranged almost in the same order except *cem*A and *pet*A, which were inverted at the LSC region of two *Stemona* species. Gene inversions at LSC were also reported in other angiosperm, such as *Silene* [24], *Cymbidium* [19] and *Acacia dealbata* [25]. The 114 unique genes included 80 protein-coding genes, 30 tRNA genes and 4 rRNA genes. In *Croomia* species, the overall GC content was 38.3%, and the GC contents of the LSC, SSC and IR regions were 36.6%, 32.3–32.5% and 42.8–42.9%, respectively, while those of *Stemona* were 38.0%, 36.2%, 32.1% and 42.7% (Table S1). In all five genomes, nine of the protein-coding genes and six of the tRNA genes possessed a single intron, while three genes (*rps*12, *clp*P and *ycf*3) contained two introns (Table 1). The *rps*12 gene was trans-spliced; the 5′ end exon was located in the LSC region, and the 3′ end exon and intron were located in the IR regions. Compared to many other species, such as *Salvia miltiorrhiza* [26] and *Cornales* [27], the SSC region of the five studied species was found to have a different (reverse) orientation. The reverse orientation of the SSC region has also been reported in a wide variety of plant species [28–30]. This phenomenon is sometimes interpreted as a major inversion existing within the species [29,31,32]. In fact, the two orientations of the SSC region have been found to occur regularly during the course of chloroplast DNA replication within individual plant cells [33,34]. Thus, the reverse orientation of the SSC region found in the five Stemonaceae cp genomes may represent a form of plastid heteroplasmy [30,35].

Table 1. List of genes in Stemonaceae chloroplast genomes.

Category of Genes	Groups of Genes	Names of Genes
Self-replication	rRNA genes	*rrn16*(×2), *rrn23*(×2), *rrn4.5*(×2), *rrn 5*(×2)
	tRNA genes	*trnA-UGC* *(×2), *trnC-GCA*, *trnD-GUC*, *trnE-UUC*, *trnF-GAA*, *trnfM-CAU*, *trnG-GCC*, *trnG-UCC* *, *trnH-GUG*(×2), *trnI-CAU*(×2), *trnI-GAU* *(×2), *trnK-UUU* *, *trnL-CAA*(×2), *trnL-UAA* *, *trnL-UAG*, *trnM-CAU*, *trnN-GUU*(×2), *trnP-UGG*, *trnQ-UUG*, *trnR-ACG*(×2), *trnR-UCU*, *trnS-GCU*, *trnS-GGA*, *trnS-UGA*, *trnT-GGU*, *trnT-UGU*, *trnV-GAC*(×2), *trnV-UAC* *, *trnW-CCA*, *trnY-FUA*
	Small subunit of ribosome	*rps2*, *rps3*, *rps4*, *rps7*(×2), *rps8*, *rps11*, *rps12* **(×2), *rps14*, *rps15*, *rps16* *, *rps18*, *rps19*(×2)
	Large subunit of ribosome	*rpl2* *(×2), *rpl14*, *rpl16* *, *rpl20*, *rpl22*(×2), *rpl23*(×2), *rpl32*, *rpl33*, *rpl36*
	DNA-dependent RNA polymerase	*rpoA*, *rpoB*, *rpoC1* *, *rpoC2*
Genes for photosynthesis	Subunit of NADH-dehydrogenase	*ndhA* *, *ndhB* *(×2), *ndhC*, *ndhD*, *ndhE*, *ndhF*, *ndhG*, *ndhI*, *ndhH*, *ndhJ*, *hdhK*
	Subunit of Photosystem 1	*psaA*, *psaB*, *psaC*, *psaI*, *psaJ*, *ycf3* **
	Subunit of Photosystem 2	*psbA*, *psbB*, *psbC*, *psbD*, *psbE*, *psbF*, *psbH*, *psbI*, *psbJ*, *psbK*, *psbL*, *psbM*, *psbN*, *psbT*
	Subunits of cytochrome b/f complex	*petA*, *petB* *, *petD* *, *petG*, *petL*, *petN*
	Subunits of ATP synthase	*atpA*, *atpB*, *atpE*, *atpF* *, *atpH*, *atpI*
	Large subunit of rubisco	*rbcL*
Other genes	Maturase	*matK*,
	Protease	*clpP* **
	Envelope membrane protein	*cemA*
	Subunit of Acetyl-CoA-carboxylase	*accD*
	c-type cytochrome synthesis gene	*ccsA*
	Translation initiation factor IF-1	*infA*
Genes of unknown function	Open reading frames (ORF, ycf)	*ycf1*, *ycf2*(×2), *ycf4*, *lhbA*

* Gene with one intron, ** gene with two introns; (×2) indicates genes duplicated in the IR region.

2.2. Contraction and Expansion of Inverted Repeats

Length variation in angiosperm cp genomes is due most typically to the expansion or contraction of the IR into or out of adjacent single-copy regions and/or changes in sequence complexity due to insertions or deletions of novel sequences [36,37]. Compared to reference cp genome *C. palmata*, all five species exhibited IR expansion at the IRb/LSC border, leading to entire *rpl22* duplication. In a previous study, a partial duplication of the *rpl22* gene was reported in some monocot species of Asparagales and Commelinales [38]. Although the gene number and gene order were conserved across these five Stemonaceae species, minor differences were still observed at the boundaries (Figure 2). At the IRa/LSC border, the spacer from *rpl22* to this border of *Stemona* (65 bp) was longer than that of *Croomia* (24–25 bp), except *C. pauciflora*. As for the *ycf1* gene, there were 4580–4662-bp sequences located at SSC in *Croomia* and 4374–4383 bp in *Stemona*, while the pseudogene fragment duplications in IRb were 970 bp and 1206 bp in *Croomia* and *Stemona*, respectively. The *ndhF* gene exhibited variable sequences in SSC of *Croomia* (2215–2226 bp), while invariable in *Stemona* with a 2190-bp length. At the border of

IRb/LSC, the spacer from *psb*A to this border of *Croomia* ranged from 91 bp–99 bp, while it ranged from 94 bp–104 bp in *Stemona*. These differences between the five cp genomes led to the length variation of their whole genome sequences.

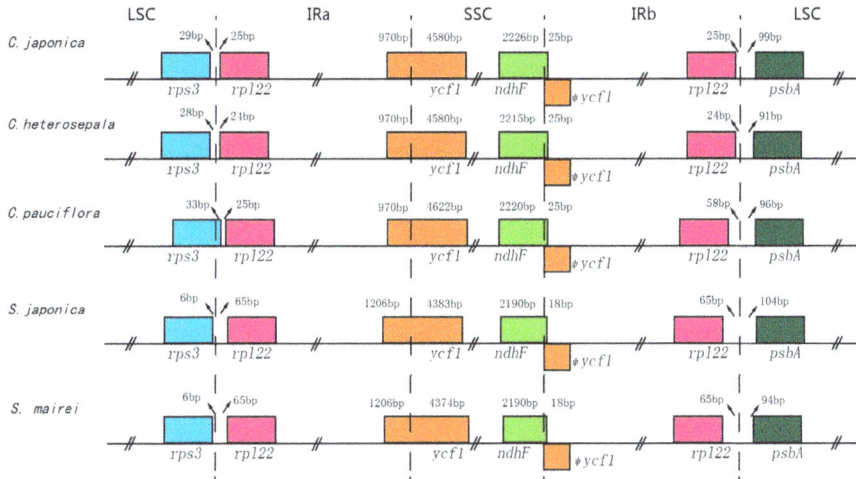

Figure 2. Comparison of LSC, IR and SSC junction positions among five Stemonaceae chloroplast genomes.

2.3. Divergence Hotspot Regions

To elucidate the level of sequence divergence, the three *Croomia* and two *Stemona* cp genome sequences were compared and plotted using the mVISTA program (Figure 3). Like most angiosperms, the sequence divergence of IR regions was lower than that of the LSC and SSC region [39,40], which may involve copy correction of IRs as a mechanism [41]. We identified 140 regions in total with more than a 200-bp length (68 protein-coding regions (CDS), 53 Intergenic Spacers (IGS) and 19 introns). The nucleotide variability (Pi) of these 140 regions ranged from 0.080% (*rrn*16) to 9.565% (IGS *petN–psbM*) among the five cp genomes. The average Pi of the non-coding region was 3.644%, much higher than coding regions (1.587%), as found in most angiosperms [42,43]. For the 68 CDS, the Pi values for each region ranged from 0.231% (*rpl*2 CDS1) to 4.047% (*ycf*1), whereby 10 regions (i.e., *matK, rpl*33, *rps*15, *psbH, rps*18, *rps*3, *rpl*20, *ccsA, ndhF, accD*) had remarkably high values (*pi* > 2.5%). For the 53 IGS regions, Pi values ranged from 0.185% (*trnN–ycf*1) to 9.565% (*petN–psbM*). Again, ten of those regions showed considerably high values (*pi* > 5.7%; i.e., *rpl*32–*ndhF, trnS–trnG, ndhE–psaC, ndhD–ccsA, atpF–atpH, psbM–trnD, trnE–trnT, petL–petG, rps*16–*trnQ, accD–psaI*; see Figure 4). A comparison of DNA sequence divergence revealed that three of these ten noncoding regions, *ndhF–rpl*32 (PICs = 96), *psbM–trnD* (PICs = 73) and *trnS–trnG* (PICs = 49), are the most variable regions across Stemonaceae (Figure A1). Thus, these three regions may be good candidates for resolving future low-level phylogeny and phylogeography in Stemonaceae. In a previous study, the availability of plastid noncoding regions was compared across 10 major lineages of angiosperms (such as Nymphaeales, monocots, eurosids) [44]. However, only five families of monocots represented by five species pairs were included, without Stemonaceae. The three variable regions predicted here are among the top 13 regions of monocots in the research by Shaw et al. [44], with *ndhF–rpl*32, *psbM–trnD* and *trnS–trnG* ranked first, third and 11th, respectively. Of these regions, *ndhF–rpl*32 has long been a popular region in phylogenetic studies of angiosperms [44,45]. Meanwhile, *psbM–trnD* and *trnS–trnG* are also noted as highly variable in Liliaceae [46] and occasionally used in low-level phylogenetic analyses (*Scabiosa*: [47]; *Solms-laubachia*: [48]). The resolution of recent divergences in monocots would benefit considerably by the inclusion of any or all of these highly variable regions.

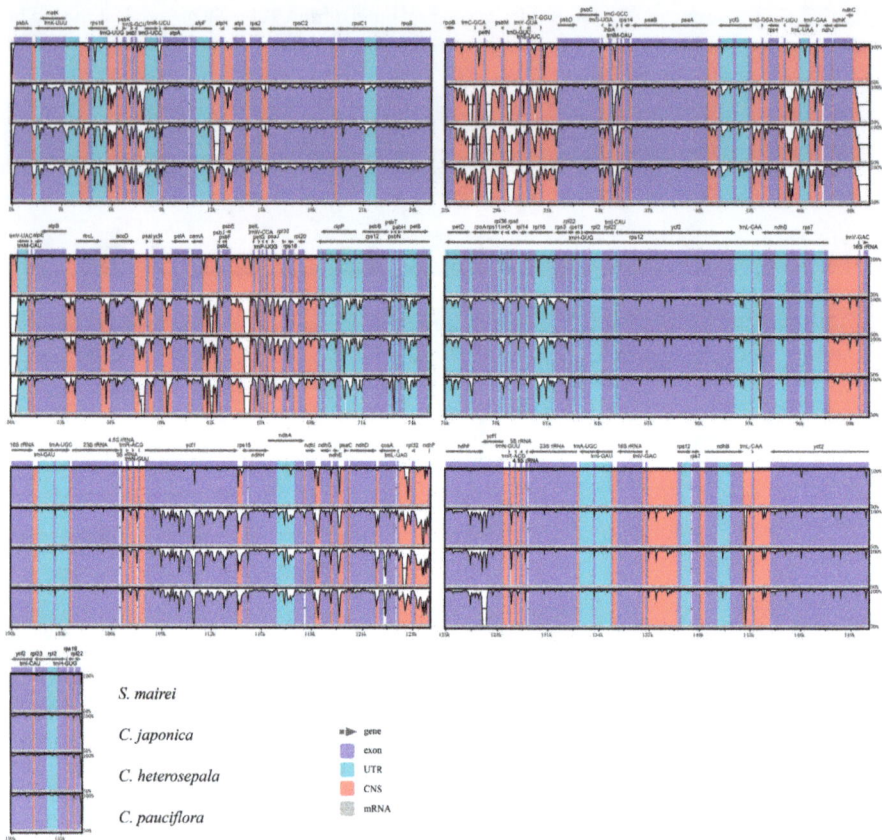

Figure 3. Sequence identity plots among five Stemonaceae chloroplast genomes, with *Stemona japonica* as a reference. CNS: conserved non-coding sequences; UTR: untranslated region.

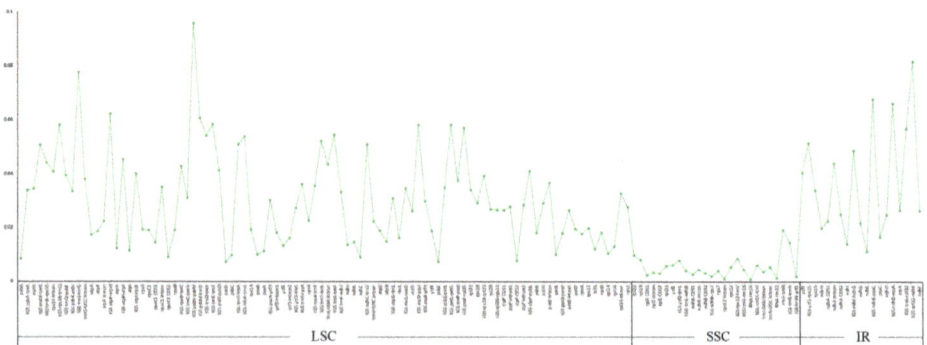

Figure 4. The nucleotide variability (Pi) values were compared among five Stemonaceae species.

2.4. Repetitive Sequences and SSR Polymorphisms

With the criterion of a copy size of 30 bp or longer and a sequence identity >90%, REPUTER [49] identified 47, 49, 48, 61 and 60 repeats (including forward, palindromic, complement and reverse repeats) in five cp genome sequences of *C. japonica*, *C. heterosepala*, *C. pauciflora*, *S. japonica* and *S. mairei*, respectively (Figure 5A). *C. japonica* contained 21 forward repeats, 24 palindromic repeats, 1 complement repeat and 1 reverse repeat, and *S. japonica* contained 27, 25, 1 and 8 repeats, correspondingly. The other two *Croomia* species and *S. mairei* contained no complement repeats. The numbers of forward repeats, palindromic and reverse repeats were, respectively, 21, 27 and 1 in *C. japonica*, 25, 21 and 2 in *C. heterosepala* and 27, 25 and 8 in *S. mairei* (Figure 5A). The lengths of majority repeats were 30, 31 and 43 bp in size (Figure 5B). For *Croomia*, the repeats were mainly located in *ycf*2 (46.8–58.3%) and non-coding regions (27.1–38.3%). As for *Stemona*, the repeats were mostly located in non-coding regions (58.3–59.0%) and *ycf*2 (33.3–34.4%). Only one repeat was across IGS (*psbC–trnS*) and CDS (*trn*S^{UGA}). The remaining repeats were found located in genes such as *ccs*A, *ycf*1, *trn*G^{UGA}, *trn*S^{GGA}, *trn*S^{GCU} and *psa*B.

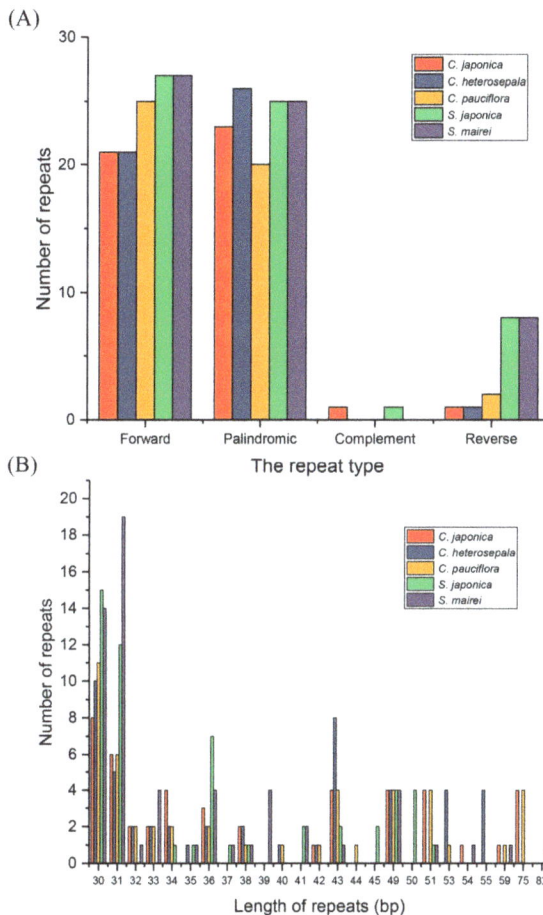

Figure 5. Analysis of repeated sequences in five Stemonaceae chloroplast genomes. (**A**) Frequency of repeats by length; (**B**) frequency of repeat types.

SSRs in the cp genome present high diversity in copy numbers, and they are important molecular markers for plant population genomics and evolutionary history [50,51]. SSRs (≥10 bp) were detected in these five Stemonaceae cp genomes by MIcroSAtellite (MISA) analysis [52], ranging from 90–116 in total. Among these SSRs, the mononucleotide repeat unit (A/T) occupied the highest proportion, with

71.2% in *C. japonica*, 70.9% in *C. heterosepala*, 63.3% in *C. pauciflora*, 64.0% in *S. japonica* and 62.4% in *S. mairei* (Figure 6A). SSR loci were mainly located in IGS (71.4%) (Figure 6B) and were also detected in introns (16.5%) and CDS (12.1%), such as *matK*, *atpA*, *rpoC2*, *rpoB*, *cemA*, *psbF*, *ycf2*, *ycf1* and *ndhD*. In general, the SSRs of these five cp genomes showed great variation, which can be used in population genetic studies of *Croomia* and *Stemona* species.

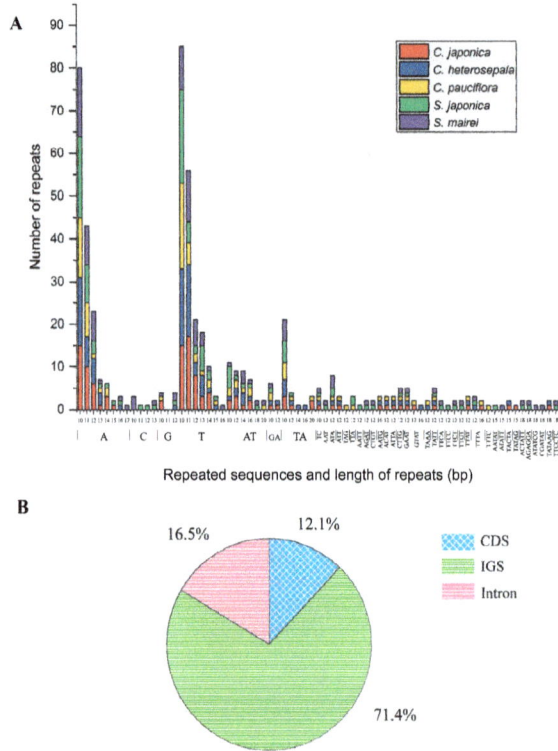

Figure 6. Simple Sequence Repeats (SSRs) in five Stemonaceae chloroplast genomes. (**A**) Numbers of SSRs by length; (**B**) distribution of SSR loci. IGS: intergenic spacer region; CDS: protein-coding regions.

2.5. Phylogenetic Analysis, Divergence Time and Ancestral Area Reconstruction

CP genome sequences have been successfully used in angiosperm phylogenetic studies [22,53]. The Maximum Likelihood (ML) and Bayesian Inference (BI) analyses of both whole sequences and protein-coding region of three *Croomia* and two *Stemona* cp genomes yielded nearly identical tree topologies, with 100% bootstrap and 1.0 Bayesian posterior probabilities at each node (Figure 7). This phylogenetic tree supports the monophyly of *Croomia*. Two Asian species *C. japonica* and *C. heterosepala* formed a clade, being strongly recovered as sisters of the North American species *C. pauciflora*. This tree topology is largely congruent with that inferred from *trnL*–F [12], but obtained much higher bootstrap support values. Using average substitution rates of whole cp genomes, the divergence time between the two Asian species, *C. japonica* and *C. heterosepala*, was estimated as ca. 1.621 Mya (1.125–2.251 Mya) and, thus, compatible with the early-Pleistocene event. By contrast, the divergence time between North American *C. pauciflora* and Asian species was estimated as ca. 4.774 Mya (3.626–6.162 Mya) (i.e., the Late Miocene). The divergence times estimated in this paper are much older than that estimated by the strict molecular clock method (*C. pauciflora*/the East Asian lineage: ca. 2.61–0.41 Mya; *C. japonica*/*C. heterosepala*: ca. 0.84–0.13 Mya) [12].

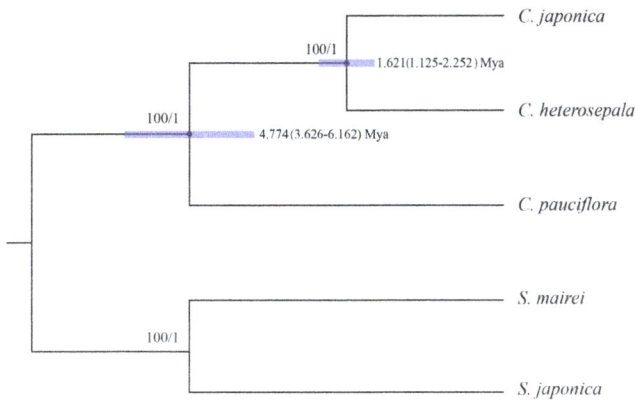

Figure 7. Phylogenetic relationships of three *Croomia* species inferred from Maximum Likelihood (ML) and Bayesian Inference (BI) and divergence time of three *Croomia* species estimated using Bayesian Evolutionary Analysis Sampling Trees (BEAST) analysis. Numbers above the lines represent ML bootstrap values and BI posterior probability. Blue bars indicate the 95% highest posterior density (HPD) credibility intervals for node ages (Mya). Numbers at the node represent divergence time (Mya) and 95% highest posterior density intervals. The phylogenetic tree based on 74 protein-coding genes is completely consistent with this topology.

The divergence between *C. pauciflora* and two Asian species coincides with the first sundering of the Bering Land Bridge (BLB) between the late Miocene and early Pliocene, most approximately at 5.4–5.5 Mya (Milne and Abbott, 2002) [54]. The Bayesian Binary MCMC (BBM) analysis of ancestral area reconstruction identified Asia as the most likely ancestral range (Node III, marginal probability: 0.93; Figure A2), indicating a possible intercontinental plant migration from Asia to North America. Indeed, the BLB served as an important route for temperate floristic exchanges between Asia and North America from the Eocene to the early Pliocene [55,56]. Subsequently, as a member of the Tertiary relict flora [15], *Croomia* species on the two continents experienced disjunct distribution and evolved separately after the Late Miocene. Thus, we conclude that the current distribution and differentiation of *Croomia* species in eastern Asia and eastern North America likely resulted from a combination of ancient migration and vicariant events. The divergence time between *C. japonica* and *C. heterosepala* fell into the Early Pleistocene. Habitat fragmentation resulting from the climatic vicissitudes of the (Late) Quaternary likely led to the speciation of *C. japonica* and *C. heterosepala* [12]. The above inferences seem to be consistent with the palaeovegetational and climatic history of eastern Asia and eastern America. However, considering that the cp genome is a haploid, uniparentaly-inherited and single locus [57], a nuclear (biparental) marker is also needed to elucidate the diversification process and demography history of *Croomia* species.

3. Materials and Methods

3.1. Sample Preparation, Sequencing, Assembly and Validation

Fresh leaves of *C. japonica* from China, *C. heterosepala* from Japan, *C. pauciflora* from North America and two outgroup species *Stemona japonica* (Bl.) Miq. and *S. mairei* (Levl.) Krause from China were sampled and dried with silica gel. The voucher specimens were deposited in the Herbarium of Zhejiang University (HZU). Total genomic DNA was extracted from ~3 mg materials using DNA Plantzol Reagent (Invitrogen, Carlsbad, CA, USA) following the manufacturer's protocol. The quality and concentration of the DNA were detected using agarose gel electrophoresis. Purified DNA was sheared into ~500-bp fragments, and the fragmentation quality was checked on a Bioanalyzer 2100

(Agilent Technologies, Santa Clara, CA, USA). Paired-end sequencing libraries were constructed according to the Illumina standard protocol (Illumina, San Diego, CA, USA). Genomic DNAs of five species were sequenced using an Illumina HiSeq^{TM} 2000 (Illumina, San Diego, CA, USA) at Beijing Genomics Institute (BGI; Shenzhen, China). Plastome sequences were assembled using a combination of de novo and reference-guided assembly [58]. Firstly, to obtain clean reads, the CLC-quality trim tool was used to remove low-quality bases ($Q < 20$, 0.01 probability error). Secondly, we assembled the clean reads into contigs on the CLC de novo assembler. Thirdly, all the contigs were aligned with the reference cp genome of *Carludovica palmate* Ruiz. & Pav. (NC_026786.1) using local BLAST (http://blast.ncbi.nlm.nih.gov/) (27 December 2016), and aligned contigs were ordered according to the reference cp genome with ≥90% similarity and query coverage. Then, to construct the draft cp genome of each species, the ordered contigs usually representing the whole reconstructed genome were imported into GENEIOUS v9.0.5 software (http://www.geneious.com) (18 March 2017), where the clean reads were remapped onto the contigs.

3.2. Genome Annotation and Whole Genome Comparison

The annotation of five species was performed using the Dual Organellar GenoMe Annotator (DOGMA) [59]. The start and stop codons and intron/exon boundaries were manually corrected by comparison to homologous genes from the reference genome of *C. palmate*. We also verified the transfer RNAs (tRNAs) using tRNAscan-SE v1.21 with default settings [60]. The circular genome maps were drawn using the OrganellarGenome DRAW tool (OGDRAW) [61], followed by manual modification.

Genome comparison among the five Stemonaceae cp genomes was analyzed using mVISTA [62] with *C. palmate* as a reference. Six genome sequences were aligned in Shuffle-LAGAN mode with default parameters, and the conservation region was visualized in an mVISTA plot. To identify the divergence hotspot regions in the five Stemonaceae cp genomes, the nucleotide variability of protein coding genes, introns and intergenic spacer sequences of five species were evaluated using DNASP v5.10 [63]. The above regions were extracted following two criteria: (a) total number of mutation (Eta) >0; and (b) the aligned length >200 bp. The inverted regions in *cem*A, *cem*A–*pet*A and *pet*A were excluded. The top ten most variable noncoding regions with a high Pi value were counted by Potentially Informative Characters (PICs) across species pair of *C. japonica* and *S. japonica* following Shaw et al. [64]. Any large structural event of the cp genome, such as gene order rearrangements or IR expansion/contractions, were recorded.

3.3. Characterization of Repeat Sequence and SSRs

REPUTER [49] was used to find the location and length of repeat sequences, including forward, palindrome, complement and reverse repeats in the five cp genomes. The minimum repeat size was set to 30 bp, and the sequence identity of repeats was no less than 90% or greater sequence identity with the Hamming distance equal to 3. The MISA perl script was used to detect simple sequence repeats (SSRs) [52] with thresholds of 10 bp in length for mono-, di-, tri, tetra-, penta- and hexa-nucleotide SSRs.

3.4. Phylogenetic Analysis, Divergence Time and Ancestral Area Reconstruction

The five cp genome sequences were aligned using MAFFT v7 [65]. Two *Stemona* species were used as outgroups. ML and BI analysis were used to reconstruct the phylogenetic trees. In order to examine the phylogenetic utility of different regions, two datasets were used: (1) the complete cp genome sequences; (2) 78 protein-coding genes shared by the five cp genomes (two inverted genes of *cem*A and *pet*A in *Stemona* species were excluded). Gaps (indels) were treated as missing data. The Akaike Information Criterion (AIC) in JMODELTEST v2.1.4 [66] was used to determine the best-fitting model of nucleotide substitutions. The GTR + I + G model was used for two datasets. The ML tree was constructed using RAXML-HPC v8.2.10 with 1000 replicates on the Cyberinfrastructure for Phylogenetic Research (CIPRES) Science Gateway website (http://www.phylo.org/) (10 May 2017) [67]. BI analysis was conducted in MRBAYES v3.2 [68]. The Markov chain Monte Carlo (MCMC) was set to run 1,000,000 generations and sampled every 1000 generations. The first 25% of generations was discarded as burn-in.

Due to the lack of fossil records, we used the average substitution rate 0.51952×10^{-9} per site per year (s/s/y) of the whole cp genome in Brassicaceae [69,70] to estimate interspecific divergence time of *Croomia*. The Bayesian analysis was implemented in BEAST v1.8.4 [71] using the GTR + I + G substitution model. MCMC analysis of 20,000,000 generations was implemented, in which every 1000 generations were sampled, under an uncorrelated lognormal relaxed clock approach using the Yule speciation tree prior with the substitution rate. TRACER v1.6 [72] was used to check the effective population size (ESS) >200. TREEANNOTATOR v.1.8.4 [73] was used to produce maximum clade credibility trees from the trees after burning-in of 25%. The final tree was visualized in FIGTREE v1.4.3 (http://tree.bio.ed.ac.uk/software/figtree/) (13 May 2017).

To reconstruct the historical biogeography of *Croomia*, we performed Bayesian Binary MCMC (BBM) analysis as implemented in RASP v3.1 [74] using trees retained from the BI analysis (see above). According to the distribution of *Croomia*, we defined the following two areas: A, Asia (East Asia/South Asia); and B, North America. Accounting for phylogenetic uncertainty, we used 500 trees randomly chosen across all post-burn-in trees generated from BEAST analysis and ran the BBM analysis. A fixed JC + G (Jukes–Cantor + Gamma) model was chosen with a null root distribution. The MCMC chains were run for 500,000 generations, and every 100 generations were sampled. The ancestral ranges obtained were projected onto the MCC tree.

4. Conclusions

Here, we sequenced the first five complete cp genomes in Stemonaceae. Each genome possesses the typical structure shared with other angiosperm species. Several highly variable noncoding cpDNA regions were identified, which should be the best choices for future phylogenetic, phylogeographic and population-level genetic studies in Stemonaceae. The phylogenomic and biogeographic analyses of *Croomia* reveal that ancient migration and vicariance-driven allopatric speciation resulting from historical climate oscillations most likely played roles in the formation of the disjunct distributions and divergence of these three *Croomia* species.

Supplementary Materials: Supplementary materials can be found at http://www.mdpi.com/1422-0067/19/8/2383/s1.

Author Contributions: Y.Q. conceived of the idea. R.L. contributed to the sampling. Q.L. performed the experiment. Q.L., W.Y., R.L. and W.X. analyzed the data. The manuscript was written by Q.L., W.Y. and Y.Q.

Funding: This research was funded by the International Cooperation and Exchange of the National Natural Science Foundation of China (Grant Nos. 31561143015, 31511140095) and the National Natural Science Foundation of China (Grant Nos. 31570214, 31700321).

Acknowledgments: The authors thank Shota Sakaguchi from Kyoto University for collecting plant materials in Japan.

Conflicts of Interest: The authors declare that the research was conducted in the absence of any commercial or financial relationships that could be construed as a potential conflict of interest.

Abbreviations

cp   Complete chloroplast
IR   Inverted Repeat
lSC   Large Single Copy
SSC   Small Single Copy
Pi   Nucleotide variability
SSR   Simple Sequence Repeat
PIC   Potentially Informative Character
ML   Maximum likelihood
BI   Bayesian Inference
MCMC   Markov chain Monte Carlo

Appendix A

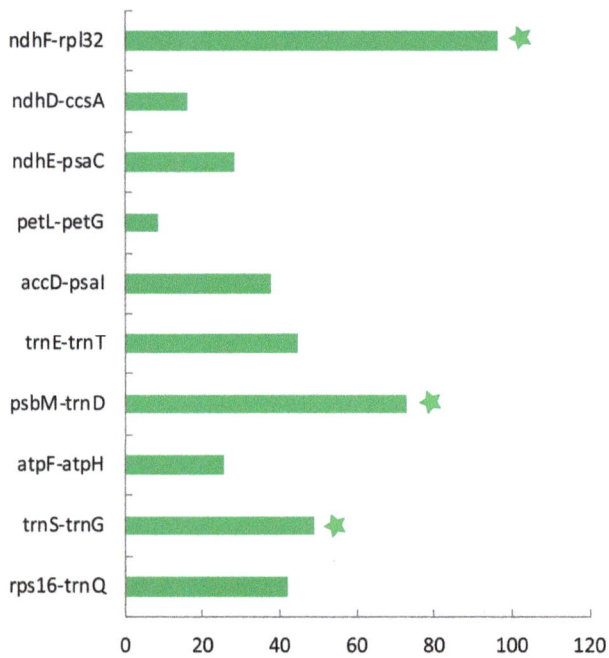

Figure A1. PIC values of the top ten most variable noncoding regions in Stemonaceae.

LEGEND

- ■ Uncertain
- ■ A (Asia)
- ■ B (North America)
- ■ AB
- → Possible dispersal event

III → (B) *C. pauciflora*

II

(A) *C. japonica*

IV (A) *C. heterosepala*

(A) *S. mairei*

(A) *S. japonica*

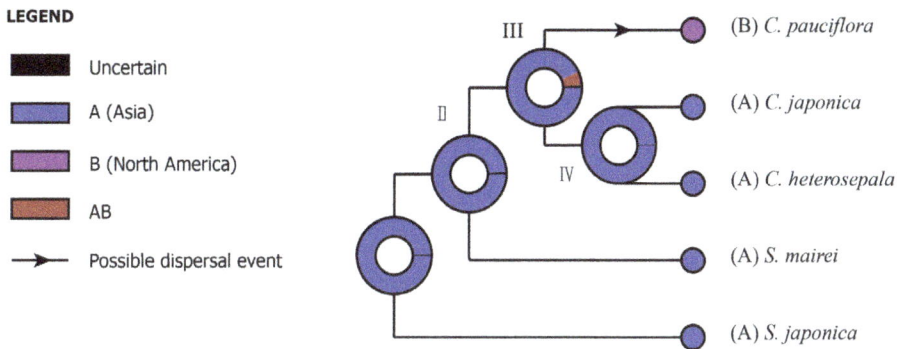

Figure A2. The Bayesian Binary MCMC (BBM) analysis of ancestral area reconstruction.

References

1. Rogers, G.K. The Stemonaceae in the southeastern United States. *J. Arnold Arboretum* **1982**, *63*, 327–336.
2. Whetstone, R.D. Notes on *Croomia* pauciflora (Stemonaceae). *Rhodora* **1984**, *25*, 131–137.
3. Li, E.X. Studies on Phylogeography of *Croomia* and Phylogeny of *Croomia* and Its Allies. Ph.D. Thesis, Zhejiang University, Hangzhou, China, 2006.
4. Okuyama, S. On the Japanese species of *Croomia. J. Jpn. Bot.* **1944**, *20*, 31–32.
5. Ohwi, J. *Croomia*. In *Flora of Japan*; Smithsonian Institution: Washington, DC, USA, 1965; p. 279.
6. Lin, W.; Cai, M.; Ying, B.; Feng, R. Studies on the chemical constituents of *Croomia japonica* Miq. *Yao Xue Xue Bao* **1993**, *28*, 202–206. [PubMed]
7. Pilli, R.A.; Ferreira, M.C. Recent progress in the chemistry of the *Stemona* alkaloids. *Nat. Prod. Rep.* **2000**, *17*, 117–127. [CrossRef] [PubMed]
8. Ji, Z.H.; Duyfjes, B.E. "Stemonaceae", in Flora of China, Flagellariaceae through Marantaceae. In *Flora of China*; Wu, Z.Y., Raven, P.H., Eds.; Missouri Botanical Garden Press: Beijing, China, 2000; Volume 24, pp. 70–72.
9. Patrick, T.S.; Allison, J.; Krakow, G. Protected plants of Georgia. Georgia Department of Natural Resources, Natural Heritage Program. *Soc. Circ.* **1995**, *25*, 36–38.
10. Estill, J.C.; Cruzan, M.B. Phytogeography of rare plant species endemic to the southeastern United States. *Castanea* **2001**, *36*, 3–23.
11. Sung, W.; Yan, X. *China Species Red List*; Higher Education Press: Beijing, China, 2004.
12. Li, E.; Yi, S.; Qiu, Y.; Guo, J.; Comes, H.P.; Fu, C. Phylogeography of two East Asian species in *Croomia* (Stemonaceae) inferred from chloroplast DNA and ISSR fingerprinting variation. *Mol. Phylogenet. Evol.* **2008**, *49*, 702–714. [CrossRef] [PubMed]
13. Xiang, Q.Y.; Soltis, D.E.; Soltis, P.S. The eastern Asian and eastern and western North American floristic disjunction: Congruent phylogenetic patterns in seven diverse genera. *Mol. Phylogenet. Evol.* **1998**, *10*, 178–190. [CrossRef] [PubMed]
14. Wen, J. Evolution of eastern Asian and eastern North American disjunct distributions in flowering plants. *Annu. Rev. Ecol. Syst.* **1999**, *30*, 421–455. [CrossRef]
15. Wolfe, J.A. Some aspects of plant geography in the Northern Hemisphere during the late Cretaceous and Tertiary. *Ann. Mo. Bot. Gard.* **1975**, *62*, 264–279. [CrossRef]
16. Tiffney, B.H.; Manchester, S.R. The use of geological and paleontological evidence in evaluating plant phylogeographic hypotheses in the Northern Hemisphere Tertiary. *Int. J. Plant Sci.* **2001**, *162*, 48–52. [CrossRef]
17. Li, H.L. Floristic relationships between eastern Asia and eastern North America. *Trans. Am. Philos. Soc.* **1952**, *42*, 371–429. [CrossRef]
18. Fukuoka, N.; Kurosaki, N. Phytogeographical notes on some species of west Honshu, Japan 5. *Shoei Jr. Col. Ann. Rep. Stud.* **1985**, *17*, 61–71.

19. Yang, J.B.; Tang, M.; Li, H.T.; Zhang, Z.R.; Li, D.Z. Complete chloroplast genome of the genus *Cymbidium*: Lights into the species identification, phylogenetic implications and population genetic analyses. *BMC Evol. Boil.* **2013**, *13*, 84–98. [CrossRef] [PubMed]

20. Ruhsam, M.; Rai, H.S.; Mathews, S.; Ross, T.G.; Graham, S.W.; Raubeson, L.A.; Mei, W.; Thomas, P.I.; Gardner, M.F.; Ennos, R.A. Does complete plastid genome sequencing improve species discrimination and phylogenetic resolution in *Araucaria*? *Mol. Ecol. Res.* **2015**, *15*, 1067–1078. [CrossRef] [PubMed]

21. Firetti, F.; Zuntini, A.R.; Gaiarsa, J.W.; Oliveira, R.S.; Lohmann, L.G.; Van Sluys, M.A. Complete chloroplast genome sequences contribute to plant species delimitation: A case study of the *Anemopaegma* species complex. *Am. J. Bot.* **2017**, *104*, 1493–1509. [CrossRef] [PubMed]

22. Jansen, R.K.; Cai, Z.; Raubeson, L.A.; Daniell, H.; Claude, W.; Leebensmack, J.; Guisingerbellian, M.; Haberle, R.C.; Hansen, A.; Chumley, T.W. Analysis of 81 genes from 64 plastid genomes resolves relationships in angiosperms and identifies genome-scale evolutionary patterns. *Proc. Natl. Acad. Sci. USA* **2007**, *104*, 19369–19374. [CrossRef] [PubMed]

23. Cai, J.; Ma, P.; Li, H.; Li, D. Complete plastid genome sequencing of four *Tilia* species (Malvaceae): A comparative analysis and phylogenetic implications. *PLoS ONE* **2015**, *10*, e0142705. [CrossRef] [PubMed]

24. Sloan, D.B.; Alverson, A.J.; Wu, M.; Palmer, J.D.; Taylor, D.R. Recent acceleration of plastid sequence and structural evolution coincides with extreme mitochondrial divergence in the angiosperm genus *Silene*. *Genome Biol. Evol.* **2012**, *4*, 294–306. [CrossRef] [PubMed]

25. Wang, Y.; Qu, X.; Chen, S.; Li, D.; Yi, T. Plastomes of Mimosoideae: Structural and size variation, sequence divergence, and phylogenetic implication. *Tree Genet. Genom.* **2017**, *13*, 41–56. [CrossRef]

26. Qian, J.; Song, J.; Gao, H.; Zhu, Y.; Xu, J.; Pang, X.; Yao, H.; Sun, C.; Li, X.E.; Li, C. The complete chloroplast genome sequence of the medicinal plant *Salvia miltiorrhiza*. *PLoS ONE* **2013**, *3*, e57607. [CrossRef] [PubMed]

27. Fu, C.; Li, H.; Milne, R.I.; Zhang, T.; Ma, P.; Yang, J.; Li, D.; Gao, L. Comparative analyses of plastid genomes from fourteen Cornales species: Inferences for phylogenetic relationships and genome evolution. *BMC Genom.* **2017**, *18*, 956–963. [CrossRef] [PubMed]

28. Hansen, D.R.; Dastidar, S.G.; Cai, Z.; Penaflor, C.; Kuehl, J.V.; Boore, J.L.; Jansen, R.K. Phylogenetic and evolutionary implications of complete chloroplast genome sequences of four early-diverging angiosperms: *Buxus* (Buxaceae), *Chloranthus* (Chloranthaceae), *Dioscorea* (Dioscoreaceae), and *Illicium* (Schisandraceae). *Mol. Phylogenet. Evol.* **2007**, *45*, 547–563. [CrossRef] [PubMed]

29. Liu, Y.; Huo, N.; Dong, L.; Wang, Y.; Zhang, S.; Young, H.A.; Feng, X.; Gu, Y.Q. Complete chloroplast genome sequences of mongolia medicine *Artemisia frigida* and phylogenetic relationships with other plants. *PLoS ONE* **2013**, *8*, e57533. [CrossRef] [PubMed]

30. Walker, J.F.; Zanis, M.J.; Emery, N.C. Comparative analysis of complete chloroplast genome sequence and inversion variation in *Lasthenia burkei* (Madieae, Asteraceae). *Am. J. Bot.* **2014**, *101*, 722–729. [CrossRef] [PubMed]

31. Yang, M.; Zhang, X.; Liu, G.; Yin, Y.; Chen, K.; Yun, Q.; Zhao, D.G.; Almssallem, I.S.; Yu, J. The complete chloroplast genome sequence of date palm (*Phoenix dactylifera* L.). *PLoS ONE* **2010**, *5*, e12762. [CrossRef] [PubMed]

32. Wang, M.; Cui, L.; Feng, K.; Deng, P.; Du, X.; Wan, F.; Weining, S.; Nie, X. Comparative analysis of asteraceae chloroplast genomes: Structural organization, RNA editing and evolution. *Plant Mol. Biol. Rep.* **2015**, *33*, 1526–1538. [CrossRef]

33. Palmer, J.D. Chloroplast DNA exists in two orientations. *Nature* **1983**, *301*, 92–93. [CrossRef]

34. Wolfe, A.D.; Randle, C.P. Recombination, heteroplasmy, haplotype polymorphism, and paralogy in plastid genes: Implications for plant molecular systematics. *Syst Bot.* **2004**, *29*, 1011–1020. [CrossRef]

35. Walker, J.F.; Jansen, R.K.; Zanis, M.J.; Emery, N.C. Sources of inversion variation in the small single copy (SSC) region of chloroplast genomes. *Am. J. Bot.* **2015**, *102*, 1751–1752. [CrossRef] [PubMed]

36. Kim, K.J.; Lee, H.L. Complete chloroplast genome sequences from Korean ginseng (*Panax schinseng* Nees) and comparative analysis of sequence evolution among 17 vascular plants. *DNA Res.* **2004**, *11*, 247–261. [CrossRef] [PubMed]

37. Downie, S.R.; Jansen, R.K. A comparative analysis of whole plastid genomes from the apiales: Expansion and contraction of the inverted repeat, mitochondrial to plastid transfer of DNA, and identification of highly divergent noncoding regions. *Sys. Bot.* **2015**, *40*, 336–351. [CrossRef]

38. Wang, R.J.; Cheng, C.L.; Chang, C.C.; Wu, C.L.; Su, T.M.; Chaw, S. Dynamics and evolution of the inverted repeat-large single copy junctions in the chloroplast genomes of monocots. *BMC Evol. Biol.* **2008**, *8*, 6–8. [CrossRef] [PubMed]

39. Nazareno, A.G.; Carlsen, M.; Lohmann, L.G. Complete chloroplast genome of tanaecium tetragonolobum: The first *Bignoniaceae plastome*. *PLoS ONE* **2015**, *10*, e0129930. [CrossRef] [PubMed]

40. Yao, X.; Tang, P.; Li, Z.; Li, D.; Liu, Y.; Huang, H. The first complete chloroplast genome sequences in Actinidiaceae: Genome structure and comparative analysis. *PLoS ONE* **2015**, *10*, e0129347. [CrossRef] [PubMed]

41. Khakhlova, O.; Bock, R. Elimination of deleterious mutations in plastid genomes by gene conversion. *Plant J.* **2006**, *46*, 85–94. [CrossRef] [PubMed]

42. Zhang, Y.; Ma, P.; Li, D. High-throughput sequencing of six bamboo chloroplast genomes: Phylogenetic implications for temperate woody bamboos (Poaceae: Bambusoideae). *PLoS ONE* **2011**, *6*, e20596. [CrossRef] [PubMed]

43. Choi, K.S.; Chung, M.G.; Park, S. The Complete Chloroplast Genome Sequences of Three Veroniceae Species (Plantaginaceae): Comparative Analysis and Highly Divergent Regions. *Front. Plant Sci.* **2016**, *7*, 355–394. [CrossRef] [PubMed]

44. Shaw, J.; Shafer, H.L.; Leonard, O.R.; Kovach, M.J.; Schorr, M.S.; Morris, A.B. Chloroplast DNA sequence utility for the lowest phylogenetic and phylogeographic inferences in angiosperms: The tortoise and the hare IV. *Am. J. Bot.* **2014**, *101*, 1987–2004. [CrossRef] [PubMed]

45. Shaw, J.; Lickey, E.B.; Schilling, E.E.; Small, R.L. Comparison of whole chloroplast genome sequences to choose noncoding regions for phylogenetic studies in angiosperms: The tortoise and the hare III. *Am. J. Bot.* **2007**, *94*, 275–288. [CrossRef] [PubMed]

46. Li, P.; Lu, R.; Xu, W.; Ohitoma, T.; Cai, M.; Qiu, Y.; Cameron, K.M.; Fu, C. Comparative genomics and phylogenomics of east Asian tulips (amana, Liliaceae). *Front. Plant Sci.* **2017**, *8*, 35–39. [CrossRef] [PubMed]

47. Carlson, S.E.; Linder, H.P.; Donoghue, M.J. The historical biogeography of *Scabiosa* (Dipsacaceae): Implications for Old World plant disjunctions. *J. Biol.* **2012**, *39*, 1086–1100. [CrossRef]

48. Yue, J.; Sun, H.; Baum, D.A.; Jianhua, L.I.; Alshehbaz, I.A.; Ree, R.H. Molecular phylogeny of *Solms-laubachia* (Brassicaceae) s.l., based on multiple nuclear and plastid DNA sequences, and its biogeographic implications. *J. Syst. Evol.* **2009**, *47*, 402–415. [CrossRef]

49. Kurtz, S.; Schleiermacher, C. REPuter: Fast computation of maximal repeats in complete genomes. *Bioinformatics* **1999**, *15*, 426–427. [CrossRef] [PubMed]

50. Huang, H.; Shi, C.; Liu, Y.; Mao, S.Y.; Gao, L.Z. Thirteen *Camellia* chloroplast genome sequences determined by high-throughput sequencing: Genome structure and phylogenetic relationships. *BMC Evol. Biol.* **2014**, *14*, 151–162. [CrossRef] [PubMed]

51. Zhao, Y.; Yin, J.; Guo, H.; Zhang, Y.; Xiao, W.; Sun, C.; Wu, J.; Qu, X.; Yu, J.; Wang, X. The complete chloroplast genome provides insight into the evolution and polymorphism of *Panax ginseng*. *Front. Plant Sci.* **2015**, *5*, 696–698. [CrossRef] [PubMed]

52. Thiel, T.; Michalek, W.; Varshney, R.; Graner, A. Exploiting EST databases for the development and characterization of gene-derived SSR-markers in barley (*Hordeum vulgare* L.). *Theor. Appl. Genet.* **2003**, *106*, 411–422. [CrossRef] [PubMed]

53. Kim, K.; Lee, S.C.; Lee, J.; Yu, Y.; Yang, K.; Choi, B.S.; Koh, H.J.; Waminal, N.E.; Choi, H.I.; Kim, N.H. Complete chloroplast and ribosomal sequences for 30 accessions elucidate evolution of Oryza AA genome species. *Sci. Rep.* **2015**, *5*, 15–16. [CrossRef] [PubMed]

54. Milne, R.I.; Abbott, R.J. The origin and evolution of tertiary relict floras. *Adv. Bot. Res* **2002**, *38*, 281–314.

55. Tiffney, B.H. Perspectives on the origin of the floristic similarity between eastern Asia and eastern North America. *J. Arn. Arb.* **1985**, *66*, 73–94. [CrossRef]

56. Xiang, Q.; Soltis, D.E.; Soltis, P.S.; Manchester, S.R.; Crawford, D.J. Timing the eastern Asian-eastern North American floristic disjunction: Molecular clock corroborates paleontological estimates. *Mol. Phylogenet. Evol.* **2000**, *15*, 462–472. [CrossRef] [PubMed]

57. Birky, C.W. Uniparental inheritance of mitochondrial and chloroplast genes: Mechanisms and evolution. *Proc. Natl. Acad. Sci. USA* **1995**, *92*, 11331–11338. [CrossRef] [PubMed]

58. Cronn, R.; Liston, A.; Parks, M.; Gernandt, D.S.; Shen, R.; Mockler, T. Multiplex sequencing of plant chloroplast genomes using Solexa sequencing-by-synthesis technology. *Nucleic Acids Res.* **2008**, *36*, 122–125. [CrossRef] [PubMed]

59. Wyman, S.K.; Jansen, R.K.; Boore, J.L. Automatic annotation of organellar genomes with DOGMA. *Bioinformatics* **2004**, *20*, 3252–3255. [CrossRef] [PubMed]

60. Schattner, P.; Brooks, A.N.; Lowe, T.M. The tRNAscan-SE, snoscan and snoGPS web servers for the detection of tRNAs and snoRNAs. *Nucleic Acids Res.* **2005**, *33*, 686–689. [CrossRef] [PubMed]

61. Lohse, M.; Drechsel, O.; Bock, R. OrganellarGenomeDRAW (OGDRAW): A tool for the easy generation of high-quality custom graphical maps of plastid and mitochondrial genomes. *Curr. Genet.* **2007**, *52*, 267–274. [CrossRef] [PubMed]

62. Frazer, K.A.; Pachter, L.; Poliakov, A.; Rubin, E.M.; Dubchak, I. VISTA: Computational tools for comparative genomics. *Nucleic Acids Res.* **2004**, *32*, 273–279. [CrossRef] [PubMed]

63. Librado, P.; Rozas, J. DnaSP v5. *Bioinformatics* **2009**, *25*, 1451–1452. [CrossRef] [PubMed]

64. Shaw, J.; Lickey, E.B.; Beck, J.T.; Farmer, S.B.; Liu, W.; Miller, J.; Siripun, K.C.; Winder, C.T.; Schilling, E.E.; Small, R.L. The tortoise and the hare II: Relative utility of 21 noncoding chloroplast DNA sequences for phylogenetic analysis. *Am. J. Bot.* **2005**, *92*, 142–166. [CrossRef] [PubMed]

65. Katoh, K.; Standley, D.M. MAFFT multiple sequence alignment software version 7: Improvements in performance and usability. *Mol. Bio. Evol.* **2013**, *30*, 772–780. [CrossRef] [PubMed]

66. Posada, D. jModelTest: Phylogenetic model averaging. *Mol. Biol. Evol.* **2008**, *25*, 1253–1256. [CrossRef] [PubMed]

67. Miller, M.A.; Pfeiffer, W.; Schwartz, T. Creating the CIPRES Science Gateway for inference of large phylogenetic trees. *Gatew. Comput. Environ. Workshop* **2010**, *3*, 1–8.

68. Ronquist, F.; Huelsenbeck, J.P. MrBayes 3: Bayesian phylogenetic inference under mixed models. *Bioinformatics* **2003**, *19*, 1572–1574. [CrossRef] [PubMed]

69. Hohmann, N.; Wolf, E.M.; Lysak, M.A.; Koch, M.A. A time-calibrated road map of Brassicaceae species radiation and evolutionary history. *Plant Cell* **2015**, *27*, 2770–2784. [CrossRef] [PubMed]

70. Hu, H.; Hu, Q.; Alshehbaz, I.A.; Luo, X.; Zeng, T.; Guo, X.; Liu, J. Species delimitation and interspecific relationships of the genus *Orychophragmus* (Brassicaceae) inferred from whole chloroplast genomes. *Front. Plant Sci.* **2016**, *7*, 1826–2884. [CrossRef] [PubMed]

71. Drummond, A.J.; Suchard, M.A.; Xie, D.; Rambaut, A. Bayesian phylogenetics with BEAUti and the BEAST 1.7. *Mol. Biol. Evol.* **2012**, *29*, 1969–1973. [CrossRef] [PubMed]

72. Rambaut, A.; Drummond, A.J. Tracer v.1.4. *Encycl. Atmos. Sci.* **2007**, *141*, 2297–2305.

73. Bouckaert, R.R.; Heled, J.; Kuhnert, D.; Vaughan, T.G.; Wu, C.H.; Xie, D.; Suchard, M.A.; Rambaut, A.; Drummond, A.J. BEAST 2: A software platform for bayesian evolutionary analysis. *PLoS Computat. Biol.* **2014**, *10*, 38–49. [CrossRef] [PubMed]

74. Yu, Y.; Harris, A.J.; Blair, C.; He, X. RASP (Reconstruct Ancestral State in Phylogenies): A tool for historical biogeography. *Mol. Phylogenet. Evol.* **2015**, *87*, 46–49. [CrossRef] [PubMed]

International Journal of
Molecular Sciences

MDPI

Article

The Complete Chloroplast Genome of *Catha edulis*: A Comparative Analysis of Genome Features with Related Species

Cuihua Gu [1,2], **Luke R. Tembrock** [2], **Shaoyu Zheng** [1] **and Zhiqiang Wu** [3,*]

[1]   School of Landscape and Architecture, Zhejiang Agriculture and Forestry University,
     Hangzhou 311300, China; gu_cuihua@126.com (C.G.); aggies.collins@gmail.com (S.Z.)
[2]   Department of Biology, Colorado State University, Fort Collins, CO 80523, USA;
     Luke.R.Tembrock@aphis.usda.gov
[3]   Department of Ecology, Evolution, and Organismal Biology, Ames, IA 50011, USA
*   Correspondence: wu.zhiqiang.1020@gmail.com; Tel.: +1-515-441-5307

Received: 18 December 2017; Accepted: 6 February 2018; Published: 9 February 2018

Abstract: Qat (*Catha edulis*, Celastraceae) is a woody evergreen species with great economic and cultural importance. It is cultivated for its stimulant alkaloids cathine and cathinone in East Africa and southwest Arabia. However, genome information, especially DNA sequence resources, for *C. edulis* are limited, hindering studies regarding interspecific and intraspecific relationships. Herein, the complete chloroplast (cp) genome of *Catha edulis* is reported. This genome is 157,960 bp in length with 37% GC content and is structurally arranged into two 26,577 bp inverted repeats and two single-copy areas. The size of the small single-copy and the large single-copy regions were 18,491 bp and 86,315 bp, respectively. The *C. edulis* cp genome consists of 129 coding genes including 37 transfer RNA (tRNA) genes, 8 ribosomal RNA (rRNA) genes, and 84 protein coding genes. For those genes, 112 are single copy genes and 17 genes are duplicated in two inverted regions with seven tRNAs, four rRNAs, and six protein coding genes. The phylogenetic relationships resolved from the cp genome of qat and 32 other species confirms the monophyly of Celastraceae. The cp genomes of *C. edulis*, *Euonymus japonicus* and seven Celastraceae species lack the *rps16* intron, which indicates an intron loss took place among an ancestor of this family. The cp genome of *C. edulis* provides a highly valuable genetic resource for further phylogenomic research, barcoding and cp transformation in Celastraceae.

Keywords: chloroplast (cp) genome; *Catha edulis*; next generation sequencing; phylogeny; repeat sequence

1. Introduction

Qat (Celastraceae: *Catha edulis* (Vahl) Forssk. ex Endl.) is a woody evergreen species of major cultural and economic importance in southwest Arabia and East Africa, which is cultivated for its stimulant alkaloids cathine and cathinone. An estimated 20 million people consume qat on a daily basis in eastern Africa [1], and its use and cultivation has been expanding in recent years [2]. Qat is the only species in Celastraceae that is cultivated on a large scale. The cultivation and/or collection (in some instances illegally from wild sources in protected areas) of qat takes place primarily in Israel, Ethiopia, Kenya, Madagascar, Rwanda, Tanzania, Somalia, Uganda, and Yemen [2–4].

The cultivation and sale of qat has become an important driver in the local and regional economies of East Africa and Yemen. In Yemen, 6% of the gross domestic product is generated from qat cultivation and sales [5]. Ethiopia has become the number one producer of qat in the world with exports in 1946 equaling only 26 tons valued at $5645, while 15,684 tons were exported in 2000 valued at $72 million [6]. A similar expansion in qat cultivation and sales has occurred in Kenya with the current trade from

Kenya to Somalia estimated at $100 million per year. Trade of qat has become international in scale with, for example, 2.26 million kilograms of qat imported into England from Ethiopian and Kenya in 2013 [7]. The biosynthesis of cathinone and similar stimulant alkaloids is rare among green plants, known only in *Catha edulis* and several Asian species of *Ephedra* [8]. In addition, Celastraceae species produce numerous unique phytochemicals of potential pharmaceutical value [9]. Chloroplast transformations of qat and related species may prove useful for the production of cathinone related alkaloids and/or novel drugs.

The phylogenetic placement of qat within the Celastraceae has been inferred from 18S, 26S, *atpB*, ITS (as Nuclear ribosomal internal transcribed spacer), *matK*, *phyB*, and *rbcL* [10]. Phylogeographic work using SSR (as simple sequence repeats) loci has been done for wild and cultivated qat in the historic areas of production—Ethiopia, Kenya, and Yemen [7,11]. Beyond these studies, no genetic resources of which we are aware have been developed for qat. In addition, no chloroplast (cp) genome has been fully sequenced and published in the genus *Catha*. Therefore, our completed cp genome will be an important genetic resource for further evolutionary studies both within the Celastrales generally and economically important qat specifically.

The cp genome in plants is noted as being highly conserved in gene content [12]. Despite the consistency between cp genomes in plants, the differences in the size of cp genomes appear to be driven by intron and gene loss, and structural changes such as loss or gain of repeat units in different types of repetitive DNA [13]. In particular, genes that straddle inversion junctions such as *ycf1* appear to be undergoing rapid evolution [14].

Contrary to the structure of most nuclear plant genomes, the cp genome is typically comprised of a highly conserved quadripartite structure which is 115 to 165 kb in length, uniparentally inherited [12,15], and with similar gene content and order shared among most land plants [16]. From the advancements made by next-generation sequencing (NGS), complete, high quality cp genomes are becoming increasingly common [17]. At present, more than 2000 completed cp genomes of angiosperm species can be downloaded in the public database of the National Center for Biotechnology Information (NCBI; [18], Available online: https://www.ncbi.nlm.nih.gov/genomes/ GenomesGroup.cgi?taxid=2759&opt=plastid). Large databases of complete cp genomes provide an indispensable resource for researchers identifying species [19], designing molecular markers for plant population studies, and for research concerning cp genome transformation [20–22]. The essentially non-recombinant structures of cp genomes make them particularly useful for the above applications. For example, cp genomes maintain a positive homologous recombination system [23–26]. Thus, in the transformation process, genes can be precisely transferred to specific genomic regions. A variety of homologous cp sites have proven useful at multiple levels of classification, including inter-specific and intra-specific [27]. In more recent years, systematic studies have employed entire cp genomes to attain high resolution phylogenies [28].

In this paper, we report the completely sequenced cp genome in the Celastrales and discuss the technical aspects of sequencing and assembly. In addition, we conduct phylogenetic analysis using other fully sequenced cp genomes from species in the closely related orders Malpighiales and Rosales. These analyses were conducted to find the top twenty loci for phylogenetic analysis and find which structural changes have taken place across cp genomes between the orders Rosales, Malpighiales, and Celastrales. The completed cp genome is a valuable resource for studying evolution and population genetics of both wild and cultivated populations of qat as well as genetic transformations related to the production of pharmaceuticals in qat or related Celastraceae species.

2. Results and Discussion

2.1. Chloroplast Assembly and Genome Features

The *C. edulis* cp genome was completely assembled into a single molecule of 157,960 bp, by combining Illumina and Sanger sequencing results. By mapping the completed genome using the paired reads, we confirm the size of our assembly for the completed cp genome with 497,848 (representing 5% of all reads) mapped pair-end reads evenly spanning the entire genome with mean read depth of 785× coverage (Figure S1). Given these quality controls and processing steps, the cp genome for qat is high quality.

Although the genome structure is highly conserved in the cp genome, several features such as the presence or lack of introns, the size of the intergenic region, gene duplication, and the length, type and number of repeat regions can vary [29]. The complete *C. edulis* cp genome has the conserved quadripartite structure and size that resembles most land plant cp genomes which are normally 115–165 kb in size including two inverted repeats (IRs) and two single-copy regions as large single copy and small single copy (LSC and SSC).

The cp genome of *C. edulis* consists of two single-copy regions isolated by two identical IRs of 26,577 bp each, one SSC region of 18,491 bp and one LSC region of 86,315 bp. The proportion of LSC, SSC, and IRs size in the entire cp genome is 54.6%, 11.7% and 33.6%, respectively (Figure 1 and Table 1). The GC contents of the LSC, IR, SSC, and the whole cp genome are 35.1%, 42.7%, 31.8%, and 37.3%, respectively, which are consistent with the published Rosid cp genomes [30].

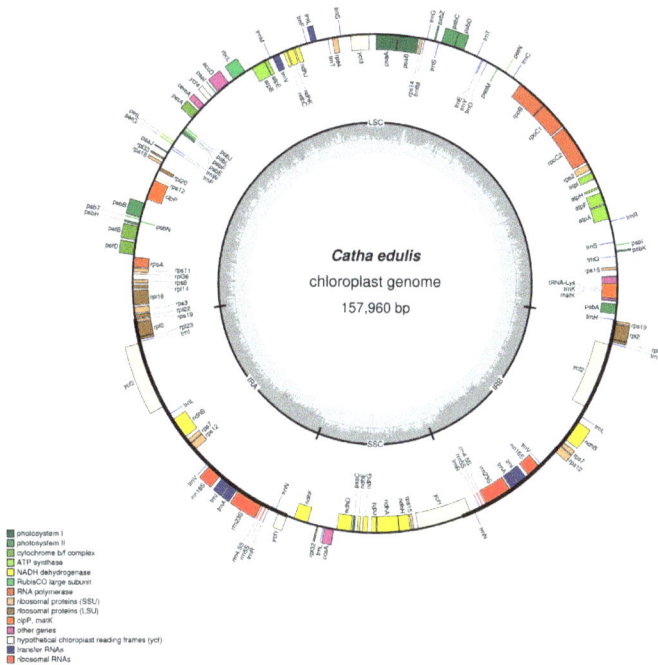

Figure 1. Circular map of the *C. edulis* cp genome. Genes shown inside and outside of the outer circle are transcribed clockwise and counterclockwise, respectively. The innermost shaded area inside the inner circle corresponds to GC content in the cp genome. Genes in different functional groups are color coded. IR, inverted repeat; LSC, large single copy region; SSC, small single copy region. The map is drawn using OGDRAW (V 1.2, Max Planck Institute of Molecular Plant Physiology, Am Mühlenberg, Germany).

The *C. edulis* cp genome is composed of tRNAs, protein coding genes and rRNAs, intergenic and intronic regions (Table 2). Non-coding DNA accounts for 67,633 bp (42.8%) of the whole *C. edulis* cp genome, protein-coding genes account for 78,471 bp (49.7%), tRNA accounts for 2806 bp (1.8%), and rRNA accounts for 9050 bp (5.7%). By comparison with seven other species, gene order, gene content, the coding genes, and non-coding region proportions are similar among these cp genomes (Table 2).

Table 1. Comparison of plastid genome size among eight species.

Region	Features	C. edulis	E. japonicus	H. brasiliensis	M. esculenta	P. euphratica	R. communis	S. purpurea	V. seoulensis
LSC	Length (bp)	86,315	85,941	89,209	89,295	84,888	89,651	84,452	85,691
	GC Content (%)	35.1	35.1	33.2	33.3	34.5	33.3	34.4	33.8
	Length Percentage (%)	54.6	54.5	55.3	55.3	54.1	54.9	54.3	54.8
SSC	Length (bp)	18,491	18,340	18,362	18,250	16,586	18,816	16,220	18,008
	GC Content (%)	31.8	31.8	29.5	29.6	30.6	29.5	31	29.6
	Length Percentage (%)	11.7	11.6	11.4	11.3	10.6	11.5	10.4	11.5
IR	Length (bp)	26,577	26,678	26,810	26,954	27,646	27,347	27,459	26,404
	GC Content (%)	42.7	42.7	42.2	42.3	41.9	41.9	41.9	42.6
	Length Percentage (%)	16.8	16.9	16.6	16.7	17.6	16.8	17.6	16.9
Total	Length (bp)	157,960	157,637	161,191	161,453	156,766	163,161	155,590	156,507
	GC Content (%)	37.3	37.3	35.7	35.9	36.7	35.7	36.7	36.3

LSC, large single copy region; SSC, small single copy region; IR, inverted repeat.

Table 2. Comparison of coding and non-coding region size among eight species.

Region	Species	C. edulis	E. japonicus	H. brasiliensis	M. esculenta	P. euphratica	R. communis	S. purpurea	V. seoulensis
Protein coding	length (bp)	78,471	77,331	78,852	79,089	78,728	78,119	77,898	78,310
	Length Percentage (%)	49.7	49.1	48.9	49.0	50.2	47.9	50.1	50.0
	GC Content (%)	38	38.2	37.1	37.2	37.6	37.5	37.6	37.2
tRNA	length (bp)	2806	2806	2798	2742	2796	2802	2792	2810
	Length Percentage (%)	1.8	1.8	1.7	1.7	1.8	1.7	1.8	1.8
	GC Content (%)	52.6	53.3	53.2	53.3	53	53.2	52.9	53
rRNA	length (bp)	9,050	9050	9050	9050	9050	9050	9,050	9050
	Length Percentage (%)	5.7	5.7	5.6	5.6	5.8	5.5	5.8	5.8
	GC Content (%)	55.2	55.4	55.4	55.5	55.5	55.5	55.4	55.4
Intron	length (bp)	18,474	19,287	18,538	18,479	18,210	18,278	17,321	18,348
	Length Percentage (%)	11.7	12.2	11.5	11.4	11.6	11.2	11.1	11.7
	GC Content (%)	37.1	36.6	36.6	36.9	36.9	37.1	37.3	36.7
Intergenic	length (bp)	49,159	49,163	51,953	52,093	47,982	54,912	48,529	47,989
	Length Percentage (%)	31.1	31.2	32.2	32.3	30.6	33.7	31.2	30.7
	GC Content (%)	31.9	31.7	29	29	31	28.7	30.7	30.1

2.2. Gene Content and Structure

The cp genome of *C. edulis* consisted of 129 coding regions made up of 37 tRNAs, 84 protein-coding genes, and eight rRNAs, of which 112 genes are unique and 17 genes were repeated in two inverted regions consisting of seven tRNAs, six protein coding genes, and four rRNAs (Figure 1 and Table 3). Among these 112 unique genes, three genes crossed different cp boundaries: $trnH^{GUG}$ crossed the IR$_B$ and LSC regions, *ycf1* crossed the IR$_B$ and SSC regions, *rps12* crossed two IR regions and the LSC region (two 3′ end exons repeated in IRs and 5′ end exon situated in LSC) (Figure 1). Of the remaining 109 genes, 80 are situated in LSC including 59 protein coding genes and 21 tRNAs, 17 in two inverted repeats (six coding genes, seven tRNAs, and four rRNAs), and 12 in the SSC including 11 coding genes and one tRNA.

Table 3. List of genes in the *C. edulis* plastid genome.

Gene Category	Groups of Genes	Name of Genes
Self-replication	Transfer RNA genes	$trnA^{UGC}$ [a,b] $trnC^{GCA}$ $trnD^{GUC}$ $trnE^{UUC}$ $trnF^{GAA}$ $trnfM^{CAU}$ $trnG^{UCC}$ $trnG^{GCC}$ $trnH^{GUG}$ $trnI^{CAU}$ [b] $trnI^{GAU}$ [a,b] $trnK^{UUU}$ [a] $trnL^{CAA}$ [b] $trnL^{UAA}$ [a] $trnL^{UAG}$ $trnM^{CAU}$ $trnN^{GUU}$ [b] $trnP^{UGG}$ $trnQ^{UUG}$ $trnR^{ACG}$ [b] $trnR^{UCU}$ $trnS^{GCU}$ $trnS^{GGA}$ $trnS^{UGA}$ $trnT^{GGU}$ $trnT^{UGU}$ $trnV^{GAC}$ [b] $trnV^{UAC}$ [a] $trnW^{CCA}$ $trnY^{GUA}$
	Small subunit of ribosome	*rps2 rps3 rps4 rps7b rps8 rps11 rps12* [a,b] *rps14 rps15 rps16 rps18 rps19*
	Ribosomal RNA genes	*rrn16* [b] *rrn23* [b] *rrn4.5* [b] *rrn5* [b]
	Large subunit of ribosome	*rpl2* [b] *rpl14 rpl16* [a] *rpl20 rpl22 rpl23* [b] *rpl32 rpl33 rpl36*
	DNA dependent RNA polymerase	*rpoA rpoB rpoC1* [a] *rpoC2*
Photosynthesis	Subunits of photosystem I	*psaA psaB psaC psaI psaJ*
	Subunits of photosystem II	*psbA psbB psbC psbD psbE psbF psbH psbI psbJ psbK psbL psbM psbN psbT psbZ*
	Subunits of cytochrome	*petA petB* [a] *petD* [a] *petG petL petN*
	Subunits of ATP synthase	*atpA atpB atpE atpF* [a] *atpH atpI*
	ATP-dependent protease subunit p gene	*clpP* [a]
	Large subunit of Rubisco	*rbcL*
	Subunits of NADH dehydrogenase	*ndhA* [a] *ndhB* [a,b] *ndhC ndhD ndhE ndhF ndhG ndhH ndhI ndhJ ndhK*
Other genes	Maturase	*matK*
	Envelop membrane protein	*cemA*
	Subunit of acetyl-CoA-carboxylase	*accD*
	c-type cytochrome synthesis gene	*ccsA*
Genes of unknown function	Conserved open reading frames	*ycf1 ycf2* [b] *ycf3* [a] *ycf4*

[a] Genes containing introns; [b] Duplicated gene (Genes present in the IR regions).

Most of the protein-coding genes contain only one exon, while 17 genes contain one intron, of which four occur in both IRs, 12 genes are distributed in LSC, and one in the SSC (Table 4), among them three genes (*rps12*, *clpP* and *ycf3*) contain two introns, while 14 genes ($trnA^{GUC}$, $trnI^{GAU}$, $trnG^{UCC}$, $trnL^{UAA}$, $trnK^{UUU}$, and $trnV^{UAC}$, *rpoC1*, *atpF*, *rpl16*, *rpl2*, *petB*, *petD*, *ndhA*, and *ndhB*) contain one intron. The longest intron of $trnK^{UUU}$ is 2495 bp including the 1533 bp encoding the *matK* gene [13]. The *rps12* gene was predicted to be trans-spliced with a repeated 3′ end duplicated in two IRs and a single 5′ end exon in LSC [31].

Table 4. Genes with intron and their length of exons and introns in plastid genome of *C. edulis*.

Gene Name	Location	Exon I (bp)	Intron I (bp)	Exon II (bp)	Intron II (bp)	Exon III (bp)
rpoC1	LSC	1632	817	441		
atpF	LSC	396	699	159		
petB	LSC	6	773	642		
petD	LSC	8	784	475		
ndhB	IR	756	687	777		
ndhA	SSC	540	1178	573		
rpl16	LSC	399	1119	9		
rpl2	IR	471	648	393		
rps12	LSC	114		27	546	231
ycf3	LSC	153	727	228	731	126
clpP	LSC	231	676	291	849	69
trnK-UUU	LSC	29	2495	37		
trnL-UAA	LSC	37	540	50		
trnV-UAC	LSC	37	663	39		
trnI-GAU	IR	42	939	35		
trnA-UGC	IR	38	801	35		
trnG-UCC	LSC	23	761	48		

2.3. Comparison of the cp Genomes

The cp genome of *C. edulis* (Celastraceae) was compared to species from 14 genera, including *Populus, Salix, Viola, Hevea, Manihot, Ricinus, Euonymus* and seven out-group species using dot-plot analysis. Besides a unique rearrangement of one 30-kb inversion in the *H. brasiliensis* cp genome [32], no other large structural differences (inversions) were detected among all compared species in the dot-plot analysis. This is consistent with the extremely conserved cp genomes in land plants [16]. The limited structural differences across the 14 species cp genomes demonstrate that gene order, gene content, and entire genome structure are conserved (Figure S3).

Based on the limited structural variation of cp genomes, we focused on seven closely related species of *C. edulis* to examine finer scale structural differences in genome length. Among these seven cp genomes, the length of genomes ranged from 155,590 bp (*S. purpurea*) to 163,161 bp (*R. communis*). The length of the LSC region varied from 84,452 bp (*S. purpurea*) to 89,651 bp (*R. communis*), and from 16,220 bp (*S. purpurea*) to 18,816 bp (*R. communis*) in SSC, and from 26,404 bp (*V. seoulensis*) to 27,646 bp (*P. euphratica*) in the IR regions (Table 2).

The entire GC content of the complete *C. edulis* cp genome is 37.3%, with 33.6% GC content in IRs, 35.1% in LSC, and 31.8% in SSC. These GC contents are consistent with other published cp genomes [33]. The whole GC content in the two Celastrales and six cp genomes of Malpighiales species ranged from 35.7% to 37.3% of the total genome, with *R. communis* having the lowest and *C. edulis* and *E. japonicus* having the highest GC content (Table 1).

These eight species have similar genetic composition at the IR-SSC and IR-LSC boundaries except *rps19*, which is not present from the border of LSC and IR$_A$ in *P. euphratica* and *R. communis* in which *rpl22* crosses the border of IR$_A$ and LSC (Figure 2).

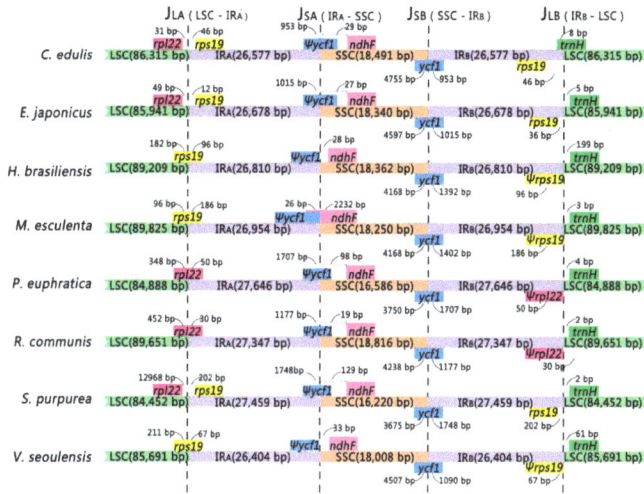

Figure 2. Comparison of junctions between the LSC, SSC, and IRs among eight species. Number above indicates the distance in bp between the ends of genes and the borders sites (distances are not to scale in this figure). The ψ symbol represents pseudogenes.

2.4. Contraction and Expansion in the Four Junction Regions

Although genomic structure including gene composition and genome size are highly conserved, expansion and contraction of IRs are common differences between plant cp genomes. Kim [34] proposed that the IRs size differ within plant cp genomes mainly results from the contraction or expansion at the junctions. Comparison of the inverted repeat-single copy (IR-SC) boundary regions of the two Celastrales and six Malpighiales species genomes showed very small differences in boundaries (Figure 2). We inspected the four boundaries (J_{LA}, J_{LB}, J_{SA}, and J_{SB}) across the two Celastrales and six Malpighiales species to detect the detailed boundary variation between the two SC regions and IRs using the methods described in [18].

The size of the IRs varied from 26,404 to 27,646 bp. The IR_A-LSC junction (J_{LA}) was situated in the *rps19* gene in *H. brasiliensis*, *M. esculenta*, and *V. seoulensis* which crossed inside the IR_A region 96 bp, 186 bp, and 67 bp, respectively, and as a result duplicated pseudogene *rps19* (ψ*rps19*) was nested within IR_B for these three species. However, in *C. edulis*, *E. japonicus* and *S. purpurea*, J_{LA} is situated in the intergenic regions between *rpl22* and *rps19* in which the distances from *rps19* to the J_{LA} were 46 bp, 12 bp and 202 bp. In two other species, *P. euphratica* and *R. communis*, J_{LA} is situated in the coding region of *rpl22* which spread into IR_A 50 bp and 30 bp, respectively, and resulted in the generation of pseudogene *rpl22* (ψ*rpl22*) in IR_B.

The IR_A-SSC junction (J_{SA}) was situated in or adjoined pseudogene *ycf1* (ψ*ycf1*) for all eight species; J_{SA} of three species (*H. brasiliensis*, *M. esculenta*, and *V. seoulensis*) were all situated just adjacent to the end of ψ*ycf1*. Overlap between *ndhF* and ψ*ycf1* was found in *M. esculenta*, in which *ndhF* expanded into the IR_A region for 26 bp. For the other five species, J_{SA} was located near ψ*ycf1*. In the other six species (*C. edulis*, *E. japonicus*, *H. brasiliensis*, *P. euphratica*, *R. communis*, *S. purpurea* and *V. seoulensis*), the distances between *ndhF* and J_{SA} were 29 bp, 27 bp, 28 bp, 98 bp, 19 bp, 129 bp and 33 bp, respectively.

The IR_B-SSC junction (J_{SB}) is situated in the *ycf1* coding region which spans into the IR_B region in all eight species. However, the length of *ycf1* in the IR region varied among the eight species from 953 bp to 1748 bp highlighting the dynamic variation of the junction regions.

The IR_B-LSC junctions (J_{LB}) were located between *rps19* and *trnH* in *E. japonicus* and *S. purpurea*; situated at the end of ψ*rps19* in *H. brasiliensis*, *M. esculenta*; and *V. seoulensis*; and at the end of ψ*rpl22* in *P. euphratica* and *R. communis*. In the J$_{LB}$ junction, the *trnH* gene is 8 bp into IR$_B$ region in *C. edulis*. In the other seven species, 2–199 bp distance is found between the *trnH* gene and the IR$_B$-SSC junction.

The variation in the IR-SC boundary area is due to the contraction or expansion of the IR observed in the IR-SSC boundaries. These expansions/contractions are likely to be mediated by molecular recombination within the two short, straight repeating sequences that occur frequently in the genes within the boundary [34].

2.5. Verification of the rps16 Intron Loss from Catha and Seven Other Celastraceae Species

The gene composition in the *C. edulis* cp genome is similar to the other angiosperm species analyzed in this study. However, we found that the *rps16* gene had no intron in the *C. edulis* cp genome. The structure and the intron size for *rps16* are conserved in the model species *Arabidopsis thaliana* and in our sampled species (NC_000932). However, it has been reported that *rps16* gene or the intron of *rps16* has been lost multiple times in numerous lineages [35,36].

To test whether the loss of the *rps16* intron is common throughout the Celastraceae family or just in certain species, two primers were designed in the flanking exons to amplify and then sequence the intron region (or lack thereof) for eight species in the Celastraceae family. Based on the PCR amplification (Figure S2), the length of this *rps16* amplicon is about 550 bp in all eight sampled Celastraceae species indicating that the intron has been lost throughout the Celastraceae family. We also conducted Sanger sequencing to verify the alignment of the *rps16* gene (Figure 3). From this alignment, all species sampled from the Celastraceae family do not contain the *rps16* intron (Figure 3A). The Sanger sequencing data provide additional evidence that all eight-species do not have this intron (Figure 3B).

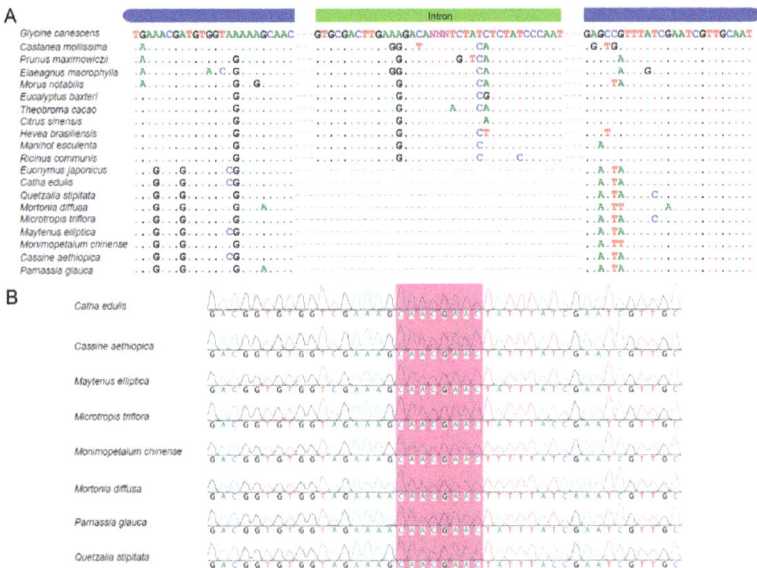

Figure 3. The sequence variation for *rps16* gene with and without intron: (**A**) The structural components of *rps16* gene in 20 species. All Species outside of Celastraceae family contained the *rps16* intron. (**B**) The purple area in all eight species from different genera of the Celastraceae family showed the connection of two exons indicating the lost intron.

Intron loss in cp genomes have been reported multiple times in different species, such as species in Desmodieae (Fabaceae) [37] and reported in both dicots and monocots. Loss of the *rps16* intron could probably be best explained by a homologous recombination and the reverse-transcriptase mediated mechanism [35]. However, intron loss from DNA fragment deletions or gene transfer between introns could be due to yet unexplained processes [37]. By increasing the sampling density within Celastraceae and its closest relatives, the timing of the *rps16* intron loss was inferred to occur between the Celastrales and Oxalidales + Malpighiales approximately 80 million years ago [38].

2.6. Identification of Long Repetitive Sequences

Long repetitive sequences play key functions in cp genome evolution, genome rearrangements and can be informative in phylogenetic studies [39]. Comparison of forward, complement, reverse, and palindromic repeats (≥30 bp) (with a sequence identity of ≥90% per repeat unit) were conducted across *C. edulis* and seven related species using REPuter (Available online: https://bibiserv.cebitec. uni-bielefeld.de/reputer/; (University of Bielefeld, Bielefeld, Germany)). *Catha edulis* had the fewest (8) repeats while its cp genome was not the shortest among those examined (157,960 bp) which is inconsistent with the general trend of shorter genomes possessing fewer repetitive regions [40].

A total of 175 unique repeats consisting of forward, reverse, complementary and palindromic were found from the eight-species examined (Figure 4A). The species *E. japonicus* included the most repeats consisting of: 14 palindromic repeats, 19 forward repeats, and eight reverse repeats, for a total of 41 repeats (Figure 4A and Table S3). In *H. brasiliensis*, *M. esculenta*, *P. euphratica*, *R. communis*, *S. purpurea* and *V. seoulensis* cp genomes, 29, 35, 20, 22, 10, and 10 total repeat pairs were found respectively (Figure 4A). Among them, 19 forward repeats were most commonly found in *E. japonicus* and *M. esculenta* and in all species the most common repeat type was forward (Figure 4A). Forward repeats are often the result of transposon activity [41], which can increase under cellular stress [42]. However, the origins and multiplication of long repetitive repeats is not fully understood [43]. Previous studies suggested that the existence of genome rearrangement could be attributed to slipped-strand mispairing and inapposite recombination of repetitive sequences [43]. Moreover, forward repeats can lead to changes in genomic structure and thus be used as markers in phylogenetic studies. The length of repeats is variable in this study, with the shortest at 30 bp and the longest at 95 bp (Table S3). The majority of repeats (82%) varied from 30 bp to 40 bp in length (Figure 4B and Table S3). Given the variability of these repeats between lineages, they can be informative regions for developing genomic markers for population and phylogenetic studies [44].

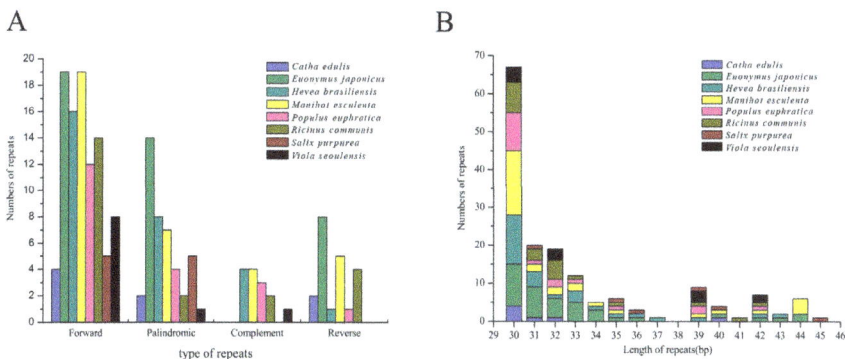

Figure 4. Analysis of repeat sequences in eight chloroplast genomes: (**A**) frequency of repeat types; and (**B**) frequency of the repeats by length ≥30 bp.

2.7. Chloroplast Genome Simple Sequence Repeats (SSRs)

Simple sequence repeats (SSRs) are sequences with motifs from 1 to 6 bp in length repeated multiple times (see methods for cutoff criteria), are found distributed throughout the cp genome, and are often used as markers for breeding studies, population genetics, and genetic linkage mapping [43,45].

A total of 278 SSRs were found in the *C. edulis* cp genome (Figure 5A and Table S4). These SSRs include 165 mononucleotide SSRs (59%), 43 dinucleotide SSRs (15%), 65 trinucleotide SSRs (23%), 3 tetranucleotide (0.01%), and 1 pentanucleotide SSR (0.003%) (Figure 5A and Table S4). Among the 165 SSRs, 98% of SSRs (161) are the AT type with copy number from 8 to 18 (Table S4). In these SSRs of the *C. edulis* cp genome, 89 SSRs were detected in protein-coding genes, 34 SSRs in introns, and 155 in intergenic regions (Figure 5B). In relation to the quadripartite, 195 SSRs were situated in the LSC, whereas 36 and 37 were identified in the SSC and IR, respectively (Figure 5C).

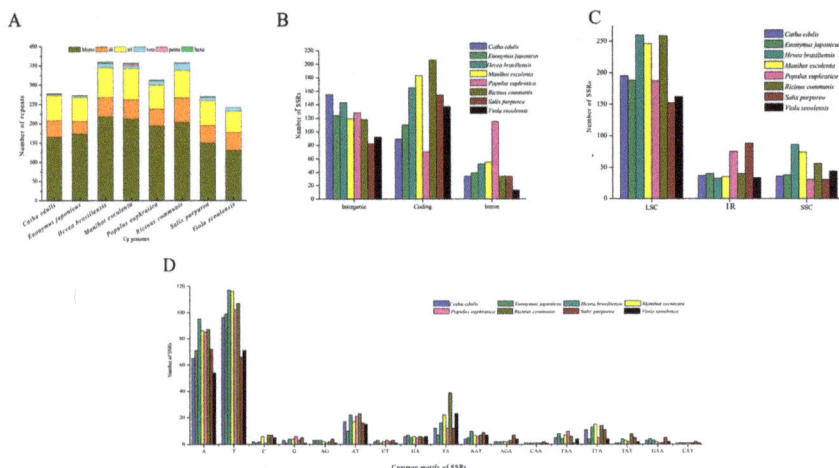

Figure 5. The distribution, type, and presence of simple sequence repeats (SSRs) in eight chloroplast genomes: (**A**) number of different SSR types detected in eight chloroplast genomes presence of SSRs at the LSC, SSC, and IR regions.; (**B**) frequency of SSRs in the protein-coding regions, intergenic spacers and intronic regions; (**C**) frequency of SSRs in the LSC, SSC, and IR regions; and (**D**) frequency of common motifs in the eight chloroplast genomes.

Among the eight species, *V. seoulensis* had the fewest SSRs (242) and *H. brasiliensis* had the most SSRs (360). *Salix purpurea* has the shortest cp genome (155,590 bp) with 270 SSRs and *R. communis* has the longest cp genome (163,161 bp) and 358 SSRs of those analyzed in this study suggesting that number of SSRs may affect genome length, but a strong correlation was not found in all species (Figure 5A). This result indicates that cp genome sizes were not obviously connected with the number of SSRs in these species. Additionally, an abundance of tetranucleotide SSRs were not found in the species studied and no pentanucleotide SSRs were found in *V. seoulensis* or hexanucleotide in *E. japonicus*, *R. communis* and *V. seoulensis* (Figure 5A). Among the eight species, most SSRs of *C. edulis* and *E. japonicus* were located in intergenic regions, most SSRs of *H. brasiliensis*, *M. esculenta*, *P. euphratica*, *R. communis*, and *V. seoulensis* in coding regions, and most SSRs of *S. purpurea* are in intronic regions (Figure 5B). Some SSRs were distributed in protein-coding regions such as *ycf1* and *rpoC2* (Table S4), which could also be employed as DNA markers for population level and genomic studies. Most SSRs in all eight-species were in the LSC region (Figure 5C). Common motifs in the eight-species studied generally consisted of polythymine (poly-T) or polyadenine (poly-A) (Figure 5D). The Euphorbiaceae species in this study all have more SSRs than the other species in this study as well as similar patterns of distribution in the genome. More work is needed to understand these patterns of SSR distribution in cp genomes. Lastly, the SSRs from this study should be valuable for phylogeographic studies and comparing phylogenetic relationships among Celastraceae species.

2.8. Highly Informative Coding Genes and Markers for Phylogenomic Analysis

Detecting highly informative and variable coding genes is important for DNA barcoding, marker development and phylogenomic analyses [46]. Coding genes such as *matK*, *rbcL* have been widely employed for barcoding applications [47,48] and phylogenetic reconstructions [49–51]. Based on compared complete cp genomes, additional informative markers were identified within the Celastraceae.

We aligned entire coding genes more than 200 bp in length to discover genes with the highest sequence identity index and the highest proportion of parsimony-informative sites, for the seven species in this study (Table 5, Table S5). In the coding regions, *matK* and *ycf1* have the largest proportion of parsimony information characters (16.83% and 16.80%, respectively). The *matK* gene is used as core DNA barcoding sequence under the suggestion of CBOL working group (CBOL is The Consortium for the Barcode of Life, an international initiative devoted to developing DNA barcoding as a global standard for the identification of biological species) and also in concert with other variable genes such as *ITS* + *psbA-trnH* + *matK* which was shown to have the highest species identification rate [52]. Given the high number of parsimony informative in *ycf1*, it may also serve as another core DNA barcode in future plant studies [14]. The coding regions identified in this analysis (Table 5) should be particularly informative for species identification and phylogenetic analyses due to the high percentage of variable sites.

Table 5. Ten highest informative sites of coding genes in eight species.

No.	Region	Length (bp) [1]	Aligned Length (bp) [2]	Conserved Sites	Parsimony Informative [3]	Parsimony Informative % [4]	CI. [5]	RI [6]	SI [7]
1	*matK*	1518	1575	1028	265	16.83	0.82	0.7	0.9
2	*ycf1*	5640	6327	3970	1063	16.80	0.82	0.6	0.8
3	*ccsA*	969	987	689	160	16.21	0.84	0.7	0.9
4	*accD*	1509	1401	242	227	16.20	0.83	0.7	0.8
5	*rps3*	648	663	467	107	16.14	0.82	0.7	0.9
6	*ndhF*	2232	2331	1606	368	15.79	0.81	0.6	0.8
7	*rps8*	405	411	294	64	15.57	0.8	0.7	0.9
8	*rpl22*	399	551	345	82	14.88	0.83	0.6	0.7
9	*petL*	96	96	70	14	14.58	0.9	0.8	0.9
10	*ndhD*	1503	1527	1116	207	13.56	0.82	0.7	0.9

[1] Length: refers to sequence length in *Catha edulis*; [2] Aligned length: refers to the alignment of seven other species considered in the comparative analysis (see Materials and Methods); [3] Number of parsimony informative sites; [4] Percentage of parsimony informative sites; [5] CI: Consistency Index; [6] RI: Retention Index; [7] SI: Sequence Identity.

2.9. Phylogenetic Analysis

Based on cp genomes, phylogenetic analyses have helped to resolve the relationships of many angiosperm lineages [53,54]. Previous phylogenetic work in Celastraceae was inferred based on nuclear (26S rDNA and ITS) together with morphological traits and chloroplast genes (*matK*, *trnL-F*) [10]. Our phylogenetic analyses included *C. edulis* and 28 species which were sampled based on relationships from NCBI database (Available online: http://www.ncbi.nlm.nih.gov/genomes/GenomesGroup.cgi?taxid=2759&opt=plastid) and the angiosperm tree of life (Available online: http://www.mobot.org/mobot/research/apweb/) with *Glycine canescens*, *Glycine falcate*, *Trifolium aureum*, and *Trifolium boissieri* from Fabaceae as outgroup taxa. The phylogenetic tree indicated that *Catha* and *Euonymus* where most closely related based on 73 common protein-coding genes (Figure 6). Most branches of the phylogenetic tree had high bootstrap support with all three methods. This suggests that the full cp genome information could be very useful in resolving phylogenetic conflicts but phylogenetic analyses with many closely related species are needed to test the resolving power of chloroplast coding genes [55].

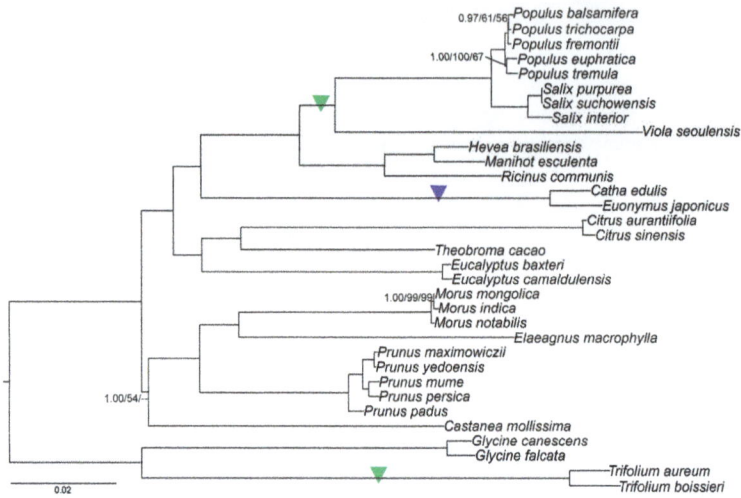

Figure 6. Phylogenetic tree based on 73 shared protein-coding genes was constructed for 33 species using three different methods, including Parsimony analysis, maximum likelihood (ML) and Bayesian inference (BI). All branches had bootstrap values or posterior probability of 100/1.00 except those labeled. The *rps16* gene losses are indicated with green triangles and the *rps16* intron loss is indicated with a purple triangle.

With a clearly resolved and strongly supported phylogeny, evolutionary patterns can be more clearly interpreted, such as gene or intron sequence loss/gain. Specifically, the intron loss of the *rps16* gene and loss of the whole *rps16* gene (Figure 6), were found in Celastraceae (*rps16* intron loss) and independently (*rps16* gene loss) in the genus *Trifolium* (Fabaceae), and the clade Salicaceae + Violaceae (Table S6). Gene and intron loss have been noted numerous times in land plant cp genomes [37]. From the phylogenetic tree, we were able to infer that the intron of *rps16* was lost in an ancestor to the Celastraceae independently from the two *rps16* gene loss events (Figure 6). Why only the *rps16* intron was lost in the Celastraceae and the entire gene in other closely related lineages is not known. Further study is needed to understand the underlying mechanisms of gene vs. intron loss in these related groups.

3. Materials and Methods

3.1. DNA Extraction and Sequencing

DNA for this project was obtained from aliquots of the extracts used in Tembrock et al., 2017. Total genomic DNA was used to build sequence libraries (Illumina Inc., San Diego, CA, USA), and was extracted from leaves using a *Catha* specific DNA extraction protocol described in Tembrock et al., 2017. At the Beijing Genomics Institute (BGI), an Illumina HiSeq 2000 sequencer was used to sequence paired-end (PE) sequencing libraries with an average 300 bp insert length. From this, over 10 million clean reads were passed through quality control with a 100 bp each read length. All other used species in this paper were listed in Table S1.

3.2. Chloroplast Genome Assembly and Sequence Analysis

The original Illumina reads were pre-processed, including the trimming and filtering of low-quality sequences with Trimmomatic v0.3 [56] in which the parameters used were as follows: minlen: 50; trailing: 3; leading: 3; and sliding window: 4:15. De novo assembly from *C. edulis* employed the default parameters (Available online: http://www.clcbio.com) in the CLC genomic workbench v7

(CLCbio, Hilden, Germany). Then, three independent de novo assemblies, which included single-end forward reads, single end reverse reads, and PE reads, were performed [18]. After that, a single assembly formed by the combination of these three separate assemblies was conducted. From the complete CLC assembly results, assembled contigs longer than 0.5 kb with over 100× coverage were compared to complete cp genomes of several species, including *Euonymus japonicus* (Celastraceae, KP189362), *Populus euphratica* (Salicaceae; NC_024747), and *Salix purpurea* (Salicaceae; NC_026722). Matching the contigs from the cp genomes was done using Local BlastN searches [57]. Using the conserved cp genome regions, the related cp genomes were matched with the mapped contigs [58] and then a single contig was connected to these contigs to create the quadripartite genome employing Contig Express 2003 (Invitrogen, Carlsbad, CA, USA). By designing primers in regions flanking gaps, PCR amplification was carried out and the gap sequences were completed by adding sequence data obtained from Sanger sequencing (Figure S2).

Additionally, primers were designed to verify de novo sequence assemblies, such as the junction regions of the cp genome (Table S2). The 40-µL PCR volume was setup as follows: 10× Taq buffer 4 µL, ddH$_2$O 33.3 µL, 10 mM dNTP 0.8 µL, 20 pmol/µL each primer 0.5 µL, 5 U/µL Taq polymerase 0.4 µL and DNA template 0.5 µL. Taq buffer, dNTP, primers were from Sangong Biotech (Shanghai, China). Cycling conditions were 94 °C for 5 min, 32 cycles 94 °C for 45 s, 54 °C for 45 s, 72 °C for 2 min and, a 10 min 72 °C final extension step. By combining the results of Sanger sequencing, the whole cp genome was used to map reference species to confirm the assembly with the uniformity of the iterative sequences.

Annotation of the transfer RNAs (tRNAs), protein-coding genes, and ribosomal RNAs (rRNAs) was first performed using DOGMA v1.2 (University of Texas at Austin, Austin, TX, USA) [59]. Then, the protein-coding gene positions in the draft annotation were verified and if necessary manually adjusted following alignment to the related species, *Euonymus japonicas* [58] to accurately determine the genes starting point, stop codons and exon borders. Finally, BLASTN searches and tRNAscan-SE v1.21 (University of California Santa Cruz, CA, USA) [60] were employed to verify both tRNA and rRNA genes.

A graphical cp genome map for *C. edulis* was completed using OGDraw (OrganellarGenomeDRAW) (V 1.2, Max Planck Institute of Molecular Plant Physiology, Am Mühlenberg, Germany) [61]. The annotated *C. edulis* cp genome reported and analyzed herein has been deposited in GenBank (KT861471).

3.3. Chloroplast Genomes Comparison

3.3.1. IR Expansion and Contraction

The changes in the size of the angiosperm cp genomes are mainly due to the contraction and expansion from the inverted repeat region, and the two single copy boundary areas. Four borders (J$_{LA}$, J$_{LB}$, J$_{SA}$, and J$_{SB}$) are present in the *C. edulis* cp genome and are situated in the middle of two IRs and two single copy regions [62]. The IR borders and neighboring genes of the two Celastrales species (*Catha edulis* and *Euonymus japonicus*) and six Malpighiales species cp genomes (*Hevea brasiliensis*, *Manihot esculea*, *Populus euphratica*, *Ricinus communis*, *Salix purpurea*, and *Viola seoulensis*) were compared in this study.

3.3.2. Repeat Analysis

Two methods were used to search repeats in *C. edulis* [63]. We identified simple sequence repeats (SSRs) using SSR Hunter v1.3 (Nanjing Agricultural University, Nanjing, China) [64] with cut-offs of eight copy number for mono-SSRs, four copy number for di-, three copy number for tri-, tetra-, penta- and hexanucleotide SSRs. To discover larger repeat regions, REPuter [65] was employed to find four possible repeats types: containing complement, forward, palindrome, and reverse repeats. Nested and low complexity repeats were not included in this study [66].

3.3.3. Dot-Plot Analysis

To identify the structural variations across all 14 genera, *Populus* (Salicaceae; Malpighiales), *Salix* (Salicaceae; Malpighiales), *Viola* (Violaceae; Malpighiales), *Hevea* (Euphorbiaceae; Malpighiales), *Manihot* (Euphorbiaceae; Malpighiales), *Ricinus* (Euphorbiaceae; Malpighiales), and *Euonymus* (Celastraceae; Celastrales), as well as outgroup genera *Prunus*, *Morus*, *Theobroma*, *Eucalyptus*, *Elaeagnus*, *Castanea*, and *Citrus*, we conducted the dot-plot analysis (based on a custom perl script) [13] between *C. edulis* and all 14 genera to visualize structural differences in two dimensional plots.

3.3.4. Verification of the *rps16* Intron Loss from Catha and Seven Other Celastraceae Genera

During annotation, the intron loss of *rps16* was found in the cp genome of *C. edulis*. To verify whether this intron loss happened throughout Celastraceae, two primers were designed (Forward-ACTTCGTTTGAGACGGTGTG, Reverse- AAAAACCCCGATTTCTTTGA) to amplify the entire *rps16* intron from *C. edulis* and seven other Celastraceae species (*Quetzalia stipitata*, *Mortonia diffusa*, *Microtropis triflora*, *Maytenus elliptica*, *Monimopetalum chinensis*, *Cassine aethiopica*, and *Parnassia glauca*). In *C. edulis*, the target *rps16* fragment without the intron is about 550 bp. Absence of the *rps16* intron was visualized on 0.8% agarose gels. The size of the fragment was determined by comparing it to a DNA size standard [67]. The *rps16* gene was sequenced using Sanger sequencing at the Beijing Genomics Institute (BGI).

3.3.5. Phylogenetic Analyses

The 73 common protein-coding genes of 26 species cp genomes, among them eight Rosales and four Fabales outgroup species, were aligned under the default parameters of Clustal X, with reading frames included by manual correction (Supplement data matrix) [68]. The phylogenetic tree based on these 73 common genes was inferred using three different methods. Implementation of Parsimony analysis, Bayesian inference (BI), and maximum likelihood (ML) were made in PAUP* 4.0b10 [69], MrBayes 3.1.2, and PHYML v 2.4.5 [70,71] respectively using the parameters from Wu et al. [18].

4. Conclusions

In this study, using next generation sequencing technology, we successfully completed the whole chloroplast genome for the economically important species *C. edulis*. In comparing the *C. edulis* cp genome with numerous closely related species, we found that it has a typical angiosperm cp genome structure and gene content. However, some unique features are reported here, such as the loss of the intron region from the *rps16* gene, and repeat structure and abundance. We also resolved the phylogenetic position of *C. edulis* with its relatives including the monophyly of Celastraceae. The whole cp genome of *C. edulis* provides a valuable genetic resource for further phylogenomic research, barcoding, and cp transformation in Celastraceae.

Supplementary Materials: Supplementary materials can be found at www.mdpi.com/1422-0067/19/2/525/s1.

Acknowledgments: This research was supported by Zhejiang Provincial Natural Science Foundation of China under Grant No. LY17C160003. The sponsors had no role in data collection, study design, data analysis, or preparing the manuscript. We also thank the editor and the constructive comments of the four anonymous reviewers who helped us to improve this manuscript. We are grateful to Nels Johnson for his kinds help on manuscript editing and improvement.

Author Contributions: Conceived and designed the experiments: Zhiqiang Wu, Cuihua Gu; Performed the experiments: Zhiqiang Wu, Cuihua Gu; Analyzed the data: Zhiqiang Wu, Cuihua Gu, Luke R. Tembrock, Shaoyu Zheng; Contributed reagents/materials/analysis tools: Zhiqiang Wu, Cuihua Gu, Luke R. Tembrock; Wrote the paper: Zhiqiang Wu, Cuihua Gu, Luke R. Tembrock, Shaoyu Zheng.

Conflicts of Interest: The authors declare no conflict of interest.

References

1. Al-Motarreb, A.; Baker, K.; Broadly, K.J. Khat: Pharmacological and medical aspects and its social use in Yemen. *Phytother. Res.* **2002**, *16*, 403–413. [CrossRef] [PubMed]
2. Anderson, D.; Beckerleg, S.; Hailu, D.; Klein, A. *The Khat Controversy: Stimulating the Debate on Drugs*; Berg: Oxford, UK, 2007.
3. Carrier, N.C.M. *The Social Life of a Stimulant*; Brill: Leiden, The Netherlands, 2007.
4. Kennedy, J.G. The flower of paradise: The Institutional Use of the Drug Qat in North Yemen. *Q. Rev. Biol.* **1988**, *63*, 364–365.
5. World Bank. *Yemen: Towards Qat Demand Reduction*; World Bank Document Report 39738-YE; World Bank: Washington, DC, USA, 2007.
6. Gebissa, E. *Leaf of Allah: Khat & Agricultural Transformation in Harerge, Ethiopia*; James Currey Ltd.: Oxford, UK, 2004.
7. Curto, M.A.; Tembrock, L.R.; Puppo, P.; Nogueira, M.; Simmons, M.P.; Meimberg, H. Evaluation of microsatellites of *Catha edulis* (qat; Celastraceae) identified using pyrosequencing. *Biochem. Syst. Ecol.* **2013**, *49*, 1–9. [CrossRef]
8. Hagel, J.M.; Krezevski, K.; Sitrit, Y.; Marsolais, F.; Facchini, J.P.; Krizevski, R.; Lewinsohn, E. Expressed sequence tag analysis of khat (*Catha edulis*) provides a putative molecular biochemical basis for the biosynthesis of phenylpropylamino alkaloids. *Genet. Mol. Biol.* **2011**, *34*, 640–646. [CrossRef] [PubMed]
9. Tembrock, L.R.; Broeckling, C.D.; Heuberger, A.L.; Simmons, M.P.; Stermitz, F.R.; Uvarov, J.M. Employing two-stage derivatisation and GC–MS to assay for cathine and related stimulant alkaloids across the Celastraceae. *Phytochem. Anal.* **2017**, *28*, 257–266. [CrossRef] [PubMed]
10. Simmons, M.P.; Cappa, J.J.; Archer, R.H.; Ford, A.J.; Eichstedt, D.; Clevinger, C.C. Phylogeny of the Celastreae (Celastraceae) and the relationships of *Catha edulis* (qat) inferred from morphological characters and nuclear and plastid genes. *Mol. Phylogenet. Evol.* **2008**, *48*, 745–757. [CrossRef] [PubMed]
11. Tembrock, L.R.; Simmons, M.P.; Richards, C.M.; Reeves, P.A.; Reilley, A.; Curto, M.A.; Al-Thobhani, M.; Varisco, D.M.; Simpson, S.; Ngugi, G.; et al. Phylogeography of the wild and cultivated stimulant plant qat (*Catha edulis*, Celastraceae) in areas of historical cultivation. *Am. J. Bot.* **2017**, *104*, 538–549. [CrossRef] [PubMed]
12. Ravi, V.; Khurana, J.P.; Tyagi, A.K.; Khurana, P. An update on chloroplast genomes. *Plant Syst. Evol.* **2008**, *271*, 101–122. [CrossRef]
13. Gu, C.H.; Tembrock, L.R.; Johnson, N.G.; Simmons, M.P.; Wu, Z.Q. The complete plastid genome of *Lagerstroemia fauriei* and loss of *rpl2* intron from *Lagerstroemia* (Lythraceae). *PLoS ONE* **2016**, *11*, e0150752. [CrossRef] [PubMed]
14. Dong, W.; Xu, C.; Li, C.; Sun, J.; Zuo, Y.; Shi, S.; Cheng, T.; Guo, J.; Zhou, S. *ycf1*, the most promising plastid DNA barcode of land plants. *Sci. Rep.* **2015**, *5*, 8348. [CrossRef] [PubMed]
15. Palmer, J.D. Comparative organization of chloroplast genomes. *Annu. Rev. Genet.* **1985**, *19*, 325–354. [CrossRef] [PubMed]
16. Wicke, S.; Schneeweiss, G.M.; DePamphilis, C.W.; Müller, K.F.; Quandt, D. The evolution of the plastid chromosome in land plants: Gene content, gene order, gene function. *Plant Mol. Biol.* **2011**, *76*, 273–297. [CrossRef] [PubMed]
17. Soltis, D.E.; Gitzendanner, M.; Stull, G.; Chester, M.; Chanderbali, A.; Jordon-Thaden, I.; Soltis, P.S.; Schnable, P.S.; Barbazuk, W.B. The potential of genomics in plant systematics. *Taxon* **2013**, *62*, 886–898. [CrossRef]
18. Wu, Z.Q.; Tembrock, L.R.; Ge, S. Are Differences in Genomic Data Sets due to True Biological Variants or Errors in Genome Assembly: An Example from Two Chloroplast Genomes. *PLoS ONE* **2015**, *10*, e0118019. [CrossRef] [PubMed]
19. CBOL. A DNA barcode for land plants. *Proc. Natl. Acad. Sci. USA* **2009**, *106*, 12794–12797.
20. Day, A.; Goldschmidt-Clermont, M. The chloroplast transformation toolbox: Selectable markers and marker removal. *Plant Biotechnol. J.* **2011**, *9*, 540–553. [CrossRef] [PubMed]

21. Shaw, J.; Lickey, E.B.; Beck, J.T.; Farmer, S.B.; Liu, W.; Miller, J.; Siripun, K.C.; Winder, C.T.; Schilling, E.E.; Small, R.L. The tortoise and the hare II: Relative utility of 21 noncoding chloroplast DNA sequences for phylogenetic analysis. *Am. J. Bot.* **2005**, *92*, 142–166. [CrossRef] [PubMed]

22. Wu, Z.Q.; Ge, S. The phylogeny of the BEP clade in grasses revisited: Evidence from the whole-genome sequences of chloroplasts. *Mol. Phylogenet. Evol.* **2012**, *62*, 573–578. [CrossRef] [PubMed]

23. Cerutti, H.; Johnson, A.M.; Boynton, J.E.; Gillham, N.W. Inhibition of chloroplast DNA recombination and repair by dominant negative mutants of Escherichia coli RecA. *Mol. Cell. Biol.* **1995**, *15*, 3003–3011. [CrossRef] [PubMed]

24. Maliga, P. Plastid transformation in higher plants. *Annu. Rev. Plant Biol.* **2004**, *55*, 289–313. [CrossRef] [PubMed]

25. Maliga, P.; Staub, J.; Carrer, H.; Kanevski, I.; Svab, Z. *Homologous Recombination and Integration of Foreign DNA in Plastids of Higher Plants*; Paszkowski, J., Ed.; Kluwer Academic: Amsterdam, The Netherlands, 1994.

26. Svab, Z.; Maliga, P. High-frequency plastid transformation in tobacco by selection for a chimeric aadA gene. *Proc. Natl. Acad. Sci. USA* **1993**, *90*, 913–917. [CrossRef] [PubMed]

27. Yang, J.B.; Li, D.Z.; Li, H.T. Highly effective sequencing whole chloroplast genomes of angiosperms by nine novel universal primer pairs. *Mol. Ecol. Resour.* **2014**, *14*, 1024–1031. [CrossRef] [PubMed]

28. O'Brien, S.J.; Stanyon, R. Phylogenomics. Ancestral primate viewed. *Nature* **1999**, *402*, 365–366. [CrossRef] [PubMed]

29. Green, B.R. Chloroplast genomes of photosynthetic eukaryotes. *Plant J.* **2011**, *66*, 34–44. [CrossRef] [PubMed]

30. Su, H.; Hogenhout, S.A.; Al-sadi, A.M.; Kuo, C. Complete chloroplast genome sequence of Omani Lime (*Citrus aurantiifolia*) and comparative analysis within the Rosids. *PLoS ONE* **2014**, *9*, e113049. [CrossRef] [PubMed]

31. Redwan, R.M.; Saidin, A.; Kumar, S.V. Complete chloroplast genome sequence of MD-2 pineapple and its comparative analysis among nine other plants from the subclass Commelinidae. *BMC Plant Biol.* **2015**, *15*, 196. [CrossRef] [PubMed]

32. Tangphatsornruang, S.; Uthaipaisanwong, P.; Sangsrakru, D.; Chanprasert, J.; Yoocha, T.; Jomchai, N.; Tragoonrung, S. Characterization of the complete chloroplast genome of *Hevea brasiliensis* reveals genome rearrangement, RNA editing sites and phylogenetic relationships. *Gene* **2011**, *475*, 104–112. [CrossRef] [PubMed]

33. Raubeson, L.A.; Peery, R.; Chumley, T.W.; Dziubek, C.; Fourcade, H.M. Comparative chloroplast genomics: Analyses including new sequences from the angiosperms Nuphar advena and Ranunculus macranthus. *BMC Genom.* **2007**, *8*, 174. [CrossRef] [PubMed]

34. Kim, K.J.; Lee, H.L. Complete chloroplast genome sequences from Korean ginseng (*Panax ginseng* Nees) and comparative analysis of sequence evolution among 17 vascular plants. *DNA Res.* **2004**, *11*, 247–261. [CrossRef] [PubMed]

35. Ryzhova, N.N.; Kholda, O.A.; Kochieva, E.Z. Structure characteristics of the chloroplast *rps16* intron in Allium sativum and related Allium species. *Mol. Biol.* **2009**, *43*, 766–775. [CrossRef]

36. Schwarz, E.N.; Ruhlman, T.A.; Sabir, J.S.; Hajrah, N.H.; Alharbi, N.S.; Al-Malki, A.L.; Bailey, C.D.; Jansen, R.K. Plastid genome sequences of legumes reveal parallel inversions and multiple losses of *rps16* in papilionoids. *J. Syst. Evol.* **2015**, *53*, 458–468. [CrossRef]

37. Downie, S.R.; Olmstead, R.G.; Zurawski, G.; Soltis, D.E.; Soltis, S.; Watson, J.C.; Palmer, J.D. Six independent losses of the Chloroplast DNA *rpl2* intron in Dicotyledons: Molecular and Phylogenetic Implications. *Evolution* **1991**, *45*, 1245–1259. [CrossRef] [PubMed]

38. Tank, D.C.; Eastman, J.M.; Pennell, M.W.; Soltis, P.S.; Soltis, D.E.; Hinchliff, C.E.; Brown, J.W.; Sessa, E.B.; Harmon, L.J. Nested radiations and the pulse of angiosperm diversification: Increased diversification rates often follow whole genome duplications. *New Phytol.* **2015**, *207*, 454–467. [CrossRef] [PubMed]

39. CavalierSmith, T. Chloroplast evolution: Secondary symbiogenesis and multiple losses. *Curr. Biol.* **2002**, *12*, 62–64. [CrossRef]

40. Rubinsztein, D.C.; Amos, W.; Leggo, J.; Goodburn, S.; Jain, S.; Li, S.H.; Margolis, R.L.; Ross, C.A.; Ferguson-Smith, M.A. Microsatellite evolution—Evidence for directionality and variation in rate between species. *Nat. Genet.* **1995**, *10*, 337–343. [CrossRef] [PubMed]

41. Gemayel, R.; Cho, J.; Boeynaems, S.; Verstrepen, K.J. Beyond junk-variable tandem repeats as facilitators of rapid evolution of regulatory and coding sequences. *Genes* **2012**, *3*, 461–480. [CrossRef] [PubMed]

42. Voronova, A.; Belevich, V.; Jansons, A.; Rungis, D. Stress-induced transcriptional activation of retrotransposon-like sequences in the Scots pine (*Pinus sylvestris* L.) genome. *Tree Genet. Genomes* **2014**, *10*, 937–951. [CrossRef]

43. Timme, R.E.; Kuehl, J.V.; Boore, J.L.; Jansen, R.K. A comparative analysis of the Lactuca and Helianthus (Asteraceae) plastid genomes: Identification of divergent regions and categorization of shared repeats. *Am. J. Bot.* **2007**, *94*, 302–312. [CrossRef] [PubMed]

44. Nie, X.; Lv, S.; Zhang, Y.; Du, X.; Wang, L.; Biradar, S.S.; Tan, X.; Wan, F.; Weining, S. Complete chloroplast genome sequence of a major invasive species, crofton weed (*Ageratina adenophora*). *PLoS ONE* **2012**, *7*, e36869. [CrossRef] [PubMed]

45. Grassi, F.; Labra, M.; Scienza, A.; Imazio, S. Chloroplast SSR markers to assess DNA diversity in wild and cultivated grapevines. *Vitis* **2002**, *41*, 157–158.

46. Dong, W.; Liu, J.; Yu, J.; Wang, L.; Zhou, S. Highly variable chloroplast markers for evaluating plant phylogeny at low taxonomic levels and for DNA barcoding. *PLoS ONE* **2012**, *7*, e35071. [CrossRef] [PubMed]

47. Kress, W.J.; Erickson, D.L. A two-locus global DNA barcode for land plants: The coding *rbcL* gene complements the non-coding *trnH-psbA* spacer region. *PLoS ONE* **2007**, *2*, e508. [CrossRef] [PubMed]

48. Li, X.; Yang, Y.; Henry, R.J.; Rossetto, M.; Wang, Y.; Chen, S. Plant DNA barcoding: From gene to genome. *Biol. Rev.* **2014**, *90*, 157–166. [CrossRef] [PubMed]

49. Hilu, K.W.; Black, C.; Diouf, D.; Burleigh, J.G. Phylogenetic signal in *matK* vs. *trnK*: A case study in early diverging eudicots (angiosperms). *Mol. Phylogenet. Evol.* **2008**, *48*, 1120–1130. [CrossRef] [PubMed]

50. Kim, K.J.; Jansen, R.K. *ndhF* sequence evolution and the major clades in the sunflower family. *Proc. Natl. Acad. Sci. USA* **1995**, *92*, 10379–10383. [CrossRef] [PubMed]

51. Li, J. Phylogeny of *Catalpa* (Bignoniaceae) inferred from sequences of chloroplast *ndhF* and nuclear ribosomal DNA. *J. Syst. Evol.* **2008**, *46*, 341–348.

52. Yan, H.F.; Liu, Y.J.; Xie, X.F.; Zhang, C.Y.; Hu, C.M.; Hao, G.; Ge, X.J. DNA barcoding evaluation and its taxonomic implications in the species-rich genus *Primula* L. in China. *PLoS ONE* **2015**, *10*, e0122903. [CrossRef] [PubMed]

53. Jansen, R.K.; Cai, Z.; Raubeson, L.A.; Daniell, H.; Depamphilis, C.W.; Leebens-Mack, J.; Müller, K.F.; Guisinger-Bellian, M.; Haberle, R.C.; Hansen, A.K.; et al. Analysis of 81 genes from 64 plastid genomes resolves relationships in angiosperms and identifies genome-scale evolutionary patterns. *Proc. Natl. Acad. Sci. USA* **2007**, *104*, 19369–19374. [CrossRef] [PubMed]

54. Moore, M.J.; Bell, C.D.; Soltis, P.S.; Soltis, D.E. Using plastid genome-scale data to resolve enigmatic relationships among basal angiosperms. *Proc. Natl. Acad. Sci. USA* **2007**, *104*, 19363–19368. [CrossRef] [PubMed]

55. Gao, L.; Su, Y.J.; Wang, T. Plastid genome sequencing, comparative genomics, and phylogenomics: Current status and prospects. *J. Syst. Evol.* **2010**, *48*, 77–93. [CrossRef]

56. Bolger, A.M.; Lohse, M.; Usadel, B. Trimmomatic: A flexible trimmer for Illumina sequence data. *Bioinformatics* **2014**, *30*, 2114–2120. [CrossRef] [PubMed]

57. Camacho, C.; Coulouris, G.; Avagyan, V.; Ma, N.; Papadopoulos, J.; Bealer, K.; Madden, T.L. BLAST+: Architecture and applications. *BMC Bioinform.* **2009**, *10*, 421. [CrossRef] [PubMed]

58. Choi, K.S.; Park, S. The complete chloroplast genome sequence of *Euonymus japonicus* (Celastraceae). *Mitochondrial DNA* **2015**, *1736*, 1–2.

59. Wyman, S.K.; Jansen, R.K.; Boore, J.L. Automatic annotation of organellar genomes with DOGMA. *Bioinformatics* **2004**, *20*, 3252–3255. [CrossRef] [PubMed]

60. Schattner, P.; Brooks, A.N.; Lowe, T.M. The tRNAscan-SE, snoscan and snoGPS web servers for the detection of tRNAs and snoRNAs. *Nucleic Acids Res.* **2005**, *33*, 686–689. [CrossRef] [PubMed]

61. Lohse, M.; Drechsel, O.; Bock, R. OrganellarGenomeDRAW (OGDRAW): A tool for the easy generation of high-quality custom graphical maps of plastid and mitochondrial genomes. *Curr. Genet.* **2007**, *52*, 267–274. [CrossRef] [PubMed]

62. Wang, R.J.; Cheng, C.L.; Chang, C.C.; Wu, C.L.; Su, T.M.; Chaw, S.M. Dynamics and evolution of the inverted repeat-large single copy junctions in the chloroplast genomes of monocots. *BMC Evol. Biol.* **2008**, *8*, 36. [CrossRef] [PubMed]

63. Huang, H.; Shi, C.; Liu, Y.; Mao, S.Y.; Gao, L.Z. Thirteen Camellia chloroplast genome sequences determined by high-throughput sequencing: Genome structure and phylogenetic relationships. *BMC Evol. Biol.* **2016**, *14*, 151. [CrossRef] [PubMed]

64. Li, Q.; Wan, J.M. SSRHunter: Development of local searching software for SSR sites. *Yi Chuan* **2005**, *27*, 808–810. [PubMed]

65. Kurtz, S.; Choudhuri, J.V.; Ohlebusch, E.; Schleiermacher, C.; Stoye, J.; Giegerich, R. REPuter: The manifold applications of repeat analysis on a genomic scale. *Nucleic Acids Res.* **2001**, *29*, 4633–4642. [CrossRef] [PubMed]

66. Yang, Y.; Dang, Y.; Li, Q.; Lu, J.J.; Li, X.W.; Wang, Y.T. Complete Chloroplast genome sequence of poisonous and medicinal plant *Datura stramonium*: Organizations and implications for genetic engineering. *PLoS ONE* **2014**, *9*, e110656. [CrossRef] [PubMed]

67. Jansen, R.K.; Wojciechowski, M.F.; Sanniyasi, E.; Lee, S.B.; Daniell, H. Complete plastid genome sequence of the chickpea (*Cicer arietinum*) and the phylogenetic distribution of *rps12* and *clpP* intron losses among legumes (Leguminosae). *Mol. Phylogenet. Evol.* **2008**, *48*, 1204–1217. [CrossRef] [PubMed]

68. Simmons, M.P. Independence of alignment and tree search. *Mol. Phylogenet. Evol.* **2004**, *31*, 874–879. [CrossRef] [PubMed]

69. Swofford, D.L. Paup*: Phylogenetic Analysis Using Parsimony (and other methods). *Mccarthy* **1993**, 1–142.

70. Guindon, S.; Dufayard, J.F.; Lefort, V.; Anisimova, M. New alogrithms and methods to estimate maximum-likelihoods phylogenies: Assessing the performance of PhyML 30. *Syst. Biol.* **2010**, *59*, 307–321. [CrossRef] [PubMed]

71. Ronquist, F.; Teslenko, M.; Van Der Mark, P.; Ayres, D.L.; Darling, A.; Höhna, S.; Larget, B.; Liu, L.; Suchard, M.A.; Huelsenbeck, J.P. Mrbayes 3.2: Efficient bayesian phylogenetic inference and model choice across a large model space. *Syst. Biol.* **2012**, *61*, 539–542. [CrossRef] [PubMed]

International Journal of
Molecular Sciences

MDPI

Article

Comparative Plastid Genomes of *Primula* Species: Sequence Divergence and Phylogenetic Relationships

Ting Ren [1], Yanci Yang [1], Tao Zhou [2] and Zhan-Lin Liu [1,*]

[1]   Key Laboratory of Resource Biology and Biotechnology in Western China (Ministry of Education),
      College of Life Sciences, Northwest University, Xi'an 710069, China; renting92@stumail.nwu.edu.cn (T.R.);
      yycjyl1@gmail.com (Y.Y.)
[2]   School of Pharmacy, Xi'an Jiaotong University, Xi'an 710061, China; zhoutao196@mail.xjtu.edu.cn
*   Correspondence: liuzl@nwu.edu.cn

Received: 13 March 2018; Accepted: 29 March 2018; Published: 1 April 2018

Abstract: Compared to traditional DNA markers, genome-scale datasets can provide mass information to effectively address historically difficult phylogenies. *Primula* is the largest genus in the family Primulaceae, with members distributed mainly throughout temperate and arctic areas of the Northern Hemisphere. The phylogenetic relationships among *Primula* taxa still maintain unresolved, mainly due to intra- and interspecific morphological variation, which was caused by frequent hybridization and introgression. In this study, we sequenced and assembled four complete plastid genomes (*Primula handeliana*, *Primula woodwardii*, *Primula knuthiana*, and *Androsace laxa*) by Illumina paired-end sequencing. A total of 10 *Primula* species (including 7 published plastid genomes) were analyzed to investigate the plastid genome sequence divergence and their inferences for the phylogeny of *Primula*. The 10 *Primula* plastid genomes were similar in terms of their gene content and order, GC content, and codon usage, but slightly different in the number of the repeat. Moderate sequence divergence was observed among *Primula* plastid genomes. Phylogenetic analysis strongly supported that *Primula* was monophyletic and more closely related to *Androsace* in the Primulaceae family. The phylogenetic relationships among the 10 *Primula* species showed that the placement of *P. knuthiana*–*P. veris* clade was uncertain in the phylogenetic tree. This study indicated that plastid genome data were highly effective to investigate the phylogeny.

Keywords: plastid genome; phylogenetic relationship; *Primula*; repeat; sequence divergence

1. Introduction

Primula is the largest genus in the family Primulaceae with approximately 500 species [1,2], where they are especially rich in the temperate and arctic areas of the Northern Hemisphere, with only a few outliers found in the Southern Hemisphere. China is the center of *Primula* diversity and speciation with over 300 species [1,3]. Many *Primula* species are grown widely as ornamental and landscape plants because of their attractive flowers and long flowering period. Therefore, *Primula* is reputed to be one of the great garden plant genera throughout the world [2,3].

As a typical cross-pollinated plant with heterostyly, *Primula* has been a particular focus of many botanists, and various studies are involved in hybridization [4], pollination biology [5,6], and distyly [7,8]. According to morphological traits, the taxonomic study of *Primula* has been revised for several times. Smith and Fletcher (1947) firstly proposed an infrageneric system with a total of 31 sections [9]. Considering some putative reticulate evolutionary relationships, Wendelbo (1961) posed a revised system with seven subgenera [10]. Richards (1993) later amended Wendelbo's version and classified six subgenera [11]. Hu and Kelso (1996) delimited the Chinese *Primula* species into 24 sections [1]. Recently, numerous molecular phylogenetic works of the genus *Primula* have also been conducted by using plastid and/or nuclear gene fragments [12–14]. These studies have greatly advanced our understanding of the evolutionary history of *Primula* species. However, the phylogenetic relationships within the genus *Primula* are still uncertain, mainly due to intra- and interspecific morphological variation, which was caused by frequent hybridization and introgression [1,2,14]. Further research has been hindered by the insufficient information of the traditional DNA markers, such as one or few chloroplast gene fragments, and by the complex evolutionary relationships in *Primula*. Therefore, more sequence resources and genome data are required in order to obtain a better understanding of the phylogeny of the genus *Primula*.

In general, the plastid genome in angiosperms is a typical quadripartite structure, where the size ranges from 115 to 165 kb, with two copies of inverted repeat (IR) regions separated by a large single copy (LSC) region and a small single copy (SSC) region [15]. Approximately 110–130 distinct genes are located along the plastid genome [16]. Most of these are protein-coding genes, the remainder being transfer RNA (*tRNA*) or ribosomal RNA (*rRNA*) genes [16]. Due to its particular advantages—such as small size, uniparental inheritance, low substitution rates, and high conservation in terms of the gene content and genome structure [17,18]—the plastid genome is considered a very promising tool for phylogenetic studies [19,20]. Significant advances in next-generation sequencing technology made it fairly inexpensive and convenient to obtain plastid genome sequences [21,22]. As a result, phylogenomic analyses have also been greatly facilitated. For example, the plastid phylogenomic analyses supported Tofieldiaceae as the most basal lineage within Alismatales [23]. The relationships between wild and domestic *Citrus* species could also be resolved with 34 plastid genomes [24]. Similarly, 142 plastid genomes were used to successfully infer deep phylogenetic relationships and the diversification history of Rosaceae [25]. These studies strongly indicate that plastid phylogenomics is helpful in determining the phylogenetic positions of various questionable lineages of angiosperms.

In the present study, we analyzed the complete plastid genomes of 10 *Primula* species including 7 published plastid genomes and 3 new data (*Primula handeliana*, *Primula woodwardii*, and *Primula knuthiana*) by using Illumina sequencing technology. Our primary aims were to: (1) compare the complete plastid genomes of 10 *Primula* species; (2) document that the extent of sequence divergence among the *Primula* plastid genomes; and (3) increase more sequence resources and genome information for investigating the phylogeny in genus *Primula*. The complete plastid genome of *Androsace laxa* from a closely related genus was used as the outgroup in the phylogenomic analysis of genus *Primula*. This study will not only contribute to further studies on the phylogeny, taxonomy, and evolutionary history of the genus *Primula*, but also provide insight into the plastid genome evolution of *Primula*.

2. Results

2.1. Genome Features

The sizes of the plastid genomes of the 10 *Primula* species ranged from 150,856 bp to 153,757 bp, where they had a typical quadripartite structure, including a LSC region (82,048–84,479 bp) and a SSC region (17,568–17,896 bp) separated by a pair of IR regions (25,182–25,855 bp) (Table 1). In the 10 *Primula* plastid genomes, gene content was similar and gene order was identical. The *Primula* plastid genomes contained about 130–132 genes, including 85–86 protein-coding genes, 37 *tRNA* genes, and 8 *rRNA* genes (Tables 1 and S4). The *accD* gene was a pseudogene in *P. sinensis*, whereas it was missing in *P. persimilis* and *P. kwangtungensis*. The *P. poissonii* plastid genome contained a pseudogene (*infA*). Among these genes, 15 genes harbored a single intron (*trnA-UGC*, *trnG-UCC*, *trnI-GAU*, *trnK-UUU*, *trnL-UAA*, *trnV-UAC*, *atpF*, *ndhA*, *ndhB*, *petB*, *petD*, *rpoC1*, *rpl2*, *rpl16*, and *rps16*) and three genes (*pafI*, *clpP*, and *rps12*) harbored two introns. Seven *tRNA* genes, seven protein-coding genes, and all four *rRNA* genes were completely duplicated in the IR regions (Table 1). *trnk-UUU* had the largest intron (2487–2568 bp) containing the *matK* gene. The GC contents of the LSC, SSC, and IR regions, as well as those of the whole plastid genomes, were nearly identical in the 10 *Primula* plastid genomes (Table 1). The complete plastid genome of *A. laxa* was 151,942 bp in length and contained 132 genes (Table 1). The overall GC content of the *A. laxa* plastid genome was 37.3%, and the corresponding values for the LSC, SSC, and IR regions were 35.2, 30.9, and 42.7%, respectively (Table 1).

Table 1. Plastid genomic characteristics of the 10 *Primula* species and *A. laxa*.

Taxa	A. laxa *	P. handeliana *	P. woodwardii *	P. knuthiana *	P. poissonii	P. sinensis	P. veris
Assembly reads	16,137,534	12,884,542	25,149,710	15,928,364	/	/	/
Mean coverage	293.4×	482.4×	508.3×	405.3×	/	/	/
GenBank numbers	MG181220	MG181221	MG181222	MG181223	NC_024543	NC_030609	NC_031428
Total genome size (bp)	151,942	151,081	151,666	152,502	151,664	150,859	150,856
LSC (bp)	83,078	82,785	83,325	83,446	83,444	82,064	82,048
IRs (bp)	25,970	25,200	25,290	25,604	25,199	25,535	25,524
SSC (bp)	16,924	17,896	17,761	17,848	17,822	17,725	17,760
Total GC content (%)	37.3	37	37	37	37	37.2	37.1
LSC (%)	35.2	34.9	34.9	34.9	34.9	35.2	35.1
IRs (%)	42.7	42.9	42.8	42.7	42.9	42.8	42.7
SSC (%)	30.9	30.2	30.2	30.3	30.1	30.5	30.2
Total number of genes	132	131	131	131	132	131	131
Protein-coding	87 (7)	86 (7)	86 (7)	86 (7)	86 (7)	85 (7)	86 (7)
tRNA	37 (7)	37 (7)	37 (7)	37 (7)	37 (7)	37 (7)	37 (7)
rRNA	8 (4)	8 (4)	8 (4)	8 (4)	8 (4)	8 (4)	8 (4)
Pseudogenes	/	/	/	/	infA	accD	/

Taxa	P. kwangtungensis	P. chrysochlora	P. stenodonta	P. persimilis
Raw Base (G)	/	/	/	/
Mean coverage	/	/	/	/
GenBank numbers	NC_034371	KX668178	KX668176	KX641757
Total genome size (bp)	153,757	151,944	150,785	152,756
LSC (bp)	84,479	83,953	82,682	83,537
IRs (bp)	25,855	25,460	25,182	25,753
SSC (bp)	17,568	17,801	17,739	17,713
Total GC content (%)	37.1	37	37.1	37.2
LSC (%)	35	35	35	35.2
IRs (%)	42.7	42.8	43	42.8
SSC (%)	30.4	30.2	30.2	30.6
Total number of genes	130	131	131	130
Protein-coding	85 (7)	86 (7)	86 (7)	85 (7)
tRNA	37 (7)	37 (7)	37 (7)	37 (7)
rRNA	8 (4)	8 (4)	8 (4)	8 (4)
Pseudogenes	/	/	/	/

*, The four newly generated plastid genomes. LSC, large single copy region, IR, inverted repeat regions, and SSC, small single copy region.

2.2. Codon Usage Analysis

Codon usage plays a crucial role in evolution of plastid genome. Here, we first analyzed codons of the protein-coding genes in the 10 *Primula* plastid genomes. The number of encoded codons ranged from 25,781 (*P. sinensis*) to 26,505 (*P. knuthiana*) (Table S5). Detailed codon analysis showed that the

10 *Primula* species had similar codon usage and relative synonymous codon usage (RSCU) values (Table S5). Leucine and Cysteine were the highest (2743–2823 codons) and lowest (280–298 codons) frequent amino acids in these species, respectively (Table S5). RSCU > 1 denotes that the codon is biased and used more frequently, RSCU = 1 shows that the codon has no bias, and RSCU < 1 indicates that the codon is used less frequently. All 10 *Primula* plastid genomes had 30 biased codons with RSCU > 1 (Table S5). The biased codons had higher representation rates for A or T at the third codon position in a similar manner to the majority of angiosperm plastid genomes. Except for TTG, all of the types of biased codons (RSCU > 1) ended with A or T. The GC% was quite different at the three codon positions (Table S6). The average values of GC% for the first, second, and third codon positions of 10 *Primula* species were 45.3, 37.9, and 29.2%, respectively (Table S6). The observation of GC% level also indicated that plastid genome in *Primula* was a strong bias toward A or T at the third codon position.

2.3. Analysis of Repeat Elements

Three categories of repeats (dispersed, palindromic, and tandem repeats) were identified in the 10 *Primula* plastid genomes. We detected 326 repeats in total comprising 144 dispersed, 123 palindromic, and 59 tandem repeats (Figure 1A and Table S7). Among them, repeats of *P. sinensis* (45) were the greatest and that of *P. woodwardii* (26) were the lowest (Figure 1A and Table S7). The majority of the repeats (95.4%) ranged in size from 14 to 62 bp (Figure 1B and Table S7). Repeats located in intergenic spacer (IGS) and intron regions comprised 44.2% (144 repeats) of the total repeats and 47.8% (156 repeats) were located in *ycf2* gene, whereas only a minority were located in other coding DNA sequence (CDS) regions, such as *psaB*, *trnS-GCU*, *ycf1*, *rpoB*, *ndhF*, etc. (Table S7).

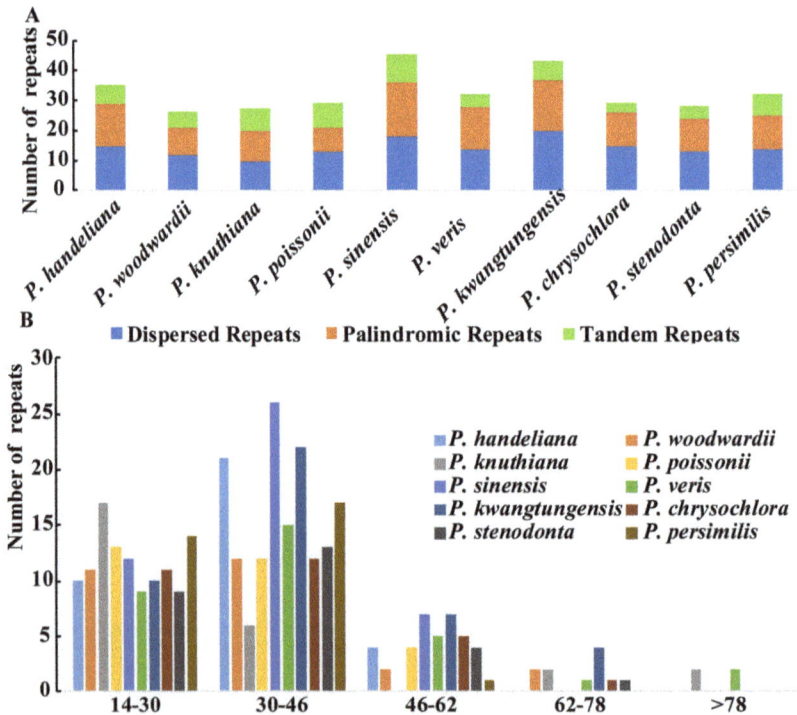

Figure 1. The type of repeated sequences in the 10 *Primula* plastid genomes. (**A**) Number of three repeat types; (**B**) number of repeat sequences by length.

Int. J. Mol. Sci. **2018**, 19, 1050

A total of 496 simple sequence repeats (SSRs) measuring at least 10 bp in length were also analyzed (Figure 2A and Table S8). Among these SSRs, the mononucleotide, dinucleotide, trinucleotide, tetranucleotide, pentanucleotide, and hexanucleotide SSRs were all detected. The mononucleotide SSR were the richest with a proportion of 76.6%, followed by dinucleotide SSR (11.7%), tetranucleotide SSR (7.7%), and trinucleotide SSR (2.8%) (Figure 2A and Table S8). We only detected six pentanucleotide and hexanucleotide SSRs in the 10 *Primula* cp genomes (Figure 2A and Table S8). Unsurprisingly, the mononucleotide A/T SSR occupied the highest portion (368; 74.2%) (Table S8). The number of mononucleotide A/T SSR was significantly higher than that of the mononucleotide G/C SSR (Figure 2B and Table S8). Furthermore, most of the SSRs were found in IGS regions (56.9%), followed by CDS regions (25%), and intron regions (17.9%) (Table S8). SSRs located in the CDS region were mainly found in the *ycf1* gene.

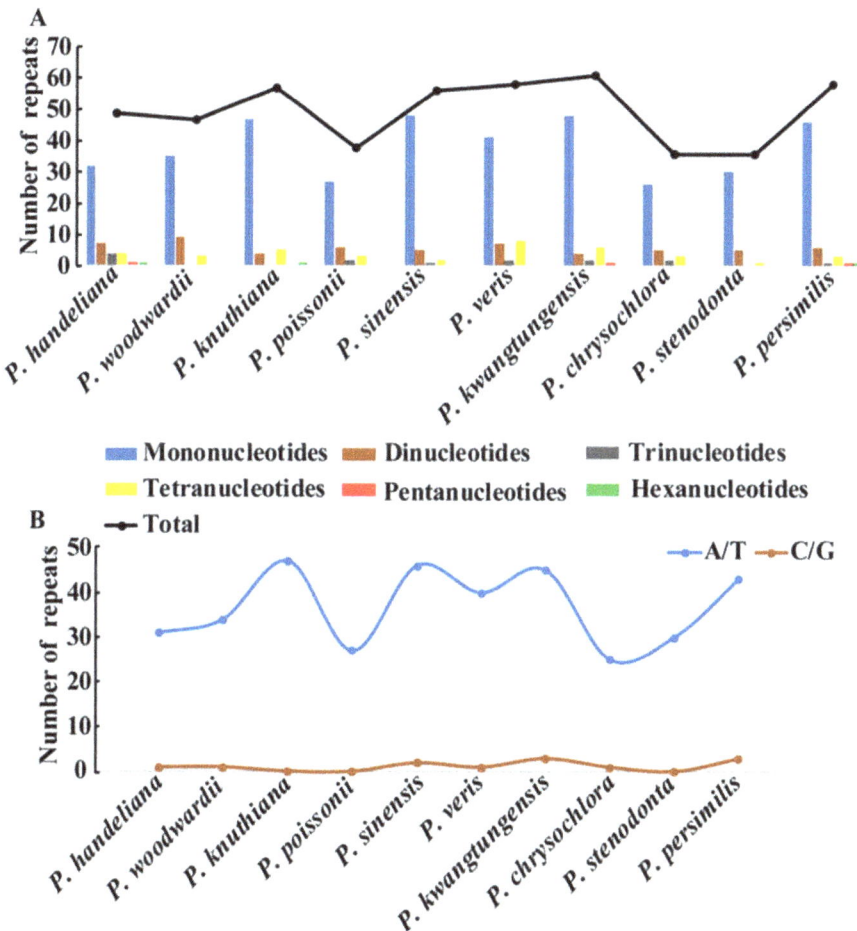

Figure 2. Simple sequence repeats (SSRs) in the 10 *Primula* plastid genomes. (**A**) Number of SSR types; (**B**) number of mononucleotide A/T and G/C SSRs.

2.4. IR/SC Boundary and Genome Rearrangement

The IR/SC boundary contents of 10 *Primula* plastid genomes were compared (Figure 3). The gene content and gene order were conserved at the IR/SC boundary, but the *Primula* plastid genomes exhibited more obvious differences. In the *P. kwangtungensis* plastid genome, the *rps19* gene was located entirely in the LSC, whereas IRb extended in a variable manner 7–175 bp into the *rps19* gene in all the other species. In the *P. chrysochlora* plastid genome, IRb even crossed completely into the *rps19* gene. IRb extended 7–74 bp in a variable manner into the *ndhF* genes, except in the *P. handeliana* and *P. poissonii* plastid genomes. In all of the *Primula* plastid genomes, IRa extended into the *ycf1* genes, where the smallest and largest extensions occurred in the *P. handeliana* (888 bp) and *P. kwangtungensis* (1048 bp) plastid genomes. The whole-genome alignment of the 10 *Primula* plastid genomes showed no rearrangement events in *Primula* (Figure S1).

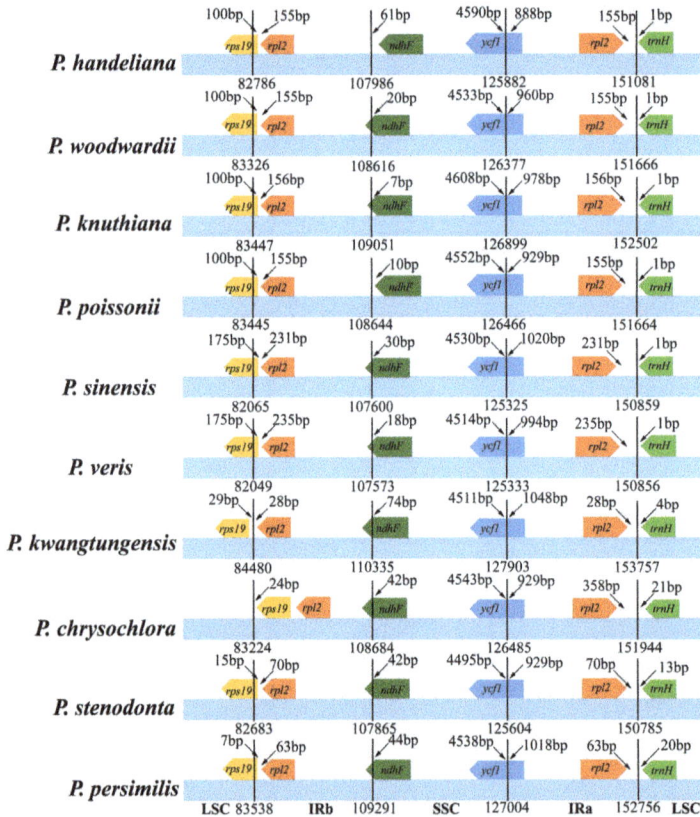

Figure 3. Comparison of the LSC, IR, and SSC border regions among the 10 *Primula* plastid genomes. Number above the gene features means the distance between the ends of genes and the borders sites. These features are not to scale.

2.5. Sequence Divergence

To investigate the levels of sequence divergence, the 10 *Primula* plastid genomes were plotted using mVISTA with *P. poissonii* as the reference (Figure 4). The *Primula* plastid genomes exhibited moderate sequence divergence (Figure 4). As expected, coding and IR regions exhibited more sequence conservation than non-coding and SC regions, respectively (Figure 4). We then calculated the percentage of variable characters for each coding region and non-coding regions with an alignment length of more than 200 bp (Table S9). The average percentage of variation in non-coding regions is 0.38, which was significantly higher than that in the coding regions (0.088 on average; Table S9). The *accD* gene contained various indels and it was a pseudogene in *P. sinensis* and missing in *P. persimilis* and *P. kwangtungensis*, which may have caused the most divergent coding region. In addition, 15 genes had a percentage of variation greater than 0.10 (Table S9), i.e., *ycf1* (0.23), *matK* (0.18), *ycf15* (0.17), *ndhF* (0.17), *rpl33* (0.16), *rpl22* (0.16), *rps16* (0.15), *rps8* (0.12), *ccsA* (0.12), *rps15* (0.12), *rpoC2* (0.11), *psbH* (0.11), *ndhD* (0.11), *rpoA* (0.10), and *ndhA* (0.10). Among the 16 genes with higher percentages of variation, 15 genes were found in SC regions and only one gene in IR regions (Table S9). The average percentages of variations in the LSC, SSC, and IR regions were 0.42, 0.43 and 0.15 in the non-coding regions, while the corresponding values in the coding regions were 0.09, 0.11, and 0.04, respectively (Table S9). All of the results demonstrated that the IR regions were more conserved than the SC regions. The overall sequence divergence based on the p-distance among the 10 *Primula* species was 0.028143 (Table S10). The pairwise p-distance between the 10 species ranged from 0.005857 to 0.041629 (Table S10). These results suggested that moderate sequence divergence has occurred within the genus *Primula*.

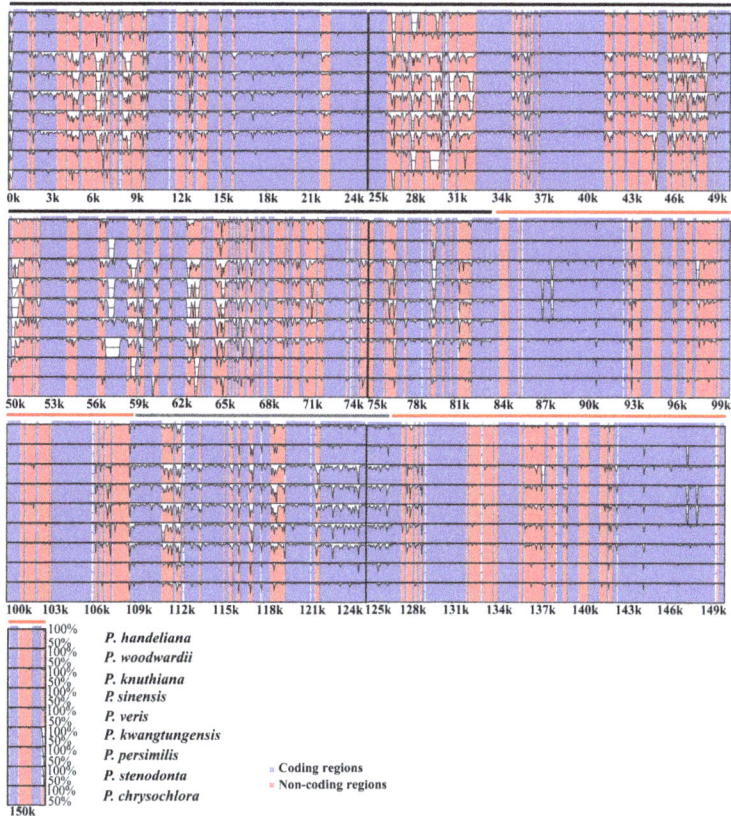

Figure 4. Sequence identity plot of the 10 *Primula* plastid genomes, with *Primula* poissonii as a reference. The *y*-axis represents % identity ranging from 50% to 100%. Coding and non-coding regions are marked in purple and pink, respectively. The red, black, and gray lines show the IRs, LSC, and SSC regions, respectively.

2.6. Phylogenomic Analysis

To investigate the phylogenetic position of *Primula*, three datasets (76 shared protein-coding genes, codon positions 1 + 2, and codon position 3) were used to conduct the BI and ML analyses (Figures 5 and S2). The selected models of each dataset were shown in Table 2. Support values were generally high for almost all relationships inferred from 76 shared protein-coding genes (the support values had a range of 78/0.91–100/1) (Figure 5). All phylogenetic trees clearly identified that *Primula* was monophyletic and more closely related to *Androsace* with high support values (Figures 5 and S2).

Table 2. Datasets and selected model in ML and BI analysis

Datasets	Best Fit Model	Model in ML	Model in BI
76 shared protein-coding genes	TVM + I + G	GTR + G	TVM + I + G
Codon positions 1 + 2	TVM + I + G	GTR + G	TVM + I + G
Codon position 3	GTR + I + G	GTR + G	GTR + I + G
Whole plastid genomes	TVM + I + G	GTR + G	TVM + I + G
Protein-coding regions	TVM + I + G	GTR + G	TVM + I + G
Introns & intergenic spacers	TVM + I + G	GTR + G	TVM + I + G
IRs	TVM + I + G	GTR + G	TVM + I + G
LSC	GTR + I + G	GTR + G	GTR + I + G
SSC	TVM + I + G	GTR + G	TVM + I + G

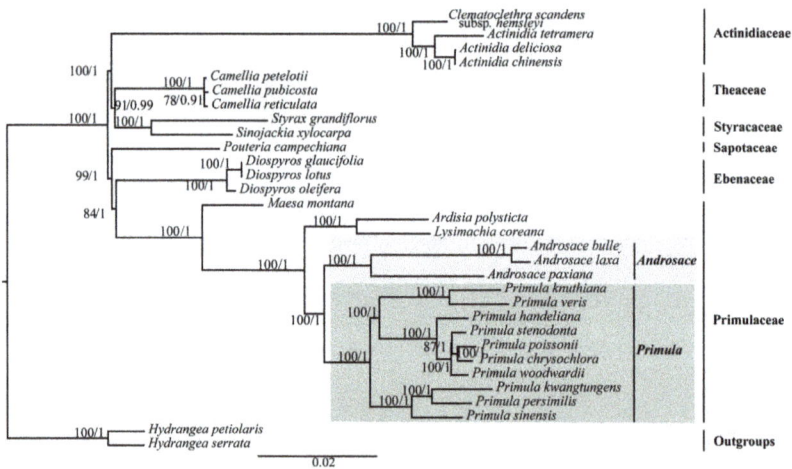

Figure 5. Phylogenetic relationship of the 31 species inferred from ML and BI analyses based on 76-shared protein-coding genes. The numbers near each node are bootstrap support values and posterior probability. *Hydrangea petiolaris* and *Hydrangea serrata* were used as the outgroups.

We then constructed six datasets (whole plastid genome, protein-coding regions, LSC, SSC, IRs, and introns & intergenic spacers) to analyze the phylogenetic relationships among the members of the genus *Primula*. The plastid genome of *A. laxa* was used as the outgroup. The selected models for each dataset used in BI and ML analyses were displayed in Table 2. The different datasets generally produced congruent phylogenetic trees (two topological structures) with moderate to high support values (Figure 6). All of the phylogenetic trees showed that *P. stenodonta*, *P. poissonii*, and *P. chrysochlora* formed a monophyletic group, where they belong to Sect. *Proliferae*. Although *P. woodwardii* and *P. handeliana* belong to Sect. *Crystallophlomis*, they were not monophyletic. *P. kwangtungensis*, *P. persimilis*, and *P. sinensis* belong to different sections, but they clustered together in the phylogenetic trees. In addition, *P. knuthiana* was more closely related to *P. veris* than other *Primula* species, but their placements varied in topological structure.

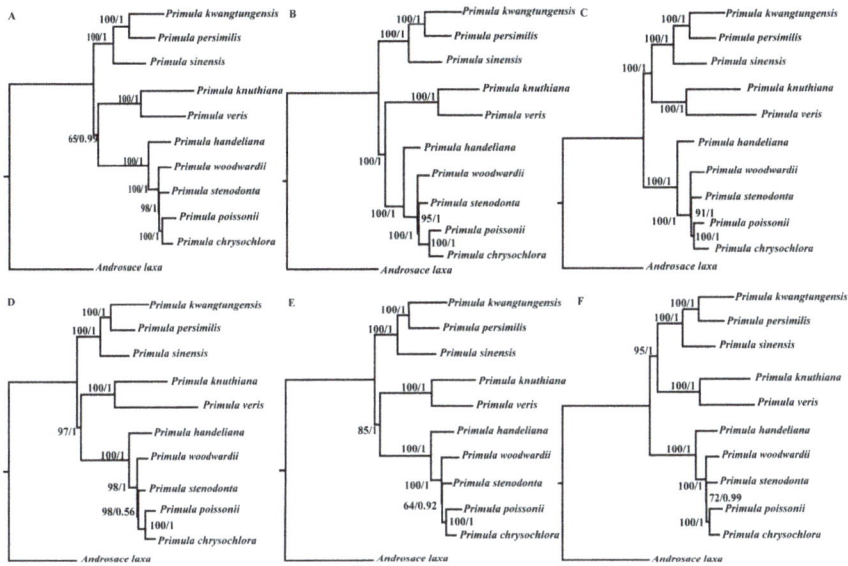

Figure 6. Phylogenetic relationships of the 10 *Primula* species and *A. laxa* inferred from ML and BI analyses. (**A**) Whole plastid genomes; (**B**) protein-coding regions; (**C**) introns and intergenic spacer regions; (**D**) IR regions; (**E**) SSC regions; and (**F**) LSC regions. The numbers near each node are bootstrap support values and posterior probability.

3. Discussion

3.1. Evolution of the Plastid Genome

Most angiosperm plastid genomes are highly conserved in terms of their gene content and order, but gene loss (deletion or production of pseudogenes) has occurred in several angiosperm lineages [26,27]. In our study, the *accD* gene was found in seven *Primula* plastid genomes, while it was a pseudogene in *P. sinensis* and was missing in *P. persimilis* and *P. kwangtungensis*. The *accD* gene encodes the acetyl-CoA carboxylase subunit D, which has been lost either partially or completely from some members of the Poales and Acoraceae [28]. The *infA* gene was a pseudogene in *P. poissonii* plastid genome, but it has been entirely lost from the other *Primula* plastid genome. The *infA* gene encodes translation initiation factor 1, which assists with the assembly of the translation initiation complex [18]. Similar events have also occurred in other angiosperm plastid genomes, such as those of *Hagenia abyssinica* [29] and *Morella rubra* [30], although the plastid genome of *A. laxa* contains the *infA* gene. The photosystem assembly factors (*ycf3* and *ycf4*) that act on photosystem I complex [31,32] should be renamed as *paf1* and *pafII* (respectively) according to recent studies [18]. Here, we use the new names of the two genes in both *Primula* and *A. laxa* plastid genomes.

IRs are the most conserved regions in the plastid genomes, where the contraction and expansion of the IR regions have occurred frequently. Our results indicated more obvious differences at the IR/SC boundaries. Particularly, in the *P. kwangtungensis* plastid genome, the *rps19* gene was located entirely in the LSC. By contrast, IRb extended into the *rps19* gene and it even completely crossed the *rps19* gene in the *P. chrysochlora* plastid genome. In addition, IRa extended into the *ycf1* genes where the smallest and largest extensions occurred in *P. handeliana* (888 bp) and *P. kwangtungensis* (1048 bp). The expansions of IRs into the *rps19* gene and *ycf1* gene have been also observed in *Cardiocrinum* [33] and *Amana* [34]. IR regions contraction and expansion events are relatively common evolutionary phenomena in plants [35]. Moreover, IR region loss was observed in some species [36,37].

Large and complex repeat sequences may play important roles in the arrangement and recombination of the plastid genome [38,39]. In all, 326 repeats were detected in the 10 *Primula* plastid genomes. Compared with other angiosperm species [40], this number is relatively small. Most of repeats ranged in size from 14 to 62 bp and almost all were not large repeats (>100 bp), which were in a similar manner to those reported in other plants [41–43]. *Pelargonium*, *Trifolium*, and *Trachelium*, the most highly rearranged plastid genomes contain a high frequency of large repeats (>100 bp) [44]. Our study revealed that no rearrangement events occurred in *Primula*, we thus deduced that may be mainly ascribed to no large repeats in these 10 *Primula* plastid genomes. Repeats located in *ycf2* gene occupied 47.8% of the total repeats. The *ycf2* gene is the largest gene in the *Primula* plastid genomes with over 6000 bp in length, and is completely duplicated in the IR regions. This phenomenon has also been reported in *Cardiocrinum* [33]. SSRs are highly polymorphic, and thus they are employed as molecular markers for population genetics and phylogenetic investigations [45,46]. Notably, the majority of the SSRs in the 10 *Primula* plastid genomes were the mononucleotide A/T SSRs (74.2%), which supports previous reports that SSRs in the plastid genome generally comprise short polyadenine (polyA) or polythymine (polyT) repeats [47,48]. Most of the SSRs were found in IGS regions (56.9%), followed by CDS regions (25%) and introns (17.9%). The CDS region with the highest number of SSRs was *ycf1*, as found in other species, such as *Cardiocrinum* [33] and *Vigna radiata* [49]. In the 10 *Primula* plastid genomes, the *ycf1* gene usually spanned the small single copy (SSC) and the inverted repeat a (IRa) region. It is very interesting that all but two of the SSRs in the *ycf1* gene are distributed in the SSC region. It is possible because the section of *ycf1* gene in the IRa region is shorter (less than one kilobase long) than these in SSC region (more than four kilobase long) [50]. The cpSSRs reported here would be potential molecular markers for future studies of *Primula* species.

According to the results obtained using mVISTA, the *Primula* plastid genomes exhibited moderate sequence divergence, especially in the non-coding regions. Our study showed that the coding regions were more conserved than the non-coding regions, as found in many plants [41–43]. Besides, the IR regions were more conserved than the SC regions as previous studies [51]. This fact that the two IR regions were less variable was attributed to the conservation of the ribosomal RNA genes, which comprised about one-third of the IR region in the plastid genomes [17]. The p-distance results also confirmed that moderate sequence divergence exists within the genus *Primula*. Compared with related herbaceous plants, trees, and shrubs generally have relatively long generation times and low rates of molecular evolution [52]. Herbs have shorter generation times and show much higher rates of molecular change and variance in rates [52]. The genetic diversity of heterotypic flower plants is higher than that of self-pollinated plants, indicating that genetic variation is easy to occur in interspecific and intraspecific species of heterotypic flower plants [53]. Therefore, the moderate sequence divergence probably be related to biology characteristics of these *Primula* species, such as perennial herbs, shorter generation times, cross-pollination, distyly, etc.

3.2. Phylogenetic Relationships

Plastid genomes have been successfully used to resolve the phylogenetic relationships in plant groups [23,25,54]. In this study, we used two methods (ML and BI) to construct the phylogenetic trees. We used three datasets to investigate the phylogenetic position of *Primula*. All of the phylogenetic trees indicated that *Primula* was monophyletic and more closely related to *Androsace* in Primulaceae family. Besides, in the genus *Primula*, all of the phylogenetic trees showed that Sect. *Proliferae* (*P. chrysochlora*, *P. poissonii*, and *P. stenodonta*) formed a monophyletic group and *P. chrysochlora* was closely related to *P. poissonii* [55]. Both *P. woodwardii* and *P. handeliana* belong to Sect. *Crystallophlomis*, but they did not have the closest relationship. The phylogenetic trees indicated that *P. woodwardii* and Sect. *Proliferae* were sister groups, then they clustered with *P. handeliana* in the same clade. Section *Crystallophlomis* and *Proliferae* were clustered into one clade in this study, which was also supported by karyotype study [56], but was inconsistent with the morphological work [9]. The placement of *P. knuthiana-P. veris* clade was uncertain in the phylogenetic tree. This was partly due to the rapid evolution of genus *Primula* [14,57]. The lack of samples might also affect the results of the phylogenetic analysis. In fact, for this large genus, our study could not fully clarify the relationships among *Primula* species due to the limited taxa sampled. Hence, more species and comprehensive analyses should be included in the future phylogenetic studies of *Primula* species. All in all, our analysis based on plastid genomes provides a valuable resource that should facilitate future phylogeny, taxonomy, and evolutionary history studies of this genus.

4. Materials and Methods

4.1. Plant Materials and DNA Extraction

The four plant materials (*Primula handeliana*, *Primula woodwardii*, *Primula knuthiana*, and *Androsace laxa*) used in this study were sampled from Taibai Mountain (Shaanxi, China; 107.77 °E, 33.95 °N). Total genomic DNA was extracted from silica-dried leaves with a modified CTAB method [58] by Biomarker Technologies Inc., Beijing, China. Voucher specimens were deposited in the Key Laboratory of Resource Biology and Biotechnology, Northwest University. All of the newly generated complete plastid genome sequences were deposited in GenBank (https://www.ncbi.nlm.nih.gov) (Table 1). The complete plastid genomes of *Primula poissonii* (NC_024543) [59], *Primula sinensis* (NC_030609) [57], *Primula veris* (NC_031428) [60], *Primula kwangtungensis* (NC_034371) [61], *Primula chrysochlora* (KX668178) [55], *Primula stenodonta* (KX668176) [62], and *Primula persimilis* (KX641757) [63] were recovered in order to conduct follow-up analysis (Table S1).

4.2. Illumina Sequencing, Assembly, and Annotation

Whole-genome sequencing was performed using the 150 bp pair-end sequencing method with the Illumina Hiseq 2500 Platform by Biomarker Technologies Inc. (Beijing, China). First, the raw Illumina reads were quality trimmed using the NGSQC Toolkit_v2.3.3 [64] with the default cutoff values. The clean reads were then subjected to reference-guided assembly with the MIRA v4.0.2 program [65] (parameters: job = genome, mapping, accurate; technology = solexa; segment_placement = FR). We used *Primula poissonii* (NC_024543) and *Androsace bulleyana* (KU513438) as reference genomes to assemble the *Primula* species and *A. laxa*, respectively. The resultant contigs were further assembled using a baiting and iteration method based on MITObim v1.8 [66] with default parameters. In addition, we also used the SPAdes v3.6.2 [67] (*k* = 33, 55, 77) to assemble the resultant clean reads of four species. We performed de novo assembly in order to verify the validity and accuracy of assembly results. Finally, a few gaps containing some ambiguous bases "N" and low-coverage regions in the assembled plastid genomes were confirmed by PCR-based Sanger sequencing. The primer pairs were designed online with the Primer3 program [68] and listed in the Supplementary Table S2. All of the genes were annotated using Dual Organellar Genome Annotator (DOGMA) software [69] with the default parameters. We then corrected the annotations with the GENEIOUS R8.0.2 program (Biomatters Ltd., Auckland, New Zealand) based on comparisons with related species. Codon usage and relative synonymous codon usage (RSCU) [70] value were estimated for all exons in the protein-coding genes with the CodonW v1.4.2 program [71].

4.3. Identification of Repeat Sequences

We used the online REPuter program [72] to identify dispersed and palindromic repeats with a minimum repeat size of 30 bp and two repeats comprising not less than 90% (Hamming distance = 3). Tandem repeats were detected using the Tandem Repeat Finder program [73] by setting two, seven, and seven as the alignment parameters for match, mismatch, and indels, respectively. The minimum alignment score and maximum period size were 80 and 500, respectively. Simple sequence repeats (SSRs) were detected using the Perl script MISA (http://pgrc.ipk-gatersleben.de/misa/) by setting the minimum number of repeats to 10, 5, 4, 3, 3, and 3 for mono-, di-, tri-, tetra-, penta-, and hexanucleotide SSRs, respectively.

4.4. Whole Plastid Genomes Comparison

Whole genome alignment with 10 *Primula* plastid genomes was run in MAUVE [74] under default settings to test rearrangement events across genomes.

4.5. Sequence Divergence Analysis

The mVISTA program [75] was used to compare the 10 *Primula* plastid genomes with *P. poissonii* as the reference. The percentages of variable characters in each coding region and non-coding region with an aligned length of more than 200 bp were calculated as described in a previous study of Poaceae species [76]. The average genetic divergences of these *Primula* plastid genomes were estimated using p-distance with MEGA6 [77]. Substitution included transition and transversion. Gaps and missing data were completely deleted.

4.6. Phylogenomic Analysis

To investigate the phylogenetic position of *Primula*, we used 31 complete plastid genomes (Table S3). Among them, 29 were from Ericales, and two *Hydrangea* species (*Hydrangea serrata* and *Hydrangea petiolaris*) were used as the outgroups. 76 shared protein-coding genes, codon positions 1 + 2, and codon position 3, were used to conduct the phylogenetic analysis.

Then, six datasets, including the whole plastid genomes, protein-coding regions, LSC, SSC, IRs, and introns & intergenic spacers were used to conduct the phylogenetic analysis among genus *Primula* with *A. laxa* as the outgroup.

All of the datasets were aligned with MAFFT [78] using the default settings. In order to examine the phylogenetic utility of different datasets, phylogenetic analyses were conducted using maximum likelihood (ML) and Bayesian inference (BI) methods. The ML analysis was conducted using RAxMLv7.2.8 [79] with 1000 bootstrap replicates. The GTRGAMMA model was used in all of the ML analyses, as suggested in the RAxML manual. For the BI analysis, the best substitution model was determined according to Akaike's information criterion (AIC) with Modeltest v3.7 [80]. The BI analysis was performed using MrBayes v3.1.2 [81]. The Markov chain Monte Carlo (MCMC) algorithm was run for two million generations and the trees were sampled very 100 generations. Convergence was determined by examining the average standard deviation of the split frequencies (<0.01). The first 25% of the trees were discarded as a burn-in and the remaining trees were used to generate the consensus tree.

Supplementary Materials: Supplementary materials can be found at http://www.mdpi.com/1422-0067/19/4/1050/s1.

Acknowledgments: This work was financially supported by the National Natural Science Foundation of China (31670219, 31370353).

Author Contributions: Zhan-Lin Liu conceived and designed the work. Ting Ren, Yanci Yang, and Tao Zhou performed the experiments and analyzed the data. Ting Ren wrote the manuscript. Yanci Yang and Zhan-Lin Liu revised the manuscript. All authors gave final approval of the paper.

Conflicts of Interest: The authors declare no conflict of interest.

References

1. Hu, C.M.; Kelso, S. *Flora of China*; Science Press: Beijing, China, 1996; Volume 15, pp. 99–185.
2. Richards, A.J. *Primula*, 2nd ed.; B. T. Batsford Ltd.: London, UK, 2002.
3. Yan, H.F.; He, C.H.; Peng, C.I.; Hu, C.M.; Hao, G. Circumscription of *Primula* subgenus *Auganthus* (Primulaceae) based on chloroplast DNA sequences. *J. Syst. Evol.* **2010**, *48*, 123–132. [CrossRef]
4. Woodell, S. Natural hybridization between the cowsip (*Primula veris* L.) and the primrose (*P. vulgaris* Huds.) in Britain. *Watsonia* **1965**, *6*, 190–202.
5. Ornduff, R. Pollen flow in a population of *Primula vulgaris* Huds. *Bot. J. Linn. Soc.* **1979**, *78*, 1–10. [CrossRef]
6. Shen, L.L. Research advances on the pollination biology of *Primula*. *J. Anhui. Agric. Sci.* **2010**, *38*, 5574–5585.

7. Li, J.H.; Webster, M.A.; Smith, M.C.; Gilmartin, P.M. Floral heteromorphy in *Primula vulgaris*: Progress towards isolation and characterization of the *S. locus. Ann. Bot.* **2011**, *108*, 715–726. [CrossRef] [PubMed]
8. Nowak, M.D.; Russo, G.; Schlapbach, R.; Huu, C.N.; Lenhard, M.; Conti, E. The draft genome of *Primula veris* yields insights into the molecular basis of heterostyly. *Genome Biol.* **2015**, *16*, 12. [CrossRef] [PubMed]
9. Smith, W.W.; Fletcher, H.R. XVII.–The genus *Primula*: Sections *Obconica, Sinenses, Reinii, Pinnatae, Malacoides, Bullatae, Carolinella, Grandis* and *Denticulata. Trans. R. Soc. Edinb.* **1947**, *61*, 415–478. [CrossRef]
10. Wendelbo, P. Studies in Primulaceae. II. An account of *Primula* subgenus *Sphondylia* (*Syn. Sect. Floribundae*) with a review of the subdivisions of the genus. *Matematisk-Naturvitenskapelig Ser.* **1961**, *11*, 1–46.
11. Richards, A.J. *Primula*; B. T. Batsford Ltd.: London, UK, 1993.
12. Conti, E.; Suring, E.; Boyd, D.; Jorgensen, J.; Grant, J.; Kelso, S. Phylogenetic relationships and character evolution in *Primula* L.: The usefulness of ITS sequence data. *Plant Biosyst.* **2000**, *134*, 385–392. [CrossRef]
13. Mast, A.R.; Kelso, S.; Richards, A.J.; Lang, D.J.; Feller, D.M.; Conti, E. Phylogenetic relationships in *Primula* L. and related genera (Primulaceae) based on noncoding chloroplast DNA. *Int. J. Plant Sci.* **2001**, *162*, 1381–1400. [CrossRef]
14. Yan, H.F.; Liu, Y.J.; Xie, X.F.; Zhang, C.Y.; Hu, C.M.; Hao, G.; Ge, X.J. DNA barcoding evaluation and its taxonomic implications in the species-rich genus *Primula* L. in China. *PLoS ONE* **2015**, *10*, e0122903. [CrossRef] [PubMed]
15. Ravi, V.; Khurana, J.P.; Tyagi, A.K.; Khurana, P. An update on chloroplast genomes. *Plant Syst. Evol.* **2008**, *271*, 101–122. [CrossRef]
16. Jansen, R.K.; Raubeson, L.A.; Boore, J.L.; Chumley, T.W.; Haberle, R.C.; Wyman, S.K. Methods for obtaining and analyzing whole chloroplast genome sequences. *Methods Enzymol.* **2005**, *395*, 348–384. [PubMed]
17. Palmer, J.D. Comparative organization of chloroplast genomes. *Annu. Rev. Genet.* **1985**, *19*, 325–354. [CrossRef] [PubMed]
18. Wicke, S.; Schneeweiss, G.M.; dePamphilis, C.W.; Müller, K.F.; Quandt, D. The evolution of the plastid chromosome in land plants: Gene content, gene order, gene function. *Plant Mol. Biol.* **2011**, *76*, 273–297. [CrossRef] [PubMed]
19. Jansen, R.K.; Cai, Z.Q.; Raubeson, L.A.; Daniell, H.; dePamphilis, C.W.; Leebens-Mack, J.; Müller, K.F.; Guisinger-Bellian, M.; Haberle, C.R.; Hansen, A.K.; et al. Analysis of 81 genes from 64 plastid genomes resolves relationships in angiosperms and identifies genome-scale evolutionary patterns. *Proc. Natl. Acad. Sci. USA* **2007**, *104*, 19369–19374. [CrossRef] [PubMed]
20. Moore, M.J.; Bell, C.D.; Soltis, P.S.; Soltis, D.E. Using plastid genome-scale data to resolve enigmatic relationships among basal angiosperms. *Proc. Natl. Acad. Sci. USA* **2007**, *104*, 19363–19368. [CrossRef] [PubMed]
21. Cronn, R.; Liston, A.; Parks, M.; Gernandt, D.S.; Shen, R.; Mockler, T. Multiplex sequencing of plant chloroplast genomes using Solexa sequencing-by-synthesis technology. *Nucleic Acids Res.* **2008**, *36*, e122. [CrossRef] [PubMed]
22. Mardis, E.R. The impact of next-generation sequencing technology on genetics. *Trends Genet.* **2008**, *24*, 133–141. [CrossRef] [PubMed]
23. Luo, Y.; Ma, P.F.; Li, H.T.; Yang, J.B.; Wang, H.; Li, D.Z. Plastid phylogenomic analyses resolve Tofieldiaceae as the root of the early diverging monocot order *Alismatales. Genome Biol. Evol.* **2016**, *8*, 932–945. [CrossRef] [PubMed]
24. Carbonell-Caballero, J.; Alonso, R.; Ibañez, V.; Terol, J.; Talon, M.; Dopazo, J. A phylogenetic analysis of 34 chloroplast genomes elucidates the relationships between wild and domestic species within the genus *Citrus. Mol. Biol. Evol.* **2015**, *32*, 2015–2035. [CrossRef] [PubMed]
25. Zhang, S.D.; Jin, J.J.; Chen, S.Y.; Chase, M.W.; Soltis, D.E.; Li, H.T.; Yang, J.B.; Li, D.Z.; Yi, T.S. Diversification of Rosaceae since the Late Cretaceous based on plastid phylogenomics. *New Phytol.* **2017**, *214*, 1355–1367. [CrossRef] [PubMed]
26. Braukmann, T.; Kuzmina, M.; Stefanović, S. Plastid genome evolution across the genus *Cuscuta* (Convolvulaceae): Two clades within subgenus *Grammica* exhibit extensive gene loss. *J. Exp. Bot.* **2013**, *64*, 977–989. [CrossRef] [PubMed]
27. Logacheva, M.D.; Schelkunov, M.I.; Nuraliev, M.S.; Samigullin, T.H.; Penin, A.A. The plastid genome of mycoheterotrophic monocot *Petrosavia stellaris* exhibits both gene losses and multiple rearrangements. *Genome Biol. Evol.* **2014**, *6*, 238–246. [CrossRef] [PubMed]

28. Katayama, H.; Ogihara, Y. Phylogenetic affinities of the grasses to other monocots as revealed by molecular analysis of chloroplast DNA. *Curr. Genet.* **1996**, *29*, 572–581. [CrossRef] [PubMed]

29. Gichira, A.W.; Li, Z.Z.; Saina, J.K.; Long, Z.C.; Hu, G.W.; Gituru, R.W.; Wang, Q.F.; Chen, J.M. The complete chloroplast genome sequence of an endemic monotypic genus *Hagenia* (Rosaceae): Structural comparative analysis, gene content and microsatellite detection. *PeerJ* **2017**, *5*, e2846. [CrossRef] [PubMed]

30. Liu, L.X.; Li, R.; Worth, J.R.; Li, X.; Li, P.; Cameron, K.M.; Fu, C.X. The complete chloroplast genome of Chinese bayberry (*Morella rubra*, Myricaceae): Implications for understanding the evolution of Fagales. *Front. Plant Sci.* **2017**, *8*. [CrossRef] [PubMed]

31. Naver, H.; Boudreau, E.; Rochaix, J.D. Functional studies of Ycf3: Its role in assembly of photosystem I and interactions with some of its subunits. *Plant Cell* **2001**, *13*, 2731–2745. [CrossRef] [PubMed]

32. Ozawa, S.I.; Nield, J.; Terao, A.; Stauber, E.J.; Hippler, M.; Koike, H.; Rochaix, J.D.; Takahashi, Y. Biochemical and structural studies of the large Ycf4-photosystem I assembly complex of the green alga *Chlamydomonas reinhardtii*. *Plant Cell* **2009**, *21*, 2424–2442. [CrossRef] [PubMed]

33. Lu, R.S.; Li, P.; Qiu, Y.X. The complete chloroplast genomes of three *Cardiocrinum* (Liliaceae) species: Comparative genomic and phylogenetic analyses. *Front. Plant Sci.* **2017**, *7*, 2054. [CrossRef] [PubMed]

34. Li, P.; Lu, R.S.; Xu, W.Q.; Ohitoma, T.; Cai, M.Q.; Qiu, Y.X.; Cameron, M.K.; Fu, C.X. Comparative genomics and phylogenomics of East Asian tulips (*Amana*, Liliaceae). *Front. Plant Sci.* **2017**, *8*, 451. [CrossRef] [PubMed]

35. Kim, K.J.; Lee, H.L. Complete chloroplast genome sequences from Korean Ginseng (*Panax schinseng* Nees) and comparative analysis of sequence evolution among 17 vascular plants. *DNA Res.* **2004**, *11*, 247–261. [CrossRef] [PubMed]

36. Perry, A.S.; Wolfe, K.H. Nucleotide substitution rates in legume chloroplast DNA depend on the presence of the inverted repeat. *J. Mol. Evol.* **2002**, *55*, 501–508. [CrossRef] [PubMed]

37. Yi, X.; Gao, L.; Wang, B.; Su, Y.J.; Wang, T. The complete chloroplast genome sequence *of Cephalotaxus oliveri* (Cephalotaxaceae): Evolutionary comparison of *Cephalotaxus* chloroplast DNAs and insights into the loss of inverted repeat copies in Gymnosperms. *Genome Biol. Evol.* **2013**, *5*, 688–698. [CrossRef] [PubMed]

38. Ogihara, Y.; Terachi, T.; Sasakuma, T. Intramolecular recombination of chloroplast genome mediated by short direct-repeat sequences in wheat species. *Proc. Natl. Acad. Sci. USA* **1988**, *85*, 8573–8577. [CrossRef] [PubMed]

39. Weng, M.L.; Blazier, J.C.; Govindu, M.; Jansen, R.K. Reconstruction of the ancestral plastid genome in Geraniaceae reveals a correlation between genome rearrangements, repeats and nucleotide substitution rates. *Mol. Biol. Evol.* **2013**, *31*, 645–659. [CrossRef] [PubMed]

40. Zhang, X.; Zhou, T.; Kanwal, N.; Zhao, Y.M.; Bai, G.Q.; Zhao, G.F. Completion of eight *Gynostemma* BL. (Cucurbitaceae) chloroplast genomes: Characterization, comparative analysis, and phylogenetic relationships. *Front. Plant Sci.* **2017**, *8*, 1583. [CrossRef] [PubMed]

41. Hu, Y.H.; Woeste, K.E.; Zhao, P. Completion of the chloroplast genomes of five Chinese *Juglans* and their contribution to chloroplast phylogeny. *Front. Plant Sci.* **2016**, *7*, 1955. [CrossRef] [PubMed]

42. Yang, Y.C.; Zhou, T.; Duan, D.; Yang, J.; Feng, L.; Zhao, G.F. Comparative analysis of the complete chloroplast genomes of five *Quercus* species. *Front. Plant Sci.* **2016**, *7*, 959. [CrossRef] [PubMed]

43. Zhou, T.; Chen, C.; Wei, Y.; Chang, Y.X.; Bai, G.Q.; Li, Z.H.; Kanwal, N.; Zhao, G.F. Comparative transcriptome and chloroplast genome analyses of two related *Dipteronia* species. *Front. Plant Sci.* **2016**, *7*, 1512. [CrossRef] [PubMed]

44. Guisinger, M.M.; Kuehl, J.V.; Boore, J.L.; Jansen, R.K. Extreme reconfiguration of plastid genomes in the angiosperm family Geraniaceae: Rearrangements, repeats, and codon usage. *Mol. Biol. Evol.* **2011**, *28*, 583–600. [CrossRef] [PubMed]

45. Powell, W.; Morgante, M.; Andre, C.; McNicol, J.W.; Machray, G.C.; Doyle, J.J. Hypervariable microsatellites provide a general source of polymorphic DNA markers for the chloroplast genome. *Curr. Biol.* **1995**, *5*, 1023–1029. [CrossRef]

46. He, S.L.; Wang, Y.S.; Volis, S.; Li, D.Z.; Yi, T.S. Genetic diversity and population structure: Implications for conservation of wild soybean (*Glycine soja* Sieb. et Zucc) based on nuclear and chloroplast microsatellite variation. *Int. J. Mol. Sci.* **2012**, *13*, 12608–12628. [CrossRef] [PubMed]

47. Kuang, D.Y.; Wu, H.; Wang, Y.L.; Gao, L.M.; Zhang, S.Z.; Lu, L. Complete chloroplast genome sequence of *Magnolia kwangsiensis* (Magnoliaceae): Implication for DNA barcoding and population genetics. *Genome* **2011**, *54*, 663–673. [CrossRef] [PubMed]

48. Martin, G.; Baurens, F.C.; Cardi, C.; Aury, J.M.; D'Hont, A. The complete chloroplast genome of banana (*Musa acuminata*, Zingiberales): Insight into plastid monocotyledon evolution. *PLoS ONE* **2013**, *8*, e67350. [CrossRef] [PubMed]

49. Tangphatsornruang, S.; Sangsrakru, D.; Chanprasert, J.; Uthaipaisanwong, P.; Yoocha, T.; Jomchai, N.; Tragoonrung, S. The chloroplast genome sequence of mungbean (*Vigna radiata*) determined by high-throughput pyrosequencing: Structural organization and phylogenetic relationships. *DNA Res.* **2009**, *17*, 11–22. [CrossRef] [PubMed]

50. Dong, W.P.; Xu, C.; Li, C.H.; Sun, J.H.; Zuo, Y.J.; Shi, S.; Cheng, T.; Guo, J.J.; Zhou, S.L. *ycf1*, the most promising plastid DNA barcode of land plants. *Sci. Rep.* **2015**, *5*. [CrossRef] [PubMed]

51. Zhu, A.D.; Guo, W.H.; Gupta, S.; Fan, W.S.; Mower, J.P. Evolutionary dynamics of the plastid inverted repeat: The effects of expansion, contraction, and loss on substitution rates. *New Phytol.* **2016**, *209*, 1747–1756. [CrossRef] [PubMed]

52. Smith, S.A.; Donoghue, M.J. Rates of molecular evolution are linked to life history in flowering plants. *Science* **2008**, *322*, 86–89. [CrossRef] [PubMed]

53. Weller, S.G.; Sakai, A.K.; Straub, C. Allozyme diversity and genetic identity in *Schiedea* and *Alsinidendron* (Caryophyllaceae: Alsinoideae) in the Hawaiian Islands. *Evolution* **1996**, *50*, 23–34. [CrossRef] [PubMed]

54. Ma, P.F.; Zhang, Y.X.; Zeng, C.X.; Guo, Z.H.; Li, D.Z. Chloroplast phylogenomic analyses resolve deep-level relationships of an intractable bamboo tribe *Arundinarieae* (Poaceae). *Syst. Biol.* **2014**, *63*, 933–950. [CrossRef] [PubMed]

55. Zhang, C.Y.; Liu, T.J.; Xu, Y.; Yan, H.F. Characterization of the whole chloroplast genome of a rare candelabra primrose *Primula chrysochlora* (Primulaceae). *Conserv. Genet. Resour.* **2017**, *9*, 361–363. [CrossRef]

56. Bruun, H.G. Cytological Studies in Primula with Special Reference to the Relation between the Karyology and Taxonomy of the Genus. Ph.D. Thesis, Acta Universitatis Upsaliensis, Uppsala, Sweden, 1932.

57. Liu, T.J.; Zhang, C.Y.; Yan, H.F.; Zhang, L.; Ge, X.J.; Hao, G. Complete plastid genome sequence of *Primula sinensis* (Primulaceae): Structure comparison, sequence variation and evidence for *accD* transfer to nucleus. *PeerJ* **2016**, *4*, e2101. [CrossRef] [PubMed]

58. Doyle, J.J. A rapid DNA isolation procedure for small quantities of fresh leaf tissue. *Phytochem. Bull.* **1987**, *19*, 11–15.

59. Yang, J.B.; Li, D.Z.; Li, H.T. Highly effective sequencing whole chloroplast genomes of angiosperms by nine novel universal primer pairs. *Mol. Ecol. Resour.* **2014**, *14*, 1024–1031. [CrossRef] [PubMed]

60. Zhou, T.; Zhao, J.X.; Chen, C.; Meng, X.; Zhao, G.F. Characterization of the complete chloroplast genome sequence of *Primula veris* (Ericales: Primulaceae). *Conserv. Genet. Resour.* **2016**, *8*, 455–458. [CrossRef]

61. Zhang, C.Y.; Liu, T.J.; Xu, Y.; Yan, H.F.; Hao, G.; Ge, X.J. Characterization of the whole chloroplast genome of an endangered species *Primula kwangtungensis* (Primulaceae). *Conserv. Genet. Resour.* **2017**, *9*, 87–89. [CrossRef]

62. Zhang, C.Y.; Liu, T.J.; Yan, H.F.; Ge, X.J.; Hao, G. The complete chloroplast genome of a rare candelabra primrose *Primula stenodonta* (Primulaceae). *Conserv. Genet. Resour.* **2017**, *9*, 123–125. [CrossRef]

63. Zhang, C.Y.; Liu, T.J.; Yan, H.F.; Xu, Y. The complete chloroplast genome of *Primula persimilis* (Primulaceae). *Conserv. Genet. Resour.* **2017**, *9*, 189–191. [CrossRef]

64. Patel, R.K.; Jain, M. NGS QC Toolkit: A toolkit for quality control of next generation sequencing data. *PLoS ONE* **2012**, *7*, e30619. [CrossRef] [PubMed]

65. Chevreux, B.; Pfisterer, T.; Drescher, B.; Driesel, A.J.; Müller, W.E.; Wetter, T.; Suhai, S. Using the miraEST assembler for reliable and automated mRNA transcript assembly and SNP detection in sequenced ESTs. *Genome Res.* **2004**, *14*, 1147–1159. [CrossRef] [PubMed]

66. Hahn, C.; Bachmann, L.; Chevreux, B. Reconstructing mitochondrial genomes directly from genomic next-generation sequencing reads-a baiting and iterative mapping approach. *Nucleic Acids Res.* **2013**, *41*, e129. [CrossRef] [PubMed]

67. Bankevich, A.; Nurk, S.; Antipov, D.; Gurevich, A.A.; Dvorkin, M.; Kulikov, A.S.; Lesin, V.M.; Nikolenko, S.I.; Pham, S.; Prjibelski, A.D.; et al. SPAdes: A new genome assembly algorithm and its applications to single-cell sequencing. *J. Comput. Biol.* **2012**, *19*, 455–477. [CrossRef] [PubMed]

68. Untergrasser, A.; Cutcutache, I.; Koressaar, T.; Ye, J.; Faircloth, B.C.; Remm, M.; Rozen, S.G. Primer3-new capabilities and interfaces. *Nucleic Acids Res.* **2012**, *40*, e115. [CrossRef] [PubMed]

69. Wyman, S.K.; Jansen, R.K.; Boore, J.L. Automatic annotation of organellar genomes with DOGMA. *Bioinformatics* **2004**, *20*, 3252–3255. [CrossRef] [PubMed]

70. Sharp, P.M.; Li, W.H. The codon adaptation index-a measure of directional synonymous codon usage bias, and its potential applications. *Nucleic Acids Res.* **1987**, *15*, 1281–1295. [CrossRef] [PubMed]

71. Peden, J.F. Analysis of codon usage. Ph.D. Thesis, University of Nottingham, University of Nottingham, UK, 1999.

72. Kurtz, S.; Choudhuri, J.V.; Ohlebusch, E.; Schleiermacher, C.; Stoye, J.; Giegerich, R. REPuter: The manifold applications of repeat analysis on a genomic scale. *Nucleic Acids Res.* **2001**, *29*, 4633–4642. [CrossRef] [PubMed]

73. Benson, G. Tandem repeats finder: A program to analyze DNA sequences. *Nucleic Acids Res.* **1999**, *27*, 573–580. [CrossRef] [PubMed]

74. Darling, A.C.E.; Mau, B.; Blattner, F.R.; Perna, N.T. Mauve: Multiple alignment of conserved genomic sequence with rearrangements. *Genome Res.* **2004**, *14*, 1394–1403. [CrossRef] [PubMed]

75. Frazer, K.A.; Pachter, L.; Poliakov, A.; Rubin, E.M.; Dubchak, I. VISTA: Computational tools for comparative genomics. *Nucleic Acids Res.* **2004**, *32*, W273–W279. [CrossRef] [PubMed]

76. Zhang, Y.J.; Ma, P.F.; Li, D.Z. High-throughput sequencing of six bamboo chloroplast genomes: Phylogenetic implications for temperate woody bamboos (*Poaceae*: *Bambusoideae*). *PLoS ONE* **2011**, *6*, e20596. [CrossRef] [PubMed]

77. Tamura, K.; Stecher, G.; Peterson, D.; Filipski, A.; Kumar, S. MEGA6: Molecular evolutionary genetics analysis version 6.0. *Mol. Biol. Evol.* **2013**, *30*, 2725–2729. [CrossRef] [PubMed]

78. Katoh, K.; Standley, D.M. MAFFT multiple sequence alignment software version 7: Improvements in performance and usability. *Mol. Biol. Evol.* **2013**, *30*, 772–780. [CrossRef] [PubMed]

79. Stamatakis, A. RAxML-VI-HPC: Maximum likelihood-based phylogenetic analysis with thousands of taxa and mixed models. *Bioinformatics* **2006**, *22*, 2688–2690. [CrossRef] [PubMed]

80. Posada, D.; Crandall, K.A. Modeltest: Testing the model of DNA substitution. *Bioinformatics* **1998**, *14*, 817–818. [CrossRef] [PubMed]

81. Ronquist, F.; Huelsenbeck, J.P. MrBayes 3: Bayesian phylogenetic inference under mixed models. *Bioinformatics* **2003**, *19*, 1572–1574. [CrossRef] [PubMed]

International Journal of
Molecular Sciences

MDPI

Article

Sequencing, Characterization, and Comparative Analyses of the Plastome of *Caragana rosea* var. *rosea*

Mei Jiang, Haimei Chen, Shuaibing He, Liqiang Wang, Amanda Juan Chen and Chang Liu *

Key Laboratory of Bioactive Substances and Resource Utilization of Chinese Herbal Medicine from Ministry of Education, Institute of Medicinal Plant Development, Chinese Academy of Medical Sciences, Peking Union Medical College, Beijing 100193, China; mjiang0502@163.com (M.J.); hmchen@implad.ac.cn (H.C.); wenyuxuan2530@163.com (S.H.); lys832000@163.com (L.W.); amanda_j_chen@163.com (A.J.C.)
* Correspondence: cliu@implad.ac.cn or cliu6688@yahoo.com; Tel.: +86-010-5783-3111

Received: 23 April 2018; Accepted: 7 May 2018; Published: 9 May 2018

Abstract: To exploit the drought-resistant *Caragana* species, we performed a comparative study of the plastomes from four species: *Caragana rosea*, *C. microphylla*, *C. kozlowii*, and *C. Korshinskii*. The complete plastome sequence of the *C. rosea* was obtained using the next generation DNA sequencing technology. The genome is a circular structure of 133,122 bases and it lacks inverted repeat. It contains 111 unique genes, including 76 protein-coding, 30 tRNA, and four rRNA genes. Repeat analyses obtained 239, 244, 258, and 246 simple sequence repeats in *C. rosea*, *C. microphylla*, *C. kozlowii*, and *C. korshinskii*, respectively. Analyses of sequence divergence found two intergenic regions: *trnI-CAU-ycf2* and *trnN-GUU-ycf1*, exhibiting a high degree of variations. Phylogenetic analyses showed that the four *Caragana* species belong to a monophyletic clade. Analyses of Ka/Ks ratios revealed that five genes: *rpl16*, *rpl20*, *rps11*, *rps7*, and *ycf1* and several sites having undergone strong positive selection in the *Caragana* branch. The results lay the foundation for the development of molecular markers and the understanding of the evolutionary process for drought-resistant characteristics.

Keywords: *Caragana*; *Caragana rosea* var. *rosea*; plastome; comparative genomics; molecular markers

---

1. Introduction

The genus *Caragana* has more than 100 species and it belongs to the family of Leguminosae. The plants mainly grow in arid and semi-arid areas of Asia and Europe. Plants from this genus are well-known in resisting drought, barren, cold, and heat, and have a strong adaptability to the sand environment to prevent wind and fixate sand [1]. Sixty two species of *Caragana* are distributed in China [2], most of them can afforest barren hills and preserve water and soil. The distributions of several *Caragana* species have been well-studied in China. *C. microphylla* and *C. korshinskii* are distributed in Northeast China, North China, and Northwest China [3]. *C. microphylla* is adapted to the typical steppe zone, forest steppe zone, and deciduous broad-leaved forest steppe zone of the Mongolian plateau. *C. korshinskii* is suitable for the fixed and semi fixed sand land in the steppe desert and the typical desert belt zone. *C kozlowii* origins in the Lancang River and Tibet, and was mostly found in riverside with 3600–4000 m altitude [4]. *C rosea* is mainly from Northeast China, North China, East China, Henan, and southern Gansu Province, growing in slopes and valleys [5]. In addition to its drought adaptability, its medicinal value, such as strengthening the spleen and tonifying the kidney, was also well known [1]. Furthermore, it was shown that chemical constituents in *C. rosea* have anti-HIV activities [6]. Until now, the plastome of *C. korshinskii*, *C. microphylla*, and *C. kozlowii* were reported, while the plastome of *C. rosea* has not been studied.

Plants live in constantly changing environments that impose many biotic stress, such as pathogen infection and herbivore attack and abiotic stress, such as drought, heat, cold, nutrient deficiency, and excess of salt or toxic metals. Through the past years, many abiotic stress signaling and response pathways in plants have been discovered, with the core pathways involve protein kinases related to the yeast SNF1 and mammalian AMPK [7]. There is also research from the evolution of chloroplast genes adapted to contrasting habitats. For example, the *Cardamine resedifolia* plastid gene has undergone a more aggressive positive selection than *Cardamine impatiens,* which is located at lower elevations, which is why it is more adapted to the plateau environment [8,9]. For *Caragana* species, dozens of studies have been carried out on the morphological changes, such as stomatal status, leaf water state, cellular carbon metabolism, and etc., in drought responses [3,10]. However, little studies have been reported on the molecular bases for the stress responses in *Caragana* species at this time.

Different species of *Caragana* exhibit significantly varied abilities in drought resistance. For example, *C. rosea, C. microphylla, C. korshinskii,* and *C. kozlowii* have been evaluated for their strength of drought resistance based on leaf microstructure analysis, the drought resistance order from large to small is *C. korshinskii* > *C. microphylla* > *C. rosea* > *C. kozlowii* [11]. Furthermore, *C. microphylla* and *C. korshinskii* are closed related phylogenatically but difficult to differentiate morphologically [1,3]. Therefore, it is important to develop molecular markers to distinguish species accurately and to promote the rational use of species.

The plastome is an ideal choice for the development of molecular markers. It has many biological characteristics when compared with the nuclear genome, such as uni-parental inheritance, simpler structure, and being easier to obtain. Moreover, it provides more genetic information than a single gene/locus, resulting in much higher resolution in distinguishing closely related species. Furthermore, the plastome contains a series of genes that are related to photosynthesis, and the photosystem II (PSII) is a key part of drought stress, high temperature, and many other stresses [12,13]. Leaf physiological characteristics, such as photosynthetic capacity, which is related to the plastome function, and stomatal conductance, are the fixed indicators of water-use efficiency [14]. The water-use efficiency is critical for plants to cope with drought stress [15]. Therefore, a comparative analysis of the plastome of *Caragana* would shed light into the molecular bases for their tolerance drought.

Here, the complete plastome of *C. rosea* was sequenced and analyzed, which complements the plastid genome database of environmental stresses-resistant plants. Comparative analyses of the plastomes from *C. rosea* and other three *Caragana* species e.g., *C. korshinskii, C. microphylla,* and *C. kozlowii* were performed. The results that were obtained here provided valuable resources to illustrate molecular mechanisms that are related to drought resistance of *C. rosea,* to carry out chloroplast genetic engineering experiments and to select for plant individuals with favorable characteristics using molecular breeding. Furthermore, the results also provide useful information for future phylogenetic and taxonomic studies in the *Caragana* species.

2. Results

2.1. General Features of the Plastome

The complete plastome of *C. rosea* var. *rosea* is 133,122 bases in length and it lacks inverted repeat (Figure 1). This genome has been deposited in GenBank (accession number: MF593790). In the legume family, the phenomenon of IR regions loss was commonly found [16,17]. The sequence of protein-coding, tRNA, and rRNA regions accounted for 49.76%, 1.77%, and 3.4% of the whole genome, respectively; and, the rest are intergenic regions (Table 1). Moreover, a total of 111 unique genes were annotated, including 77 protein-coding, 30 tRNA, and four rRNA genes. The functional classification of these genes is shown in Table S1. The *C. rosea* plastome has 16 intron-containing genes, including 10 protein-coding genes and six tRNA genes. A total of 10 genes have only one intron, and only the *ycf3* contains two introns (Table S2). The *trnK-UUU* has the largest intron (2485 bases), which contains the *matK* gene. The *rps12* gene, which has the intron in the plastomes of other legume species [18], does not have the intron in the plastome of *Caragana.* Previous studies have shown that the loss of *rps12* intron may have occurred after the loss of the IR [19].

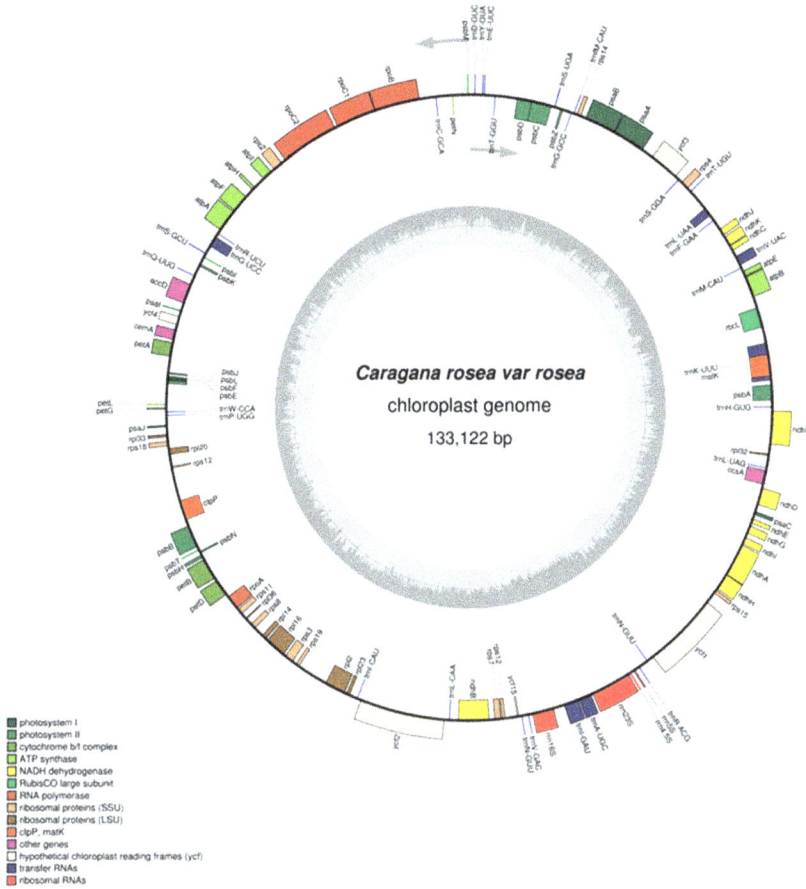

Figure 1. Circular gene map of the *C. rosea* plastome. Genes drawn inside the circle are transcribed clockwise, and those outside the circle are transcribed counterclockwise. Genes belonging to different functional groups are color codes. The inner circle shows the GC content. The two grey arrows represent the direction of transcription.

Table 1. Characteristics of *Caragana* plastome.

Plastome Characteristics	*C. rosea*	*C. microphylla*	*C. kozlowii*	*C. korshinskii*
complete genome length	133,122 bp	130,029 bp	131,274 bp	129,331 bp
No.of unique genes	110	110	110	110
No.of unique protein-coding genes	76	76	76	76
No.of unique tRNA genes	30	30	30	30
No.of unique rRNA genes	4	4	4	4
Size of protein-coding genes	66,243 bp (49.76%)	66,231 bp (50.94%)	66,234 bp (50.45%)	66,231 bp (51.21%)
Size of tRNA genes	2359 bp (1.77%)	2370 bp (1.82%)	2285 bp (1.74%)	2370 bp (1.83)
Size of rRNA genes	4537 bp (3.4%)	4520 bp (3.48%)	4521 bp (3.44%)	4520 bp (3.49)
Overall GC contents	34.84%	34.26%	34.50%	34.36%
GC contents of protein-coding genes	Coding GC 37.13% #1st position 45.36% #2nd position 37.62% #3rd position 28.41%	Coding GC 36.88% #1st position 44.98% #2nd position 37.67% #3rd position 27.99%	Coding GC 37.03% #1st position 45.30% #2nd position 37.58% #3rd position 28.21%	Coding GC 36.88% #1st position 44.96% #2nd position 37.67% #3rd position 27.99%
GC contents of tRNA genes	52.73%	53.14%	53.15%	53.05%
GC contents of rRNA genes	54.77%	54.82%	54.75%	54.82%

The overall GC content of the *C. rosea* plastome is 34.84%, whereas that for the protein-coding regions is 37.13%. The GC contents for the first, second, and third codon position with the protein-coding regions are 45.36%, 37.62%, and 28.41%, respectively. A bias towards a higher AT representation at the third codon position has also been observed in other land plant plastomes [20–22]. The 77 protein-coding genes comprise 66,243 bases coding for 22,081 codons. Among these codons, 2336 (10.58%) encode leucine, whereas just 258 (1.17%) encode cysteine, which are the most and least frequently amino acid in *C. rosea* plastome, respectively. The 30 unique tRNA genes include all the 20 amino acids required for protein biosynthesis (Table S3). However, there are 61 codons (excluding the three stop codons) that are found in the coding sequence (CDs) of the plastome, Since 31 of them do not have the corresponding tRNAs, the translation of their amino acids have to depend on tRNAs encoded in the nuclear genome.

The basic characteristics of cp genome from *C. rosea* and other three *Caragana* species (*C. microphylla, C. kozlowii,* and *C. korshinskii*) are shown in Table 1. As shown, the lengths of the four genomes are quite different, ranging from 129,331 bp to 133,122 bp. However, the length of the coding sequence differs by only 12 bp; and, the sizes of the tRNA genes and rRNA genes are also very similar. This suggests that the difference in the length of the cp genomes is caused by those of the intergenic spacers (IGS). In addition, the gene numbers, gene types, and GC contents in the four cp genomes are very similar.

2.2. Gene Loss Analysis

Chloroplast gene losses were analyzed between IRLC (inverted-repeat-lacking clade) of Papilionoideae in 34 species in detail (Table 2). The species are arranged in the same order as they are shown in the phylogenetic tree (see below). The *rpl22* gene and the *rps16* gene were both absent in all plastome. Moreover, the gene *rpl22* and *rps16* have been lost in most members of the angiosperm [23,24]. The two genes, which are essential for plant survival, have been transferred to the nucleus to maintain the plant's photosynthetic capacity [25,26]. In addition, only two (*Wisteria floribunda* and *Wisteria sinensis*) plastomes possess the gene *ycf15*. The *ycf15* gene belongs to the PFAM protein family PF10705 and its function is unknown. In fact, in some plant species, the *ycf15* gene may not produce any protein because of a premature stop codons in the coding sequences of these species [27]. Because the *ycf15* genes in the GenBank are highly variable in terms of gene length and sequence, it has been difficult to annotate this gene. The absence of the *ycf4* gene occurs in many species of Papilionatae [23,28,29], whereas the plastome of four *Caragana* species contains this gene. *PsaI, ycf1, rpl23, rps18,* and *ndhB* genes were absent in 5, 4, 3, 3, 2 species, respectively, and the *rps32* gene was also found to be transferred to nuclear genomes from plastomes in several species [30,31]. The losses of *accD, atpE, ndhA, psbJ, psbL, psbZ,* and *rps2* were only found in *Trifolium boissieri, Astragalus mongholicus* var. *nakaianus, Medicago falcate, Lathyrus littoralis, Medicago falcate, Lens culinaris,* and *Medicago falcate,* respectively; the deletion of these genes rarely occurs in the plastome of angiosperms [23]. Overall, the losses of genes are monophyletic. However, exceptions can be found in the *Medicago falcate* and several species in the genus *Lathyrus,* including *L. sativus, L. odoratus, L. inconspicuus, L. tingitanus, L. davidii,* and *L. pubescens.*

2.3. Repeat Analysis

Repeated units play an important role in genome evolution, such as structural rearrangements and size evolution [32,33]. We analyzed the content and distribution of repeated sequences in the *C. rosea* plastomes. A total of 19 repeated elements that were longer than 30 bases were identified. The similarities between all the repeated elements were greater than 90%. Tandem, forward, and palindromic repeats presented a similar pattern of distribution. Most of them were found in the intergenic spacer region (IGS), around 31.6% were found in the protein-coding region, and only one repeat was found in the tRNA gene. However, the length of repeats among *C. rosea* plastome was obviously longer than those in other legumes [24,34], the largest repeat unit was 291 base long and was located in the spacer between the genes *rps12* and *clpP*. The type, location, and sequence of the repeat units are shown in Table S4.

Table 2. Gene losses in the plastomes from the inverted-repeat-lacking clade (IRLC) of Papilionoideae.

Category	Name of Species	rps16	rpl22	ycf15	ycf4	psaI	ycf1	rpl23	rps18	ndhB
Millettieae	*W. floribunda*	−	−	+	+	+	−	+	+	+
	W. sinensis	−	−	+	+	+	−	+	+	−
Galegeae	*G. glabra*	−	−	−	+	+	+	+	+	+
	G. lepidota	−	−	−	+	+	+	+	+	+
	A. mongholicus	−	−	−	+	+	+	+	+	+
	A. mongholicus var. nakaianus	−	−	−	+	+	+	+	+	+
Caraganeae	*C. kozlowii*	−	−	−	+	+	+	+	+	+
	C. korshinskii	−	−	−	+	+	+	+	+	+
	C. microphylla	−	−	−	+	+	+	+	+	+
	C. rosea var. rosea	−	−	−	+	+	+	+	+	+
Cicereae	*C. arietinum*	−	−	−	−	+	+	+	+	+
Trifolieae	*M. truncatula*	−	−	−	−	+	+	+	+	+
	M. papillosa	−	−	−	−	+	+	+	+	+
	M. hybrida	−	−	−	−	+	+	+	+	+
	M. falcata	−	−	−	+	+	−	+	+	−
	T. boissieri	−	−	−	−	+	+	+	+	+
	T. glanduliferum	−	−	−	−	+	+	+	+	+
Fabeae	*T. strictum*	−	−	−	−	+	+	+	+	+
	L. sativus	−	−	−	+	−	+	−	+	+
	L. odoratus	−	−	−	−	+	+	+	−	+
	L. inconspicuus	−	−	−	−	−	+	+	+	+
	L. ochroleucus	−	−	−	−	+	+	+	+	+
	L. venosus	−	−	−	−	+	+	+	+	+
	L. palustris	−	−	−	−	+	+	+	+	+
	L. tingitanus	−	−	−	−	+	+	+	−	+
	L. davidii	−	−	−	−	−	+	+	+	+
	L. graminifolius	−	−	−	−	+	+	+	+	+
	L. littoralis	−	−	−	−	+	+	+	+	+
	L. japonicus	−	−	−	−	+	+	+	+	+
	L. pubescens	−	−	−	−	−	+	+	+	+
	L. clymenum	−	−	−	−	+	+	+	+	+
	P. sativum	−	−	−	−	+	+	−	+	+
	V. sativa	−	−	−	−	+	+	−	+	+
	L. culinaris	−	−	−	−	+	−	+	−	+
	Number [a]	34	34	32	22	5	4	3	3	2

[a] The number refers to the total number of species that do not have the gene. "+": presence; "−" absence.

More remarkably, we found that 31.6% and 21.1% of these repeats were located in the IGS(*rps12-clpP*) and IGS(*rps19-rpl2*) regions, with the total length of the IGS regions being 2139 bases and 4060 bases, respectively. By contrast, the other three *Caragana* species do not have so many repeat units in the same regions. For example, the length of the corresponding IGS(*rps12-clpP*) regions in *C. microphylla* and *C. korshinskii* are less than 1 kb. Repeats are known to play a major role in plastome size evolution in angiosperm [35], the presence of abundant repeat sequences may be the reason why the genome of *C. rosea* is larger than the other three *Caragana* species.

Simple sequence repeat (SSR) loci are effective molecular markers because of their high variability that are wide distribution throughout the whole plastome [36]. SSR can provide useful information in polymorphism investigations and population genetics [37,38]. We identified SSRs in the plastome of four *Caragana* species. The number of SSRs ranges from 239 to 258 (Figure 2). In *Caragana rosea*, 66.9% are mono-nucleotide repeats, as compared with only 26.4%, 2.5%, and 4.2%, of di-, tri-, and tetra-nucleotide repeats, respectively. Of these SSR loci, 159 contained A or T, whereas only one had G or C; similarly, most dinucleotide repeat sequences were composed of AT/AT repeats. This result is consistent with previous reports that most of the SSR in plastomes are composed of short polyA or polyT repeats, while tandem repeats of G or C are rare [39]. We also analyzed the occurrence of SSRs in the CDs and found that there are fewer SSRs in the protein-coding regions than in the non-coding regions.

The plastomes of the other three *Caragana* species are similar to that of *C. rosea* in terms of the number, distribution, and the GC contents of the SSRs.

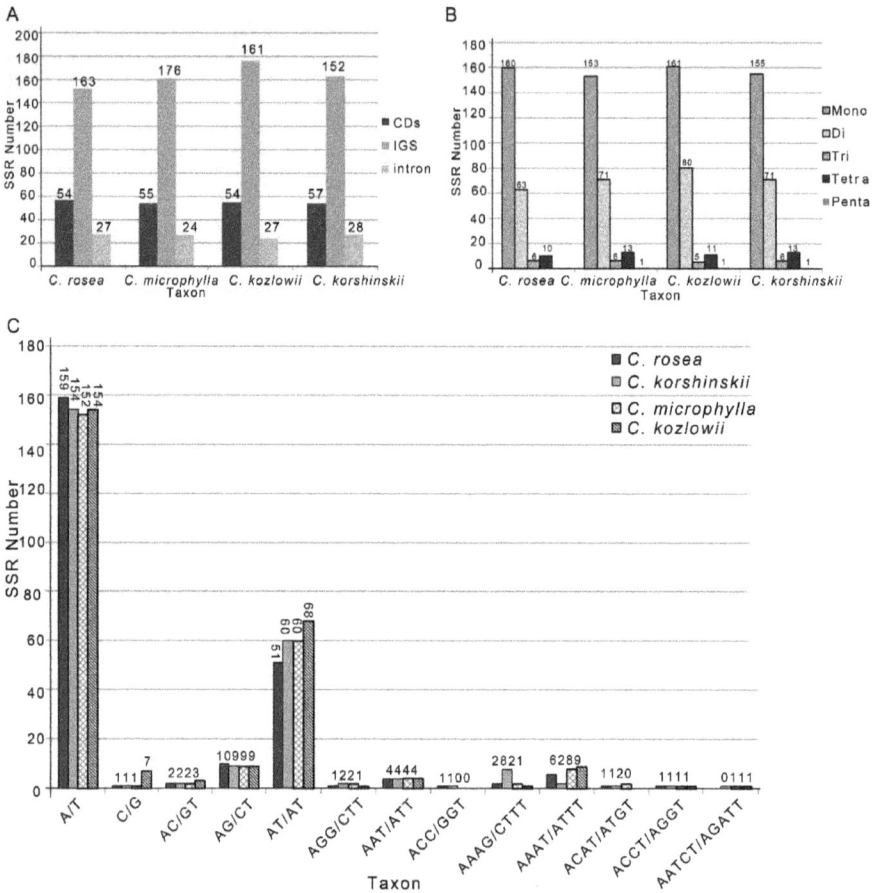

Figure 2. Statistics of simple sequence repeat (SSRs) detected in the plastome of four *Caragana* species. (**A**) Numbers of SSRs found in the coding (CDS), intergenic (IGS), and intronic regions, respectively; (**B**) number of different SSR types identified in the four genomes; and, (**C**) number of identified SSR motifs in different repeat class types.

2.4. Sequence Divergence Analysis among Caragana Species

To elucidate the level of sequence similarities between the *C. rosea* and the other three *Caragana*, the plastome sequences were compared while using the annotated *C. rosea* plastome as the reference (Figure 3). As shown, the four plastome sequences are highly similar. However, there are significant differences between *C. rosea* and other three *Caragana* species in some IGS, such as IGS(*rps12-clpP*) (square A), IGS(*rps19-rpl2*) (square B), and etc., which may be related to the unique large repeat fragments in the *C. rosea* plastome. For the regions IGS(*psaC-ndhD*) (square C), the sequence of *C. rosea* plastome is highly similar to those of *C. microphylla* and *C. korshinskii*, but it is quite different from that of *C. kozlowii*. Overall, the protein-coding regions are highly conserved, while the non-coding regions have different degrees of divergence between the *C. rosea* plastome and those of the other three. This suggests that the IGS of *Caragana* species has evolved rapidly.

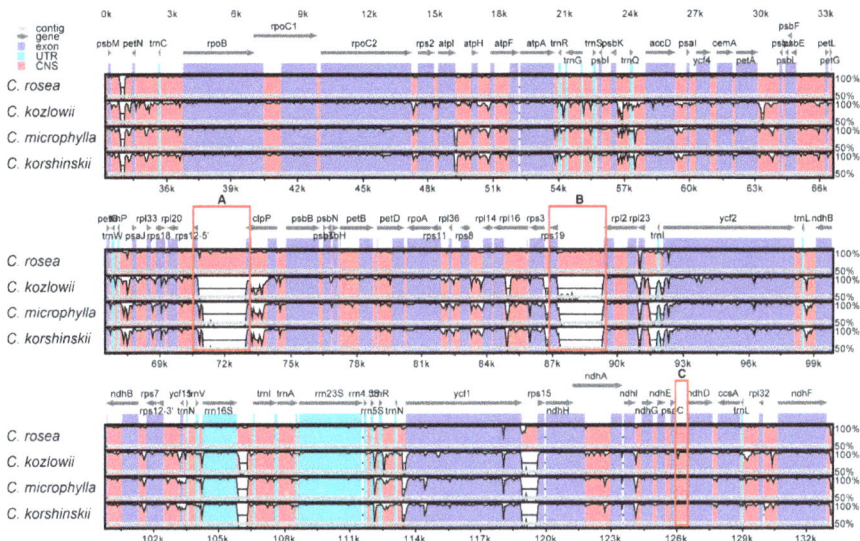

Figure 3. Structure comparison of the four plastomes by using the mVISTA program. Gray arrows and thick black lines above the alignment indicate genes with their orientation and the position of the IRs, respectively. A cut-off value of 70% identity was used for the plots, and the Y-scale represents the percent identity between 50% and 100%. UTR: Untranlated Regions; CNS: Conserved Non-coding Sequences. A: IGS(*rps12-clpP*); B: IGS(*rps19-rp12*); C: IGS(*psaC-ndhD*).

Highly variable sites in the genome can be used to develop molecular markers. We set out to identify the highly variable sites. We conducted pairwise distance comparison analysis for each non-coding region, including the intergenic region and the intron region using Kimura 2-parameter (K2p) model to identify divergence hotspot regions among the *Caragana* species. As expected, the variation of the intron sequence is relatively low, and the K2p distances ranged from 0.000 to 0.0269 (Figure 4, Table S5). The *clpP*, *ndhA* introns and the second intron of *ycf3* show the highest K2p values among the four *Caragana* species.

For IGSs, the K2p values ranged from 0 to 0.6207 (Figure 5). As described in the method section, we calculatecd the mean+2*STD (Standard Deviation) as the threshold for a IGS to be highly variable. It is 0.1649 in this case. The sequence divergences of the IGS regions ranged from 0 to 0.395 between *C. microphylla* and *C. kozlowii*, ranged from 0 to 0.6207 between *C. microphylla* and *C. rosea*, ranged from 0 to 0.5641 between *C. kozlowii* and *C. rosea*, ranged from 0 to 0.068 between *C. korshinskii* and *C. microphylla*, ranged from 0 to 0.3094 between *C. korshinskii* vs C. kozlowii, and ranged from 0 to 0.5176 between *C. korshinskii* vs C. rosea. The seven IGS regions (*atpB-atpE*, *ndhC-ndhK*, *ndhH-ndhA*, *psaA-psaB*, *psbC-psaD*, *psbT-psbN*, and *rpoB-rpoC1*) are 100% identical with the K2p values of 0. Meanwhile, we found that the K2p values were particularly high for the five IGS regions: *rpl23-trnI-CAU*, *rps12-clpP*, *rps19-rpl2*, *trnI-CAU-ycf2*, and *ycf1-rps15*.

The high degree of similarity among the plastomes of *C. korshinskii* and *C. microphylla* is consistent with their high degree of similarity in morphological characteristics. To determine whether these species can be distinguished with common molecular markers, such as ITS, *rbcL*, and *matK*, we compared the sequences of these markers from these species. It is found that these regions could not be used to distinguish the two confusable species with 99%, 100%, and 100% identities among these marker sequences, respectively. Interestingly, two regions IGS(*trnI-CAU-ycf2*) and IGS(*trnN-GUU-ycf1*) from *C. korshinskii* and *C. microphylla* have large K2p distances, indicating a high degree of sequence divergences. Moreover, these two regions have relatively high K2p values in the pairwise distance comparison analysis of these four species, which can be used to develop novel molecular markers to distinguish the four *Caragana* species accurately.

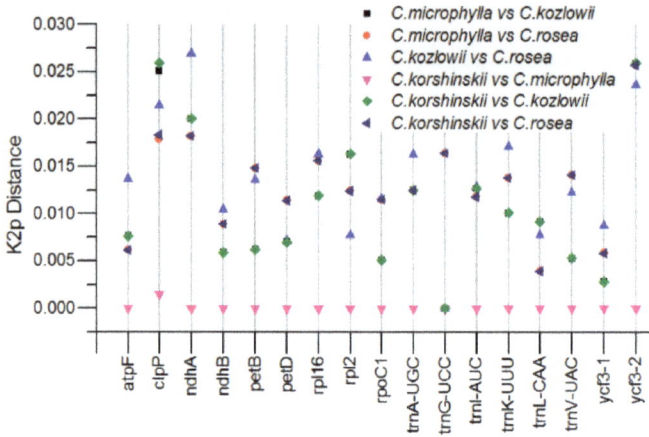

Figure 4. K2p distances for introns among *C. rosea*, *C. microphylla*, *C. kozlowii*, and *C. korshinskii*.

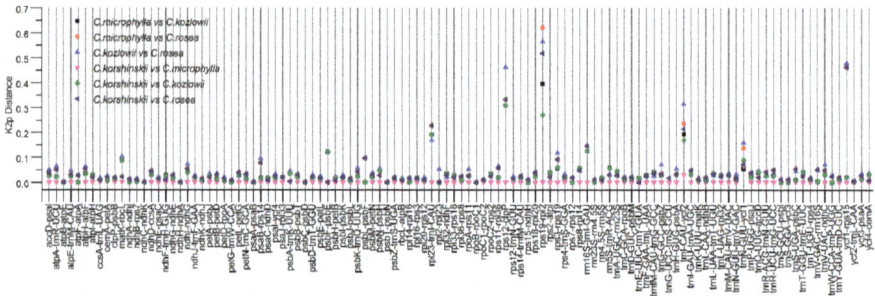

Figure 5. K2p distances for IGS regions among *C. rosea*, *C. microphylla*, *C. kozlowii*, and *C. korshinskii*.

2.5. Phylogenetic Analysis

The plastome sequence is an important resource for studying phylogenetic relationships and taxonomic status in the angiosperm [28]. In order to determine the phylogenetic position of *Caragana* in the Papilionoideae, we conducted multiple sequence alignments while using 63 common protein sequences from the plastomes of 36 species. *Arabidopsis thaliana* and *Nicotiana tabacum* were set as outgroup. The other 34 species contained six plant families that belong to the IRLC of Papilionoideae, including *Galegeae* (4), *Caraganeae* (4), *Cicereae* (1), *Fabeae* (16), *Trifolieae* (7), and *Millettieae* (2). The numbers in the parentheses represent the number of species in the corresponding taxa. The final dataset comprised 18745 positions and were subjected to phylogentic analysis using RaxML. Without surprise, *C. rosea* is found to locate in the same branch as the other three *Caragana* species, with 100% bootstrap values (Figure 6).

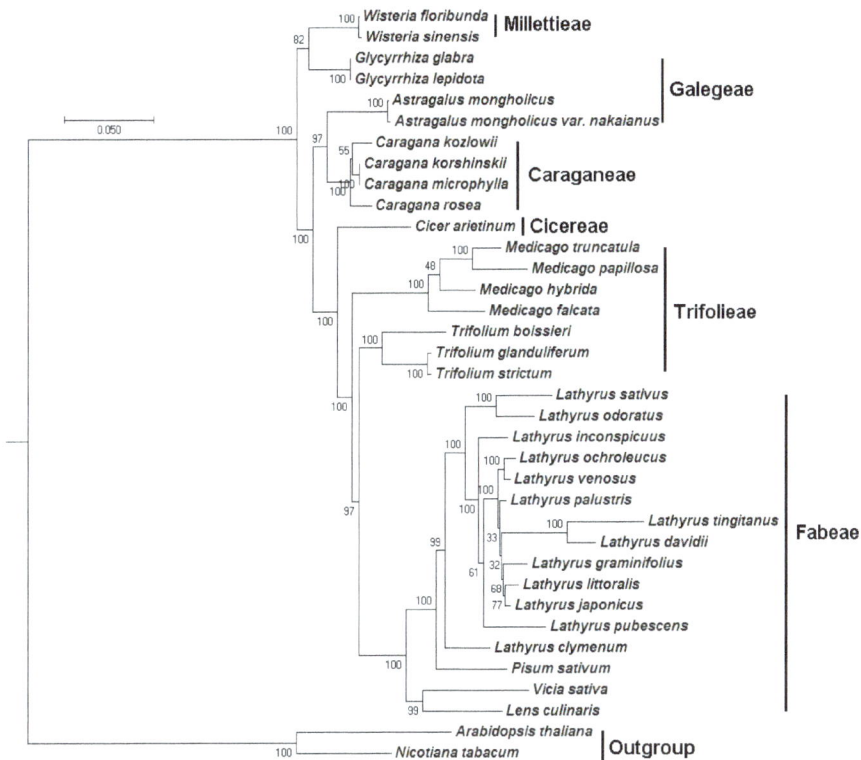

Figure 6. Molecular phylogenetic analyses of plastomes in the inverted-repeat-lacking clade of Papilionoideae. The tree was constructed with the sequences of 63 proteins present in all 36 species by using the maximum likelihood method implemented in RAxML. Bootstrap supports were calculated from 1000 replicates. *Nicotiana tabacum* and *Arabidopsis thaliana* were set as outgroups.

2.6. Selective Pressure Analysis

As synonymous substitutions accumulate nearly neutrally, non-synonyous substitutions are subject to selective pressures of varying degree and direction (positive or negative). In general, the ratio of nonsynonymous to synonymous substitution (ω) measures the levels of selective pressure operating in a protein coding gene. To test which genes were subject to positive selection at the *Caragana* branch, we conducted the selection analysis of the exons of each protein-encoding gene using the adaptive Branch-Site Random Effects Likelihood (aBSREL) model. A total of 69 branches among 36 species (listed in Section 2.5) were tested for diversifying selection. Significance was assessed using the Likelihood Ratio Test at a threshold of $p \leq 0.05$, after correcting for multiple testing.

Five genes (*rpl16*, *rpl20*, *rps11*, *rps7*, and *ycf1*) were found to have evolved under positive selection in *Caragana* branch in the phylogeny, the significance and number of rate categories inferred at the *Caragana* branch are provided in Table 3. The optimized branch length of the *Caragana* branch is 0.0128. The genes can be classified into two groups. The first group contains four ribosomal protein-coding genes. There is very strong evidence in selective pressure for the *rpl20* and *rps11* genes: positive selection was detected in 16 branches for both genes. In contrast, four branches (*Medicago hybrida*, *Medicago papillosa*, *Medicago falcata*, and *Caragana*) have experienced positive selection for the *rpl16* gene. Two branches (*Lathyrus clymenum*, *Caragana*) excluding those for the outgroup have experienced positive selection for the *rps7* gene, with two rate classes per branch. The second group contain the *ycf1* gene. The analysis did not include four species (*Lens culinaris*, *Wisteria floribunda*, *Wisteria sinensis*, and *Medicago falcata*) lacking the *ycf1* gene. The aBSREL model selection procedure identifies 42 branches for *ycf1* to have been significantly selected.

Table 3. The results of positive selection genes at *Caragana* branch.

Gene	B	LRT	Test *p*-Value	Uncorrected *p*-Value	ω Distribution over Sites
rpl16	0.0128	17.4234	0.0037	0.0001	ω1 = 0.121 (98%) ω2 = 781 (1.9%)
rpl20	0.0128	31.0401	0	0	ω1 = 0.147 (99%) ω2 = 290 (0.92%)
rps11	0.0128	12.0303	0.0452	0.0008	ω1 = 0.226 (100%) ω2 = 0.824 (0.20%) ω3 = 10000 (0.14%)
rps7	0.0128	24.3289	0.0001	0	ω1 = 0.00 (100%) ω2 = 286 (0.43%)
ycf1	0.0128	34.4372	0	0	ω1 = 0.384 (99%) ω2 = 91.3 (0.74%)

B: Optimized branch length; LRT: Likelihood ratio test statistic for selection; Test *p*-value: *p*-value corrected for multiple testing; Uncorrected *p*-value: Raw *p*-value without correction for multiple testing.

To find out which sites were subject to positive selection along the *Caragana* branch, codeml from PAML (v4.9) were used to analyze the Ka/Ks for the four genes: *rpl16*, *rpl20*, *rps11*, and *rps7*, using the branch-site model. The *Caragana* branch shown in Figure 6 were set as the foreground branch and all the other branches were set as the background branches. It is found that two sites: 36Y and 67Q was potentially under positive selection for the *rpl16* gene along the *Caragana* branch. In contrast, one site: 97Y were potentially under positive selection for the *rpl20* gene. Seven sites: 32K, 72T, 86N, 92V, 97Q, 103I, 137T were potential positively selected for the *rps11* gene. Four sites: 59E, 60T, 65V, 132V were positively selected for the *rps7* gene. As no three-dimensional (3D) structures are available for *Caragana* proteins, we cannot determine the functional and evolutionary significance of these sites at this time.

3. Discussion

Caragana species are xeromorphic, heat-resistant, and cold-resistant plants. Understanding the underlying molecular mechanism of this genus is of great interest for molecular breeding. As sessile organisms, plants must cope with abiotic stresses, such as soil salinity, drought, and extreme temperatures. Stress signaling pathways in plants have been extensively studied and reviewed [7,40]. All of these signaling pathways involve protein kinases that are related to yeast SNF1 and mammalian AMPK, suggesting that stress signaling in plants evolved from energy sensing.

Signals that are caused by limited water (drought stress) or excessive salt (salt stress) can be divided into primary and secondary types. The primary signal caused by drought is also called hyperosmotic stress. Salt stress exerts both osmotic and ionic effects on cells. The secondary effects are rather complex and they include oxidative stress; such effects include damage to cellular components (such as membrane lipids, proteins, and nucleic acids), chloroplast, mitochondria, ER, and metabolic dysfunction [7].

Chloroplast is an organelle where photosynthetic electron transport and many metabolic reactions occur. Environmental stress can easily perturb the metabolic balance in chloroplasts. The disturbance in chloroplast homeostasis is then passed to the nucleus through retrograde signals; as such, all cellular activities can be adjusted and coordinated. The chloroplast is a major site for the production of reactive oxygen species (ROS), such as superoxide anion, hydrogen peroxide, hydroxyl radical, and singlet oxygen [41]. Various environmental stresses, particularly high light stress, can exacerbate ROS production, thereby disrupting ROS-managing systems and generating various secondary messengers.

The relationship between protein synthesis and stress response has been studied [7]. The accuracy of protein synthesis is critical for life because a high degree of fidelity of the translation of the genetic information is required to accomplish the needs for cellular functions and to preserve variability developed by evolution. Even in simple organisms, this process involves more than 100 macromolecules, such as ribosomal proteins, translation factors, aminoacyl-tRNA synthetases, ribosomal RNAs, and transfer RNAs. Moreover, in vivo or in vitro experiments indicated that several macromolecules that participate in translation are the targets of oxidation; hence, translation is directly targeted by oxidative species. In bacteria, target macromolecules include elongation factors, such as Tu [42,43], Ts (EF-Ts) [44], and G (EF-G) [45–48]; several ribosomal proteins [31,36,37,49–51]; tRNAs [38–42,52,53]; and, aminoacyl-tRNA synthetases (aaRS) [43,49–51,54]. In particular, the disulfide bond formation of ribosomal proteins S7 and L16 are affected by oxidative stress [51]. Furthermore, scholars have identified the covalent binding of ribosomal protein S11 to chaperon/oxido-reductase protein through cysteine bond [43].

Previous studies showed that environmental stress can cause oxidative stress, which in turn affects the translation system, particularly in prokaryote originated systems, such as chloroplast. When considering the axis of environmental stress->oxidative stress->translation system, we can speculate that ribosomal protein-coding genes, such as *rps7*, *rps11*, *rpl16*, and *rpl20*, are strongly selected to maintain the integrity of the protein synthesis machinery under various environmental stresses. This phenomenon might, at least in part, contribute to the strong environmental stress resistance characteristics of the Caragana species. Additional analyses would be needed to confirm this hypothesis.

4. Materials and Methods

4.1. Plant Material, DNA Extraction, and Sequencing

The fresh leaves of the *C. rosea* were collected from the Institute of Medicinal Plant Development, China. After washing, the leaves were kept in the −80 refrigerator until use. DNA from about 100 mg leaves was extracted using the modified CTAB (Cetyltrimethylammonium bromide) method. Subsequently, the extracted DNA integrity and concentration were detected by electrophoresis in 1% (*w/v*) agarose gel and spectrophotometer (Nanodrop 2000, Thermo Fisher Scientific, Waltham, MA, USA). The genomic DNA of *C. rosea* was subjected to high-throughput sequencing using an Illumina Hiseq2000 sequencer (Illumina Inc., San Diego, CA, USA), with insert sizes of 500 bases for the library.

A total of 18,932,846 paired-end reads were obtained with 100 bases long. The other three plastome sequences of *Caragana* e.g., *C. korshinskii* (Accession number: NC_035229), *C. kozlowii* (Accession number NC_035228), *C. microphylla* (Accession number NC_032691), and ITS (Accession number: FJ537266, FJ537264) were obtained from Genbank.

4.2. Genome Assembly and Gap Filling

In order to extract the reads belonging to the plastome from those for the total DNAs, we downloaded 1688 plastome sequences from GenBank in February 2016, which were used to search against Illumina paired-end reads using BLASTN with an E-value cutoff of 1×10^5 [55]. The genome sequence of *C. kozlowii* was found to have the highest similarity and it was chosen as the reference sequence for the following assembly.

A total of 3076 paired-end reads similar to the *C. kozlowii* plastome sequence were selected and assembled by AbySS (v1.5.2) [56] and CLC Genomics Workbench (v7) Software. Nine and seven contigs were obtained using the two software tools, respectively. Then, the 16 sequences were further assembled by the Seqman module of DNAStar (v6.10.01), resulting in three contigs. The gaps between the contigs were filled with PCR amplification and Sanger sequencing using the sequence-specific primers (Table S6) designed to cross the gaps. Finally, the draft plastome sequence was validated by mapping the raw Illumina paired-end reads against it using Bowtie2 (v2.0.1) with default settings [57].

4.3. Genome Annotation and Characteristics Analysis

The *C. rosea* plastome sequence was annotated by CpGAVAS web service [58], with the default parameter. The tRNA genes were annotated using ARAGORN [59] and tRNAscan-SE [60]; the protein sequences were verified again by BLASP against the GenBank sequences. Subsequently, the intron/exon boundaries and the start/stop codons of predicted genes are manually edited using the Apollo program (v1.11.8) [61]. The circular plastome map of *C. rosea* was drawn using OrganellarGenomeDRAW [62]. Both GC contents and codon usage were calculated using the programs Cusp and Compseq from EMBOSS (v6.3.1) [63].

4.4. Repeat and SSR Analysis

Repeats (palindrome and forward repeats) were identified by REPuter web service [64], with the settings of 3 for the Hamming Distance (sequence identity ≥ 90%) and 30 for Minimal Repeat Size, as reported previously [34,65]. The number and location of tandem repeated elements in the *Caragana* genus plastome were determined using the Tandem Repeats Finder [66], with the following parameters: matches, mismatches and indels, minimum alignment score, and the maximum period size were 2, 7, 50, and 500, respectively. We manually verified all of the detected repeats and removed nested and redundant sequences. SSR in the plastome was analyzed by MISA software with the same parameters as reported previously [67]. Briefly, the cutoff for the numbers of units for mono-, di-, tri-, tetra-, penta-, and hexa-nucleotides were 8, 4, 4, 3, 3, and 3, respectively.

4.5. Comparative Genomic Analysis

The complete plastome sequence of *C. rosea* was compared with the those of *C. korshinskii*, *C. kozlowii*, and *C. microphylla*, using the mVISTA program in a Shuffle-LAGAN mode with default parameters [68]. The annotated *C. rosea* plastome was used as the reference. In order to analyze sequence diversity and selective pressure, a total of 112 intergenic regions, 17 introns, and 76 exons were extracted from the four plastomes using custom MatLab scripts. The corresponding nucleotide sequences were aligned using the CLUSTALW2 (v2.0.12) program with options "-type = DNA -gapopen = 10 -gapext = 2" [69]. Pairwise distance were determined with the Distmat program that was implemented in EMBOSS (v6.3.1) [63] using the Kimura 2-parameters (K2p) evolution model [70] for intergenic regions and introns. To determine the threshold for the K2p distance to be highly variable, we calculated the mean and the standard deviation for all the K2p values. The mean + 2*STD were then set as the threshold. The 76 exons sequences were aligned using the RevTrans (v2.0) [71] with the option of CLUSTALW2 program. Subsequently, the selective pressure analysis were conducted using adaptive branch-site random effects likelihood (aBSREL) model [72], implemented in HyPhy (https://veg.github.io/hyphy-site/getting-started/#characterizing-selective-pressures). Finally, we analyzed which sites were subject to positive selection along *Caragana* branch using the Codeml program that was implemented in PAML (v4.9) [73].

4.6. Phylogenetic Analysis

The plastome sequences of 33 species belonging to the IRLC (inverted-repeat-lackingclade) of Papilionoideae and two outgroup species (*Arabidopsis thaliana* and *Nicotiana tabacum*) were downloaded from NCBI RefSeq database, and a total of 63 protein sequences that were present in all of the 35 species and *C. rosea* were obtained by manual detection (ATPA, ATPB, ATPF, ATPH, ATPI, CCSA, CEMA, CLPP, MATK, NDHC, NDHD, NDHE, NDHF, NDHG, NDHH, NDHI, NDHJ, NDHK, PETA, PETB, PETD, PETG, PETL, PETN, PSAA, PSAB, PSAC, PSAJ, PSBA, PSBB, PSBC, PSBD, PSBE, PSBF, PSBH, PSBI, PSBK, PSBM, PSBN, PSBT, RBCL, RPL14, RPL16, RPL2, RPL20, RPL32, RPL33, RPL36, RPOA, RPOB, RPOC1, RPOC2, RPS11, RPS12, RPS14, RPS15, RPS19, RPS3, RPS4, RPS7, RPS8, YCF2, and YCF3) (Table S7). For the phylogenetic analysis, these protein sequences were aligned using the CLUSTALW2 (v2.0.12) program with options "-gapopen = 10 -gapext = 2 -output = phylip".

The Maximum Likelihood method implemented in RaxML (v8.2.4) [69] was used to inferred the evolutionary history, using "raxmlHPC-PTHREADS-SSE3 -f a -N 1000 -m PROTGAMMACPREV -x 551314260 -p 551314260 -o A_thaliana, N_tabacum -T 20". Subsequently, the Bootstrap analysis was also performed with 1000 replicates for the phylogenetic tree.

5. Conclusions

In this study, we sequenced the plastome of *C. rosea* and carried out a comparative study with those from *C. microphylla*, *C. kozlowii*, and *C. korshinskii*. Phylogenetic analyses showed that four *Caragana* species were on a monophyletic clade, with 100% bootstrap values. Analyses of selective pressure revealed that five genes: *rpl16, rpl20, rps11, rps7*, and *ycf1* were evolved undergoing positive selection. Analyses of sequence divergence found two sites: IGS(*trnI-CAU-ycf2*) and IGS(*trnN-GUU-ycf1*) had high degree of variations and might be sources for markers that can be used to distinguish these four species. The results presented in this paper will facilitate the further investigation for these four species in terms the molecular mechanisms for drought-resistance.

Supplementary Materials: Supplementary materials can be found at http://www.mdpi.com/1422-0067/19/5/1419/s1.

Author Contributions: C.L. conceived the study; M.J. collected samples of *C. rosea* var. *rosea*, extracted DNA for next-generation sequencing, assembled the genome, performed data analysis, conducted PCR validation and drafted the manuscript; H.C. annotated the genome; L.W. wrote the matlab scripts. S.H. and A.J.C. reviewed the manuscript critically. All authors have read and agreed the contents of the manuscript.

Funding: This work was supported by the CAMS Innovation Fund for Medical Sciences (CIFMS) (2016-I2M-3-016, 2017-I2M-1-013) from Chinese Academy of Medical Science. The funders were not involved in the study design, data collection and analysis, decision to publish, or manuscript preparation.

Conflicts of Interest: The authors declare no conflict of interest.

References

1. Meng, Q.; Niu, Y.; Niu, X.; Roubin, R.H.; Hanrahan, J.R. Ethnobotany, phytochemistry and pharmacology of the genus *Caragana* used in traditional chinese medicine. *J. Ethnopharmacol.* **2009**, *124*, 350–368. [CrossRef] [PubMed]
2. Delectis Flora Reipublicae Popularis Sinicae Agendae Academiae Sinicae Edita. *Flora Reipublicae Popularis Sinicae*; Science Press: Beijing, China, 1993; Volume 42, p. 18.
3. Ma, F.; Na, X.; Xu, T. Drought responses of three closely related *Caragana* species: Implication for their vicarious distribution. *Ecol. Evol.* **2016**, *6*, 2763–2773. [CrossRef] [PubMed]
4. Delectis Flora Reipublicae Popularis Sinicae Agendae Academiae Sinicae Edita. *Flora Reipublicae Popularis Sinicae*; Science Press: Beijing, China, 1993; Volume 42, p. 31.
5. Delectis Flora Reipublicae Popularis Sinicae Agendae Academiae Sinicae Edita. *Flora Reipublicae Popularis Sinicae*; Science Press: Beijing, China, 1993; Volume 42, p. 60.
6. Yang, G.X.; Qi, J.B.; Cheng, K.J.; Hu, C.Q. Anti-HIV chemical constituents of aerial parts of *Caragana rosea*. *Yao Xue Xue Bao (Acta Pharm. Sin.)* **2007**, *42*, 179–182.
7. Zhu, J.K. Abiotic stress signaling and responses in plants. *Cell* **2016**, *167*, 313–324. [CrossRef] [PubMed]
8. Hu, S.; Sablok, G.; Wang, B.; Qu, D.; Barbaro, E.; Viola, R.; Li, M.; Varotto, C. Plastome organization and evolution of chloroplast genes in *Cardamine* species adapted to contrasting habitats. *BMC Genom.* **2015**, *16*, 306. [CrossRef] [PubMed]
9. Ometto, L.; Li, M.; Bresadola, L.; Varotto, C. Rates of evolution in stress-related genes are associated with habitat preference in two *Cardamine* lineages. *BMC Evol. Biol.* **2012**, *12*, 7. [CrossRef] [PubMed]
10. Gong, C.; Bai, J.; Wang, J.; Zhou, Y.; Kang, T.; Wang, J.; Hu, C.; Guo, H.; Chen, P.; Xie, P.; et al. Carbon storage patterns of *Caragana korshinskii* in areas of reduced environmental moisture on the loess plateau, China. *Sci. Rep.* **2016**, *6*, 28883. [CrossRef] [PubMed]
11. Li, M.; Liu, D.; Liu, Y. Evaluation on drought-resistant characteristics of ten *Caragana* species based on leaf micromorphological structure. *J. Desert Res.* **2016**, *3*, 708–717. [CrossRef]

12. Yamamoto, Y.; Aminaka, R.; Yoshioka, M.; Khatoon, M.; Komayama, K.; Takenaka, D.; Yamashita, A.; Nijo, N.; Inagawa, K.; Morita, N.; et al. Quality control of photosystem II: Impact of light and heat stresses. *Photosynth. Res.* **2008**, *98*, 589–608. [CrossRef] [PubMed]

13. Mulo, P.; Sakurai, I.; Aro, E.M. Strategies for psba gene expression in cyanobacteria, green algae and higher plants: From transcription to psii repair. *Biochim. Biophys. Acta* **2012**, *1817*, 247–257. [CrossRef] [PubMed]

14. Wright, I.J.; Reich, P.B.; Westoby, M.; Ackerly, D.D.; Baruch, Z.; Bongers, F.; Cavender-Bares, J.; Chapin, T.; Cornelissen, J.H.; Diemer, M.; et al. The worldwide leaf economics spectrum. *Nature* **2004**, *428*, 821–827. [CrossRef] [PubMed]

15. Neufeld, H.S. Plant physiological ecology. *Photosynthetica* **1999**, *80*, 1785–1787. [CrossRef]

16. Sabir, J.; Schwarz, E.; Ellison, N.; Zhang, J.; Baeshen, N.A.; Mutwakil, M.; Jansen, R.; Ruhlman, T. Evolutionary and biotechnology implications of plastid genome variation in the inverted-repeat-lacking clade of legumes. *Plant Biotechnol. J.* **2014**, *12*, 743–754. [CrossRef] [PubMed]

17. Cardoso, D.; de Queiroz, L.P.; Pennington, R.T.; de Lima, H.C.; Fonty, E.; Wojciechowski, M.F.; Lavin, M. Revisiting the phylogeny of papilionoid legumes: New insights from comprehensively sampled early-branching lineages. *Am. J. Bot.* **2012**, *99*, 1991–2013. [CrossRef] [PubMed]

18. Dugas, D.V.; Hernandez, D.; Koenen, E.J.; Schwarz, E.; Straub, S.; Hughes, C.E.; Jansen, R.K.; Nageswara-Rao, M.; Staats, M.; Trujillo, J.T.; et al. Mimosoid legume plastome evolution: Ir expansion, tandem repeat expansions, and accelerated rate of evolution in *clpP. Sci. Rep.* **2015**, *5*, 16958. [CrossRef] [PubMed]

19. Jansen, R.K.; Wojciechowski, M.F.; Sanniyasi, E.; Lee, S.B.; Daniell, H. Complete plastid genome sequence of the chickpea (*Cicer arietinum*) and the phylogenetic distribution of *rps12* and *clpP* intron losses among legumes (leguminosae). *Mol. Phylogenet. Evol.* **2008**, *48*, 1204–1217. [CrossRef] [PubMed]

20. Qian, J.; Song, J.; Gao, H.; Zhu, Y.; Xu, J.; Pang, X.; Yao, H.; Sun, C.; Li, X.; Li, C.; et al. The complete chloroplast genome sequence of the medicinal plant *Salvia miltiorrhiza. PLoS ONE* **2013**, *8*, e57607. [CrossRef] [PubMed]

21. Shen, X.; Wu, M.; Liao, B.; Liu, Z.; Bai, R.; Xiao, S.; Li, X.; Zhang, B.; Xu, J.; Chen, S. Complete chloroplast genome sequence and phylogenetic analysis of the medicinal plant *Artemisia annua. Molecules* **2017**, *22*, 1330. [CrossRef] [PubMed]

22. He, L.; Qian, J.; Li, X.; Sun, Z.; Xu, X.; Chen, S. Complete chloroplast genome of medicinal plant *Lonicera japonica*: Genome rearrangement, intron gain and loss, and implications for phylogenetic studies. *Molecules* **2017**, *22*, 249. [CrossRef] [PubMed]

23. Daniell, H.; Lin, C.S.; Yu, M.; Chang, W.J. Chloroplast genomes: Diversity, evolution, and applications in genetic engineering. *Genome Biol.* **2016**, *17*, 134. [CrossRef] [PubMed]

24. Keller, J.; Rousseau-Gueutin, M.; Martin, G.E.; Morice, J.; Boutte, J.; Coissac, E.; Ourari, M.; Ainouche, M.; Salmon, A.; Cabello-Hurtado, F.; et al. The evolutionary fate of the chloroplast and nuclear *rps16* genes as revealed through the sequencing and comparative analyses of four novel legume chloroplast genomes from *Lupinus. DNA Res. Int. J. Rapid Publ. Rep. Genes Genomes* **2017**, *24*, 343–358. [CrossRef] [PubMed]

25. Jansen, R.K.; Saski, C.; Lee, S.B.; Hansen, A.K.; Daniell, H. Complete plastid genome sequences of three rosids (*Castanea, Prunus, Theobroma*): Evidence for at least two independent transfers of *rpl22* to the nucleus. *Mol. Biol. Evol.* **2011**, *28*, 835–847. [CrossRef] [PubMed]

26. Gantt, J.S.; Baldauf, S.L.; Calie, P.J.; Weeden, N.F.; Palmer, J.D. Transfer of *rpl22* to the nucleus greatly preceded its loss from the chloroplast and involved the gain of an intron. *EMBO J.* **1991**, *10*, 3073–3078. [PubMed]

27. Steane, D.A. Complete nucleotide sequence of the chloroplast genome from the tasmanian blue gum, *Eucalyptus globulus* (Myrtaceae). *DNA Res. Int. J. Rapid Publ. Rep. Genes Genomes* **2005**, *12*, 215–220. [CrossRef] [PubMed]

28. Jansen, R.K.; Cai, Z.; Raubeson, L.A.; Daniell, H.; Depamphilis, C.W.; Leebens-Mack, J.; Muller, K.F.; Guisinger-Bellian, M.; Haberle, R.C.; Hansen, A.K.; et al. Analysis of 81 genes from 64 plastid genomes resolves relationships in angiosperms and identifies genome-scale evolutionary patterns. *Proc. Nat. Acad. Sci. USA* **2007**, *104*, 19369–19374. [CrossRef] [PubMed]

29. Magee, A.M.; Aspinall, S.; Rice, D.W.; Cusack, B.P.; Semon, M.; Perry, A.S.; Stefanovic, S.; Milbourne, D.; Barth, S.; Palmer, J.D.; et al. Localized hypermutation and associated gene losses in legume chloroplast genomes. *Genome Res.* **2010**, *20*, 1700–1710. [CrossRef] [PubMed]

30. Park, S.; Jansen, R.K.; Park, S. Complete plastome sequence of *Thalictrum coreanum* (ranunculaceae) and transfer of the *rpl32* gene to the nucleus in the ancestor of the subfamily thalictroideae. *BMC Plant Biol.* **2015**, *15*, 40. [CrossRef] [PubMed]

31. Ueda, M.; Fujimoto, M.; Arimura, S.; Murata, J.; Tsutsumi, N.; Kadowaki, K. Loss of the *rpl32* gene from the chloroplast genome and subsequent acquisition of a preexisting transit peptide within the nuclear gene in *Populus*. *Gene* **2007**, *402*, 51–56. [CrossRef] [PubMed]

32. Jo, Y.D.; Park, J.; Kim, J.; Song, W.; Hur, C.G.; Lee, Y.H.; Kang, B.C. Complete sequencing and comparative analyses of the pepper (*Capsicum annuum* L.) plastome revealed high frequency of tandem repeats and large insertion/deletions on pepper plastome. *Plant Cell Rep.* **2011**, *30*, 217–229. [CrossRef] [PubMed]

33. Sloan, D.B.; Triant, D.A.; Forrester, N.J.; Bergner, L.M.; Wu, M.; Taylor, D.R. A recurring syndrome of accelerated plastid genome evolution in the angiosperm tribe *Sileneae* (caryophyllaceae). *Mol. Phylogenet. Evol.* **2014**, *72*, 82–89. [CrossRef] [PubMed]

34. Martin, G.E.; Rousseau-Gueutin, M.; Cordonnier, S.; Lima, O.; Michon-Coudouel, S.; Naquin, D.; de Carvalho, J.F.; Ainouche, M.; Salmon, A.; Ainouche, A. The first complete chloroplast genome of the genistoid legume *Lupinus luteus*: Evidence for a novel major lineage-specific rearrangement and new insights regarding plastome evolution in the legume family. *Annu. Bot.* **2014**, *113*, 1197–1210. [CrossRef] [PubMed]

35. Haberle, R.C.; Fourcade, H.M.; Boore, J.L.; Jansen, R.K. Extensive rearrangements in the chloroplast genome of *Trachelium caeruleum* are associated with repeats and tRNA genes. *J. Mol. Evol.* **2008**, *66*, 350–361. [CrossRef] [PubMed]

36. Provan, J.; Corbett, G.; McNicol, J.W.; Powell, W. Chloroplast DNA variability in wild and cultivated rice (*Oryza* spp.) revealed by polymorphic chloroplast simple sequence repeats. *Genome* **1997**, *40*, 104–110. [CrossRef] [PubMed]

37. Xue, J.; Wang, S.; Zhou, S.L. Polymorphic chloroplast microsatellite loci in *Nelumbo* (nelumbonaceae). *Am. J. Bot.* **2012**, *99*, e240–e244. [CrossRef] [PubMed]

38. Pauwels, M.; Vekemans, X.; Gode, C.; Frerot, H.; Castric, V.; Saumitou-Laprade, P. Nuclear and chloroplast DNA phylogeography reveals vicariance among european populations of the model species for the study of metal tolerance, *Arabidopsis halleri* (Brassicaceae). *New Phytol.* **2012**, *193*, 916–928. [CrossRef] [PubMed]

39. Kuang, D.Y.; Wu, H.; Wang, Y.L.; Gao, L.M.; Zhang, S.Z.; Lu, L. Complete chloroplast genome sequence of *Magnolia kwangsiensis* (Magnoliaceae): Implication for DNA barcoding and population genetics. *Genome* **2011**, *54*, 663–673. [CrossRef] [PubMed]

40. Zhu, J.K. Salt and drought stress signal transduction in plants. *Annu. Rev. Plant Biol.* **2002**, *53*, 247–273. [CrossRef] [PubMed]

41. Mignolet-Spruyt, L.; Xu, E.; Idanheimo, N.; Hoeberichts, F.A.; Muhlenbock, P.; Brosche, M.; van Breusegem, F.; Kangasjarvi, J. Spreading the news: Subcellular and organellar reactive oxygen species production and signalling. *J. Exp. Bot.* **2016**, *67*, 3831–3844. [CrossRef] [PubMed]

42. Ichimura, K.; Mizoguchi, T.; Yoshida, R.; Yuasa, T.; Shinozaki, K. Various abiotic stresses rapidly activate *Arabidopsis* MAP kinases ATMPK4 and ATMPK6. *Plant J. Cell Mol. Biol.* **2000**, *24*, 655–665. [CrossRef]

43. Iuchi, S.; Kobayashi, M.; Taji, T.; Naramoto, M.; Seki, M.; Kato, T.; Tabata, S.; Kakubari, Y.; Yamaguchi-Shinozaki, K.; Shinozaki, K. Regulation of drought tolerance by gene manipulation of 9-*cis*-epoxycarotenoid dioxygenase, a key enzyme in abscisic acid biosynthesis in *Arabidopsis*. *Plant J. Cell Mol. Biol.* **2001**, *27*, 325–333. [CrossRef]

44. Ishitani, M.; Liu, J.; Halfter, U.; Kim, C.S.; Shi, W.; Zhu, J.K. Sos3 function in plant salt tolerance requires n-myristoylation and calcium binding. *Plant Cell* **2000**, *12*, 1667–1678. [CrossRef] [PubMed]

45. Chen, H.H.; Li, P.H.; Brenner, M.L. Involvement of abscisic acid in potato cold acclimation. *Plant Physiol.* **1983**, *71*, 362–365. [CrossRef] [PubMed]

46. Choi, H.; Hong, J.; Ha, J.; Kang, J.; Kim, S.Y. ABFs, a family of ABA-responsive element binding factors. *J. Biol. Chem.* **2000**, *275*, 1723–1730. [CrossRef] [PubMed]

47. Gustin, M.C.; Albertyn, J.; Alexander, M.; Davenport, K. MAP kinase pathways in the yeast *Saccharomyces cerevisiae*. *Microbiol. Mol. Biol. Rev. (MMBR)* **1998**, *62*, 1264–1300. [PubMed]

48. Jacob, T.; Ritchie, S.; Assmann, S.M.; Gilroy, S. Abscisic acid signal transduction in guard cells is mediated by phospholipase D activity. *Proc. Nat. Acad. Sci. USA* **1999**, *96*, 12192–12197. [CrossRef] [PubMed]

49. Ingram, J.; Bartels, D. The molecular basis of dehydration tolerance in plants. *Annu. Rev. Plant Physiol. Plant Mol. Biol.* **1996**, *47*, 377–403. [CrossRef] [PubMed]

50. Jaglo-Ottosen, K.R.; Gilmour, S.J.; Zarka, D.G.; Schabenberger, O.; Thomashow, M.F. *Arabidopsis CBF1* overexpression induces *COR* genes and enhances freezing tolerance. *Science* **1998**, *280*, 104–106. [CrossRef] [PubMed]

51. Jonak, C.; Kiegerl, S.; Ligterink, W.; Barker, P.J.; Huskisson, N.S.; Hirt, H. Stress signaling in plants: A mitogen-activated protein kinase pathway is activated by cold and drought. *Proc. Nat. Acad. Sci. USA* **1996**, *93*, 11274–11279. [CrossRef] [PubMed]

52. Knight, H.; Trewavas, A.J.; Knight, M.R. Calcium signalling in *Arabidopsis thaliana* responding to drought and salinity. *Plant J. Cell Mol. Biol.* **1997**, *12*, 1067–1078. [CrossRef]

53. Katagiri, T.; Takahashi, S.; Shinozaki, K. Involvement of a novel *Arabidopsis* phospholipase D, AtPLDδ, in dehydration-inducible accumulation of phosphatidic acid in stress signalling. *Plant J. Cell Mol. Biol.* **2001**, *26*, 595–605. [CrossRef]

54. Kovtun, Y.; Chiu, W.L.; Tena, G.; Sheen, J. Functional analysis of oxidative stress-activated mitogen-activated protein kinase cascade in plants. *Proc. Nat. Acad. Sci. USA* **2000**, *97*, 2940–2945. [CrossRef] [PubMed]

55. Camacho, C.; Coulouris, G.; Avagyan, V.; Ma, N.; Papadopoulos, J.; Bealer, K.; Madden, T.L. Blast+: Architecture and applications. *BMC Bioinform.* **2009**, *10*, 421. [CrossRef] [PubMed]

56. Simpson, J.T.; Wong, K.; Jackman, S.D.; Schein, J.E.; Jones, S.J.; Birol, I. ABySS: A parallel assembler for short read sequence data. *Genome Res.* **2009**, *19*, 1117–1123. [CrossRef] [PubMed]

57. Langmead, B.; Trapnell, C.; Pop, M.; Salzberg, S.L. Ultrafast and memory-efficient alignment of short DNA sequences to the human genome. *Genome Biol.* **2009**, *10*, R25. [CrossRef] [PubMed]

58. Liu, C.; Shi, L.; Zhu, Y.; Chen, H.; Zhang, J.; Lin, X.; Guan, X. CpGAVAS, an integrated web server for the annotation, visualization, analysis, and genbank submission of completely sequenced chloroplast genome sequences. *BMC Genom.* **2012**, *13*, 715. [CrossRef] [PubMed]

59. Laslett, D.; Canback, B. ARAGORN, a program to detect tRNA genes and tmRNA genes in nucleotide sequences. *Nucleic Acids Res.* **2004**, *32*, 11–16. [CrossRef] [PubMed]

60. Schattner, P.; Brooks, A.N.; Lowe, T.M. The tRNAscan-SE, snoscan and snoGPS web servers for the detection of tRNAs and snoRNAs. *Nucleic Acids Res.* **2005**, *33*, W686–W689. [CrossRef] [PubMed]

61. Misra, S.; Harris, N. Using apollo to browse and edit genome annotations. *Curr. Protocol. Bioinform.* **2006**. Chapter 9, Unit 9 5. [CrossRef]

62. Lohse, M.; Drechsel, O.; Bock, R. Organellargenomedraw (OGDRAW): A tool for the easy generation of high-quality custom graphical maps of plastid and mitochondrial genomes. *Curr. Genet.* **2007**, *52*, 267–274. [CrossRef] [PubMed]

63. Rice, P.; Longden, I.; Bleasby, A. Emboss: The European molecular biology open software suite. *Trends Genet. (TIG)* **2000**, *16*, 276–277. [CrossRef]

64. Kurtz, S.; Choudhuri, J.V.; Ohlebusch, E.; Schleiermacher, C.; Stoye, J.; Giegerich, R. REPuter: The manifold applications of repeat analysis on a genomic scale. *Nucleic Acids Res.* **2001**, *29*, 4633–4642. [CrossRef] [PubMed]

65. Tangphatsornruang, S.; Sangsrakru, D.; Chanprasert, J.; Uthaipaisanwong, P.; Yoocha, T.; Jomchai, N.; Tragoonrung, S. The chloroplast genome sequence of mungbean (*Vigna radiata*) determined by high-throughput pyrosequencing: Structural organization and phylogenetic relationships. *DNA Res. Int. J. Rapid Publ. Rep. Genes Genomes* **2010**, *17*, 11–22. [CrossRef] [PubMed]

66. Benson, G. Tandem repeats finder: A program to analyze DNA sequences. *Nucleic Acids Res.* **1999**, *27*, 573–580. [CrossRef] [PubMed]

67. Lei, W.; Ni, D.; Wang, Y.; Shao, J.; Wang, X.; Yang, D.; Wang, J.; Chen, H.; Liu, C.; Lei, W. Intraspecific and heteroplasmic variations, gene losses and inversions in the chloroplast genome of *Astragalus membranaceus*. *Sci. Rep.* **2016**, *6*, 21669. [CrossRef] [PubMed]

68. Frazer, K.A.; Pachter, L.; Poliakov, A.; Rubin, E.M.; Dubchak, I. Vista: Computational tools for comparative genomics. *Nucleic Acids Res.* **2004**, *32*, W273–W279. [CrossRef] [PubMed]

69. Stamatakis, A. Raxml version 8: A tool for phylogenetic analysis and post-analysis of large phylogenies. *Bioinformatics* **2014**, *30*, 1312–1313. [CrossRef] [PubMed]

70. Kimura, M. A simple method for estimating evolutionary rates of base substitutions through comparative studies of nucleotide sequences. *J. Mol. Evol.* **1980**, *16*, 111–120. [CrossRef] [PubMed]

71. Wernersson, R.; Pedersen, A.G. RevTrans: Multiple alignment of coding DNA from aligned amino acid sequences. *Nucleic Acids Res.* **2003**, *31*, 3537–3539. [CrossRef] [PubMed]

72. Smith, M.D.; Wertheim, J.O.; Weaver, S.; Murrell, B.; Scheffler, K.; Kosakovsky Pond, S.L. Less is more: An adaptive branch-site random effects model for efficient detection of episodic diversifying selection. *Mol. Biol. Evol.* **2015**, *32*, 1342–1353. [CrossRef] [PubMed]

73. Yang, Z.; Nielsen, R. Codon-substitution models for detecting molecular adaptation at individual sites along specific lineages. *Mol. Biol. Evol.* **2002**, *19*, 908–917. [CrossRef] [PubMed]

Article

Comparative Chloroplast Genome Analyses of Species in *Gentiana* section *Cruciata* (Gentianaceae) and the Development of Authentication Markers

Tao Zhou [1], Jian Wang [1], Yun Jia [2], Wenli Li [1], Fusheng Xu [1] and Xumei Wang [1,*]

[1] School of Pharmacy, Xi'an Jiaotong University, Xi'an 710061, China; zhoutao196@mail.xjtu.edu.cn (T.Z.); wangjian6318@126.com (J.W.); lwl3659003@stu.xjtu.edu.cn (W.L.); xfs19940903@stu.xjtu.edu.cn (F.X.)
[2] Key Laboratory of Resource Biology and Biotechnology in Western China (Ministry of Education), School of Life Sciences, Northwest University, Xi'an 710069, China; jy878683@163.com
*   Correspondence: wangxumei@mail.xjtu.edu.cn; Tel.: +86-29-8265-5424

Received: 16 May 2018; Accepted: 3 July 2018; Published: 5 July 2018

Abstract: *Gentiana* section *Cruciata* is widely distributed across Eurasia at high altitudes, and some species in this section are used as traditional Chinese medicine. Accurate identification of these species is important for their utilization and conservation. Due to similar morphological and chemical characteristics, correct discrimination of these species still remains problematic. Here, we sequenced three complete chloroplast (cp) genomes (*G. dahurica*, *G. siphonantha* and *G. officinalis*). We further compared them with the previously published plastomes from sect. *Cruciata* and developed highly polymorphic molecular markers for species authentication. The eight cp genomes shared the highly conserved structure and contained 112 unique genes arranged in the same order, including 78 protein-coding genes, 30 tRNAs, and 4 rRNAs. We analyzed the repeats and nucleotide substitutions in these plastomes and detected several highly variable regions. We found that four genes (*accD*, *clpP*, *matK* and *ycf1*) were subject to positive selection, and sixteen InDel-variable loci with high discriminatory powers were selected as candidate barcodes. Our phylogenetic analyses based on plastomes further confirmed the monophyly of sect. *Cruciata* and primarily elucidated the phylogeny of Gentianales. This study indicated that cp genomes can provide more integrated information for better elucidating the phylogenetic pattern and improving discriminatory power during species authentication.

Keywords: *Gentiana* section *Cruciata*; chloroplast genome; molecular markers; species authentication

1. Introduction

Gentiana is the largest genus in the family Gentianaceae and widely distributed throughout the northern Hemisphere [1]. Approximately 362 species are recognized in genus *Gentiana* which have been divided into 15 sections [2]. Section *Cruciata* contains 21 species which are mainly distributed in eastern Eurasia [3]. Most species of this section are restricted to alpine regions, although some of them could be found at altitudes below 1000 m at higher latitudes [1]. Four species (*G. macrophylla*, *G. crassicaulis*, *G. straminea*, and *G. dahurica*) in sect. *Cruciata* are used as the original plants of traditional Chinese medicine named Qin-jiao [4]. The roots of these plants contain abundant secoiridoid active compounds which could be used for the treatment of diabetes, apoplexy, paralysis, and rheumatism [5–8].

Recently, the wild resources of some *Gentiana* species are dramatically declined due to overexploitation and some of them have been listed in the National Key Protected Wild Herbs in China [5,7]. However, the demand of natural sources for these plants remains high due to the high pharmacological and economical values. Therefore, many economically motivated adulterants of Qin-jiao products with similar morphological characters have been developed to substitute the genuine medicinal materials. Generally, the authentication of herbs was based on the morphological and histological inspection. But these methods may not be suitable for authenticating some species in sect. *Cruciata* due to the following reasons. Firstly, most species of sect. *Cruciata* shared the similar morphological characters especially in terms of leaf shape. Secondly, some species in this section are usually located in the sympatric distributions, thus intermediate morphology could be detected due to interspecific hybridization [9,10]. Thirdly, pharmacognostical studies showed that some species such as *G. siphonantha* and *G. straminea* usually shared similar chemical profiles [11]. Some other factors, such as growth conditions, developmental stage, and internal metabolism may affect the secondary metabolite accumulation in Qin-jiao and limit the application of such chemical analyses for authenticating the species of sect. *Cruciata*. In addition, chemical methods for identifying the medicinal plants are also expensive and not suitable for high-throughput analysis [12]. Therefore, reliable and cost-efficient methods are needed to authenticate the medical plants of sect. *Cruciata*.

Chloroplast (cp) genome of angiosperm is characterized by a typical quadripartite structure that contains a pair of inverted repeat (IR) regions separated by a large single-copy (LSC) and a small single-copy (SSC) region [13], and it is highly conserved compared to nuclear and mitochondrial genomes. Although chloroplast genomes are highly conserved, some hotspot regions with single nucleotide polymorphisms and insertion/deletions could be found and these regions may provide enough information for species identification [14,15]. Due to low recombination, uniparental inheritance, and low nucleotide substitution rates, many cp genetic markers have been used for plant phylogenetic, phylogeographic, and population genetic analyses [16]. It has been proven that some chloroplast sequences such as *trnH-psbA*, *rbcL*, and *matK* were commonly used as DNA barcodes for plants discrimination [17]. But in some cases, above commonly used DNA barcodes were not suitable to distinguish closely related plants due to limited variation loci [16,18]. Recently, it has been proposed that the complete cp genome could be used as a plant barcode, and various research have demonstrated that complete cp genome can greatly increase resolution for resolving difficult phylogenetic relationships at lower taxonomic levels [16,19–22]. In addition, using the cp genome as a genetic marker for identifying the plant will avoid the problems such as gene deletion and low Polymerase Chain Reaction (PCR) efficiency [23].

Most species in section *Cruciata* were recently diverged and originated from a common radiation in the Qinghai-Tibet Plateau (QTP) before the Pleistocene [1,10], therefore these species were usually closely related and showed parallel evolutionary relationships [1]. Previous research showed that commonly used DNA barcodes in some cases may not be suitable to identify the medicinal plant of this section [24,25]. Therefore, more specific barcodes with enough variation are needed to discriminate closely related species belong to sect. *Cruciata*. Nowadays, with the improvement of sequencing and assembly technologies, it is comparatively simple to obtain comprehensive chloroplast sequences for identifying *Gentiana* species. By utilizing the variable information provided from cp genomes, we can not only obtain more specific barcodes for species authentication in sect. *Cruciata*, but also shed light on the complex evolutionary relationships of the species in this section.

In the present study, we obtained the chloroplast genome sequences of *G. dahurica*, *G. siphonantha* and *G. officinalis* by using de novo assembly of whole-genome sequencing (WGS) data derived from high throughput sequencing technology. We also comparatively analyzed the chloroplast genomes of eight species in sect. *Cruciata* and developed credible cp genome derived InDel markers to authenticate these species. These markers are not only valuable tools for further evolutionary and population genetic studies on *Gentiana*, but also could be used as standardized barcodes for authenticating the original plants of Qin-jiao.

2. Results

2.1. Complete Chloroplast Genome Features of Sect. Cruciata

The chloroplast genomes of *G. dahurica*, *G. siphonantha*, and *G. officinalis* were sequenced with approximately 5.2, 5.8, and 5.6 Gb of paired-end reads, respectively. The raw reads with a sequence length of 125 bp were trimmed to generate the clean reads for the next assembly. After quality filtering, 10,114,902, 11,405,694, and 11,288,676 clean reads were recovered for *G. dahurica*, *G. siphonantha* and *G. officinalis*, respectively. Combined with the de novo and reference guided assembly, the cp genomes were obtained. The four junction regions between the IRs and SSC/LSC regions were confirmed by PCR amplification and Sanger sequencing. We mapped the obtained sequences to the new assembled genomes and no mismatch or InDel was observed. We compared the basic genome features of three newly sequenced cp genomes with five previously published cp genomes [26–28] and found that all the chloroplast genomes possessed the typical quadripartite structure with the length range from 148,765 to 149,916 bp (Table 1, Figure 1). The whole cp genome contained a pair of inverted repeat regions (IRs: 24,955–25,337 bp) which were separated by a small single copy region (SSC: 17,070–17,095 bp) and a large single copy region (LSC: 81,119–82,911 bp) (Table 1). Although genomic structure and size were highly conserved in eight cp genomes, the IR/SC boundary regions still varied slightly (Figure 2). All the eight chloroplast genomes contained 112 unique genes arranged in the same order, including 78 protein-coding genes, 30 tRNA genes, and 4 rRNA genes. Two genes (*rps16*, *infA*) were inferred to be pseudogenes (Figure S1). The overall guanine and cytosine (GC) content in each chloroplast genome is identically 37.7% (Table 1).

Table 1. Summary of complete chloroplast genomes for eight *Gentiana* species.

Name of Taxon	*G. dahurica*	*G. siphonantha*	*G. officinalis*	*G. straminea*
Genome length	148,803	148,908	148,879	148,991
LSC length	81,154	81,121	81,119	81,240
SSC length	17,093	17,113	17,088	17,085
IR length	25,278	25,337	25,336	25,333
Total gene number	112	112	112	112
No. of protein coding genes	78	78	78	78
No. of tRNA genes	30	30	30	30
No. of rRNA genes	4	4	4	4
GC content in genome (%)	37.7	37.7	37.7	37.7
Name of Taxon	***G. crassicaulis***	***G. robusta***	***G. tibetica***	***G. macrophylla***
Genome length	148,776	148,911	148,765	149,916
LSC length	81,164	81,164	81,163	82,911
SSC length	17,071	17,085	17,070	17,095
IR length	25,271	25,333	25,266	24,955
Total gene number	112	112	112	112
No. of protein coding genes	78	78	78	78
No. of tRNA genes	30	30	30	30
No. of rRNA genes	4	4	4	4
GC content in genome (%)	37.7	37.7	37.7	37.7

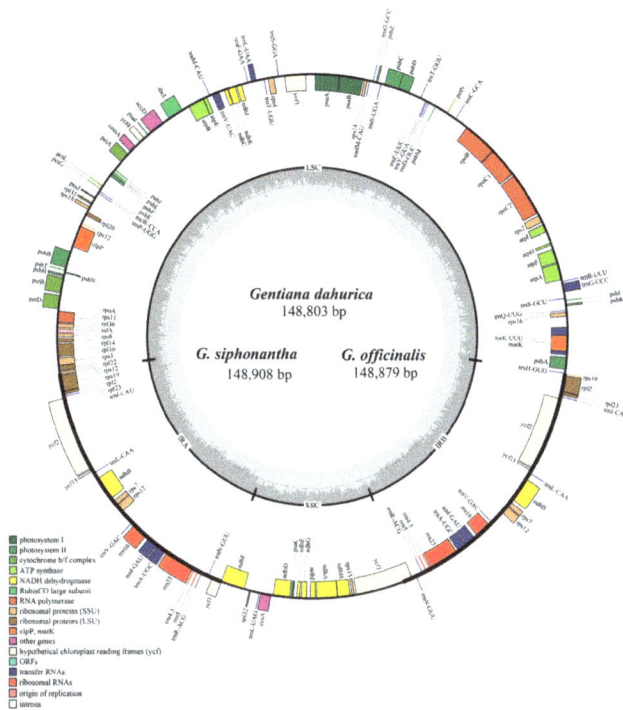

Figure 1. Merged gene map of the complete chloroplast genomes of three *Gentiana* species. Genes belonging to different functional groups are classified by different colors. The genes drawn outside of the circle are transcribed counterclockwise, while those inside are clockwise. Dashed area in the inner circle represent GC content of chloroplast genome.

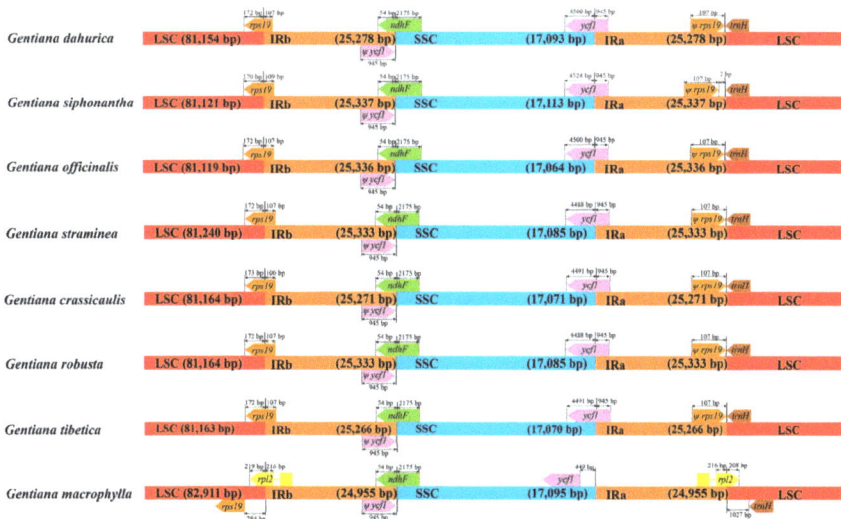

Figure 2. Comparison of chloroplast genome borders of LSC, SSC, and IRs among eight species in *Gentiana* sect. *Cruciata*. Ψ indicates a pseudogene.

2.2. Comparative Analyses of the Chloroplast Genomes of Species of Sect. Cruciata

Repeat analyses of three newly sequenced cp genomes showed 13/13/13 (*G. siphonantha/ G. officinalis/G. dahurica*) palindromic repeats, 12/11/11 dispersed repeats, and 7/6/6 tandem repeats (Figure 3A,B) with the repeat length range from 15 to 38 bp (Tables S1 and S2). The numbers and distribution of all repeat types were similar and conserved in these three cp genomes. Overall, 32/30/30 repeats were detected in three cp genomes. Similarly, 37, 34, 34, and 37 repeats were found in previously reported *G. crassicaulis*, *G. robusta*, *G. straminea*, and *G. tibetica* cp genomes (Figure 3A,B). Unexpectedly, 61 repeats, including 28 dispersed repeats, 18 palindromic repeats and 15 tandem repeats, were found in the cp genome of *G. macrophylla*. We found most of repeats in eight cp genomes were located in the intergenic or intron regions, and only a few repeats were distributed in protein-coding regions (*ycf1*, *ycf2*, and *psaA*) (Tables S1 and S2). Simple sequence repeats (SSRs) consisting of 1–6 bp repeat unit are distributed throughout the genome. In our study, perfect SSRs in eight *Gentiana* cp genomes were detected. The results showed that Mono-nucleotide repeats were most abundant type, followed by Tetra-nucleotides, Di-nucleotides and Tri-nucleotides. The penta- and hexa-nucleotides were very rare across the cp genomes (Figure 3C,D). Most SSRs are located in intergenic regions, but some were found in *rpoC2*, *rpoC1*, *atpB*, *ndhF*, and *ycf1* coding genes (Table S3). To investigate the evolutionary characteristics of cpDNA genes in eight *Gentiana* cp genomes and estimate selection pressures, nonsynonymous (dN), synonymous substitution rates (dS), and the ratio of dN/dS were calculated for 78 protein-coding genes (Table S4). We obtained 771 pairwise comparison results of dN/dS values and the remaining could not be calculated due to dS = 0. Only four genes (*accD*, *clpP*, *matK*, and *ycf1*) had dN/dS values ≥ 1 indicating that they had undergone positive selection.

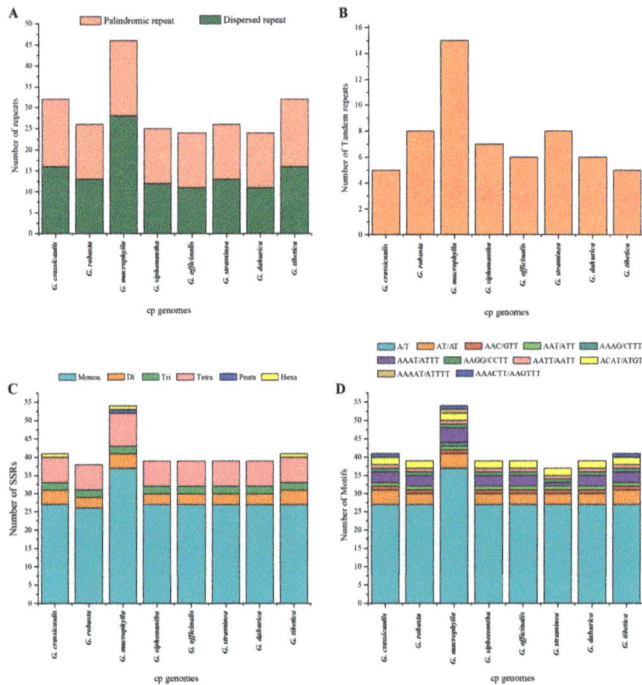

Figure 3. Analysis of different repeats in eight chloroplast genomes of *Gentiana* sect. *Cruciata*. (**A**) Histogram showing the number of palindromic repeats and dispersed repeats; (**B**) histogram showing the number of tandem repeats; (**C**) number of different simple sequence repeat (SSR) types detected in eight chloroplast genomes; (**D**) total numbers of different SSR motifs in eight chloroplast genomes.

To understand the level of sequence divergence, comparative analysis among eight *Gentiana* cp genomes was performed using mVISTA with the annotation of *G. crassicaulis* as a reference. The cp genomes within sect. *Cruciata* showed high sequence similarities with identities of only a few regions below 90%, indicating a high conservatism of these chloroplast genomes (Figure 4). The single-copy regions and intergenic regions were more divergent than the IR regions and genic regions (Figure 5). According to the comparative analyses, some hotspot regions for genome divergence that could be utilized as potential genetic markers to elucidate the phylogenies and to discriminate the species in sect. *Cruciata*. These regions were *psbA-trnH*, *trnK-rps16*, *rps16-trnQ*, *trnS-trnG*, *trnE-trnT*, *psbM-trnD*, *trnT-psbD*, *trnS-psbZ*, *ndhC-trnV*, *atpB-rbcL*, *rbcL-accD*, *accD-psbI*, *rpl33-rps18*, *trnR-trnA*, and *trnV-rps7* (Figure 4).

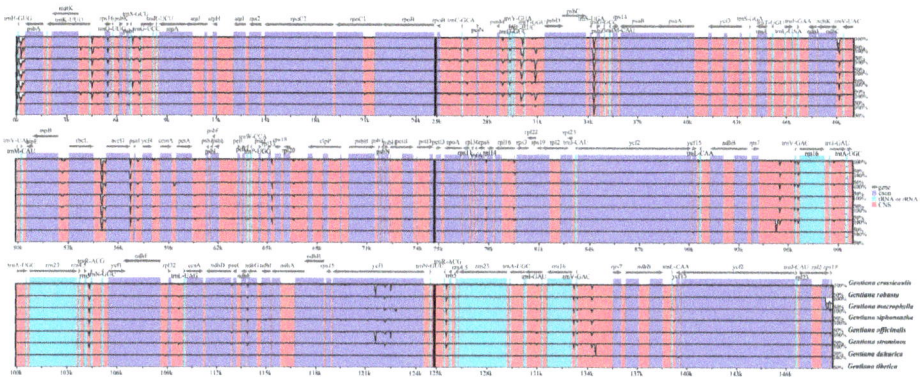

Figure 4. mVISTA percent identity plot comparing the eight chloroplast genomes of *Gentiana* sect. *Cruciata* with *G. crassicaulis* as a reference. The *y*-axis represents the percent identity within 50–100%. Genome regions are color-coded as protein coding (purple), rRNA, or tRNA coding genes (blue), and noncoding sequences (pink).

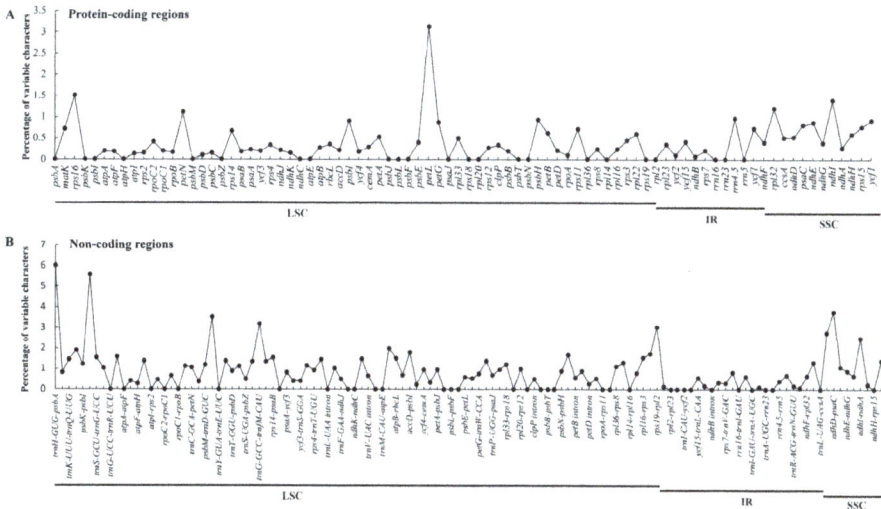

Figure 5. Percentage of variable characters in eight aligned chloroplast genomes of *Gentiana* sect. *Cruciata*. (**A**) Coding region; (**B**) Noncoding region.

2.3. Development of InDel Markers to Discriminate Species of Sect. Cruciata

Based on the alignment of complete cp genome sequences, the 16 most InDel-variable loci were selected as candidate DNA markers for authentication (Table S5). After PCR amplification, these 16 markers could successfully amplify the expected polymorphic band sizes (Figure 6). Some of these 16 markers had unique amplicon sizes specific to different *Gentiana* species (Figure 6). Especially five markers (QJcpm9, QJcpm12, QJcpm14, QJcpm15, and QJcpm16) were specific to *G. crassicaulis*, which all derived from long InDels in the intergenic regions including *rps16-trnQ*, *psbM-trnD*, *trnS-psbZ*, *accD-psbI*, and *trnK-rps16*. The marker QJcpm1 was specific to *G. robusta* and *G. crassicaulis* and was derived from a 54 and 64 bp InDel in the *ndhC-trnV* region. The QJcpm2 marker derived from 14 bp tandem repeat (TR) in *cemA-petA* region was specific to *G. siphonantha* and *G. crassicaulis*. QJcpm3 marker, which was specific to *G. officinalis* and *G. crassicaulis*, was derived from 72, 14 bp InDels, and 7 bp TR in *rbcL-accD* region. Three markers (QJcpm4, QJcpm10, and QJcpm11) were specific to *G. straminea*, *G. robusta*, and *G. crassicaulis*. QJcpm4 marker was derived from 12 bp InDels and 6 bp TR in the *rpl33-rps18* region; QJcpm10 marker was derived from 9 bp TR and 33 bp InDel in the *trnT-psbD*; QJcpm11 marker was derived from 18 bp InDel in *rrn5-trnA* region. The QJcpm5 marker, which was derived from 14, 4, and 7 bp TR in *atpB-rbcL*, was specific to *G. macrophylla*, *G. robusta*, and *G. crassicaulis*. Three markers QJcpm6, QJcpm8, and QJcpm13 were derived from a 42 bp InDel in *ycf1*, 9 bp InDel in *rps8-rpl14* region, and 24 bp TR in the *trnS-trnG* region, respectively, and were specific to *G. straminea* and *G. robusta*. The marker QJcpm7, which was specific *G. dahurica* and *G. siphonantha*, was also derived from 24 InDel in *ycf1* CDS region. Our validation results indicated all these markers can be used to identify species in sect. *Cruciata*.

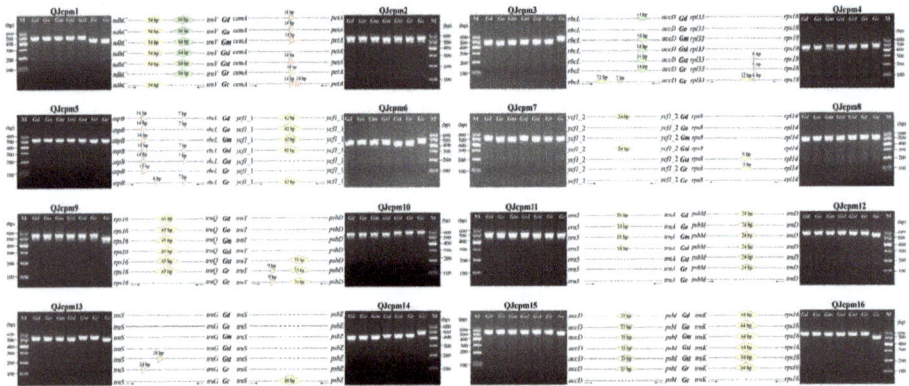

Figure 6. Validation of 16 molecular markers derived from InDel regions of eight chloroplast genomes of *Gentiana* sect. *Cruciata*. Inserted sequences and tandem repeats are designated by diamonds and triangle, respectively. Solid and dotted lines indicate conserved and deleted sequences, respectively. Left and right black arrows indicate forward and reverse primers, respectively. Abbreviated species names were shown on schematic diagrams: Gd, *G. dahurica*; Go, *G. officinalis*; Gm, *G. macrophylla*; Gsi, *G. siphonantha*; Gst, *G. straminea*; Gr, *G. robusta*; Gc, *G. crassicaulis*; M, D600 DNA ladder.

2.4. Phylogenetic Relationships of Species Belong to Sect. Cruciata

Here, 27 cp genomes were retrieved to infer the interspecific relationships of eight species in sect. *Cruciata* as well as to clarify the phylogenetic relationships of some Gentianales species (Table S6). Phylogenetic analyses were performed using Maximum parsimony (MP), Maximum likelihood (ML) and Bayesian inference (BI) methods, and *Arabidopsis thaliana* was set as outgroup. Three different datasets including complete cp genomes, 70 shared protein-coding genes (PCGs) and the most conserved regions (TMCRs) of cp genomes were used to construct the phylogenetic trees. The results showed the same phylogenetic signals for these three datasets and the phylogenetic trees inferred from MP/ML/BI methods also shared identical topologies (Figure 7, Figures S2 and S3). In these phylogenetic trees, we found all the species of sect. *Cruciata* formed a monophyletic clade a with high bootstrap and BI support values and clustered with another two Gentianaceae species (*G. lawrencei* and *Swertia mussotii*) in the same clade [29,30]. Of these species, *G. macrophylla*, *G. officinalis*, and *G. siphonantha* showed paraphyletic relationships with each other and formed a monophyletic clade with *G. dahurica*. *G. tibetica* and *G. crassicaulis* formed a monophyletic clade and located in the basal position of these eight species in sect. *Cruciata*. Interestingly, *G. robusta* and *G. straminea* with similar morphological characteristics were clustered in a monophyletic clade with a high resolution value. In addition, our phylogenetic results supported the monophyly of two families, including Apocynaceae and Rubiaceae, in the order Gentianales. Unexpectedly, *Gynochthodes nanlingensis* (*Morinda nanlingensis*) belongs to Rubiaceae was embed in the Apocynaceae species.

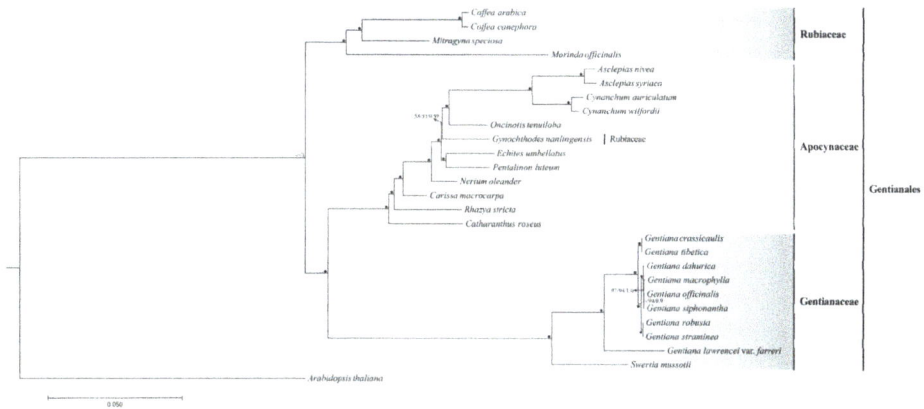

Figure 7. Phylogenetic relationships of species belong to *Gentiana* sect. *Cruciata* inferred from MP/ML/BI analysis based on complete chloroplast genome sequences. The numbers associated with each node are bootstrap support and posterior probability values, and the symbol ★ in the phylogenetic tree indicated that the support value of branch is 100/100/1.0.

3. Discussion

Three cp genomes of sect. *Cruciata* were sequenced using Illumina Hiseq platform, which provided more resources for evolutionary and genetics studies of *Gentiana*. The cp genomic information presented in this study will also contribute to the conservation and management of wild resources of sect. *Cruciata*. Although a recent research reported that 11 ndh genes had been lost in the cp genomes of *Gentiana* sect. *Kudoa* [31], eight cp genomes of sect. *Cruciata* analyzed in present study are rather conserved in gene structures, contents and arrangement, and no significant structural rearrangements, such as inversions or gene relocations, were detected. Of these eight species, *G. macrophylla* has the largest cp genome size and other species showed minor differences in genome size. The length variations of these cp genomes may result from the length of intergenic regions, similar result has been reported for *Paris* (Melanthiaceae) cp genomes [18].

All the eight cp genomes of sect. *Cruciata* had the same protein-coding genes, tRNA and rRNA genes. We found that exon2 of *rps16* gene was lost in three newly sequenced cp genomes, and *rps16* in other cp genomes also showed same structure. Therefore, *rps16* pseudogene may commonly exist in the genus *Gentiana* [26]. And *infA* gene, which contains internal stop codons, was also inferred as pseudogene in these species. This pseudogene had been reported in many species [32–35]. Except for cp genome of *G. macrophylla*, the remained cp genomes showed minor variations in the junctions between the SSC and IRs regions. As most species of sect. *Cruciata* derived from a common radiation and usually showed closely interspecific relationships, we thus speculated that highly conserved nature of cp genomes resulted in the similar gene distributions at SC/IR boundaries.

Repeat structure plays an important role in genomic rearrangement, recombination, and sequence divergence in plastomes [36–38]. In the present study, cp genome of *G. macrophylla* has the largest number of repeats, while the number of repeats was similar in other cp genomes. Most of the repeated regions in different species showed similar characteristics and most repeats were located in intergenic regions or in *ycf1/pasA*. Repeats in these genes are commonly observed in other angiosperm lineages [22,32,39]. Cp microsatellites (cpSSR) usually showed high polymorphism within the same species and which are potentially useful markers for population genetics [40]. Here, 326 SSRs varying in number and type between eight major *Gentiana* species, and the most abundant repeat type was found to be stretches of mononucleotides (A/T). Similar to the distribution status of dispersed and tandem repeats, most cpSSRs were observed in noncoding regions, and only small proportion were found in coding regions. CpSSRs located in noncoding regions of the cp genome are generally short mononucleotide tandem repeats and commonly showed intraspecific variation in repeat number [15]. Therefore, cpSSRs derived from eight *Gentiana* species in this study are expected to be useful for the genetic diversity studies in *Gentiana*. As the wild resources of some *Gentiana* species were dramatically declined due to overexploitation, we thought these species need to transplant or cultivate in order to preserve their germplasm resources. We believe the obtained SSRs among these chloroplast genomes will also be useful for the domestication and breeding of *Gentiana* species.

Sequence divergence of the coding genes was observed between different species. Our analyses indicated that all of cp genes showed a low sequence divergence (dS < 0.1) and most cp genes were under purifying selection (dN/dS < 1); similar results were reported for other cp genomes [32,41,42]. Only four genes (*accD*, *clpP*, *matK*, and *ycf1*) were under positive selection. Previous research reported that *accD* and *clpP* genes had a high evolution rate in *Fagopyrum* species [43,44], we thus presumed that these genes may have a high evolution rate in *Gentiana* species. One other gene (*matK*) was highly divergent in Caryophyllaceae, and comparative cp genomes analyses of Myrtaceae also indicated *matK* was under positive pressure [45,46]. The *ycf1* gene with unknown functions showed a biased higher value for dN/dS ratio compared to dS value indicating that this gene evolved at a faster rate. It has also been shown to be subject to positive selection in many angiosperms [20,22,32,44,45].

DNA barcodes are defined as the short DNA sequences with a sufficiently high mutation rate to discriminate a species within a given taxonomic group and are confirmed as reliable tools for the identification of plant species [16,47]. Previously, *rbcL*, *trnH-psbA*, and *matK* were considered as "core" plant barcodes for species identification, but they often have limited resolutions at species level [18]. Previous research showed that three commonly used barcodes in some cases may not be suitable to authenticate the medicinal plant in section *Cruciata* [24,25]. Therefore, seeking for more effective DNA barcodes with high evolutionary rates is very important for the molecular identification of species in *Gentiana* sect. *Cruciata*. The complete cp genome has a conserved sequence from 110k to 160k bp, which far exceeds the length of commonly used molecular markers and provides more variation to distinguish closely related species [12,16]. Therefore, some mutation hotspot regions, including *trnK-rps16*, *rps16-trnQ*, *trnS-trnG*, *trnE-trnT*, *trnT-psbD*, *trnS-psbZ*, *ndhC-trnV*, *rbcL-accD*, *accD-psbI*, *trnR-trnA*, *trnV-rps7*, and *ycf1*, detected from the cp genomes can provide more specific DNA barcodes for the authentication of medicinal materials of sect. *Cruciata* and also provide sufficient genetic markers for resolving the phylogeny of Gentianaceae.

We developed the specific markers for species authentication of sect. *Cruciata* based on the hotspot regions derived from cp genomes. Most of these markers were derived from the intergenic regions of cp genomes and showed high interspecific polymorphism. Previous molecular identification of *Panax*, *Zanthoxylum*, and *Eclipta* species also indicated that chloroplast-derived genetic markers had high discriminatory powers [12,14,48]. Therefore, specific markers developed from the comparative cp genomes were superior than the commonly used markers for identifying the closely related species. Especially for medicinal plants, these specific genetic markers are more effective in the authentication of their source plants. We found two InDels (42 and 24 bp) in the *ycf1* gene, which can be used to distinguish species in sect. *Cruciata*. *Ycf1*, which encodes a protein of approximately 1800 amino acids with unknown function, is the second largest gene in the cp genome. Because the sequence of *ycf1* is too long and too variable for designing universal primers, it has received little attention for DNA barcodes at low taxonomy [18,49]. But two markers derived from *ycf1* gene showed high PCR efficiency and polymorphism in species of sect. *Cruciata*, and could be used as specific barcodes for the authentication of *Gentiana* species. Although our study provided 16 genetic markers which had enough interspecies polymorphism for species identification, some of the markers were usually specific to two species. We thus suggest a combination of several markers should be considered for credible authentication between different species in genus *Gentiana*.

We inferred the phylogenetic relationships of sect. *Cruciata* using complete cp genomes. Three different methods (MP/ML/BI) were used to rebuilt the phylogenetic trees based on different datasets (cp genomes, 70 shared PCGs, and TMCRs), and the derived phylogenetic trees shared identical topology. All the species of sect. *Cruciata* formed a monophyletic clade with high bootstrap and BI support values. This result is comparable with the previous phylogenetic research based on four cpDNA fragments [1]. Four species, including *G. dahurica*, *G. macrophylla*, *G. siphonantha*, and *G. officinalis*, were clustered in the same clade with high support values. Although the flower color of *G. officinalis* was different from other three species, it shared similar morphological and chemical characters with *G. macrophylla* [50]. We found that *G. straminea* was closely related to *G. robusta*. *G. robusta* may have originated from introgression between *G. straminea* and another relative species, and these two species are usually closer to each other [26,51]. Two species, *G. tibetica* and *G. crassicaulis* were clustered in the same clade and located in the basal position in the clade of sect. *Cruciata*. However, a previous phylogenetic result indicated that *G. tibetica* was closely related to *G. straminea* and *G. robusta* [1]. As *G. tibetica* and *G. crassicaulis* distributed sympatrically in Tibet and intermediate types were produced by introgression between these two species [52], we thus inferred these two species should be closely related. In addition, based on the phylogenetic results, we found that the family Gentianaceae was closer to family Apocynaceae than to family Rubiaceae in order Gentianales. Previous phylogenetic studies of order Gentianales resulted in similar findings, but with relatively low support values [53,54]. Although our result confirmed the monophyly of section *Cruciata* and primarily elucidated the phylogeny of Gentianales based on available cp genomes, more complete cp genome sequences are needed to resolve the comprehensive phylogenies of this section, especially since limited taxon sampling may produce discrepancies in tree topologies [15,55].

4. Materials and Methods

4.1. Plant Materials and DNA Isolation

Samples of *G. dahurica*, *G. siphonantha* and *G. officinalis* were collected from Tianzhu (102.54° E, 37.01° N), Sunan (98.05° E, 39.55° N) and Yuzhong (104.05° E, 35.78° N) Counties in Gansu Province, China. Young leaves of three species were collected and immediately dried with silica gel for further DNA isolation. Total genomic DNA was isolated from each sample using the modified Cetyl Trimethyl Ammonium Bromide (CTAB) method [56]. The quantity and quality of extracted genomic DNA was determined by gel electrophoresis and NanoDrop 2000 Spectrophotometer (Thermo Scientific, Carlsbad, CA, USA).

4.2. Chloroplast Genome Sequencing, Assembly and Annotation

The DNA Library with insert size of 200 bp was prepared according to the description by Zhou et al. [32], and sequenced using Illumina Hiseq™ 2500 platform (Illumina Inc., San Diego, CA, USA) with the average read length of 125 bp. The obtained raw reads were filtered with the NGS QC Toolkit_v2.3.3 (National Institute of Plant Genome Research, New Delhi, India) [57]. Adapter sequences and low-quality reads with Q-value $\leq$ 20 were removed. Filtered paired-end reads were firstly mapped to the chloroplast genome of *Gentiana straminea* (KJ657732) by using the Bowtie 2-2.2.6 (University of Maryland,

College Park, MD, USA.) with default parameter [58]. And then the matched paired-end reads were de novo assembled using SPAdes-3.6.0 (St. Petersburg Academic University, St. Petersburg, Russia) [59]. After de novo assembly, the resultant scaffolds were further assembled using a baiting and iteration method based on Perl script MITObim_1.9.pl (University of Oslo, Oslo, Norway) [60]. Finally, all obtained reads were mapped to the spliced cp genome sequence using Geneious 10.1 (Biomatters Ltd., Auckland, New Zealand) in order to avoid assembly errors. The four junction regions between the IRs and SSC/LSC were confirmed by PCR amplification and Sanger sequencing (Primers and sequencing results are listed in Table S7). The cp genome genes were annotated with the online program Organellar Genome Annotator (DOGMA) [61], and the primary annotated results were manually verified according to the annotation information from other closely related species. The circular plastid genome maps were drawn using the online program OrganellarGenome DRAW (Max planck Institute of Molecular Plant Physiology, Potsdam, Germany) [62] and three newly sequenced cp genome were deposited in GenBank (MH261259–MH261261).

4.3. Repeat Structure, Genome Comparison and Sequence Divergence

Dispersed and palindromic repeats within the cp genomes were identified using REPuter (University of Bielefeld, Bielefeld, Germany) with a minimum repeat size of 30 bp and a sequence identity > 90% [63]. Tandem repeat sequences were searched using the Tandem Repeats Finder program (Mount Sinai School of Medicine, New York, NY, USA) with the following parameters: 2 for alignment parameters match, 7 for mismatch and InDel, respectively [64]. Simple sequence repeats (SSRs) were predicted using MISA perl script (Institute of Plant Genetics and Crop Plant Research, Gatersleben, Germany) with the parameters of ten for mono, five for di-, four for tri-, and three for tetra-, penta, and hexa-nucleotide motifs [65]. The nonsynonymous (dN), synonymous (dS), and dN/dS values of each protein coding gene were calculated using PAML packages 4.0 (University College London, London, UK) with Yang and Nielsen (YN) algorithm to detect whether selective pressure exists for plastid genes [66]. The cp genome gene distribution of eight *Gentiana* species was compared and visualized using mVISTA software with the annotation of *G. crassicaulis* as a reference [67]. To examine mutation hotspot regions of the cp genomes of eight *Gentiana* species, the percentages of variable characters for each coding and noncoding regions were analyzed using the method described by Zhang et al. [68].

4.4. Development and Validation of the InDel Molecular Marker

In order to validate interspecies polymorphisms within the chloroplast genomes and develop DNA genetic markers for identifying species belong to sect. *Cruciata*, specific primers were designed using Primer 3 based on the mutational hotspot regions found in these *Gentiana* chloroplast genomes [69]. PCR amplifications were performed in a reaction volume of 25 μL with 12.5 μL 2× Taq PCR Master Mix, 0.4 μM of each primer, 2 μL template DNA and 10.1 μL ddH$_2$O. All amplifications were carried out in SimpliAmp™ Thermal Cycler (Applied Biosystems, Carlsbad, CA, USA) as follow: denaturation at 94 °C for 5 min, followed by 30 cycles of 94 °C for 50 s, at specific annealing temperature (Tm) for 40 s, 72 °C for 90 s and 72 °C for 7 min as final extension. PCR products were visualized on 2% agarose gels after staining with ethidium bromide and then the DNA fragments were sequenced by Sangon Biotech (Shanghai, China) (Sequencing results are listed in Table S8).

4.5. Phylogenetic Analysis

The complete chloroplast genomes of 26 Gentianales species were recovered to clarify the phylogenetic relationships of sect. *Cruciata* and the cp genome of *Arabidopsis thaliana* was set as outgroup. In order to obtain a reliable result, phylogenetic analyses were implemented based on different cp genome datasets. On the one hand, whole cp genome sequences and 70 common cp protein-coding genes (PCGs) were separately used to infer the phylogenetic relationships of these species. On the other hand, multi-gene alignment matrix, which contained the most conserved regions (TMCRs) of cp genome was generated using HomBlocks (Ocean University of China, Qingdao, China) [70], was used to understand the phylogenetic relationships at cp genome level. Alignments were constructed using

MAFFT v7.308 (Osaka University, Suita, Japan) with default parameters and the best-fit nucleotide substitution model (General Time Reversible + Invariant + Gamma, GTR + I + G) was determined with Modeltest 3.7 (Brigham Young University, Provo, UT, USA) [71,72]. Maximum parsimony (MP) analyses of the resulting alignments from different datasets were performed using PAUP 4.0b10 (Smithsonian Institution, Washington, DC, USA) [73]. Maximum likelihood (ML) analyses were performed using RAxML 8.1.24 (Heidelberg Institute for Theoretical Studies, Heidelberg, Germany) with GTR + I + G nucleotide substitution model [74]. The reliability of each tree node was tested by bootstrap analysis with 1000 replicates. Bayesian analyses were also conducted with MrBayes v3.2.6 (Swedish Museum of Natural History, Stockholm, Sweden) [75] under the same substitution model (GTR + I + G). The Markov chain Monte Carlo (MCMC) algorithm was run for one million generations, with one tree sampled every 100 generations. The first 25% of trees were discarded as burn-in to construct majority-rule consensus tree and estimate posterior probabilities (PP) for each node.

Supplementary Materials: Supplementary materials can be found at http://www.mdpi.com/1422-0067/19/7/1962/s1.

Author Contributions: T.Z. and X.W. conceived and designed the work; T.Z. and J.W. collected samples; T.Z., J.W., Y.J., W.L., and F.X. performed the experiments and analyzed the data; T.Z. wrote the manuscript; X.W. revised the manuscript. All authors gave final approval of the paper.

Acknowledgments: This work was financially co-supported by the National Natural Science Foundation of China (31770364) and Scientific Research Supporting Project for New Teacher of Xi'an Jiaotong University (YX1K105).

Conflicts of Interest: The authors declare no conflict of interest.

References

1. Zhang, X.L.; Wang, Y.J.; Ge, X.J.; Yuan, Y.M.; Yang, H.L.; Liu, J.Q. Molecular phylogeny and biogeography of *Gentiana* sect. *Cruciata* (Gentianaceae) based on four chloroplast DNA datasets. *Taxon* **2009**, *58*, 862–870.
2. Ho, T.N.; Liu, S.W. *A Worldwide Monograph of Gentiana*; Science Press: Beijing, China, 2001.
3. Ho, T.N.; Pringle, S.J. *"Gentianaceae," Flora of China*; Science Press: Beijing, China, 1995; Volume 16, pp. 1–140.
4. State Pharmacopoeia Commission of the PRC. *Pharmacopoeia of P.R. China, Part 1*; Chemical Industry Publishing House: Beijing, China, 2015; pp. 270–271.
5. Hua, W.; Zheng, P.; He, Y.; Cui, L.; Kong, W.; Wang, Z. An insight into the genes involved in secoiridoid biosynthesis in *Gentiana macrophylla* by RNA-seq. *Mol. Biol. Rep.* **2014**, *41*, 4817–4825. [CrossRef] [PubMed]
6. Chang-Liao, W.-L.; Chien, C.-F.; Lin, L.-C.; Tsai, T.-H. Isolation of gentiopicroside from *Gentianae* Radix and its pharmacokinetics on liver ischemia/reperfusion rats. *J. Ethnopharmacol.* **2012**, *141*, 668–673. [CrossRef] [PubMed]
7. Yin, H.; Zhao, Q.; Sun, F.-M.; An, T. Gentiopicrin-producing endophytic fungus isolated from *Gentiana macrophylla*. *Phytomedicine* **2009**, *16*, 793–797. [CrossRef] [PubMed]
8. Yu, F.; Yu, F.; Li, R.; Wang, R. Inhibitory effects of the *Gentiana macrophylla* (Gentianaceae) extract on rheumatoid arthritis of rats. *J. Ethnopharmacol.* **2004**, *95*, 77–81. [CrossRef] [PubMed]
9. Li, X.; Wang, L.; Yang, H.; Liu, J. Confirmation of natural hybrids between *Gentiana straminea* and *G. siphonantha* (Gentianaceae) based on molecular evidence. *Front. Biol. China* **2008**, *3*, 470–476. [CrossRef]
10. Hu, Q.; Peng, H.; Bi, H.; Lu, Z.; Wan, D.; Wang, Q.; Mao, K. Genetic homogenization of the nuclear ITS loci across two morphologically distinct gentians in their overlapping distributions in the Qinghai-Tibet Plateau. *Sci. Rep.* **2016**, *6*, 34244. [CrossRef] [PubMed]
11. Zhao, Z.; Su, J.; Wang, Z. Pharmacognostical studies on root of *Gentiana siphonantha*. *Chin. Tradit. Herbal Drugs* **2006**, *37*, 1875–1878.
12. Nguyen, V.B.; Park, H.-S.; Lee, S.-C.; Lee, J.; Park, J.Y.; Yang, T.-J. Authentication markers for five major *Panax* species developed via comparative analysis of complete chloroplast genome sequences. *J. Agric. Food Chem.* **2017**, *65*, 6298–6306. [CrossRef] [PubMed]
13. Bendich, A.J. Circular chloroplast chromosomes: The grand illusion. *Plant Cell* **2004**, *16*, 1661–1666. [CrossRef] [PubMed]
14. Lee, H.J.; Koo, H.J.; Lee, J.; Lee, S.-C.; Lee, D.Y.; Giang, V.N.L.; Kim, M.; Shim, H.; Park, J.Y.; Yoo, K.-O.; et al. Authentication of *Zanthoxylum* species based on integrated analysis of complete chloroplast genome sequences and metabolite profiles. *J. Agric. Food Chem.* **2017**, *65*, 10350–10359. [CrossRef] [PubMed]

15. Eguiluz, M.; Rodrigues, N.F.; Guzman, F.; Yuyama, P.; Margis, R. The chloroplast genome sequence from *Eugenia uniflora*, a Myrtaceae from Neotropics. *Plant Syst. Evol.* **2017**, *303*, 1199–1212. [CrossRef]

16. Li, X.; Yang, Y.; Henry, R.J.; Rossetto, M.; Wang, Y.; Chen, S. Plant DNA barcoding: From gene to genome. *Biol. Rev.* **2015**, *90*, 157–166. [CrossRef] [PubMed]

17. Hollingsworth, P.M.; Forrest, L.L.; Spouge, J.L.; Hajibabaei, M.; Ratnasingham, S.; van der Bank, M.; Chase, M.W.; Cowan, R.S.; Erickson, D.L.; Fazekas, A.J.; et al. A DNA barcode for land plants. *Proc. Natl. Acad. Sci. USA* **2009**, *106*, 12794–12797.

18. Song, Y.; Wang, S.; Ding, Y.; Xu, J.; Li, M.F.; Zhu, S.; Chen, N. Chloroplast genomic resource of *Paris* for species discrimination. *Sci. Rep.* **2017**, *7*, 3427. [CrossRef] [PubMed]

19. Ma, P.-F.; Zhang, Y.-X.; Zeng, C.-X.; Guo, Z.-H.; Li, D.-Z. Chloroplast phylogenomic analyses resolve deep-level relationships of an intractable bamboo tribe Arundinarieae (Poaceae). *Systematic Biol.* **2014**, *63*, 933–950. [CrossRef] [PubMed]

20. Carbonell-Caballero, J.; Alonso, R.; Ibañez, V.; Terol, J.; Talon, M.; Dopazo, J. A phylogenetic analysis of 34 chloroplast genomes elucidates the relationships between wild and domestic species within the genus *Citrus*. *Mol. Biol. Evol.* **2015**, *32*, 2015–2035. [CrossRef] [PubMed]

21. Dong, W.; Xu, C.; Li, W.; Xie, X.; Lu, Y.; Liu, Y.; Jin, X.; Suo, Z. Phylogenetic resolution in *Juglans* based on complete chloroplast genomes and nuclear DNA sequences. *Front. Plant Sci.* **2017**, *8*, 1148. [CrossRef] [PubMed]

22. Yang, Y.; Zhou, T.; Duan, D.; Yang, J.; Feng, L.; Zhao, G. Comparative analysis of the complete chloroplast genomes of five *Quercus* species. *Front. Plant Sci.* **2016**, *7*, 959. [CrossRef] [PubMed]

23. Huang, C.-Y.; Gruenheit, N.; Ahmadinejad, N.; Timmis, J.; Martin, W. Mutational decay and age of chloroplast and mitochondrial genomes transferred recently to angiosperm nuclear chromosomes. *Plant Physiol.* **2005**, *138*, 1723–1733. [CrossRef] [PubMed]

24. Liu, J.; Yan, H.-F.; Ge, X.-J. The use of DNA barcoding on recently diverged species in the genus *Gentiana* (Gentianaceae) in China. *PLoS ONE* **2016**, *11*, e0153008. [CrossRef] [PubMed]

25. Zhang, D.; Gao, Q.; Li, F.; Li, Y. DNA molecular identification of botanical origin in Chinese herb Qingjiao. *J. Anhui Agric. Sci.* **2011**, *39*, 14609–14612.

26. Ni, L.; Zhao, Z.; Xu, H.; Chen, S.; Dorje, G. Chloroplast genome structures in *Gentiana* (Gentianaceae), based on three medicinal alpine plants used in Tibetan herbal medicine. *Curr. Genet.* **2017**, *63*, 241–252. [CrossRef] [PubMed]

27. Ni, L.; Zhao, Z.; Xu, H.; Chen, S.; Dorje, G. The complete chloroplast genome of *Gentiana straminea* (Gentianaceae), an endemic species to the Sino-Himalayan subregion. *Gene* **2016**, *577*, 281–288. [CrossRef] [PubMed]

28. Wang, X.; Yang, N.; Su, J.; Zhang, H.; Cao, X. The complete chloroplast genome of *Gentiana macrophylla*. *Mitochondrial DNA B* **2017**, *2*, 395–396. [CrossRef]

29. Xiang, B.; Li, X.; Qian, J.; Wang, L.; Ma, L.; Tian, X.; Wang, Y. The complete chloroplast genome sequence of the medicinal plant *Swertia mussotii* using the PacBio RS II platform. *Molecules* **2016**, *21*, 1029. [CrossRef] [PubMed]

30. Fu, P.-C.; Zhang, Y.-Z.; Geng, H.-M.; Chen, S.-L. The complete chloroplast genome sequence of *Gentiana lawrencei* var. *farreri* (Gentianaceae) and comparative analysis with its congeneric species. *PeerJ* **2016**, *4*, e2540. [CrossRef] [PubMed]

31. Sun, S.-S.; Fu, P.-C.; Zhou, X.-J.; Cheng, Y.-W.; Zhang, F.-Q.; Chen, S.-L.; Gao, Q.-B. The complete plastome sequences of seven species in *Gentiana* sect. *Kudoa* (Gentianaceae): Insights into plastid gene loss and molecular evolution. *Front. Plant Sci.* **2018**, *9*, 493. [CrossRef] [PubMed]

32. Zhou, T.; Chen, C.; Wei, Y.; Chang, Y.; Bai, G.; Li, Z.; Kanwal, N.; Zhao, G. Comparative transcriptome and chloroplast genome analyses of two related *Dipteronia* Species. *Front. Plant Sci.* **2016**, *7*, 1512. [CrossRef] [PubMed]

33. Yang, J.-B.; Li, D.-Z.; Li, H.-T. Highly effective sequencing whole chloroplast genomes of angiosperms by nine novel universal primer pairs. *Mol. Ecol. Resour.* **2014**, *14*, 1024–1031. [CrossRef] [PubMed]

34. Hu, Y.; Woeste, K.E.; Zhao, P. Completion of the chloroplast genomes of five Chinese *Juglans* and their contribution to chloroplast phylogeny. *Front. Plant Sci.* **2016**, *7*, 1955. [CrossRef] [PubMed]

35. Sun, Y.; Moore, M.J.; Zhang, S.; Soltis, P.S.; Soltis, D.E.; Zhao, T.; Meng, A.; Li, X.; Li, J.; Wang, H. Phylogenomic and structural analyses of 18 complete plastomes across nearly all families of early-diverging eudicots, including an angiosperm-wide analysis of IR gene content evolution. *Mol. Phylogenet. Evol.* **2016**, *96*, 93–101. [CrossRef] [PubMed]

36. Weng, M.-L.; Blazier, J.C.; Govindu, M.; Jansen, R.K. Reconstruction of the ancestral plastid genome in Geraniaceae reveals a correlation between genome rearrangements, repeats and nucleotide substitution rates. *Mol. Biol. Evol.* **2013**, *31*, 645–659. [CrossRef] [PubMed]

37. Lu, L.; Li, X.; Hao, Z.; Yang, L.; Zhang, J.; Peng, Y.; Xu, H.; Lu, Y.; Zhang, J.; Shi, J.; et al. Phylogenetic studies and comparative chloroplast genome analyses elucidate the basal position of halophyte *Nitraria sibirica* (Nitrariaceae) in the Sapindales. *Mitochondrial DNA A* **2017**, 1–11. [CrossRef] [PubMed]

38. Asano, T.; Tsudzuki, T.; Takahashi, S.; Shimada, H.; Kadowaki, K. Complete nucleotide sequence of the sugarcane (*Saccharum officinarum*) chloroplast genome: A comparative analysis of four monocot chloroplast genomes. *DNA Res.* **2004**, *11*, 93–99. [CrossRef] [PubMed]

39. Curci, P.L.; De Paola, D.; Danzi, D.; Vendramin, G.G.; Sonnante, G. Complete chloroplast genome of the multifunctional crop globe artichoke and comparison with other asteraceae. *PLoS ONE* **2015**, *10*, e0120589. [CrossRef] [PubMed]

40. Provan, J.; Powell, W.; Hollingsworth, P.M. Chloroplast microsatellites: New tools for studies in plant ecology and evolution. *Trends Ecol. Evol.* **2001**, *16*, 142–147. [CrossRef]

41. Rousseau-Gueutin, M.; Bellot, S.; Martin, G.E.; Boutte, J.; Chelaifa, H.; Lima, O.; Michon-Coudouel, S.; Naquin, D.; Salmon, A.; Ainouche, K. The chloroplast genome of the hexaploid *Spartina maritima* (Poaceae, Chloridoideae): Comparative analyses and molecular dating. *Mol. Phylogenet. Evol.* **2015**, *93*, 5–16. [CrossRef] [PubMed]

42. Xu, J.-H.; Liu, Q.; Hu, W.; Wang, T.; Xue, Q.; Messing, J. Dynamics of chloroplast genomes in green plants. *Genomics* **2015**, *106*, 221–231. [CrossRef] [PubMed]

43. Yamane, K.; Yasui, Y.; Ohnishi, O. Intraspecific cpDNA variations of diploid and tetraploid perennial buckwheat, *Fagopyrum cymosum* (Polygonaceae). *Am. J. Bot.* **2003**, *90*, 339–346. [CrossRef] [PubMed]

44. Cho, K.-S.; Yun, B.-K.; Yoon, Y.-H.; Hong, S.-Y.; Mekapogu, M.; Kim, K.-H.; Yang, T.-J. Complete chloroplast genome sequence of tartary buckwheat (*Fagopyrum tataricum*) and comparative analysis with common buckwheat (*F. esculentum*). *PLoS ONE* **2015**, *10*, e0125332. [CrossRef] [PubMed]

45. Machado, L.D.O.; Vieira, L.D.N.; Stefenon, V.M.; Oliveira Pedrosa, F.D.; Souza, E.M.D.; Guerra, M.P.; Nodari, R.O. Phylogenomic relationship of feijoa (*Acca sellowiana* (O.Berg) Burret) with other Myrtaceae based on complete chloroplast genome sequences. *Genetica* **2017**, *145*, 163–174. [CrossRef] [PubMed]

46. Cuenoud, P.; Savolainen, V.; Chatrou, L.W.; Powell, M.; Grayer, R.J.; Chase, M.W. Molecular phylogenetics of Caryophyllales based on nuclear 18S rDNA and plastid *rbcL*, *atpB*, and *matK* DNA sequences. *Am. J. Bot.* **2002**, *89*, 132–144. [CrossRef] [PubMed]

47. Techen, N.; Parveen, I.; Pan, Z.; Khan, I.A. DNA barcoding of medicinal plant material for identification. *Curr. Opin. Biotech.* **2014**, *25*, 103–110. [CrossRef] [PubMed]

48. Kim, I.; Young Park, J.; Sun Lee, Y.; Lee, H.; Park, H.-S.; Jayakodi, M.; Waminal, N.; Hwa Kang, J.; Joo Lee, T.; Sung, S.; et al. Discrimination and authentication of *Eclipta prostrata* and *E. alba* based on the complete chloroplast genomes. *Plant Breed. Biotech.* **2017**, *5*, 334–343. [CrossRef]

49. Dong, W.; Xu, C.; Li, C.; Sun, J.; Zuo, Y.; Shi, S.; Cheng, T.; Guo, J.; Zhou, S. *ycf1*, the most promising plastid DNA barcode of land plants. *Sci. Rep.* **2015**, *5*, 8348. [CrossRef] [PubMed]

50. Liu, L.; Wu, D.; Zhang, X. Pharmacognostical studies on root of *Gentiana officinalis*. *J. Chin. Med. Mater.* **2008**, *31*, 1635–1638.

51. Xiong, B.; Zhao, Z.; Ni, L.; Gaawe, D.; Mi, M. DNA-based identification of *Gentiana robusta* and related species. *Chin. Med. Mater.* **2015**, *40*, 4680–4685.

52. Zhang, X.; Ge, X.; Liu, J.; Yuan, Y. Morphological, karyological and molecular delimitation of two gentians: *Gentiana crassicaulis* versus *G. tibetica* (Gentianaceae). *Acta Phytotaxon. Sin.* **2006**, *44*, 627–640. [CrossRef]

53. Maria, B.; Bengt, O.; Birgitta, B. Phylogenetic relationships within the Gentianales based on *ndhF* and *rbcL* sequences, with particular reference to the Loganiaceae. *Am. J. Bot.* **2000**, *87*, 1029–1043.

54. Yang, L.L.; Li, H.L.; Wei, L.; Yang, T.; Kuang, D.Y.; Li, M.H.; Liao, Y.Y.; Chen, Z.D.; Wu, H.; Zhang, S.Z. A supermatrix approach provides a comprehensive genus-level phylogeny for Gentianales. *J. Syst. Evol.* **2016**, *54*, 400–415. [CrossRef]

55. Leebens-Mack, J.; Raubeson, L.A.; Cui, L.; Kuehl, J.V.; Fourcade, M.H.; Chumley, T.W.; Boore, J.L.; Jansen, R.K.; Depamphilis, C.W. Identifying the basal angiosperm node in chloroplast genome phylogenies: Sampling one's way out of the Felsenstein zone. *Mol. Biol. Evol.* **2005**, *22*, 1948–1963. [CrossRef] [PubMed]

56. Doyle, J.J. A rapid DNA isolation procedure for small quantities of fresh leaf tissue. *Phytochem. Bull.* **1987**, *19*, 11–15.

57. Patel, R.K.; Jain, M. NGS QC Toolkit: A toolkit for quality control of next generation sequencing data. *PLoS ONE* **2012**, *7*, e30619. [CrossRef] [PubMed]

58. Langmead, B.; Salzberg, S.L. Fast gapped-read alignment with Bowtie 2. *Nat. Methods* **2012**, *9*, 357–359. [CrossRef] [PubMed]

59. Bankevich, A.; Nurk, S.; Antipov, D.; Gurevich, A.A.; Dvorkin, M.; Kulikov, A.S.; Lesin, V.M.; Nikolenko, S.I.; Pham, S.; Prjibelski, A.D. SPAdes: A new genome assembly algorithm and its applications to single-cell sequencing. *J. Comput. Biol.* **2012**, *19*, 455–477. [CrossRef] [PubMed]

60. Hahn, C.; Bachmann, L.; Chevreux, B. Reconstructing mitochondrial genomes directly from genomic next-generation sequencing reads—A baiting and iterative mapping approach. *Nucleic Acids Res.* **2013**, *41*, e129. [CrossRef] [PubMed]

61. Wyman, S.K.; Jansen, R.K.; Boore, J.L. Automatic annotation of organellar genomes with DOGMA. *Bioinformatics* **2004**, *20*, 3252–3255. [CrossRef] [PubMed]

62. Lohse, M.; Drechsel, O.; Kahlau, S.; Bock, R. OrganellarGenomeDRAW—A suite of tools for generating physical maps of plastid and mitochondrial genomes and visualizing expression data sets. *Nucleic Acids Res.* **2013**, *41*, W575–W581. [CrossRef] [PubMed]

63. Kurtz, S.; Choudhuri, J.V.; Ohlebusch, E.; Schleiermacher, C.; Stoye, J.; Giegerich, R. REPuter: The manifold applications of repeat analysis on a genomic scale. *Nucleic Acids Res.* **2001**, *29*, 4633–4642. [CrossRef] [PubMed]

64. Benson, G. Tandem repeats finder: A program to analyze DNA sequences. *Nucleic Acids Res.* **1999**, *27*, 573. [CrossRef] [PubMed]

65. Thiel, T.; Michalek, W.; Varshney, R.; Graner, A. Exploiting EST databases for the development and characterization of gene-derived SSR-markers in barley (*Hordeum vulgare* L.). *Theor. Appl. Genet.* **2003**, *106*, 411–422. [CrossRef] [PubMed]

66. Yang, Z. PAML 4: Phylogenetic analysis by maximum likelihood. *Mol. Biol. Evol.* **2007**, *24*, 1586–1591. [CrossRef] [PubMed]

67. Frazer, K.A.; Pachter, L.; Poliakov, A.; Rubin, E.M.; Dubchak, I. VISTA: Computational tools for comparative genomics. *Nucleic Acids Res.* **2004**, *32* (Suppl. 2), W273–W279. [CrossRef] [PubMed]

68. Zhang, Y.-J.; Ma, P.-F.; Li, D.-Z. High-throughput sequencing of six bamboo chloroplast genomes: Phylogenetic implications for temperate woody bamboos (Poaceae: Bambusoideae). *PLoS ONE* **2011**, *6*, e20596. [CrossRef] [PubMed]

69. Koressaar, T.; Remm, M. Enhancements and modifications of primer design program Primer3. *Bioinformatics* **2007**, *23*, 1289–1291. [CrossRef] [PubMed]

70. Bi, G.; Mao, Y.; Xing, Q.; Cao, M. HomBlocks: A multiple-alignment construction pipeline for organelle phylogenomics based on locally collinear block searching. *Genomics* **2018**, *110*, 18–22. [CrossRef] [PubMed]

71. Katoh, K.; Standley, D.M. MAFFT multiple sequence alignment software version 7: Improvements in performance and usability. *Mol. Biol. Evol.* **2013**, *30*, 772–780. [CrossRef] [PubMed]

72. Posada, D.; Crandall, K.A. Modeltest: Testing the model of DNA substitution. *Bioinformatics* **1998**, *14*, 817–818. [CrossRef] [PubMed]

73. Swofford, D.L. *Commands Used in the PAUP Block in PAUP 4.0: Phylogenetic Analysis Using Parsimony 132–135*; Smithsonian Institution: Washington, DC, USA, 1998.

74. Stamatakis, A. RAxML version 8: A tool for phylogenetic analysis and post-analysis of large phylogenies. *Bioinformatics* **2014**, *30*, 1312–1313. [CrossRef] [PubMed]

75. Ronquist, F.; Teslenko, M.; van der Mark, P.; Ayres, D.L.; Darling, A.; Höhna, S.; Larget, B.; Liu, L.; Suchard, M.A.; Huelsenbeck, J.P. MrBayes 3.2: Efficient Bayesian phylogenetic inference and model choice across a large model space. *Syst. Biol.* **2012**, *61*, 539–542. [CrossRef] [PubMed]

International Journal of
Molecular Sciences

MDPI

Article

Complete Chloroplast Genome Sequence and Phylogenetic Analysis of *Quercus acutissima*

Xuan Li [1], Yongfu Li [1], Mingyue Zang [1], Mingzhi Li [2] and Yanming Fang [1,*]

[1] Co-Innovation Center for Sustainable Forestry in Southern China, College of Biology and the Environment, Key Laboratory of State Forestry Administration on Subtropical Forest Biodiversity Conservation, Nanjing Forestry University, 159 Longpan Road, Nanjing 210037, China; xuanli18851128817@163.com (X.L.); liyongfu199417@gmail.com (Y.L.); sanskritm@163.com (M.Z.)

[2] Genepioneer Biotechnologies Co. Ltd., Nanjing 210014, China; limzhi87@foxmail.com

[*] Correspondence: jwu4@njfu.edu.cn; Tel.: +86-25-8542-7428

Received: 13 July 2018; Accepted: 16 August 2018; Published: 18 August 2018

Abstract: *Quercus acutissima*, an important endemic and ecological plant of the *Quercus* genus, is widely distributed throughout China. However, there have been few studies on its chloroplast genome. In this study, the complete chloroplast (cp) genome of *Q. acutissima* was sequenced, analyzed, and compared to four species in the Fagaceae family. The size of the *Q. acutissima* chloroplast genome is 161,124 bp, including one large single copy (LSC) region of 90,423 bp and one small single copy (SSC) region of 19,068 bp, separated by two inverted repeat (IR) regions of 51,632 bp. The GC content of the whole genome is 36.08%, while those of LSC, SSC, and IR are 34.62%, 30.84%, and 42.78%, respectively. The *Q. acutissima* chloroplast genome encodes 136 genes, including 88 protein-coding genes, four ribosomal RNA genes, and 40 transfer RNA genes. In the repeat structure analysis, 31 forward and 22 inverted long repeats and 65 simple-sequence repeat loci were detected in the *Q. acutissima* cp genome. The existence of abundant simple-sequence repeat loci in the genome suggests the potential for future population genetic work. The genome comparison revealed that the LSC region is more divergent than the SSC and IR regions, and there is higher divergence in noncoding regions than in coding regions. The phylogenetic relationships of 25 species inferred that members of the *Quercus* genus do not form a clade and that *Q. acutissima* is closely related to *Q. variabilis*. This study identified the unique characteristics of the *Q. acutissima* cp genome, which will provide a theoretical basis for species identification and biological research.

Keywords: *Quercus*; chloroplast genome; phylogenetic relationship

1. Introduction

Oak trees provide humans with materials used in food, clothing, and houses, while oak forests supply living organisms and animals with comfortable habitats, good air, and sufficient and pure moisture. Oak trees are linked to Chinese culture, and are also often called eucalyptus or pecking trees. In China, eucalyptus is regarded as a mysterious tree, growing silently, watching its ancestors forge ahead, and passing through generation to generation. Many countries regard oaks as sacred trees, and consider them to be magical and a symbol of longevity, strength, and pride.

The genus *Quercus* L. (Oak) contains more than 400 species that are widespread in the northern hemisphere [1]. These species play important roles in China's forest ecosystem. *Quercus* L. (Oak)'s taxonomy, genetic structure, and breeding is complicated because of its wide variety of species, diverse forms, complex habitat conditions, and gene exchanges between species. Many studies have used nuclear simple sequence repeat (SSR) chloroplast DNA makers to study phylogeny and population variation [2,3]. Previously, studies found a conflict (inconsistency) between the phylogeny of plastid data and nuclear data in Senecioneae and Neotropical Catasetinae [4,5]. Therefore, it is not sufficient to study *Quercus* simply by using plastid regions. With the rapid development of next-generation sequencing, genome acquisition is now cheaper and faster than traditional Sanger sequencing. Complete chloroplast (cp) genome size data will be necessarily used to infer the phylogenetic relationship of *Quercus* or Fagaceae in future studies.

The genus is characterized by a high variability of morphological and ecological traits, the occurrence of mixed stands, the presence of large population sizes, and high levels of gene flow within the *Quercus* complex [6–11]. A new classification of *Quercus* L. was proposed by Denk with eight sections: *Cyclobalanopsis, Cerris, Ilex, Lobatae, Quercus, Ponticae, Protobalanus,* and *Virentes* [12]. In China, *Quercus* is divided into five morphology-based sections: *Quercus, Aegilops, Heterobalanus, Engleriana,* and *Echinolepides* [13–15]. Due to incomplete sampling and the use of markers with insufficient phylogenetic signals and complex evolutionary problems, the relationships among *Quercus* species are not fully understood.

Q. acutissima is an ecological and economic tree species in deciduous broad-leaved forests in the temperate zone of East Asia, widely distributed on the Hu Huanyong line or in Southeast China (latitude from 18° to 41° N and longitude from 91° to 123° E) [16]. This line from Heilongjiang Province to Tengchong, Yunnan Province, is roughly inclined in a 45° straight line. The development, origin, and reproduction of China are linked with *Q. acutissima*. Therefore, we need to protect, cultivate, and utilize *Q. acutissima*, and this has received substantial attention in phylogeny and biogeography studies. Most previous studies have focused on its population structure [17], breeding [18], forest management [19], and physiology [20]. Studies on the genetic variation of *Q. acutissima* using simple sequence repeat (SSR) and cpDNA makers have been carried out in China and South Korea [16,21]. According to this research, the distribution of *Q. acutissima* often overlaps with other oak trees, i.e., *Q. variabilis* and *Q. chenii* [22]. There is often a variety of species found in the population, although this has usually been determined from a comparison of morphology, rather than at a molecular level. Therefore, an analysis of the complete cp genome of *Q. acutissima* will help to identify the species further.

In the present study, we constructed the whole chloroplast genome of *Q. acutissima* by using next-generation sequencing and applying a combination of de novo and reference-guided assembly. Here, we describe the whole chloroplast genome sequence of *Q. acutissima* and the characterization of long repeats and simple sequence repeats (SSRs). We compare and analyze the chloroplast genome of *Q. acutissima* and the chloroplast genome of other members of Fagaceae. It is expected that the results will provide a theoretical basis for the determination of phylogenetic status and future scientific research.

2. Results and Discussion

2.1. Features of Q. Acutissima cpDNA

A total number of 63 million pair-end reads were produced with 9.82 Gb of clean data. Data from all of the reads were deposited in the NCBI Sequence Read Archive (SRA) under accession number MH607377. The size of the complete cp genome is 161,124 bp (Figure 1). The cp genome displayed a typical quadripartite structure, including a pair of IR (25,816 bp) separated by the large single copy (LSC; 90,423 bp) and small single copy (SSC; 19,069 bp) regions (Figure 1 and Table 1). The DNA G + C contents of the LSC, SSC, and IR regions, and the whole genome are 34.62, 30.84, 42.78, and 36.08 mol %, respectively, which is also similar to the chloroplast genomes of other *Quercus* species (Figure A1; Table 2). The DNA G + C content is a very important indicator of species affinity [23]. It is obvious that the DNA G + C content of the IR region is higher than that of other regions (LSC, SSC). This phenomenon is very common in other plants [23,24]. GC skewness has been shown to be an indicator of DNA lead chains, lag chains, replication origin, and replication terminals [25–27].

Int. J. Mol. Sci. **2018**, *19*, 2443

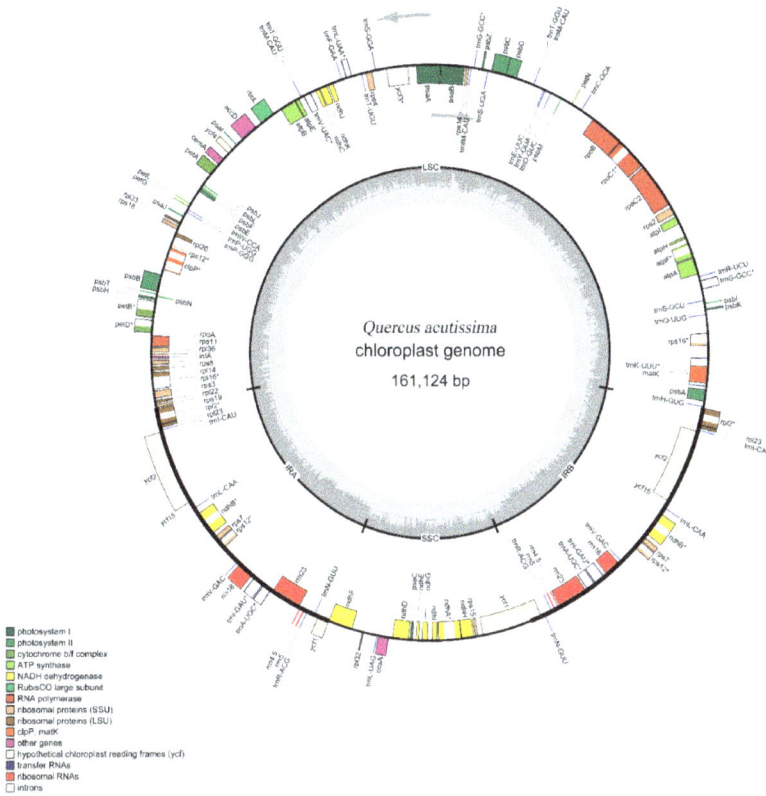

Figure 1. Chloroplast genome map of *Q. acutissima*. Genes inside the circle are transcribed clockwise, and those outside are transcribed counterclockwise. Genes of different functions are color-coded. The darker gray in the inner circle shows the GC content, while the lighter gray shows the AT content.

Table 1. Summary of five *Quercus* chloroplast genome features.

Genome Features	Q. acutissima	Q. variabilis	Q. dolicholepis	C. mollissima	L. balansae	F. engleriana
Genome size (bp)	161,124	161,077	161,237	160,799	161,020	158,346
LSC length (bp)	90,423	90,387	90,461	90,432	90,596	87,667
SSC length (bp)	19,068	19,056	19,048	18,995	19,160	18,895
IR length (bp)	51,632	51,634	51,728	51,372	51,264	51,784
Number of genes	136	134	134	130	134	131
Number of protein–coding genes	88	86	86	83	87	83
Number of tRNA genes	40	40	40	37	39	40
Number of rRNA genes	8	8	8	8	8	8

Plant chloroplast genomes may have 63–209 genes, but most are concentrated between 110 and 130, with a highly conserved composition and arrangement, including photosynthetic genes, chloroplast transcriptional expression-related genes, and some other protein-coding genes [28]. In the *Q. acutissima* chloroplast genome, 136 functional genes were predicted and divided into six groups, including eight rRNA genes, 40 tRNA genes, and 88 protein-coding genes (Tables 1 and 3). In addition, 14 tRNA genes, eight rRNA genes, and 15 protein-coding genes are duplicated in the IR regions (Figure 1). The LSC region includes 62 protein-coding and 25 tRNA genes, while the SSC region includes 13 protein-coding genes (Table A1).

Based on the protein-coding sequences and tRNA genes, the frequency of codon usage was estimated for the *Q. acutissima* cp genome and is summarized in Table A2. In total, all genes are encoded by 6311 codons. Among these, leucine, with 2824 (44.4%) codons, is the most frequent amino acid in the cp genome, and cysteine, with 293 (1.1%), is the least frequent (Table 3). A- and U-ending codons are common. The most preferred synonymous codons (relative synonymous codon usage values (RSCU) > 1) end with A or U [23,29].

Table 2. Base composition of the *Q. acutissima* chloroplast genome.

Region	A (%)	T (U) (%)	C (%)	G (%)	A + T (%)	G + C (%)
LSC	31.99	33.4	17.74	16.88	65.39	34.62
SSC	34.46	34.71	16.24	14.6	69.17	30.84
IR	28.61	28.61	21.39	21.39	57.22	42.78
Total	31.69	32.24	18.46	17.62	63.93	36.08

Table 3. List of genes annotated in the cp genomes of *Q. acutissima* sequenced in this study.

Function	Genes
RNAs, transfer	*trnH-GUG, trnK-UUU, trnQ-UUG, trnS-GCU, trnG-GCC, trnR-UCU, trnC-GCA, trnD-GUC, trnY-GUA, trnE-UUC, trnT-GGU, trnM-CAU, trnS-UGA, trnG-GCC, trnfM-CAU, trnS-GGA, trnT-UGU, trnL-UAA, trnF-GAA, trnV-UAC, trnM-CAU, trnT-GGU, trnW-CCA, trnP-UGG, trnP-GGG, trnI *-CAU, trnL-CAA *, trnV-GAC, trnI-GAU *, trnA-UGC, trnR-ACG, trnN-GUU, trnL-UAG, trnN-GUU, trnR-ACG, trnA-UGC, trnV-GAC*
RNAs, ribosomal	*rrn23 *, rrn16 *, rrn5 *, rrn4.5 **
Transcription and splicing	*rpoC1 *, rpoC2, rpoA, rpoB*
Translation, ribosomal proteins	
Small subunit	*rps2, rps3, rps4, rps7, rps8, rps11, rps12 **, rps14, rps15, rps16 *, rps18, rps19*
Large subunit	*rpl2 *, rpl14, rpl16 *, rpl20, rpl22, rpl23, rpl32, rpl33, rpl36*
Photosynthesis	
ATP synthase	*atpE, atpB, atpA, atpF *, atpH, atpI*
Photosystem I	*psaI, psaB, psaA, psaC, psaJ, ycf3 *, ycf4*
Photosystem II	*psbD, psbC, psbZ, psbT, psbH, psbK, psbI, psbJ, psbF, psbE, psbM, psbN, psbL, psbA, psbB*
Calvin cycle	*rbcL*
Cytochrome complex	*petN, petA, petL, petG, petB *, petD **
NADH dehydrogenase	*ndhB *, ndhI, ndhK, ndhC, ndhF, ndhD, ndhG, ndhE, ndhA, ndhH, ndhJ*
Others	*inFA, ycf15 *, ycf1 *, ycf2 *, accD, cemA, ccsA, clpP ***

* Genes containing one intron; ** genes containing two introns.

In total, we found 23 intron-containing genes, including 15 protein-coding genes, and eight tRNA genes (Table 4). 21 genes (13 protein-coding and eight tRNA genes) contain one intron, and two genes (*ycf3* and *clpP*) contain two introns. The *trnK-UUU* has the largest intron (2505 bp), and the *trnL-UAA* has the smallest intron (483bp). Studies have shown that *ycf3* is required for stable accumulation of photosystem I complexes [30]. Therefore, we speculate that the *ycf3* intron gain of *Q. acutissima* may be helpful for further study of the mechanism of photosynthesis evolution.

Table 4. The lengths of exons and introns in genes with introns in the *Q. acutissima* chloroplast genome.

Gene	Location	Exon I (bp)	Intron I (bp)	Exon II (bp)	Intron II (bp)	Exon III (bp)
rps16	LSC	42	898	195		
atpF	LSC	144	780	411		
rpoC1	LSC	432	827	1626		
ycf3	LSC	127	718	228	778	155
clpP	LSC	69	844	294	649	228
petB	LSC	6	841	642		
petD	LSC	9	640	474		
rpl16	LSC	9	1102	399		
rpl2	RepeatA	390	628	471		
ndhB	RepeatA	777	680	756		
rps12	RepeatA	10	537	231		
ndhA	SSC	551	1040	541		
rps12	RepeatB			232	536	26
ndhB	RepeatB	777	680	756		
rpl2	RepeatB	390	628	471		
trnG-GCC	LSC	23	734	37		
trnK-UUU	LSC	37	2505	35		
trnL-UAA	LSC	35	483	50		
trnV-UAC	LSC	36	630	37		
trnI-GAU	RepeatA	42	950	35		
trnA-UGC	RepeatA	38	800	35		
TRNA-UGC	RepeatB	38	800	35		
trnI-GAU	RepeatB	42	950	35		

2.2. Comparative Analysis of Genomic Structure

The chloroplast sequence are often used to measure the genetic diversity within a species, the gene flow between species, and the size of ancestral populations of separated sister species [31]. Thus, it is necessary to understand the chloroplast differences between species. The complete cp genome sequence of *Q. acutissima* was compared to those of *Q. variabilis*, *Q. dolicholepis*, *Castanea mollissima*, *Lithocarpus balansae*, and *Fagus engleriana*. *F. engleriana* has the smallest cp genome with the largest IR region (51,784 bp), and *Q. dolicholepis* has the largest cp genome (Table 1). We assumed that the different lengths of the SSC and IR regions is the main reason for variety in sequence lengths. To verify the possibility of genome divergence, sequence identity was calculated for six species' chloroplast DNA using the program mVISTA with *Q. variabilis* as a reference (Figure 2). The results of this comparison revealed that LSC regions are more divergent than SSC and IR regions and that higher divergence is found in noncoding than in coding regions. The complete cp genome sequence of *F. engleriana* is quite different from the five other plants. There was no significant difference between the chloroplast genome sequences of evergreen and deciduous trees. At the same time, the results of the sliding window indicated that the location of the variation in the cp genome among the six species occurred in the LSC and SSC regions (Figure A2). Significant variation was found in coding regions of some genes, including *psbI*, *rpl33*, *petB*, *rpl2*, *rps16*, *rpoC2*, *ndhK*, *ycf2*, *ycf1*, and *ndhI*. The highest divergence in noncoding regions was found in the intergenic regions of *trnK-rps16*, *rps 16-trnQ*, *psbK-psbI*, *trnS-trnG*, *atpH-atpI*, *atpI-rps2*, *rpoB-trnC*, *trnC-petN*, *psbM-trnD*, *trnD-trnY*, *trnE-trnM*, *trnT-petD*, *psbZ-trnG*, *trnT-trnL*, *trnF-ndhJ*, *rbcL-accD*, *psaI-ycf4*, *ycf4-cemA*, *petA-psbL*, *psaJ-rpl33*, *clpP-psbB*, *rpl14-rpl16*, *ndhF-rpl32*, *ccsA-ndhD*, *ndhD-psaC*, and *rps15-ycf1*.

The contraction and expansion of the IR region at the borders play important roles in evolution. They are common evolutionary events and a major cause of changes in the size of the chloroplast genome. They may also cause variation in the length of angiosperm plastid genome [32–34]. Detailed comparisons of the IR–SSC and IR–LSC boundaries among the cp genomes of the above six Fagaceae species were presented in Figure 3. The IR regions are relatively highly conserved in the *Quercus* genus—the *rpl2* gene in the *Quercus* cp genome is shifted by 62 bp from IRb to LSC at the LSC/IRb border, and by 62 bp from IRa to LSC at the IRa/LSC border. Compared to other species in the genus, the range of the IRa/SSC regions changes greatly. Compared with evergreen and deciduous species, we found significant differences in IRb/SSC. Some reports showed that *ycf1* is necessary for plant viability and encodes *Tic214*, an important component of the *Arabidopsis TIC* complex [35,36]. The *ycf1* gene crossed the SSC/IRb region, with 1041bp of *ycf1_like* within IRb (incompletely duplicated in IRb). The SSC/IRa junction is located in the *ycf1* region in all Fagaceae species chloroplast genomes and extends into the SSC region by different lengths depending on the genome (*Q. acutissima*, 4619 bp; *Q. variabilis*, 4620 bp; *Q. dolicholepis*, 4611 bp; *C. mollissima*, 4623 bp; *L. balansae*, 4626 bp; *F. engleriana*, 4633 bp); the IRa region includes 1041, 1041, 1068, 1059, 828, and 1049 bp of the *ycf1* gene.

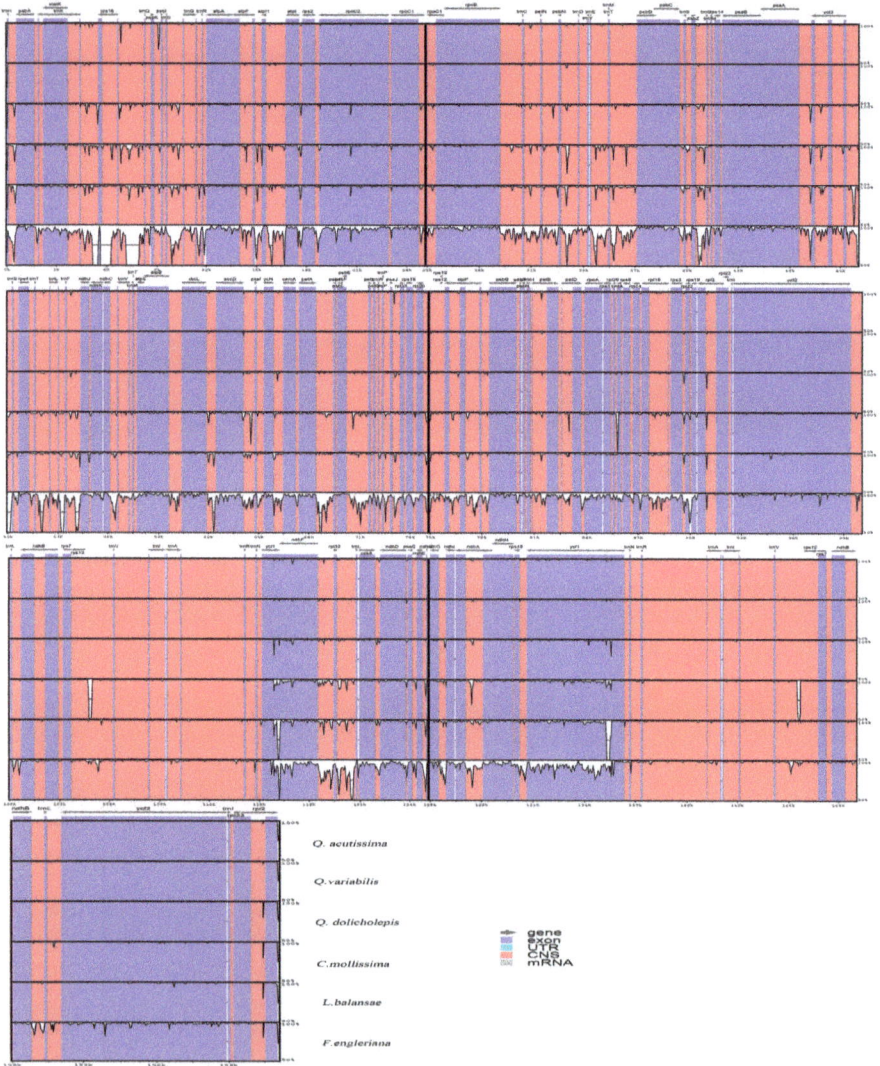

Figure 2. Complete chloroplast genome comparison of six species using the chloroplast genome of *Q. variabilis* as a reference. The grey arrows and thick black lines above the alignment indicate the genes' orientations. The Y-axis represents the identity from 50% to 100%.

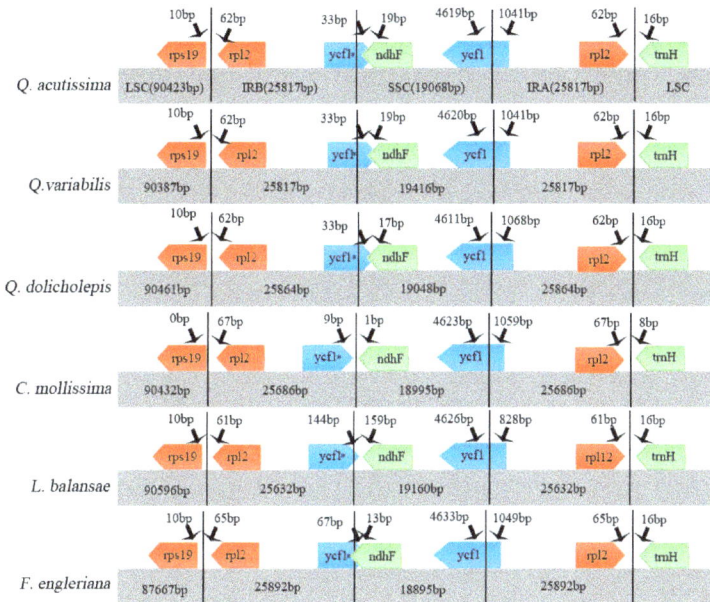

Figure 3. Comparison of the large single copy (LSC), small single copy (SSC), and inverted repeat (IR) regions in chloroplast genomes of four species. Genes are denoted by colored boxes. The gaps between the genes and the boundaries are indicated by the base lengths (bp). Extensions of the genes are indicated above the boxes.

2.3. Long-Repeat and SSR Analysis

For the repeat structure analysis (Table 5), 31 forward and 22 inverted repeats were detected in the *Q. acutissima* cp genome. Most of these repeats are between 19 and 46 bp. The longest forward repeat is 46 bp in length and is located in the LSC region. A total of 35, 18, and eight repeats were found in the LSC, SSC, IR regions, respectively. Seven forward repeats were located in IR, including one repeat associated with *ycf1* genes and one repeat related to the *trnV-UAC* and *trnA-UGC* genes. Most repeats in the intergenic spacers are distributed in the LSC region. Ten repeats are distributed in the SSC region, and only four of them are in the intergenic spacers.

As chloroplast-specific SSRs are uniparentally inherited and are inclined to undergo slipped-strand mispairing, they are often used in population genetics, species identification, and evolutionary process research of wild plants [37,38]. In addition, chloroplast genome sequences are highly conserved, and the SSR primer for chloroplast genomes can be transferred across species and genera. Yoko et al. used six maternally inherited chloroplast (cpDNA) simple sequence repeat (SSR) markers to study the genetic variation in *Q. acutissima* [39]. In this study, a total of 65 SSRs were found in *Q. acutissima*, most of them distributed in LSC and SSC and partly distributed in IR. These included 61 mononucleotide SSRs (93.85%) and four dinucleotide SSRs (6.15%) (Table 6). Compared with other *Quercus* species, fewer types of SSRs were identified in *Q. acutissima* [40]. Among them, two SSRs belonged to the C type, and the others all belonged to the A/T types. These results are consistent with the hypothesis that cpSSRs are generally composed of short polyadenine (polyA) or polythymine (polyT) repeats and rarely contain tandem guanine (G) or cytosine (C) repeats [41]. We also found that 12 SSRs were located in genes, and the remaining were all located in intergenic regions. These cpSSR markers could be used to examine the genetic structure, diversity, differentiation, and maternity in *Q. acutissima* and its relative species in future studies.

Table 5. Long repeat sequence in the *Q. acutissima* chloroplast genome.

ID	Repeat Start I	Type	Size (bp)	Repeat Start 2	Mismatch (bp)	E-Value	Gene	Region
1	6831	F	46	6853	0	1.47×10^{-18}	IGS	LSC
2	11,847	R	31	11,847	0	1.58×10^{-9}	IGS	LSC
3	6818	R	26	6818	0	1.62×10^{-6}	*rps16*	LSC
4	47,242	F	25	47,264	0	6.49×10^{-6}	IGS	LSC
5	6831	F	24	6875	0	2.59×10^{-5}	IGS	LSC
6	115,801	F	24	135,722	0	2.59×10^{-5}	*ycf1*	IRA; IRB
7	113,545	F	23	113,576	0	1.04×10^{-4}	IGS	IRA
8	118,844	R	23	118,844	0	1.04×10^{-4}	IGS	IRA
9	137,948	F	23	137,979	0	1.04×10^{-4}	IGS	IRB
10	11,371	F	22	41,193	0	4.15×10^{-4}	*trnG-GCC* (exon), *trnG-GCC*	LSC
11	9536	F	21	39,849	0	1.66×10^{-3}	*trnS-UGA, trnS-GCU*	LSC
12	10,319	F	21	18,682	0	1.66×10^{-3}	IGS	LSC
13	117,049	R	21	117,049	0	1.66×10^{-3}	*ndhF*	SSC
14	36,478	F	20	53,719	0	6.64×10^{-3}	IGS	LSC
15	53,720	F	20	130,481	0	6.64×10^{-3}	IGS	LSC; SSC
16	55,907	R	20	55,907	0	6.64×10^{-3}	*atpB*	LSC
17	57,271	F	20	142,064	0	6.64×10^{-3}	*trnV-UAC, trnA-UGC*	LSC; IRB
18	105,331	F	20	105,349	0	6.64×10^{-3}	IGS	IRA
19	146,178	F	20	146,196	0	6.64×10^{-3}	IGS	IRB
20	4930	F	19	36,476	0	2.66×10^{-2}	IGS	LSC
21	8915	R	19	8915	0	2.66×10^{-2}	IGS	LSC
22	13,541	R	19	76,642	0	2.66×10^{-2}	*atpA*	LSC
23	18,685	R	19	118,842	0	2.66×10^{-2}	*clpP*	LSC; SSC
24	21,297	R	19	54,183	0	2.66×10^{-2}	*rpoC2*	LSC
25	36,479	F	19	130,481	0	2.66×10^{-2}	IGS	LSC; SSC
26	39,957	R	19	39,957	0	2.66×10^{-2}	IGS	LSC
27	62,040	R	19	62,040	0	2.66×10^{-2}	IGS	LSC
28	64,751	R	19	64,751	0	2.66×10^{-2}	IGS	LSC
29	69,026	R	19	69,026	0	2.66×10^{-2}	IGS	LSC
30	71,277	R	19	71,277	0	2.66×10^{-2}	IGS	LSC
31	72,561	R	19	72,561	0	2.66×10^{-2}	IGS	LSC
32	4430	R	18	4430	0	1.06×10^{-1}	IGS	LSC
33	4437	F	18	24,828	0	1.06×10^{-1}	*rpoC1* (intron)	SSC

Int. J. Mol. Sci. **2018**, *19*, 2443

Table 5. *Cont.*

ID	Repeat Start 1	Type	Size (bp)	Repeat Start 2	Mismatch (bp)	E-Value	Gene	Region
34	4935	F	18	52,105	0	1.06×10^{-1}	IGS	LSC
35	4938	F	18	118,695	0	1.06×10^{-1}	IGS	LSC
36	6813	F	18	6847	0	1.06×10^{-1}	IGS	LSC
37	6813	F	18	6869	0	1.06×10^{-1}	IGS	LSC
38	6817	F	18	127,945	0	1.06×10^{-1}	*ndhA* (intron)	LSC
39	7369	F	18	7387	0	1.06×10^{-1}	IGS	LSC; SSC
40	7465	R	18	7465	0	1.06×10^{-1}	IGS	LSC; SSC
41	8589	R	18	34,768	0	1.06×10^{-1}	IGS	LSC; SSC
42	9996	R	18	9996	0	1.06×10^{-1}	IGS	LSC
43	10,283	F	18	31,730	0	1.06×10^{-1}	IGS	LSC
44	10,322	R	18	118,843	0	1.06×10^{-1}	IGS	LSC; IRA
45	10,548	F	18	133,365	0	1.06×10^{-1}	*ycf1*	LSC
46	31,728	F	18	125,951	0	1.06×10^{-1}	IGS	LSC
47	39,812	F	18	40,698	0	1.06×10^{-1}	*trnS-UGA*	LSC; SSC
48	40,022	R	18	69,093	0	1.06×10^{-1}	IGS	LSC
49	40,700	F	18	123,827	0	1.06×10^{-1}	IGS	LSC
50	43,446	F	18	45,670	0	1.06×10^{-1}	*psaB*	SSC
51	40,022	R	18	69,093	0	1.06×10^{-1}	IGS	LSC
52	40,700	F	18	123,827	0	1.06×10^{-1}	IGS	LSC
53	43,446	F	18	45,670	0	1.06×10^{-1}	*psaB, psaA*	LSC

F: forward; I: inverted; IGS: intergenic space.

Table 6. Simple sequence repeats (SSRs) in the *Q. acutissima* chloroplast genome.

ID	Repeat Motif	Length (bp)	Start	End	Region	Gene
1	(A)10	9	1809	1818	LSC	
2	(C)14	13	4433	4446	LSC	
3	(T)11	10	4697	4707	LSC	
4	(A)10	9	4939	4948	LSC	trnK-UUU
5	(T)11	10	7001	7011	LSC	
6	(T)10	9	7746	7755	LSC	
7	(A)10	9	8174	8183	LSC	
8	(A)12	11	8590	8601	LSC	
9	(A)11	10	8920	8930	LSC	psbK
10	(A)10	9	9465	9474	LSC	
11	(A)10	9	10,161	10,170	LSC	
12	(A)11	10	13,547	13,557	LSC	
13	(T)12	11	15,345	15,356	LSC	
14	(T)10	9	16,160	16,169	LSC	
15	(A)12	11	18,692	18,703	LSC	rpoC2
16	(T)12	11	21,295	21,306	LSC	rpoC2
17	(T)14	13	25,299	25,312	LSC	
18	(T)10	9	28,563	28,572	LSC	
19	(T)10	9	29,651	29,660	LSC	
20	(T)11	10	30,275	30,285	LSC	
21	(C)14	13	30,428	30,441	LSC	
22	(T)11	10	31,731	31,741	LSC	
23	(A)10	9	32,094	32,103	LSC	
24	(A)10	9	33,986	33,995	LSC	
25	(A)13	12	34,775	34,787	LSC	
26	(A)10	9	34,955	34,964	LSC	
27	(A)10	9	36,485	36,494	LSC	
28	(AT)6	11	39,819	39,830	LSC	trnfM-CAU
29	(T)10	9	41,238	41,247	LSC	
30	(T)11	10	53,217	53,227	LSC	
31	(A)10	9	53,726	53,735	LSC	
32	(T)15	14	54,110	54,124	LSC	
33	(A)11	10	54,990	55,000	LSC	
34	(T)10	9	55,713	55,722	LSC	
35	(T)10	9	59,591	59,600	LSC	
36	(T)10	9	60,063	60,072	LSC	
37	(T)10	9	64,092	64,101	LSC	accD
38	(A)11	10	64,266	64,276	LSC	
39	(AT)7	13	64,570	64,583	LSC	
40	(T)14	13	64,945	64,958	LSC	
41	(T)13	12	66,170	66,182	LSC	
42	(T)11	10	68,616	68,626	LSC	petA
43	(T)11	10	70,730	70,740	LSC	
44	(T)11	10	71,398	71,408	LSC	
45	(T)11	10	73,389	73,399	LSC	
46	(AT)6	11	77,274	77,285	LSC	clpP
47	(TA)7	13	82,928	82,941	LSC	petD
48	(A)11	10	85,781	85,791	LSC	
49	(T)10	9	86,100	86,109	LSC	
50	(T)10	9	88,820	88,829	LSC	
51	(T)11	10	114,070	114,080	IRA	
52	(T)12	11	118,582	118,593	SSC	
53	(A)11	10	118,695	118,705	SSC	
54	(T)11	10	119,000	119,010	SSC	
55	(A)10	9	119,794	119,803	SSC	ndhD
56	(T)11	10	122,199	122,209	SSC	
57	(A)10	9	122,546	122,555	SSC	
58	(AT)8	15	123,832	123,847	SSC	
59	(T)11	10	125,812	125,822	SSC	
60	(T)11	10	125,954	125,964	SSC	
61	(T)11	10	130,262	130,272	SSC	
62	(A)10	9	130,487	130,496	SSC	
63	(T)10	9	133,465	133,474	SSC	
64	(T)13	12	134,042	134,054	SSC	ycf1
65	(A)11	10	137,468	137,478	SSC	ycf1

2.4. Phylogenetic Analysis

Phylogenetic analysis was completed on an alignment of concatenated nucleotide sequences of all chloroplast genomes from 25 angiosperm species (Figure 4). We used the Bayesian inference (BI) method based on RAxML to build a phylogenetic tree, and *Malus prunifolia* and *Ulmus gaussenii* were used as the outgroup. Support is generally high for almost all relationships inferred from all chloroplast genome data based on BI methods (the support values have a range of 0.8956 to 1). It is noteworthy that the species in genus *Quercus* do not form a clade. Several evergreen tree species gather together to form one clade. *Q. acutissima* and *Q. variabilis* are sister species and are frequently mixed in Chinese endemic species; the second clade splits into two subclades. *F. engleriana* is in the top position, while *Q. acutissima* appears to be more closely related to *Q. variabilis*, *Q. dolicholepis*, and *Q. baronii*. In general, the topologies of the other branches (genus *Fagus, Trigonobalanus, Lithocarpus,* and *Castanopsis*) are almost the same based on two nuclear loci (ITS and CRC) [3].

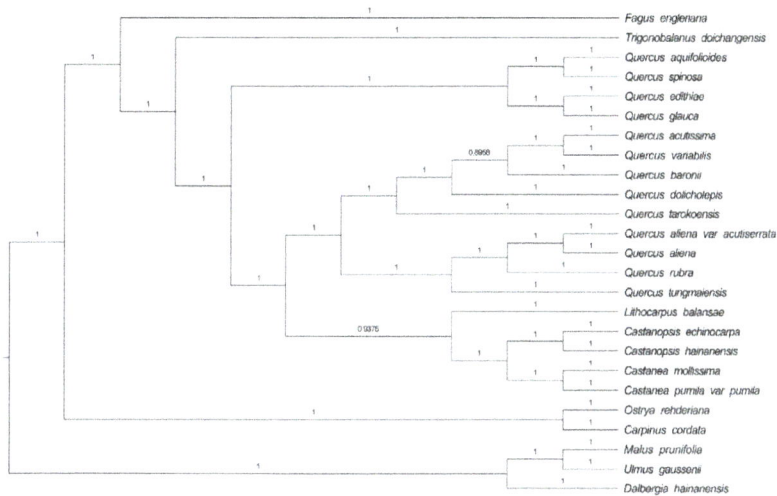

Figure 4. Bayesian inference (BI) phylogenetic tree reconstruction including 25 species based on all chloroplast genomes. *Malus prunifolia* and *Ulmus gaussenii* were used as the outgroup.

3. Materials and Methods

3.1. Sampling, DNA Extraction, Sequencing, and Assembly

Q. acutissima was planted in Nanjing Forestry University and Zijin Mountain in Nanjing, China (32°04′ N, 118°48′ E; 32°04′ N, 118°50′ E), respectively. Fresh leaves were collected and wrapped in ice and immediately stored at −80 °C until analysis. Genomic DNA was isolated by the modified method CTAB [42]. Agarose gel electrophoresis and one drop spectrophotometer (OD-1000, Shanghai Cytoeasy Biotech Co., Ltd., Shanghai, China) were used to detect DNA integrity and quality. Shotgun libraries (250 bp) were constructed using pure DNA according to the manufacturer's instructions. Sequencing was performed with an Illumina Hiseq 2500 platform (Nanjing, China), yielding at least 9.82 GB of clean data for *Q. acutissima*. Firstly, all of the raw reads were trimmed by Fastqc. Next, we performed a BLAST analysis between trimmed reads and references (*Q. variabilis* and *Q. dolicholepis*) to extract cp-like reads. Finally, we used the chloroplast-like reads to assemble sequences using NOVOPlasty [43]. NOVOPlasty assembled part reads and stretched as far as possible until a circular genome formed. When the assembly result was within the expected range, the overlap was larger than 200 bp, and the assembly formed a ring.

3.2. Annotation and Analysis of the cpDNA Sequences

CpGAVAS was used to annotate the sequences; DOGMA (http://dogma.ccbb.utexas.edu/) and BLAST were used to check the results of the annotation [44,45]. tRNAscanSE was used to identify the tRNAs [46]. The circular gene maps of the species of *Q. acutissima* were drawn using the OGDRAWv1.2 program [47] (http://ogdraw.mpimp-golm.mpg.de/). An analysis of variation in synonymous codon usage, relative synonymous codon usage values (RSCU), codon usage, and the GC content of the complete plastid genomes and commonly analyzed CDS was conducted. MISA(available online: http://pgrc.ipk-gatersleben.de/misa/misa.html) [48] and REPuter (available online: https://bibiserv.cebitec.uni-bielefeld.de/reputer/) [49] was used to visualize the SSRs and long repeats, respectively.

3.3. Genome Comparison

MUMmer [50] was used for pairing sequence alignment of the cp genome. The mVISTA [51] program was applied to compare the complete cp genome of *Q. acutissima* to the other published cp genomes of its related species, i.e., *Q. variabilis* (KU240009), *Q. dolicholepis* (KU240010), *C. mollissima* (HQ336406), *L. balansae* (KP299291), and *F. engleriana* (KX852398) with the shuffle-LAGAN mode [52], using the annotation of *Q. variabilis* as a reference.

3.4. Phylogenetic Analysis

Phylogenies were constructed by Bayesian inference (BI) analysis using the 25 cp genome of the Fagaceae species sequences from the NCBI Organelle Genome and Nucleotide Resources database. The sequences were initially aligned using MAFFT [53]. Then, the visualization and manual adjustment of multiple sequence alignment were conducted in BioEdit [54]. An IQ-tree was used to select the best-fitting evaluation of models of nucleotide sequences [55]. TVM + F + R4 and GTR + G were selected as the best substitution models for the BI analyses. BI analyses were conducted using Mrbayes [56]. *Malus prunifolia* (NC_031163), and the *Ulmus gaussenii* (NC_037840) were used as the outgroups.

4. Conclusions

In this study, we reported and analyzed the complete cp genome of *Q. acutissima*, an endemic and ecological tree species in China. The chloroplast genome was shown to be more conservative with similar characteristics to other genus *Quercus* species. Compared to the cp genomes of five other oak species, its LSC were shown to be more divergent among the four regions, and noncoding regions showed higher divergence. An analysis of the phylogenetic relationships among six species found *Q. acutissima* to be closely related to *Q. variabilis*. The developmental position of the tree in the Fagaceae family is consistent with previous studies. The results of this study provide an assembly of a whole chloroplast genome of *Q. acutissima* which might facilitate genetics, breeding, and biological discoveries in the future.

Author Contributions: X.L. performed most of the experiments, data analysis, and the writing of the manuscript; Y.L. participated in the data analysis; M.Z. and M.L. participated in the preprocessing of data; and Y.F. supervised the project and provided suggestions for the manuscript.

Acknowledgments: This research was supported by the National Natural Science Foundation of China (31770699, 31370666), the Priority Academic Program Development of Jiangsu Higher Education Institutions (PAPD), and the Nanjing Forestry University Excellent Doctoral Thesis Fund.

Conflicts of Interest: The authors declare no conflict of interest.

Abbreviations

LSC	Large single copy
SSC	Small single copy
IR	Inverted repeat
Cp	Chloroplast
BI	Bayesian inference
A	Adenine
T	Thymine
G	Guanine
C	Cytosine

Appendix A

Table A1. The number of genes in the *Q. acutissima* cp genome.

Region	Number of CDS	Number of tRNA	Number of rRNA	Total
LSC region	62	25	0	87
SSC region	13	1	0	14
IRA region	6	7	4	17
IRB region	7	7	4	18

Table A2. Codon-anticodon recognition patterns and codon usage of the *Q. acutissima* chloroplast genome.

Amino Acid	Codon	No.	RSCU	tRNA	Amino Acid	Codon	No.	RSCU	tRNA
Ala	GCG	164	0.47		Pro	CCA	313	1.13	trnP-TGG
Ala	GCC	224	0.64		Pro	CCC	226	0.82	
Ala	GCU	630	1.79		Pro	CCU	409	1.48	
Ala	GCA	388	1.1		Pro	CCG	161	0.58	
Cys	UGU	221	1.44		Gln	CAG	215	0.45	
Cys	UGC	86	0.56	trnC-GCA	Gln	CAA	731	1.55	trnQ-TTG
Asp	GAC	209	0.39	trnD-GTC	Arg	CGU	337	1.26	trnR-ACG
Asp	GAU	870	1.61		Arg	AGA	500	1.87	trnR-TCT
Glu	GAA	1064	1.5	trnE-TTC	Arg	CGA	358	1.34	
Glu	GAG	357	0.5		Arg	AGG	183	0.68	
Phe	UUU	983	1.3		Arg	CGG	118	0.44	
Phe	UUC	535	0.7	trnF-GAA	Arg	CGC	109	0.41	
Gly	GGU	580	1.27		Ser	AGC	125	0.37	trnS-GCT
Gly	GGG	330	0.72		Ser	UCU	557	1.66	
Gly	GGA	706	1.55		Ser	UCA	397	1.18	trnS-TGA
Gly	GGC	206	0.45	trnG-GCC	Ser	UCC	349	1.04	trnS-GGA
His	CAU	486	1.54		Ser	AGU	391	1.17	
His	CAC	145	0.46	trnH-GTG	Ser	UCG	193	0.58	
Ile	AUC	458	0.58		Thr	ACU	538	1.6	
Ile	AUA	758	0.97		Thr	ACG	160	0.48	
Ile	AUU	1139	1.45		Thr	ACC	247	0.73	trnT-GGT
Lys	AAG	379	0.5		Thr	ACA	402	1.19	trnT-TGT
Lys	AAA	1062	1.4		Val	GUU	508	1.41	
Leu	UUG	572	1.22	trnL-CAA	Val	GUC	181	0.5	trnV-GAC
Leu	UUA	894	1.9		Val	GUA	547	1.52	
Leu	CUU	583	1.24		Val	GUG	207	0.57	
Leu	CUA	373	0.79	trnL-TAG	Trp	UGG	462	1	trnW-CCA
Leu	CUC	204	0.43		Tyr	UAC	212	0.42	trnY-GTA
Leu	CUG	198	0.42		Tyr	UAU	792	1.58	
Met	AUG	620	1	trnI-CAT	Stop	UAA	47	1.6	
Asn	AAU	1004	1.5		Stop	UAG	22	0.75	
Asn	AAC	304	0.46		Stop	UGA	19	0.65	

RSCU: Relative Synonymous Codon Usage.

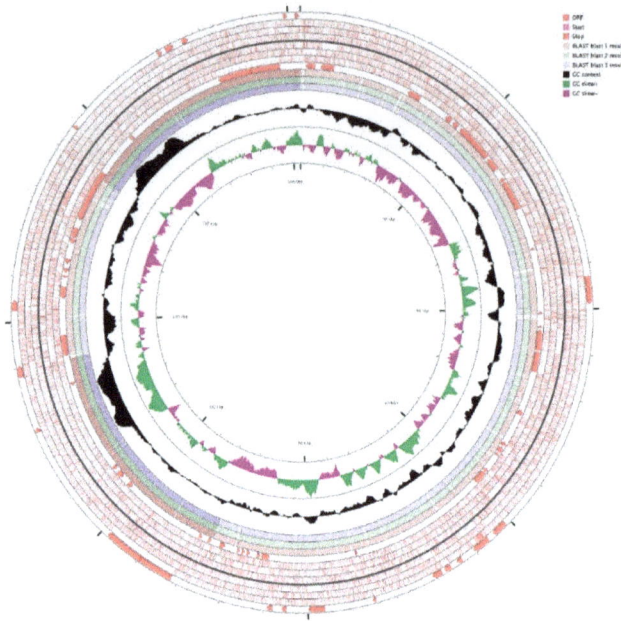

Figure A1. BLAST result of the chloroplast genome and the GC stew of *Q. acutissima.* BlAST 1 represents *L. balansae*; BlAST 2 represents *Q. variabilis*; BlAST 3 represents *Q. dolicholepis*.

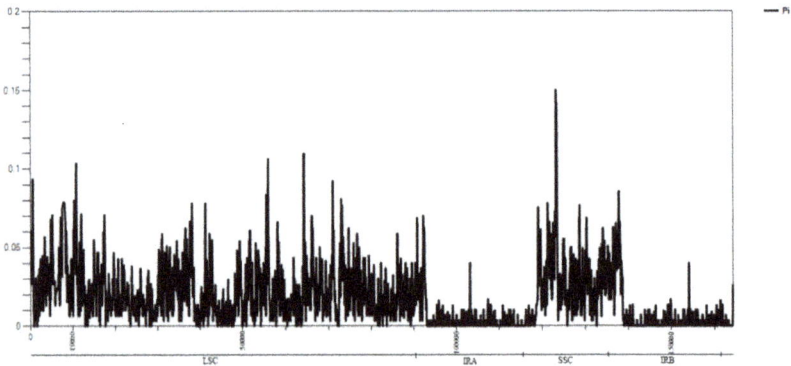

Figure A2. Percentage of variation in the complete cp genomes of the six species. The regions are oriented according to their locations in the genome.

References

1. Aldrich, P.R.; Cavender-Bares, J. Quercus. *Wild Crop Relat. Genom. Breed. Resour.* **2011**, 89–129. [CrossRef]
2. Manos, P.S.; Cannon, C.H.; Oh, S.H. Phylogenetic relationships and taxonomic status of the paleoendemic Fagaceae of western North America: Recognition of a new genus, *Notholithocarpus. Madroño* **2008**, *55*, 181–190. [CrossRef]
3. Oh, S.H.; Manos, P.S. Molecular phylogenetics and cupule evolution in Fagaceae as inferred from nuclear crabs claw sequences. *Taxon* **2008**, *57*, 434–451.

4. Pelser, P.B.; Kennedy, A.H.; Tepe, E.J.; Shidler, J.B.; Nordenstam, B.; Kadereit, J.W.; Watson, L.E. Patterns and causes of incongruence between plastid and nuclear *Senecioneae* (Asteraceae) phylogenies. *Am. J. Bot.* **2010**, *97*, 856–873. [CrossRef] [PubMed]

5. Pérezescobar, O.A.; Balbuena, J.A.; Gottschling, M. Rumbling orchids: How to assess divergent evolution between chloroplast endosymbionts and the nuclear host. *Syst. Biol.* **2016**, *65*, 51. [CrossRef] [PubMed]

6. Curtu, A.L.; Gailing, O.; Leinemann, L.; Finkeldey, R. Genetic variation and differentiation within a natural community of five oak species (*Quercus* spp.). *Plant Biol.* **2006**, *9*, 116–126. [CrossRef] [PubMed]

7. Kleinschmit, J.; Kleinschmit, J.G.R.; Vukelic, J.; Anic, I. *Quercus robur-Quercus petraea*: A critical review of the species concept. *Glasnik Za Šumske Pokuse* **2000**, *37*, 441–452.

8. Denk, T.; Grimm, G.W. The oaks of western Eurasia: Traditional classifications and evidence from two nuclear markers. *Taxon* **2010**, *59*, 351–366.

9. Kremer, A.; Abbott, A.G.; Carlson, J.E.; Manos, P.S.; Plomion, C.; Sisco, P.; Staton, M.E.; Ueno, S.; Vendramin, G.G. Genomics of Fagaceae. *Tree Genet. Genomes* **2012**, *8*, 583–610. [CrossRef]

10. Simeone, M.C.; Piredda, R.; Papini, A.; Vessella, F.; Schirone, B. Application of plastid and nuclear markers to DNA barcoding of Euro-Mediterranean oaks (*Quercus*, Fagaceae): Problems, prospects and phylogenetic implications. *Bot. J. Linn. Soc.* **2013**, *172*, 478–499. [CrossRef]

11. Hipp, A.L. Should hybridization make us skeptical of the oak phylogeny? *Int. Oaks* **2015**, *26*, 9–17.

12. Denk, T.; Grimm, G.W.; Manos, P.S.; Deng, M.; Hipp, A.L. An updated infrageneric classification of the oaks: Review of previous taxonomic schemes and synthesis of evolutionary patterns. In *Oaks Physiological Ecology. Exploring the Functional Diversity of Genus Quercus* L.; Springer: Cham, Switzerland, 2017; pp. 13–38.

13. Zhou, Z.; Wilkinson, H.; Wu, Z. Taxonomical and evolutionary implications of the leaf anatomy and architecture of *Quercus* L. Subgenus *Quercus* from China. *Cathaya* **1995**, *7*, 1–34.

14. Pu, C.; Zhou, Z.; Luo, Y. A cladistic analysis of *Quercus* (Fagaceae) in China based on leaf epidermic and architecture. *Acta Bot. Yunnanica* **2002**, *24*, 689–698.

15. Peng, Y.S.; Chen, L.; Li, J.Q. Study on Numerical Taxonomy of *Quercus* L. (Fagaceae) in China. *J. Plant Sci.* **2007**, *25*, 149–157.

16. Zhang, X.; Yao, L.I.; Fang, Y. Geographical distribution and prediction of potential ranges of *Quercus acutissima* in China. *Acta Bot. Boreali-Occident. Sin.* **2014**, *34*, 1685–1692.

17. Zhang, X.; Li, Y.; Liu, C.; Xia, T.; Zhang, Q.; Fang, Y. Phylogeography of the temperate tree species *Quercus acutissima* in China: Inferences from chloroplast DNA variations. *Biochem. Syst. Ecol.* **2015**, *63*, 190–197. [CrossRef]

18. Hui, L.; Xie, H.; Jiang, Z.; Li, C.; Zhang, G. Photosynthetic response of potted *Quercus acutissima* Carruth seedlings under different soil moisture conditions. *Sci. Soil Water Conserv.* **2013**, *11*, 93–97.

19. Fang, S.; Liu, Z.; Cao, Y.; Liu, D.; Yu, M.; Tang, L. Sprout development, biomass accumulation and fuelwood characteristics from coppiced plantations of *Quercus acutissima*. *Biomass Bioenergy* **2011**, *35*, 3104–3114. [CrossRef]

20. Wu, T.; Wang, G.G.; Wu, Q.; Cheng, X.; Yu, M.; Wang, W.; Yu, X. Patterns of leaf nitrogen and phosphorus stoichiometry among *Quercus acutissima* provenances across China. *Ecol Complex.* **2014**, *17*, 32–39. [CrossRef]

21. Choi, H.S.; Kim, Y.Y.; Hong, K.N.; Hong, Y.P.; Hyun, J.O. Genetic structure of a population of *Quercus acutissima* in Korea revealed by microsatellite markers. *Korean J. Genet.* **2005**, *27*, 267–271.

22. Huang, L.; Xiao, L.I.; Yan, J. Studies on Introduction of North American Oaks. China Forestry Science and Technology. 2005. Available online: http://xueshu.baidu.com/s?wd=paperuri%3A%2866d7b49f4975cf2de13aa699e48387b1%29&filter=sc_long_sign&tn=SE_xueshusource_2kduw22v&sc_vurl=http%3A%2F%2Fen.cnki.com.cn%2FArticle_en%2FCJFDTOTAL-LKKF200501009.htm&ie=utf-8&sc_us=11198188077522908127 (accessed on 16 August 2018).

23. Shen, X.; Wu, M.; Liao, B.; Liu, Z.; Bai, R.; Xiao, S.; Li, X.; Zhang, B.; Xu, J.; Chen, S. Complete chloroplast genome sequence and phylogenetic analysis of the medicinal plant *Artemisia annua*. *Molecules* **2017**, *22*, 1330. [CrossRef] [PubMed]

24. Guo, S.; Guo, L.; Zhao, W.; Xu, J.; Li, Y.; Zhang, X.; Shen, X.; Wu, M.; Hou, X. Complete chloroplast genome sequence and phylogenetic analysis of *Paeonia ostii*. *Molecules* **2018**, *23*, 246. [CrossRef] [PubMed]

25. Lobry, J.R. Asymmetric substitution patterns in the two DNA strands of bacteria. *Mol. Biol. Evol.* **1996**, *13*, 660–665. [CrossRef] [PubMed]

26. Necsulea, A.; Lobry, J. A new method for assessing the effect of replication on DNA base composition asymmetry. *Mol. Biol. Evol.* **2007**, *24*, 2169–2179. [CrossRef] [PubMed]

27. Tillier, E.R.; Collins, R.A. The contributions of replication orientation, gene direction, and signal sequences to base-composition asymmetries in bacterial genomes. *J. Mol. Evol.* **2000**, *50*, 249–257. [CrossRef] [PubMed]

28. Jansen, R.K.; Raubeson, L.A.; Boore, J.L.; Depamphilis, C.W.; Chumley, T.W.; Haberle, R.C.; Wyman, S.K.; Alverson, A.J.; Peery, R.; Herman, S.J. Methods for obtaining and analyzing whole chloroplast genome sequences. *Method Enzymol.* **2005**, *395*, 348.

29. Shetty, S.M.; Md Shah, M.U.; Makale, K.; Mohd-Yusuf, Y.; Khalid, N.; Othman, R.Y. Complete chloroplast genome sequence of *Musa balbisiana* corroborates structural heterogeneity of inverted repeats in wild progenitors of cultivated bananas and plantains. *Plant Genome* **2016**, *9*. [CrossRef] [PubMed]

30. Boudreau, E.; Takahashi, Y.; Lemieux, C.; Turmel, M.; Rochaix, J.D. The chloroplast *ycf3* and *ycf4* open reading frames of Chlamydomonas reinhardtii are required for the accumulation of the photosystem I complex. *Embo J.* **1997**, *16*, 6095–6104. [CrossRef] [PubMed]

31. Cavender Bares, J.; González Rodríguez, A.; Eaton, D.A.R.; Hipp, A.A.L.; Beulke, A.; Manos, P.S. Phylogeny and biogeography of the American live oaks (*Quercus* subsection *Virentes*): A genomic and population genetics approach. *Mol Ecol.* **2015**, *24*, 3668–3687. [CrossRef] [PubMed]

32. Kode, V.; Mudd, E.A.; Iamtham, S.; Day, A. The tobacco plastid *accD* gene is essential and is required for leaf development. *Plant J.* **2005**, *44*, 237–244. [CrossRef] [PubMed]

33. Raubeson, L.A.; Peery, R.; Chumley, T.W.; Dziubek, C.; Fourcade, H.M.; Boore, J.L.; Jansen, R.K. Comparative chloroplast genomics: Analyses including new sequences from the angiosperms *Nuphar advena* and *Ranunculus macranthus*. *BMC Genom.* **2007**, *8*, 174. [CrossRef] [PubMed]

34. Yao, X.; Tang, P.; Li, Z.; Li, D.; Liu, Y.; Huang, H. The first complete chloroplast genome sequences in *Actinidiaceae*: Genome structure and comparative analysis. *PLoS ONE* **2015**, *10*, e129347. [CrossRef] [PubMed]

35. Dong, W.; Xu, C.; Li, C.; Sun, J.; Zuo, Y.; Shi, S.; Cheng, T.; Guo, J.; Zhou, S. Ycf1, the most promising plastid DNA barcode of land plants. *Sci. Rep.* **2015**, *5*, 8348. [CrossRef] [PubMed]

36. Kikuchi, S.; Bédard, J.; Hirano, M.; Hirabayashi, Y.; Oishi, M.; Imai, M.; Takase, M.; Ide, T.; Nakai, M. Uncovering the protein translocon at the chloroplast inner envelope membrane. *Science* **2013**, *339*, 571. [CrossRef] [PubMed]

37. Provan, J. Novel chloroplast microsatellites reveal cytoplasmic variation in *Arabidopsis thaliana*. *Mol. Ecol.* **2000**, *9*, 2183–2185. [CrossRef] [PubMed]

38. Flannery, M.L.; Mitchell, F.J.; Coyne, S.; Kavanagh, T.A.; Burke, J.I.; Salamin, N.; Dowding, P.; Hodkinson, T.R. Plastid genome characterisation in *Brassica* and Brassicaceae using a new set of nine SSRs. *Theor. Appl. Genet.* **2006**, *113*, 1221–1231. [CrossRef] [PubMed]

39. Saito, Y.; Tsuda, Y.; Uchiyama, K.; Saito, Y.; Tsuda, Y.; Uchiyama, K.; Fukuda, T.; Seto, Y.; Kim, P.G.; Shen, H.L.; et al. Genetic Variation in *Quercus acutissima* Carruth., in Traditional Japanese Rural Forests and Agricultural Landscapes, Revealed by Chloroplast Microsatellite Markers. *Forests* **2017**, *8*, 451. [CrossRef]

40. Yang, Y.; Zhu, J.; Feng, L.; Zhou, T.; Bai, G.; Yang, J.; Zhao, G. Plastid genome comparative and phylogenetic analyses of the key genera in Fagaceae: Highlighting the effect of codon composition bias in phylogenetic inference. *Front. Plant Sci.* **2018**, *9*, 82. [CrossRef] [PubMed]

41. Wang, L.; Wuyun, T.N.; Du, H.; Wang, D.; Cao, D. Complete chloroplast genome sequences of *Eucommia ulmoides*: Genome structure and evolution. *Tree Genet. Genomes* **2016**, *12*, 12. [CrossRef]

42. Doyle, J.J. A rapid DNA isolation procedure for small quantities of fresh leaf tissue. *Phytochem. Bull.* **1987**, *19*, 11–15.

43. Dierckxsens, N.; Mardulyn, P.; Smits, G. Novoplasty: *De novo* assembly of organelle genomes from whole genome DNA. *Nucleic Acids Res.* **2017**, *45*, e18. [PubMed]

44. Chang, L.; Shi, L.; Zhu, Y.; Chen, H.; Zhang, J.; Lin, X.; Guan, X. CpGAVAS, an integrated web server for the annotation, visualization, analysis, and GenBank submission of completely sequenced chloroplast genome sequences. *BMC Genom.* **2012**, *13*, 715.

45. Wyman, S.K.; Jansen, R.K.; Boore, J.L. Automatic annotation of organellar genomes with DOGMA. *Bioinformatics* **2004**, *20*, 3252–3255. [CrossRef] [PubMed]

46. Schattner, P.; Brooks, A.N.; Lowe, T.M. The tRNAscan-SE, snoscan and snoGPS web servers for the detection of tRNAs and snoRNAs. *Nucleic Acids Res.* **2005**, *33*, W686. [CrossRef] [PubMed]

47. Lohse, M.; Drechsel, O.; Bock, R. Organellar Genome DRAW (OGDRAW): A tool for the easy generation of high-quality custom graphical maps of plastid and mitochondrial genomes. *Curr. Genet.* **2007**, *52*, 267–274. [CrossRef] [PubMed]

48. Mudunuri, S.B.; Nagarajaram, H.A. IMEx: Imperfect Microsatellite Extractor. *Bioinformatics* **2007**, *23*, 1181–1187. [CrossRef] [PubMed]

49. Kurtz, S.; Choudhuri, J.V.; Ohlebusch, E.; Schleiermacher, C.; Stoye, J.; Giegerich, R. REPuter: The manifold applications of repeat analysis on a genomic scale. *Nucleic Acids Res.* **2001**, *29*, 4633–4642. [CrossRef] [PubMed]

50. Kurtz, S.; Phillippy, A.; Delcher, A.L.; Smoot, M.; Shumway, M.; Antonescu, C.; Salzberg, S.L. Versatile and open software for comparing large genomes. *Genome Biol.* **2004**, *5*, R12. [CrossRef] [PubMed]

51. Mayor, C.; Brudno, M.; Schwartz, J.R.; Poliakov, A.; Rubin, E.M.; Frazer, K.A.; Pachter, L.S.; Dubchak, I. VISTA: Visualizing global DNA sequence alignments of arbitrary length. *Bioinformatics* **2000**, *16*, 1046–1047. [CrossRef] [PubMed]

52. Frazer, K.A.; Pachter, L.; Poliakov, A.; Rubin, E.M.; Dubchak, I. VISTA: Computational tools for comparative genomics. *Nucleic Acids Res.* **2004**, *32*, W273. [CrossRef] [PubMed]

53. Katoh, K.; Kuma, K.; Toh, H.; Miyata, T. MAFFT version 5: Improvement in accuracy of multiple sequence alignment. *Nucleic Acids Res.* **2005**, *33*, 511–518. [CrossRef] [PubMed]

54. Hall, T.A. BioEdit: A user-friendly biological sequence alignment editor and analysis program for windows 95/98/NT. *Nucleic Acids Symp. Ser.* **1999**, *41*, 95–98.

55. Lam-Tung, N.; Schmidt, H.A.; Arndt, V.H.; Quang, M.B. IQ-TREE: A fast and effective stochastic algorithm for estimating maximum-likelihood phylogenies. *Mol. Biol. Evol.* **2015**, *32*, 268–274.

56. Huelsenbeck, J.P.; Ronquist, F. MRBAYES: Bayesian inference of phylogenetic trees. *Bioinformatics* **2001**, *17*, 754–755. [CrossRef] [PubMed]

International Journal of
Molecular Sciences

MDPI

Article

Comparative Genomics of the Balsaminaceae Sister Genera *Hydrocera triflora* and *Impatiens pinfanensis*

Zhi-Zhong Li [1,2,†], Josphat K. Saina [1,2,3,†], Andrew W. Gichira [1,2,3], Cornelius M. Kyalo [1,2,3], Qing-Feng Wang [1,3,*] and Jin-Ming Chen [1,3,*]

[1] Key Laboratory of Aquatic Botany and Watershed Ecology, Wuhan Botanical Garden, Chinese Academy of Sciences, Wuhan 430074, China; wbg_georgelee@163.com (Z.-Z.L.); jksaina@wbgcas.cn (J.K.S.); gichira@wbgcas.cn (A.W.G.); cmulili90@gmail.com (C.M.K.)
[2] University of Chinese Academy of Sciences, Beijing 100049, China
[3] Sino-African Joint Research Center, Chinese Academy of Sciences, Wuhan 430074, China
* Correspondence: qfwang@wbgcas.cn (Q.-F.W.); jmchen@wbgcas.cn (J.-M.C.); Tel.: +86-27-8751-0526 (Q.-F.W.); +86-27-8761-7212 (J.-M.C.)
† These authors contributed equally to this work.

Received: 21 December 2017; Accepted: 15 January 2018; Published: 22 January 2018

Abstract: The family Balsaminaceae, which consists of the economically important genus *Impatiens* and the monotypic genus *Hydrocera*, lacks a reported or published complete chloroplast genome sequence. Therefore, chloroplast genome sequences of the two sister genera are significant to give insight into the phylogenetic position and understanding the evolution of the Balsaminaceae family among the Ericales. In this study, complete chloroplast (cp) genomes of *Impatiens pinfanensis* and *Hydrocera triflora* were characterized and assembled using a high-throughput sequencing method. The complete cp genomes were found to possess the typical quadripartite structure of land plants chloroplast genomes with double-stranded molecules of 154,189 bp (*Impatiens pinfanensis*) and 152,238 bp (*Hydrocera triflora*) in length. A total of 115 unique genes were identified in both genomes, of which 80 are protein-coding genes, 31 are distinct transfer RNA (tRNA) and four distinct ribosomal RNA (rRNA). Thirty codons, of which 29 had A/T ending codons, revealed relative synonymous codon usage values of >1, whereas those with G/C ending codons displayed values of <1. The simple sequence repeats comprise mostly the mononucleotide repeats A/T in all examined cp genomes. Phylogenetic analysis based on 51 common protein-coding genes indicated that the Balsaminaceae family formed a lineage with Ebenaceae together with all the other Ericales.

Keywords: Balsaminaceae; chloroplast genome; *Hydrocera triflora*; *Impatiens pinfanensis*; phylogenetic analyses

1. Introduction

The family Balsaminaceae of the order Ericales contains only two genera, *Impatiens* Linnaeus (1753:937) and *Hydrocera* Wight and Arnott (1834:140) and are predominantly perennial and annual herbs [1]. The monotypic genus *Hydrocera*, with a single species *Hydrocera triflora*, is characterized by actinomorphic flowers, a pentamerous calyx and corolla without any fusion between perianth parts, contrary to highly similar sister genus *Impatiens* whose flowers are highly zygomorphic [2]. *Impatiens*, one of the largest genera in angiosperms, consists of over 1000 species [3–6] primarily distributed in the Old World tropics, subtropics and temperate regions, but also in Europe, and central and North America [5,7]. In contrast, the sister *Hydrocera*, which is a semi-aquatic plant, is restricted to the lowlands of Indo-Malaysia [1]. Besides, the geographical regions, including south-east Asia, the eastern Himalayas, tropical Africa, Madagascar, southern India and Sri Lanka occupied by *Impatiens*, have been identified as diversity hotspots [7,8]. Recently, numerous new species have been recorded within these regions each year [9–14].

The controversial nature of classification of the genus *Impatiens* [1,15], for example different floral characters, its hybridization nature and species radiation, has made it under-studied. The species in prolific genus *Impatiens* are economically used as ornamentals, medicinal, as well as experimental research plant materials [16]. Additionally, previous studies have shown the genus *Impatiens* to possess potential anticancer compounds by decreasing patients' cancer cell count and increasing their life span and body weight [17]. The glanduliferins A and B isolated from the stem act to inhibit the growth of human cancer cells for growth inhibitory activity of human cancer cells [18]. As well, some polyphenols from *Impatiens* stems have showed antioxidant and antimicrobial activities [19].

In angiosperms, the chloroplast genome (cp) typically has a quadripartite organization consisting of a small single copy (SSC, 16–27 kb) and one large single copy (LSC) of about 80–90 kb long separated by two identical copies of inverted repeats (IRs) of about 20–88 kb with the total complete chloroplast genome size ranging from 72 to 217 kb [20–22]. Most of the complete cp genomes contains 110–130 distinct genes, with approximately 80 genes coding for proteins, 30 tRNA and 4 rRNA genes [21]. In addition, due to the highly conserved gene order and gene content, they have been used in plant evolution and systematic studies [23], determining evolutionary patterns of the cp genomes [24], phylogenetic analysis [25,26], and comparisons of angiosperm, gymnosperm, and fern families [27]. Moreover, the cp genomes are useful in genetic engineering [28], phylogenetics and phylogeography of angiosperms [29], and estimation of the diversification pattern and ancestral state of the vegetation within the family [30].

The Ericales (Bercht and Presl) form a well-supported clade (Asterid) containing more than 20 families [31]. Up to now, complete cp genomes representing approximately half of the families in the order Ericales have been sequenced including: Actinidiaceae [32,33], Ericaceae [34,35], Ebenaceae [36], Sapotaceae [37], Primulaceae [38,39] Styracaceae [40], and Theaceae, Pentaphylacaceae, Sladeniaceae, Symplocaceae, Lecythidaceae [30]. In addition the *Impatiens* and *Hydrocera* intergeneric phylogenetic relationship has been done using chloroplast *atpB-rbcL* spacer sequences [4]. However, there are no reports of complete chloroplast genomes in the family Balsaminaceae to date. This limitation of genetic information has hindered the progress and understanding in taxonomy, phylogeny, evolution and genetic diversity of Balsaminaceae. Analyses of more cp genomes are needed to provide a robust picture of generic and familial relationships of families in order Ericales.

This study aims to determine the complete sequences of the chloroplast genomes of *I. pinfanensis* (Hook. f.) and *H. triflora* using a high-throughput sequencing method. Additionally, comparisons with other published cp genomes in the order Ericales will be made in order to determine phylogenetic relationships among the representatives of Ericales.

2. Results and Discussion

2.1. The I. pinfanensis and H. triflora Chloroplast Genome Structure and Gene Content

The complete chloroplast genomes of *I. pinfanensis* and *H. triflora* share the common feature of possessing a typical quadripartite structure composed of a pair of inverted repeats (IRs) separating a large single copy (LSC) and a small single copy (SSC), similar to other angiosperm cp genomes [23]. The cp genome size of *I. pinfanensis* is 154,189 bp, with a pair of inverted repeats (IRs) of 17,611 bp long that divide LSC of 83,117 bp long and SSC of 25,755 bp long (Table 1). On the other hand, the *H. triflora* complete cp genome is 152,238 bp in length comprising a LSC region of 84,865 bp in size, a SSC of 25,622 bp size, and a pair of IR region 18,082 bp each in size. The overall guanine-cytosine (GC) contents of *I. pinfanensis* and *H. triflora* genomes are 36.8% and 36.9% respectively. Meanwhile, the GC contents in the LSC, SSC, and IR regions are 34.5%/34.7%, 29.3%/29.9%, and 43.1%/43.1% respectively.

Table 1. Comparison of the chloroplast genomes of *Impatiens pinfanensis* and *Hydrocera triflora*.

Species	*Impatiens pinfanensis*	*Hydrocera triflora*
Total Genome length (bp)	154,189	152,238
Overall G/C content (%)	36.8	36.9
Large single copy region	83,117	84,865
GC content (%)	34.5	34.7
Short single copy region	25,755	25,622
GC content (%)	29.3	29.9
Inverted repeat region	17,611	18,082
GC content (%)	43.1	43.1
Protein-Coding Genes	80	80
tRNAs	31	31
rRNAs	4	4
Genes with introns	17	17
Genes duplicated by IR	18	18

Like in typical angiosperms, both *I. pinfanensis* and *H. triflora* cp genomes encode 115 total distinct genes of which 80 are protein coding, 31 distinct tRNA and four distinct rRNA genes. Of these 62 genes coding for proteins and 23 tRNA genes were located in the LSC region, seven protein-coding genes, all the four rRNA genes and seven tRNA genes were replicated in the IR regions, while the SSC region was occupied by 11 protein-coding genes and one tRNA gene. The *ycf1* gene was located at the IR and SSC boundary region (Figures 1 and 2).

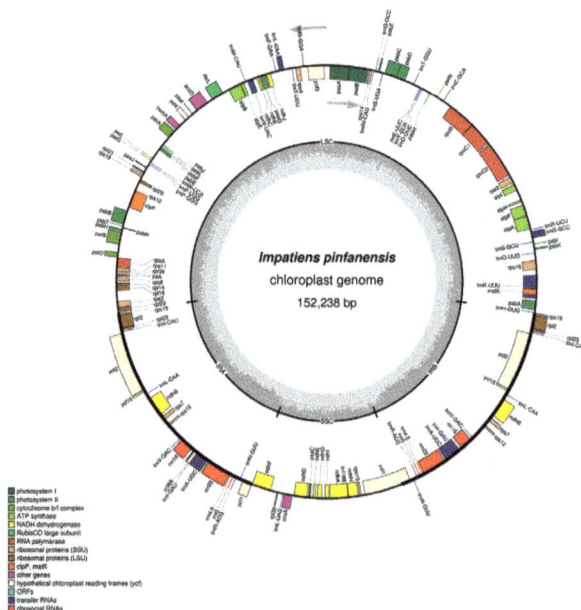

Figure 1. Gene map of the *Impatiens pinfanensis* chloroplast genome. Genes lying outside of the circle are transcribed clockwise, while genes inside the circle are transcribed counterclockwise. The colored bars indicate different functional groups. The dark gray area in the inner circle corresponds to GC content while the light gray corresponds to the adenine-thymine (AT) content of the genome.

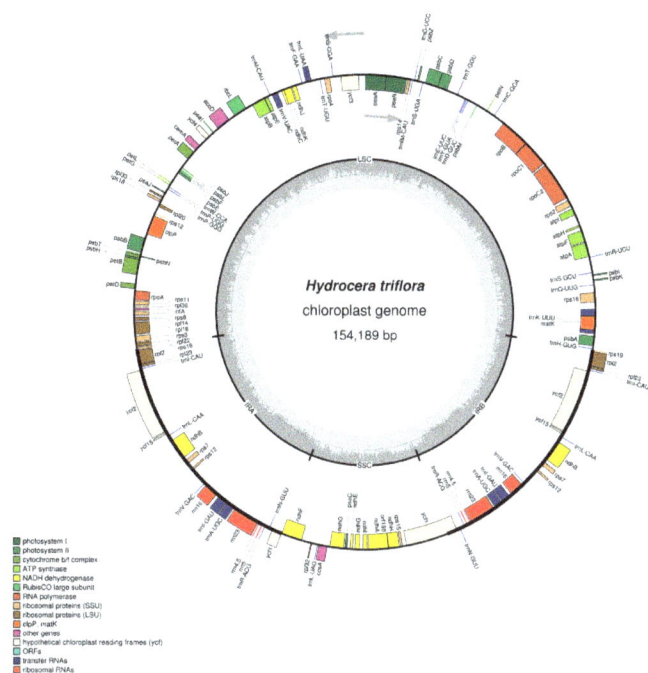

Figure 2. Gene map of the *Hydrocera triflora* chloroplast genome. Genes lying outside of the circle are transcribed clockwise, while genes inside the circle are transcribed counterclockwise. The colored bars indicate different functional groups. The dark gray area in the inner circle corresponds to (guanine cytosine) GC content while the light gray corresponds to the AT content of the genome.

Among the 115 unique genes in *I. pinfanensis* and *H. triflora* cp genomes, 14 genes contain one intron, comprised of eight genes coding for proteins (*atpF, rpoC1, rpl2, petB, rps16, ndhA, ndhB, ndhK*) and six tRNAs (*trnL-UAA, trnV-UAC, trnK-UUU, trnI-GAU, trnG-GCC* and *trnA-UGC*) (Table 2), while *ycf3, clpP* and *rps12* genes each contain two introns. These genes have maintained intron content in other angiosperms. The trans-splicing gene *rps12* has its 5′exon located in LSC, whereas the 3′exon is located in the IRs, which is similar to that in *Diospyros* species (Ebenaceae) [36,41] and *Actinidia chinensis* (Actinidiaceae) [41]. Oddly, *rps19* and *ndhD* genes in both species begin with uncommon start codons GTG and ACG respectively, which is consistent with previous reports in other plants [36]. However, the standard start codon can be restored through RNA editing process [42,43].

The complete cp genome of *I. pinfanensis* and *H. triflora* were found to be similar, although some slight variations such as genome size, gene loss and IR expansion and contraction factors were detected, despite the two species being from the same family Balsaminaceae. For instance, *H. triflora* cp genome is 1951 bp smaller than that of sister species *I. pinfanensis*. The SSC region of *I. pinfanensis* is shorter (17,611 bp) compared to that of *H. triflora*, which is 18,082 bp long. The GC content of *H. triflora* is slightly higher (36.9%) than that of *I. pinfanensis* (36.8%). Both species possess highest GC values in the IR regions (43.1%) compared to LSC and SSC region showing the lowest values (34.5%/34.7% and 29.3%/29.9%) respectively. The IR region is more conserved than the single copy region (SSC) in both species, due to presence of conserved rRNA genes in the IR region, which is also the reason for its high GC content. Both cp genomes are AT-rich with the genome organization and content of the two species almost the same and highly conserved, these results are similar to those of other recently published Ericales chloroplast genomes [34,36].

Table 2. Genes encoded in the *Impatiens pinfanensis* and *Hydrocera triflora* Chloroplast genomes.

Group of Genes	Gene Name
rRNA genes	*rrn*16(×2), *rrn*23(×2), *rrn*4.5(×2), *rrn*5(×2),
tRNA genes	*trn*A-UGC * (×2), *trn*C-GCA, *trn*D-GUC, *trn*E-UUC, *trn*F-GAA, *trn*G-GCC *, *trn*G-UCC, *trn*H-GUG, *trn*I-CAU(×2), *trn*I-GAU * (×2), *trn*K-UUU *, *trn*L-CAA(×2), *trn*L-UAA *, *trn*L-UAG, *trn*fM-CAU, *trn*M-CAU, *trn*N-GUU(×2), *trn*P-GGG *trn*P-UGG, *trn*Q-UUG, *trn*R-ACG(×2), *trn*R-UCU, *trn*S-GCU, *trn*S-GGA, *trn*S-UGA, *trn*T-GGU, *trn*T-UGU, *trn*V-GAC(×2), *trn*V-UAC *, *trn*W-CCA, *trn*Y-GUA
Ribosomal small subunit	*rps*2, *rps*3, *rps*4, *rps*7(×2), *rps*8, *rps*11, *rps*12_5′end, *rps*12_3′end * (×2), *rps*14, *rps*15, *rps*16 *, *rps*18, *rps*19
Ribosomal large subunit	*rpl*2 * (×2), *rpl*14, *rpl*16, *rpl*20, *rpl*22, *rpl*23(×2), *rpl*32, *rpl*33, *rpl*36
DNA-dependent RNA polymerase	*rpo*A, *rpo*B, *rpo*C1 *, *rpo*C2
Large subunit of rubisco	*rbc*L
Photosystem I	*psa*A, *psa*B, *psa*C, *psa*I, *psa*J, *ycf*3 **
Photosystem II	*psb*A, *psb*B, *psb*C, *psb*D, *psb*E, *psb*F, *psb*H, *psb*I, *psb*J, *psb*K, *psb*L, *psb*M, *psb*N, *psb*T, *psb*Z
NADH dehydrogenase	*ndh*A *, *ndh*B * (×2), *ndh*C, *ndh*D, *ndh*E, *ndh*F, *ndh*G, *ndh*H, *ndh*I, *ndh*J, *ndh*K
Cytochrome b/f complex	*pet*A, *pet*B *, *pet*D, *pet*G, *pet*L, *pet*N
ATP synthase	*atp*A, *atp*B, *atp*E, *atp*F *, *atp*H, *atp*I
Maturase	*mat*K
Subunit of acetyl-CoA carboxylase	*acc*D
Envelope membrane protein	*cem*A
Protease	*clp*P **
Translational initiation factor	*inf*A
c-type cytochrome synthesis	*ccs*A
Conserved open reading frames (*ycf*)	*ycf*1, *ycf*2(×2), *ycf*4, *ycf*15(×2)

Genes with one or two introns are indicated by one (*) or two asterisks (**), respectively. Genes in the IR regions are followed by the (×2) symbol.

2.2. Codon Usage

The relative synonymous codon usage (RSCU) has been divided into four models, i.e., RSCU value of less than 1.0 (lack of bias), RSCU value between 1.0 and 1.2 (low bias), RSCU value between 1.2 and 1.3 (moderately bias) and RSCU value greater than 1.3 (highly bias) [44,45]. To determine codon usage, we selected 52 shared protein-coding genes between *I. pinfanensis* and *H. triflora* with length of >300 bp for calculating the effective number of codons. As shown in (Table 3), the relative synonymous codon usage (RSCU) and codon usage revealed biased codon usage in both species with values of 30 codons showing preferences (<1) except tryptophan and methionine, with 29 having A/T ending codons. The TAA stop codon was found to be preferred. All the protein-coding genes contained 22,900 and 22,995 codons in *I. pinfanensis* and *H. triflora* cp genomes respectively. In addition, our results indicated that 2408 and 2439 codons encode leucine while 253 and 259 encode cysteine in *I. pinfanensis* and *H. triflora* cp genomes as the most and least frequently universal amino acids respectively. The Number of codons (Nc) of the individual PCGs varied from *petD* (37.10) to *ycf3* (54.84) and *rps18* (32.11) to *rpl2* (54.24) in *I. pinfanensis* and *H. triflora* respectively (Table S1). Like recently reported in cp genomes of higher plants, our study showed that there was bias in the usage of synonymous codons except tryptophan and methionine. Our result is in line with previous findings of codon usage preference for A/T ending in other land plants [46,47].

2.3. SSR Analysis Results

Analysis of SSR occurrence using the microsatellite identification tool (MISA) detected Mono-, di-, tri-, tetra-, penta- and hexa-nucleotides categories of SSRs in the cp genomes of eight Ericales. A total of 197 and 159 SSRs were found in the *I. pinfanensis* and *H. triflora* cp genomes respectively. Not all the SSR types were identified in all the species, Penta and hexanucleotide repeats were not found in *I. pinfanensis*, *Diospyros lotus*, and *Pouteria campechiana*, while only hexanucleotides were not identified in *Ardisia polysticta* and *Barringtonia fusicarpa* (Table 4). Among the SSR types discovered mononucleotide repeat units were highly represented, which were found 180 and 141 times in *I. pinfanensis* and *H. triflora* respectively. Most of the mononucleotide repeats consisting of A or T were most common (117–176 times), whereas C/G were less in number (1–8 times), and all the dinucleotide repeat sequences in all the species were AT repeats. This result is consistent with previous reports, which showed most angiosperm cp genome to be AT-rich [36,38,48].

Table 3. Codon usage in *Impatiens pinfanensis* and *Hydrocera triflora* chloroplast genomes.

Amino Acid	Codon	Number I. pinfanensis	Number H. triflora	RSCU I. pinfanensis	RSCU H. triflora
Phe	UUU	913	908	**1.40**	**1.38**
	UUC	387	406	0.60	0.62
Leu	UUA	854	842	**2.11**	**2.07**
	UUG	468	486	**1.16**	**1.20**
	CUU	517	503	**1.28**	**1.24**
	CUC	160	162	0.40	0.40
	CUA	310	315	0.77	0.78
	CUG	121	128	0.30	0.32
Ile	AUU	1035	1020	**1.54**	**1.52**
	AUC	359	376	0.53	0.56
	AUA	624	611	0.93	0.91
Met	AUG	547	548	**1.00**	**1.00**
Val	GUU	482	469	**1.55**	**1.52**
	GUC	134	135	0.43	0.44
	GUA	457	457	**1.47**	**1.48**
	GUG	167	174	0.54	0.56
Tyr	UAU	704	697	**1.64**	**1.65**
	UAC	155	146	0.36	0.35
TER	UAA	41	44	**1.50**	**1.63**
	UAG	23	19	0.84	0.70
His	CAU	405	421	**1.54**	**1.57**
	CAC	121	114	0.46	0.43
Gln	CAA	627	626	**1.54**	**1.53**
	CAG	186	192	0.46	0.47
Asn	AAU	885	868	**1.59**	**1.57**
	AAC	231	238	0.41	0.43
Lys	AAA	976	978	**1.55**	**1.54**
	AAG	284	289	0.45	0.46
Asp	GAU	720	737	**1.64**	**1.64**
	GAC	159	160	0.36	0.36
Glu	GAA	914	929	**1.55**	**1.55**
	GAG	264	272	0.45	0.45
Ser	UCU	482	482	**1.69**	**1.67**
	UCC	252	264	0.88	0.92
	UCA	360	324	**1.26**	**1.12**
	UCG	142	181	0.50	0.63
Pro	CCU	376	371	**1.59**	**1.58**
	CCC	175	167	0.74	0.71
	CCA	294	290	**1.24**	**1.23**
	CCG	103	112	0.43	0.48
Thr	ACU	493	500	**1.70**	**1.74**
	ACC	198	180	0.68	0.63
	ACA	358	368	**1.24**	**1.28**
	ACG	108	104	0.37	0.36
Ala	GCU	580	593	**1.86**	**1.85**
	GCC	183	191	0.59	0.60
	GCA	346	353	**1.11**	**1.10**
	GCG	141	143	0.45	0.45
Cys	UGU	191	196	**1.53**	**1.51**
	UGC	58	63	0.47	0.49
TER	UGA	18	18	0.66	0.67
Trp	UGG	412	412	**1.00**	**1.00**
Arg	AGA	406	407	**1.81**	**1.77**
	AGG	134	143	0.60	0.62
Arg	CGU	302	299	**1.35**	**1.30**
	CGC	88	95	0.39	0.41
	CGA	317	333	**1.41**	**1.45**
	CGG	98	103	0.44	0.45
Ser	AGU	363	72	**1.27**	**1.29**
	AGC	110	108	0.39	0.37
Gly	GGU	525	525	**1.33**	**1.35**
	GGC	160	165	0.40	0.42
	GGA	639	625	**1.62**	**1.61**
	GGG	258	238	0.65	0.61

RSCU: Relative synonymous Codon Usage. RSCU > 1 are highlighted in bold.

Table 4. SSR types and amount in the *Impatiens pinfanensis* and *Hydrocera triflora* Chloroplast genomes.

SSR Type	Repeat Unit	Amount							
		Impatiens pinfanensis	*Hydrocera triflora*	*Actinidia kolomikta*	*Ardisia polysticta*	*Diospyros lotus*	*Barringtonia fusicarpa*	*Pouteria campechiana*	*Primula persimilis*
Mono	A/T	176	139	117	153	146	154	161	134
	C/G	4	2	4	4	4	8	1	4
Di	AT/AT	8	9	8	5	3	13	11	6
Tri	AAG/CTT	1	0	0	0	0	0	1	1
	AAT/ATT	3	3	2	1	1	2	4	2
	AGC/CTG	0	0	0	0	1	0	0	0
Tetra	AAAG/CTTT	1	0	3	2	1	3	1	1
	AAAT/ATTT	2	3	3	3	4	3	6	2
	AATG/ATTC	1	0	0	1	0	0	0	0
	AATT/AATT	1	0	1	0	0	0	1	0
	AGAT/ATCT	1	0	0	0	0	0	0	0
	AAGT/ACTT	0	1	0	0	0	1	0	0
	AACT/AGTT	0	0	0	1	0	1	0	0
	AATC/ATTG	0	0	2	0	1	1	0	0
	AAAC/GTTT	0	0	0	0	0	0	1	0
	AAGG/CCTT	0	0	0	0	0	0	1	0
Penta	AATAC/ATTGT	0	1	0	0	0	0	0	0
	AAAAT/ATTTT	0	0	1	0	0	0	0	0
	AAATT/AATTT	0	0	0	1	0	0	0	0
	AATGT/ACATT	0	0	0	0	0	1	0	0
	AATAT/ATATT	0	0	0	0	0	0	0	1
Hexa	AATCCC/ATTGGG	0	1	0	0	0	0	0	0
	AGATAT/ATATCT	0	0	0	0	0	0	0	1
	AAGATG/ATCTTC	0	0	1	0	0	0	0	0
Total		197	159	143	171	161	187	188	150

2.4. Selection Pressure Analysis of Evolution

The ratio of Synonymous (Ks) and non-synonymous (Ka) Substitution can determine whether the selection pressure has acted on a particular protein-coding sequence. Eighty common protein-coding genes shared by *I. pinfanensis* and *H. triflora* genomes were used. As suggested by Makałowski and Boguski [49] the Ka/Ks values are less than one in protein-coding genes as a result of less frequent non-synonymous (Ka) nucleotide Substitutions than the Synonymous (Ks) substitutions (Table S2). We found that the Ka/Ks values of the two species were low (<1) approaching zero, except for one gene *psbK* found in the LSC region, which has a ratio of 1.0259 (Figure 3). This indicates a negative selection all genes except *psbK* gene and shows that the protein-coding genes in both species are quite highly conserved (Table S2). The LSC, SSC, and IR regions average Ks values between the two species were 0.0995, 0.0314, and 0.1334 respectively. Based on Ka/Ks comparison among the regions, only *ycf1* gene in IR region and most of the genes in the LSC and SSC regions revealed higher Ks values. The higher Ks values signaled that on average more genes found in the SSC region have experienced higher selection pressures in contrast to other cp genome regions (LSC and IR). The non-synonymous (Ka) value varied from 0.005 (*psbE*) to 0.0927 (*ycf1*) while Ks ranged from 0.058 (*psbN*) to 0.2944 (*ndhE*). Based on sequence similarity among the IR, SSC and LSC regions, the IR region was more conserved. This is in agreement with previous reports that found out that IR region diverged at a slower rate than the LSC and SSC regions as a result of frequent recombinant events taking place in IR region leading to selective constraints on sequence homogeneity [50,51].

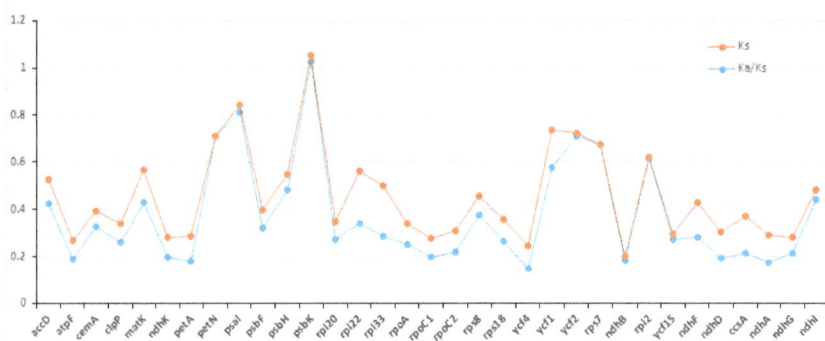

Figure 3. Non-synonymous (Ka) and synonymous (Ks) substitution rates and Ka/Ks ratio between *I. pinfanensis* and *H. triflora*. One gene *psbK* had Ka/Ks ratio greater than 1.0, whereas all the other genes were less than 1.0.

2.5. IR Expansion and Contraction

Despite of the highly conserved nature of the angiosperms inverted repeat (IRa/b) regions, the contraction or expansion at the IR junction are the usual evolutionary events resulting in varying cp genome sizes [52,53]. In our study, the IR/SSC and IR/LSC borders of *I. pinfanensis* and *H. triflora* were compared to those of the other six Ericales representatives (*P. persimilis, P. campechiana, D. lotus, B. fusicarpa, A. kolomikta* and *A. polysticta*) to identify the IR expansion or contraction (Figure 4). The IRb/SSC boundary expansions in all the eight species extended into the *ycf1* genes creating long *ᵠycf1* pseudogene fragments with varying length. The *ycf1* pseudogene length in *I. pinfanensis* is 1101 bp, 1095 bp in *H. triflora*, 394 bp in *A. kolomikta*, 974 bp in *A. polysticta*, 1058 bp in *B. fusicarpa*, 1203 bp in *D. lotus*, 1078 bp in *P. campechiana* and 1018 bp in *P. persimilis*. Additionally, the *ndhF* gene is situated in the SSC region in *I. pinfanensis, H. triflora, A. kolomikta, D. lotus,* and *P. persimilis*, and it ranges from 32 bp, 9 bp, 71 bp, 10 bp and 44 bp away from the IRb/SSC boundary region respectively, but this gene formed an overlap with the *ycf1* pseudogene in *A. polystica, B. fusicarpa* and *P. campechiana* cp genomes sharing some nucleotides of 3 bp, 1 bp and 1 bp in that order. The *rps19* gene is located at the /IRb/LSC junction, of *I. pinfanensis, H. triflora* and of the other five cp genomes, apart from *A. kolomikta* in which this gene is found in the LSC region, 151 bp gap from the LSC/IRb junction. Moreover, the occurrence of *rps19* gene at the LSC/IRb junction resulted in partial duplication of this gene at the corresponding region (IRa/LSC border) in *I. pinfanensis, H. triflora,* and *A. polysticta* cp genomes. The *trnH* gene is detected in the LSC region in *I. pinfanensis* and *H. triflora*. However, complete gene rearrangement of this *trnH* gene was observed resulting in complete duplication in the IR in the *A. kolomikta* chloroplast genome, 630 bp apart from the IR/LSC junction with *psbA* gene extending towards LSC/IRa border, however this gene is found in the LSC regions of the other five chloroplast genomes.

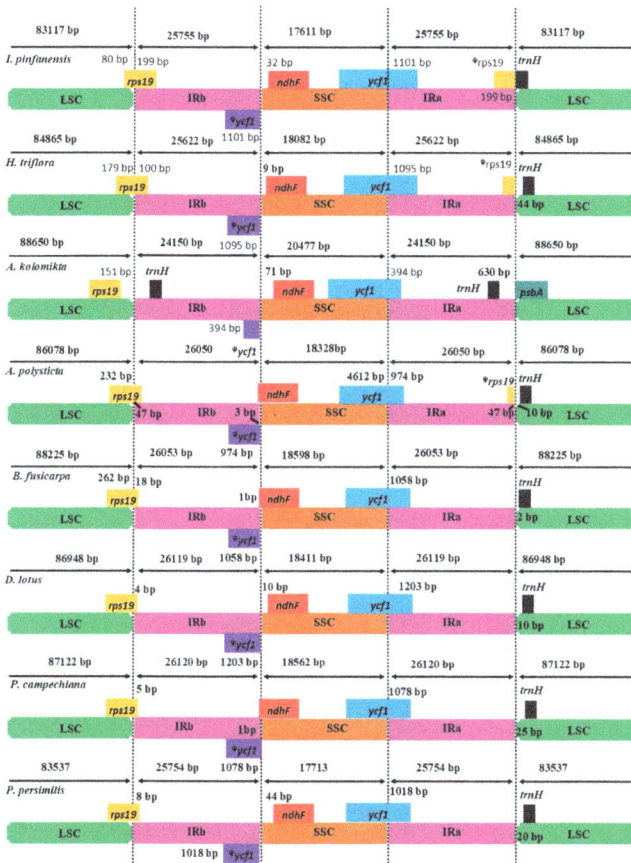

Figure 4. Comparison of IR, LSC and SSC border regions among eight Ericales cp genomes. The IRb/SSC junction extended into the *ycf1* genes creating various lengths of *ycf1* pseudogenes among the eight cp genomes. The numbers above, below or adjacent to genes shows the distance between the ends of genes and the boundary sites. The figure features are not to scale. $^{\varphi}$ indicates a pseudogene.

The border regions of the Ericales revealed that the *I. pinfanensis* and *H. triflora* cp genomes varied a little compared to other analyzed cp genomes. As shown in Figure 4, our analyses confirmed the

IR evolution as revealed by the incomplete *rps19* gene, which was duplicated in the IR region in *I. pinfanensis*, *H. triflora*, and *A. polysticta*. Conversely, this *rps19* gene was not duplicated among the remaining representatives of Ericales cp genomes. In a recent study [36,54] found that the *trnH* gene duplication occurs in Actinidiaceae, and Ericaceae. This duplication of genes in the LSC/IRb junction and the IRa/LSC junction would be of great importance in systematic studies. Furthermore, the *rps19* gene at the LSC/IRb in *I. pinfanensis* and *H. triflora* is largely extended into the IRb region (199 bp and 100 bp) respectively. The SSC region of *I. pinfanensis* is 471 bp smaller than that of sister species *H. triflora*, but also smallest among the other species used in this study. Additionally, the *I. pinfanensis* LSC region is smaller than that of other species. Previous studies have shown that there is expansion of single copy (SC) and IR regions of angiosperms cp genomes during evolution [50,55], the *I. pinfanensis* and *H. triflora* cp genomes revealed that the border areas were highly conserved despite of slight genome size differences between the two species.

2.6. Phylogenetic Analysis

Phylogenetic relationships within the order Ericales have been resolved in recent published reports but the position of Balsaminaceae still remains controversial [33,35–40]. In our study, the phylogenetic relationship of *I. pinfanensis*, and *H. triflora* and 38 other species of Ericales downloaded from GenBank (Table S3) was determined, with four cp genomes sequences belonging to Cornales being used as Outgroup species. Fifty-one common protein-coding sequences in all the selected cp genomes employed a single alignment data matrix of a total 35,548 characters (Supplementary Materials File S4). Almost all the nodes in the phylogenetic tree showed a strong bootstrap support. Though, Sapotaceae and Ebenaceae had low support (bootstrap < 70), this could be as a result of fewer samples in these families (Figure 5). *I. pinfanensis* and *H. triflora* as sister taxa (Balsaminaceae) formed the basal family of Ericales with intensive support. In general, all the 38 species together with the two Balsaminaceae family species formed a lineage (Ericales) recognizably discrete from the four outgroup species (Cornales). All the species grouped together into 10 clades corresponding to the 10 families in order Ericales according to APGIV system [31]. This study will provide resources for species identification and resolution of deeper phylogenetic branches among *Impatiens* and *Hydrocera* genera.

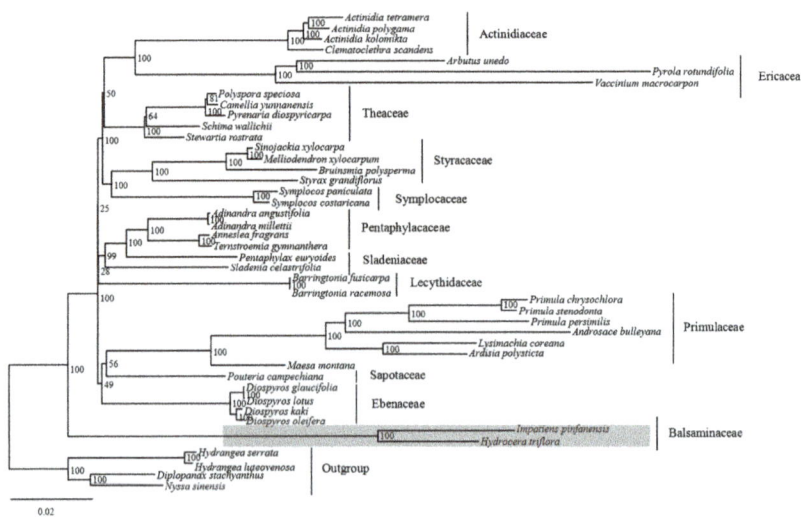

Figure 5. Phylogenetic relationships based on 51 common protein-coding genes of 38 representative species from order Ericales and four Cornales as Outgroup species with maximum likelihood. The numbers associated with the nodes indicate bootstrap values tested with 1000 replicates.

3. Materials and Methods

3.1. Plant Materials and DNA Extraction

Total genomic DNA was extracted from fresh leaves of the *I. pinfanensis* and *H. triflora* collected from Hubei province (108°42′19″ E, 30°12′33″ N) and Hainan province (110°18′57″ E, 19°23′10″ N) in China using a modified cetyltrimethylammonium bromide (CTAB) method [56]. The DNA quality was checked using spectrophotometry and their integrity examined by electrophoresis in 2% agarose gel. The voucher specimens (HIB-lzz07, HIB-lzz18) were deposited at the Wuhan Botanical Garden herbarium (HIB).

3.2. Chloroplast Genome Sequence Assembly and Annotation

The pair-end libraries were constructed using the Illumina Hiseq 2500 platform at NOVOgene Company (Beijing, China) with an average insert size of approximately 150 bp for each genome. The high-quality reads were filtered from Illumina raw reads using the PRINSEQ lite v0.20.4 (San Diego State University, San Diego, CA, USA) [57] (phredQ $\geq$ 20, Length $\geq$ 50), then assembled with closely related species cp genome using a BLASTn (with E value of 10^{-6}) with *Primula chrysochlora* (NC_034678) and *Diospyros lotus* (NC_030786) as reference species. In addition, the software Velvet v1.2.10 (Wellcome Trust Genome Campus, Hinxton, Cambridge, UK) [58] was used to assemble the obtained reads with K-mer length of 99–119. Then, consensus sequences with reference chloroplast genome was mapped using GENEIOUS 8.0.2 (Biomatters Ltd., Auckland, New Zealand) [59]. We used the online software local blast to verify the single copy (SC) and inverted repeat (IR) boundary regions of the assembled sequences.

The annotations of the complete cp genomes were performed using DOGMA (Dual Organellar GenoMe Annotator, University of Texas at Austin, Austin, TX, USA) [60]. The start and stop codons positions were further checked by local blast searches. Further, the tRNAs locations were confirmed with tRNAscan-SE v1.23 (http://lowelab.ucsc.edu/tRNAscan-SE/) [61]. The circular cp genome maps were generated using an online program (OGDrawV1.2, Max planck Institute of Molecular Plant Physiology, Potsdam, Germany) OrganellarGenomeDraw [62] with default settings plus manual corrections. Putative tRNAs, rRNAs and protein-coding genes were corrected by comparing them with the more similar reference species *Primula chrysochlora* (NC_034678) and *Diospyros lotus* (NC_030786) resulting from BLASTN and BLASTX searches against the nucleotide database NCBI (https://blast.ncbi.nlm.nih.gov/). The cp genome sequences were submitted to GenBank database, accession numbers *I. pinfanensis* (MG162586) and *H. triflora* (MG162585).

3.3. Genome Comparison and Structure Analyses

The IR and SC boundary regions of *I. pinfanensis* and *H. triflora*, and the other six Ericales species were compared and examined. For synonymous codon usage analysis, about 52 protein-coding genes of length > 300 bp were chosen. Online program CodonW1.4.2 (http://downloads.fyxm.net/CodonW-76666.html) was used to investigate the Nc and RSCU parameters. The simple sequence repeats (SSRs) of the two study species and other Ericales representatives were detected using MISA software [63] with SSR search parameters set same as Gichira et al. [48].

3.4. Substitution Rate Analysis—Synonymous (Ks) and Non-Synonymous (Ka)

We examined substitution rates synonymous (Ks) and non-synonymous (Ka) using Model Averaging in the KaKs_Cal-culator program (Institute of Genomics, Chinese Academy of Sciences, Beijing, China) [64]. Eighty common protein-coding genes shared by the *I. pinfanensis* and *H. triflora* were aligned separately using Geneious software v5.6.4 (Biomatters Ltd., Auckland, New Zealand) [59].

3.5. Phylogenetic Analyses

To locate the phylogenetic positions of *I. pinfanensis* and *H. triflora* (Balsaminaceae) within order Ericales, the chloroplast genome sequences of 38 species belonging to order Ericales and four Cornales species as outgroups, were used to reconstruct a phylogenetic relationships tree. The Phylogenetic tree was performed based on maximum likelihood (ML) analysis using RAxMLversion 8.0.20 (Scientific Computing Group, Heidelberg Institute for Theoretical Studies, Institute of Theoretical Informatics, Karlsruhe Institute of Technology, Karlsruhe, Germany) [65]. Consequently, based on the Akaike information criterion (AIC), the best-fitting substitution models (GTR + I + G) were selected (p-inv = 0.47, and gamma shape = 0.93) from jModelTest v2.1.7 [66]. The bootstrap test was performed in algorithm of RAxML with 1000 replicates.

4. Conclusions

The cp genomes of *I. pinfanensis*, and *H. triflora* from the family Balsaminaceae provide novel genome sequences and will be of benefit as a reference for further complete chloroplast genome sequencing within the family. The genome organization and gene content are well conserved typical of most angiosperms. Fifty protein-coding sequences, shared by selected species from Ericales as well as our study species, were used to construct the phylogenetic tree using the maximum likelihood (ML). Majority of the nodes showed strong bootstrap support values, and the few nodes with low support, should be solved using other methods (e.g., restriction-site-associated DNA sequencing). The two species (*I. pinfanensis*, and *H. triflora*) were placed close to each other. These findings strongly support Balsaminaceae as a basal family of the order Ericales. Lastly, the Balsaminaceae (*I. pinfanensis*, and *H. triflora*) has a relationship with the other 38 species, which are all grouped into one Clade (Ericales). This study will be of value in determining genome evolution and understanding phylogenomic relationships within Ericales and give precious resources for the evolutionary study of Balsaminaceae.

Supplementary Materials: Supplementary materials can be found at http://www.mdpi.com/1422-0067/19/2/319/s1.

Acknowledgments: This study was supported by the Special Funds for the Young Scholars of Taxonomy of Chinese Academy of Sciences grants to Liao Kuo (Grant number; ZSBR-013), the Special Foundation for State Basic Working Program of China (2013FY112300) and the National Natural Science Foundation of China (grant no. 31570220).

Author Contributions: Qing-Feng Wang and Jin-Ming Chen conceived and designed the experiment; Zhi-Zhong Li, Josphat K. Saina, Andrew W. Gichira and Cornelius M. Kyalo assembled sequences and revised the manuscript; Zhi-Zhong Li and Josphat K. Saina performed the experiments, analyzed the data and wrote the paper; Jin-Ming Chen and Zhi-Zhong Li collected the plant materials. All authors have read and approved the final version of the manuscript.

Conflicts of Interest: The authors declare no conflict of interest.

Abbreviations

IR	Inverted repeat
LSC	Large single copy
SSC	Small single copy
SSR	Simple sequence repeats
RSCU	Relative synonymous codon usage

References

1. Grey-Wilson, C. *Impatiens of Africa; Morphology, Pollination and Pollinators, Ecology, Phytogeography, Hybridization, Keys and a Systematics of All African Species with a Note on Collecting and Cultivation*; AA Balkema: Rotterdam, The Netherlands, 1980.

2. Janssens, S.B.; Smets, E.F.; Vrijdaghs, A. Floral development of *Hydrocera* and *Impatiens* reveals evolutionary trends in the most early diverged lineages of the Balsaminaceae. *Ann. Bot.* **2012**, *109*, 1285–1296. [CrossRef] [PubMed]

3. Fischer, E.; Rahelivololona, M.E. New taxa of *Impatiens* (Balsaminaceae) from Madagascar iii. *Adansonia* **2004**, *26*, 37–52.

4. Janssens, S.; Geuten, K.; Yuan, Y.-M.; Song, Y.; Küpfer, P.; Smets, E. Phylogenetics of *Impatiens* and *Hydrocera* (Balsaminaceae) using chloroplast atpb-rbcl spacer sequences. *Syst. Bot.* **2006**, *31*, 171–180. [CrossRef]

5. Janssens, S.B.; Knox, E.B.; Huysmans, S.; Smets, E.F.; Merckx, V.S. Rapid radiation of *Impatiens* (Balsaminaceae) during pliocene and pleistocene: Result of a global climate change. *Mol. Phylogenet. Evol.* **2009**, *52*, 806–824. [CrossRef] [PubMed]

6. Janssens, S.B.; Viaene, T.; Huysmans, S.; Smets, E.F.; Geuten, K.P. Selection on length mutations after frameshift can explain the origin and retention of the AP3/DEF-like paralogues in *Impatiens*. *J. Mol. Evol.* **2008**, *66*, 424–435. [CrossRef] [PubMed]

7. Yuan, Y.-M.; Song, Y.; Geuten, K.; Rahelivololona, E.; Wohlhauser, S.; Fischer, E.; Smets, E.; Küpfer, P. Phylogeny and biogeography of Balsaminaceae inferred from its sequences. *Taxon* **2004**, *53*, 391. [CrossRef]

8. Song, Y.; Yuan, Y.-M.; Küpfer, P. Chromosomal evolution in Balsaminaceae, with cytological observations on 45 species from Southeast Asia. *Caryologia* **2003**, *56*, 463–481. [CrossRef]

9. Tan, Y.-H.; Liu, Y.-N.; Jiang, H.; Zhu, X.-X.; Zhang, W.; Yu, S.-X. Impatiens pandurata (Balsaminaceae), a new species from Yunnan, China. *Bot. Stud.* **2015**, *56*, 29. [CrossRef] [PubMed]

10. Zeng, L.; Liu, Y.-N.; Gogoi, R.; Zhang, L.-J.; Yu, S.-X. Impatiens tianlinensis (Balsaminaceae), a new species from Guangxi, China. *Phytotaxa* **2015**, *227*, 253–260. [CrossRef]

11. Raju, R.; Dhanraj, F.I.; Arumugam, M.; Pandurangan, A. Impatiens matthewiana, a new scapigerous balsam from Western Ghats, India. *Phytotaxa* **2015**, *227*, 268–274. [CrossRef]

12. Guo, H.; Wei, L.; Hao, J.-C.; Du, Y.-F.; Zhang, L.-J.; Yu, S.-X. Impatiens occultans (Balsaminaceae), a newly recorded species from Xizang, China, and its phylogenetic position. *Phytotaxa* **2016**, *275*, 62–68. [CrossRef]

13. Cho, S.-H.; Kim, B.-Y.; Park, H.-S.; Phourin, C.; Kim, Y.-D. Impatiens bokorensis (Balsaminaceae), a new species from Cambodia. *PhytoKeys* **2017**, *77*, 33. [CrossRef] [PubMed]

14. Yang, B.; Zhou, S.-S.; Maung, K.W.; Tan, Y.-H. Two new species of *Impatiens* (Balsaminaceae) from Putao, Kachin state, northern Myanmar. *Phytotaxa* **2017**, *321*, 103–113. [CrossRef]

15. Hooker, J.D. Les Espèces du Genre "*Impatiens*" dans l'herbier du Museum de Paris. *Nov. Arch. Mus. Nat. Hist. Paris Ser.* **1908**, *10*, 233–272.

16. Bhaskar, V. *Taxonomic Monograph on 'Impatiens' ('Balsaminaceae') of Western Ghats, South India: The Key Genus for Endemism*; Centre for Plant Taxonomic Studies: Bengaluru, India, 2012.

17. Baskar, N.; Devi, B.P.; Jayakar, B. Anticancer studies on ethanol extract of *Impatiens balsamina*. *Int. J. Res. Ayurveda Pharm.* **2012**, *3*, 631–633.

18. Cimmino, A.; Mathieu, V.; Evidente, M.; Ferderin, M.; Banuls, L.M.Y.; Masi, M.; De Carvalho, A.; Kiss, R.; Evidente, A. Glanduliferins A and B, two new glucosylated steroids from *Impatiens glandulifera*, with in vitro growth inhibitory activity in human cancer cells. *Fitoterapia* **2016**, *109*, 138–145. [CrossRef] [PubMed]

19. Szewczyk, K.; Zidorn, C.; Biernasiuk, A.; Komsta, Ł.; Granica, S. Polyphenols from *Impatiens* (Balsaminaceae) and their antioxidant and antimicrobial activities. *Ind. Crops Prod.* **2016**, *86*, 262–272. [CrossRef]

20. Sugiura, M. The chloroplast genome. In *10 Years Plant Molecular Biology*; Springer: Berlin, Germany, 1992; pp. 149–168.

21. Chumley, T.W.; Palmer, J.D.; Mower, J.P.; Fourcade, H.M.; Calie, P.J.; Boore, J.L.; Jansen, R.K. The complete chloroplast genome sequence of *Pelargonium × hortorum*: Organization and evolution of the largest and most highly rearranged chloroplast genome of land plants. *Mol. Biol. Evol.* **2006**, *23*, 2175–2190. [CrossRef] [PubMed]

22. Tangphatsornruang, S.; Sangsrakru, D.; Chanprasert, J.; Uthaipaisanwong, P.; Yoocha, T.; Jomchai, N.; Tragoonrung, S. The chloroplast genome sequence of mungbean (*Vigna radiata*) determined by high-throughput pyrosequencing: Structural organization and phylogenetic relationships. *DNA Res.* **2009**, *17*, 11–22. [CrossRef] [PubMed]

23. Wicke, S.; Schneeweiss, G.M.; Müller, K.F.; Quandt, D. The evolution of the plastid chromosome in land plants: Gene content, gene order, gene function. *Plant Mol. Biol.* **2011**, *76*, 273–297. [CrossRef] [PubMed]

24. Jansen, R.K.; Cai, Z.; Raubeson, L.A.; Daniell, H.; Leebens-Mack, J.; Müller, K.F.; Guisinger-Bellian, M.; Haberle, R.C.; Hansen, A.K.; Chumley, T.W. Analysis of 81 genes from 64 plastid genomes resolves relationships in angiosperms and identifies genome-scale evolutionary patterns. *Proc. Natl. Acad. Sci. USA* **2007**, *104*, 19369–19374. [CrossRef] [PubMed]

25. Parks, M.; Cronn, R.; Liston, A. Increasing phylogenetic resolution at low taxonomic levels using massively parallel sequencing of chloroplast genomes. *BMC Biol.* **2009**, *7*, 84. [CrossRef] [PubMed]

26. Moore, M.J.; Soltis, P.S.; Bell, C.D.; Burleigh, J.G.; Soltis, D.E. Phylogenetic analysis of 83 plastid genes further resolves the early diversification of eudicots. *Proc. Natl. Acad. Sci. USA* **2010**, *107*, 4623–4628. [CrossRef] [PubMed]

27. Zhu, A.; Guo, W.; Gupta, S.; Fan, W.; Mower, J.P. Evolutionary dynamics of the plastid inverted repeat: The effects of expansion, contraction, and loss on substitution rates. *New Phytol.* **2016**, *209*, 1747–1756. [CrossRef] [PubMed]

28. Maliga, P. Engineering the plastid genome of higher plants. *Curr. Opin. Plant Biol.* **2002**, *5*, 164–172. [CrossRef]

29. Shaw, J.; Shafer, H.L.; Leonard, O.R.; Kovach, M.J.; Schorr, M.; Morris, A.B. Chloroplast DNA sequence utility for the lowest phylogenetic and phylogeographic inferences in angiosperms: The tortoise and the hare iv. *Am. J. Bot.* **2014**, *101*, 1987–2004. [CrossRef] [PubMed]

30. Yu, X.Q.; Gao, L.M.; Soltis, D.E.; Soltis, P.S.; Yang, J.B.; Fang, L.; Yang, S.X.; Li, D.Z. Insights into the historical assembly of East Asian subtropical evergreen broadleaved forests revealed by the temporal history of the tea family. *New Phytol.* **2017**, *215*, 1235–1248. [CrossRef] [PubMed]

31. Allantospermum, A.; Apodanthaceae, A.; Boraginales, B.; Buxaceae, C.; Centrolepidaceae, C.; Cynomoriaceae, D.; Dilleniales, D.; Dipterocarpaceae, E.; Forchhammeria, F.; Gesneriaceae, H. An update of the angiosperm phylogeny group classification for the orders and families of flowering plants: APG IV. *Bot. J. Linn. Soc.* **2016**, *181*, 1–20. [CrossRef]

32. Lan, Y.; Cheng, L.; Huang, W.; Cao, Q.; Zhou, Z.; Luo, A.; Hu, G. The complete chloroplast genome sequence of *Actinidia kolomikta* from north China. *Conserv. Genet. Resour.* **2017**, 1–3. [CrossRef]

33. Wang, W.-C.; Chen, S.-Y.; Zhang, X.-Z. Chloroplast genome evolution in Actinidiaceae: Clpp loss, heterogenous divergence and phylogenomic practice. *PLoS ONE* **2016**, *11*, e0162324. [CrossRef] [PubMed]

34. Logacheva, M.D.; Schelkunov, M.I.; Shtratnikova, V.Y.; Matveeva, M.V.; Penin, A.A. Comparative analysis of plastid genomes of non-photosynthetic Ericaceae and their photosynthetic relatives. *Sci. Rep.* **2016**, *6*, 30042. [CrossRef] [PubMed]

35. Fajardo, D.; Senalik, D.; Ames, M.; Zhu, H.; Steffan, S.A.; Harbut, R.; Polashock, J.; Vorsa, N.; Gillespie, E.; Kron, K. Complete plastid genome sequence of *Vaccinium macrocarpon*: Structure, gene content, and rearrangements revealed by next generation sequencing. *Tree Genet. Genomes* **2013**, *9*, 489–498. [CrossRef]

36. Fu, J.; Liu, H.; Hu, J.; Liang, Y.; Liang, J.; Wuyun, T.; Tan, X. Five complete chloroplast genome sequences from *Diospyros*: Genome organization and comparative analysis. *PLoS ONE* **2016**, *11*, e0159566. [CrossRef] [PubMed]

37. Jo, S.; Kim, H.-W.; Kim, Y.-K.; Cheon, S.-H.; Kim, K.-J. The first complete plastome sequence from the family Sapotaceae, *Pouteria campechiana* (kunth) baehni. *Mitochondr. DNA Part B* **2016**, *1*, 734–736. [CrossRef]

38. Ku, C.; Hu, J.-M.; Kuo, C.-H. Complete plastid genome sequence of the basal Asterid *Ardisia polysticta* miq. and comparative analyses of Asterid plastid genomes. *PLoS ONE* **2013**, *8*, e62548. [CrossRef]

39. Zhang, C.-Y.; Liu, T.-J.; Yan, H.-F.; Ge, X.-J.; Hao, G. The complete chloroplast genome of a rare candelabra primrose *Primula stenodonta* (Primulaceae). *Conserv. Genet. Resour.* **2017**, *9*, 123–125. [CrossRef]

40. Wang, L.-L.; Zhang, Y.; Yang, Y.-C.; Du, X.-M.; Ren, X.-L.; Liu, W.-Z. The complete chloroplast genome of *Sinojackia xylocarpa* (Ericales: Styracaceae), an endangered plant species endemic to China. *Conserv. Genet. Resour.* **2017**. [CrossRef]

41. Yao, X.; Tang, P.; Li, Z.; Li, D.; Liu, Y.; Huang, H. The first complete chloroplast genome sequences in Actinidiaceae: Genome structure and comparative analysis. *PLoS ONE* **2015**, *10*, e0129347. [CrossRef] [PubMed]

42. Kuroda, H.; Suzuki, H.; Kusumegi, T.; Hirose, T.; Yukawa, Y.; Sugiura, M. Translation of *psbC* mRNAs starts from the downstream GUG, not the upstream AUG, and requires the extended shine–dalgarno sequence in tobacco chloroplasts. *Plant Cell Physiol.* **2007**, *48*, 1374–1378. [CrossRef] [PubMed]

43. Takenaka, M.; Zehrmann, A.; Verbitskiy, D.; Härtel, B.; Brennicke, A. RNA editing in plants and its evolution. *Annu. Rev. Genet.* **2013**, *47*, 335–352. [CrossRef] [PubMed]

44. Zhao, J.; Qi, B.; Ding, L.; Tang, X. Based on RSCU and QRSCU research codon bias of F/10 and G/11 xylanase. *J. Food Sci. Biotechnol.* **2010**, *29*, 755–764.

45. Zuo, L.-H.; Shang, A.-Q.; Zhang, S.; Yu, X.-Y.; Ren, Y.-C.; Yang, M.-S.; Wang, J.-M. The first complete chloroplast genome sequences of *Ulmus* species by de novo sequencing: Genome comparative and taxonomic position analysis. *PLoS ONE* **2017**, *12*, e0171264. [CrossRef] [PubMed]

46. Zhou, J.; Chen, X.; Cui, Y.; Sun, W.; Li, Y.; Wang, Y.; Song, J.; Yao, H. Molecular structure and phylogenetic analyses of complete chloroplast genomes of two *Aristolochia* medicinal species. *Int. J. Mol. Sci.* **2017**, *18*, 1839. [CrossRef] [PubMed]

47. Wang, W.; Yu, H.; Wang, J.; Lei, W.; Gao, J.; Qiu, X.; Wang, J. The complete chloroplast genome sequences of the medicinal plant *Forsythia suspensa* (Oleaceae). *Int. J. Mol. Sci.* **2017**, *18*, 2288. [CrossRef] [PubMed]

48. Gichira, A.W.; Li, Z.; Saina, J.K.; Long, Z.; Hu, G.; Gituru, R.W.; Wang, Q.; Chen, J. The complete chloroplast genome sequence of an endemic monotypic genus *Hagenia* (Rosaceae): Structural comparative analysis, gene content and microsatellite detection. *PeerJ* **2017**, *5*, e2846. [CrossRef] [PubMed]

49. Makałowski, W.; Boguski, M.S. Evolutionary parameters of the transcribed mammalian genome: An analysis of 2,820 orthologous rodent and human sequences. *Proc. Natl. Acad. Sci. USA* **1998**, *95*, 9407–9412. [CrossRef] [PubMed]

50. Hong, S.-Y.; Cheon, K.-S.; Yoo, K.-O.; Lee, H.-O.; Cho, K.-S.; Suh, J.-T.; Kim, S.-J.; Nam, J.-H.; Sohn, H.-B.; Kim, Y.-H. Complete chloroplast genome sequences and comparative analysis of *Chenopodium quinoa* and *C. album*. *Front. Plant Sci.* **2017**, *8*, 1696. [CrossRef] [PubMed]

51. Saina, J.K.; Gichira, A.W.; Li, Z.-Z.; Hu, G.-W.; Wang, Q.-F.; Liao, K. The complete chloroplast genome sequence of *Dodonaea viscosa*: Comparative and phylogenetic analyses. *Genetica* **2017**, 1–13. [CrossRef] [PubMed]

52. Raubeson, L.A.; Peery, R.; Chumley, T.W.; Dziubek, C.; Fourcade, H.M.; Boore, J.L.; Jansen, R.K. Comparative chloroplast genomics: Analyses including new sequences from the angiosperms *Nuphar advena* and *Ranunculus macranthus*. *BMC Genom.* **2007**, *8*, 174. [CrossRef] [PubMed]

53. Wang, R.-J.; Cheng, C.-L.; Chang, C.-C.; Wu, C.-L.; Su, T.-M.; Chaw, S.-M. Dynamics and evolution of the inverted repeat-large single copy junctions in the chloroplast genomes of monocots. *BMC Evol. Biol.* **2008**, *8*, 36. [CrossRef] [PubMed]

54. Huotari, T.; Korpelainen, H. Complete chloroplast genome sequence of *Elodea canadensis* and comparative analyses with other monocot plastid genomes. *Gene* **2012**, *508*, 96–105. [CrossRef] [PubMed]

55. Choi, K.S.; Chung, M.G.; Park, S. The complete chloroplast genome sequences of three Veroniceae species (Plantaginaceae): Comparative analysis and highly divergent regions. *Front. Plant Sci.* **2016**, *7*, 355. [CrossRef] [PubMed]

56. Doyle, J. DNA protocols for plants. In *Molecular Techniques in Taxonomy*; Springer: Berlin, Germany, 1991; pp. 283–293.

57. Schmieder, R.; Edwards, R. Quality control and preprocessing of metagenomic datasets. *Bioinformatics* **2011**, *27*, 863–864. [CrossRef] [PubMed]

58. Zerbino, D.R.; Birney, E. Velvet: Algorithms for de novo short read assembly using de bruijn graphs. *Genome Res.* **2008**, *18*, 821–829. [CrossRef] [PubMed]

59. Kearse, M.; Moir, R.; Wilson, A.; Stones-Havas, S.; Cheung, M.; Sturrock, S.; Buxton, S.; Cooper, A.; Markowitz, S.; Duran, C. Geneious basic: An integrated and extendable desktop software platform for the organization and analysis of sequence data. *Bioinformatics* **2012**, *28*, 1647–1649. [CrossRef] [PubMed]

60. Wyman, S.K.; Jansen, R.K.; Boore, J.L. Automatic annotation of organellar genomes with DOGMA. *Bioinformatics* **2004**, *20*, 3252–3255. [CrossRef] [PubMed]

61. Schattner, P.; Brooks, A.N.; Lowe, T.M. The tRNAscan-SE, snoscan and snoGPS web servers for the detection of tRNAs and snoRNAs. *Nucleic Acids Res.* **2005**, *33*, W686–W689. [CrossRef] [PubMed]

62. Lohse, M.; Drechsel, O.; Bock, R. OrganellarGenomeDRAW (OGDRAW): A tool for the easy generation of high-quality custom graphical maps of plastid and mitochondrial genomes. *Curr. Genet.* **2007**, *52*, 267–274. [CrossRef] [PubMed]

63. Thiel, T.; Michalek, W.; Varshney, R.; Graner, A. Exploiting EST databases for the development and characterization of gene-derived SSR-markers in barley (*Hordeum vulgare* L.). *TAG Theor. Appl. Genet.* **2003**, *106*, 411–422. [CrossRef] [PubMed]

64. Wang, D.; Liu, F.; Wang, L.; Huang, S.; Yu, J. Nonsynonymous substitution rate (Ka) is a relatively consistent parameter for defining fast-evolving and slow-evolving protein-coding genes. *Biol. Direct* **2011**, *6*, 13. [CrossRef] [PubMed]
65. Stamatakis, A. RAxML version 8: A tool for phylogenetic analysis and post-analysis of large phylogenies. *Bioinformatics* **2014**, *30*, 1312–1313. [CrossRef] [PubMed]
66. Posada, D. Jmodeltest: Phylogenetic model averaging. *Mol. Biol. Evol.* **2008**, *25*, 1253–1256. [CrossRef] [PubMed]

International Journal of
Molecular Sciences

MDPI

Article

Complete Chloroplast Genome Sequences of Four Meliaceae Species and Comparative Analyses

Malte Mader [1], Birte Pakull [1], Céline Blanc-Jolivet [1], Maike Paulini-Drewes [1],
Zoéwindé Henri-Noël Bouda [1], Bernd Degen [1], Ian Small [2] and Birgit Kersten [1,*]

[1]   Thünen Institute of Forest Genetics, Sieker Landstrasse 2, D-22927 Grosshansdorf, Germany;
      malte.mader@thuenen.de (M.M.); birte.pakull@thuenen.de (B.P.); celine.blanc-jolivet@thuenen.de (C.B.-J.);
      maike.paulini@thuenen.de (M.P.-D.); henri.bouda@thuenen.de (Z.H.-N.B.); bernd.degen@thuenen.de (B.D.)
[2]   Australian Research Centre of Excellence in Plant Energy Biology, School of Molecular Sciences,
      The University of Western Australia, 35 Stirling Highway, Crawley, WA 6009, Australia;
      ian.small@uwa.edu.au
*   Correspondence: birgit.kersten@thuenen.de; Tel.: +49-4102-696105

Received: 31 January 2018; Accepted: 26 February 2018; Published: 1 March 2018

Abstract: The Meliaceae family mainly consists of trees and shrubs with a pantropical distribution. In this study, the complete chloroplast genomes of four Meliaceae species were sequenced and compared with each other and with the previously published *Azadirachta indica* plastome. The five plastomes are circular and exhibit a quadripartite structure with high conservation of gene content and order. They include 130 genes encoding 85 proteins, 37 tRNAs and 8 rRNAs. Inverted repeat expansion resulted in a duplication of *rps19* in the five Meliaceae species, which is consistent with that in many other Sapindales, but different from many other rosids. Compared to *Azadirachta indica*, the four newly sequenced Meliaceae individuals share several large deletions, which mainly contribute to the decreased genome sizes. A whole-plastome phylogeny supports previous findings that the four species form a monophyletic sister clade to *Azadirachta indica* within the Meliaceae. SNPs and indels identified in all complete Meliaceae plastomes might be suitable targets for the future development of genetic markers at different taxonomic levels. The extended analysis of SNPs in the *matK* gene led to the identification of four potential Meliaceae-specific SNPs as a basis for future validation and marker development.

Keywords: chloroplast genome; Next Generation Sequencing; genome skimming; Meliaceae; DNA marker; SNP; indel; *matK*; *rps19*

1. Introduction

The Meliaceae or mahogany family is a flowering plant family of mainly trees and shrubs (and a few mangroves and herbaceous plants) in the order Sapindales. The species of the Meliaceae family, which are contained in The Plant List [1] belong to 52 plant genera, all showing a pantropical distribution. The Plant List includes 3198 scientific plant names of species rank for the family Meliaceae. Of these, 669 are accepted species names.

Cedrela odorata L., a fast-growing deciduous tree species, is the most commercially important and widely distributed species in the genus *Cedrela*, and one of the world's most important timber species. It is found from Mexico southwards throughout central America to northern Argentina, as well as in the Caribbean. The aromatic wood is in high demand in the American tropics because it is naturally termite- and rot-resistant. It contains an aromatic and insect-repelling resin that is the source of one of its popular names, Spanish-cedar (it resembles the aroma of true cedars, *Cedrus* spp.). Other common names include Cuban cedar or cedro in Spanish. It is used for a wide range of purposes, including the production of furniture and craft items as well as different medicinal uses of the bark. It is an excellent choice for use in reforestation, because it is considered a pioneer species.

Entandrophragma cylindricum (Sprague) Sprague, commonly known as sapele or sapelli, is a large tree native to tropical Africa. The commercially important wood is reminiscent of mahogany (*Swietenia macrophylla*), a member of the same family. Demand for sapelli as a mahogany substitute, often traded as "African mahogany", has increased sharply in recent years. It is sold both in lumber and veneer form. Among other applications is its use in musical instruments.

Khaya senegalensis (Desv.) A. Juss. represents a *Khaya* species of the African riparian woodlands and savanna zone. Common names include African mahogany, dry zone mahogany, Gambia mahogany, or Senegal mahogany, among others. The wood is used for a variety of purposes, such as carpentry, interior trim, and construction. The bitter tasting bark is utilized for a variety of medical purposes.

Carapa guianensis Aubl. is a tree from the tropical regions of Southern Central America, the Amazon region, and the Caribbean. The wood resembles mahogany and is used in quality furniture. The seed oil is used in traditional medicine and as an insect repellent.

All four Meliaceae species analyzed in this study (*Cedrela odorata*, *Entandrophragma cylindricum*, *Khaya senegalensis*, *Carapa guianensis*) are listed as vulnerable on the IUCN red list of threatened species [2]. Furthermore, *Cedrela odorata* is one of the six Meliaceae species that are included in the list of CITES-protected species to ensure that international trade will not threaten the survival of this species [3]. Detecting violations of CITES regulations in the tropical timber trade requires accurate genus and species identification of wood, which is often difficult or even impossible using anatomical methods, especially if the wood is processed. Thus, there is a high demand for the development of genetic markers for these purposes.

For genus and species identification, chloroplast DNA (cpDNA) markers are often useful. In contrast to nuclear genomes, chloroplast genomes are inherited uniparentally (maternally in most seed plants) [4], show a dense gene content and a slower evolutionary rate of change (e.g., [5]). The double-stranded cpDNA is present in many copies per cell, leading to the convenient situation that cpDNA can be retrieved relatively easily from low-quantity and/or degraded DNA samples, including wood samples. These advantages of cpDNA are reflected in the recommendation of the Barcode of Life Consortia to apply molecular markers, mainly based on organelle DNA, to genetically differentiate all eukaryotic species [6]. For the genetic differentiation of vascular plants, molecular markers which are mainly based on DNA variations in two chloroplast regions (*rbcL* and *matK*) are used as a two-locus barcode [7].

Chloroplast-derived DNA sequences have been widely used for taxonomic purposes and phylogenetic studies (e.g., [8,9]). Complete cp genome sequences provide valuable data sets to resolve complex evolutionary relationships [10] and have been shown to improve resolution at lower taxonomic levels (e.g., [11]). The application of Next-Generation Sequencing (NGS) technologies [12], especially using genome-skimming strategies [13,14] has made it relatively easy to obtain complete cpDNA sequences for low cost. The high abundance of cpDNA compared to nuclear DNA allows the use of total DNA for genome skimming without prior purification of chloroplasts or cpDNA [15,16].

The comparative analysis of complete plastomes allows the extension of two-locus barcoding to next-generation barcoding (whole-plastome barcoding), gene-based phylogenetics to genome-based phylogenomics, and the development of molecular markers for taxonomic and phylogeographic purposes [15–18].

In Meliaceae, the development of cpDNA markers has so far been restricted to specific loci [19,20]. Although the Meliaceae form a large family, only the plastome sequence of one species, *Azadirachta indica*, has been previously published in GenBank (KF986530.1). In this study, we sequenced the plastomes of four additional species and compared them to that of *Azadirachta indica* as well as those of other species in the order Sapindales to analyze interspecific variation within the Meliaceae and uncover Meliaceae-specific genome features.

2. Results

2.1. The Structure of Chloroplast Genomes from Four Meliaceae Species

The complete cp genomes of *Cedrela odorata*, *Entandrophragma cylindricum*, *Khaya senegalensis* and *Carapa guianensis* were sequenced using Illumina sequencing technology in a genome-skimming approach (Table 1) and compared to the previously published plastome of *Azadirachta indica* (GenBank KF986530.1).

Table 1. Information on samples and NGS data of four Meliaceae individuals sequenced in this study.

Species	Individual (Thuenen-ID)	Origin/Location	Longitude	Latitude	Trimmed Reads	Coverage *
Cedrela odorata	CEODO_205_2	Cuba, population Guisa	−76.68	20.16	254214	101×
Entandrophragma cylindricum	c-5-ENTC-46	Cameroon, FBR, Parc National de Lobeke	15.6442	2.26286	2206300	165×
Khaya senegalensis	KS	Unknown/Green house, Thünen Institute Hamburg-Lohbrügge			2783117	346×
Carapa guianensis	CAGUI_332_1	French Guiana, region Rorota	−52.262392	4.87761	422759	135×

* The coverage is based on a final mapping of the trimmed reads to the assembled cpDNA sequence.

The plastomes of these four Meliaceae species (and of *Azadirachta indica*) are small circular DNA molecules of sizes in the range of 158,558 bp to 160,737 bp, with the typical quadripartite structure of land plant cp genomes consisting of two inverted repeats (IRa and IRb) separated by large (LSC) and small (SSC) single copy regions, respectively (Table 2).

Table 2. Summary of Meliaceae chloroplast genome features.

	Cedrela Odorata (MG724915)	*Entandrophragma Cylindricum* (KY923074.1)	*Khaya Senegalensis* (KX364458.1)	*Carapa Guianensis* (MF401522.1)	*Azadirachta Indica* * (KF986530.1)
Genome size (bp)	158,558	159,609	159,787	159,483	160,737
LSC length (bp)	86,390	87,117	87,404	87,054	88,137
SSC length (bp)	18,380	18,532	18,311	18,277	18,636
IR length (bp)	26,894	26,980	27,036	27,076	26,982
Number of genes	130	130	130	130	131

* Not sequenced in this study. Identifiers (in parenthesis) under the species name refer to GenBank accession numbers.

In each of the cp genomes of the four newly sequenced Meliaceae species (Table 1), 112 different genes were annotated (78 protein-coding, 30 tRNA and 4 rRNA genes), 18 of which were duplicated in the IR regions, giving a total of 130 genes encoding 85 proteins, 37 tRNAs and 8 rRNAs (Tables 2 and 3). Among the 112 unique genes, 18 included one or two introns. The intron sizes are ranging from 532 bp for trnL-UAA to 2535 bp for trnK-UUU when considering the plastome of *Cedrela odorata*. One gene, *rps12*, is presumed to require trans-splicing (Table 3; [21]).

Table 3. List of genes annotated in the cp genomes of four Meliaceae species sequenced in this study (Table 2).

Function	Genes
RNAs, ribosomal	*rrn23, rrn16, rrn5, rrn4.5*
RNAs, transfer	*trnA-UGC *, trnC-GCA, trnD-GUC, trnE-UUC, trnF-GAA, trnG-GCC, trnG-UCC *, trnH-GUG, trnI-CAU, trnI-GAU *, trnK-UUU *, trnL-CAA, trnL-UAA *, trnL-UAG, trnM-CAU, trnfM-CAU, trnN-GUU, trnP-UGG, trnQ-UUG, trnR-ACG, trnR-UCU, trnS-GCU, trnS-GGA, trnS-UGA, trnT-GGU, trnT-UGU, trnV-GAC, trnV-UAC *, trnW-CCA, trnY-GUA*
Transcription and splicing	*rpoA, rpoB, rpoC1 *, rpoC2, matK*
Translation, ribosomal proteins	
Small subunit	*rps2, rps3, rps4, rps7, rps8, rps11, rps12 **,^{T}, rps14, rps15, rps16 *, rps18, rps19*
Large subunit	*rpl2 *, rpl14, rpl16 *, rpl20, rpl22, rpl23, rpl32, rpl33, rpl36*
Photosynthesis	
ATP synthase	*atpA, atpB, atpE, atpF *, atpH, atpI*
Photosystem I	*psaA, psaB, psaC, psaI, psaJ, ycf3 **, ycf4*
Photosystem II	*psbA, psbB, psbC, psbD, psbE, psbF, psbH, psbI, psbJ, psbK, psbL, psbM, psbN, psbT, psbZ*
Calvin cycle	*rbcL*
Cytochrome complex	*petA, petB *, petD *, petG, petL, petN*
NADH dehydrogenase	*NdhA *, ndhB *, ndhC, ndhD, ndhE, ndhF, ndhG, ndhH, ndhI, ndhJ, ndhK*
Others	*clpP1 **, accD, cemA, ccsA, ycf1, ycf2*

* Genes containing one intron; ** genes containing two introns; ^{T} trans-splicing of the related gene. Genes in bold are located in inverted repeats (two gene copies in the genome).

The gene maps of the newly sequenced Meliaceae species are provided in Figure 1 (*Cedrela odorata*), and in Figures S1–S3 (*Entandrophragma cylindricum, Khaya senegalensis* and *Carapa guianensis*). The gene content and gene order in the cp genomes of the four species are nearly identical to the previously published plastome of *Azadirachta indica* (GenBank KF986530.1) with the exception that the *ycf1* gene at the IRa/SSC border was not annotated as a pseudogene in *Azadirachta indica* resulting in a total number of 131 genes in this species compared to 130 genes in the other four Meliaceae species (Table 2). Another exception is that the gene names of trnG-UCC and trnC-GCC are swapped due to misannotation in *Azadirachta indica*.

As mentioned above, 18 unique genes were annotated to include introns in the four newly sequenced Meliaceae species (Table 3), whereas introns are missing in the annotations of two of these genes in *Azadirachta indica* (GenBank KF986530.1), namely *petD* and *rps12* (intron in the 3′ part of the gene). The annotations of these genes are probably not correct in *Azadirachta indica*; both exons are relatively short (exon 1 of *petD*: 8 bp; exon 3 of *rps12*: 26 bp), thus making their annotation difficult.

The gene *rps19* is included in the inverted repeats (close to the IRa/LSC or IRb/LSC border, respectively; Figure 1, Figures S1–S3, Table 3) in the four Meliaceae species sequenced in this study as well as in *Azadirachta indica* (KF986530.1) thus resulting in a duplication of *rps19*.

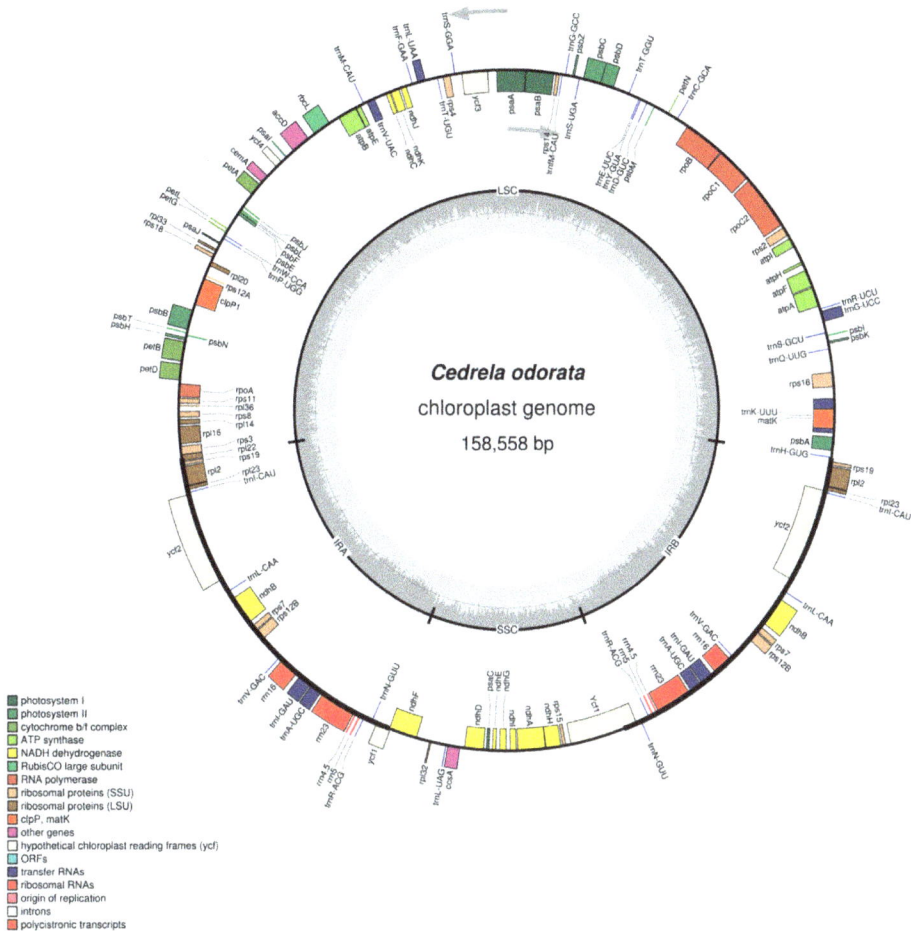

Figure 1. Gene map of the complete chloroplast genome of *Cedrela odorata* (GenBank MG724915). The grey arrows indicate the direction of transcription of the two DNA strands. A GC-content graph is depicted within the inner circle. The circle inside the GC content graph marks the 50% threshold. The maps were created using OrganellarGenomeDRAW [22].

2.2. Diversity of the Meliaceae Plastome Sequences

The newly sequenced individuals of the four different Meliaceae species share several large regions, located in intergenic regions and in the *rpl16* intron (LSC), which show low similarity to *Azadirachta indica* (Figure 2). These regions are related to large deletions which become obvious in a multiple alignment of related whole plastomes (Figure S4). The largest deletion which is in the *psbE-petL* linker is of a size of about 199 bp (Figure S4; at position 69531 bp of the *Azadirachta indica* sequence). These deletions mainly contribute to the smaller genome sizes of the four individuals compared to *Azadirachta indica* (Table 2). The *Cedrela odorata* individual, which is the individual with the smallest cp genome size (158,558 bp; Table 2), shows, compared to the four other individuals, additional large deletions in different intergenic linkers, e.g., the *ycf3-rps4* linker (Figure 2). A large exclusive deletion was also detected for the *Carapa guianensis* individual in the *psbM-psbD* linker region in the LSC (Figure 2, Figure S4).

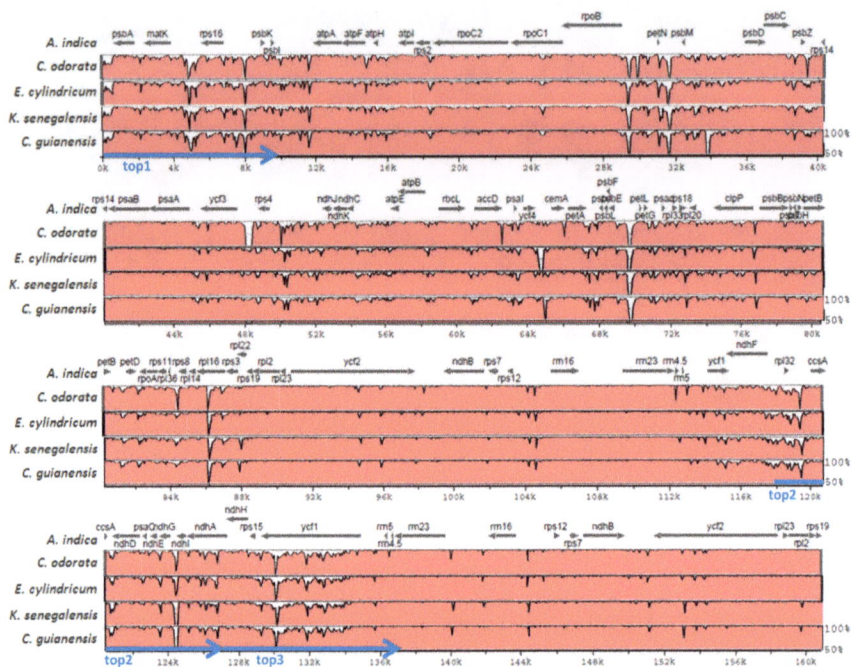

Figure 2. Visualization of pairwise alignments of complete cpDNA sequences of four Meliaceae species each with the cpDNA sequence of *Azadirachta indica* (reference). VISTA-based similarity plots portraying the sequence identity of each of the four Meliacea species with the reference *Azadirachta indica* are shown. The annotation (protein-encoding genes) is provided for *Azadirachta indica* on top (based on the related GenBank file; KF986530.1). Plastome regions with the highest diversity between the 5 Meliaceae individuals are marked by blue arrows (top1–3). Further details are provided in the text below.

Based on the multiple whole-plastome alignment of the five Meliaceae individuals (Figure S4), 7635 positions that showed DNA sequence variations (SNPs or indels) in at least one of the plastomes (compared to the consensus sequence), were called by the SNiPlay tool (SNiPlay genotyping table in Table S5). The following regions of the consensus sequence showed the highest frequencies of variations (considering intervals of 10,000 bp): 1–10,000 bp (1–9681 bp in *Azadirachta indica*) with 923 variable positions (top1); 120,000–130,000 bp (117,660–127,364 bp in *A. indica*) with 771 positions (top2); and 130,000–140,000 bp (127,365–137,199 bp in *A. indica*) with 735 positions (top3). The top1-region is located at the 5-prime part of the LSC including *psbA*, *matK*, *rps16*, *psbK* and *psbI* (Figure 2). The top2- and top3-region are connected and represent the SSC downstream of *ndhF*, the SSC/IRb border and parts of the rRNA cluster of the IRb (Figure 2).

A phylogenetic tree based on a multiple alignment of the five complete Meliaceae cpDNA sequences (Table 2) and *Acer buergerianum* (family Sapindaceae in the Sapindales; NC_034744.1) as an outgroup, shows that the analyzed Meliaceae form a monophyletic sister clade to *Azadirachta indica* within the Meliaceae. Within this clade, *Cedrela odorata* and *Entandrophragma cylindricum* group together, as do *Khaya senegalensis* and *Carapa guianensis* (Figure 3).

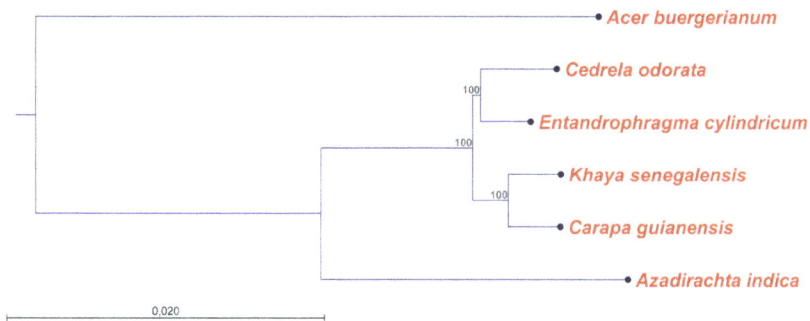

Figure 3. Phylogenetic tree (maximum likelihood) based on whole-plastome sequences of five Meliaceae species and *Acer buergerianum* (outgroup). Bootstrap values (%) are shown above branches. The bootstrap value on the branch separating *Azadirachta indica* from the other Meliaceae is below 70% and was not shown for this reason. GenBank accession numbers of the plastomes are given in Table 2.

2.3. Comparative Analyses for the Identification of Potential Meliaceae-Specific Plastome DNA Variations in One Barcoding Region

The *matK* gene, one of the cpDNA barcoding regions (see Introduction), was selected for comparative analyses aiming at the identification of potential Meliaceae-specific cpDNA variations. Based on a multiple alignment (Figure S5) of the extracted *matK* gene sequences of the five Meliaceae species listed in Table 2 together with five non-Meliaceae species of the order Sapindales (*Boswellia sacra* Flueck (Burseraceae), NC_029420.1; *Anacardium occidentale* L. (Anacardiaceae), NC_035235.1; *Leitneria floridana* Chapm. (Simaroubaceae), NC_030482.1; *Citrus sinensis* (L.) Osbeck (Rutaceae), NC_008334.1 [23]; *Acer buergerianum* Miq. (Aceraceae), NC_034744.1), 16 SNP positions were identified where the five Meliaceae individuals showed the same nucleotide, which, however, differed from the nucleotide(s) of the non-Meliaceae species at the same position (SNP positions summarized in Table 4).

These 16 SNP positions were further analyzed in a multiple alignment of *matK* gene sequences from 100 Meliaceae individuals downloaded from GenBank (Figure S6). Only SNPs where all the downloaded Meliaceae sequences showed the same nucleotide at the given position were further validated with more *matK* sequences from members of other taxonomic groups (other Sapindales, other Rosids, and other land plants; Table 4).

The SNPs at the following four positions of the *matK* gene were selected as potential Meliaceae-specific: 346 bp/328 bp (consensus sequence/*Cedrela odorata* sequence), 1318 bp/1270 bp, 1478 bp/1430 bp and 1494 bp/1446 bp (Table 4). At position 1318 bp/1270 bp, e.g., all Meliaceae individuals considered include a G (Table 4, Figure S6) in contrast to members of other Sapindales families, other Rosids, as well as other land plants, where the considered individuals show an A (Table 4). This SNP is at the first position of the codon encoding for MatK amino acid 424 in *Cedrela odorata* and will result in an amino acid exchange (Meliaceae analyzed: Gly (Figure S6); other Sapindales, Rosids and land plants analyzed: Ser or Asp).

The other barcoding region, the *rbcL* gene (see Introduction), showed less nucleotide variation compared to *matK* when analyzed in a similar way (Figure S7). In a multiple alignment of the *rbcL* sequences of five Meliaceae species and five non-Meliaceae species of the order Sapindales (Figure S7), only 2 SNP positions were identified where the five Meliaceae individuals showed another nucleotide than the non-Meliaceae species at the same position (SNP positions 225 and 582 related to the consensus sequence). However, at these two positions, some other Meliaceae individuals showed differing nucleotides at GenBank; thus, these positions were not further analyzed.

Table 4. Identification of potential Meliaceae-specific SNPs (highlighted in grey) in the *matK* gene.

Position (Consensus)	Position (*Cedrela odorata*)	Meliaceae	Sapindales without Meliaceae	Rosids without Sapindales	All without Rosids
208	208	T	A or C or T	C or A	C or A
280	262	T or C [1]			
346	328	G	C or T	C or A	C or A
402	378	A	C or T or A or G	A or C	A or C
574	550	C or T [3]			
618	588	C or A [1]			
639	609	G	T or G or A or C	T or G or A or C	G or A or T
861	819	T	C	C or T or G	C
995	953	C or T [4]			
1194	1146	T	C or T	C or T	C
1237	1189	G or A [2] or C [1]			
1239	1191	G or A [3]			
1318	1270	G	A	A	A
1389	1341	C or T [1]			
1478	1430	C	G or A	G or A or T	A or G
1494	1446	T	A or G	G or A	G or C

The SNP positions listed were validated in different multiple alignments of *matK* sequences from member individuals of different taxonomic groups downloaded from GenBank (100 sequences each). The "position (consensus)" refers to the position in the consensus sequence in the alignment in Figure S5. [1,2,3,4] Only 1/2/3/4 individual(s) show the indicated nucleotide.

3. Discussion

In this study, the complete chloroplast genomes of four Meliaceae species (order Sapindales) were sequenced by NGS (genome-skimming), annotated (Figure 1, Figures S1–S3, Table 3) and compared with each other and with the only previously published Meliaceae plastome sequence, the sequence of *Azadirachta indica* (KF986530.1; Table 2). Across eight other families within the Sapindales, there are currently 38 complete cpDNA sequences available at the Organellar Genome Resource at NCBI [24].

Gene and intron content are highly conserved among land plant plastomes, although losses have been identified in several angiosperm lineages (reviewed, e.g., in [25,26]). The following genes known to be lost in some other Angiosperm species [25] are also missing in the four newly sequenced Meliaceae plastomes (Table 3) as well as in *Azadirachta indica* (KF986530.1): *chlB, chlL, chlN, infA, trnP-GGG*. The losses of *chlB, chlL, chlN*, and *trnP-GGG* represent synapomorphies for flowering plants [25]. The most common gene loss involves *infA* [25,27]. There is evidence that *infA* has been transferred to the nucleus in some species [27], which has not yet been investigated in Meliaceae. The low depth of coverage (related to the nuclear genome) of the genome skimming data used in this study did not allow to perform such an investigation.

Intron content is also highly conserved across angiosperms with most genomes containing 18 genes with introns [25]. In the four Meliaceae species sequenced in this study, 18 unique genes with one or two introns were annotated (Table 3). The apparent lack of the *petD* intron and the intron in the 3'-*rps12* locus in *Azadirachta indica* (KF986530.1) are annotation errors in our opinion.

Hotspots for structural rearrangements within plastomes include the IRs, which are frequently subjected to expansion, contraction, or even complete loss (summarized in [21]). Inverted repeat expansion resulted in a duplication of *rps19* in the five Meliaceae plastomes analyzed so far (Figure 1, Figures S1–S3, Table 3, KF986530.1). This is consistent with research considering the plastomes of *Nitraria sibirica* (Nitrariaceae [28]) and 38 other Sapindales species [24] not belonging to the Meliaceae. However, some of these plastomes have only incomplete second *rps19* copies (*Anacardium occidentale*, NC_035235.1; *Acer miaotaiense*, NC_030343.1; *Acer buergerianum*, NC_034744.1; *Mangifera indica*, NC_035239.1) or the second copy is even completely missing, such as in *Aesculus wangii* (NC_035955.1) or *Pistacia vera* (NC_034998.1), where the first *rps19* copy is in addition incomplete. The *rps19* gene is completely absent in *Rhus chinensis* (NC_033535.1). In contrast to the Sapindales, many other rosid species include only one *rps19* gene in the LSC of the cpDNA sequence [28]. To answer the question if the duplication of *rps19* is a general plastome feature in Meliaceae, the plastomes of other member species of the 52 plant genera within the Meliaceae must be sequenced and annotated.

Complete plastome sequences are valuable for deciphering phylogenetic relationships especially between closely related taxa, or where recent divergence, rapid speciation or slow genome evolution has resulted in limited sequence variation [18,29,30]. Considering that most species-level diversity of Meliaceae in rainforests is recent [31], a whole-plastome phylogeny is highly desired for this family. The whole-plastome phylogeny that was generated in this study (Figure 3) based on the whole-plastome alignment for five Meliaceae species (Figure S4), is an initial step in this direction. The result that *Azadirachta indica* (member of the subfamily Melioidaeae) belongs to another subclade than the four other Meliaceae species (members of the subfamily Swietenioideae) as well as the further sub-grouping of the four Meliaceae species, agrees with previous studies (subfamilies according to [31–34]). In the future, more complete cpDNA sequences of member individuals of other Meliaceae species are needed to exploit the power of the whole-plastome phylogenies, especially for differentiation at lower taxonomic levels. Additional integration of complete cpDNA sequences with small amplicon datasets could further improve the phylogenetic resolution, as recently shown in *Acacia*, where the greatest support has been achieved when using a whole-plastome phylogeny as a constraint on the amplicon-derived phylogeny [30].

Whole-plastome alignments are also very useful to develop cpDNA markers for the genetic identification of species or other taxonomic categories [17]. Especially, large indels can be easily identified in whole-plastome alignments and used, e.g., for the development of robust PCR-based markers, as recently shown for a *Populus tremula*-specific marker that has been developed based on a 96 bp-indel [17,35]. In the present study, we identified and compactly visualized large indels between the analyzed Meliaceae individuals based on pairwise alignments using VISTA (Figure 2). Exclusive large deletions were identified for the *Cedrela odorata* and the *Carapa guianensis* individuals, respectively (Figure 2). Before effective markers can be developed, it must be further validated with more Meliaceae species and individuals whether these deletions are genus-, species- or individual-specific. PCR-marker development based on indels in small intergenic regions might be particularly robust, because the primers could be placed into conserved genic regions adjacent to the indel. The SNPs and indels identified between the five Meliaceae plastomes (Table S5), based on a multiple alignment, may also serve—after further validation—as targets for future marker development at different taxonomic levels.

Aiming at the identification of potential Meliaceae-specific SNPs, the *matK* gene, one of the barcoding regions [7]—was selected and the SNPs identified in this gene were analyzed in extended sets of previously published *matK* sequences. In general, nucleotide variation per site in *matK* is three times higher than in *rbcL* (the large subunit of RUBISCO; [36]), and the amino acid substitution rate is six times that of *rbcL* [37]. Here, four potential Meliaceae-specific SNPs were identified (Table 4), three of which are in the 3′-terminus of the *matK* gene encoding for the C-terminus of MatK in a region that is homolog to domain X of mitochondrial group II intron maturases [38] and has a significantly higher amount of basic amino acids compared to the N-terminal region [39]. The potential Meliaceae-specific SNPs—once further validated—should be attractive for the taxonomic differentiation of samples from wood or wood products.

4. Materials and Methods

4.1. Sampling, DNA Extraction and Sequencing

The individual trees analyzed are described in detail in Table 1. Genomic DNA was isolated from leaves or cambium according to Dumolin et al. 1995. Genomic library generation and sequencing on the Illumina MiSeq v3 (2 × 300 bp paired-end reads) was done by Eurofins Genomics (Ebersberg, Germany).

4.2. Assemblies of Chloroplast Genome Sequences and Annotation

If not otherwise stated, CLC Genomics Workbench (CLC-GWB; v8.5.1 and v9.5.3; CLC bio, A QIAGEN company; Aarhus, Denmark) was used for data processing. Initial quality control of the NGS reads was done with FastQC [40].

4.2.1. Assembly of cpDNA Sequences of *Khaya Senegalensis* and *Entandrophragma Cylindricum*

All reads were trimmed with CLC-GWB including adapter trimming, quality trimming (quality limit of 0.01), trimming of 10 nucleotides at the 5′-end and removing reads of less than 50 bp in length. All other options were set to default. Potential chloroplast reads were extracted by mapping all trimmed reads against the chloroplast sequence of *Azadirachta indica* (KF986530.1) using the tool MITObim [41]. Multiple de novo assemblies based on different word sizes were performed on extracted chloroplast reads. Overlapping contigs of one or more assemblies were used to combine the contigs to complete cpDNA sequences.

4.2.2. Assembly of cpDNA Sequences of *Carapa Guianensis* and *Cedrela Odorata*

Adapter sequences were removed by the trimming software Trimmomatic [42] using simple and palindromic trim. Further trimming was done by CLC-GWB including quality trimming (quality limit of 0.01), trimming of 10 nucleotides at the 3′- and 15 nucleotides 5′-end and removing reads shorter than 50 bp. All other options were set to default. A first set of contigs was generated by de novo assembly of all trimmed reads, using a length fraction of 0.9 and a similarity fraction of 0.95. All resulting contigs were blasted with the command line tool *blastn* [43] against the nucleotide BLAST database downloaded from GenBank [44] to identify and select chloroplast contigs. All resulting Blast hits were filtered for the keyword "chloroplast" and finally validated with the web Blast tool at NCBI [45]. Trimmed reads were mapped to the chloroplast contigs, mapped reads were extracted and stored in a separate read set as chloroplast reads. Multiple de novo assemblies based on different word sizes were performed on these chloroplast reads. Overlapping contigs of one or more assemblies were used to assemble the complete cpDNA sequence.

4.3. Annotation of the cpDNA Sequences and Preparation of GenBank Submission Files

Draft annotations were generated with the web-based software CPGAVAS [46,47] to check gene content and order. These draft annotations were corrected where necessary, guided by alignments to other well-characterized eudicot plastomes including those of *Arabidopsis thaliana* (NC_000932), *Nicotiana tabacum* (NC_001879) and *Spinacia oleracea* (NC_002202).

The file resulting from the fine annotation of the plastome of *Khaya senegalensis* (GB-format) was transferred to a draft SQN-file using the CHLOROBOX-GenBank2Sequin-tool [48] and edited using the Sequin tool (v13.05; [49]). The error-corrected SQN-file was submitted to GenBank. Because gene content and order were the same in all species analyzed (according to the fine annotation), the GenBank submission files (SQN-format) for the other species were created by updating the sqn-file of *Khaya senegalensis* with the sequences of the other three species (using the "update sequence" function in Sequin) and subsequent manual editing of shifted annotations in Sequin.

The circular gene maps of the four Meliaceae plastomes (Figure 1; Figures S1–S3) were obtained using the OrganellarGenomeDRAW software (OGDRAW v1.2; [22,50].

4.4. Alignments and Construction of a Phylogenetic Tree

Pairwise alignments of complete cpDNA sequences were performed using the VISTA tool mVISTA ("AVID" as alignment program) at VISTA [51,52]. Identity plots of each pairwise alignment were downloaded from the related VISTA-point results page.

Multiple alignments of complete cpDNA sequences were run using CLC-GWB (v. 8.5.1.; parameters: gap open cost = 10.0; gap extension cost = 1.0; end gap cost = as any other; alignment mode = very accurate; redo alignments = no; use fixpoints = no).

The phylogenetic tree was constructed based on the alignment of five Meliaceae species together with one outgroup (*Acer buergerianum*, NC_034744.1; family Sapindaceae in the order Sapindales) using the "Maximum likelihood phylogeny" tool of CLC-GWB including bootstrap analysis with 100 replicates (other parameters: construction method for the start tree = UPGMA; existing start tree = not set; nucleotide substitution model = Jukes Cantor; protein substitution model = WAG; transition/transversion ratio = 2.0; include rate variation = No; number of substitution rate categories = 4; gamma distribution parameter = 1.0; estimate substitution rate parameter(s) = Yes; estimate topology = Yes; estimate gamma distribution parameter = no).

4.5. SNP and Indel Detection in Multiple Alignments Using SNiPlay

To identify SNPs between the complete cpDNA sequences of the 5 Meliaceae species (Table 2), the alignment-FASTA of the related multiple alignment was exported from CLC-GWB and further edited (replacement of alignment spaces "-"—if present—by "?"at the 5'- or 3'- end of the sequences). The edited alignment-FASTA was used as an input for the web tool SNiPlay (pipeline v2; [53]) to run SNP/indel discovery (default parameters) [54].

4.6. NCBI-Blast Analyses of matK and Download of matK Gene Sequences of Different Taxonomic Groups for Multiple Alignments

The *matK* gene sequence of *Cedrela odorata* (MG724915) was used as a query in different BlastN searches (parameters: optimized for highly similar sequences, maximal target sequences: 100) at GenBank (NCBI; [45]): (i) restrict to Meliaceae (taxid:43707); (ii) restrict to Sapindales (taxid:41937) and exclude Meliaceae; (iii) restrict to Rosids (taxid:71275) and exclude Sapindales; and (iv) no restriction, but exclude Rosids. After each Blast analysis, all 100 hits were selected and a FASTA was downloaded with the aligned sequences. Each FASTA file was used as input for a multiple alignment together with the *matK* sequence of *Cedrela odorata* (used as reference). In the case of the Meliaceae alignment, sequences of wrong orientation in the alignment were reverse complemented and the alignment was repeated. In the case of other alignments only sequences in the right orientation (CDS orientation) were considered in the further analyses.

Supplementary Materials: Supplementary materials can be found at www.mdpi.com/1422-0067/19/3/701/s1. Figure S1. Gene map of the complete cpDNA sequence of Entandrophragma cylindricum (KY923074). Figure S2. Gene map of the complete cpDNA sequence of Khaya senegalensis (KX364458). Figure S3. Gene map of the complete cpDNA sequence of Carapa guianensis (MF401522). Figure S4. Whole-plastome alignment of five Meliaceae species. Figure S5. Alignment of the matK gene sequences of 5 Meliaceae species (Table 2) and 5 non-Meliaceae members of the order Sapindales (Boswellia sacra, NC_029420.1; Anacardium occidentale, NC_035235.1; Leitneria floridana, NC_030482.1; Citrus sinensis, NC_008334.1; Acer buergerianum, NC_034744.1). Figure S6. Alignment of the matK gene sequences of 100 Meliaceae individuals (GenBank) and Cedrela odorata (MG724915; reference). Figure S7. Alignment of the rbcL gene sequences of 5 Meliaceae species (Table 2) and 5 non-Meliaceae members of the order Sapindales (Boswellia sacra, NC_029420.1; Anacardium occidentale, NC_035235.1; Leitneria floridana, NC_030482.1; Citrus sinensis, NC_008334.1; Acer buergerianum, NC_034744.1). Table S1. SNPs and Indels identified by SNIplay in the whole-plastome alignment of 5 Meliaceae species.

Acknowledgments: This work was funded by the Federal Ministry of Food and Agriculture (BMEL), Germany in the scope of the "Large scale project on genetic timber verification" and by the International Tropical Timber Organization (project PD 620/11 Rev.1). We would like to thank Susanne Bein for technical assistance in the laboratory, Stephen Cavers and Niklas Tysklind for providing samples from South America. Open access costs were covered by the Thuenen Institute, Braunschweig, Germany.

Author Contributions: B.D. initiated the project; B.P., M.M., C.B.-J., B.D. and B.K. conceived and designed the experiments; B.P. and M.P.-D. performed the experiments; M.M., B.K. and I.S. analyzed the data; Z.H.-N.B. collected and contributed sample material; B.K., M.M. and I.S. wrote the manuscript; and all authors edited and approved the final manuscript.

Conflicts of Interest: The authors declare no conflict of interest. The funding sponsors had no role in the design of the study; in the collection, analyses, or interpretation of data; in the writing of the manuscript, and in the decision to publish the results.

Abbreviations

cp	chloroplast
cpDNA	chloroplast DNA
CITES	Convention on International Trade in Endangered Species of Wild Fauna and Flora
IRa	Inverted Repeat a
IRb	Inverted Repeat b
IUCN	International Union for Conservation of Nature
LSC	Large Single-Copy Region
NGS	Next-Generation Sequencing
SSC	Small Single-Copy Region

References

1. The Plant List. Available online: http://www.theplantlist.org/ (accessed on 15 January 2018).
2. The IUCN Red List of Threatened Species. Available online: http://www.iucnredlist.org/ (accessed on 11 January 2018).
3. CITES—Convention on International Trade in Endangered Species of Wild Fauna and Flora/the Cites Species. Available online: https://cites.org/eng/disc/species.php (accessed on 11 January 2018).
4. Birky, C.W. Uniparental inheritance of mitochondrial and chloroplast genes—Mechanisms and evolution. *Proc. Natl. Acad. Sci. USA* **1995**, *92*, 11331–11338. [CrossRef] [PubMed]
5. Drouin, G.; Daoud, H.; Xia, J. Relative rates of synonymous substitutions in the mitochondrial, chloroplast and nuclear genomes of seed plants. *Mol. Phylogenet. Evol.* **2008**, *49*, 827–831. [CrossRef] [PubMed]
6. Barcode of Life. Available online: http://www.barcodeoflife.org/ (accessed on 1 January 2018).
7. Hollingsworth, P.M.; Forrest, L.L.; Spouge, J.L.; Hajibabaei, M.; Ratnasingham, S.; van der Bank, M.; Chase, M.W.; Cowan, R.S.; Erickson, D.L.; Fazekas, A.J.; et al. A DNA barcode for land plants. *Proc. Natl. Acad. Sci. USA* **2009**, *106*, 12794–12797. [CrossRef]
8. Shaw, J.; Shafer, H.L.; Leonard, O.R.; Kovach, M.J.; Schorr, M.; Morris, A.B. Chloroplast DNA sequence utility for the lowest phylogenetic and phylogeographic inferences in angiosperms: The tortoise and the hare IV. *Am. J. Bot.* **2014**, *101*, 1987–2004. [CrossRef] [PubMed]
9. Wang, W.; Li, H.L.; Chen, Z.D. Analysis of plastid and nuclear DNA data in plant phylogenetics-evaluation and improvement. *Sci. China Life Sci.* **2014**, *57*, 280–286. [CrossRef] [PubMed]
10. Moore, M.J.; Soltis, P.S.; Bell, C.D.; Burleigh, J.G.; Soltis, D.E. Phylogenetic analysis of 83 plastid genes further resolves the early diversification of eudicots. *Proc. Natl. Acad. Sci. USA* **2010**, *107*, 4623–4628. [CrossRef] [PubMed]
11. Carbonell-Caballero, J.; Alonso, R.; Ibanez, V.; Terol, J.; Talon, M.; Dopazo, J. A phylogenetic analysis of 34 chloroplast genomes elucidates the relationships between wild and domestic species within the genus Citrus. *Mol. Biol. Evol.* **2015**, *32*, 2015–2035. [CrossRef] [PubMed]
12. Goodwin, S.; McPherson, J.D.; McCombie, W.R. Coming of age: Ten years of next-generation sequencing technologies. *Nat. Rev. Genet.* **2016**, *17*, 333–351. [CrossRef] [PubMed]
13. Dodsworth, S. Genome skimming for next-generation biodiversity analysis. *Trends Plant Sci.* **2015**, *20*, 525–527. [CrossRef] [PubMed]
14. Straub, S.C.K.; Parks, M.; Weitemier, K.; Fishbein, M.; Cronn, R.C.; Liston, A. Navigating the tip of the genomic iceberg: Next-generation sequencing for plant systematics. *Am. J. Bot.* **2012**, *99*, 349–364. [CrossRef] [PubMed]
15. Pakull, B.; Mader, M.; Kersten, B.; Ekue, M.R.M.; Dipelet, U.G.B.; Paulini, M.; Bouda, Z.H.N.; Degen, B. Development of nuclear, chloroplast and mitochondrial SNP markers for *Khaya* sp. *Conserv. Genet. Resour.* **2016**, *8*, 283–297. [CrossRef]

16. Schroeder, H.; Cronn, R.; Yanbaev, Y.; Jennings, T.; Mader, M.; Degen, B.; Kersten, B. Development of molecular markers for determining continental origin of wood from white oaks (*Quercus* l. Sect. Quercus). *PLoS ONE* **2016**, *11*, e0158221. [CrossRef] [PubMed]

17. Kersten, B.; Rampant, P.F.; Mader, M.; Le Paslier, M.C.; Bounon, R.; Berard, A.; Vettori, C.; Schroeder, H.; Leple, J.C.; Fladung, M. Genome sequences of *Populus tremula* chloroplast and mitochondrion: Implications for holistic poplar breeding. *PLoS ONE* **2016**, *11*, e0147209. [CrossRef] [PubMed]

18. Tonti-Filippini, J.; Nevill, P.G.; Dixon, K.; Small, I. What can we do with 1000 plastid genomes? *Plant J.* **2017**, *90*, 808–818. [CrossRef] [PubMed]

19. Duminil, J.; Kenfack, D.; Viscosi, V.; Grumiau, L.; Hardy, O.J. Testing species delimitation in sympatric species complexes: The case of an african tropical tree, *Carapa* spp. (Meliaceae). *Mol. Phylogenet. Evol.* **2012**, *62*, 275–285. [CrossRef] [PubMed]

20. Holtken, A.M.; Schroder, H.; Wischnewski, N.; Degen, B.; Magel, E.; Fladung, M. Development of DNA-based methods to identify cites-protected timber species: A case study in the Meliaceae family. *Holzforschung* **2012**, *66*, 97–104. [CrossRef]

21. Wicke, S.; Schneeweiss, G.M.; dePamphilis, C.W.; Muller, K.F.; Quandt, D. The evolution of the plastid chromosome in land plants: Gene content, gene order, gene function. *Plant Mol. Biol.* **2011**, *76*, 273–297. [CrossRef] [PubMed]

22. Lohse, M.; Drechsel, O.; Kahlau, S.; Bock, R. Organellargenomedraw-a suite of tools for generating physical maps of plastid and mitochondrial genomes and visualizing expression data sets. *Nucleic Acids Res.* **2013**, *41*, W575–W581. [CrossRef] [PubMed]

23. Bausher, M.G.; Singh, N.D.; Lee, S.B.; Jansen, R.K.; Daniell, H. The complete chloroplast genome sequence of *Citrus sinensis* (L.) Osbeck var 'ridge pineapple': Organization and phylogenetic relationships to other angiosperms. *BMC Plant Biol.* **2006**, *6*, 21. [CrossRef] [PubMed]

24. Organelle Resources at NCBI. Available online: http://www.ncbi.nlm.nih.gov/genome/organelle/ (accessed on 1 February 2018).

25. Jansen, R.K.; Cai, Z.; Raubeson, L.A.; Daniell, H.; Depamphilis, C.W.; Leebens-Mack, J.; Muller, K.F.; Guisinger-Bellian, M.; Haberle, R.C.; Hansen, A.K.; et al. Analysis of 81 genes from 64 plastid genomes resolves relationships in angiosperms and identifies genome-scale evolutionary patterns. *Proc. Natl. Acad. Sci. USA* **2007**, *104*, 19369–19374. [CrossRef] [PubMed]

26. Green, B.R. Chloroplast genomes of photosynthetic eukaryotes. *Plant J.* **2011**, *66*, 34–44. [CrossRef] [PubMed]

27. Millen, R.S.; Olmstead, R.G.; Adams, K.L.; Palmer, J.D.; Lao, N.T.; Heggie, L.; Kavanagh, T.A.; Hibberd, J.M.; Giray, J.C.; Morden, C.W.; et al. Many parallel losses of infa from chloroplast DNA during angiosperm evolution with multiple independent transfers to the nucleus. *Plant Cell* **2001**, *13*, 645–658. [CrossRef] [PubMed]

28. Lu, L.; Li, X.; Hao, Z.; Yang, L.; Zhang, J.; Peng, Y.; Xu, H.; Lu, Y.; Zhang, J.; Shi, J.; et al. Phylogenetic studies and comparative chloroplast genome analyses elucidate the basal position of halophyte *Nitraria sibirica* (Nitrariaceae) in the sapindales. *Mitochondrial DNA Part A DNA Mapp. Seq. Anal.* **2017**, 1–11. [CrossRef] [PubMed]

29. Daniell, H.; Lin, C.S.; Yu, M.; Chang, W.J. Chloroplast genomes: Diversity, evolution, and applications in genetic engineering. *Genome Biol.* **2016**, *17*, 134. [CrossRef] [PubMed]

30. Williams, A.V.; Miller, J.T.; Small, I.; Nevill, P.G.; Boykin, L.M. Integration of complete chloroplast genome sequences with small amplicon datasets improves phylogenetic resolution in Acacia. *Mol. Phylogenet. Evol.* **2016**, *96*, 1–8. [CrossRef] [PubMed]

31. Koenen, E.J.M.; Clarkson, J.J.; Pennington, T.D.; Chatrou, L.W. Recently evolved diversity and convergent radiations of rainforest mahoganies (Meliaceae) shed new light on the origins of rainforest hyperdiversity. *New Phytol.* **2015**, *207*, 327–339. [CrossRef] [PubMed]

32. Pennington, T.D.; Styles, B.T. A generic monograph of the meliaceae. *Blumea* **1975**, *22*, 419–540.

33. Muellner, A.N.; Samuel, R.; Johnson, S.A.; Cheek, M.; Pennington, T.D.; Chase, M.W. Molecular phylogenetics of Meliaceae (Sapindales) based on nuclear and plastid DNA sequences. *Am. J. Bot.* **2003**, *90*, 471–480. [CrossRef] [PubMed]

34. Muellner, A.N.; Pennington, T.D.; Chase, M.W. Molecular phylogenetics of neotropical Cedreleae (mahogany family, Meliaceae) based on nuclear and plastid DNA sequences reveal multiple origins of "cedrela odorata". *Mol. Phylogenet. Evol.* **2009**, *52*, 461–469. [CrossRef] [PubMed]

35. Schroeder, H.; Kersten, B.; Fladung, M. Development of multiplexed marker sets to identify the most relevant poplar species for breeding. *Forests* **2017**, *8*, 492. [CrossRef]
36. Soltis, D.E.; Soltis, P.S. Choosing an approach and an appropriate gene for phylogenetic analysis. In *Molecular Systematics of Plants II*; Soltis, D.E., Soltis, P.S., Doyle, J.J., Eds.; Kluwer Academic Publishers: Boston, MA, USA, 1998; pp. 1–42.
37. Olmstead, R.G.; Palmer, J.D. Chloroplast DNA systematics—A review of methods and data-analysis. *Am. J. Bot.* **1994**, *81*, 1205–1224. [CrossRef]
38. Neuhaus, H.; Link, G. The chloroplast tRNALys(UUU) gene from mustard (*Sinapis alba*) contains a class-II intron potentially coding for a maturase-related polypeptide. *Curr. Genet.* **1987**, *11*, 251–257. [CrossRef] [PubMed]
39. Barthet, M.M.; Hilu, K.W. Evaluating evolutionary constraint on the rapidly evolving gene *matK* using protein composition. *J. Mol. Evol.* **2008**, *66*, 85–97. [CrossRef] [PubMed]
40. Braham Bioinformatics/Fastqc. Available online: http://www.bioinformatics.babraham.ac.uk/projects/fastqc/ (accessed on 1 December 2017).
41. Hahn, C.; Bachmann, L.; Chevreux, B. Reconstructing mitochondrial genomes directly from genomic next-generation sequencing reads-a baiting and iterative mapping approach. *Nucleic Acids Res.* **2013**, *41*. [CrossRef] [PubMed]
42. Bolger, A.M.; Lohse, M.; Usadel, B. Trimmomatic: A flexible trimmer for illumina sequence data. *Bioinformatics* **2014**, *30*, 2114–2120. [CrossRef] [PubMed]
43. Camacho, C.; Coulouris, G.; Avagyan, V.; Ma, N.; Papadopoulos, J.; Bealer, K.; Madden, T.L. Blast+: Architecture and applications. *BMC Bioinf.* **2009**, *10*, 421. [CrossRef] [PubMed]
44. Genbank. Available online: http://www.ncbi.nlm.nih.gov/genbank/ (accessed on 1 January 2018).
45. Ncbi/Blast. Available online: https://blast.ncbi.nlm.nih.gov/Blast.cgi?PROGRAM=blastn&PAGE_TYPE=BlastSearch&LINK_LOC=blasthome (accessed on 19 January 2018).
46. Liu, C.; Shi, L.C.; Zhu, Y.J.; Chen, H.M.; Zhang, J.H.; Lin, X.H.; Guan, X.J. Cpgavas, an integrated web server for the annotation, visualization, analysis, and genbank submission of completely sequenced chloroplast genome sequences. *BMC Genom.* **2012**, *13*, 715. [CrossRef] [PubMed]
47. CpGAVAS: Chloroplast Genome Annotation, Visualization, Analysis and Genbank. Available online: http://www.herbalgenomics.org/0506/cpgavas (accessed on 1 February 2018).
48. Chlorobox/genbank2sequin. Available online: https://chlorobox.mpimp-golm.mpg.de/GenBank2Sequin.html (accessed on 1 February 2018).
49. NCBI/sequin. Available online: https://www.ncbi.nlm.nih.gov/Sequin/ (accessed on 1 February 2018).
50. OrganellarGenomeDRAW. Available online: http://ogdraw.mpimp-golm.mpg.de/ (accessed on 1 February 2018).
51. VISTA. Available online: http://genome.lbl.gov/vista/index.shtml (accessed on 1 February 2018).
52. Mayor, C.; Brudno, M.; Schwartz, J.R.; Poliakov, A.; Rubin, E.M.; Frazer, K.A.; Pachter, L.S.; Dubchak, I. VISTA: Visualizing global DNA sequence alignments of arbitrary length. *Bioinformatics* **2000**, *16*, 1046–1047. [CrossRef] [PubMed]
53. SNiPlay Pipeline v2. Available online: http://sniplay.cirad.fr/cgi-bin/analysis.cgi (accessed on 1 February 2018).
54. Dereeper, A.; Homa, F.; Andres, G.; Sempere, G.; Sarah, G.; Hueber, Y.; Dufayard, J.F.; Ruiz, M. SNiPlay3: A web-based application for exploration and large scale analyses of genomic variations. *Nucleic Acids Res.* **2015**, *43*, W295–W300. [CrossRef] [PubMed]

International Journal of
Molecular Sciences

MDPI

Article

Molecular Evolution of Chloroplast Genomes of Orchid Species: Insights into Phylogenetic Relationship and Adaptive Evolution

Wan-Lin Dong [†], Ruo-Nan Wang [†], Na-Yao Zhang, Wei-Bing Fan, Min-Feng Fang and Zhong-Hu Li *

Key Laboratory of Resource Biology and Biotechnology in Western China, Ministry of Education, College of Life Sciences, Northwest University, Xi'an 710069, China; dongwl@stumail.nwu.edu.cn (W.-L.D.); wangruonan@stumail.nwu.edu.cn (R.-N.W.); 201620812@stumail.nwu.edu.cn (N.-Y.Z.); 201631689@stumail.nwu.edu.cn (W.-B.F.); fangmf@nwu.edu.cn (M.-F.F.)
* Correspondence: lizhonghu@nwu.edu.cn; Tel./Fax: +86-029-8830-2411
† These two authors contributed equally to this study.

Received: 24 January 2018; Accepted: 27 February 2018; Published: 2 March 2018

Abstract: Orchidaceae is the 3rd largest family of angiosperms, an evolved young branch of monocotyledons. This family contains a number of economically-important horticulture and flowering plants. However, the limited availability of genomic information largely hindered the study of molecular evolution and phylogeny of Orchidaceae. In this study, we determined the evolutionary characteristics of whole chloroplast (cp) genomes and the phylogenetic relationships of the family Orchidaceae. We firstly characterized the cp genomes of four orchid species: *Cremastra appendiculata*, *Calanthe davidii*, *Epipactis mairei*, and *Platanthera japonica*. The size of the chloroplast genome ranged from 153,629 bp (*C. davidi*) to 160,427 bp (*E. mairei*). The gene order, GC content, and gene compositions are similar to those of other previously-reported angiosperms. We identified that the genes of *ndhC*, *ndhI*, and *ndhK* were lost in *C. appendiculata*, in that the *ndh I* gene was lost in *P. japonica* and *E. mairei*. In addition, the four types of repeats (forward, palindromic, reverse, and complement repeats) were examined in orchid species. *E. mairei* had the highest number of repeats (81), while *C. davidii* had the lowest number (57). The total number of Simple Sequence Repeats is at least 50 in *C. davidii*, and, at most, 78 in *P. japonica*. Interestingly, we identified 16 genes with positive selection sites (the *psbH*, *petD*, *petL*, *rpl22*, *rpl32*, *rpoC1*, *rpoC2*, *rps12*, *rps15*, *rps16*, *accD*, *ccsA*, *rbcL*, *ycf1*, *ycf2*, and *ycf4* genes), which might play an important role in the orchid species' adaptation to diverse environments. Additionally, 11 mutational hotspot regions were determined, including five non-coding regions (*ndhB* intron, *ccsA-ndhD*, *rpl33-rps18*, *ndhE-ndhG*, and *ndhF-rpl32*) and six coding regions (*rps16*, *ndhC*, *rpl32*, *ndhI*, *ndhK*, and *ndhF*). The phylogenetic analysis based on whole cp genomes showed that *C. appendiculata* was closely related to *C. striata* var. *vreelandii*, while *C. davidii* and *C. triplicate* formed a small monophyletic evolutionary clade with a high bootstrap support. In addition, five subfamilies of Orchidaceae, Apostasioideae, Cypripedioideae, Epidendroideae, Orchidoideae, and Vanilloideae, formed a nested evolutionary relationship in the phylogenetic tree. These results provide important insights into the adaptive evolution and phylogeny of Orchidaceae.

Keywords: adaptive variation; chloroplast genome; molecular evolution; Orchidaceae; phylogenetic relationship

1. Introduction

Orchidaceae is the biggest family of monocotyledons and the third largest angiosperm family, containing about five recognized subfamilies (Apostasioideae, Cypripedioideae, Epidendroideae,

Orchidoideae, and Vanilloideae) [1], with over 700 genera and 25,000 species [2–4]. The orchid species are generally distributed in tropical and subtropical regions in the world, while a few species are found in temperate zones. Many orchid species have important ornamental and flowering values, e.g., their flowers are characterized by labella and a column, and they are attractive to humans [5,6]. In recent years, due to the overexploitation and habitat destruction of orchid species, many wild population resources have become rare and endangered [7]. Presently, some scholars have mainly concentrated on the study of Orchidaceae for their morphology and medicinal value, and research on genomes has been relatively scarce [8,9]. Some studies showed that the two subfamilies, Apostasioideae and Cypripedioideae, were clustered into the two respective genetic clades based partial on chloroplast DNA regions and nuclear markers [4,10]. However, the major phylogenetic relationships among the five orchid subfamilies remain unresolved [11].

In recent years, the fast progress of next-generation sequencing technology has provided a good opportunity for the study of genomic evolution and interspecific relationships of organisms based on large-scale genomic dataset resources, such as complete plastid sequences [12,13]. The chloroplast (cp) is made up of multifunctional organelles, playing a critical role in photosynthesis and carbon fixation [5,14–16]. The majority of the cp genomes of angiosperms are circular DNA molecules, ranging from 120 to 160 kb in length, with highly-conserved compositions, in terms of gene content and gene order [17–20]. Generally, the typical cp genome is composed of a large single copy (LSC) region and a small single copy (SSC) region, which are separated by two copies of inverted repeats (IRa/b) [21–23]. Due to its maternal inheritance and conserved structure characteristics [24–27], the cp genomes can provide abundant genetic information for studying species divergence and the interspecific relationships of plants [28–31]. For example, based on complete cp genomes, some studies suggested that *Dactylorhiza viridis* diverged earlier than *Dactylorhiza incarnate* [12]; *Lepanthes* is was distinct from *Pleurothallis* and *Salpistele* [13]. In addition, some researchs based on one nuclear region (*ITS-1*) and five chloroplast DNA fragment variations revealed that *Bolusiella talbotii* and the congeneric *B. iridifolia* were clustered into an earlier diverged lineage [10]. However, up to now, the phylogenetic relationships of some major taxons (e.g., *Cremastra* and *Epipactis*) in the Orchidaceae family remain unclear.

In this study, the complete cp genomes of four orchid species (*Cremastra appendiculata*, *Calanthe davidii*, *Epipactis mairei*, and *Platanthera japonica*) were first assembled and annotated. Following this, we analyzed the differences in genome size, content, and structure, and the inverted repeats (IR) contraction and expansion, identifying the sequence divergence, along with variant hotspot regions and adaptive evolution through combination with other available orchid cp genomes. In addition, we also constructed the evolutionary relationships of the Orchidaceae family, based on the large number of cp genome datasets.

2. Results

2.1. The Chloroplast Genome Structures

In this study, the cp genomes of four species displayed a typical quadripartite structure and similar lengths, containing a pair of inverted repeats IR regions (IRa and IRb), one large single-copy (LSC) region, and one small single-copy (SSC) region (Figure 1, Table 2). The cp genome size ranged from 153,629 bp in *C. davidii* to 160,427 bp in *E. mairei*, with *P. japonica* at 154,995 bp and *C. appendiculata* at 155,320 bp. The length of LSC ranged from 85,979 bp (*P. japonica*) to 88,328 bp (*E. mairei*), while the SSC length and IR length ranged from 13,664 bp (*P. japonica*) to 18,513 bp (*E. mairei*), and from 25,956 bp (*C. davidii*) to 27,676 bp (*P. japonica*). In the four species, the GC contents of the LSC and SSC regions (about 34% and 40%) were lower than those of the IR regions (about 43%) (Table 1). There were 37 tRNA genes and eight rRNA genes that were identified in each orchid cp genome, but there were some differences in terms of protein-coding genes. In *C. davidii*, we annotated 86 protein-coding genes. There were no *ndhC*, *ndhI*, and *ndhK* genes in *C. appendiculata*. In *P. japonica* and *E. mairei*, the *ndhI* gene was lost (Tables 1 and 2). Fourteen out of the seventeen genes contained a single intron, while three (*clpP*, *ycf3*, and *rps12*) had two introns (Table 2).

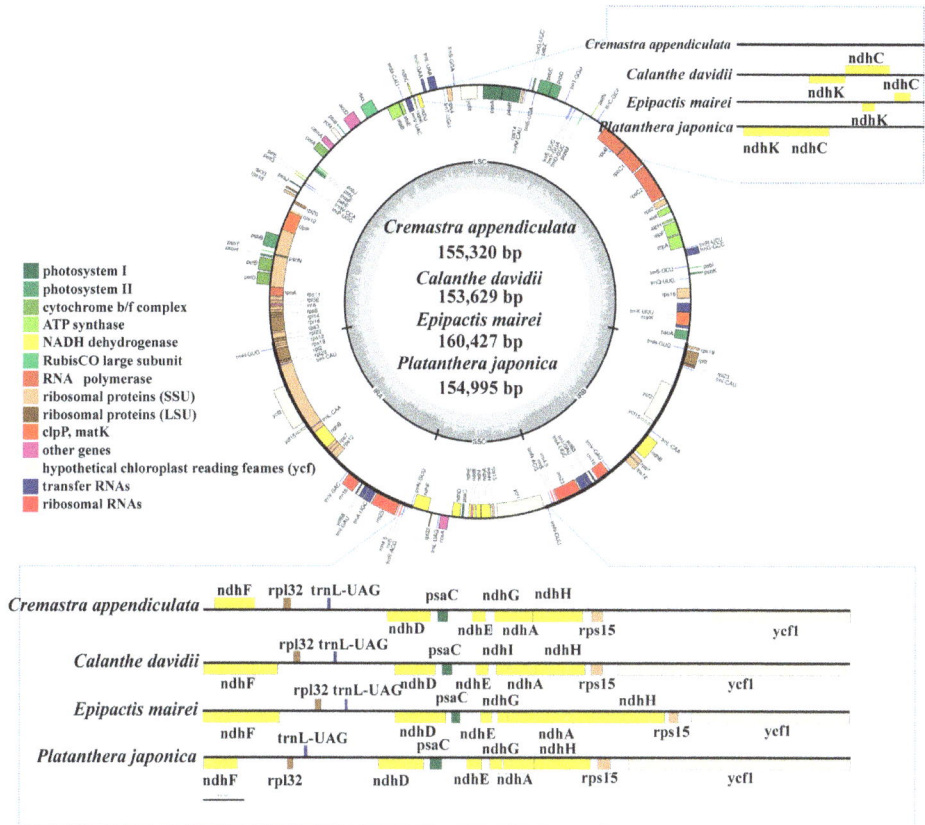

Figure 1. Chloroplast genome maps of the four orchid species. Gene locations outside of the outer rim are transcribed in the counter clockwise direction, whereas genes inside are transcribed in the clockwise direction. The colored bars indicate known different functional groups. The dashed gray area in the inner circle shows the proportional GC content of the corresponding genes. LSC, SSC and IR are large single-copy region, small single-copy region, and inverted repeat region, respectively.

Table 1. Comparison of chloroplast genome features in four orchid species.

Species	Cremastra appendiculata	Calanthe davidii	Epipactis mairei	Platanthera japonica
Accession number	MG925366	MG925365	MG925367	MG925368
Genome size (bp)	155,320	153,629	160,427	154,995
LSC length (bp)	87,098	86,045	88,328	85,979
SSC length (bp)	15,478	15,672	18,513	13,664
IR length (bp)	26,372	25,956	26,790	27,676
Coding (bp)	100,018	104,531	113,915	107,028
Non-coding (bp)	55,302	49,098	46,512	47,967
Number of genes	130 (0)	132 (19)	131 (19)	128 (17)
Number of protein-coding genes	83 (7)	86 (7)	85 (7)	85 (7)
Number of tRNA genes	38 (8)	38 (8)	38 (8)	38 (8)
Number of rRNA genes	8 (4)	8 (4)	8 (4)	8 (4)
GC content (%)	37.2	36.9	37.2	37
GC content in LSC (%)	34.5	34.5	34.9	34.2
GC content in SSC (%)	30.4	30.2	31.0	29
GC content in IR (%)	43.5	43.1	43.1	43.2
Mapped read number	551,680	324,741	230,968	322,259
Chloroplast coverage	544.9	217.4	216	313.6

The numbers in parenthesis indicate the genes duplicated in the IR regions.

Table 2. List of genes present in four orchid chloroplast genomes.

Category of Genes	Group of Gene	Name of Gene	Name of Gene	Name of Gene	Name of Gene	Name of Gene
Self-replication	Ribosomal RNA genes	rrn16 (×2)	rrn2 (×2)	rrn4.5 (×2)	rrn5 (×2)	
	Transfer RNA genes	trnA-UGC *(×2)	trnC-GCA	trnD-GUC	trnE-UUC	trnF-GAA
		trnfM-CAU	trnG-GCC *	trnG-UCC	trnH-GUG (×2)	trnI-CAU (×2)
		trnL-GAU *(×2)	trnK-UUU *(×2)	trnL-CAA (×2)	trnL-UAA *	trnL-UAG
		trnM-CAU	trnN-GUU (×2)	trnP-UGG	trnQ-UUG	trnR-ACG (×2)
		trnR-UCU	trnS-GCU	trnS-GGA	trnS-UGA	trnT-GGU
		trnT-UGU	trnV-GAC (×2)	trnV-UAC (×2)	trnW-CCA	trnY-GUA
	Small subunit of ribosome	rps2	rps3	rps4	rps7 (×2)	rps8
		rps11	rps12 **(×2)	rps14	rps15	rps16 *
		rps18	rps19 (×2)			
	Large subunit of ribosome	rpl2 *(×2)	rpl14	rpl16 *	rpl20	rpl22
		rpl23 (×2)	rpl32	rpl33	rpl36	
	DNA-dependent RNA polymerase	rpoA	rpoB	rpoC1 *	rpoC2	
	Translational initiation factor	infA				
Genes for photosynthesis	Subunits of NADH-dehydrogenase	ndhA *	ndhB *(×2)	ndhC [a]	ndhD	ndhE
		ndhF	ndhG	ndhH	ndhI [a,c,d]	ndhJ
		ndhK [a]				
	Subunits of photosystem I	psaA	psaB	psaC	psaI	psaJ
		ycf3 **	ycf4			
	Subunits of photosystem II	psbA	psbB	psbC	psbD	psbE
		psbF	psbH	psbI	psbJ	psbK
		psbL	psbM	psbN	psbT	psbZ
	Subunits of cytochrome b/f complex	petA	petB*	petD *	petG	petL
		petN				
	Subunits of ATP synthase	atpA	atpB	atpE	atpF *	atpH
		atpI				
	Subunits of rubisco	rbcL				
Other genes	Maturase	matK				
	Protease	clpP **				
	Envelope membrane protein	cemA				
	Subunit of acetyl-CoA carboxylase	accD				
	C-type cytochrome synthesis gene	ccsA				
Genes of unknown function	Conserved open reading frames	ycf1	ycf2 (×2)			

[a] gene is no in *Cremastra appendiculata*; [c] gene is not in *Epipactis mairei*; [d] gene is not in *Platanthera japonica*; * Gene contains one intron; ** gene contains two introns; (×2) indicates that the number of the repeat unit is 2.

2.2. Repeat Structure and Simple Sequence Repeats

Repeats in cp genomes were analyzed using REPuter (Figure 2a and Table S2). *E. mairei* had the greatest number, including 46 forward, 31 palindromic, three reverse repeats, and 1 complement repeat. This was followed by *C. appendiculata* with 43 forward, 33 palindromic, and 2 reverse repeats. *P. japonica* had 42, 21, 1, and 1 forward, palindromic, reverse, and complement repeats. *C. davidii* had the least number, with only 30 forward and 27 palindromic repeats. The comparison analyses revealed that most of the repeats were 30–90 bp, and that the longest repeats, with a length of 309 bp, were detected in the *E. mairei* cp genome (Figure 2b). Most of the repeats were distributed in non-coding regions. There were 9% repeats in coding sequence and intergenic spacer parts (CDS-IGS) in *E. mairei*, but none in *C. appendiculata* (Figure 2c). The highest number of tandem repeats was 53 in *E. mairei*, and the lowest was 29 in *C. davidii* (Table S3). The total number of SSRs was 51 in *C. appendiculata*, 50 in *C. davidii*, 58 in *E. mairei*, and 78 in *P. japonica* (Table S4). Only one six compound, SSR, was found in *C. appendiculata* (Figure 3a). A large proportion of SSRs were found in the LSC region, and we did not identify C/G mononucleotide repeats, while the majority of the dinucleotide repeat sequences were comprised of AT/TA repeats (Figure 3b).

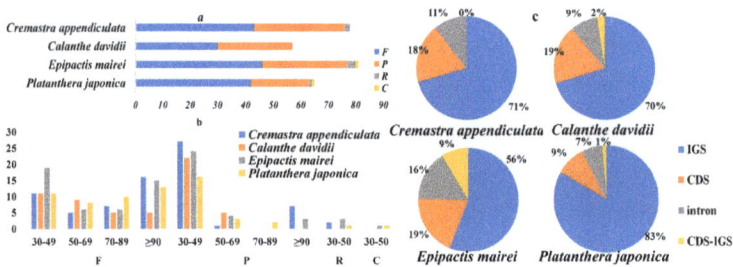

Figure 2. Maps of repeat sequence analyses. Repeat sequences in *C. appendiculata*, *C. davidii*, *E. mairei*, and *P. japonica* chloroplast genomes. (**a**) Number of the four repeat types, F, P, R, and C, indicate the repeat type (F: forward, P: palindrome, R: reverse, and C: complement, respectively). (**b**) Frequency of the four repeat types by length. (**c**) Repeat distribution among four different regions: IGS: intergenic spacer, CDS: coding sequence, intron and CDS-IGS part in CDS and part in IGS.

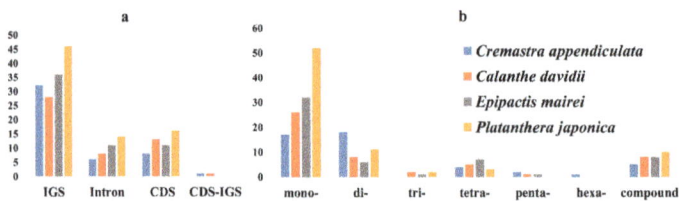

Figure 3. The distribution maps of simple sequence repeats (SSR) in *C. appendiculata*, *C. davidii*, *E. mairei*, and *P. japonica* chloroplast genomes. (**a**) Classification of SSRs in four orchid species. IGS, intergenic spacer; CDS, coding sequence, CDS-IGS, part in CDS and part in IGS. (**b**) Classification of SSRs by repeat type. mono-, mononucleotides; di-, dinucleotides; tri-, trinucleotides; tetra-, tetranucleotides; penta-, pentanucleotides; and hexa-, hexanucleotides.

2.3. IR Contraction and Expansion

We examined the differences between inverted repeat and single-copy (IR/SC) boundary regions among 20 orchid genera, which were classified into several different types (Figure 4). First, the *rps19* gene crossed the large single-copy and inverted repeat b (LSC/IRb) regions within the two parts for eighteen Orchidaceae genera. In *C. crispate* and *C. appendiculata*, the *rps19* gene existed only in the IRb region. Second, in 12 genera, the *ndhF* gene and the *ycf1* pseudogene overlapped in the IRb/SSC region. In *C. appendiculata* and *Dendrobium strongylanthum*, the *ndhF* gene was complete in the SSC region, 8–35 bp away from the IRb region. In *C. crispate*, *E. pusilla*, and *Phalaenopsis equestris*, the *rpl32* gene was in the SSC region instead of the *ndhF* gene, 280–464 bp away from the IRb region. For the 17 genera mentioned above, the *ycf1* gene crossed the SSC/IRa region. In *C. edavidii* and *Bletilla ochracea*, the *ndhF* gene crossed the IRb/SSC region, and the *ycf1* gene was complete in the SSC region, 101 and 4 bp away from the IRa region. The *trnH-GUG* genes were all located in the LSC region, which was 231 to 1390 bp away from the LSC/IRa boundary. Most specifically, in *Vanilla planifolia*, the *ccsA* gene crossed the IRb/SSC region, as we did not find the *ndhF* and *ycf1* genes where they should be. The SSC/IRa borders were located between the *rpl32* and *ycf1* genes. Thirdly, all 20 genera had the same IRa/LSC borders: the *rps19* gene in the IRa region and the *psbA* gene in the LSC region.

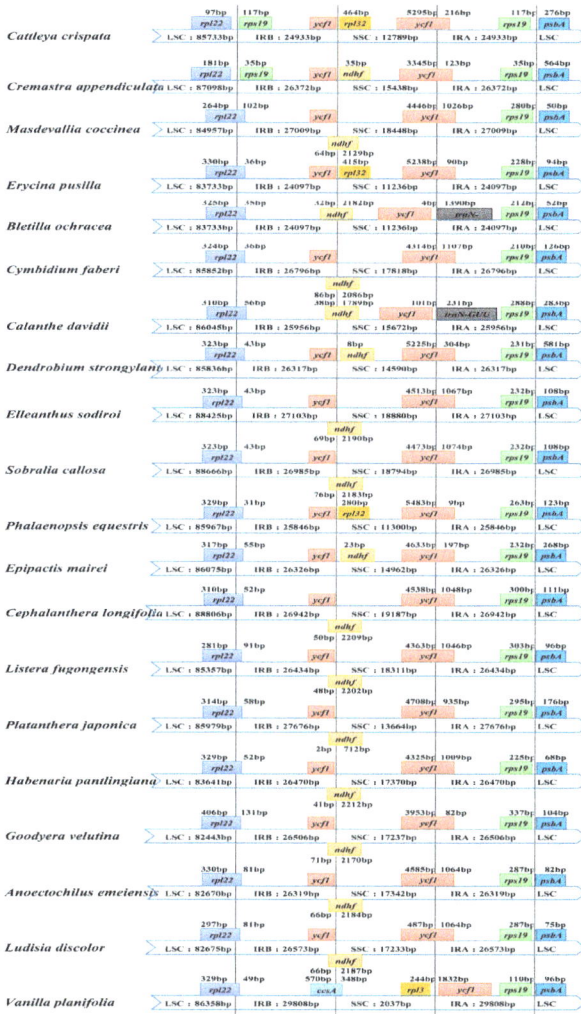

Figure 4. Comparison of the borders of LSC, SSC, and IR regions in 20 orchid complete chloroplast genomes.

2.4. Sequence Divergence and Mutational Hotspot

The whole chloroplast genome sequences of *C. appendiculata*, *C. davidii*, *E. mairei*, and *P. japonica* were compared to 16 other species, using mVISTA [32] (Figures 5 and 6, and Table S5). The comparison analyses showed a high sequence similarity across the cp genomes, with a sequence identity of 82.0%. Interestingly, the proportions of variability in the non-coding regions (introns and intergenic spacers) ranged from 6.77% to 100% with a mean value of 45.97%, i.e., values that are twice as high as in the coding regions (where the range was from 5.80% to 61.76% with a mean value of 24.68%). Five regions within the non-coding regions (*ndhB* intron, *ccsA-ndhD*, *rpl33-rps18*, *ndhE-ndhG*, and *ndhF-rpl32*) and six regions within the coding parts (*rps16*, *ndhC*, *rpl32*, *ndhI*, *ndhK*, and *ndhF*) showed greater levels of variations (percentage of variability >80% and 50%, respectively). In particular, the *ndhB* intron and *ccsA-ndhD* showed a variable percentage of 100%.

In addition, we performed a MAUVE [33] alignment of the 20 orchid chloroplast genomes. The *C. appendiculata* genome is shown at the top as the reference genome (Figure 7). These species maintained a consistent sequence order in most of the genes. However, in *B. ochracea* and *C. faberi*, the *psbM* gene was in front of the *petN*, while the others were upside-down. *Bletilla* and *Cymbidium* actually had the nearest relationship.

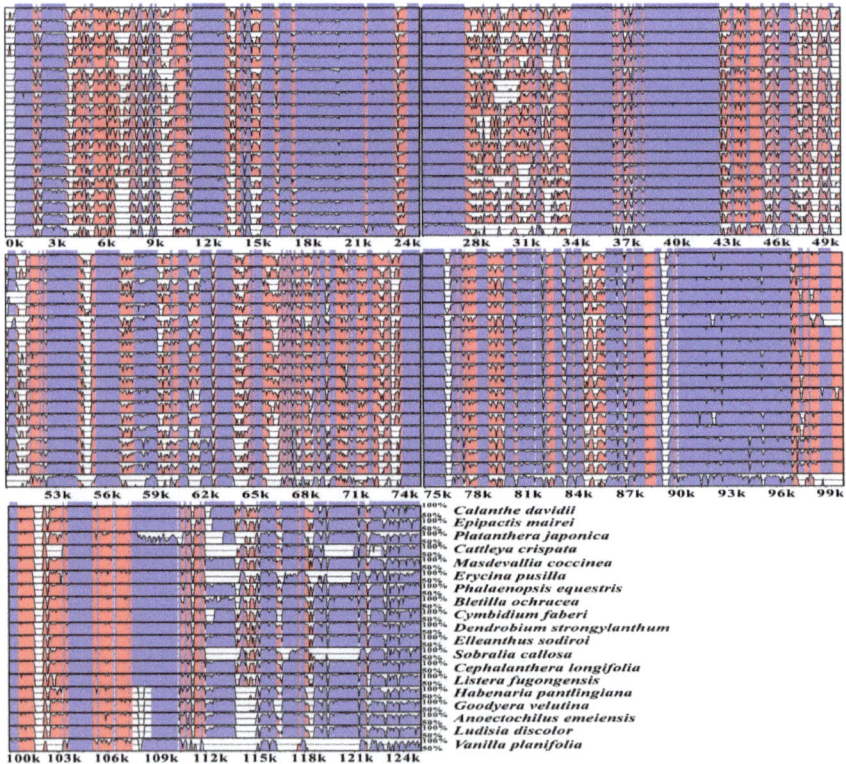

Figure 5. Sequence alignment of chloroplast genomes of 20 orchid species. Sequence identity plot comparing the chloroplast genomes with *C. appendiculata* as a reference using mVISTA. The red color-coded as intergenic spacer regions. The blue color-coded as gene regions. A cut-off of 70% identity was used for the plots, and the Y-scale represents the percent identity between 50% and 100%.

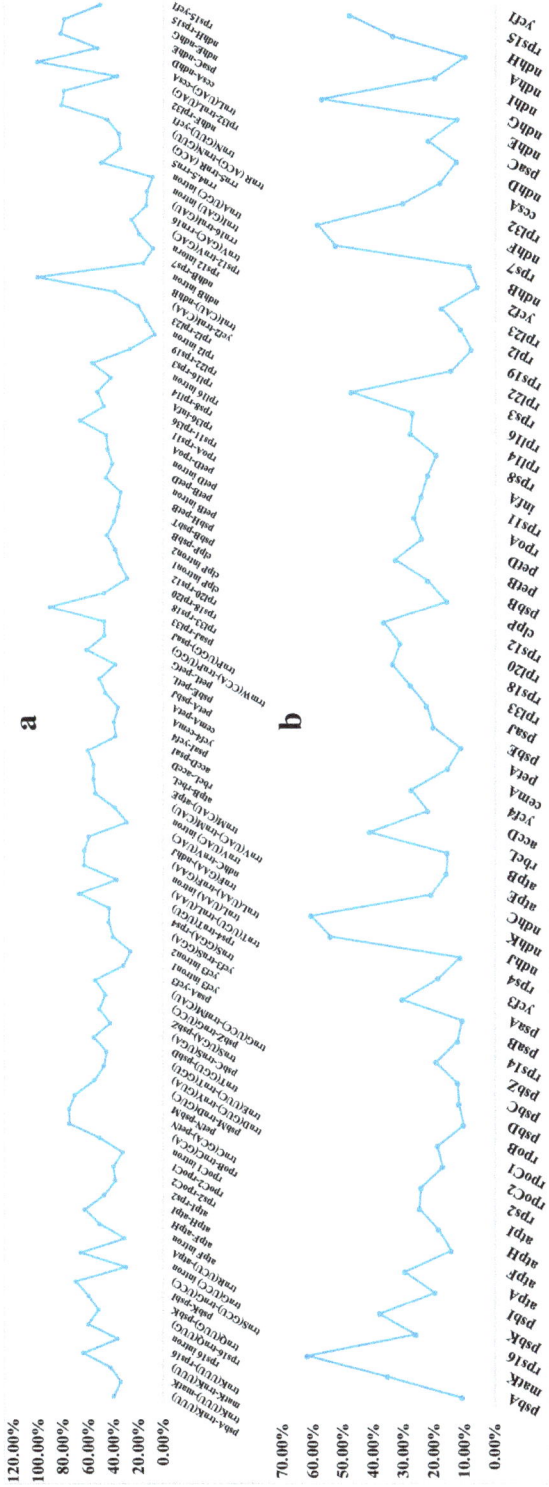

Figure 6. Percentages of variable sites in homologous regions across the 20 orchids with complete chloroplast genomes. (**a**) The introns and spacers (IGS); and (**b**) protein coding sequences (CDS).

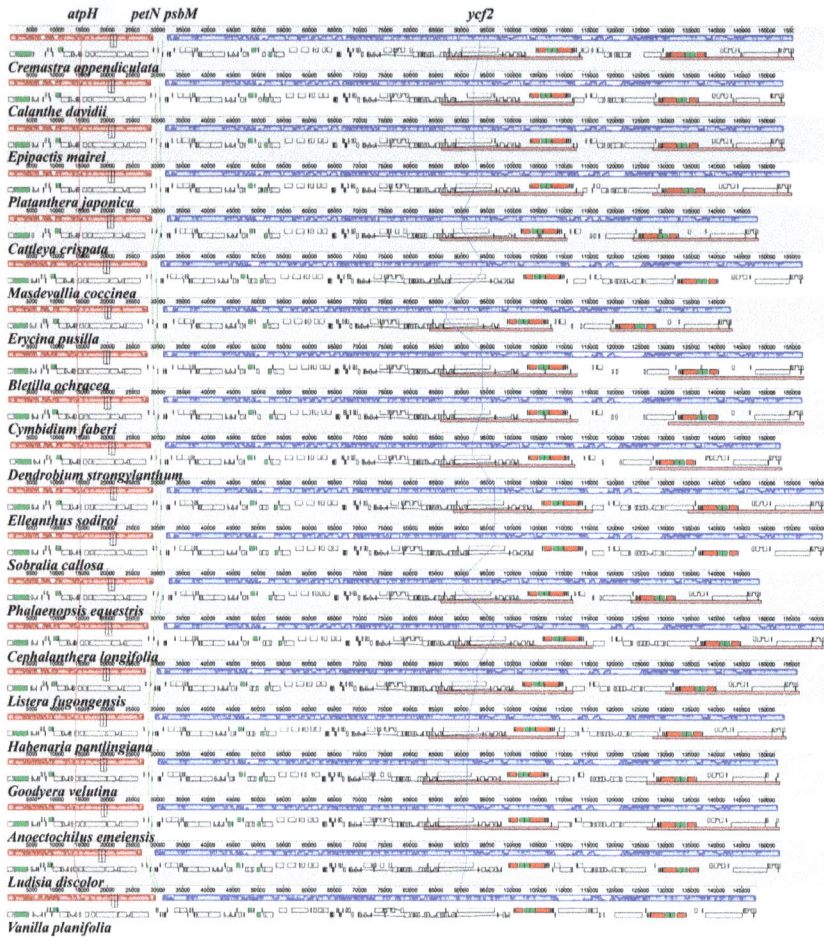

Figure 7. MAUVE genome alignments of the 20 orchid chloroplast genomes, with *C. appendiculata* set as a reference genome. The corresponding colored boxes indicate locally-collinear blocks, which present homologous gene clusters. The red vertical line is the location of *atpH* gene. The yellow vertical line is the location of *petN* gene. The green vertical line is the location of *psbM* gene. The blue vertical line is the location of *ycf2* gene.

2.5. Gene Selective Analysis

We compared the rate of nonsynonymous (dN) and synonymous (dS) substitutions for 68 common protein-coding genes between *C. appendiculata*, *C. davidii*, *E. mairei*, and *P. japonica* with 16 other Orchidaceae species (Table S6). Sixteen genes with positive selection sites were identified (Table S7). These genes included one subunit of the photosystem II gene (*psbH*), two genes for cytochrome b/f complex subunit proteins (*petD* and *petL*), two genes for ribosome large subunit proteins (*rpl22* and *rpl32*), two DNA-dependent RNA polymerase genes (*rpoC1* and *rpoC2*), three genes for ribosome small subunit proteins (*rps12*, *rps15*, and *rps16*), and *accD*, *ccsA*, *rbcL*, *ycf1*, *ycf2*, and *ycf4* genes. Interestingly, the *ycf1* gene possesses 13 and 15 positive selective sites, followed by *accD* (8, 10), *rbcL* (4, 7), *ycf2* (2, 3), *rpoC1* (2, 4), *rpoC2* (1, 2), *rpl22* (1, 2), *rps16* (1, 2), *rpl32* (1, 1), *rps12* (1, 1), *ccsA* (0, 2), *petD* (0, 1), *petL* (0, 1), *psbH* (0, 1), and *ycf4* (0, 1). What is more, the likelihood ratio tests (LRTs) of variables under different models were compared in the site-specific models, M0 vs. M3, M1 vs. M2 and M7 vs. M8, in order to support the sites under positive selection ($p < 0.01$) (Table S7).

2.6. Phylogenetic Relationship

In this study, the maximum likelihood (ML) analysis suggested that *C. appendiculata* and the congeneric *C. davidii* clustered into the Epidendroideae subfamily clade with high bootstrap support, and that *E. mairei* and *P. japonica* clustered into Orchidoideae subfamily (Figure 8). Interestingly, five subfamilies of Orchidaceae, Apostasioideae, Cypripedioideae, Epidendroideae, Orchidoideae, and Vanilloideae have a nested evolutionary relationship in the ML tree. Meanwhile, *C. appendiculata* was closely-related to *C. striata* var. *vreelandii*, *C. davidii*, and *C. triplicate*, which formed a small evolutionary clade with a high bootstrap. *P. japonica* and *Habenaria pantlingiana* had a relatively-closer affinity in the Orchidoideae subfamily.

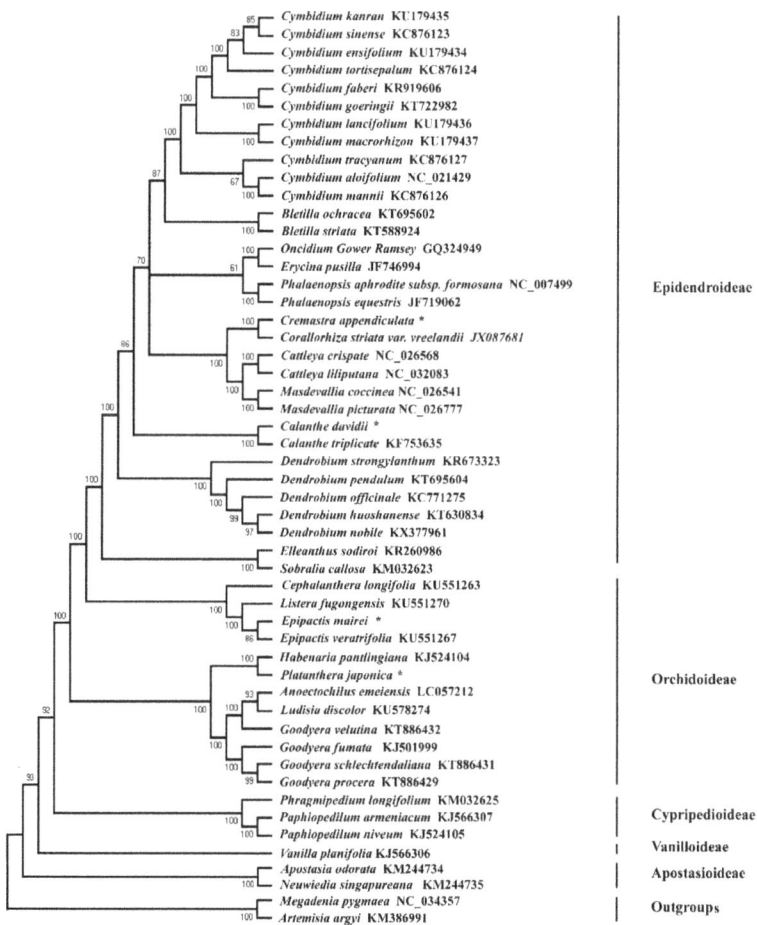

Figure 8. Cladogram of 54 nucleotide sequences of complete chloroplast genomes of orchid species based on the GTRGAMMA model with maximum likelihood (ML) analysis. * The newly generated chloroplast genomes of orchid species.

3. Discussion

3.1. Sequence Variation

In this study, we first determined the whole chloroplast genomes of four orchid species. The cp genome size of *C. davidii* was shorter than that of others, which might be the result of the expansion and contraction of the border positions between the IR and SC regions [21–23]. In addition, the GC contents of the LSC and SSC regions in all the orchid species were much lower than those of the IR regions, which possibly resulted from four rRNA genes (*rrn16*, *rrn23*, *rrn4.5*, and *rrn5*) sequences in the IR regions. In addition, we identified some obvious differences in the protein-coding genes for the orchid chloroplast genomes, despite that the cp genomes of land plants are generally considered to be highly conserved [34]. Interestingly, there were no *ndhC*, *ndhI*, and *ndhK* genes in *C. appendiculata*. In *P. japonica* and *E. mairei*, the *ndhI* gene was lost. Previous studies also found that some orchid species had lost the *ndh* gene, which encodes the subunits of the nicotinamide-adenine dinucleotid (NADH) dehydrogenase-like complex proteins [35–37]. The loss of this gene might have hindered cyclic electron flow around photosystem I and affected the plant photosynthesizing [35–38]. In addition, some studies suggested that the different Orchidaceae species harbored a variable loss or retention of *ndh* genes [35]. For example, *Cymbidium* has the *ndhE*, *ndhJ*, and *ndhC* genes [39], and *Oncidium* has the *ndhB* gene [40]. Nevertheless, the mechanisms that underlie the variable loss or retention of *ndh* genes in orchids remain unclear [11,41].

In addition, we identified 233 SSRs in four orchid species (*C. appendiculata*, *C. davidii*, *E. mairei*, and *P. japonica*); 77.68% of SSRs were distributed in the IGS and intron regions. Generally, microsatellites consist of 1–6 nucleotide repeat units, which are widely distributed across the entire genome and have a great influence on genome recombination and rearrangement [42,43]. The large amount of SSRs also have been identified in *Forsythia suspense* [44], *Dendrobium nobile*, *Dendrobium officinale*, and so on [45]. The majority of these SSRs consisted of mono- and di-nucleotide repeats. Tri-, tetra-, and penta-nucleotide repeat sequences were detected at a much lower frequency in these orchid species and in other organisms [46,47].

Meanwhile, our analyses revealed that the mutational hotspots among orchid genera were highly variable. A diversity of IR contraction and expansion, along with the high level of mutational hotspots, revealed that Orchidaceae had experienced a complex evolution process. Interestingly, in the orchid species, the two IR regions were less divergent than the LSC and SSC regions. Five regions within the non-coding regions (*ndhB* intron, *ccsA-ndhD*, *rpl33-rps18*, *ndhE-ndhG*, and *ndhF-rpl32*) and six regions within the coding regions (*rps16*, *ndhC*, *rpl32*, *ndhI*, *ndhK*, and *ndhF*) showed greater levels of variations (percentage of variability >80% and 50%, respectively). These regions can be used as potential DNA barcodes for the further study of phylogenetic relationships, species identification, and population genetics.

3.2. Adaptive Evolution

We used the site-specific model (seqtype = 1, model = 0, NSsites = 0, 1, 2, 3, 7, 8), one of the codon substitution models, to estimate the selection pressure [48]. Sixteen genes with positive selection sites were identified in these orchid species. These genes included one subunit of the photosystem II gene (*psbH*), two genes for cytochrome b/f complex subunit proteins (*petD* and *petL*), two genes for ribosome large subunit proteins (*rpl22* and *rpl32*), two DNA-dependent RNA polymerase genes (*rpoC1* and *rpoC2*), three genes for ribosome small subunit proteins (*rps12*, *rps15*, and *rps16*), and *accD*, *ccsA*, *rbcL*, *ycf1*, *ycf2*, and *ycf4* genes. We found that the genes with positive selection sites can be divided into four categories: Subunits of photosystem (*psbH* and *ycf4*), subunits of cytochrome (*petD*, *petL*, and *ccsA*), subunits of ribosome (*rpl22*, *rpl32*, *rps12*, *rps15*, and *rps16*) and others (*rpoC1*, *rpoC2*, *accD*, *rbcL*, *ycf1*, and *ycf2*). The plastid *accD* gene, which encodes the β-carboxyl transferase subunit of acetyl-CoA carboxylase, is an essential and required component for plant leaf development [49–53]. In this study, 10 positively-selected sites were identified in *accD* genes for orchid species, suggesting that the *accD* gene played a possible pivotal role in the adaptive evolution of orchids. What is more, the *ycf1* gene is also essential for almost all plant lineages [5,54], except for Gramineae, which lost the *ycf1* gene in its cp genomes [55]. Additionally, *ycf1* is one of the largest cp genes, encoding a component of the chloroplast's inner envelope membrane protein translocon [56]. This gene, which is also highly variable in terms of phylogenetic information at the level of species, has also been shown to be subject to positive selection with 15 sites, as has also been identified in many plant lineages [57–59]. In addition, we found that the *rbcL* gene possessed seven sites under positive selection in orchid species. Generally, *rbcL* is the gene for the Rubisco large subunit protein, and as the result of enzymatic activity of Rubisco, which is an important component as a modulator of photosynthetic electron transport [60,61]. Current research has revealed that positive selection of the *rbcL* gene in land plants may be a common phenomenon [62]. Additionally, the *rbcL* gene is also widely used in the phylogenetic analysis of land plants [63]. In conclusion, these results showed that multiple factors, several of them interconnected (positive selection, heterogeneity environments), have possibly contributed to orchid diversification and adaptation. For example, some positively-selected sites that were identified (e.g., *rbcL*, *ycf1*, and *accD*) were associated in a significant manner with environment adaptation, including factors such as temperature, light, humidity, and atmosphere [49]. Additionally, epiphytism in orchid species is a key innovation which should help generate and maintain high levels of plant diversity. On the other hand, the tropical distributions of orchid species might have increased the rates of speciation relative to those outside of the tropics as a result of more stable climates (e.g., the lack of glaciation and suitable temperatures), the greater habitat area, and together, this possibly provided a greater opportunity for the co-evolution of plants and their mutualists, and for greater adaptation [49,58,59].

3.3. Phylogenetic Relationship

In this study, the maximum likelihood (ML) tree obtained high bootstrap support values, which had 33 nodes with 100% bootstrap support, with 36 of the 46 nodes having values ≥95%. The phylogenetic analyses based on complete cp genomes, suggested that five subfamilies of Orchidaceae (Apostasioideae, Cypripedioideae, Epidendroideae, Orchidoideae, and Vanilloideae) have a nested evolutionary relationship (Figure 8). Apostasioideae is the earliest diverging subfamily of orchids. Some recent molecular studies have shown that the five subfamilies had formed their respective five monophylies [11,41,49]. The generic relationships of the five subfamilies found in our analyses are basically congruent with those of recent studies. However, our finding that Orchidoideae is a nested subfamily is different from the studies of Kim et al. [41] and Givnish et al. [49]. They reconstructed ML trees using the concatenated coding sequences of plastid genes, resulting in large amounts of missing data for these orchid taxa. In this study, we sampled these newly orchid species (*C. longifolia*, *L. fugongensis*, *E. mairei*, and *E. veratrifolia*) to construct a more widespread Orchidaceae phylogenetic tree, through which we obtained the different species relationships. However, some molecular phylogenetic studies, to date, have failed to identify the placement of Cypripedioideae and Vanilloideae [8,25,64,65]. Recently, Givnish et al. [49] and Niu et al. [11] reconstructed ML trees from 39 and 53 orchids species, respectively, using the sequence variations in 75 genes and 67 genes from the plastid genomes. Their results showed that five orchid subfamilies clustered into the five monophyletic clades: Epidendroideae–Orchidoideae–Cypripedioideae–Vanilloideae–Apostasioideae. However, the current study found that *C. appendiculata* and the congeneric *C. davidii* clustered into the Epidendroideae subfamily clade, and that *E. mairei* and *P. japonica* clustered into the Orchidoideae subfamily. These results were largely consistent with traditional morphological evidence [66–68]. However, the inconsistent phylogenetic relationships for the five subfamilies may be due to the differences in the collected samples used in different studies [11,49,64,65], which need to be further explored by sampling a much higher number of orchid species.

4. Materials and Methods

4.1. Plant Material, DNA Extraction, Library Construction, and Sequencing

Fresh leaf tissues were collected from *Cremastra appendiculata*, *Calanthe davidii*, *Epipactis mairei*, and *Platanthera japonica* in the Qinling Mountains, Shaanxi Province, China. The leaves were cleaned and preserved in a −80 °C refrigerator at Northwest University. The voucher specimens of the four species materials were deposited into the Northwest University Herbarium (NUH). The total genomic DNA was isolated using the modified Cetyltrimethyl Ammonium Bromide (CTAB) method [69], which added the EDTA buffer (Amresco, Washington, DC, USA) (1.0 mol/L Tris-HCl (Amresco, Washington, DC, USA) (pH 8.0), 0.5 mol/L EDTA-Na$_2$ (Amresco, Washington, DC, USA), 5.0 mol/L NaCl) solution before isolating the high-quality DNA with the CTAB solution (1.0 mol/L Tris-HCl (pH 8.0), 0.5 mol/L EDTA-Na$_2$, 2% CTAB). Following this, we constructed a pair-end (PE) library with 350 bp insert size fragments using TruSeq DNA sample preparation kits (Sangon, Shanghai, China). Subsequently, we sequenced at least 4.5 GB of clean data for each orchid species. The detailed next-generation sequencing was conducted on the Illumina Hiseq 2500 platform by Sangon Biotech (Shanghai, China).

4.2. Chloroplast Genome Assembly and Annotation

First, we used the software, NGSQCToolkit v2.3.3 [70], to trim the low-quality reads. After removing the low-quality sequences, the clean reads were assembled using MIRA v4.0.2 [70] and MITObim v1.8 [71] with the cp genome of a closely-related species, *Dendrobium nobile* (KX377961), as reference. The programs, DOGMA (http://dogma.ccbb.utexas.edu/) [72] and Geneious v8.0.2 [73] were used to annotate the chloroplast genome. Finally, we obtained four high-quality, complete chloroplast genome sequences. The Circle maps of the four species were drawn using OGDRAW v1.1 [74].

4.3. Repeat Sequence Analyses

The REPuter program (Available online: https://bibiserv.cebitec.uni-bielefeld.de/reputer/manual.html) was used to identify repeats, including forward, reverse, palindrome, and complement sequences. The maximum computed repeats and the minimal repeat size were limited to 50 and 30, respectively, with a Hamming distance equal to 3 [75]. The tandem repeats finder welcome page (http://tandem.bu.edu/trf/trf.html) was used to identify tandem repeats sequences [76]. The alignment parameters match, mismatch, and indels, were 2, 7, and 7, respectively. The minimum alignment score to report repeat, maximum period size and maximum TR array Size (bp, millions) are limited to 80, 500, and 2, respectively. A Perl script MISA (MIcroSAtellite identification tool, http://pgrc.ipk-gatersleben.de/misa/) was used to search for simple sequence repeat (SSR or microsatellite) loci in the chloroplast genomes [77]. Tandem repeats of 1–6 nucleotides were viewed as microsatellites. The minimum number of repeats were set to 10, 5, 4, 3, 3, and 3, for mono-, di-, tri-, tetra-, penta-, and hexanucleotides, respectively.

4.4. Genome Structure and Mutational Hotspot

In order to compare the genome structures and divergence hotspots in a broad manner, we used 16 cp genomes (available in Genbank https://www.ncbi.nlm.nih.gov/) representing each orchid genus, and added the four newly-sequenced ones (Table 3). The boundaries between the IR and SC regions of *C. appendiculata*, *C. davidii*, *E. mairei*, and *P. japonica* and other 16 sequences were compared and analyzed. Meanwhile, the whole-genome alignment of the chloroplast genomes of the 20 species were performed and plotted using the mVISTA program [32]. Following this, we selected the regions within non-coding and coding regions that had a greater level of variation (percentage of variability >80% and 50%, respectively) as mutational hotspots. The formula was as follows: percentage of variable = (number of nucleotide substitutions + the number of indels)/(the length of aligned sites−the length of indels + the number of indels) × 100%.

Table 3. List of taxa sampled in the study and species accessions numbers (GenBank).

Subfamily	Species	Accession Number
Orchidaceae subfamily. Epidendroideae	*Cattleya crispata*	KP168671
	Cremastra appendiculata	MG925366
	Masdevallia coccinea	KP205432
	Erycina pusilla	JF746994
	Phalaenopsis equestris	JF719062
	Bletilla ochracea	KT695602
	Cymbidium faberi	KR919606
	Calanthe davidii	MG925365
	Dendrobium strongylanthum	KR673323
	Elleanthus sodiroi	KR260986
	Sobralia callosa	KM032623
Orchidaceae subfamily. Orchidoideae	*Epipactis mairei*	MG925367
	Cephalanthera longifolia	KU551263
	Listera fugongensis	KU551270
	Platanthera japonica	MG925368
	Habenaria pantlingiana	KJ524104
	Goodyera velutina	KT886432
	Anoectochilus emeiensis	LC057212
	Ludisia discolor	KU578274
Orchidaceae subfamily. Vanilloideae	*Vanilla planifolia*	KJ566306

4.5. Gene Selective Pressure Analysis

The codon substitution models in the Codeml program, PAML3.15 [46] were used for calculating the non-synonymous (dN) and synonymous (dS) substitution rates, along with their ratios (ω = dN/dS). We analyzed all CDS gene regions, except *ndh*, due to there being too many losses there. These unique CDS gene sequences were separately extracted and aligned using Geneious v8.0.2 [73]. A maximum likelihood phylogenetic tree was built based on the complete cp genomes of the 20 species using RAxML [78]. We used the site-specific model (seqtype = 1, model = 0, NSsites = 0, 1, 2, 3, 7, 8) to estimate the selection pressure [79]. This model allowed the ω ratio to vary among sites, with a fixed ω ratio in all the branches. Comparing the site-specific model, M1 (nearly neutral) vs. M2 (positive selection), M7 (β) vs. M8 (β and ω) and M0 (one-ratio) vs. M3 (discrete) were calculated in order to detect positive selection [79].

4.6. Phylogenetic Analysis

In order to deeply detect the evolutionary relationship of the Orchidaceae family, 50 available complete chloroplast genomes were downloaded from the NCBI Organelle Genome Resources database (Table S1). In addition, *Artemisia argyi* and *Megadenia pygmaea* were used as outgroups. In total, 54 nucleotide sequences of complete chloroplast genomes were aligned using MAFFT [73]; the detailed parameters were as follows: 200 PAM/K = 2 and 1.53 gap open penalty [73]. The choice of the best nucleotide sequence substitution model (GTRGAMMA model) was determined using the Modeltest v3.7 [80]. We constructed a maximum likelihood phylogenetic tree based on these complete plastomes using MAGA7 [34] with 1000 bootstrap replicates under the GTRGAMMA model [80].

Supplementary Materials: The following are available online at http://www.mdpi.com/1422-0067/19/3/716/s1.

Acknowledgments: This research was co-supported by the National Natural Science Foundation of China (31470400) and Shaanxi Provincial Key Laboratory Project of Department of Education (grant no. 17JS135).

Author Contributions: Zhong-Hu Li conceived the work. Wan-Lin Dong and Ruo-Nan Wang performed the experiments. Zhong-Hu Li, Wan-Lin Dong, Ruo-Nan Wang, Na-Yao Zhang, Wei-Bing Fan and Min-Feng Fang contributed materials/analysis tools. Wan-Lin Dong and Zhong-Hu Li wrote the paper. Zhong-Hu Li and Wan-Lin Dong revised the paper. All authors approved the final paper.

Conflicts of Interest: The authors declare no conflict of interest.

References

1. Chase, M.W.; Cameron, K.M.; Barrett, R.L.; Freudenstein, J.V. DNA data and Orchidaceae systematics: A new phylogenetic classification. In *Orchid Conservation*; Dixon, K.W., Kell, S.P., Barrett, R.L., Cribb, P.J., Eds.; Natural History Publications: Kota Kinabalu, Malaysia, 2003; pp. 69–89.
2. Dressler, R.L. *The Orchids: Natural History and Classification*; Harvard University Press: Cambridge, MA, USA, 1990.
3. Chase, M.W. Classification of Orchidaceae in the age of DNA data. *Curtis's Bot. Mag.* **2005**, *22*, 2–7. [CrossRef]
4. Luo, J.; Hou, B.W.; Niu, Z.T.; Liu, W.; Xue, Q.Y.; Ding, X.Y. Comparative chloroplast genomes of photosynthetic orchids: Insights into evolution of the Orchidaceae and development of molecular markers for phylogenetic applications. *PLoS ONE* **2014**, *9*, e99016. [CrossRef] [PubMed]
5. Raubeson, L.A.; Jansen, R.K. Chloroplast genomes of plants. In *Plant Diversity and Evolution: Genotypic and Phenotypic Variation in Higher Plants*; Henry, R.J., Ed.; CAB International: Wallingford, UK, 2005; pp. 45–68.
6. Van den Berg, C.; Goldman, D.H.; Freudenstein, J.V.; Pridgeon, A.M.; Cameron, K.M.; Chase, M.W. An overview of the phylogenetic relationships within Epidendroideae inferred from multiple DNA regions and recircumscription of Epidendreae and Arethuseae (Orchidaceae). *Am. J. Bot.* **2005**, *92*, 13–24. [CrossRef] [PubMed]
7. Mendonca, M.P.; Lins, L.V. *Revisao das Listas das Especies da Flora eda Fauna Ameaçadas de Extincao do Estado de Minas Gerais*; Fundacao Biodiversitas: BeloHorizonte, Brazil, 2007.
8. Cameron, K.M.; Chase, M.W.; Whitten, W.M.; Kores, P.J.; Jarrell, D.C.; Albert, V.A.; Yukawa, T.; Hills, H.G.; Goldman, D.H. A phylogenetic analysis of the Orchidaceae: Evidence from *rbcL* nucleotide. *Am. J. Bot.* **1999**, *86*, 8–24. [CrossRef]
9. Van den Berg, C.; Higgins, W.E.; Dressler, R.L.; Whitten, W.M.; Soto-Arenas, M.; Chase, M.W. A phylogenetic study of *Laeliinae* (Orchidaceae) based on combined nuclear and plastid DNA sequences. *Ann. Bot.* **2009**, *104*, 17–30. [CrossRef] [PubMed]
10. Verlynde, S.; D'Haese, C.A.; Plunkett, G.M.; Simo-Droissart, M.; Edwards, M.; Droissart, V.; Stévart, T. Molecular phylogeny of the genus *Bolusiella* (Orchidaceae, Angraecinae). *Plant Syst. Evol.* **2017**, *304*, 269–279. [CrossRef]
11. Niu, Z.T.; Xue, Q.Y.; Zhu, S.Y.; Sun, J.; Liu, W.; Ding, X.Y. The complete plastome sequences of four orchid species: Insights into the evolution of the Orchidaceae and the utility of plastomic mutational hotspots. *Front. Plant. Sci.* **2017**, *8*, 1–11. [CrossRef] [PubMed]
12. Bateman, R.M.; Rudall, P.J. Clarified relationship between *Dactylorhiza viridis* and *Dactylorhiza iberica* renders obsolete the former genus *Coeloglossum* (Orchidaceae: Orchidinae). *Kew Bull.* **2018**, *73*, 1–17. [CrossRef]
13. Wilson, M.; Frank, G.S.; Lou, J.; Pridgeon, A.M.; Vieira-Uribe, S.; Karremans, A.P. Phylogenetic analysis of *Andinia* (Pleurothallidinae; Orchidaceae) and a systematic re-circumscription of the genus. *Phytotaxa* **2017**, *295*, 101–131. [CrossRef]
14. Neuhaus, H.E.; Emes, M.J. Nonphotosynthetic metabolism in plastids. *Annu. Rev. Plant Biol.* **2000**, *51*, 111–140. [CrossRef] [PubMed]
15. Rodríguezezpeleta, N.; Brinkmann, H.; Burey, S.C.; Roure, B.; Burger, G.; Löffelhardt, W.; Bohnert, H.J.; Philippe, H.; Lang, B.F. Monophyly of primary photosynthetic eukaryotes: Green plants, red algae, and glaucophytes. *Curr. Biol.* **2005**, *15*, 1325–1330. [CrossRef] [PubMed]
16. Yap, J.Y.; Rohner, T.; Greenfield, A.; Van Der Merwe, M.; McPherson, H.; Glenn, W.; Kornfeld, G.; Marendy, E.; Pan, A.Y.; Wilton, A.; et al. Complete chloroplast genome of the Wollemi pine (*Wollemia nobilis*): Structure and evolution. *PLoS ONE* **2015**, *106*, 126–128. [CrossRef] [PubMed]
17. Wicke, S.; Schneeweiss, G.M.; Müller, K.F.; Quandt, D. The evolution of the plastid chromosome in land plants: Gene content, gene order, gene function. *Plant Mol. Biol.* **2011**, *76*, 273–297. [CrossRef] [PubMed]

18. Dong, W.; Liu, H.; Xu, C.; Zuo, Y.J.; Chen, Z.J.; Zhou, S.L. A chloroplast genomic strategy for designing taxon specific DNA mini-barcodes: A case study on ginsengs. *BMC Genet.* **2014**, *15*, 138–145. [CrossRef] [PubMed]

19. Zhang, Y.; Li, L.; Yan, T.L.; Liu, Q. Complete chloroplast genome sequences of Praxelis (*Eupatorium catarium* Veldkamp), an important invasive species. *Gene* **2014**, *549*, 58–69. [CrossRef] [PubMed]

20. Xu, C.; Dong, W.P.; Li, W.Q.; Lu, Y.Z.; Xie, X.M.; Jin, X.B.; Shi, J.; He, K.; Suo, Z. Comparative analysis of six *Lagerstroemia* complete chloroplast genomes. *Front. Plant Sci.* **2017**, *8*, 15–26. [CrossRef] [PubMed]

21. Jer, J.D. Plastid chromosomes: Structure and evolution. In *Cell Culture and Somatic Cell Genetics in Plants, the Molecular Biology of Plastids 7A*; Vasil, I.K., Bogorad, L., Eds.; Academic Press: San Diego, CA, USA, 1991; pp. 5–53.

22. Bendich, A.J. Circular chloroplast chromosomes: The grand illusion. *Plant Cell* **2004**, *16*, 1661–1666. [CrossRef] [PubMed]

23. Jansen, R.K.; Raubeson, L.A.; Boore, J.L.; Pamphilis, C.W.; Chumley, T.W.; Haberle, R.C.; Wyman, S.K.; Alverson, A.J.; Peery, R.; Herman, S.J.; et al. Methods for obtaining and analyzing whole chloroplast genome sequences. *Methods Enzymol.* **2015**, *395*, 348–384.

24. Burke, S.V.; Grennan, C.P.; Duvall, M.R. Plastome sequences of two new world bamboos-*Arundinaria gigantea* and *Cryptochloa strictiflora* (Poaceae)-extend phylogenomic understanding of Bambusoideae. *Am. J. Bot.* **2012**, *99*, 1951–1961. [CrossRef] [PubMed]

25. Civan, P.; Foster, P.G.; Embley, M.T.; Séneca, A.; Cox, C.J. Analyses of charophyte chloroplast genomes help characterize the ancestral chloroplast genome of land plants. *Genome Biol. Evol.* **2014**, *6*, 897–911. [CrossRef] [PubMed]

26. Guo, W.; Grewe, F.; Cobo-Clark, A.; Fan, W.; Duan, Z.; Adams, R.P.; Schwarzbach, A.E.; Mower, J.P. Predominant and substoichiometric isomers of the plastid genome coexist within *Juniperus* plants and have shifted multiple times during cupressophyte evolution. *Genome Biol. Evol.* **2014**, *6*, 580–590. [CrossRef] [PubMed]

27. Ruhfel, B.R.; Gitzendanner, M.A.; Soltis, P.S.; Soltis, D.E.; Burleigh, J.G. From algae to angiosperms-inferring the phylogeny of green plants (Viridiplantae) from 360 plastid genomes. *BMC Evol. Biol.* **2014**, *14*, 385–399. [CrossRef] [PubMed]

28. Moore, M.J.; Bell, C.D.; Soltis, P.S.; Soltis, D.E. Using plastid genome-scale data to resolve enigmatic relation-ships among basal angiosperms. *Proc. Natl. Acad. Sci. USA* **2007**, *104*, 19363–19368. [CrossRef] [PubMed]

29. Huang, H.; Shi, C.; Liu, Y.; Mao, S.Y.; Gao, L.Z. Thirteen Camellia chloroplast genome sequences determined by high-throughput sequencing: Genome structure and phylogenetic relationships. *BMC Evol. Biol.* **2014**, *14*, 4302–4315. [CrossRef] [PubMed]

30. Walker, J.F.; Zanis, M.J.; Emery, N.C. Comparative analysis of complete chloroplast genome sequence and inversion variation in *Lasthenia burkei* (Madieae, Asteraceae). *Am. J. Bot.* **2014**, *101*, 722–729. [CrossRef] [PubMed]

31. Oldenburg, D.J.; Bendich, A.J. The linear plastid chromosomes of maize: Terminal sequences, structures, and implications for DNA replication. *Curr. Genet.* **2015**, *62*, 1–12. [CrossRef] [PubMed]

32. Frazer, K.A.; Pachter, L.; Poliakov, A.; Rubin, E.M.; Dubchak, I. VISTA: Computational tools for comparative genomics. *Nucleic Acids Res.* **2004**, *32*, 273–279. [CrossRef] [PubMed]

33. Doose, D.; Grand, C.; Lesire, C. MAUVE Runtime: A Component-Based Middleware to Reconfigure Software Architectures in Real-Time. In Proceedings of the IEEE International Conference on Robotic Computing (IRC), Taichung, Taiwan, 10–12 April 2017; pp. 208–211.

34. Kumar, S.; Stecher, G.; Tamura, K. Mega7: Molecular evolutionary genetics analysis version 7.0 for bigger datasets. *Mol. Biol. Evol.* **2016**, *33*, 1870–1874. [CrossRef] [PubMed]

35. Clegg, M.T.; Gaut, B.S.; Learn, G.H., Jr.; Morton, B.R. Rates and patterns of chloroplast DNA evolution. *Proc. Natl. Acad. Sci. USA* **1994**, *91*, 6795–6801. [CrossRef] [PubMed]

36. Delannoy, E.; Fujii, S.; Colas des Francs-Small, C.; Brundrett, M.S. Rampant gene loss in the underground orchid *Rhizanthella gardneri* highlights evolutionary constraints on plastid genomes. *Mol. Biol. Evol.* **2011**, *28*, 2077–2086. [CrossRef] [PubMed]

37. Logacheva, M.D.; Schelkunov, M.I.; Penin, A.A. Sequencing and analysis of plastid genome in mycoheterotrophic orchid Neottia nidus-avis. *Genome Biol. Evol.* **2011**, *3*, 1296–1303. [CrossRef] [PubMed]

38. Barrett, C.F.; Davis, J.I. The plastid genome of the mycoheterotrophic *Corallorhiza striata* (Orchidaceae) is in the relatively early stages of degradation. *Am. J. Bot.* **2012**, *99*, 1513–1523. [CrossRef] [PubMed]

39. Yang, J.B.; Tang, M.; Li, H.T.; Zhang, Z.R.; Li, D.Z. Complete chloroplast genome of the genus *Cymbidium*: Lights into the species identification, phylogenetic implications and population genetic analyses. *BMC Evol. Biol.* **2013**, *13*, 84. [CrossRef] [PubMed]

40. Wu, F.H.; Chan, M.T.; Liao, D.C.; Hsu, C.T.; Lee, Y.W.; Daniell, H.; Duvall, M.R.; Lin, C.S. Complete chloroplast genome of *Oncidium Gower Ramsey* and evaluation of molecular markers for identification and breeding in Oncidiinae. *BMC Plant Biol.* **2010**, *10*, 68. [CrossRef] [PubMed]

41. Kim, H.T.; Kim, J.S.; Moore, M.J.; Neubig, K.M.; Williams, N.H.; Whitten, W.M.; Kim, J.H. Seven new complete plastome sequences reveal rampant independent loss of the *ndh* gene family across orchids and associatedinstability of the inverted repeat/small single-copy region boundaries. *PLoS ONE* **2015**, *10*, e0142215.

42. Ni, L.; Zhao, Z.; Xu, H.; Chen, S.; Dorje, G. The complete chloroplast genome of *Gentiana straminea* (Gentianaceae), an endemic species to the Sino-Himalayan subregion. *Gene* **2016**, *577*, 281–288. [CrossRef] [PubMed]

43. Ni, L.; Zhao, Z.; Xu, H.; Chen, S.; Dorje, G. Chloroplast genome structures in *Gentiana* (Gentianaceae), based on three medicinal alpine plants used in Tibetan herbal medicine. *Curr. Genet.* **2017**, *63*, 241–252. [CrossRef] [PubMed]

44. Wang, W.B.; Yu, H.; Wang, J.H.; Lei, W.J.; Gao, J.H.; Qiu, X.P.; Wang, J.S. The complete chloroplast genome sequences of the medicinal plant *Forsythia suspensa* (Oleaceae). *Int. J. Mol. Sci.* **2017**, *18*, 2288. [CrossRef] [PubMed]

45. Kanga, J.Y.; Lua, J.J.; Qiua, S.; Chen, Z.; Liu, J.J.; Wang, H.Z. *Dendrobium* SSR markers play a good role in genetic diversity and phylogenetic analysis of Orchidaceae species. *Sci. Hortic.* **2015**, *183*, 160–166. [CrossRef]

46. Song, Y.; Wang, S.; Ding, Y.; Xu, J.; Li, M.F.; Zhu, S.; Chen, N. Chloroplast genomic resource of Paris for species discrimination. *Sci. Rep.* **2017**, *7*, 3427–3434. [CrossRef] [PubMed]

47. Yu, X.Q.; Drew, B.T.; Yang, J.B.; Gao, L.M.; Li, D.Z. Comparative chloroplast genomes of eleven *Schima* (Theaceae) species: Insights into DNA barcoding and phylogeny. *PLoS ONE* **2017**, *12*, e0178026. [CrossRef] [PubMed]

48. Wang, B.; Jiang, B.; Zhou, Y.; Su, Y.; Wang, T. Higher substitution rates and lower dN/dS for the plastid genes in Gnetales than other gymnosperms. *Biochem. Syst. Ecol.* **2015**, *59*, 278–287. [CrossRef]

49. Givnish, T.J.; Spalink, D.; Ames, M.; Lyon, S.P.; Hunter, S.J.; Zuluaga, A.; Iles, W.J.; Clements, M.A.; Arroyo, M.T.; Leebens-Mack, J.; et al. Orchid phylogenomics and multiple drivers of their extraordinary diversification. *Proc. Biol. Sci. B* **2015**, *282*, 2108–2111. [CrossRef] [PubMed]

50. Sasaki, Y.; Hakamada, K.; Suama, Y.; Nagano, Y.; Furusawa, I.; Matsuno, R. Chloroplast encoded protein as a subunit of acetyl-COA carboxylase in pea plant. *J. Biol. Chem.* **1993**, *268*, 25118–25123. [PubMed]

51. Konishi, T.; Shinohara, K.; Yamada, K.; Sasaki, Y. Acetyl-CoA carboxylase in higher plants: Most plants other than Gramineae have both the prokaryotic and the eukaryotic forms of this enzyme. *Plant Cell Physiol.* **1996**, *37*, 117–122. [CrossRef] [PubMed]

52. Kode, V.; Mudd, E.A.; Iamtham, S.; Day, A. The tobacco plastid *accD* gene is essential and is required for leaf development. *Plant J.* **2005**, *44*, 237–244. [CrossRef] [PubMed]

53. Nakkaew, A.; Chotigeat, W.; Eksomtramage, T.; Phongdara, A. Cloning and expression of a plastid-encoded subunit, beta-carboxyltransferase gene (*accD*) and a nuclear-encoded subunit, biotin carboxylase of acetyl-CoA carboxylase from oil palm (*Elaeis guineensis* Jacq.). *Plant Sci.* **2008**, *175*, 497–504. [CrossRef]

54. Drescher, A.; Ruf, S.; Calsa, T.J.; Carrer, H.; Bock, R. The two largest chloroplast genome-encoded open reading frames of higher plants are essential genes. *Plant J.* **2000**, *22*, 97–104. [CrossRef] [PubMed]

55. Asano, T.; Tsudzuki, T.; Takahashi, S.; Shimada, H.; Kadowaki, K. Complete nucleotide sequence of the sugarcane (*Saccharum officinarum*) chloroplast genome: A comparative analysis of four monocot chloroplast genomes. *DNA Res.* **2004**, *11*, 93–99. [CrossRef] [PubMed]

56. Kikuchi, S.; Bédard, J.; Hirano, M.; Hirabayashi, Y.; Oishi, M.; Imai, M.; Takase, M.; Ide, T.; Nakai, M. Uncovering the protein translocon at the chloroplast inner envelope membrane. *Science* **2013**, *339*, 571–574. [CrossRef] [PubMed]

57. Greiner, S.; Wang, X.; Herrmann, R.G.; Rauwolf, U.; Mayer, K.; Haberer, G.; Meurer, J. The complete nucleotide sequences of the 5 genetically distinct plastid genomes of *Oenothera*, subsection *Oenothera*: II. A microevolutionary view using bioinformatics and formal genetic data. *Mol. Biol. Evol.* **2008**, *25*, 2019–2030. [CrossRef] [PubMed]

58. Carbonell-Caballero, J.; Alonso, R.; Ibañez, V.; Terol, J.; Talon, M.; Dopazo, J. A phylogenetic analysis of 34 chloroplast genomes elucidates the relationships between wild and domestic species within the genus *Citrus*. *Mol. Biol. Evol.* **2015**, *32*, 2015–2035. [CrossRef] [PubMed]

59. Hu, S.; Sablok, G.; Wang, B.; Qu, D.; Barbaro, E.; Viola, R.; Li, M.; Varotto, C. Plastome organization and evolution of chloroplast genes in *Cardamine* species adapted to contrasting habitats. *BMC Genom.* **2015**, *16*, 1. [CrossRef] [PubMed]

60. Allahverdiyeva, Y.; Mamedov, F.; Mäenpää, P.; Vass, I.; Aro, E.M. Modulation of photosynthetic electron transport in the absence of terminal electron acceptors: Characterization of the *rbcL* deletion mutant of tobacco. *Biochim. Biophys. Acta Bioenerg.* **2005**, *1709*, 69–83. [CrossRef] [PubMed]

61. Piot, A.; Hackel, J.; Christin, P.A.; Besnard, G. One-third of the plastid genes evolved under positive selection in PACMAD grasses. *Planta* **2018**, *247*, 255–266. [CrossRef] [PubMed]

62. Kapralov, M.V.; Filatov, D.A. Widespread positive selection in the photosynthetic Rubisco enzyme. *BMC Evol. Biol.* **2007**, *7*, 73–82. [CrossRef] [PubMed]

63. Ivanova, Z.; Sablok, G.; Daskalova, E.; Zahmanova, G.; Apostolova, E.; Yahubyan, G.; Baev, V. Chloroplast genome analysis of resurrection tertiary relict *Haberlea rhodopensis* highlights genes important for desiccation stress response. *Front. Plant Sci.* **2017**, *8*, 1–15. [CrossRef] [PubMed]

64. Shi, Y.; Yang, L.F.; Yang, Z.Y.; Ji, Y.H. The complete chloroplast genome of *Pleione bulbocodioides* (Orchidaceae). *Conserv. Genet. Resour.* **2017**, 1–5. [CrossRef]

65. Górniak, M.; Paun, O.; Chase, M.W. Phylogenetic relationships with Orchidaceae based on a low-copy nuclear-coding gene, Xdh: Congruence with organellar and nuclear ribosomal DNA results. *Mol. Phylogenet. Evol.* **2010**, *56*, 784–795. [CrossRef] [PubMed]

66. Lin, C.S.; Chen, J.J.; Huang, Y.T.; Chan, M.T.; Daniell, H.; Chang, W.J.; Hsu, C.T.; Liao, D.C.; Wu, F.H.; Lin, S.Y.; et al. The location and translocation of *ndh* genes of chloroplast origin in the Orchidaceae family. *Sci. Rep.* **2015**, *5*, 9040. [CrossRef] [PubMed]

67. Rasmussen, F.N. The families of the monocotyledones—Structure, evolution and taxonomy. In *Orchids*; Dahlgren, R., Cliford, H.T., Yeo, P.F., Eds.; Springer: Berlin/Heidelberg, Germany, 1985; pp. 249–274.

68. Szlachetko, D.L. Systema orchidalium. *Fragm. Florist. Geobot. Pol.* **1995**, *3*, 1–152.

69. Doyle, J.J. A rapid DNA isolation procedure from small quantities of fresh leaf tissues. *Phytochem. Bull.* **1987**, *19*, 11–15.

70. Chevreux, B.; Pfisterer, T.; Drescher, B.; Driesel, A.J.; Müller, W.E.; Wetter, T.; Suhai, S. Using the mira EST assembler for reliable and automated mRNA transcript assembly and SNP detection in sequenced ESTs. *Genome Res.* **2004**, *14*, 1147–1159. [CrossRef] [PubMed]

71. Hahn, C.; Bachmann, L.; Chevreux, B. Reconstructing mitochondrial genomes directly from genomic next-generation sequencing reads-abaiting and iterative mapping approach. *Nucleic Acids Res.* **2013**, *41*, e129. [CrossRef] [PubMed]

72. Wyman, S.K.; Jansen, R.K.; Boore, J.L. Automatic annotation of organellar genomes with DOGMA. *Bioinformatics* **2004**, *20*, 3252–3255. [CrossRef] [PubMed]

73. Kearse, M.; Moir, R.; Wilson, A.; Steven, S.H.; Matthew, C.; Shane, S.; Simon, B.; Alex, C.; Markowitz, S.; Duran, C.; et al. Geneious Basic: An integrated and extendable desktop software platform for the organization and analysis of sequence data. *Bioinformatics* **2012**, *12*, 1647–1649. [CrossRef] [PubMed]

74. Lohse, M.; Drechsel, O.; Kahlau, S.; Bock, R. Organellar Genome DRAW-a suite of tools for generating physical maps of plastid and mitochondrial genomes and visualizing expression data sets. *Nucleic Acids Res.* **2013**, *41*, 575–581. [CrossRef] [PubMed]

75. Kurtz, S.; Schleiermacher, C. REPuter: Fast computation of maximal repeats incomplete genomes. *Bioinformatics* **1999**, *15*, 426–427. [CrossRef] [PubMed]

76. Benson, G. Tandem repeats finder: A program to analyze DNA sequences. *Nucleic Acids Res.* **1999**, *27*, 573. [CrossRef] [PubMed]

77. Thiel, T.; Michalek, W.; Varshney, R.K.; Graner, A. Exploiting EST databases for the development and characterization of gene-derived SSR-markers in barley (*Hordeum vulgare* L.). *Theor. Appl. Genet.* **2003**, *106*, 411–422. [CrossRef] [PubMed]

78. Stamatakis, A. RAxML-VI-HPC: Maximum likelihood-based phylogenetic analyses with thousands of taxa and mixed models. *Bioinformatics* **2006**, *22*, 2688–2690. [CrossRef] [PubMed]

79. Yang, Z.; Nielsen, R. Codon-substitution models for detecting molecular adaptation at individual sites along specific lineages. *Mol. Biol. Evol.* **2002**, *19*, 908–917. [CrossRef] [PubMed]

80. Posada, D.; Crandall, K.A. Modeltest: Testing the model of DNA substitution. *Bioinformatics* **1998**, *14*, 817–818. [CrossRef] [PubMed]

International Journal of
Molecular Sciences

MDPI

Article

Complete Chloroplast Genome of *Cercis chuniana* (Fabaceae) with Structural and Genetic Comparison to Six Species in Caesalpinioideae

Wanzhen Liu [1], Hanghui Kong [2,3], Juan Zhou [1], Peter W. Fritsch [4], Gang Hao [1,*] and Wei Gong [1,*]

[1] College of Life Sciences, South China Agricultural University, Guangzhou 510614, China;
    xixuegui17@163.com (W.L.); zj13808847221@126.com (J.Z.)
[2] Key Laboratory of Plant Resources Conservation and Sustainable Utilization, South China Botanical Garden,
    Chinese Academy of Sciences, Guangzhou 510650, China; konghh@scbg.ac.cn
[3] Guangdong Provincial Key Laboratory of Applied Botany, South China Botanical Garden,
    Chinese Academy of Sciences, Guangzhou 510650, China
[4] Botanical Research Institute of Texas, 1700 University Drive, Fort Worth, TX 76107, USA; pfritsch@brit.org
*   Correspondence: haogang@scau.edu.cn (G.H.); wgong@scau.edu.cn (W.G.); Tel.: +86-3829-7712 (G.H.)

Received: 3 April 2018; Accepted: 19 April 2018; Published: 25 April 2018

Abstract: The subfamily Caesalpinioideae of the Fabaceae has long been recognized as non-monophyletic due to its controversial phylogenetic relationships. *Cercis chuniana*, endemic to China, is a representative species of *Cercis* L. placed within Caesalpinioideae in the older sense. Here, we report the whole chloroplast (cp) genome of *C. chuniana* and compare it to six other species from the Caesalpinioideae. Comparative analyses of gene synteny and simple sequence repeats (SSRs), as well as estimation of nucleotide diversity, the relative ratios of synonymous and nonsynonymous substitutions (dn/ds), and Kimura 2-parameter (K2P) interspecific genetic distances, were all conducted. The whole cp genome of *C. chuniana* was found to be 158,433 bp long with a total of 114 genes, 81 of which code for proteins. Nucleotide substitutions and length variation are present, particularly at the boundaries among large single copy (LSC), inverted repeat (IR) and small single copy (SSC) regions. Nucleotide diversity among all species was estimated to be 0.03, the average dn/ds ratio 0.3177, and the average K2P value 0.0372. Ninety-one SSRs were identified in *C. chuniana*, with the highest proportion in the LSC region. Ninety-seven species from the old Caesalpinioideae were selected for phylogenetic reconstruction, the analysis of which strongly supports the monophyly of Cercidoideae based on the new classification of the Fabaceae. Our study provides genomic information for further phylogenetic reconstruction and biogeographic inference of *Cercis* and other legume species.

Keywords: *Cercis chuniana*; Cercidoideae; Caesalpinioideae; chloroplast genome; legume; next-generation sequencing

1. Introduction

The chloroplast (cp) is widely present in algae and plants with important functions in photosynthesis, carbon fixation, and stress response [1,2]. The cp genome in most angiosperms is a circular molecule with a typically quadripartite structure, comprising a large single copy (LSC) region and a small single copy (SSC) region separated by two copies of a large inverted repeat (IR) region [3–6]. Although the cp genome is highly conserved, some differences in gene synteny, simple sequence repeats (SSRs) and pseudogenes have been observed [7–9] and an accelerated rate of evolution has been observed in some cp regions at different taxonomic levels [10,11]. A complete cp genome is a valuable resource of information for studying plant taxonomy, phylogenetic reconstruction, and historical biogeographic inference. Next-generation sequencing (NGS) technologies have enabled a rapid expansion in the database of whole cp genomes [12,13].

Fabaceae (legumes) are the third largest angiosperm family, with an estimated 727 genera and 20,000 species [14]. The family has been traditionally classified into three well-known and widely accepted subfamilies, i.e., Caesalpinioideae DC., Mimosoideae DC. and Papilionoideae DC. However, the subfamily Caesalpinioideae has been long considered to be non-monophyletic and not reflective of accurate phylogenetic relationships among the species [15–20]. As based on recent phylogenetic analyses, a new classification of six subfamilies has been recognized in Leguminosae: Cercidoideae, Detarioideae, Dialioideae, Duparquetioideae, Papilionoideae and a recircumscribed Caesalpinioideae [21].

The genus *Cercis* L. is removed from Caesalpinioideae and currently placed within Cercidoideae [21]. This genus comprises a clade of about nine species, with a disjunct distribution across the warm temperate zones of Eastern Asia, Europe and North America [22–26]. In China, five species are recognized, i.e., *C. chinensis*, *C. chingii*, *C. chuniana*, *C. glabra* and *C. racemosa* [26,27]. *Cercis chuniana*, a small tree or shrub, occurs mainly in subtropical evergreen broadleaf forest with a relatively narrow geographic distribution in southern China. Unique among *Cercis* species, it has an asymmetrical leaf blade [27,28].

Previous research has been focused on plant anatomy, phylogenetic reconstruction, and historical biogeography of *Cercis* [24–26,29,30]. However, *C. chuniana* has frequently failed to be analyzed in most phylogenetic research, resulting in an unclear phylogenetic position within the genus. Because *Cercis* has been removed to Cercidoideae, it would also be useful to detect additional genomic evidence that might support the new classification system of Fabaceae. Moreover, Sanger-based and whole-cp genome DNA barcoding can been used for phylogenetic reconstruction. Here we present and characterize the complete cp genome of *C. chuniana*. The structural variation, gene arrangement, and distribution of SSRs are compared with previously published cp genome of *C. canadensis* and five species from various genera in Caesalpinioideae. Our results provide cp information for *Cercis* and other legumes for use in comparative genomics, phylogenetic reconstruction, and biogeographic inference.

2. Results and Discussion

2.1. Genome Organization and Features of C. chuniana

A total number of 2×250 bp pair-end reads of 1,917,920 were produced with 1.17 Gb of clean data. All reads data were deposited in the NCBI Sequence Read Archive (SRA) under accession number SRP118607. In total, 102 contigs (N50 = 8438 bp) were generated for *C. chuniana*. The size of the complete cp genome is 158,433 bp (Figure 1; Table 1). The cp genome displays a typical quadripartite structure, including a pair of IR regions (25,505 bp) separated by the LSC (88,063 bp) and SSC (19,360 bp) regions (Figure 1 and Table 1). The G + C content of the cp genome is 36.10% for *C. chuniana*, demonstrating congruence with that of *C. canadensis* (36.20%) (Table 1). When duplicated genes in the IR regions were counted only once, the cp genome of *C. chuniana* were found to encode 114 predicted functional genes, including 81 protein-coding genes (PCGs), 29 tRNA genes, and four rRNA genes, all of which are comparable to the numbers in *C. canadensis* and other related species (Table 1). The remaining non-coding regions include introns, intergenic spacers, and pseudogenes. Nineteen genes are duplicated in the IR regions, including eight PCGs, seven tRNA genes, and four rRNA genes (Figure 1 and Table S1). Fifteen genes (nine PCGs and six tRNA genes) contain one intron, and two PCGs (*clpP* and *ycf3*) have two introns each (Table S1). The maturase K (*matK*) gene in the cp genome is located within the *trnK* intron, consistent with the location in *C. canadensis* and similar to most other plant species [31]. In the IR regions of *C. chuniana*, the four rRNA genes and two tRNA genes (*trnE* and *trnA*) are clustered as 16S-*trnE*-*trnA*-23S-4.5S-5S. This differs from the cp genomes of *C. canadensis* and most legumes, which show a cluster of 16S-*trnI*-*trnA*-23S-4.5S-5S [32–37].

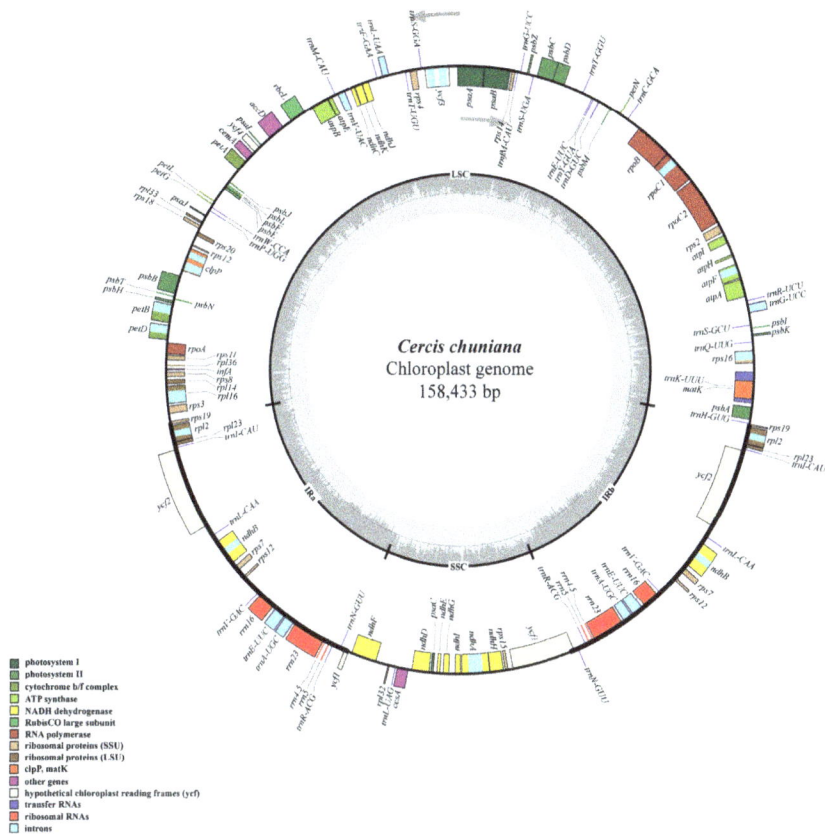

Figure 1. Gene map of the *Cercis chuniana* cp genome. The genes lying inside and outside the outer circle are transcribed in clockwise and counterclockwise direction, respectively (as indicated by arrows). Colors denote the genes belonging to different functional groups. The hatch marks on the inner circle indicate the extent of the inverted repeats (IRa and IRb) that separate the small single copy (SSC) region from the large single copy (LSC) region. The dark gray and light gray shading within the inner circle correspond to percentage G + C and A + T content, respectively.

Table 1. Summary of characteristics in cp genome sequences of *Cercis chuniana* and six other species of caesalpinioid legumes compared in this study.

Genome Features	C. chuniana	C. canadensis	T. indica	Cera. siliqua	L. coriaria	M. cucullatum	H. brasiletto
GenBank Accession No.	MF741770	KF856619	KJ468103	KJ468096	KJ468095	KU569489	KJ468097
Size (bp)	158,433	158,995	159,551	156,367	158,045	158,357	157,728
LSC length (bp)	88,063	88,118	87,967	85,801	87,581	87,663	87,465
SSC length (bp)	19,360	19,621	19,546	18,492	18,160	18,091	18,185
IR length (bp)	25,505	25,628	26,019	26,037	26,152	26,294	26,039
Number of genes	114	113	113	112	113	114	113
PCGs	81	79	79	78	80	80	79
tRNA genes	29	30	30	30	29	30	30
rRNA genes	4	4	4	4	4	4	4
G + C content (%)	36.10	36.20	36.20	36.70	36.50	36.40	36.70

2.2. Comparative Analysis of Genomic Structure

Synteny analysis identified a lack of genome rearrangement and inversions in the cp genome sequences among the seven species (Figure S1). Therefore, genomic structure, including gene number and gene order, is highly conserved among the seven species. However, some nucleotide substitutions and indels as well as length variation are still present, particularly in the LSC/IR/SSC boundaries (Figure 2 and Figure S2).

Pseudogenes are frequently identified in cp genomes [38,39]. Four pseudogenes were identified in the current study, i.e., Ψrps19, Ψycf1, ΨinfA and ΨaccD (Table 2). Ψrps19 and ycf1 are partially repeated in the IR regions and were generally found to be pseudogenized. The *rps19* gene is 279 bp in all species (Figure 2) with length variation in the IR regions, from 73 bp in *Tamarindus indica* to 107 bp in *Libidibia coriaria*. It has the same length (152 bp) in both *C. chuniana* and *C. canadensis* in the IR regions (Figure 2). Because it is partially duplicated in the IR regions, the Ψrps19 gene has lost its protein-coding ability, thus producing the pseudogenized Ψrps 19 gene. Two nonsynonymous substitutions were detected in the Ψrps19 gene between *C. chuniana* and *C. canadensis*. Among the seven species, 28 substitutions (seven in the IRb region and 21 in the LSC region, respectively) and 4 indels with length variation from 4 to 47 bp were identified (Figure 2; Table 2). The same was found with the Ψycf1 gene, as the IRb/SSC junction region is located within the Ψycf1 CDS region and only a partial gene is duplicated in the IRa region, thus producing the pseudogene Ψycf1. This is generally the case in the dicots. The length of the Ψycf1 pseudogene in the IR regions ranges from 385 bp in *C. chuniana* to 899 bp in *Mezoneuron cucullatum*. Four nonsynonymous substitutions were detected between *C. chuniana* and *C. canadensis*. Altogether 20 substitutions (19 in the IRa region and one in the SSC region) and 7 indels with length variation ranging from 1 to 33 bp are present among the seven species (Figure 2; Table 2). The ΨinfA gene is pseudogenized in all species except *Ceratonia siliqua*, with a length of 135 bp in both *C. chuniana* and *C. canadensis* and with length ranging from 192 to 252 bp among the other four species. A total of 23 substitutions and 6 indels ranging from 1 to 13 bp in length occurs in ΨinfA (Figure 2; Table 2). The pseudogenized ΨinfA gene has also been frequently found in other angiosperm chloroplast genomes as well [40–42]. The pseudogenized ΨaccD gene is present in all species except *T. indica* and *M. cucullatum*, with a length of 1473 bp in both *C. chuniana* and *C. canadensis* and with length ranging from 1395 to 1500 bp in the other three species. Six indels ranging from 3 to 36 bp in length, and 101 substitutions were detected in ΨaccD (Table 2).

Table 2. The location and characteristics of the four pseudogenes in the seven species of caesalpinioid legumes compared in this study.

Species	IRa Ψycf1	IRb Ψrps19	LSC ΨinfA	LSC ΨaccD
C. chuniana	385 bp *	152 bp *	1 indel, 91-bp SV *	5 indels *
C. canadensis	418 bp *	152 bp *	1 indel, 91-bp SV *	5 indels *
T. indica	644 bp *	73 bp *	71-bp SV *	-
Cera. siliqua	776 bp *	85 bp *	-	4 indels, 63-bp SV *
L. coriaria	819 bp *	107 bp *	-	4 indels *
M. cucullatum	899 bp *	103 bp *	4 indels *	3 indels *
H. brasiletto	697 bp *	96 bp *	4 indels *	4 indels *

* Pseudogene present; SV: structural variation with indels ≥ 50 bp.

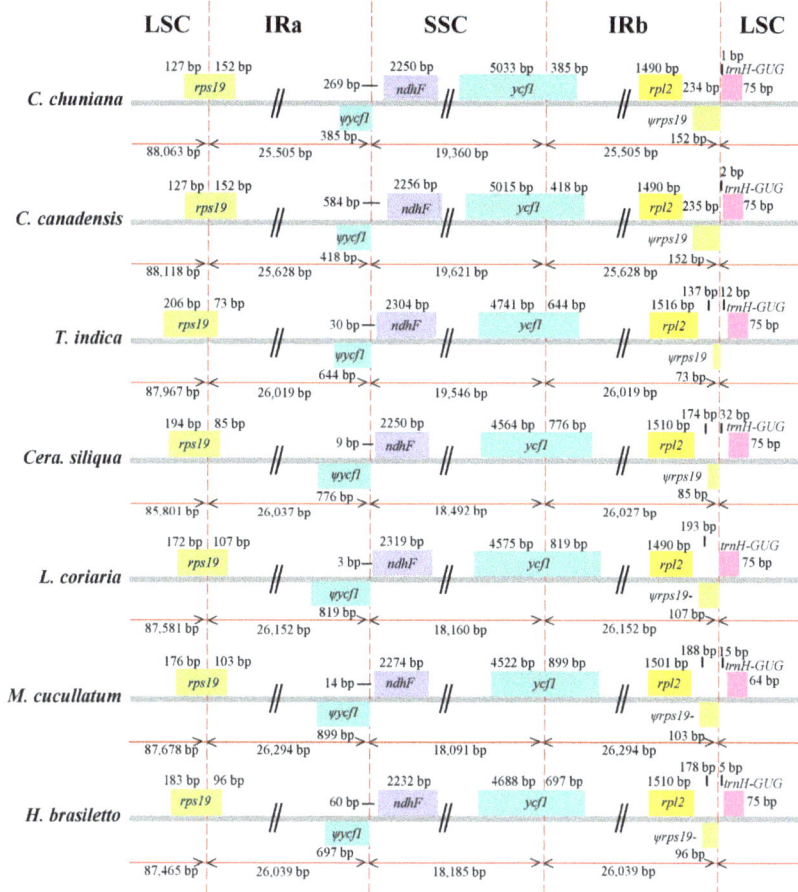

Figure 2. Comparison of the border positions of LSC, SSC and IR regions among the seven species of caesalpinioid legumes compared in this study. Genes are denoted by colored boxes. The gaps between the genes and the boundaries are indicated by the base lengths (bp). Extensions of the genes are indicated above the boxes.

2.3. Characterization of Simple Sequence Repeats

Variable copy numbers and resulting length variation have impelled the wide use of cp SSRs in plant population genetics and biogeographic studies, especially at lower taxonomic levels [43,44]. A total of 91 SSRs of ≥10 bp in length were found in both *C. chuniana* and *C. canadensis*. These two species exhibit the highest number of SSRs among the seven species (Table 3). The lowest number of SSRs was detected in *Haematoxylum brasiletto*, with only 38 SSRs in total (Table 3). Most SSRs are present in the LSC regions, accounting for an average of 75.00% of the total SSRs in each species. Among all of the SSRs, the mononucleotide A or T repeat units were found in highest proportion, with an average of 78.10% of the total SSRs in each species. The SSRs have a remarkably high A + T content, with only 15 compound SSRs containing the nucleotides C or G in *C. chuniana* (Table S2). The lengths of SSRs in the seven species range from 10 to 20 bp, whereas the compound SSRs range from 21 to 275 bp. The copy lengths of 10 to 13 bp are most common, with an average of 77.00% among all species (Figure 3). No pentanucleotide or hexanucleotide SSRs were detected among the seven species.

The shared interspecific SSRs were identified among species, with identical repeats and locations in homologous regions (Table 4). *Cercis chuniana* and *C. canadensis* demonstrated the highest number of 19 common SSRs. Conversely, *Tamarindus indica* has the lowest number of shared SSRs (≤3). Altogether 13 SSRs were isolated and corresponding primer pairs were designed for each di-, tri- and tetranucleotide SSRs of *C. chuniana* (Table S3). These SSRs are expected to be useful in the assessment of genetic diversity and population structure as well as the investigations of biogeographic patterns among the species of *Cercis*.

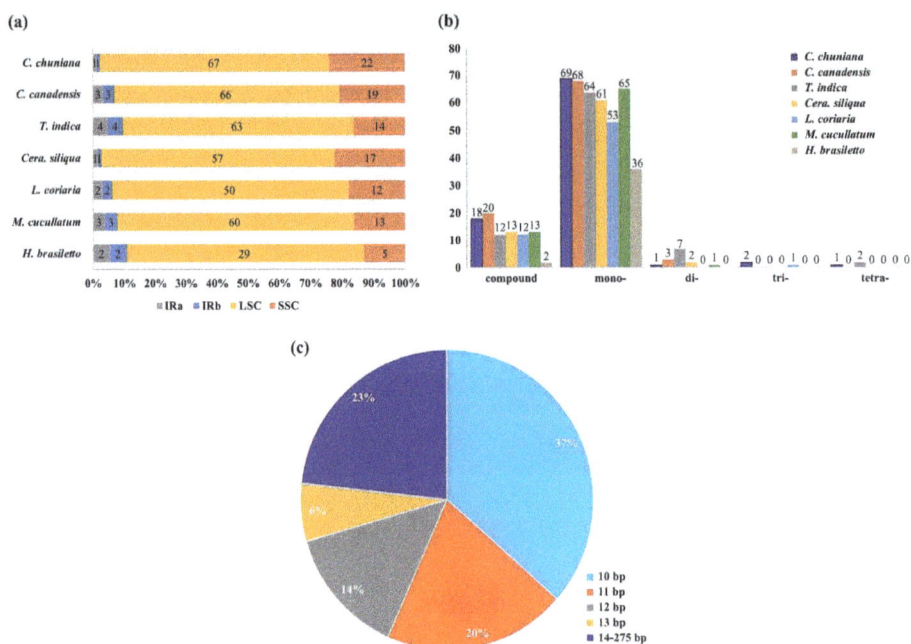

Figure 3. Analysis of repeated sequences of the seven species compared in this study. (**a**) The number of SSRs distributed in different regions; (**b**) The number of SSRs with different types, including compound, mono-, di-, tri-, and tetranucleotides; (**c**) The proportion of SSRs with different lengths.

2.4. Sequence Divergence and Nucleotide Diversity

A complete cp genome is valuable for plant taxonomic analyses, phylogenetic reconstruction, speciation processes, and biogeographical inferences at different taxonomic levels [45–49]. Highly variable regions among cp genomes can provide useful data for phylogenetic reconstruction. In the current study, the average nucleotide variability (*Pi*) was estimated to be 0.006 between *C. chuniana* and *C. canadensis* as based on the comparative analysis with DnaSP (Figure 4a). The highest variation was found in the LSC and SSC regions. The IR regions had a much lower nucleotide diversity with *Pi* < 0.006. Eight regions (*trnS-trnT, atpF-atpH, trnT-psbD, trnL-trnF-ndhJ, accD-psaI, rps3-rps19, ycf1-ndhF* and the *ndhA* intron) were highly variable, with *Pi* values >0.030. The first five loci are present in the LSC, whereas the remaining two are present in the SSC region. In contrast, much higher nucleotide diversity with *Pi* = 0.038 was detected among the seven species (Figure 4b). Five regions (*psbZ-trnG, trnT-trnL, rps3-rps19, rpl32,* and *ycf1*) exhibit the highest nucleotide diversity, all with *Pi* >0.12. These loci are thus suggested as useful regions for phylogenetic analysis at higher taxonomic levels in the Fabaceae.

Int. J. Mol. Sci. **2018**, *19*, 1286

Table 3. Number of chloroplast SSRs in different regions or different types present in the seven species from caesalpinioid legumes.

Species	N	LSC	SSC	IRa	IRb	Compound	Mono-(≥10)	Di-(≥6)	Tri-(≥5)	Tetra-(≥5)
C. chuniana	91	67 (73.63)	22 (24.18)	1 (1.10)	1 (1.10)	18 (19.78)	69 (75.82)	1 (1.10)	2 (2.20)	1 (1.10)
C. canadensis	91	66 (72.53)	19 (20.88)	3 (3.30)	3 (3.30)	20 (21.98)	68 (74.73)	3 (3.30)	0 (0.00)	0 (0.00)
T. indica	85	63 (74.12)	14 (16.47)	4 (4.71)	4 (4.71)	12 (14.12)	64 (75.29)	7 (8.24)	0 (0.00)	2 (2.35)
Cera. siliqua	76	57 (75.00)	17 (22.37)	1 (1.32)	1 (1.32)	13 (17.11)	61 (80.26)	2 (2.63)	0 (0.00)	0 (0.00)
L. coriaria	66	50 (75.76)	12 (18.18)	2 (3.03)	2 (3.03)	12 (18.18)	53 (80.30)	0 (0.00)	1 (1.52)	0 (0.00)
M. cucullatum	79	60 (75.95)	13 (16.46)	3 (3.80)	3 (3.80)	13 (16.46)	65 (82.28)	1 (1.27)	0 (0.00)	0 (0.00)
H. brasiletto	38	29 (76.32)	5 (13.16)	2 (5.26)	2 (5.26)	2 (5.26)	36 (94.74)	0 (0.00)	0 (0.00)	0 (0.00)
Average	75	56 (74.76)	15 (18.81)	2 (3.22)	2 (3.22)	13 (17.14)	59 (79.24)	2 (2.67)	0.43 (0.57)	0.43 (0.57)

Note: The numbers in parentheses are the percentage of each region or SSR types.

Table 4. Shared SSRs among the seven species in caesalpinioid legumes.

Species	C. chuniana	C. canadensis	T. indica	Cera. siliqua	L. coriaria	M. cucullatum	H. brasiletto
C. chuniana	-						
C. canadensis	19	-					
T. indica	1	0	-				
Cera. siliqua	6	4	1	-			
L. coriaria	4	6	3	7	-		
M. cucullatum	6	5	1	1	8	-	
H. brasiletto	4	3	2	6	6	4	-

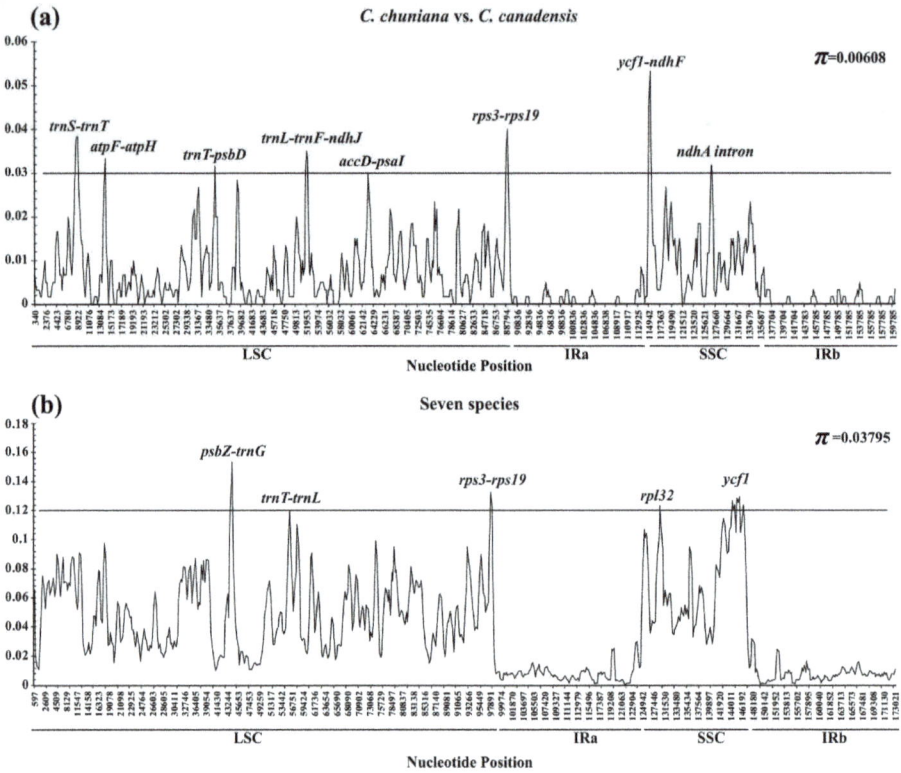

Figure 4. Sliding window analysis of the whole cp genome. (**a**) C. *chuniana* and C. *canadensis*; (**b**) All seven species. X-axis: position of the midpoint of a window; Y-axis: nucleotide diversity (π) of each window.

2.5. dn/ds Ratio and Kimura 2-Parameter (K2P) Genetic Distance

A total of 76 PCGs in all seven species was used to estimate dn/ds ratios. The dn and ds values range from 0 to 0.1713 and 0.0046 to 0.5330, respectively. If dn or ds is 0, the dn/ds ratio cannot be calculated. Among all genes, 67 proteins possess dn/ds ratios <0.5, indicating purifying selection (Figure 5a). In *ndhD*, Ψ*ycf1*, Ψ*infA* and *rpl23* the dn/ds ratios were >1, indicating positive selection (Figure 5a). Among the different regions, the dn/ds ratio was the highest in the IR regions (0.9022) and the lowest in the LSC region (0.2205). Based on the K2P model, we calculated the interspecific genetic distance among the seven species using 80 PCGs. The average K2P interspecific genetic distance was found to be 0.0373 (Figure 5b). The minimum K2P values were identified in *ndhB* and *rps7* (0.0030) and the maximum in *psaB* (0.2020).

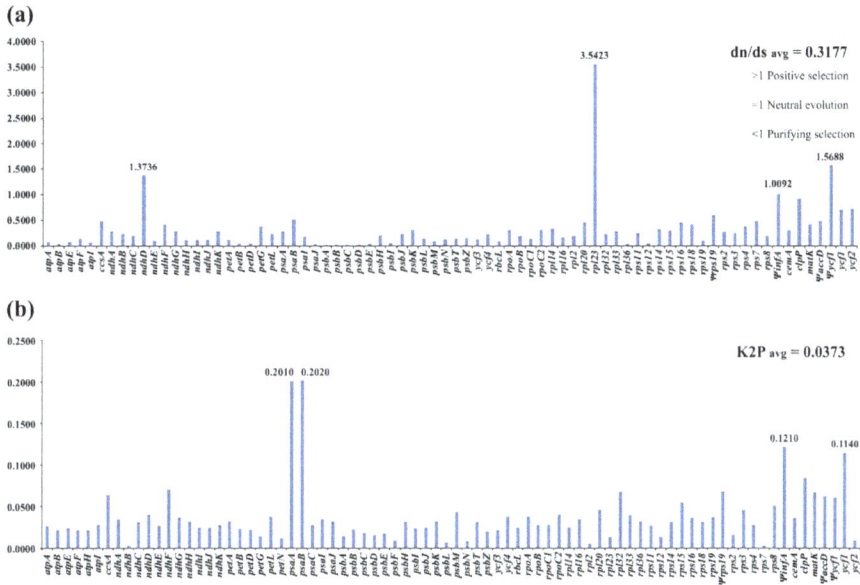

Figure 5. Evolutionary dynamics of genes in the cp genomes. (**a**) The dn/ds ratios for individual genes; (**b**) The K2P values for individual genes.

2.6. Phylogenetic Analyses

A total of 97 representative species from the old Caesalpinioideae and Mimosoideae were selected to reconstruct phylogenetic relationships (Table S4). *Cucumis sativus* (DQ119058) was used as the outgroup. Two phylogenetic methods of Bayesian inference (BI) and maximum likelihood (ML) resulted in highly similar phylogenetic trees based on the complete cp genome sequences and 61 protein-coding genes (PCGs) (Figure 6). The total aligned length was 302,882 bp for the complete cp genome sequences and 69,253 bp for the PCGs, and the number of parsimony-informative sites was 163,470 bp and 25,698 bp, respectively. The trees based on ML exhibit completely congruent topologies with higher bootstrap support values in the tree based on complete cp genome sequences than those based on the PCGs (Figure 6a). The relationship between subfamilies Cercidoideae and Detarioideae was not stable in the BI analysis, but otherwise high posterior probability values were detected in both the ML and BI analyses based on the two data sets (Figure 6b). All analyses recover the monophyly of both the Cercidoideae and Detarioideae with strong support. Our results are consistent with [50] and strongly support the new classification system of the Fabaceae [21].

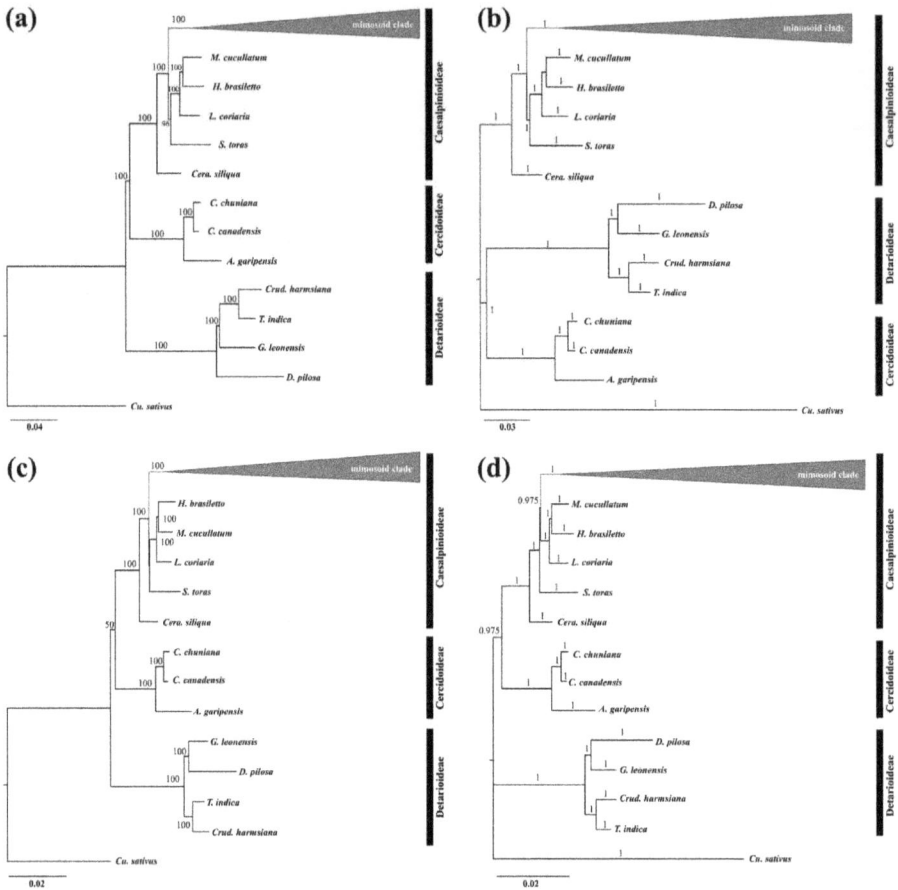

Figure 6. Phylogenetic trees of sampled species inferred from the concatenated whole cp genome sequences and 61 protein-coding genes (PCGs) in the cp genome based on maximum likelihood (ML) and Bayesian inference (BI). (**a**) ML analysis based on whole cp genome sequences; (**b**) BI analysis based on whole cp genome sequences; (**c**) ML analysis based on PCGs; (**d**) BI analysis based on PCGs. Numbers in bold above branches are bootstrap values ≥50% and Bayesian posterior probability values ≥90%.

3. Materials and Methods

3.1. Ethics Statement

Sample collection and transplanting were carried out for scientific purposes. *Cercis chuniana* was collected from the field in Dadongshan Natural Reserve in Guangdong Province, China. One individual seedling was permitted by the management of the reserve to be transplanted and grown in the greenhouse at the College of Life Sciences, South China Agricultural University (SCAU, Guangzhou, China).

3.2. Plant Samples

Fresh leaves were collected from *C. chuniana* growing at SCAU. The voucher (LWZ109) is deposited in the herbarium of SCAU (CANT). The cp genome of *C. canadensis* (KF856619) was downloaded from NCBI and used as the reference sequence in the assembly of *C. chuniana*. Five additional species from the old

Caesalpinioideae were used for comparison, i.e., *Tamarindus indica* (KJ468103), *Ceratonia siliqua* (KJ468096), *Libidibia coriaria* (KJ468095), *Mezoneuron cucullatum* (KU569489) and *Haematoxylum brasiletto* (KJ468097).

3.3. DNA Extraction and PCR Amplification

Total genomic DNA was extracted with the modified Cetyl Trimethyl Ammonium Bromide (CTAB) method [51]. The DNA concentration was quantified with a Nanodrop spectrophotometer (Thermo Scientific, Carlsbad, CA, USA), and a final DNA concentration of >30 ng/µL was used. Sequences of complete cp genome of *C. chuniana* were amplified with fifteen universal primer pairs developed by Zhang et al. [52]. The PCR amplification was performed in a total volume of 25 µL, containing 1 ng of template DNA, 0.12 U of Primerstar GXL DNA Polymerase, 0.2 µM of each primer, 200 µM of each dNTP, 5 µL of 5× PCR Buffer and 13.5 µL of sterilized double-distilled water. Thermocycling conditions were 95 °C (1 min), followed by 32 cycles of denaturation at 94 °C (15 s), annealing at 58 °C (30 s), and extension at 68 °C (10 min), and a final extension of 68 °C (10 min).

3.4. Chloroplast Genome Sequencing, Assembly and Annotation

A paired-end library was constructed with the Nextera XT DNA Library Prep Kit (Illumina Inc., San Diego, CA, USA). The genomic DNA mixture was fragmented into ~300 bp size by the Nextera XT transposome. Library Sequencing acquired 2 × 250 bp paired reads with Illumina MiSeq Desktop Sequencer at South China Botanical Garden, Chinese Academy of Sciences. Reads of the *C. chuniana* cp genome were initially filtered for quality, and then adapters were removed, errors were checked, and contigs and scaffolds generated, all with the A5-miseq pipeline [53]. Scaffolds from the assembly with *k*-mer values of 35 to 145 were matched to reference cp genome sequences, and were used to determine the relative position and direction respectively. We assembled the cp genome using Geneious 9.1.4 (Biomatters Ltd., Auckland, New Zealand) [54] with BLAST 2.0.3+ (National Institutes of Health, Bethesda, MD, USA) [55] and map reference tools. DOGMA (available online: http://dogma.ccbb.utexas.edu/) [56] and Geneious (Biomatters Ltd., Auckland, New Zealand) were used for annotating the cp genome in comparison with that of *C. canadensis* (KF856619) [57]. The annotation of tRNA genes were confirmed with the ARAGORN program (Lund University, Lund, Sweden) [58] and then manually adjusted with Geneious. Contigs with BLAST hits to the consensus sequence from the "map to reference function" were assembled manually to construct the complete cp genome. Finally, the circular genome map of *C. chuniana* was illustrated with the Organellar Genome DRAW tool (OGDRAW, available online: http://ogdraw.mpimp-golm.mpg.de/) [59]. To further refine the draft genome, the quality and coverage of was confirmed by remapping reads. The Sequence Read Archive (SRA) can be found in GenBank under an accession number of SRP118607. The annotated cp genomic sequence of *C. chuniana* was deposited in GenBank (Accession Number: MF741770).

3.5. Genome Comparison

The cp genome sequences from the finalized data set were aligned with MAFFT v7.0.0 (Osaka University, Suita, Japan) [60] and adjusted manually when necessary. The expansion/contraction of the IR regions can lead to changes in the structure of the cp genome, resulting in the length variation of angiosperm cp genomes and contributing to the formation of pseudogenes [9,61,62]. Therefore, we conducted a comparative analysis to detect the variation in the LSC/IR/SSC boundaries among the seven species included in comparisons. Gene synteny analysis was performed with MAUVE (University of Wisconsin, Madison, WI, USA) [63] as implemented in Geneious with default settings. To elucidate the level of sequence divergence, the complete cp genomes were compared and plotted with the mVISTA program in Shuffle-LAGAN mode [64–66].

3.6. Simple Sequence Repeats Analysis

MISA (available online: http://pgrc.ipk-gatersleben.de/misa/misa.html) [67] is a tool for the identification and location of perfect simple sequence repeat loci (SSRs) and compound SSRs (the latter being two individual SSRs that are disrupted by a certain number of bases). We used MISA to search for potential SSRs in the cp genomes of the seven species. The minimum number (thresholds) of SSRs was set as 10, 6, 5, 5, and 5 for mono-, di-, tri-, tetra-, and pentanucleotide SSRs, respectively. All SSRs, motif types and length variants were manually verified and the redundant ones removed. We investigated the shared repeats among the cp genomes of the seven species, based on the criterion that identical lengths located in homologous regions are considered to be shared repeats. Using the program Primer 3-1.1.1 (Premier Biosoft International, Palo Alto, CA, USA) [68], we developed SSR primers specific for *C. chuniana* for potential application in further analysis.

3.7. Sequence Divergence, dn/ds Ratio and K2P Genetic Distance

Comparative analyses of the nucleotide diversity (*Pi*) among the complete cp genomes of the seven species were performed with DnaSP 6 (Universitat de Barcelona, Barcelona, Spain) [69,70], as based on a sliding window analysis. The window length was 600 bp and step size was 200 bp. The 80 PCGs were extracted and aligned with MAFFT. We estimated the dn/ds ratio for each PCG as well as the interspecific genetic distance with DnaSP 6 and MEGA 6.0 (Tokyo Metropolitan University, Hachioji, Tokyo, Japan) [71], as based on the Kimura 2-parameter (K2P) model.

3.8. Phylogenetic Analysis

Altogether 97 representative species from the old Caesalpinioideae and Mimosoideae were selected for phylogenetic analyses (Table S4). *Cucumis sativus* (DQ119058) was used as the outgroup. Two data sets of the complete cp genome sequences and PCGs were used for phylogenetic reconstruction based on two methods of Bayesian inference (BI) and maximum likelihood (ML), respectively. All analyses were performed on the high-performance computer cluster available in the CIPRES Science Gateway 3.3 (available online: www.phylo.org) [72]. Gaps were treated as missing data. BI was performed by using MrBayes v. 3.2.6 (Swedish Museum of Natural History, Stockholm, Sweden) [73] with base frequencies estimated from the data. We ran four Markov Chains Monte Carlo (MCMC) for 50 million generations using default settings for priors and saved one tree every 1000 generations. The first 10% of the trees were discarded, as determined with the aid of the program Tracer version 1.6 (University of Auckland, Auckland, New Zealand) [74]. The posterior probability (PP) of each clade (i.e., the "clade credibility value") was estimated with 50% majority-rule consensus trees. We conducted ML using RAxML 8.2.10 (Heidelberg Institute for Theoretical Studies, Heidelberg, Germany) [75] and the RAxML graphical interface (rxmlGUI v. 1.3) (Research Institute Senckenberg, Frankfurt, Germany) [76]. RaxML was conducted by using Python v.2.7.6 (available online: http://www.python.org/ftp/python/2.7.6/python-2.7.6.msi) with 1000 rapid bootstrap replicates. The general time-reversible (GTR) model was chosen with a gamma model for the rate of heterogeneity.

4. Conclusions

We report the complete cp genome of *C. chuniana* endemic to China, which belongs to *Cercis* L., an intercontinentally disjunct genus. Using a high-throughput sequencing method, we sequenced and annotated the whole genome, detected the arrangement of the genes, and identified SSRs in *C. chuniana*. We compared the cp genomic characteristics of *C. chuniana* to its congener *C. canadensis* and five other species from the old Caesalpinioideae. The current study is the first structural and gene comparison among the cp genomes of seven species from three subfamilies of legumes, including Cercidoideae, Detarioideae and Caesalpinioideae at the genomic level. Nearly 100 representative species from the old Caesalpinioideae and Mimosoideae were used for phylogenetic reconstruction, strongly corroborating the monophyly of Cercidoideae and Detarioideae in the sense of the new classification of Fabaceae. Our study contributes to the taxonomy, phylogenetic reconstruction and biogeographical research of *Cercis* and other legume species.

Supplementary Materials: Supplementary materials can be found at http://www.mdpi.com/1422-0067/19/5/1286/s1.

Author Contributions: Wanzhen Liu performed most of the experiments and data analyses; Hanghui Kong participated in data analyses and writing the manuscript; Juan Zhou participated in sample collection; Peter W. Fritsch participated in writing the manuscript; Gang Hao supervised the project and provided suggestions for the manuscript; Wei Gong conceived and designed the experiment and research, supervised the project and contributed to the writing of the manuscript.

Acknowledgments: The authors thank Tongjian Liu and Gang Yao lab assistance and data analyses; El Mahdi Bendif for linguistic assistance; and the National Natural Science Foundation of China (31470312; 31470319) and Science and Technology Planning Project of Guangdong Province, China (2016A030303048) for financial support.

Conflicts of Interest: The authors declare no conflict of interest.

Abbreviations

LSC	Large single copy
SSC	Small single copy
IR	Inverted repeat
Cp	Chloroplast
ML	Maximum likelihood
BI	Bayesian inference
A	Adenine
T	Thymine
G	Guanine
C	Cytosine

References

1. Neuhaus, H.E.; Emes, M.J. Nonphotosynthetic metabolism in plastids. *Annu. Rev. Plant Biol.* **2000**, *51*, 111–140. [CrossRef] [PubMed]
2. Inoue, K. Emerging roles of the chloroplast outer envelope membrane. *Trends Plant Sci.* **2011**, *16*, 550–557. [CrossRef] [PubMed]
3. Raubeson, L.A.; Jansen, R.K. Chloroplast genomes of plants. In *Plant Diversity and Evolution: Genotypic and Phenotypic Variation in Higher Plants*; Henry, R.J., Ed.; CABI Publishing: Cambridge, MA, USA, 2005; pp. 45–68.
4. Yang, M.; Zhang, X.; Liu, G.; Yin, Y.; Chen, K.; Yun, Q.; Zhao, D.; Al-Mssallem, I.S.; Yu, J. The complete chloroplast genome sequence of date palm (*Phoenix dactylifera* L.). *PLoS ONE* **2010**, *5*, e12762. [CrossRef] [PubMed]
5. Green, B.R. Chloroplast genomes of photosynthetic eukaryotes. *Plant J.* **2011**, *66*, 34–44. [CrossRef] [PubMed]
6. Wicke, S.; Schneeweiss, G.M.; Müller, K.F.; Quandt, D. The evolution of the plastid chromosome in land plants: Gene content, gene order, gene function. *Plant Mol. Biol.* **2011**, *76*, 273–297. [CrossRef] [PubMed]
7. Roy, S.; Ueda, M.; Kadowaki, K.; Tsutsumi, N. Different status of the gene for ribosomal protein S16 in the chloroplast genome during evolution of the genus *Arabidopsis* and closely related species. *Genes Genet. Syst.* **2010**, *85*, 319–326. [CrossRef] [PubMed]
8. Lei, W.; Ni, D.; Wang, Y.; Shao, J.; Wang, X.; Yang, D.; Wang, J.; Chen, H.; Liu, C. Intraspecific and heteroplasmic variations, gene losses and inversions in the chloroplast genome of *Astragalus membranaceus*. *Sci. Rep.* **2016**, *6*, 21669. [CrossRef] [PubMed]
9. Ivanova, Z.; Sablok, G.; Daskalova, E.; Zahmanova, G.; Apostolova, E.; Yahubyan, G.; Baev, V. Chloroplast genome analysis of resurrection Tertiary relict *Haberlea rhodopensis* highlights genes important for desiccation stress response. *Front. Plant Sci.* **2017**, *8*, 204. [CrossRef] [PubMed]
10. Gaut, B.; Yang, L.; Takuno, S.; Eguiarte, L.E. The patterns and causes of variation in plant nucleotide substitution rates. *Annu. Rev. Ecol. Evol. Syst.* **2011**, *42*, 245–266. [CrossRef]

11. Dong, W.; Xu, C.; Cheng, T.; Zhou, S. Complete chloroplast genome of *Sedum sarmentosum* and chloroplast genome evolution in Saxifragales. *PLoS ONE* **2013**, *8*, e77965. [CrossRef] [PubMed]

12. Duan, Y.; Shen, Y.; Kang, F.; Wang, J. Characterization of the complete chloroplast genomes of the endangered shrub species *Prunus mongolica* and *Prunus pedunculata* (Rosales: Rosaceae). *Conserv. Genet. Resour.* **2018**, 1–4. [CrossRef]

13. Wang, H.; Park, S.; Lee, A.; Jang, S.; Im, D.; Jun, T.; Lee, J.; Chung, J.; Ham, T.; Kwon, S. Next-generation sequencing yields the complete chloroplast genome of *C. goeringii* acc. smg222 and phylogenetic analysis. *Mitochondrial DNA Part B* **2018**, *3*, 215–216. [CrossRef]

14. Lewis, G.P.; Schrire, B.D.; Mackinder, B.A.; Lock, J.M. (Eds.) *Legumes of the World*; Royal Botanic Gardens, Kew: Richmond, UK, 2005.

15. Käss, E.; Wink, M. Molecular evolution of the Leguminosae: Phylogeny of the three subfamilies based on rbcL-sequences. *Biochem. Syst. Ecol.* **1996**, *24*, 365–378. [CrossRef]

16. Doyle, J.; Ballenger, J.; Dickson, E.; Kajita, T.; Ohashi, H. A phylogeny of the chloroplast gene *rbcL* in the Leguminosae: Taxonomic correlations and insights into the evolution of nodulation. *Am. J. Bot.* **1997**, *84*, 541–554. [CrossRef] [PubMed]

17. Doyle, J.J.; Chappill, J.A.; Bailey, C.D.; Kajita, T. Towards a comprehensive phylogeny of legumes: Evidence from *rbcL* sequences and non-molecular data. In *Advances in Legume Systematics*; Herendeen, P.S., Bruneau, A., Eds.; Royal Botanic Gardens, Kew: Richmond, UK, 2000; pp. 1–20.

18. Wojciechowski, M.F.; Lavin, M.; Sanderson, M.J. A phylogeny of legumes (Leguminosae) based on analysis of the plastid *matK* gene resolves many well-supported subclades within the family. *Am. J. Bot.* **2004**, *91*, 1846–1862. [CrossRef] [PubMed]

19. Lavin, M.; Herendeen, P.S.; Wojciechowski, M.F. Evolutionary rates analysis of Leguminosae implicates a rapid diversification of lineages during the Tertiary. *Syst. Biol.* **2005**, *54*, 575–594. [CrossRef] [PubMed]

20. The Legume phylogeny Working Group. Legume phylogeny and classification in the 21st century: Progress, prospects and lessons for other species-rich clades. *Taxon* **2013**, *62*, 217–248.

21. The Legume Phylogeny Working Group. A new subfamily classification of the Leguminosae based on a taxonomically comprehensive phylogeny. *Taxon* **2017**, *66*, 44–77.

22. Li, H. Taxonomy and distribution of the genus *Cercis* in China. *Bull. Torrey Bot. Club* **1944**, *71*, 419–425. [CrossRef]

23. Robertson, K.R. *Cercis*: The redbuds. *Arnoldia* **1976**, *36*, 37–49.

24. Hao, G.; Zhang, D.; Guo, L.; Zhang, M.; Deng, Y.; Wen, X. A phylogenetic and biogeographic study of *Cercis* (Leguminosae). *Acta Bot. Sin.* **2001**, *43*, 1275–1278.

25. Davis, C.C.; Fritsch, P.W.; Li, J.; Donoghue, M.J. Phylogeny and biogeography of *Cercis* (Fabaceae): Evidence from nuclear ribosomal ITS and chloroplast *ndhF* sequence data. *Syst. Bot.* **2002**, *27*, 289–302.

26. Fritsch, P.W.; Cruz, B.C. Phylogeny of *Cercis* based on DNA sequences of nuclear ITS and four plastid regions: Implications for transatlantic historical biogeography. *Mol. Phylogenet. Evol.* **2012**, *62*, 816–825. [CrossRef] [PubMed]

27. Dezhao, C.; Dianxiang, Z.; Larsen, S.S.; Vincent, M.A. *Cercis*. In *Flora of China*; Wu, Z.Y., Raven, P.H., Eds.; Science Press: Beijing, China; Missouri Botanical Garden: St. Louis, MO, USA, 2010; Volume 10, pp. 5–6.

28. Metcalf, F.P. Eight new species of Leguminosae from Southeastern China. *Lingnan Sci. J.* **1940**, *19*, 549–563.

29. Coşkun, F.; Parks, C.R. A molecular phylogenetic study of red buds (*Cercis* L., Fabaceae) based on ITS nrDNA sequences. *Pak. J. Bot.* **2009**, *41*, 1577–1586.

30. Coşkun, F.; Parks, C.R. A molecular phylogeny of *Cercis* L. (Fabaceae) using the chloroplast *trnL-F* DNA sequences. *Pak. J. Bot.* **2009**, *41*, 1587–1592.

31. Kong, W.Q.; Yang, J.H. The complete chloroplast genome sequence of *Morus cathayana* and *Morus multicaulis*, and comparative analysis within genus *Morus* L. *PeerJ* **2017**, *5*, e3037. [CrossRef] [PubMed]

32. Guo, X.; Castillo-Ramírez, S.; González, V.; Bustos, P.; Fernández-Vázquez, J.L.; Santamaría, R.I.; Arellano, J.; Cevallos, M.A.; Dávila, G. Rapid evolutionary change of common bean (*Phaseolus vulgaris* L.) plastome, and the genomic diversification of legume chloroplasts. *BMC Genom.* **2007**, *8*, 228. [CrossRef] [PubMed]

33. Williams, A.V.; Boykin, L.M.; Howell, K.A.; Nevill, P.G.; Small, I. Correction: The complete sequence of the *Acacia ligulata* chloroplast genome reveals a highly divergent *clpP1* gene. *PLoS ONE* **2015**, *10*, e138367. [CrossRef] [PubMed]

34. Kaila, T.; Chaduvla, P.K.; Saxena, S.; Bahadur, K.; Gahukar, S.J.; Chaudhury, A.; Sharma, T.R.; Singh, N.K.; Gaikwad, K. Chloroplast genome sequence of Pigeonpea (*Cajanus cajan* (L.) Millspaugh) and *Cajanus scarabaeoides* (L.) Thouars: Genome organization and comparison with other legumes. *Front. Plant Sci.* **2016**, *7*, 1847. [CrossRef] [PubMed]

35. Wang, Y.; Qu, X.; Chen, S.; Li, D.; Yi, T. Plastomes of Mimosoideae: Structural and size variation, sequence divergence, and phylogenetic implication. *Tree Genet. Genomes* **2017**, *13*, 41. [CrossRef]

36. Wang, Y.; Wang, H.; Yi, T.; Wang, Y. The complete chloroplast genomes of *Adenolobus garipensis* and *Cercis glabra* (Cercidoideae, Fabaceae). *Conserv. Genet. Resour.* **2017**, *9*, 635–638. [CrossRef]

37. Choi, I.; Choi, B. The distinct plastid genome structure of *Maackia fauriei* (Fabaceae: Papilionoideae) and its systematic implications for genistoids and tribe Sophoreae. *PLoS ONE* **2017**, *12*, e173766. [CrossRef] [PubMed]

38. Xiang, B.; Li, X.; Qian, J.; Wang, L.; Ma, L.; Tian, X.; Wang, Y. The complete chloroplast genome sequence of the medicinal plant *Swertia mussotii* using the PacBio RS II platform. *Molecules* **2016**, *21*, 1029. [CrossRef] [PubMed]

39. Raman, G.; Park, V.; Kwak, M.; Lee, B.; Park, S. Characterization of the complete chloroplast genome of *Arabis stellari* and comparisons with related species. *PLoS ONE* **2017**, *12*, e183197. [CrossRef] [PubMed]

40. Park, S.; Jansen, R.K.; Park, S.J. Complete plastome sequence of *Thalictrum coreanum* (Ranunculaceae) and transfer of the *rpl32* gene to the nucleus in the ancestor of the subfamily Thalictroideae. *BMC Plant Biol.* **2015**, *15*, 40. [CrossRef] [PubMed]

41. Lu, R.; Li, P.; Qiu, Y. The complete chloroplast genomes of three *Cardiocrinum* (Liliaceae) species: Comparative genomic and phylogenetic analyses. *Front. Plant Sci.* **2017**, *7*, 2054. [CrossRef] [PubMed]

42. Kong, H.; Liu, W.; Yao, G.; Gong, W. A comparison of chloroplast genome sequences in *Aconitum* (Ranunculaceae): A traditional herbal medicinal genus. *PeerJ* **2017**, *5*, e4018. [CrossRef] [PubMed]

43. Piovani, P.; Leonardi, S.; Piotti, A.; Menozzi, P. Conservation genetics of small relic populations of silver fir (*Abies alba* Mill.) in the northern Apennines. *Plant Biosyst.* **2010**, *144*, 683–691. [CrossRef]

44. Wang, T.; Wang, Z.; Chen, G.; Wang, C.; Su, Y. Invasive chloroplast population genetics of *Mikania micrantha* in China: No local adaptation and negative correlation between diversity and geographic distance. *Front. Plant Sci.* **2016**, *7*, 1426. [CrossRef] [PubMed]

45. Der, J.P.; Thomson, J.A.; Stratford, J.K.; Wolf, P.G. Global chloroplast phylogeny and biogeography of bracken (*Pteridium*; Dennstaedtiaceae). *Am. J. Bot.* **2009**, *96*, 1041–1049. [CrossRef] [PubMed]

46. Greiner, S.; Rauwolf, U.; Meurer, J.; Herrmann, R.G. The role of plastids in plant speciation. *Mol. Ecol.* **2011**, *20*, 671–691. [CrossRef] [PubMed]

47. Zhang, Y.J.; Ma, P.F.; Li, D.Z. High-throughput sequencing of six bamboo chloroplast genomes: Phylogenetic implications for temperate woody bamboos (*Poaceae*: Bambusoideae). *PLoS ONE* **2011**, *6*, e20596. [CrossRef] [PubMed]

48. Zhang, Y.; Du, L.; Liu, A.; Chen, J.; Wu, L.; Hu, W.; Zhang, W.; Kim, K.; Lee, S.; Yang, T. The complete chloroplast genome sequences of five *Epimedium* species: Lights into phylogenetic and taxonomic analyses. *Front. Plant Sci.* **2016**, *7*, 306. [CrossRef] [PubMed]

49. Myszczyński, K.; Bączkiewicz, A.; Buczkowska, K.; Ślipiko, M.; Szczecińska, M.; Sawicki, J. The extraordinary variation of the organellar genomes of the *Aneura pinguis* revealed advanced cryptic speciation of the early land plants. *Sci. Rep.* **2017**, *7*, 9804. [CrossRef] [PubMed]

50. Wang, Y.; Wicke, S.; Wang, H.; Jin, J.; Chen, S.; Zhang, S.; Li, D.; Yi, T. Plastid genome evolution in the early-diverging legume subfamily Cercidoideae (Fabaceae). *Front. Plant Sci.* **2018**, *9*, 138. [CrossRef] [PubMed]

51. Doyle, J.J. A rapid DNA isolation procedure for small quantities of fresh leaf tissue. *Phytochem. Bull.* **1987**, *19*, 11–15.

52. Zhang, T.; Zeng, C.X.; Yang, J.B.; Li, H.T.; Li, D.Z. Fifteen novel universal primer pairs for sequencing whole chloroplast genomes and a primer pair for nuclear ribosomal DNAs. *J. Syst. Evol.* **2016**, *54*, 219–227. [CrossRef]

53. Coil, D.; Jospin, G.; Darling, A.E. A5-miseq: An updated pipeline to assemble microbial genomes from Illumina MiSeq data. *Bioinformatics* **2014**, *31*, 587–589. [CrossRef] [PubMed]

54. Kearse, M.; Moir, R.; Wilson, A.; Stones-Havas, S.; Cheung, M.; Sturrock, S.; Buxton, S.; Cooper, A.; Markowitz, S.; Duran, C. Geneious Basic: An integrated and extendable desktop software platform for the organization and analysis of sequence data. *Bioinformatics* **2012**, *28*, 1647–1649. [CrossRef] [PubMed]

55. Altschul, S.F.; Gish, W.; Miller, W.; Myers, E.W.; Lipman, D.J. Basic local alignment search tool. *J. Mol. Biol.* **1990**, *215*, 403–410. [CrossRef]

56. Wyman, S.K.; Jansen, R.K.; Boore, J.L. Automatic annotation of organellar genomes with DOGMA. *Bioinformatics* **2004**, *20*, 3252–3255. [CrossRef] [PubMed]

57. Schwarz, E.N.; Ruhlman, T.A.; Sabir, J.S.M.; Hajrah, N.H.; Alharbi, N.S.; Al Malki, A.L.; Bailey, C.D.; Jansen, R.K. Plastid genome sequences of legumes reveal parallel inversions and multiple losses of *rps16* in papilionoids. *J. Syst. Evol.* **2015**, *53*, 458–468. [CrossRef]

58. Laslett, D.; Canback, B. ARAGORN, a program to detect tRNA genes and tmRNA genes in nucleotide sequences. *Nucleic Acids Res.* **2004**, *32*, 11–16. [CrossRef] [PubMed]

59. Lohse, M.; Drechsel, O.; Kahlau, S.; Bock, R. OrganellarGenomeDRAW—A suite of tools for generating physical maps of plastid and mitochondrial genomes and visualizing expression data sets. *Nucleic Acids Res.* **2013**, *41*, W575–W581. [CrossRef] [PubMed]

60. Katoh, K.; Standley, D.M. MAFFT multiple sequence alignment software version 7: Improvements in performance and usability. *Mol. Biol. Evol.* **2013**, *30*, 772–780. [CrossRef] [PubMed]

61. Kim, K.; Lee, H. Complete chloroplast genome sequences from Korean ginseng (*Panax schinseng* Nees) and comparative analysis of sequence evolution among 17 vascular plants. *DNA Res.* **2004**, *11*, 247–261. [CrossRef] [PubMed]

62. Nazareno, A.G.; Carlsen, M.; Lohmann, L.G. Complete chloroplast genome of *Tanaecium tetragonolobum*: The first Bignoniaceae plastome. *PLoS ONE* **2015**, *10*, e129930. [CrossRef] [PubMed]

63. Darling, A.E.; Mau, B.; Perna, N.T. progressiveMauve: Multiple genome alignment with gene gain, loss and rearrangement. *PLoS ONE* **2010**, *5*, e11147. [CrossRef] [PubMed]

64. Mayor, C.; Brudno, M.; Schwartz, J.R.; Poliakov, A.; Rubin, E.M.; Frazer, K.A.; Pachter, L.S.; Dubchak, I. VISTA: Visualizing global DNA sequence alignments of arbitrary length. *Bioinformatics* **2000**, *16*, 1046–1047. [CrossRef] [PubMed]

65. Brudno, M.; Malde, S.; Poliakov, A.; Do, C.B.; Couronne, O.; Dubchak, I.; Batzoglou, S. Glocal alignment: Finding rearrangements during alignment. *Bioinformatics* **2003**, *19*, i54–i62. [CrossRef] [PubMed]

66. Frazer, K.A.; Pachter, L.; Poliakov, A.; Rubin, E.M.; Dubchak, I. VISTA: Computational tools for comparative genomics. *Nucleic Acids Res.* **2004**, *32*, W273–W279. [CrossRef] [PubMed]

67. Thiel, T.; Michalek, W.; Varshney, R.; Graner, A. Exploiting EST databases for the development and characterization of gene-derived SSR-markers in barley (*Hordeum vulgare* L.). *Theor. Appl. Genet.* **2003**, *106*, 411–422. [CrossRef] [PubMed]

68. Rozen, S.; Skaletsky, H. *Primer3 on the WWW for General Users and for Biologist Programmers; Bioinformatics Methods and Protocols*; Humana Press: Totowa, NJ, USA, 2000; pp. 365–386.

69. Librado, P.; Rozas, J. DnaSP v5: A software for comprehensive analysis of DNA polymorphism data. *Bioinformatics* **2009**, *25*, 1451–1452. [CrossRef] [PubMed]

70. Rozas, J.; Ferrer-Mata, A.; Sánchez-DelBarrio, J.C.; Guirao-Rico, S.; Librado, P.; Ramos-Onsins, S.E.; Sánchez-Gracia, A. DnaSP 6: DNA sequence polymorphism analysis of large data sets. *Mol. Biol. Evol.* **2017**, *34*, 3299–3302. [CrossRef] [PubMed]

71. Tamura, K.; Stecher, G.; Peterson, D.; Filipski, A.; Kumar, S. MEGA6: Molecular evolutionary genetics analysis version 6.0. *Mol. Biol. Evol.* **2013**, *30*, 2725–2729. [CrossRef] [PubMed]

72. Miller, M.A.; Schwartz, T.; Pickett, B.E.; He, S.; Klem, E.B.; Scheuermann, R.H.; Passarotti, M.; Kaufman, S.; O'Leary, M.A. A RESTful API for access to phylogenetic tools via the CIPRES science gateway. *Evol. Bioinform.* **2015**, *11*, S21501. [CrossRef] [PubMed]

73. Ronquist, F.; Teslenko, M.; Van Der Mark, P.; Ayres, D.L.; Darling, A.; Höhna, S.; Larget, B.; Liu, L.; Suchard, M.A.; Huelsenbeck, J.P. MrBayes 3.2: Efficient Bayesian phylogenetic inference and model choice across a large model space. *Syst. Biol.* **2012**, *61*, 539–542. [CrossRef] [PubMed]

74. Drummond, A.J.; Suchard, M.A.; Xie, D.; Rambaut, A. Bayesian phylogenetics with BEAUti and the BEAST 1.7. *Mol. Biol. Evol.* **2012**, *29*, 1969–1973. [CrossRef] [PubMed]

75. Stamatakis, A. RAxML version 8: A tool for phylogenetic analysis and post-analysis of large phylogenies. *Bioinformatics* **2014**, *30*, 1312–1313. [CrossRef] [PubMed]
76. Silvestro, D.; Michalak, I. raxmlGUI: A graphical front-end for RAxML. *Org. Divers. Evol.* **2012**, *12*, 335–337. [CrossRef]

International Journal of

Molecular Sciences

MDPI

Article

The Complete Plastome Sequence of an Antarctic Bryophyte *Sanionia uncinata* (Hedw.) Loeske

Mira Park [1,2], Hyun Park [1,3], Hyoungseok Lee [1,3,*], Byeong-ha Lee [2,*] and Jungeun Lee [1,3,*]

[1] Unit of Polar Genomics, Korea Polar Research Institute, Incheon 21990, Korea; mira0295@kopri.re.kr (M.P.); hpark@kopri.re.kr (H.P.)

[2] Department of Life Science, Sogang University, Seoul 04107, Korea

[3] Polar Science, University of Science & Technology, Daejeon 34113, Korea

* Correspondence: soulaid@kopri.re.kr (H.L.); byeongha@sogang.ac.kr (B.-h.L.); jelee@kopri.re.kr (J.L.); Tel.: +82-32-760-5571 (H.L.); +82-2-705-8794 (B.-h.L.); +82-32-760-5576 (J.L.)

Received: 31 January 2018; Accepted: 27 February 2018; Published: 1 March 2018

Abstract: Organellar genomes of bryophytes are poorly represented with chloroplast genomes of only four mosses, four liverworts and two hornworts having been sequenced and annotated. Moreover, while Antarctic vegetation is dominated by the bryophytes, there are few reports on the plastid genomes for the Antarctic bryophytes. *Sanionia uncinata* (Hedw.) Loeske is one of the most dominant moss species in the maritime Antarctic. It has been researched as an important marker for ecological studies and as an extremophile plant for studies on stress tolerance. Here, we report the complete plastome sequence of *S. uncinata*, which can be exploited in comparative studies to identify the lineage-specific divergence across different species. The complete plastome of *S. uncinata* is 124,374 bp in length with a typical quadripartite structure of 114 unique genes including 82 unique protein-coding genes, 37 tRNA genes and four rRNA genes. However, two genes encoding the α subunit of RNA polymerase (*rpoA*) and encoding the cytochrome $b_{6/f}$ complex subunit VIII (*petN*) were absent. We could identify nuclear genes homologous to those genes, which suggests that *rpoA* and *petN* might have been relocated from the chloroplast genome to the nuclear genome.

Keywords: *Sanionia uncinata*; chloroplast genome; Antarctic bryophyte; moss; plastome

1. Introduction

Antarctic terrestrial ecosystems are dominated by lichens and bryophytes (including mosses, liverworts and hornworts), encompassing more than 200 lichens and 109 mosses species [1]. Only two vascular plant species have survived and adapted to these extreme environments, with a very limited distribution restricted to the maritime Antarctic, while mosses are common plants on extensive ice-free areas of Antarctica.

Sanionia uncinata is one of the most dominant moss species in Antarctica and is mainly distributed over coastal areas [2,3]. Moreover, *S. uncinata* is distributed across multiple geographic regions, ranging from Northern Hemisphere (Europe, Asia, North America and the Arctic) to Southern Hemisphere (Africa, South America and the Antarctica) and also found at high-altitude mountains in tropical and subtropical areas [4]. A recent phylogeographic study has established the haplotype networks of *S. uncinata* populations by identifying their genetic diversity with massive molecular marker datasets of more than 200 specimens collected from various regions around the world [4].

S. uncinata is a pleurocarpous moss species that form dense and extensive carpets on terrestrial habitats over a wide range of water regimes, from dry rock surfaces to wet areas at the edges of streams or melt pools [5–7]. *S. uncinata* has been extensively used as an experimental model for the study of environmental impacts on plants [8–10]. This species is known to tolerate dehydration by retaining moisture in their tissues for a long period of time by forming a carpet-like community shape that helps to avoid water loss [11]. However, the molecular mechanism and molecular ecology underlying stress tolerance have yet to be elucidated. The dehydration process, although it prevents the moss from freezing, directly affects cell metabolism and as a result, photosynthetic capacity decreases when the water content of moss drops below the optimum level [12]. On the other hand, photosynthesis is essential for the production of energy needed for plant growth and takes place in the chloroplasts. Thus, the integrity and metabolic performance of chloroplasts are very important for photosynthetic activities [13].

Chloroplasts are unique organelles, derived from cyanobacteria through endosymbiosis, that provide essential energy for plants and algae through photosynthesis [14,15]. They contain their own genomes that have a unique mechanism of RNA transcription and are inherited maternally. Chloroplasts are known to play an important role in the synthesis of pigments, starch, fatty acids and amino acids as well as the photosynthesis process [16,17]. In general, chloroplast genomes—namely, plastomes—are highly conserved with regards to gene sequences and gene content in terrestrial plants. Their highly conservative nature is sufficient to perform comparative studies on different species to discuss evolutionary relationships between species in terms of molecular phylogeny and molecular ecology [18]. For instance, plastome sequences provide species-specific information that includes genome size, gene order, genome rearrangements, patterns of base pair composition, codon usage, massive plastid gene losses and various type of nucleotide polymorphism [19,20].

Despite their genetic diversity and evolutionary significance, genetic resources for bryophytes are very limited when compared to angiosperms. Of the 2352 records for chloroplast genome sequences of green plants, only 15 plastomes of bryophytes comprised of 2 from hornworts, 5 from liverworts and 8 from mosses (including *Sanonia uncinata* NC_025668 which was directly submitted by the authors of this study) are available in public repositories [21] (http://www.ncbi.nlm.nih.gov). There are currently only four complete chloroplast genomes fully published for mosses, for example, *Physcomitrella patens* [22], *Tortula ruralis* [13], *Tetraphis pellucida* [23] and *Tetraplodon fuegianus* [24]. The plastome information of more bryophytes and comparative genomic studies are necessary to better understand the molecular evolutionary events or functions of chloroplast genes. In this regard, the plastome information of *S. uncinata* provided in this study will be a very useful resource for future research on the ecology, physiology and molecular evolution of bryophytes.

2. Results and Discussion

2.1. Overall Genome Organization

Illumina MiSeq sequencing produced 4,993,466 raw reads with an average read length of 301 bp and a total number of 1,503,033,266 base pairs. A total of 46,573 chloroplast-related reads were obtained as a result of alignment of quality trimmed reads against other chloroplast genomes publically available in NCBI. Assembly of the nucleotide sequence reads was performed to obtain non-redundant contigs and singletons using CLC Genomics Workbench V7.5 (CLC bio, Aarhus, Denmark). The final *S. uncinata* plastome sequence has been submitted to GenBank (Accession: NC_025668).

The gene map for the *S. uncinata* plastome is shown in Figure 1. The complete plastome of *S. uncinata* is 124,374 base pairs (bp) in length with a typical quadripartite structure including large and small single-copy regions (LSC of 86,570 bp and SSC of 18,430 bp) separated by a pair of identical inverted repeats (IRA and IRB) of 9687 bp each (Figure 1). Most of the chloroplast DNA had a well preserved quadripartite structure in bryophytes [23,25,26] and vascular plants [27,28]. The genome contained 114 unique genes including 82 unique protein-coding genes, 37 tRNA genes and 4 rRNA genes (Table 1). The gene content of the IR regions was conserved among *S. uncinata, T. ruralis* and *P. patens*.

The size of the plastome of *S. uncinata* is very similar to those of liverworts (*Marchantia polymorpha* 121,024 bp NC_001319, *Pellia endiviifolia* 120,546 bp NC_019628, *Aneura mirabilis* 108,007 bp NC_010359, *Ptilidium pulcherrimum* 119,007 bp NC_015402) and mosses (*Physcomitrella patens* 122,890 bp NC_005087, *Tortula ruralis* 124,374 bp NC_012052, *Tetraphis pellucida* 127,489 bp NC_024291) and *Tetraplodon fuegianus*

123,670 bp, KU_095851(unverified) but much smaller than hornworts (*Anthoceros formosae* 161,162 bp NC_004543 and *Nothocerosaenigmaticus* 153,208 bp NC_020259), which have an increased length of intragenic spacers in the LSC region or in the identical IR regions [18,29,30].

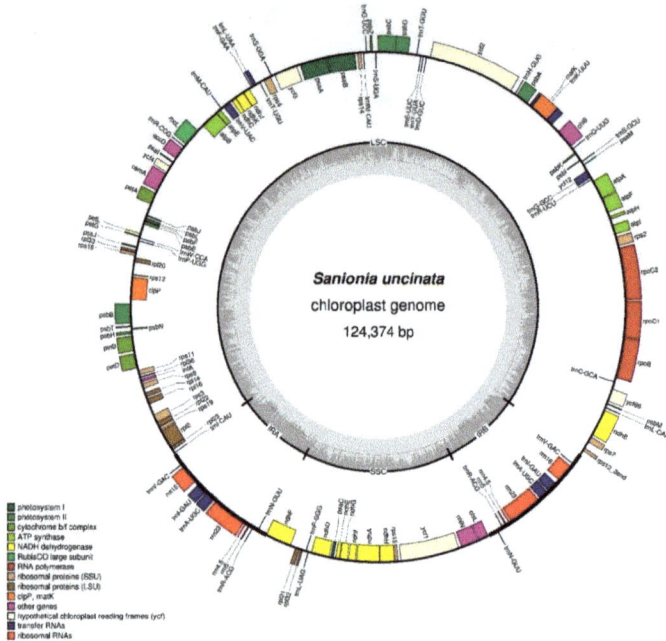

Figure 1. Map of the *Sanionia uncinata* plastome. Complete plastome sequences were obtained from the de novo assembly of Illumina paired-end reads. Genes are color coded by functional group, which are located in the left box. The inner darker gray circle indicates the GC content while the lighter gray corresponds to AT content. IR, inverted repeat; LSC, large single copy region; SSC, small single copy region. Genes shown on the outside of the outer circle are transcribed clockwise and those on the inside counter clockwise. The map was made with OGDraw [31].

Table 1. Genes present in the *S. uncinata* plastome.

Gene Products	Genes
Photosystem I	*psaA, B, C, I, J, M*
Photosystem II	*psbA, B, C, D, E, F, H, I, J, K, L, M, N, T, Z*
Cytochrome b6/f	*petA, B [a], D [a], G, L*
ATP synthase	*atpA, B, E, F [a], H, I*
Translation factor	*infA*
Chlorophyll biosynthesis	*chlB, L, N*
Rubisco	*rbcL*
NADH oxidoreductase	*ndhA [a], B [a], C, D, E, F, G, H, I, J, K*
Large subunit ribosomal proteins	*rpl2 [a], 14, 16 [a], 20, 21, 22, 23, 32, 33, 36*
Small subunit ribosomal proteins	*rps2, 3, 4, 7, 8, 11, 12 [a,b], 14, 15, 18, 19*
RNAP	*rpoB, C1 [a], C2*
Other proteins	*accD, cemA, clpP [c], matK*
Proteins of unknown function	*ycf1, 2, 3 [c], 4, 12, 66 [a]*
Ribosomal	*rrn4.5 [d], 5 [d], 16 [d], 23 [d]*
Transfer RNAs	*trnA(UGC) [a,d], C(GCA), D(GUC), E(UUC), F(GAA), G(UCC) [a], G(UCC), H(GUG), I(CAU), I(GAU) [a,d], K(UUU) [a], L(CAA), L(UAA) [a], L(UAG), fM(CAU), M(CAU), N(GUU) [d], P(UGG), P(GGG), Q(UUG), R(ACG) [d], R(CCG), R(UCU), S(GCU), S(GGA), S(UGA), T(GGU), T(UGU), V(GAC) [d], V(UAC) [a], W(CCA), Y(GUA)*

[a] Gene containing a single intron; [b] Gene divided into two independent transcription units; [c] Gene containing two introns; [d] Two gene copies in the IRs.

The overall G/C content was 29.3% for *S. uncinata*, similar to other known bryophyte plastomes (*P. patens* (28.5%) [22], *T. pellucida* (29.4%) [23], *T. fuegianus* (28.7%) [24]), as well as the liverwort *M. polymorpha* (28.8%) [26], hornwort *A. formosae* (32.9%) [25], the charophyte *Chaetosphaeridium* (29.6%) [32] and algae (30–33%) [22] but significantly less than the 34~40% found in seed plants [33].

The chloroplast genes found in the complete plastome are represented in Table 2. There were 14 intron-containing genes including 5 tRNA genes and 9 protein-coding genes and almost all of which were single-intron genes except for *ycf3* and *clpP*, which each had two introns. Two exons of trans-spliced gene *rps12* are located in the LSC 71 kb apart from each other. The *trnK-UUU* gene has the largest intron (2272 bp), which has *matK* ORF (1548 bp) encoding a maturase involved in splicing type II introns [34].

Table 2. The genes with introns in the *S. uncinata* plastome and the length of the exons and introns.

Gene	Location	Length (bp)				
		Exon I	**Intron I**	**Exon II**	**Intron II**	**Exon III**
rps12	LSC	114	-	270		
ndhB	LSC	729	629	780		
ycf66	LSC	106	591	320		
rpoC1	LSC	423	789	1614		
atpF	LSC	411	654	135		
ycf3	LSC	126	684	228	739	153
clpP	LSC	69	687	291	483	234
rpl2	LSC	396	637	438		
ndhA	SSC	556	731	551		
trnK-UUU	LSC	37	2272	42		
trnL-UAA	LSC	38	262	50		
trnV-UAC	LSC	37	542	37		
trnI-GAU	IR	42	769	35		
trnA-UGC	IR	38	763	35		

The 82 protein-coding genes in this genome represented nucleotide coding for 40,330 codons. On the basis of the sequences of protein-coding genes and tRNA genes within the plastome, the frequency of codon usage was deduced (Table 3). Among these codons, 4343 (10.85%) encoded for leucine and 455 (1.14%) for Tryptophan, which were the most and the least amino acids, respectively. The codon usage was biased towards a high representation at the third codon position. A biased frequency of codons included the levels of available tRNA, functionally related genes, evolutionary pressures and the rate of gene evolution [35].

Table 3. The codon-anticodon recognition pattern and codon usage for *S. uncinata* plastome.

Amino Acid	Codon	No. *	tRNA	Amino Acid	Codon	No. *	tRNA
Phe	UUU	2862		Tyr	UAU	1492	
Phe	UUC	916	*trnF-GAA*	Tyr	UAC	578	*trnY-GUA*
Leu	UUG	688	*trnL-UAA*	Stop	UAA	1772	
Leu	UUA	1766	*trnL-CAA*	Stop	UAG	577	
Leu	CUG	252		His	CAU	550	
Leu	CUA	591	*trnL-UAG*	His	CAC	224	*trnH-GUG*
Leu	CUU	741		Gln	CAA	685	*trnQ-UUG*
Leu	CUC	305		Gln	CAG	273	
Ile	AUG	1582	*trnI-CAU*	Asn	AAU	1878	
Ile	AUU	1943		Asn	AAC	665	*trnN-GUU*
Ile	AUC	629	*trnI-GAU*	Lys	AAA	2942	*trnK-UUU*
Met	AUG	515	*trnfM-CAU*	Lys	AAG	768	
Val	GUG	244		Asp	GAU	619	

Table 3. *Cont.*

Amino Acid	Codon	No. *	tRNA	Amino Acid	Codon	No. *	tRNA
Val	GUA	588	*trnV-UAC*	Asp	GAC	210	*trnD-GUC*
Val	GUU	652		Glu	GAA	845	*trnE-UUC*
Val	GUC	237	*trnV-GAC*	Glu	GAG	286	
Ser	AGU	579		Cys	UGU	515	
Ser	AGC	470	*trnS-GCU*	Cys	UGC	362	*trnC-GCA*
Ser	UCG	272		Stop	UGA	657	
Ser	UCA	643	*trnS-UGA*	Trp	UGG	455	*trnW-CCA*
Pro	CCG	169		Arg	AGG	387	
Pro	CCA	429	*trnP-UGG*	Arg	AGA	732	*trnR-UCU*
Pro	CCU	423		Arg	CGG	164	*trnR-CCG*
Pro	CCC	230	*trnP-GGG*	Arg	CGA	322	
Thr	ACG	228		Arg	CGU	270	*trnR-ACG*
Thr	ACA	508	*trnT-UGU*	Arg	CGC	143	
Thr	ACU	561		Ser	UCU	735	
Thr	ACC	363	*trnT-GGU*	Ser	UCC	419	*trnS-GGA*
Ala	GCG	155		Gly	GGG	426	
Ala	GCA	367	*trnA-UGC*	Gly	GGA	465	*trnG-UCC*
Ala	GCU	413		Gly	GGU	653	
Ala	GCC	514		Gly	GGC	237	

* Numerals indicate the frequency of usage of each codon in 40,330 in codons in 82 potential protein-coding genes.

2.2. Comparison with Other Bryophyte Plastomes

Multiple complete bryophyte plastomes available provide an opportunity to compare the sequence variation at the genome-level. We therefore compared the whole plastome sequence of *S. uncinata* with those of mosses *T. ruralis*, *P. patens*, *T. pellucida*, the liverwort *M. polymorpha* and hornwort *A. formosae*. The sequence identity between all five bryophyte plastomes was plotted using the mVISTA program with the annotation of *S. uncinata* as a reference (Figure 2).

Sequence similarities of the genes between *S. uncinata* and other bryophytes (mosses, liverwort and hornwort) were compared (Table S1). The rRNA genes (*rrn5*, *rrn16*, *rrn4.5* and *rrn23*) in the IRs region showed the highest sequence similarity (average 98.3–96.1%) and PSII-associated genes such as *psbL*, *psbA*, *psbN*, *psbZ*, *psbE*, *psbH*, *psbK*, *psbF*, *psbB*, *psbD* and *psbJ*, which also displayed high levels of sequence similarity (average 93.9–89.8%). Genes for large subunit ribosomal proteins (*rpl32*, *rpl20*, *rpl33* and *rpl23*) and a small subunit ribosomal protein (*rps12*) were relatively more conserved than other coding genes. Notably, the highest sequence variation occurred in the *matK* gene (average 79.4%), widely known to be evolved rapidly and thus often used as a barcoding marker in phylogenetic and evolutionary studies [34,36], suggesting that this gene has also undergone evolutionary pressure within bryophytes. Following this, genes such as *rpoC2*, *petB atpE*, *atpF* and *rpl22* showed lower similarity (average 80.2–83.5%) than other plastid genes in order (Table S1). Those genes with large sequence variations had evolutionary significance in inferring divergence times and branching patterns among early land plant lineages, while relatively less varied genes such as rRNA and PSII-associated genes were well conserved during the evolution of bryophytes.

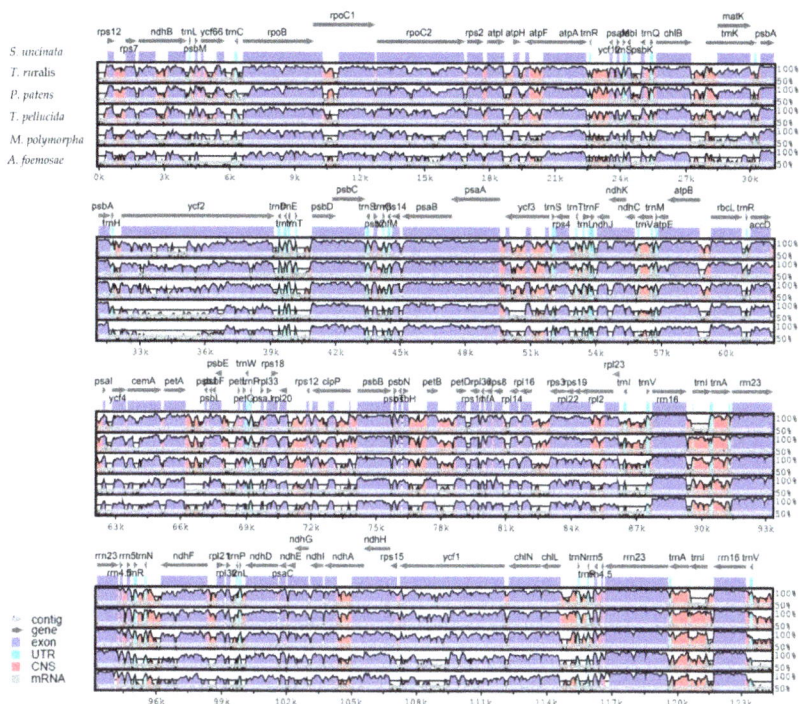

Figure 2. Alignment of complete plastome sequences from six species. Alignment and comparison were performed using mVISTA and the percentage of identity between the plastomes was visualized in the form of an mVISTA plot. The sequence similarity of the aligned regions between *S. uncinata* and other five species is shown as horizontal bars indicating average percent identity between 50–100% (shown on the y-axis of graph). The x-axis represents the coordinate in the plastome. Genome regions are color-coded for protein-coding (exon), rRNA, tRNA and conserved non-coding sequences (CNS) as the guide at the bottom-left.

2.3. Phylogenomic Analysis

Plastome information has provided an important resource for uncovering evolutionary relationships between various plant lineages [20,37]. The whole plastomes and protein-coding genes have been widely used for the reconstruction of phylogenetic relationships among different plant species [38]. The availability of completed *S. uncinata* plastome provided us with the sequence information to study the molecular evolution and phylogeny of *S. uncinata* with closely related species.

Phylogenomic analysis of representatives from the bryophyte subfamily including *S. uncinata*, produced a single, well-supported tree using maximum parsimony (MP) (Figure 3). To do this, a set of 40 protein-coding genes in plastome analyzed in other species were concatenated and these concatenated sequences were used to infer the phylogenetic relationships of 23 taxa including *S. uncinata* using MEGA7. Phylogenetic analysis based on the multigene dataset revealed that mosses, liverwort and hornwort have been resolved as monophyletic in MP tree (Figure 3). Based on molecular phylogeny results, liverworts are placed in a basal position representing the earliest diverging lineage, while hornworts are the closest relatives of extant vascular plants, corroborating a previously reported branch order of "liverworts (mosses(hornworts(vascular plants)))" [37]. *S. uncinata* was most related to *T. ruralis* and formed a sister group with other moss species, which was supported by bootstrap values (100 for both ML and MP). In addition, it provides convincing support for many traditionally recognized genera and identifies higher level phylogenetic structure of mosses [39]. This result suggests that plastome information can effectively resolve phylogenetic positions and evolutionary relationships between different Bryophyte lineages.

The *T. ruralis* and *S. uncinata*, two mosses inhabiting extreme environments of polar and alpine regions, share common features in their plastomes–lack of *petN* and *rpoA* and presence of *trnP-GGG*–which is not the case for other two species of *P. patens* and *T. pellucid* (Table 4). Due to very scarce taxon sampling and the limitation of available plastome data, it is not clear whether the presence or absence of those genes is related to the resistance or adaptation to the extreme environments where they inhabit. To address this, synapomorphic characteristics developed during adaptation should be investigated together with genome evolution.

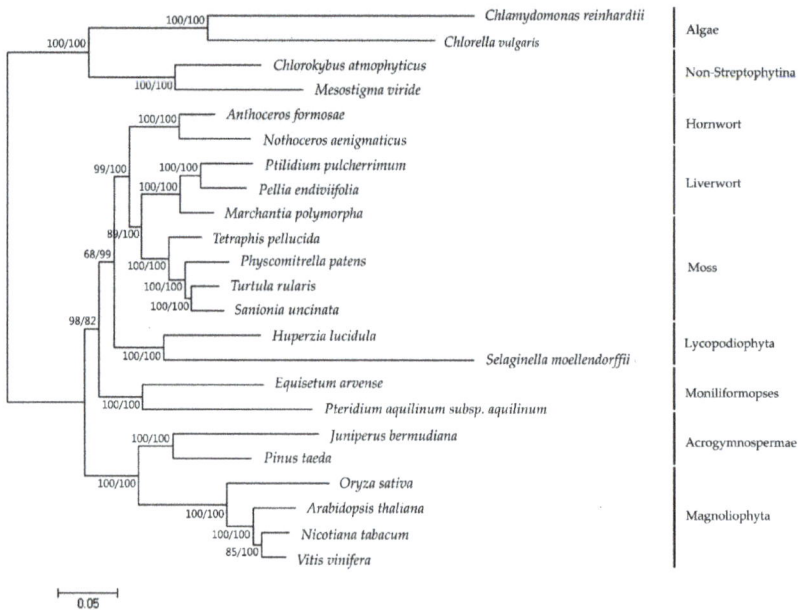

Figure 3. Phylogenetic tree reconstruction of 23 taxa using MEGA7 based on concatenated sequences of 40 protein-coding genes in the plastome. Maximum likelihood (ML) topology is shown with the bootstrap support values (MP/ML) given at nodes. Forty protein-coding sequences were extracted from annotated plastomes found in GenBank [21] (http://www.ncbi.nlm.nih.gov) (Table S2). The nucleotide sequences for each gene were translated into amino acids, aligned in MEGA7 and manually adjusted. Nucleotide sequences were aligned by constraining them to the amino acid sequence alignment. Individual gene alignments were then assembled into a single dataset.

2.4. Loss of rpoA and petN in the S. uncinata Plastome

Previous studies on bryophyte plastomes have shown that the overall structure and gene contents of plastomes are highly conserved among different bryophyte lineages [22,23,25,26]. The plastome of *S. uncinata* also showed a similar level of size and structure when compared to those of other three moss plastomes, *P. patens*, *T. ruralis* and *T. pellucida*. The plastome of *S. uncinata* did not show a large inversion of the ~71 kb fragment found in the LSC region of *P. patens* plastome, which follows the cases of *T. ruralis* and *T. pellucida*, suggesting that this large inversion is limited to the order *Funariales* in Bryophyte [22,40] (Figure 4).

While most mosses and plant groups are conserved in the contents of chloroplast genes, certain lineages undergo apparent and frequent gene loss (e.g., *rpoA* transfer from the chloroplast to the nucleus: [13,22,23]). Two genes encoding α subunit of RNA polymerase (*rpoA*) and encoding cytochrome $b_{6/f}$ complex subunit VIII (*petN*) were absent in the *S. uncinata* plastome (Figure 4). As discussed in previous moss plastome studies, the most prominent and typical feature of the plastomes of moss is the absence of *rpoA*, which is thought to have disappeared together with *ycf5*, *cysA*, *cysT* after the divergence of the mosses from the hepatic bryophytes [22,41]. We could not identify *rpoA* with *cysT* and *cysA* in the *S. uncinata* plastome, which means that the *S. uncinata* plastome follows the typical characteristics of moss plastomes (Table 4, Figure 4). The loss of *petN* reported in other mosses, except in *P. patens* [22], was also observed in the *S. uncinata* plastome (Table 4, Figure 4). The presence of *trnP-GGG* also varied depending on species [13,23], the gene was present in the plastomes of *S. uncinata* and *T. ruralis*, while it was absent in the plastomes of *P. patens* and *T. pellucida* (Table 4).

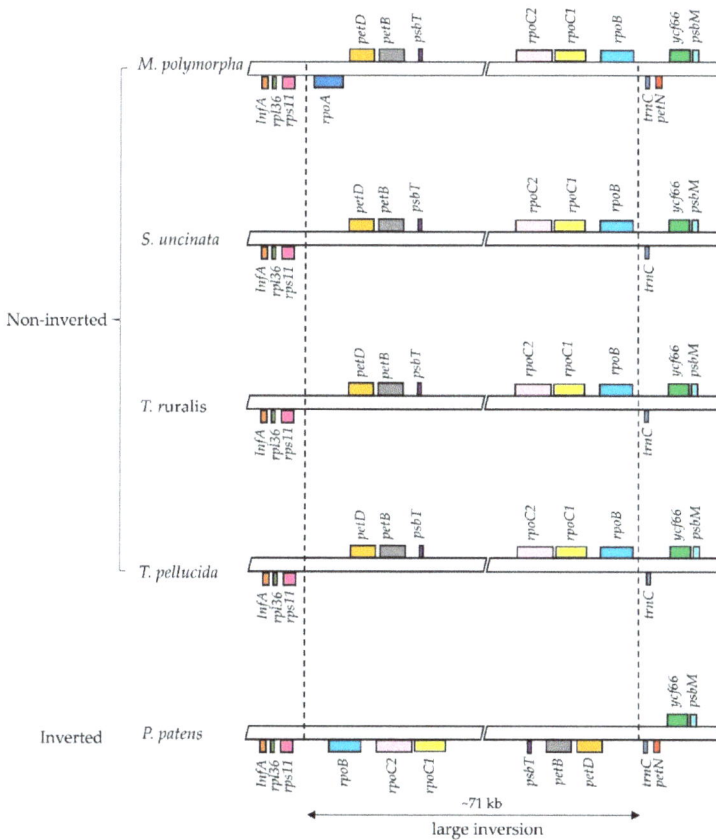

Figure 4. Comparison of the large inversion in the LSC region among six bryophytes plastomes. In comparative LSC region alignment of *rpoA*, *petN* coding regions from *M. polymorpha*, *S. uncinata*, *T. ruralis*, *T. pellucida* and *P. patens*. The inverted-arrangement of 71 kb fragment was only detected for *P. patens*.

Table 4. Gene contents of plastomes from green alga, bryophytes and land plants.

	Plants *	Genome Size (bp)	petN	rpoA	ccsA	cysA	cysT	ycf66	matK	rps15	trnP-GGG
Alga	*Chlorella vulgaris*	150,613	-	+	+	+	+	-	-	-	-
Hornwort	*Anthoceros formosae*	161,162	+	+	+	+	+	-	Ψ	Ψ	+
Liverwort	*Marchantia polymorpha*	121,024	+	+	+	+	+	+	+	+	Ψ
Moss	*Tetraphis pellucida*	127,489	-	-	-	-	-	+	+	+	-
Moss	*Physcomitrella patens*	122,890	+	-	-	-	-	+	+	+	-
Moss	*Tortula ruralis*	122,630	-	-	-	-	-	+	+	+	+
Moss	*Sanionia uncinata*	124,374	-	-	-	-	-	+	+	+	+
Lycopodiophyta	*Huperzia lucidula*	154,373	+	+	+	-	-	+	+	+	+
Moniliformopses	*Equisetum arvense*	133,309	+	+	+	-	-	+	+	+	+
Acrogymnospermae	*Pinus thunbergii*	116,635	+	+	+	-	-	-	+	+	+
Magnoliophyta	*Arabidopsis thaliana*	154,515	+	+	+	-	-	-	+	+	-

* The annotated plastomes are listed in Table S2. The presence '+' or absence '-' of each molecular character and pseudogenes are marked 'Ψ'.

2.5. Identification of Nuclear Genes Encoding rpoA and petN

Many cyanobacterial genes have been lost or transferred to the nuclear genomes during the transition from endosymbiont to organelle and the chloroplast genome might have been decreased in size [42,43]. Most gene loss occurred early in the endosymbiotic process; however, some losses have occurred in subsequent evolutionary processes [43]. The gene loss and transfer to the nucleus have been especially frequent, sporadic and temporary during the evolution of embryophytes and bryophytes [44]. Revealing why certain genes lose their functions is of great importance in the genome evolution of chloroplast, enabling the reconstruction of genomic events that occurred after the split of vascular plants and moss.

Therefore, to verify the chance of the nuclear relocation of the lost two chloroplast genes encoding the α subunit of RNA polymerase (*rpoA*) and cytochrome $b_{6/f}$ complex subunit VIII (*petN*), we used blast to search for putative homologues against the draft sequences of the *S. uncinata* nuclear genome (unpublished results), as a result, we found one target sequence for each gene and then verified the sequences of PCR amplified fragments (nucleotide sequences are listed in Table S3), respectively, which might function as the replaced gene for *rpoA* or *petN* in *S. uncinata*. To check whether those nuclear genes would have a conserved function with chloroplastic genes, we performed a comparative alignment of nuc-*rpoA* (nuclear-*rpoA)* and nuc-*petN* (nuclear-*petN*) of *S. uncinata* with the homologous corresponding genes of *P. patens*, *M. polymorpha*, *A. formosae* and *A. thaliana* (Figure 5). Particularly, for *P. patens*, the nuc-*rpoA* sequence (*Pp* nuc-*rpoA*) was used in the alignment [20]. The nuc-*rpoA* sequence showed 41% identity with the nuc-*rpoA* sequence of *P. patens*, 76% of *M. polymorpha*, 66% of *A. formosae* and 67% of *A. thaliana* of the other cp-*rpoA* sequence from other compared species. The nuclear *petN* sequence showed 59–79% identified with the other chloroplastic *petN* sequence from other compared species *P. patens* (79%), *A. formosae* (59%) and *A. thaliana* (76%), respectively. When the sequences were scanned in the signalP 4.1 program [45] (SignalP 4.1 server; http://www.cbs.dtu.dk/services/SignalP), we could identify the signal peptide which targets chloroplast at the 5' terminus of the nuc-*rpoA* gene of *S. uncinata* (Figure 5), suggesting that those genes would be localized in the chloroplast after being translated and functioned as other cp-*rpoA* or cp-*petN* genes in chloroplasts. This is very similar to the case of nuc-*rpoA* of *P. patens*, which has been shown to target the chloroplast organelle by the fluorescence-labelled protein method [22]. It would therefore be very interesting to prove that the proteins encoded by these nuclear genes actually constitute the chloroplastic-RNA polymerase complex or the Cyt $b_{6/f}$ complex in the moss protein expression system.

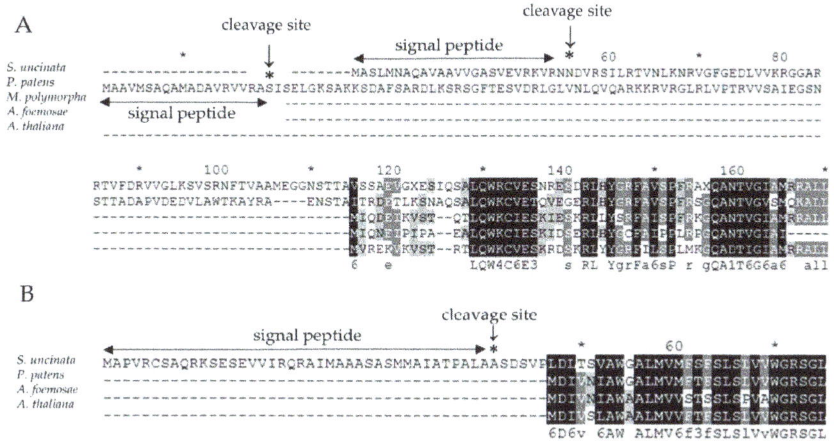

Figure 5. Amino acid alignment of (**A**) nuc-*rpoA* and (**B**) nuc-*petN* genes of *S. uncinata* with the nuc-*rpoA* or cp-*rpoA* and cp-*petN* genes from other green plants. Identical amino acid residues are boxed in black, other residues are printed in grey. Signal peptide sequences were predicted using SignalP [45] and shown as double arrow lines and the asterisk.

2.6. Prediction of RNA Editing Sites of Chloroplast Genes

Genetic information on DNA is sometimes altered at the transcript level by a process known as RNA editing, a sequence-specific post transcriptional modification resulting in the conversion, insertion, or deletion of nucleotides in a precursor RNA [46,47]. Such modifications are observed across organisms and have been reported with the discovery of C to U as well as U to C conversions in mitochondria and chloroplast in plants [48–50]. Since the first evidence of RNA editing in chloroplasts was found 24 years ago [51], RNA editing has been found in chloroplast transcripts from all major lineages of land plants [52]. In general, it is known that extensive RNA-editing occurs in hornworts and ferns when compared to seed plants [53]. The number of shared editing sites increases in closely related taxa, implying that RNA editing sites are evolutionarily conserved.

We predicted 72 editing sites in 22 protein coding genes by analyzing chloroplast coding sequences of *S. uncinata* using the PREP [54] and PREPACT 2.0-chloroplast program (http://www.prepact.de) [55] (Table S4). Most RNA editings are known to occur at the second codon position [56], with a frequency from 58.6% in hornwort to 68% in fern, 80% or 73.1% in *Cycad taitungensis* and *Pinus thunbergii*, respectively, 85.3%, 91.9% and 92% in *Arabidopsis thaliana*, *Nicotiana tabacum* and *Zea mays*, respectively, and up to 95.2% in *Oryza sativa* [57]. The RNA-editing of *S. uncinata* were mostly C-U editing events enriched at the second positions (64/72, 89%) as well, which allowed for the identification of the putative amino acid conversion by RNA editing (A-V, P-L, S-L, S-F, T-M and T-I).

3. Materials and Methods

3.1. Ethics Statement

Sample collection and field activities were carried out for scientific purposes in accordance with the Protocol on Environmental Protection to the Antarctic Treaty and approved by the Ministry of Foreign Affairs and Trade of the Republic of Korea (International Legal Affairs Division document No. 4029, approved on 18 December 2014). Sampling sites were not located within Antarctic Specially Protected Areas and no protected species were sampled in this study.

3.2. Plant Culture Conditions

Sanionia uncinata (Hedw.) Loeske plants growing under natural conditions were collected in the vicinity of the Korean King Sejong Antarctic Station (62°14′29″ S; 58°44′18″ W) on the Barton Peninsula of King George Island during the austral summer (mainly in January 2012) and then transferred to the lab and grown hydroponically in BCDAT solid media under a 16:8 h light:dark cycle at 15 °C.

3.3. Library Preparation and Sequencing

The total genomic DNA was extracted from tissues of plants using the DNeasy Plant Mini Kit (Qiagen). The quality of DNA was checked by Bioanalyzer 2100 (Aligent, Santa Clara, CA, USA). Library quantification was performed using the KAPA Library Quantification Kit (KAPA Biosystems, Boston, MA, USA). Paired-end cluster generation and sequencing were performed on the Illumina MiSeq, with each library being allocated to one lane of a flow cell. For the DNA library, TruSeq DNA sample preparation kits were used and sequenced in one lane of Illumina MiSeq 300 X 2 bp. The files containing the sequences and quality scores of reads were deposited in the NCBI Short Read Archive and the accession numbers are SRR6440975 genomic DNA-Seq (BioProject PRJNA428497).

3.4. Assembly and Annotation

After performing read preprocessing including adapter removal and quality filtering, the high-quality raw reads were aligned to 3 publicly available moss plastomes (*Physcomitrella patens* NC_005087, *Tortula ruralis* NC_012052 and *Tetraphis pellucida* NC_024291) downloaded from the NCBI organelle genome resources. The chloroplast reads were recovered from whole genome sequences by identifying them based on the alignment results and then de novo assembled using CLC Genomics Workbench V7.5 (CLC bio, Aarhus, Denmark). The assembled contigs were ordered based on their position in the reference plastome of *T. ruralis* because *T. ruralis* was identified as the top-hit species when BLAST searches were performed against the nr database. All gaps and junction regions between contigs and highly variable regions were validated by Sanger sequencing. The complete plastome was annotated using the program online software DOGMA [58] with default parameters, tRNAscan-SE [59] and BLAST similarity search tools from NCBI. OGDraw was used for map drawing [31]. The final *S. uncinata* plastome sequence has been submitted to GenBank (Accession: NC_025668).

3.5. Genome Alignment

The complete plastome sequences of 7 species were aligned using the mVISTA online suite [60]. A comparison of *Sanionia uncinata* (NC_025668) plastome structures with *Physcomiterella patens* (NC_005087), *Tortula ruralis* (NC_012052), *Tetraphis geniculate* (NC_024291), *Marchantia polymorpha* (NC_001319) and *Anthoceros formosae* (NC_004543), which are all in the bryophyte, was performed using the mVISTA program in Shuffle-LAGAN mode [60]. Default parameters were applied and the sequence annotation information of *S. uncinata* was used. Percentage identity between each plastome, all relative to that of *S. uncinata*, was subsequently visualized through an mVISTA plot. The mVISTA program only compared the sequence similarity by aligning the entire plastomes of all 6 taxa and the ~71 Kb inversion region of *P. patens* was reversed to match with other plastomes.

3.6. Phylogenetic Analysis

A set of 40 protein-coding genes, which have been analyzed in other species (accession numbers are listed in Table S2 and 40 protein-coding genes sequence of plastomes derived from 23 plants in Table S5), were used to infer the phylogenetic relationships among *S. uncinata*. Sequences were aligned using ClustalW. Phylogenetic analyses using maximum parsimony (MP) and maximum likelihood (ML) were performed, with MEGA7 [61]. For the MP analyses, the Subtree-Pruning-Regrafting algorithm was used with search level 1 in which the initial trees were obtained by the random addition of sequences (10 replicates). Sites with gaps or missing data were excluded from the analysis and statistical

Int. J. Mol. Sci. **2018**, *19*, 709

support was achieved through bootstrapping using 1000 replicates. For the ML analyses, the "Models" function of MEGA7 was used to find the best model for ML analysis. The Jones-Taylor-Thornton (JTT) + G + I + F model which was estimated to be the Best-Fit substitution model showing the lowest Bayesian Information Criterion, was employed in subsequent analyses. All positions containing gaps and missing data were eliminated. Bootstrap support was estimated with 1000 bootstrap replicates.

3.7. Prediction of RNA-Editing Sites

The RNA editing sites (C-to-U) in protein-coding genes were predicted by the online program Plant RNA Editing Prediction and Analysis Computer Tool (PREPACT 2.0) (http://www.prepact.de) [55] and Predictive RNA Editor for Plants (PREP) suite (http://prep.unl.edu/) [54] with a cutoff value of 0.8.

4. Conclusions

This study provided the whole sequence of the *S. uncinata* plastome, the dominant species of the Antarctic Peninsula, with information on the sequence and regulatory regions of chloroplast genes. We completed the *S. uncinata* plastome using a high-throughput sequencing method. The sequence and structure of *S. uncinata* were well conserved with the moss plastome sequences and was the most similar to the plastome sequence of *T. ruralis*. We confirmed that the two genes coding for *rpoA* and *petN* were lost in the *S. uncinata* plastome through comparative analysis of the plastome contents of the representative species of land plants and the sequence analysis results suggested the possibility of their relocation from plastome to nuclear genome. Our results also suggested that the possibility of post-transcriptional regulation of the chloroplast genes of *S. uncinata* by predicting the RNA-editing site. These results will contribute not only to the functional utilization of chloroplast genes but also to systematic phylogenetic analysis of land plants using whole plastome sequences.

Supplementary Materials: Supplementary materials can be found at www.mdpi.com/1422-0067/19/3/709/s1.

Acknowledgments: This work was supported by "Polar Genome 101 Project: Genome Analysis of Polar Organisms and Establishment of Application Platform (PE18080)" and "Modeling responses of terrestrial organisms to environmental changes on King George Island (PE18090)" funded by the Korea Polar Research Institute.

Author Contributions: Hyoungseok Lee, Hyun Park and Jungeun Lee conceived and designed the experiment; Mira Park performed the experiments; Mira Park, Hyoungseok Lee, Byeong-ha Lee and Jungeun Lee analyzed the data and wrote the paper. All authors read and approved the final version of the manuscript.

Conflicts of Interest: The authors declare no conflict of interest.

Abbreviations

LSC	Large single copy
SSC	Small single copy
IR	Inverted repeat
Cp	Chloroplast
nuc	Nucleus
MP	Maximum parsimony
ML	Maximum likelihood
A	Adenine
T	Thymine
G	Guanine
C	cytosine

References

1. Bramley-Alves, J.; King, D.H.; Robinson, S.A.; Miller, R.E. Dominating the Antarctic environment: Bryophytes in a time of change. In *Photosynthesis in Bryophytes and Early Land Plants*; Hanson, D., Rice, S., Eds.; Springer Netherlands: Dordrecht, The Netherlands, 2014; pp. 309–324.

2. Smith, R.L. Introduced plants in Antarctica: Potential impacts and conservation issues. *Biol. Conserv.* **1996**, *76*, 135–146. [CrossRef]

3. Victoria, F.D.C.; Pereira, A.B.; da Costa, D.P. Composition and distribution of moss formations in the ice-free areas adjoining the Arctowski region, admiralty bay, King George Island, Antarctica. *Iheringia Ser. Bot.* **2009**, *64*, 81–91.

4. Hedenäs, L. Global phylogeography in *Sanionia uncinata* (amblystegiaceae: Bryophyta). *Bot. J. Linn. Soc.* **2011**, *168*, 19–42. [CrossRef]

5. Nakatsubo, T. Predicting the impact of climatic warming on the carbon balance of the moss *Sanionia uncinata* on a maritime Antarctic island. *J. Plant Res.* **2002**, *115*, 99–106. [CrossRef] [PubMed]

6. Hokkanen, P.J. Environmental patterns and gradients in the vascular plants and bryophytes of eastern Fennoscandian herb-rich forests. *For. Ecol. Manag.* **2006**, *229*, 73–87. [CrossRef]

7. Kushnevskaya, H.; Mirin, D.; Shorohova, E. Patterns of epixylic vegetation on spruce logs in late-successional boreal forests. *For. Ecol. Manag.* **2007**, *250*, 25–33. [CrossRef]

8. Samecka-Cymerman, A.; Wojtuń, B.; Kolon, K.; Kempers, A. *Sanionia uncinata* (Hedw.) loeske as bioindicator of metal pollution in polar regions. *Polar Biol.* **2011**, *34*, 381–388. [CrossRef]

9. Lud, D.; Moerdijk, T.; Van de Poll, W.; Buma, A.; Huiskes, A. DNA damage and photosynthesis in Antarctic and Arctic *Sanionia uncinata* (hedw.) Loeske under ambient and enhanced levels of UV-B radiation. *Plant Cell Environ.* **2002**, *25*, 1579–1589. [CrossRef]

10. Lud, D.; Schlensog, M.; Schroeter, B.; Huiskes, A. The influence of UV-B radiation on light-dependent photosynthetic performance in *Sanionia uncinata* (Hedw.) Loeske in Antarctica. *Polar Biol.* **2003**, *26*, 225–232.

11. Zúñiga-González, P.; Zúñiga, G.E.; Pizarro, M.; Casanova-Katny, A. Soluble carbohydrate content variation in *Sanionia uncinata* and polytrichastrum alpinum, two Antarctic mosses with contrasting desiccation capacities. *Biol. Res.* **2016**, *49*, 6. [CrossRef] [PubMed]

12. Van Gaalen, K.E.; Flanagan, L.B.; Peddle, D.R. Photosynthesis, chlorophyll fluorescence and spectral reflectance in sphagnum moss at varying water contents. *Oecologia* **2007**, *153*, 19–28. [CrossRef] [PubMed]

13. Oliver, M.J.; Murdock, A.G.; Mishler, B.D.; Kuehl, J.V.; Boore, J.L.; Mandoli, D.F.; Everett, K.D.; Wolf, P.G.; Duffy, A.M.; Karol, K.G. Chloroplast genome sequence of the moss Tortula ruralis: Gene content, polymorphism and structural arrangement relative to other green plant chloroplast genomes. *BMC Genom.* **2010**, *11*, 143. [CrossRef] [PubMed]

14. Gray, M.W. The evolutionary origins of organelles. *Trends Genet.* **1989**, *5*, 294–299. [CrossRef]

15. Howe, C.J.; Barbrook, A.C.; Koumandou, V.L.; Nisbet, R.E.R.; Symington, H.A.; Wightman, T.F. Evolution of the chloroplast genome. *Philos. Trans. R. Soc. Lond. B Biol. Sci.* **2003**, *358*, 99–107. [CrossRef] [PubMed]

16. Sugiura, M. The chloroplast chromosomes in land plants. *Annu. Rev. Cell Biol.* **1989**, *5*, 51–70. [CrossRef] [PubMed]

17. Neuhaus, H.; Emes, M. Nonphotosynthetic metabolism in plastids. *Annu. Rev. Plant Biol.* **2000**, *51*, 111–140. [CrossRef] [PubMed]

18. Daniell, H.; Lin, C.-S.; Yu, M.; Chang, W.-J. Chloroplast genomes: Diversity, evolution and applications in genetic engineering. *Genome Biol.* **2016**, *17*, 134. [CrossRef] [PubMed]

19. Chumley, T.W.; Palmer, J.D.; Mower, J.P.; Fourcade, H.M.; Calie, P.J.; Boore, J.L.; Jansen, R.K. The complete chloroplast genome sequence of pelargonium × hortorum: Organization and evolution of the largest and most highly rearranged chloroplast genome of land plants. *Mol. Biol. Evol.* **2006**, *23*, 2175–2190. [CrossRef] [PubMed]

20. Karol, K.G.; Arumuganathan, K.; Boore, J.L.; Duffy, A.M.; Everett, K.D.; Hall, J.D.; Hansen, S.K.; Kuehl, J.V.; Mandoli, D.F.; Mishler, B.D. Complete plastome sequences of *Equisetum arvense* and *isoetes flaccida*: Implications for phylogeny and plastid genome evolution of early land plant lineages. *BMC Evol. Biol.* **2010**, *10*, 321. [CrossRef] [PubMed]

21. Coordinators, N.R. Database resources of the National Center for Biotechnology Information. *Nucleic Acids Res.* **2016**, *44*, D7–D19.

22. Sugiura, C.; Kobayashi, Y.; Aoki, S.; Sugita, C.; Sugita, M. Complete chloroplast DNA sequence of the moss Physcomitrella patens: Evidence for the loss and relocation of rpoA from the chloroplast to the nucleus. *Nucleic Acids Res.* **2003**, *31*, 5324–5331. [CrossRef] [PubMed]

23. Bell, N.E.; Boore, J.L.; Mishler, B.D.; Hyvönen, J. Organellar genomes of the four-toothed moss, *Tetraphis pellucida*. *BMC Genom.* **2014**, *15*, 383. [CrossRef] [PubMed]

24. Lewis, L.R.; Liu, Y.; Rozzi, R.; Goffinet, B. Infraspecific variation within and across complete organellar genomes and nuclear ribosomal repeats in a moss. *Mol. Phylogenet. Evol.* **2016**, *96*, 195–199. [CrossRef] [PubMed]

25. Kugita, M.; Kaneko, A.; Yamamoto, Y.; Takeya, Y.; Matsumoto, T.; Yoshinaga, K. The complete nucleotide sequence of the hornwort (*Anthoceros formosae*) chloroplast genome: Insight into the earliest land plants. *Nucleic Acids Res.* **2003**, *31*, 716–721. [CrossRef] [PubMed]

26. Ohyama, K.; Fukuzawa, H.; Kohchi, T.; Shirai, H.; Sano, T.; Sano, S.; Umesono, K.; Shiki, Y.; Takeuchi, M.; Chang, Z. Chloroplast gene organization deduced from complete sequence of liverwort Marchantia polymorpha chloroplast DNA. *Nature* **1986**, *322*, 572–574. [CrossRef]

27. Wakasugi, T.; Tsudzuki, T.; Sugiura, M. The genomics of land plant chloroplasts: Gene content and alteration of genomic information by RNA editing. *Photosynth. Res.* **2001**, *70*, 107–118. [CrossRef] [PubMed]

28. Lee, J.; Kang, Y.; Shin, S.C.; Park, H.; Lee, H. Combined analysis of the chloroplast genome and transcriptome of the Antarctic vascular plant Deschampsia antarctica Desv. *PLoS ONE* **2014**, *9*, e92501. [CrossRef] [PubMed]

29. Maul, J.E.; Lilly, J.W.; Cui, L.; Miller, W.; Harris, E.H.; Stern, D.B. The Chlamydomonas reinhardtii plastid chromosome islands of genes in a sea of repeats. *Plant Cell* **2002**, *14*, 2659–2679. [CrossRef] [PubMed]

30. Turmel, M.; Pombert, J.-F.; Charlebois, P.; Otis, C.; Lemieux, C. The green algal ancestry of land plants as revealed by the chloroplast genome. *Int. J. Plant Sci.* **2007**, *168*, 679–689. [CrossRef]

31. Lohse, M.; Drechsel, O.; Kahlau, S.; Bock, R. OrganellarGenomeDRAW—A suite of tools for generating physical maps of plastid and mitochondrial genomes and visualizing expression data sets. *Nucleic Acids Res.* **2013**, *41*, W575–W581. [CrossRef] [PubMed]

32. Turmel, M.; Otis, C.; Lemieux, C. The chloroplast and mitochondrial genome sequences of the charophyte Chaetosphaeridium globosum: Insights into the timing of the events that restructured organelle DNAs within the green algal lineage that led to land plants. *Proc. Natl. Acad. Sci. USA* **2002**, *99*, 11275–11280. [CrossRef] [PubMed]

33. Jansen, R.K.; Ruhlman, T.A. Plastid genomes of seed plants. In *Genomics of Chloroplasts and Mitochondria*; Bock, R., Knoop, V., Eds.; Springer Netherlands: Dordrecht, The Netherlands, 2012; pp. 103–126.

34. Barthet, M.M.; Hilu, K.W. Expression of matK: Functional and evolutionary implications. *Am. J. Bot.* **2007**, *94*, 1402–1412. [CrossRef] [PubMed]

35. Quax, T.E.; Claassens, N.J.; Söll, D.; van der Oost, J. Codon bias as a means to fine-tune gene expression. *Mol. Cell* **2015**, *59*, 149–161. [CrossRef] [PubMed]

36. Hausner, G.; Olson, R.; Simon, D.; Johnson, I.; Sanders, E.R.; Karol, K.G.; McCourt, R.M.; Zimmerly, S. Origin and evolution of the chloroplast trnK (matK) intron: A model for evolution of group II intron RNA structures. *Mol. Biol. Evol.* **2005**, *23*, 380–391. [CrossRef] [PubMed]

37. Gao, L.; Su, Y.J.; Wang, T. Plastid genome sequencing, comparative genomics and phylogenomics: Current status and prospects. *J. Syst. Evol.* **2010**, *48*, 77–93. [CrossRef]

38. Shaw, J.; Lickey, E.B.; Schilling, E.E.; Small, R.L. Comparison of whole chloroplast genome sequences to choose noncoding regions for phylogenetic studies in angiosperms: The tortoise and the hare III. *Am. J. Bot.* **2007**, *94*, 275–288. [CrossRef] [PubMed]

39. Cox, C.J.; Goffinet, B.; Wickett, N.J.; Boles, S.B.; Shaw, A.J. Moss diversity: A molecular phylogenetic analysis of genera. *Phytotaxa* **2014**, *9*, 175–195. [CrossRef]

40. Goffinet, B.; Wickett, N.J.; Werner, O.; Ros, R.M.; Shaw, A.J.; Cox, C.J. Distribution and phylogenetic significance of the 71-kb inversion in the plastid genome in Funariidae (Bryophyta). *Am. J. Bot.* **2007**, *99*, 747–753. [CrossRef] [PubMed]

41. Wicke, S.; Schneeweiss, G.M.; Müller, K.F.; Quandt, D. The evolution of the plastid chromosome in land plants: Gene content, gene order, gene function. *Plant Mol. Biol.* **2011**, *76*, 273–297. [CrossRef] [PubMed]

42. Martin, W.; Rujan, T.; Richly, E.; Hansen, A.; Cornelsen, S.; Lins, T.; Leister, D.; Stoebe, B.; Hasegawa, M.; Penny, D. Evolutionary analysis of *Arabidopsis*, cyanobacterial and chloroplast genomes reveals plastid

phylogeny and thousands of cyanobacterial genes in the nucleus. *Proc. Natl. Acad. Sci. USA* **2002**, *99*, 12246–12251. [CrossRef] [PubMed]

43. Martin, W.; Stoebe, B.; Goremykin, V.; Hansmann, S.; Hasegawa, M.; Kowallik, K.V. Gene transfer to the nucleus and the evolution of chloroplasts. *Nature* **1998**, *393*, 162–165. [CrossRef] [PubMed]

44. Rensing, S.A.; Lang, D.; Zimmer, A.D.; Terry, A.; Salamov, A.; Shapiro, H.; Nishiyama, T.; Perroud, P.-F.; Lindquist, E.A.; Kamisugi, Y. The Physcomitrella genome reveals evolutionary insights into the conquest of land by plants. *Science* **2008**, *319*, 64–69. [CrossRef] [PubMed]

45. Petersen, T.N.; Brunak, S.; von Heijne, G.; Nielsen, H. SignalP 4.0: Discriminating signal peptides from transmembrane regions. *Nat. Methods* **2011**, *8*, 785–786. [CrossRef] [PubMed]

46. Wakasugi, T.; Hirose, T.; Horihata, M.; Tsudzuki, T.; Kössel, H.; Sugiura, M. Creation of a novel protein-coding region at the RNA level in black pine chloroplasts: The pattern of RNA editing in the gymnosperm chloroplast is different from that in angiosperms. *Proc. Natl. Acad. Sci. USA* **1996**, *93*, 8766–8770. [CrossRef] [PubMed]

47. Zandueta-Criado, A.; Bock, R. Surprising features of plastid *ndhD* transcripts: Addition of non-encoded nucleotides and polysome association of mRNAs with an unedited start codon. *Nucleic Acids Res.* **2004**, *32*, 542–550. [CrossRef] [PubMed]

48. Covello, P.S.; Gray, M.W. RNA editing in plant mitochondria. *Nature* **1989**, *341*, 662–666. [CrossRef] [PubMed]

49. Gualberto, J.M.; Lamattina, L.; Bonnard, G.; Weil, J.-H.; Grienenberger, J.-M. RNA editing in wheat mitochondria results in the conservation of protein sequences. *Nature* **1989**, *341*, 660–662. [CrossRef] [PubMed]

50. Hiesel, R.; Wissinger, B.; Schuster, W.; Brennicke, A. RNA editing in plant mitochondria. *Science* **1989**, *246*, 1632–1634. [CrossRef] [PubMed]

51. Hoch, B.; Maier, R.M.; Appel, K.; Igloi, G.L.; Kössel, H. Editing of a chloroplast mRNA by creation of an initiation codon. *Nature* **1991**, *353*, 178–180. [CrossRef] [PubMed]

52. Freyer, R.; Kiefer-Meyer, M.-C.; Kössel, H. Occurrence of plastid RNA editing in all major lineages of land plants. *Proc. Natl. Acad. Sci. USA* **1997**, *94*, 6285–6290. [CrossRef] [PubMed]

53. Stern, D.B.; Goldschmidt-Clermont, M.; Hanson, M.R. Chloroplast RNA metabolism. *Annu. Rev. Plant Biol.* **2010**, *61*, 125–155. [CrossRef] [PubMed]

54. Mower, J.P. The prep suite: Predictive RNA editors for plant mitochondrial genes, chloroplast genes and user-defined alignments. *Nucleic Acids Res.* **2009**, *37*, W253–W259. [CrossRef] [PubMed]

55. Lenz, H.; Knoop, V. PREPACT 2.0: Predicting C-to-U and U-to-C RNA editing in organelle genome sequences with multiple references and curated RNA editing annotation. *Bioinform. Biol. Insights* **2013**, *7*, 1–19. [CrossRef] [PubMed]

56. Bock, R. Sense from nonsense: How the genetic information of chloroplastsis altered by RNA editing. *Biochimie* **2000**, *82*, 549–557. [CrossRef]

57. Chen, H.; Deng, L.; Jiang, Y.; Lu, P.; Yu, J. RNA editing sites exist in protein-coding genes in the chloroplast genome of Cycas taitungensis. *J. Integr. Plant Biol.* **2011**, *53*, 961–970. [CrossRef] [PubMed]

58. Wyman, S.K.; Jansen, R.K.; Boore, J.L. Automatic annotation of organellar genomes with DOGMA. *Bioinformatics* **2004**, *20*, 3252–3255. [CrossRef] [PubMed]

59. Schattner, P.; Brooks, A.N.; Lowe, T.M. The tRNAscan-SE, snoscan and snoGPS web servers for the detection of tRNAs and snoRNAs. *Nucleic Acids Res.* **2005**, *33*, W686–W689. [CrossRef] [PubMed]

60. Frazer, K.A.; Pachter, L.; Poliakov, A.; Rubin, E.M.; Dubchak, I. Vista: Computational tools for comparative genomics. *Nucleic Acids Res.* **2004**, *32*, W273–W279. [CrossRef] [PubMed]

61. Kumar, S.; Stecher, G.; Tamura, K. MEGA7: Molecular evolutionary genetics analysis version 7.0 for bigger datasets. *Mol. Biol. Evol.* **2016**, *33*, 1870–1874. [CrossRef] [PubMed]

International Journal of
Molecular Sciences

MDPI

Article

The Complete Chloroplast Genome Sequence of Tree of Heaven (*Ailanthus altissima* (Mill.) (Sapindales: Simaroubaceae), an Important Pantropical Tree

Josphat K. Saina [1,2,3,4], Zhi-Zhong Li [2,3], Andrew W. Gichira [2,3,4] and Yi-Ying Liao [1,*]

[1] Fairy Lake Botanical Garden, Shenzhen & Chinese Academy of Sciences, Shenzhen 518004, China; jksaina@wbgcas.cn
[2] Key Laboratory of Aquatic Botany and Watershed Ecology, Wuhan Botanical Garden, Chinese Academy of Sciences, Wuhan 430074, China; wbg_georgelee@163.com (Z.-Z.L.); gichira@wbgcas.cn (A.W.G.)
[3] University of Chinese Academy of Sciences, Beijing 100049, China
[4] Sino-African Joint Research Center, Chinese Academy of Sciences, Wuhan 430074, China
* Correspondence: liaoyiying666@163.com; Tel.: +86-150-1949-8243

Received: 31 January 2018; Accepted: 16 March 2018; Published: 21 March 2018

Abstract: *Ailanthus altissima* (Mill.) Swingle (Simaroubaceae) is a deciduous tree widely distributed throughout temperate regions in China, hence suitable for genetic diversity and evolutionary studies. Previous studies in *A. altissima* have mainly focused on its biological activities, genetic diversity and genetic structure. However, until now there is no published report regarding genome of this plant species or Simaroubaceae family. Therefore, in this paper, we first characterized *A. altissima* complete chloroplast genome sequence. The tree of heaven chloroplast genome was found to be a circular molecule 160,815 base pairs (bp) in size and possess a quadripartite structure. The *A. altissima* chloroplast genome contains 113 unique genes of which 79 and 30 are protein coding and transfer RNA (tRNA) genes respectively and also 4 ribosomal RNA genes (rRNA) with overall GC content of 37.6%. Microsatellite marker detection identified A/T mononucleotides as majority SSRs in all the seven analyzed genomes. Repeat analyses of seven Sapindales revealed a total of 49 repeats in *A. altissima, Rhus chinensis, Dodonaea viscosa, Leitneria floridana*, while *Azadirachta indica, Boswellia sacra*, and *Citrus aurantiifolia* had a total of 48 repeats. The phylogenetic analysis using protein coding genes revealed that *A. altissima* is a sister to *Leitneria floridana* and also suggested that Simaroubaceae is a sister to Rutaceae family. The genome information reported here could be further applied for evolution and invasion, population genetics, and molecular studies in this plant species and family.

Keywords: *Ailanthus altissima*; chloroplast genome; microsatellites; Simaroubaceae; Sapindales

1. Introduction

Ailanthus altissima (Mill.) Swingle, a deciduous tree in the Simaroubaceae family, is widely distributed throughout temperate regions in China. It grows rapidly reaching heights of 15 m (49ft) in 25 years and can tolerate various levels of extreme environments (e.g., low temperatures, sterile soils, arid land). Besides, it reproduces through sexual (seeds disperse by wind) or asexual (sprouts) methods [1]. Two hundred years ago it was brought to Europe and North America. *A. altissima* being an early colonizer can survive high levels of natural or human disturbance [2]. Therefore, in recent years, it is commonly known as an exotic invasive tree developed into an invasive species expanding on all continents except Antarctica [1]. While previous studies in *A. altissima* have mainly focused on discovering the biological features of this plant to prevent its expansion, there is limited information to understand the impact of genetic diversity and evolution. Thus, genomic information of *A. altissima* is essential for further molecular studies, identification, and evolutionary studies.

Many studies have analyzed the genetic diversity of *A. altissima* using various markers, for example, chloroplast DNA [2,3], microsatellite primer [4,5]. These studies provided a detailed series of information about genetic structure and genetic diversity of *A. altissima* in native and invasive area. However, to understand the genetic diversity and population structure within *A. altissima* natural populations, more genetic resources are required.

It is well known that chloroplasts (cp) are key organelles in plants, with crucial functions in the photosynthesis and biosynthesis [6]. Current research shows that chloroplast genomes in angiosperms have highly conserved structure, gene content, organization, compared with either nuclear or mitochondrial genomes [7,8]. In general, cp genomes in angiosperms have circular structure consisting of two inverted repeat regions (IRa and IRb) that divides a large–single-copy (LSC) and a small-single-copy (SSC) regions [9]. Nevertheless, mutations, duplications, arrangements and gene loss have been observed, including the loss of the inverted repeat region in leguminous plants [7,10–12]. Some studies have applied plant plastomes to study population genetic analyses and basal phylogenetic relationships at different taxonomic levels [13], also to investigate the functional and structural evolution in plants [14–16]. At present, more cp genomes have been sequenced as a result of next-generation sequencing technologies advancement resulting in low sequencing costs.

More than 800 sequenced plastomes from various land plants have boosted our understanding of intracellular gene transfer, conservation, diversity, and genetic basis [17]. Although cp genomes have been sequenced in many trees such as *Castanea mollissima* [18]), *Liriodendron tulipifera* [19], *Eucalyptus globules* [20], and *Larix deciduas* [21], the plastome of *Leitneria floridana* (GenBank NC_030482) a member of Simaroubaceae has been sequenced but no analysis has been published at present despite the family containing many high economic value trees. Regardless of its potential use in crop or tree species improvement, studies on invasive species such as *A. altissima* which is also an important economic tree in the North China are too few. Chloroplast genome sequencing in invasive species could bring insights into evolutionary aspects in stress-tolerance related trait and genetic variation.

Simple sequence repeat (SSR) also called microsatellite markers are known to be more informative and versatile DNA-based markers used in plant genetic research [22]. These DNA markers are reliable molecular tools that can be used to examine plants genetic variation. SSR loci are evenly distributed and very abundant in angiosperms plastomes [23,24]. Chloroplast microsatellites are typically mononucleotide tandem repeats, and SSR in the fully sequenced genome could be used in plant species identification and diversity analysis. CpSSR in the fully sequenced plants plastomes such as; orchid genus *Chiloglottis* [25], *Cryptomeria japonica* [26], *Podocarpus lambertii* [27], *Actinidia chinensis* [28], have proven to be useful genetic tools in determining gene flow and population genetics of cp genomes. However, the lack of published plastome of *A. altissima* has limited the development of suitable SSR markers.

Here, we sequenced the complete chloroplast genome of *A. altissima*, and characterized its organization and structure. Furthermore, phylogenetic relationship using protein coding genes from selected species, consisting of 31 species from five families was uncovered for the Simaroubaceae family within the order Sapindales. Lastly, this resource will be used to develop SSR markers for analyzing genetic diversity and structure of several wild populations of *A. altissima*.

2. Results and Discussion

2.1. Ailanthus altissima Genome Size and Features

The *A. altissima* chloroplast genome has a quadripartite organizational structure with overall size of 160,815 base pairs (bp) including two copies of Inverted repeats (IRa and IRb) (27,051 bp each) separating the Large Single Copy (LSC) (88,382 bp) and Small Single Copy Region (SSC) (18,332 bp) (Figure 1). Notably, the genome content; gene order, orientation and organization of *A. altissima* is similar to the reference genome and other sequenced Sapindales plastomes [29,30] with genome size of about 160 kb. The overall guanine-cytosine (GC) content of the whole genome is 37.6%, while the average adenine-thymine A + T content is 62.36%. The relatively higher IR GC content and A + T bias in this chloroplast have been previously reported in genomes of relative species in order Sapindales [31]. The GC content of the LSC, SSC and IR regions are 35.7, 32.2 and 42.6% respectively. Moreover, the protein coding sequences had 38.3% GC content.

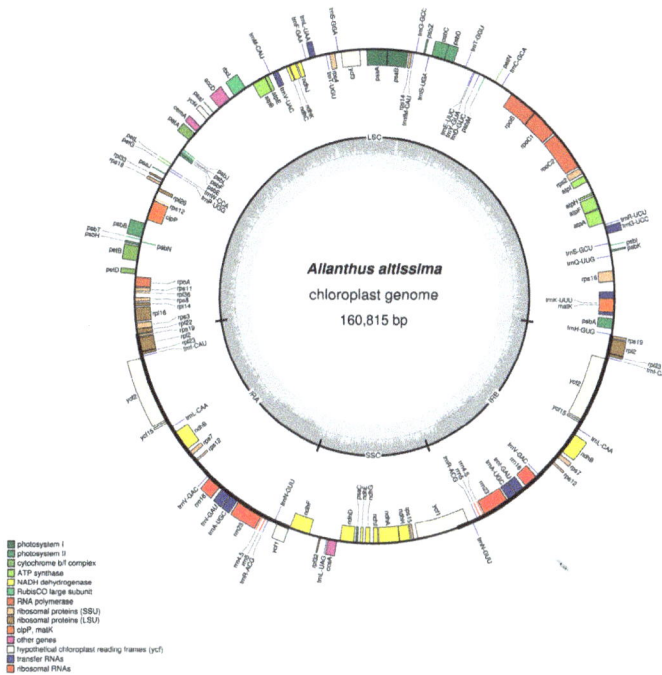

Figure 1. Circular gene map of *A. altissima* complete chloroplast genome. Genes drawn on the outside of the circle are transcribed clockwise, whereas those inside are transcribed clockwise. The light gray in the inner circle corresponds to AT content, while the darker gray corresponds to the GC content. Large single copy (LSC), Inverted repeats (IRa and IRb), and Small single copy (SSC) are indicated.

The tree of heaven (*A. altissima*) chloroplast genome harbored a total of 113 different genes, comprising 79 protein coding genes (PCGS), 30 transfer RNA (tRNA) genes, and four ribosomal RNA (rRNA) genes (Table 1). All the 77 PCGS started with the common initiation codon (ATG), but *rps19* and *ndhD* genes started with alternative codons GTG and ACG respectively, this unusual initiator codons have been observed to be common in other angiosperm chloroplast genomes [32–34]. Of the 79 protein coding sequences, 60 are located in the LSC, 11 in the SSC and eight genes were duplicated the IR region, while 22 tRNA genes were found in LSC, seven replicated in the IR region and one located in the SSC region.

Similar to some closely related plant species in order Sapindales, the chloroplast genome of *A. altissima* has maintained intron content [35]. Among the 113 unique genes, the *rps16, rpoC1, petB, rpl2, ycf15, ndhB, ndhA, atpF*, six tRNA genes (*trnG-GCC*, trnA-UGC*, trnL-UAA*, trnI-GAU*, trnK-UUU*, trnV-UAC**) possess one intron, and *ycf3* and *clpP* genes harbored two introns. The *rps12* trans-splicing gene has two 3′ end exons repeated in the IRs and the 5′ end exon located in the LSC region, which is similar to that in *C. platymamma* [30], *C. aurantiifolia* [29], *Dipteronia* species [35]. The *ycf1* gene crossed the IR/SSC junction forming a pseudogene *ycf1* on the corresponding IR region. The *rps19* gene in *A. altissima* was completely duplicated in the inverted repeat (IR) region, which most other chloroplast genomes have presented [29,30].

Table 1. List of genes found in *Ailanthus altissima* Chloroplast genome.

Functional Category	Group of Genes	Gene Name	Number
Self-replication	rRNA genes	*rrn16*(×2), *rrn23*(×2), *rrn4.5*(×2), *rrn5*(×2),	4
	tRNA genes	*trnA-UGC**(×2), *trnC-GCA*, *trnD-GUC*, *trnE-UUC*, *trnF-GAA*, *trnG-UCC*, *trnH-GUG*, *trnI-CAU*(×2), *trnI-GAU**(×2), *trnK-UUU**, *trnL-CAA*(×2), *trnL-UAA**, *trnL-UAG*, *trnG-GCC**, *trnM-CAU*, *trnN-GUU*(×2), *trnP-GGG*, *trnP-UGG*, *trnQ-UUG*, *trnR-ACG*(×2), *trnR-UCU*, *trnS-GCU*, *trnS-GGA*, *trnS-UGA*, *trnT-GGU*, *trnT-UGU*, *trnV-GAC*(×2), *trnV-UAC**, *trnW-CCA*, *trnY-GUA*	30
	Ribosomal small subunit	*rps2*, *rps3*, *rps4*, *rps7*(×2), *rps8*, *rps11*, *rps12*, *rps14*, *rps15*, *rps16**, *rps18*, *rps19*	12
	Ribosomal large subunit	*rpl2**(×2), *rpl14*, *rpl16*, *rpl20*, *rpl22*, *rpl23*(×2), *rpl32*, *rpl33*, *rpl36*	9
	DNA-dependent RNA polymerase	*rpoA*, *rpoB*, *rpoC1**, *rpoC2*	4
Photosynthesis	Large subunit of rubisco	*rbcL*	1
	Photosystem I	*psaA*, *psaB*, *psaC*, *psaI*, *psaJ*, *ycf3***	6
	Photosystem II	*psbA*, *psbB*, *psbC*, *psbD*, *psbE*, *psbF*, *psbH*, *psbI*, *psbJ*, *psbK*, *psbL*, *psbM*, *psbN*, *psbT*, *psbZ*	15
	NADH dehydrogenase	*ndhA**, *ndhB**(×2), *ndhC*, *ndhD*, *ndhE*, *ndhF*, *ndhG*, *ndhH*, *ndhI*, *ndhJ*, *ndhK*	11
	Cytochrome b/f complex	*petA*, *petB**, *petD*, *petG*, *petL*, *petN*	6
	ATP synthase	*atpA*, *atpB*, *atpE*, *atpF**, *atpH*, *atpI*	6
Other	Maturase	*matK*	1
	Subunit of acetyl-CoA carboxylase	*accD*	1
	Envelope membrane protein	*cemA*	1
	Protease	*clpP***	1
	c-type cytochrome synthesis	*ccsA*	1
Functions unknown	Conserved open reading frames (*ycf*)	*ycf1*, *ycf2*(×2), *ycf4*, *ycf15*(×2)	4
Total			113

Note: * Gene with one intron, ** Genes with two introns. (×2) Genes with two copies.

2.2. IR Expansion and Contraction and Genome Rearrangement

The angiosperms chloroplast genomes are highly conserved, but slightly vary as a result of either expansion or contraction of the single-copy (SC) and IR boundary regions [36]. The expansion and contraction of the IR causes size variations and rearrangements in the SSC/IRa/IRb/LSC junctions [37]. Therefore, in this study, exact IR boundary positions and their adjacent genes of seven representative species from different families in order Sapindales were compared (Figure 2). The functional *ycf1* gene crossed the IRa/SSC boundary creating *ycf1* pseudogene fragment at the IRb region in all the genomes. Besides, *ycf1* pseudogene overlapped with the *ndhF* gene in the SSC and IRa junctions in four genomes with a stretch of 9 to 85 bp, but *ndhF* gene is located in SSC region in *L. floridana*, *R. chinensis* and *A. altissima*.

The *rpl22* gene crossed the LSC/IRb junction in all the chloroplast genomes except in *R. chinensis*. Furthermore, this gene was partially duplicated forming a pseudogene fragment at the corresponding IRA/LSC junction in *L. floridana* and *B. sacra*, but completely duplicated in *D. viscosa*. In all the seven chloroplast genomes, the *trnH-GUG* gene was located in the LSC regions, however this gene overlapped with *rpl22* gene in *D. viscosa*. The results reported here are congruent with the recent studies which showed that the *trnH-GUG* gene was situated in the LSC region in some species from order Sapindales, while the SSC/IRa border extends into the protein coding gene *ycf1* with subsequent formation of a *ycf1* pseudogene [29,30]. Despite the seven chloroplast genomes of Sapindales having well-conserved genomic structure in terms of gene order and number, length variation of the whole chloroplast genome sequences and LSC, SSC and IR regions was detected among these genomes. This sequence variation might have been as a result of boundary expansion and contraction between the single copy and IR boundary regions among plant lineages as suggested by Wang and Messing 2011 [38].

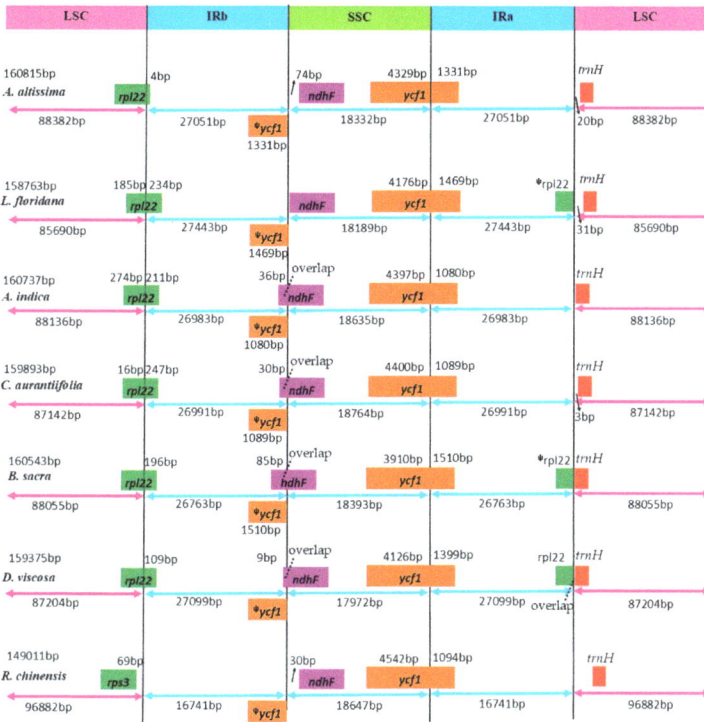

Figure 2. Comparison of IR, LSC and SSC junction positions among seven Chloroplast genomes. The features drawn are not to scale. The symbol φ means pseudogene created by IRb/SSC border extension into *ycf1* genes. Colored boxes for genes represent the gene position.

The mauve alignment for seven species revealed that all the genomes formed collinear blocks (LCBs). In particular, all the seven species; *A. altissima*, *Leitneria floridana*, *Azadirachta indica*, *Citrus aurantiifolia*, *Boswellia Sacra*, *Spondias bahiensis* and *Dodonaea viscosa* reveal a syntenic structure, however block two was inverted (from *rpl20* to *rbcL* genes) compared to the reference genome (*Aquilaria sinensis*). The collinear blocks of the genes including ribosomal RNA, tRNA, and protein coding genes revealed that all the seven genomes were relatively conserved with no gene rearrangement (Figure 3). Some other studies have revealed homology in genome organization and no gene rearrangements, thus our findings support their conclusions [31,39,40].

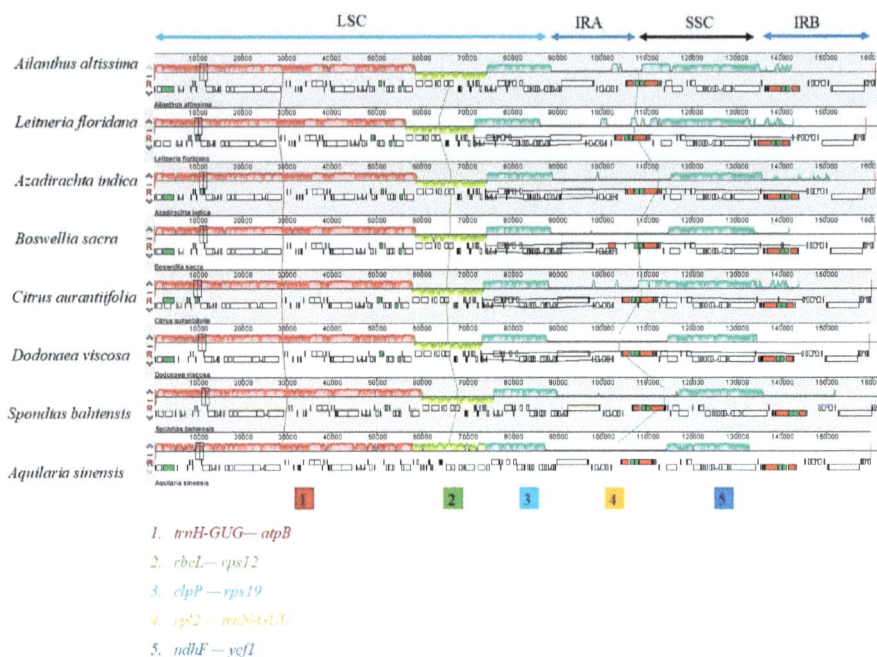

Figure 3. Gene arrangement map of seven chloroplast genomes representing families from Sapindales, and one reference species (*Aquilaria sinensis*) aligned using Mauve software Local collinear blocks within each alignment are represented in as blocks of similar color connected with lines. Annotations of rRNA, protein coding and tRNA genes are shown in red, white and green boxes respectively.

2.3. Codon Usage and Putative RNA Editing Sites in Chloroplast Genes of A. altissima

In this study, we analyzed codon usage frequency and the relative synonymous codon usage (RSCU) in the *A. altissima* plastome. All the protein coding genes presented a total of 68,952 bp and 22,964 codons in *A. altissima* chloroplast genome. Of 22,964 codons, leucine (Leu) being the most abundant amino acid had a frequency of 10.56%, then isoleucine (Ile) with 8.54%, while cysteine (Cys) was rare with a proportion of 1.12% (Tables S1 and S2, Figure 4). Our study species genome is like other previously reported genomes which showed that leucine and isoleucine are more common [41–45]. Furthermore, comparable to other angiosperm chloroplast genomes, our results followed the trend of codon preference towards A/T ending which was observed in plastomes of two *Aristolochia* species [46], *Scutellaria baicalensis* [47], *Decaisnea insignis* [34], *Papaver rhoeas* and *Papaver orientale* [48] *Cinnamomum camphora* [49], and *Forsythia suspensa* [41]. All the twenty-eight A/U—ending codons had RSCU values of more than one (RSCU > 1), whereas the C/G—ending codons had RSCU values of less than one. Two amino acids, Methionine (Met) and tryptophan (Trp) showed no codon bias. The results for number of codons (Nc) of each protein coding gene ranged from 38.94 (*rps14* gene) to 58.37 (*clpP* gene).

The potential RNA editing sites in tree of heaven chloroplast genome was done using PREP program which revealed that most conversions at the codon positions change from serine (S) to leucine (L) (Table 2). In addition, 15 (27.78%), 39 (72.22%), and 0 editing locations were used in the first, second and third codons respectively. One RNA editing site converted the amino acid from apolar to polar (proline (P) to serine (S)). Overall, the PREP program identified a total of 54 editing sites in 21 protein coding genes, with *ndhB* and *ndhD* genes predicted to have the highest number of editing sites (9). Followed by *ndhA*, *matK*, *rpoC2*, and *rpoB* with four editing sites, whereas *ndhF* had three sites. Interestingly, fifty three of fifty four RNA editing conversions in the *A. altissima* chloroplast genome resulted into hydrophobic products comprising; isoleucine, leucine, tryptophan, tyrosine valine, methionine, and phenylalanine. In general our results are congruent with the preceding reports which also found that most RNA editing sites led to amino acid change from polar to apolar, resulting in increase in protein hydrophobicity [41,46,50].

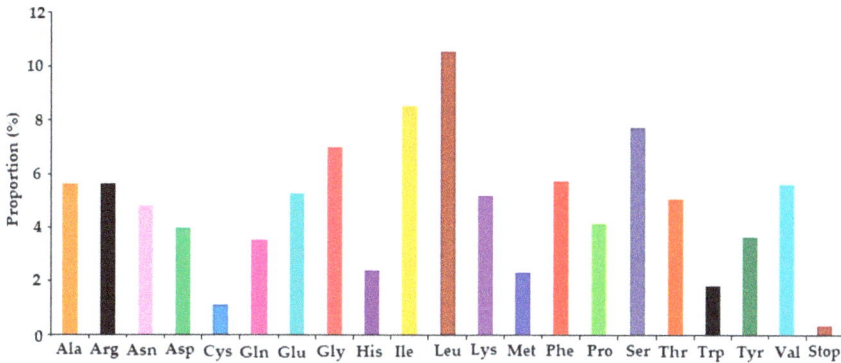

Figure 4. Amino acid frequencies in *A. altissima* chloroplast genome protein coding sequences.

Table 2. Predicted RNA editing site in the *A. altissima* chloroplast genome.

Gene	Nucleotide Position	Amino Acid Position	Codon Conversion	Amino Acid Conversion	Score
accD	818	273	TCG ≥ TTG	S ≥ L	0.80
atpF	92	31	CCA ≥ CTA	P ≥ L	0.86
	353	118	TCA ≥ TTA	S ≥ L	1.00
atpB	403	135	CCA ≥ TCA	P ≥ S	0.86
rps14	80	27	TCA ≥ TTA	S ≥ L	1.00
	149	50	TCA ≥ TTA	S ≥ L	1.00
ccsA	145	49	CTT ≥ TTT	L ≥ F	1.00
clpP	556	186	CAT ≥ TAT	H ≥ Y	1.00
MatK	319	107	CTT ≥ TTT	L ≥ F	0.86
	457	153	CAC ≥ TAC	H ≥ Y	1.00
	643	215	CAT ≥ TAT	H ≥ Y	1.00
	1246	416	CAC ≥ TAC	H ≥ Y	1.00
ndhA	107	36	CCT ≥ CTT	P ≥ L	1.00
	341	114	TCA ≥ TTA	S ≥ L	1.00
	566	189	TCA ≥ TTA	S ≥ L	1.00
	1073	358	TCC ≥ TTC	S ≥ F	1.00
ndhB	149	50	TCA ≥ TTA	S ≥ L	1.00
	467	156	CCA ≥ CTA	P ≥ L	1.00
	586	196	CAT ≥ TAT	H ≥ Y	1.00
	611	204	TCA ≥ TTA	S ≥ L	0.80
	746	249	TCT ≥ TTT	S ≥ F	1.00
	830	277	TCA ≥ TTA	S ≥ L	1.00
	836	279	TCA ≥ TTA	S ≥ L	1.00
	1255	419	CAT ≥ TAT	H ≥ Y	1.00
	1481	494	CCA ≥ CTA	P ≥ L	1.00

Table 2. *Cont.*

Gene	Nucleotide Position	Amino Acid Position	Codon Conversion	Amino Acid Conversion	Score
	2	1	ACG ≥ ATG	T ≥ M	1.00
	313	105	CGG ≥ TGG	R ≥ W	0.80
	383	128	TCA ≥ TTA	S ≥ L	1.00
	674	225	TCA ≥ TTA	S ≥ L	1.00
ndhD	878	293	TCA ≥ TTA	S ≥ L	1.00
	887	296	CCT ≥ CTT	P ≥ L	1.00
	1076	359	GCT ≥ GTT	A ≥ V	1.00
	1298	433	TCA ≥ TTA	S ≥ L	0.80
	1310	437	TCA ≥ TTA	S ≥ L	0.80
	290	97	TCA ≥ TTA	S ≥ L	1.00
ndhF	586	196	CTT ≥ TTT	L ≥ F	0.80
	1919	640	GCT ≥ GTT	A ≥ V	0.80
ndhG	166	56	CAT ≥ TAT	H ≥ Y	0.80
	320	107	ACA ≥ ATA	T ≥ I	0.80
petL	119	40	CCT ≥ CTT	P ≥ L	0.86
psbF	77	26	TCT ≥ TTT	S ≥ F	1.00
rpl20	308	103	TCA ≥ TTA	S ≥ L	0.86
rpoA	830	277	TCA ≥ TTA	S ≥ L	1.00
	338	113	TCT ≥ TTT	S ≥ F	1.00
rpoB	551	184	TCA ≥ TTA	S ≥ L	1.00
	566	189	TCG ≥ TTG	S ≥ L	1.00
	2426	809	TCA ≥ TTA	S ≥ L	0.86
rpoC1	41	14	TCA ≥ TTA	S ≥ L	1.00
	1681	561	CAT ≥ TAT	H ≥ Y	0.86
rpoC2	2030	677	ACT ≥ ATT	T ≥ I	1.00
	2314	772	CGG ≥ TGG	R ≥ W	1.00
	4183	1395	CTT ≥ TTT	L ≥ F	0.80
rps2	248	83	TCA ≥ TTA	S ≥ L	1.00
rps16	209	70	TCA ≥ TTA	S ≥ L	0.83

The cytidines marked are putatively edited to uredines.

Comparisons of RNA editing sites with other six species from other families revealed that *R. chinensis* and *D. viscosa* have high RNA editing sites (61 each distributed in 20 and 17 genes respectively) followed by *B. sacra* (57 in 20 genes), *A. altissima* (54 in 21 genes), *A. indica* (53 in 21 genes), *C. aurantiifolia* (52 in 21 genes), and *L. floridana* 48 in 20 genes. As shown in Table S3, these results are consistent with several studies in that all the RNA editing sites predicted among the seven species are cytidine (C) to uridine (U) conversions [41,50–52]. Majority of RNA editing occurred at the second positions of the codons with a frequency from 62.30% (38/61) in *D. viscosa* to 81.28% (39/48) in *L. floridana*, which concurs with previous plastid genome studies in other land plants [53,54]. All the species shared 19 editing sites distributed in twelve genes (Table 3), whereas the two species from Simaroubaceae family (*L. floridana* and *A. altissima*) shared 33 editing sites in 16 genes this implies that the RNA editing sites in these two species are highly conserved (Table S4). Like previous studies [41,51,55], the *ndhB* gene in most of species analyzed here have the highest number of editing sites. Notably, a RNA editing event was detected at the initiator codon (ACG) resulting in ATG translational start codon in the *ndhD* gene.

Table 3. List of RNA editing sites shared by the seven plastomes predicted by PREP program.

Gene	A.A Position	Citrus aurantiifolia	Rhus chinensis	Dodonaea viscosa	Boswellia Sacra	Leitneria floridana	Azadirachta indica	Ailanthus altissima
		Codon (A.A) Conversion						
atpF	31	CCA (P) ≥ CTA (L)	CCA (P) ≥ CTA (L)	CCA (P) ≥ CTA (L)	CCA (P) ≥ CTA (L)	CCA (P) ≥ CTA (L)	CCA (P) ≥ CTA (L)	CCA (P) ≥ CTA (L)
clpP	187	CAT (H) ≥ TAT (Y)	CAT (H) ≥ TAT (Y)	CAT (H) ≥ TAT (Y)	CAT (H) ≥ TAT (Y)	CAT (H) ≥ TAT (Y)	CAT (H) ≥ TAT (Y)	CAT (H) ≥ TAT (Y)
MatK		CAT (H) ≥ TAT (Y)	CAT (H) ≥ TAT (Y)	CAT (H) ≥ TAT (Y)	CAT (H) ≥ TAT (Y)	CAT (H) ≥ TAT (Y)	CAT (H) ≥ TAT (Y)	CAT (H) ≥ TAT (Y)
ndhA	358	TCA (S) ≥ TTA (L)	TCA (S) ≥ TTA (L)	TCA (S) ≥ TTA (L)	TCA (S) ≥ TTA (L)	TCA (S) ≥ TTA (L)	TCA (S) ≥ TTA (L)	TCA (S) ≥ TTA (L)
ndhB	50	TCA (S) ≥ TTA (L)	TCA (S) ≥ TTA (L)	TCA (S) ≥ TTA (L)	TCA (S) ≥ TTA (L)	TCA (S) ≥ TTA (L)	TCA (S) ≥ TTA (L)	TCA (S) ≥ TTA (L)
	156	CCA (P) ≥ CTA (L)	CCA (P) ≥ CTA (L)	CCA (P) ≥ CTA (L)	CCA (P) ≥ CTA (L)	CCA (P) ≥ CTA (L)	CCA (P) ≥ CTA (L)	CCA (P) ≥ CTA (L)
	196	CAT (H) ≥ TAT (Y)	CAT (H) ≥ TAT (Y)	CAT (H) ≥ TAT (Y)	CAT (H) ≥ TAT (Y)	CAT (H) ≥ TAT (Y)	CAT (H) ≥ TAT (Y)	CAT (H) ≥ TAT (Y)
	249	TCT (S) ≥ TTT (F)	TCT (S) ≥ TTT (F)	TCT (S) ≥ TTT (F)	TCT (S) ≥ TTT (F)	TCT (S) ≥ TTT (F)	TCT (S) ≥ TTT (F)	TCT (S) ≥ TTT (F)
	419	CAT (H) ≥ TAT (Y)	CAT (H) ≥ TAT (Y)	CAT (H) ≥ TAT (Y)	CAT (H) ≥ TAT (Y)	CAT (H) ≥ TAT (Y)	CAT (H) ≥ TAT (Y)	CAT (H) ≥ TAT (Y)
ndhD	1	ACG (T) ≥ ATG (M)	ACG (T) ≥ ATG (M)	ACG (T) ≥ ATG (M)	ACG (T) ≥ ATG (M)	ACG (T) ≥ ATG (M)	ACG (T) ≥ ATG (M)	ACG (T) ≥ ATG (M)
	128	TCA (S) ≥ TTA (L)	TCA (S) ≥ TTA (L)	TCA (S) ≥ TTA (L)	TCA (S) ≥ TTA (L)	TCA (S) ≥ TTA (L)	TCA (S) ≥ TTA (L)	TCA (S) ≥ TTA (L)
ndhG	107	ACA (T) ≥ ATA (I)	ACA (T) ≥ ATA (I)	ACA (T) ≥ ATA (I)	ACA (T) ≥ ATA (I)	ACA (T) ≥ ATA (I)	ACA (T) ≥ ATA (I)	ACA (T) ≥ ATA (I)
rpoA	278	TCA (S) ≥ TTA (L)	TCA (S) ≥ TTA (L)	TCA (S) ≥ TTA (L)	TCA (S) ≥ TTA (L)	TCA (S) ≥ TTA (L)	TCA (S) ≥ TTA (L)	TCA (S) ≥ TTA (L)
rpoB	113	TCT (S) ≥ TTT (F)	TCT (S) ≥ TTT (F)	TCT (S) ≥ TTT (F)	TCT (S) ≥ TTT (F)	TCT (S) ≥ TTT (F)	TCT (S) ≥ TTT (F)	TCT (S) ≥ TTT (F)
	184	TCA (S) ≥ TTA (L)	TCA (S) ≥ TTA (L)	TCA (S) ≥ TTA (L)	TCA (S) ≥ TTA (L)	TCA (S) ≥ TTA (L)	TCA (S) ≥ TTA (L)	TCA (S) ≥ TTA (L)
	809	TCG (S) ≥ TTG (L)	TCG (S) ≥ TTG (L)	TCG (S) ≥ TTG (L)	TCG (S) ≥ TTG (L)	TCG (S) ≥ TTG (L)	TCG (S) ≥ TTG (L)	TCG (S) ≥ TTG (L)
rpoC1	14	TCA (S) ≥ TTA (L)	TCA (S) ≥ TTA (L)	TCA (S) ≥ TTA (L)	TCA (S) ≥ TTA (L)	TCA (S) ≥ TTA (L)	TCA (S) ≥ TTA (L)	TCA (S) ≥ TTA (L)
rpoC2	563	CAT (H) ≥ TAT (Y)	CAT (H) ≥ TAT (Y)	CAT (H) ≥ TAT (Y)	CAT (H) ≥ TAT (Y)	CAT (H) ≥ TAT (Y)	CAT (H) ≥ TAT (Y)	CAT (H) ≥ TAT (Y)
rps14	27	TCA (S) ≥ TTA (L)	TCA (S) ≥ TTA (L)	TCA (S) ≥ TTA (L)	TCA (S) ≥ TTA (L)	TCA (S) ≥ TTA (L)	TCA (S) ≥ TTA (L)	TCA (S) ≥ TTA (L)

2.4. Repeat Sequence Analysis

Microsatellites are usually 1–6 bp tandem repeat DNA sequences and are distributed throughout the genome. The presence of microsatellites was detected in the chloroplast genome of *A. altissima* (Figure 5). A total of 219 simple sequence repeats (SSRs) loci were detected, of which mononucleotide repeats occurred with high frequency constituting 190 (86.76%) of all the SSRs. Majority of mononucleotides composed of poly A (polyadenine) (39.27%) and poly T (polythymine) (47.49%) repeats, whereas poly G (polyguanine)or polyC (polycytosine) repeats were rather rare (2.74%). Among the dinucleotide repeat motifs AT/AT were more abundant, while AG/CT were less frequent. One trinucleotide motif (AAT/ATT), five tetra-(AAAG/CTTT, AAAT/ATTT, AACT/AGTT, AATC/ATTG and AGAT/ATCT) and two pentanucleotide repeats (AAAAG/CTTTT and AATAG/ATTCT) were identified. Hexanucleotide repeats were not detected in the *A. altissima* chloroplast genome.

As shown in Figure 5, the SSR analysis for seven species showed that *Leitneria floridana* had the highest number of SSRs (256) while *Dodonaea viscosa* and *Rhus chinensis* had the lowest (186). In all the seven species, mononucleotide repeats were more abundant with A/T repeats being the most common repeats. This result is consistent with earlier studies in [31,35,46] which revealed that many angiosperm chloroplast genomes are rich in poly A and poly T. Moreover, in the seven analyzed species, hexanucleotide repeats were not detected, whereas *Azadirachta indica*, *Dodonaea viscosa* and *Leitneria floridana* had no pentanucleotide repeats.

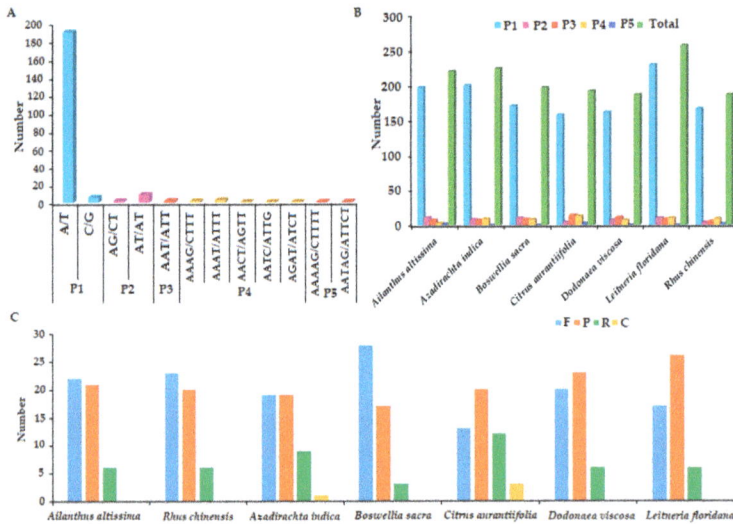

Figure 5. Simple sequence repeat (SSRs) type, distribution and presence in *A. altissima* and other representative species from Sapindales. (**A**) Number of detected SSR motifs in different repeat types in *A. altissima* Chloroplast genome. (**B**) Number of identified repeat sequences in seven chloroplast genomes. (**C**) Number of different SSR types in seven representative species. F, indicate (forward), P (palindromic), R (reverse), and C (complement), while P1, P2, P3, P4, P5 indicates Mono-, di-, tri-, tetra-, and penta-nucleotides respectively. F: forward; P: palindromic, R: reverse; C: complement.

The REPuter program revealed that *A. altissima* chloroplast genome contains 21 palindromic (p), 22 forward (f) and six reverse (r) repeats, however the complement repeats were not detected (Table 4). We notice that all the identified tandem repeats in *A. altissima* were more than 20 bp, while thirteen had length of more than 30 bp. Repeat analyses of seven Sapindales revealed a total of 48 or 49 repeats for each species, with all species containing forward, palindromic and reverse repeats

(Figure 5). Compliment repeats were not identified in other species except for *Azadirachta indica* and *Citrus aurantiifolia* which had one and three repeats respectively. *Citrus aurantiifolia* had the highest number of reverse repeats but also lowest number of forward repeats. Most of the repeat lengths were less than 50 bp, however *Boswellia sacra* chloroplast had seven forward repeats with length of between 65 to 251 bp. Besides, we found out that almost all the repeat sequences were located in either IR or LSC region.

Table 4. Distribution and localization of repetitive sequences F, forward: P, palindromic, R; reverse in *A. altissima* chloroplast genome.

Number	Size	Position 1	Type	Position 2	Location 1 (2)	Region
1	48	95,957	F	95,975	*ycf2*	IRa
2	48	153,174	F	153,192	*ycf2*	IRb
3	37	103,326	F	125,821	*rps12/trnV-GAC(ndhA*)*	IRa/SSC
4	30	95,957	F	95,993	*ycf2*	IRa
5	30	153,174	F	153,210	*ycf2*	IRb
6	29	50,944	F	50,972	*trnL-UAA**	LSC
7	29	58,040	F	58,078	*rbcL*	LSC
8	28	115,434	F	115,460	*ycf1*	SSC
9	26	39,399	F	39,625	*psbZ/trnG-UCC*	LSC
10	25	71,153	F	71,178	*trnP-GGG/psaJ*	LSC
11	23	47,036	F	103,323	*ycf3**(rps12/trnV-GAC)*	LSC/IRa
12	23	112,456	F	112,488	*rrn4.5/rrn5*	IRa
13	23	136,686	F	136,718	*rrn5/rrn4.5*	IRb
14	22	11,749	F	11,771	*trnR-UCU/atpA*	LSC
15	21	248	F	270	*trnH-GUG/psbA*	LSC
16	21	9541	F	38,293	*trnS-GCU (trnS-UGA)*	LSC
17	21	41,956	F	44,180	*psaB(psaA)*	LSC
18	21	49,678	F	49,699	*trnL-UAA**	LSC
19	20	1945	F	1965	*trnK-UUU*	LSC
20	20	15,166	F	92,503	*atpH/atpI(ycf2)*	LSC
21	20	47,039	F	125,821	*ycf3**(rps15)*	LSC/IRa
22	20	88,907	F	160,270	*rpl2*	IRa/IRb
25	48	31,790	P	31,790	*petN/psbM*	LSC
26	48	95,957	P	153,174	*ycf2*	IRa/IRb
27	48	95,975	P	153,192	*ycf2*	IRa/IRb
28	37	125,821	P	145,834	*ndhA*(trnV-GAC/rps12)*	SSC/IRb
29	36	30,970	P	30,970	*petN/psbM*	LSC
30	30	72,117	P	72,117	*rpl33/rps18*	LSC
31	30	95,957	P	153,174	*ycf2*	IRa/IRb
32	30	95,993	P	153,210	*ycf2*	IRa/IRb
33	27	542	P	571	*trnH-GUG/psbA*	LSC
34	25	11,403	P	11,430	*trnS-GCU/trnR-UCU*	LSC
35	24	4867	P	4897	*trnK-UUU/rps16*	LSC
36	24	9535	P	48,164	*trnS-GCU(psaA/ycf3)*	LSC
37	23	47,036	P	145,851	*ycf3**(trnV-GAC/rps12)*	LSC/IRb
38	23	51,804	P	119,066	*trnF-GAA/ndhJ(rpl32/trnL-UAG)*	LSC/SSC
39	23	112,456	P	136,686	*rrn4.5/rrn5*	IRa/IRb
40	23	112,488	P	136,718	*rrn4.5/rrn5*	IRa/IRb
41	22	39,195	P	39,195	*psbZ/trnG-UCC*	LSC
42	20	15,166	P	156,674	*atpH(ycf2)*	LSC/IRb
43	20	38,361	P	48,100	*trnS-UGA(trnS-GGA)*	LSC
44	20	88,907	P	88,907	*rpl2*	IRa
45	20	107,097	P	107,130	*rrn16/trnI-GAU*	IRa
46	23	39,184	R	39,184	*psbZ/trnG-UCC*	LSC
47	21	9751	R	9751	*trnS-GCU/trnR-UCU*	LSC
48	21	51,281	R	51,281	*trnL-UAA/trnF-GAA*	LSC
49	21	85,055	R	85,055	*rps8/rpl14*	LSC
50	20	53,712	R	53,712	*ndhC*	LSC
51	20	9385	R	13,356	*psbI(atpA/atpF)*	LSC

F: forward; P: palindrome; R; reverse* intron or ** introns.

2.5. Phylogenetic Tree

The phylogenetic position of *A. altissima* within Sapindales was carried out using 75 protein coding sequences shared by thirty-one taxa from Sapindales (Table S5). Three remaining species were from Thymelaeaceae family (*Aquilaria sinensis*) and Malvaceae (*Theobroma cacao* and *Abelmoschus esculentus*) from order Malvales selected as outgroups (Figure 6). The maximum likelihood (ML) analysis produced a phylogenetic tree which fully supported *A. altissima* to be closely related with *Leitneria floridana* with 100% bootstrap value. The ML resolved 26 nodes with high branch support (over 98% bootstrap values), however six nodes were moderately supported perhaps as a result of less samples use (59 to 95%). Concerning relationships among families within Sapindales order, family Simaroubaceae early diverged and formed a sister clade/relationship with a 95% bootstrap support to Rutaceae family. Interestingly, the placement of families within Sapindales in our phylogenetic tree supports the one reported by previous studies [30,56,57] based on some chloroplast and nuclear markers. The families Anacardiaceae and Burseraceae formed a sister clade/ group, this clade further branched forming a sister clade with families Sapindaceae, Meliaceae, Simaroubaceae and Rutaceae analyzed in our study. Therefore, it is crucial to use more species for better understanding of Simaroubaceae phylogeny and evolution. This study provides a basis for future phylogenetic of Simaroubaceae species.

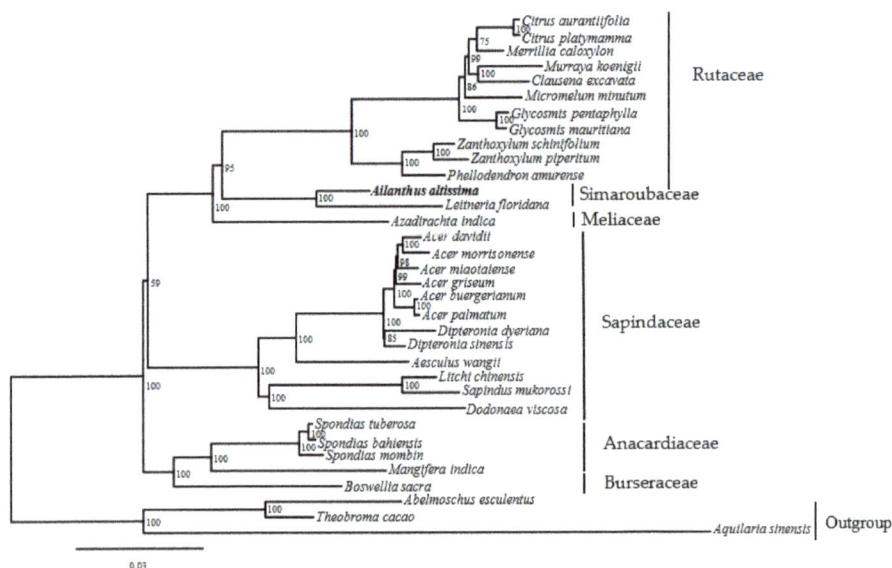

Figure 6. Phylogenetic tree of 31 Sapindales species with three outgroup Malvales species inferred from ML (Maximum likelihood) based on common protein coding genes. The position of *A. altissima* is shown in bold, while bootstrap support values are shown at each node.

3. Materials and Methods

3.1. Plant Materials and DNA Extraction

Fresh leaves of *Ailanthus altissima* were collected in Wuhan Botanical Garden, Chinese Academy of Sciences in China. Total genomic DNA isolation was carried out using MagicMag Genomic DNA Micro Kit (Sangon Biotech Co., Shanghai, China) based on the manufactures protocol. The quality and integrity of DNA were checked and inspected using spectrophotometry and agarose gel electrophoresis respectively. The voucher specimen (HIB-LZZ-CC003) has been deposited at the Wuhan Botanical Garden herbarium (HIB) Wuhan, China.

3.2. The Tree of Heaven Plastome Sequence Assembly and Annotation

Library preparation was constructed using the Illumina Hiseq 2500 platform at NOVOgene Company (Beijing, China) with an average insert size of approximately 350 bp. The high-quality data (5 Gb) were filtered from raw sequence data (5.2 Gb) using the PRINSEQ lite v0.20.4 (San Diego State University, San Diego, CA, USA) [58] (phredQ $\geq$ 20, Length $\geq$ 50), followed by de novo assembling using NOVOPlasty [59] with default sets (K-mer = 31). The seeds and reference plastome used was from the closely related species *Leitneria floridana* (NC_030482) with high coverage of chloroplast reads ~1500×. Lastly, one contig of *Ailanthus altissima* was generated and mapped with reference plastome using GENEIOUS 8.1 (Biomatters Ltd., Auckland, New Zealand) [60]. Finally, online web-based server local blast was used to verify the inverted repeat (IR) single copy (SC) junctions.

Preliminary gene annotation of assembled genome was done using the program DOGMA (Dual Organellar GenoMe Annotator, University of Texas at Austin, Austin, TX, USA) [61], and BLAST (http://blast.ncbi.nlm.nih.gov/). The positions of start and stop codons together with position of introns were confirmed by comparing with homologous genes of other chloroplast genomes available at the GenBank database. Moreover, tRNA genes were verified with tRNAscan-SE server (http://lowelab.ucsc.edu/tRNAscan-SE/) [62]. The chloroplast genome physical circular map was drawn using program OGDRAW (Organellar Genome DRAW) [63] Max planck Institute of Molecular Plant Physiology, Potsdam, Germany) accompanied by manual corrections. The chloroplast genome sequence of *A. altissima* was deposited in the GenBank database, accession (MG799542).

3.3. Genome Comparison and Gene Rearrangement

The border region between Inverted repeat (IR) and large single copy (LSC), also between inverted repeats and small single copy (SSC) junction were compared among seven representative species from Sapindales order. Additionally, alignments of seven chloroplast with one reference genome to determined gene rearrangements was carried out using Mauve v.4.0 [64].

3.4. Repeat Analysis in A. altissima Chloroplast Genome

Microsatellites were identified in the tree of heaven chloroplast genome and other selected representative genomes belonging to order Sapindales using an online software MIcroSAtellite (MISA) [65].The minimum number of repetitions were set to eight repeat units for mononucleotide SSR motifs, five repeat units for dinucleotide SSRs, four for trinucleotide SSRs and three repeat units for tetra-, penta-, and hexanucleotide motifs. The REPuter (https://bibiserv.cebitec.uni-bielefeld.de/reputer) program [66] (University of Bielefeld, Bielefeld, Germany) with default parameters was used to identify the location and sizes of forward, palindromic, complement and reverse repeat sequences in *A. altissima* chloroplast genome.

3.5. Codon Usage and RNA Editing Sites

CodonW1.4.2 (http://downloads.fyxm.net/CodonW-76666.html) was used to analyze codon usage. Subsequently, possible RNA editing sites in *A. altissima* protein coding genes were predicted using the program predictive RNA editor for plants (PREP) suite [67] with the cutoff value set to 0.8. PREP server uses 35 genes as reference for potential RNA editing sites prediction by comparing the predicted protein genes to homologous proteins from other plants.

3.6. Phylogenetic Analysis

Seventy five protein coding sequences present in 31 species from order Sapindales and three species from Thymelaeaceae (*Aquilaria sinensis*), Malvaceae (*Theobroma cacao* and *Abelmoschus esculentus*) as outgroups were used for the phylogenetic reconstruction. These species chloroplast genomes were downloaded from GenBank (Table S5). The protein coding sequences alignment was done using GENEIOUS v8.0.2 (Biomatters Ltd., Auckland, New Zealand) [60]. Maximum likelihood (ML) analysis was carried out using RAxMLversion 8.0.20 (Scientific Computing Group, Heidelberg Institute for Theoretical Studies, Institute of Theoretical Informatics, Karlsruhe Institute of Technology, Karlsruhe, Germany) [68] with 1000 replicates for bootstrap test. Lastly, the jModelTest v2.1.7 [69] was used to select the best substitution model (GTR + I + G).

4. Conclusions

In this study, we present the plastome of tree of heaven from family Simaroubaceae which contains about 22 genera and over 100 species. *A. altissima* chloroplast genome possess circular and quadripartite structure which is well conserved similar to other plants chloroplast genomes. Nonetheless, the plastome showed slight variations at the four boundary junctions due to expansion and contraction in SC and IR borders. About 219 SSR loci and 49 repeats sequences were identified in *A. altissima* genome, this provides genetic information for designing DNA molecular markers for analyzing gene pool dynamics and genetic diversity of *A. altissima* natural populations aiming dispersal mechanism of this invasive tree. The phylogenetic analysis performed using 75 protein coding genes of 34 species available at the GenBank database, comprising 3 outgroup species from Malvales and 31 species representing families from order Sapindales. The two species from family Simaroubaceae formed a cluster and were group together with other families to form a single clade (Sapindales). In addition, the RNA editing analysis in *A. altissima* genome identified a total of 54 possible editing sites in 21 chloroplast genes with C-to-U transitions being the most. The availability of this chloroplast genome provides a tool to advance the study of evolution and invasion in *A. altissima* in order to address present evolutionary, ecological and genetic questions regarding this species.

Supplementary Materials: Supplementary materials can be found at http://www.mdpi.com/1422-0067/19/4/929/s1.

Acknowledgments: This work was supported by the National Natural Scientific Foundation of China (Grant No. 31500457) and CAS-TWAS President's Fellowship for International PhD Students.

Author Contributions: Yi-Ying Liao conceived and designed the experiment; Josphat K. Saina, Zhi-Zhong Li and Andrew W. Gichira assembled plastome sequence and revised the manuscript. Josphat K. Saina performed the experiments, analyzed data and wrote the manuscript. Yi-Ying Liao and Zhi-Zhong Li collected the plant material. All the authors read and approved submission of the final manuscript.

Conflicts of Interest: The authors declare that they have no conflict of interest.

Abbreviations

SC	Single copy
LSC	Large single copy
SSC	Small single copy
IR	Inverted repeat

References

1. Kowarik, I.; Säumel, I. Biological flora of central Europe: *Ailanthus altissima* (Mill.) swingle. *Perspect. Plant Ecol. Evol. Syst.* **2007**, *8*, 207–237. [CrossRef]
2. Liao, Y.Y.; Guo, Y.H.; Chen, J.M.; Wang, Q.F. Phylogeography of the widespread plant *Ailanthus altissima* (Simaroubaceae) in China indicated by three chloroplast DNA regions. *J. Syst. Evol.* **2014**, *52*, 175–185. [CrossRef]
3. Kurokochi, H.; Saito, Y.; Ide, Y. Genetic structure of the introduced heaven tree (*Ailanthus altissima*) in Japan: Evidence for two distinct origins with limited admixture. *Botany* **2014**, *93*, 133–139. [CrossRef]
4. Dallas, J.F.; Leitch, M.J.; Hulme, P.E. Microsatellites for tree of heaven (*Ailanthus altissima*). *Mol. Ecol. Resour.* **2005**, *5*, 340–342. [CrossRef]
5. Aldrich, P.R.; Briguglio, J.S.; Kapadia, S.N.; Morker, M.U.; Rawal, A.; Kalra, P.; Huebner, C.D.; Greer, G.K. Genetic structure of the invasive tree *Ailanthus altissima* in eastern United States cities. *J. Bot.* **2010**, *2010*, 795735.
6. Neuhaus, H.; Emes, M. Nonphotosynthetic metabolism in plastids. *Annu. Rev. Plant Biol.* **2000**, *51*, 111–140. [CrossRef] [PubMed]

7. Henry, R.J. *Plant Diversity and Evolution: Genotypic and Phenotypic Variation in Higher Plants*; CABI Publishing: Cambridge, MA, USA, 2005.

8. Raubeson, L.A.; Jansen, R.K. Chloroplast genomes of plants. In *Plant Diversity and Evolution: Genotypic and Phenotypic Variation in Higher Plants*; CABI Publishing: Cambridge, MA, USA, 2005; pp. 45–68.

9. Wicke, S.; Schneeweiss, G.M.; Müller, K.F.; Quandt, D. The evolution of the plastid chromosome in land plants: Gene content, gene order, gene function. *Plant Mol. Biol.* **2011**, *76*, 273–297. [CrossRef] [PubMed]

10. Yue, F.; Cui, L.; Moret, B.M.; Tang, J. Gene rearrangement analysis and ancestral order inference from chloroplast genomes with inverted repeat. *BMC Genom.* **2008**, *9*, S25. [CrossRef] [PubMed]

11. Jansen, R.K.; Wojciechowski, M.F.; Sanniyasi, E.; Lee, S.-B.; Daniell, H. Complete plastid genome sequence of the chickpea (*Cicer arietinum*) and the phylogenetic distribution of *rps12* and *clpP* intron losses among legumes (Leguminosae). *Mol. Phylogen. Evol.* **2008**, *48*, 1204–1217. [CrossRef] [PubMed]

12. Lee, H.-L.; Jansen, R.K.; Chumley, T.W.; Kim, K.-J. Gene relocations within chloroplast genomes of *Jasminum* and *Menodora* (Oleaceae) are due to multiple, overlapping inversions. *Mol. Biol. Evol.* **2007**, *24*, 1161–1180. [CrossRef] [PubMed]

13. Parks, M.; Cronn, R.; Liston, A. Increasing phylogenetic resolution at low taxonomic levels using massively parallel sequencing of chloroplast genomes. *BMC Biol.* **2009**, *7*, 84. [CrossRef] [PubMed]

14. Yi, X.; Gao, L.; Wang, B.; Su, Y.-J.; Wang, T. The complete chloroplast genome sequence of *Cephalotaxus oliveri* (Cephalotaxaceae): Evolutionary comparison of *Cephalotaxus* chloroplast DNAs and insights into the loss of inverted repeat copies in gymnosperms. *Genome Biol. Evol.* **2013**, *5*, 688–698. [CrossRef] [PubMed]

15. Moore, M.J.; Bell, C.D.; Soltis, P.S.; Soltis, D.E. Using plastid genome-scale data to resolve enigmatic relationships among basal angiosperms. *Proc. Natl. Acad. Sci. USA* **2007**, *104*, 19363–19368. [CrossRef] [PubMed]

16. Wu, C.-S.; Wang, Y.-N.; Hsu, C.-Y.; Lin, C.-P.; Chaw, S.-M. Loss of different inverted repeat copies from the chloroplast genomes of Pinaceae and Cupressophytes and influence of heterotachy on the evaluation of gymnosperm phylogeny. *Genome Biol. Evol.* **2011**, *3*, 1284–1295. [CrossRef] [PubMed]

17. Daniell, H.; Lin, C.-S.; Yu, M.; Chang, W.-J. Chloroplast genomes: Diversity, evolution, and applications in genetic engineering. *Genome Biol.* **2016**, *17*, 134. [CrossRef] [PubMed]

18. Jansen, R.K.; Saski, C.; Lee, S.-B.; Hansen, A.K.; Daniell, H. Complete plastid genome sequences of three rosids (*Castanea*, *Prunus*, *Theobroma*): Evidence for at least two independent transfers of *rpl22* to the nucleus. *Mol. Biol. Evol.* **2010**, *28*, 835–847. [CrossRef] [PubMed]

19. Cai, Z.; Penaflor, C.; Kuehl, J.V.; Leebens-Mack, J.; Carlson, J.E.; Boore, J.L.; Jansen, R.K. Complete plastid genome sequences of drimys, *Liriodendron*, and *Piper*: Implications for the phylogenetic relationships of magnoliids. *BMC Evol. Biol.* **2006**, *6*, 77. [CrossRef] [PubMed]

20. Steane, D.A. Complete nucleotide sequence of the chloroplast genome from the tasmanian blue gum, *Eucalyptus globulus* (myrtaceae). *DNA Res.* **2005**, *12*, 215–220. [CrossRef] [PubMed]

21. Wu, C.-S.; Lin, C.-P.; Hsu, C.-Y.; Wang, R.-J.; Chaw, S.-M. Comparative chloroplast genomes of Pinaceae: Insights into the mechanism of diversified genomic organizations. *Genome Biol. Evol.* **2011**, *3*, 309–319. [CrossRef] [PubMed]

22. Zalapa, J.E.; Cuevas, H.; Zhu, H.; Steffan, S.; Senalik, D.; Zeldin, E.; McCown, B.; Harbut, R.; Simon, P. Using next-generation sequencing approaches to isolate simple sequence repeat (SSR) loci in the plant sciences. *Am. J. Bot.* **2012**, *99*, 193–208. [CrossRef] [PubMed]

23. Buschiazzo, E.; Gemmell, N.J. The rise, fall and renaissance of microsatellites in eukaryotic genomes. *Bioessays* **2006**, *28*, 1040–1050. [CrossRef] [PubMed]

24. Kelkar, Y.D.; Tyekucheva, S.; Chiaromonte, F.; Makova, K.D. The genome-wide determinants of human and chimpanzee microsatellite evolution. *Genome Res.* **2008**, *18*, 30–38. [CrossRef] [PubMed]

25. Ebert, D.; Peakall, R. A new set of universal de novo sequencing primers for extensive coverage of noncoding chloroplast DNA: New opportunities for phylogenetic studies and CPSSR discovery. *Mol. Ecol. Resour.* **2009**, *9*, 777–783. [CrossRef] [PubMed]

26. Hirao, T.; Watanabe, A.; Miyamoto, N.; Takata, K. Development and characterization of chloroplast microsatellite markers for *Cryptomeria japonica* D. Don. *Mol. Ecol. Resour.* **2009**, *9*, 122–124. [CrossRef] [PubMed]

27. Do Nascimento Vieira, L.; Faoro, H.; Rogalski, M.; de Freitas Fraga, H.P.; Cardoso, R.L.A.; de Souza, E.M.; de Oliveira Pedrosa, F.; Nodari, R.O.; Guerra, M.P. The complete chloroplast genome sequence of *Podocarpus lambertii*: Genome structure, evolutionary aspects, gene content and SSR detection. *PLoS ONE* **2014**, *9*, e90618.

28. Yao, X.; Tang, P.; Li, Z.; Li, D.; Liu, Y.; Huang, H. The first complete chloroplast genome sequences in Actinidiaceae: Genome structure and comparative analysis. *PLoS ONE* **2015**, *10*, e0129347. [CrossRef] [PubMed]

29. Su, H.-J.; Hogenhout, S.A.; Al-Sadi, A.M.; Kuo, C.-H. Complete chloroplast genome sequence of Omani lime (*Citrus aurantiifolia*) and comparative analysis within the rosids. *PLoS ONE* **2014**, *9*, e113049. [CrossRef] [PubMed]

30. Lee, M.; Park, J.; Lee, H.; Sohn, S.-H.; Lee, J. Complete chloroplast genomic sequence of *Citrus platymamma* determined by combined analysis of sanger and NGS data. *Hortic. Environ. Biotechnol.* **2015**, *56*, 704–711. [CrossRef]

31. Saina, J.K.; Gichira, A.W.; Li, Z.-Z.; Hu, G.-W.; Wang, Q.-F.; Liao, K. The complete chloroplast genome sequence of *Dodonaea viscosa*: Comparative and phylogenetic analyses. *Genetica* **2018**, *146*, 101–113. [CrossRef] [PubMed]

32. Raman, G.; Park, S. The complete chloroplast genome sequence of *Ampelopsis*: Gene organization, comparative analysis, and phylogenetic relationships to other angiosperms. *Front. Plant Sci.* **2016**, *7*, 341. [CrossRef] [PubMed]

33. Park, I.; Kim, W.J.; Yeo, S.-M.; Choi, G.; Kang, Y.-M.; Piao, R.; Moon, B.C. The complete chloroplast genome sequences of *Fritillaria ussuriensis* maxim. In addition, *Fritillaria cirrhosa* D. Don, and comparative analysis with other *Fritillaria* species. *Molecules* **2017**, *22*, 982. [CrossRef] [PubMed]

34. Li, B.; Lin, F.; Huang, P.; Guo, W.; Zheng, Y. Complete chloroplast genome sequence of *Decaisnea insignis*: Genome organization, genomic resources and comparative analysis. *Sci. Rep.* **2017**, *7*, 10073. [CrossRef] [PubMed]

35. Zhou, T.; Chen, C.; Wei, Y.; Chang, Y.; Bai, G.; Li, Z.; Kanwal, N.; Zhao, G. Comparative transcriptome and chloroplast genome analyses of two related *Dipteronia* species. *Front. Plant Sci.* **2016**, *7*, 1512. [CrossRef] [PubMed]

36. Kim, K.-J.; Lee, H.-L. Complete chloroplast genome sequences from Korean ginseng (*Panax schinseng* nees) and comparative analysis of sequence evolution among 17 vascular plants. *DNA Res.* **2004**, *11*, 247–261. [CrossRef] [PubMed]

37. Wang, R.-J.; Cheng, C.-L.; Chang, C.-C.; Wu, C.-L.; Su, T.-M.; Chaw, S.-M. Dynamics and evolution of the inverted repeat-large single copy junctions in the chloroplast genomes of monocots. *BMC Evol. Biol.* **2008**, *8*, 36. [CrossRef] [PubMed]

38. Wang, W.; Messing, J. High-throughput sequencing of three Lemnoideae (duckweeds) chloroplast genomes from total DNA. *PLoS ONE* **2011**, *6*, e24670. [CrossRef] [PubMed]

39. Khan, A.L.; Al-Harrasi, A.; Asaf, S.; Park, C.E.; Park, G.-S.; Khan, A.R.; Lee, I.-J.; Al-Rawahi, A.; Shin, J.-H. The first chloroplast genome sequence of *Boswellia sacra*, a resin-producing plant in Oman. *PLoS ONE* **2017**, *12*, e0169794. [CrossRef] [PubMed]

40. Yang, J.; Yue, M.; Niu, C.; Ma, X.-F.; Li, Z.-H. Comparative analysis of the complete chloroplast genome of four endangered herbals of *Notopterygium*. *Genes* **2017**, *8*, 124. [CrossRef] [PubMed]

41. Wang, W.; Yu, H.; Wang, J.; Lei, W.; Gao, J.; Qiu, X.; Wang, J. The complete chloroplast genome sequences of the medicinal plant *Forsythia suspensa* (oleaceae). *Int. J. Mol. Sci.* **2017**, *18*, 2288. [CrossRef] [PubMed]

42. Yang, Y.; Zhu, J.; Feng, L.; Zhou, T.; Bai, G.; Yang, J.; Zhao, G. Plastid genome comparative and phylogenetic analyses of the key genera in Fagaceae: Highlighting the effect of codon composition bias in phylogenetic inference. *Front. Plant Sci.* **2018**, *9*, 82. [CrossRef] [PubMed]

43. Guo, S.; Guo, L.; Zhao, W.; Xu, J.; Li, Y.; Zhang, X.; Shen, X.; Wu, M.; Hou, X. Complete chloroplast genome sequence and phylogenetic analysis of *Paeonia ostii*. *Molecules* **2018**, *23*, 246. [CrossRef] [PubMed]

44. Shen, X.; Wu, M.; Liao, B.; Liu, Z.; Bai, R.; Xiao, S.; Li, X.; Zhang, B.; Xu, J.; Chen, S. Complete chloroplast genome sequence and phylogenetic analysis of the medicinal plant *Artemisia annua*. *Molecules* **2017**, *22*, 1330. [CrossRef] [PubMed]

45. Li, Z.-Z.; Saina, J.K.; Gichira, A.W.; Kyalo, C.M.; Wang, Q.-F.; Chen, J.-M. Comparative genomics of the Balsaminaceae sister genera *Hydrocera triflora* and *Impatiens pinfanensis*. *Int. J. Mol. Sci.* **2018**, *19*, 319. [CrossRef] [PubMed]

46. Zhou, J.; Chen, X.; Cui, Y.; Sun, W.; Li, Y.; Wang, Y.; Song, J.; Yao, H. Molecular structure and phylogenetic analyses of complete chloroplast genomes of two *Aristolochia* medicinal species. *Int. J. Mol. Sci.* **2017**, *18*, 1839. [CrossRef] [PubMed]

47. Jiang, D.; Zhao, Z.; Zhang, T.; Zhong, W.; Liu, C.; Yuan, Q.; Huang, L. The chloroplast genome sequence of *Scutellaria baicalensis* provides insight into intraspecific and interspecific chloroplast genome diversity in *Scutellaria*. *Genes* **2017**, *8*, 227. [CrossRef] [PubMed]

48. Zhou, J.; Cui, Y.; Chen, X.; Li, Y.; Xu, Z.; Duan, B.; Li, Y.; Song, J.; Yao, H. Complete chloroplast genomes of *Papaver rhoeas* and *Papaver orientale*: Molecular structures, comparative analysis, and phylogenetic analysis. *Molecules* **2018**, *23*, 437. [CrossRef] [PubMed]

49. Chen, C.; Zheng, Y.; Liu, S.; Zhong, Y.; Wu, Y.; Li, J.; Xu, L.-A.; Xu, M. The complete chloroplast genome of *Cinnamomum camphora* and its comparison with related *Lauraceae* species. *PeerJ* **2017**, *5*, e3820. [CrossRef] [PubMed]

50. De Santana Lopes, A.; Pacheco, T.G.; Nimz, T.; do Nascimento Vieira, L.; Guerra, M.P.; Nodari, R.O.; de Souza, E.M.; de Oliveira Pedrosa, F.; Rogalski, M. The complete plastome of macaw palm [*Acrocomia aculeata* (jacq.) lodd. Ex mart.] and extensive molecular analyses of the evolution of plastid genes in arecaceae. *Planta* **2018**, 1–20. [CrossRef] [PubMed]

51. Kumbhar, F.; Nie, X.; Xing, G.; Zhao, X.; Lin, Y.; Wang, S.; Weining, S. Identification and characterisation of rna editing sites in chloroplast transcripts of einkorn wheat (*Triticum monococcum*). *Ann. Appl. Biol.* **2018**, *172*, 197–207. [CrossRef]

52. Huang, Y.-Y.; Cho, S.-T.; Haryono, M.; Kuo, C.-H. Complete chloroplast genome sequence of common Bermuda grass (*Cynodon dactylon* (L.) pers.) and comparative analysis within the family poaceae. *PLoS ONE* **2017**, *12*, e0179055.

53. Park, M.; Park, H.; Lee, H.; Lee, B.-H.; Lee, J. The complete plastome sequence of an antarctic bryophyte *Sanionia uncinata* (Hedw.) loeske. *Int. J. Mol. Sci.* **2018**, *19*, 709. [CrossRef] [PubMed]

54. Chen, H.; Deng, L.; Jiang, Y.; Lu, P.; Yu, J. RNA editing sites exist in protein-coding genes in the chloroplast genome of *Cycas taitungensis*. *J. Integr. Plant Biol.* **2011**, *53*, 961–970. [CrossRef] [PubMed]

55. De Santana Lopes, A.; Pacheco, T.G.; dos Santos, K.G.; do Nascimento Vieira, L.; Guerra, M.P.; Nodari, R.O.; de Souza, E.M.; de Oliveira Pedrosa, F.; Rogalski, M. The *Linum usitatissimum* L. Plastome reveals atypical structural evolution, new editing sites, and the phylogenetic position of Linaceae within malpighiales. *Plant Cell Rep.* **2018**, *37*, 307–328. [CrossRef] [PubMed]

56. Clayton, J.W.; Fernando, E.S.; Soltis, P.S.; Soltis, D.E. Molecular phylogeny of the tree-of-heaven family (Simaroubaceae) based on chloroplast and nuclear markers. *Int. J. Plant Sci.* **2007**, *168*, 1325–1339. [CrossRef]

57. Lee, Y.S.; Kim, I.; Kim, J.-K.; Park, J.Y.; Joh, H.J.; Park, H.-S.; Lee, H.O.; Lee, S.-C.; Hur, Y.-J.; Yang, T.-J. The complete chloroplast genome sequence of *Rhus chinensis* mill (Anacardiaceae). *Mitochondrial DNA Part B* **2016**, *1*, 696–697. [CrossRef]

58. Schmieder, R.; Edwards, R. Quality control and preprocessing of metagenomic datasets. *Bioinformatics* **2011**, *27*, 863–864. [CrossRef] [PubMed]

59. Dierckxsens, N.; Mardulyn, P.; Smits, G. NOVOPlasty: De novo assembly of organelle genomes from whole genome data. *Nucleic Acids Res.* **2016**, *45*, e18.

60. Kearse, M.; Moir, R.; Wilson, A.; Stones-Havas, S.; Cheung, M.; Sturrock, S.; Buxton, S.; Cooper, A.; Markowitz, S.; Duran, C. Geneious basic: An integrated and extendable desktop software platform for the organization and analysis of sequence data. *Bioinformatics* **2012**, *28*, 1647–1649. [CrossRef] [PubMed]

61. Wyman, S.K.; Jansen, R.K.; Boore, J.L. Automatic annotation of organellar genomes with DOGMA. *Bioinformatics* **2004**, *20*, 3252–3255. [CrossRef] [PubMed]

62. Schattner, P.; Brooks, A.N.; Lowe, T.M. The tRNAscan-SE, snoscan and snoGPS web servers for the detection of tRNAs and snoRNAs. *Nucleic Acids Res.* **2005**, *33*, W686–W689. [CrossRef] [PubMed]

63. Lohse, M.; Drechsel, O.; Bock, R. OrganellarGenomeDRAW (OGDRAW): A tool for the easy generation of high-quality custom graphical maps of plastid and mitochondrial genomes. *Curr. Genet.* **2007**, *52*, 267–274. [CrossRef] [PubMed]

64. Darling, A.C.; Mau, B.; Blattner, F.R.; Perna, N.T. Mauve: Multiple alignment of conserved genomic sequence with rearrangements. *Genome Res.* **2004**, *14*, 1394–1403. [CrossRef] [PubMed]

65. Thiel, T.; Michalek, W.; Varshney, R.; Graner, A. Exploiting EST databases for the development and characterization of gene-derived SSR-markers in barley (*Hordeum vulgare* L.). *Theor. Appl. Genet.* **2003**, *106*, 411–422. [CrossRef] [PubMed]

66. Kurtz, S.; Choudhuri, J.V.; Ohlebusch, E.; Schleiermacher, C.; Stoye, J.; Giegerich, R. Reputer: The manifold applications of repeat analysis on a genomic scale. *Nucleic Acids Res.* **2001**, *29*, 4633–4642. [CrossRef] [PubMed]

67. Mower, J.P. The prep suite: Predictive RNA editors for plant mitochondrial genes, chloroplast genes and user-defined alignments. *Nucleic Acids Res.* **2009**, *37*, W253–W259. [CrossRef] [PubMed]

68. Stamatakis, A. Raxml version 8: A tool for phylogenetic analysis and post-analysis of large phylogenies. *Bioinformatics* **2014**, *30*, 1312–1313. [CrossRef] [PubMed]

69. Darriba, D.; Taboada, G.L.; Doallo, R.; Posada, D. jModelTest 2: More models, new heuristics and parallel computing. *Nat. Methods* **2012**, *9*, 772. [CrossRef] [PubMed]

International Journal of
Molecular Sciences

MDPI

Article

Comparative Analysis of the Chloroplast Genomes of the Chinese Endemic Genus *Urophysa* and Their Contribution to Chloroplast Phylogeny and Adaptive Evolution

Deng-Feng Xie, Yan Yu, Yi-Qi Deng, Juan Li, Hai-Ying Liu, Song-Dong Zhou and Xing-Jin He *

Key Laboratory of Bio-Resources and Eco-Environment of Ministry of Education, College of Life Sciences, Sichuan University, Chengdu 610065, Sichuan, China; df_xie2017@163.com (D.-F.X.); yyu@scu.edu.cn (Y.Y.); yiqiden@gmail.com (Y.-Q.D.); lijuanxxn@163.com (J.L.); lhy921180@163.com (H.-Y.L.); songdongzhou@aliyun.com (S.-D.Z.)
* Correspondence: xjhe@scu.edu.cn; Tel./Fax: +86-28-85415006

Received: 13 April 2018; Accepted: 19 June 2018; Published: 22 June 2018

Abstract: *Urophysa* is a Chinese endemic genus comprising two species, *Urophysa rockii* and *Urophysa henryi*. In this study, we sequenced the complete chloroplast (cp) genomes of these two species and of their relative *Semiquilegia adoxoides*. Illumina sequencing technology was used to compare sequences, elucidate the intra- and interspecies variations, and infer the phylogeny relationship with other Ranunculaceae family species. A typical quadripartite structure was detected, with a genome size from 158,473 to 158,512 bp, consisting of a pair of inverted repeats separated by a small single-copy region and a large single-copy region. We analyzed the nucleotide diversity and repeated sequences components and conducted a positive selection analysis by the codon-based substitution on single-copy coding sequence (CDS). Seven regions were found to possess relatively high nucleotide diversity, and numerous variable repeats and simple sequence repeats (SSR) markers were detected. Six single-copy genes (*atpA*, *rpl20*, *psaA*, *atpB*, *ndhI*, and *rbcL*) resulted to have high posterior probabilities of codon sites in the positive selection analysis, which means that the six genes may be under a great selection pressure. The visualization results of the six genes showed that the amino acid properties across each column of all species are variable in different genera. All these regions with high nucleotide diversity, abundant repeats, and under positive selection will provide potential plastid markers for further taxonomic, phylogenetic, and population genetics studies in *Urophysa* and its relatives. Phylogenetic analyses based on the 79 single-copy genes, the whole complete genome sequences, and all CDS sequences showed same topologies with high support, and *U. rockii* was closely clustered with *U. henryi* within the *Urophysa* genus, with *S. adoxoides* as their closest relative. Therefore, the complete cp genomes in *Urophysa* species provide interesting insights and valuable information that can be used to identify related species and reconstruct their phylogeny.

Keywords: *Urophysa*; *Semiaquilegia adoxoides*; cp genome; repeat analysis; SSRs; positive selection analysis; phylogeny

---

1. Introduction

The genus *Urophysa* (Ranunculaceae) is a Chinese endemic genus with only two species, *Urophysa rockii* Ulbr. and *Urophysa henryi* (Oliv.) Ulbr. *U. rockii* is an extremely rare species with fewer than 2000 individuals living in Jiangyou, a Sichuan province of China, and *U. henryi* is distributed in Guizhou, south Chongqing, north Hunan, and west Hubei [1]. The two species' natural populations are restricted to small and isolated areas separated by high mountains and deep valleys and grow in steep and karstic cliffs with dramatically shrinking and fragmenting natural distributions [2]. In addition, the plants are collected for Chinese traditional medicine for the treatment of contusions and bruises, which contributed to the decline of their populations [3]. Previous studies on the genus *Urophysa* are scarce and mainly focused on the endangered *U. rockii*, its growing environment and conservation strategies [4], its biological and ecological characteristics, and its reproductive biology [5,6]. A recent study suggested that the uplift of the Yungui Plateau played an important role in the species divergence of *Urophysa* [2]. However, the chloroplast DNA (cpDNA) phylogeny showed inconsistency with the nuclear ribosomal DNA (nrDNA). Hence, to gain a better insight into the relationship of these two species and understand their genome structure so as to facilitate their speciation process and the conservation of *U. rockii*, we assembled and characterized the complete chloroplast genome sequence of *U. rockii* and *U. henryi* using the Illumina paired-end sequencing reads.

The angiosperm cp genome is one of the three DNA genomes (the other two are nuclear and mitochondrial genome), is uniparentally inherited, and has a high conserved circular DNA arrangement [7]. It is widely considered an informative and valuable resource for investigating evolutionary biology because of its relatively stable genome structure, gene content, and gene order [8–13]. The cp genome of plants always ranges from 115 to 210 kb and has a quadripartite structure that is typically composed of two copies of inverted repeat (IR) regions, which are separated by a large single-copy (LSC) region and a small single-copy (SSC) region [14–16]. Because of its compact size, less recombination, and maternal inheritance, the cp genome has been used to generate genetic markers for phylogenetic analysis [17,18], molecular identification [19], and divergence dating [20]. Especially, the low evolutionary rate of the cp genome in taxa that are not very young makes it an ideal system for assessing plant phylogeny [21].

In the present study, we report the complete chloroplast genome sequences of these two *Urophysa* species and their relative *Semiaquilegia adoxoides* for the first time. Combining previously reported cp genome sequences, we performed phylogenetic analyses according to the whole cp genome and shared single-copy genes. Our findings will contribute to our understanding of the evolutionary history of the genus *Urophysa*. Additionally, highly variable regions and genes that were detected to be under positive selection could be employed to develop potential markers for phylogenetic analyses or candidates for DNA barcoding in future studies.

2. Results and Discussion

2.1. Complete Chloroplast Genomes of Three Species

The complete chloroplast genome of *U. rockii*, *U. henryi*, and *S. adoxoides* showed a single circular molecule with a typical quadripartite structure (Figure 1). The sizes of the *U. rockii*, *U. henry*, and *S. adoxoides* cp genomes were found to be 158,512 bp, 158,303, and 158,340 bp, respectively, which are in the range of most angiosperm plastid genomes [22]. The cp genome consists of a pair of IRs (IRa and IRb, with length 26,473–26,584 bp), separated by a LSC (87,031–87,202 bp) region and one SSC (18,192–18,220 bp) region (Table 1). The GC content of each species was very similar in the whole cp genome and the same region (LSC, SSC, and IR), but in the IR regions it was clearly higher than in the other regions, possibly because of the high GC content of the rRNA (55.8%) that was located in the IR regions (Table 2). These results are similar to a previously reported high GC percentage in IR regions [23–25].

The genomes contain 87 coding genes, 36 transfer RNA genes (tRNA), and 8 ribosomal RNA genes (rRNA) (Table 3). Most of the genes occur as a single copy in LSC or SSC regions, while 18 genes are duplicated in the IR regions, including seven protein-coding genes (*ndhB*, *rpl2*, *rpl23*, *rps7*, *rps12*, *rps19*, *ycf2*), seven tRNA species (*trnA-UGC*, *trnI-CAU*, *trnI-GAU*, *trnL-CAA*, *trnN-GUU*, *trnR-ACG*, and *trnV-GAC*) and four rRNA species (*rrn4.5*, *rrn5*, *rrn16*, and *rrn23*). The gene *ycf1* straddles the SSC and IRs, while *rps12* locates its first exon in the LSC region and two other exons in the IRs. The LSC region comprises 63 protein-coding genes and 21 tRNA genes, whereas the SSC and IR regions include 12 and 7 protein-coding genes, with one and seven tRNA, respectively. The protein-coding genes present in the *U. rockii* cp genome include 9 genes encoding large ribosomal proteins (*rpl2, rpl14, rpl16, rpl20, rpl22, rpl23, rpl32, rpl33, rpl36*) and 12 genes encoding small ribosomal proteins (*rps2, rps3, rps4, rps7, rps8, rps11, rps12, rps14, rps15, rps16, rps18, rps19*). There are 5 genes encoding phytosystem I subunits (*psaA, psaB, psaC, psaI, psaJ*), along with 15 genes related to photosystem II subunits (*psbA, psbB, psbC, psbD, psbE, psbF, psbH, psbI, psbJ, psbK, psbL, psbM, psbN, psbT, psbZ*) (Table 3). Six genes (*atpA, atpB, atpE, atpF, atpH, atpI*) encode ATP synthase and electron transport chain components (Table 3). A similar pattern of protein-coding genes is also present in *U. henryi* and *S. adoxoides*. There are eight intron-containing genes, six of which contain one intron; only the genes *clpP* and *ycf3* have two introns (Table S1). All these eight genes possess at least two exons, and *ycf3* has three exons. The *rps16* gene has the longest intron (866 bp), and *rpoC1* has the longest exon (1613 bp).

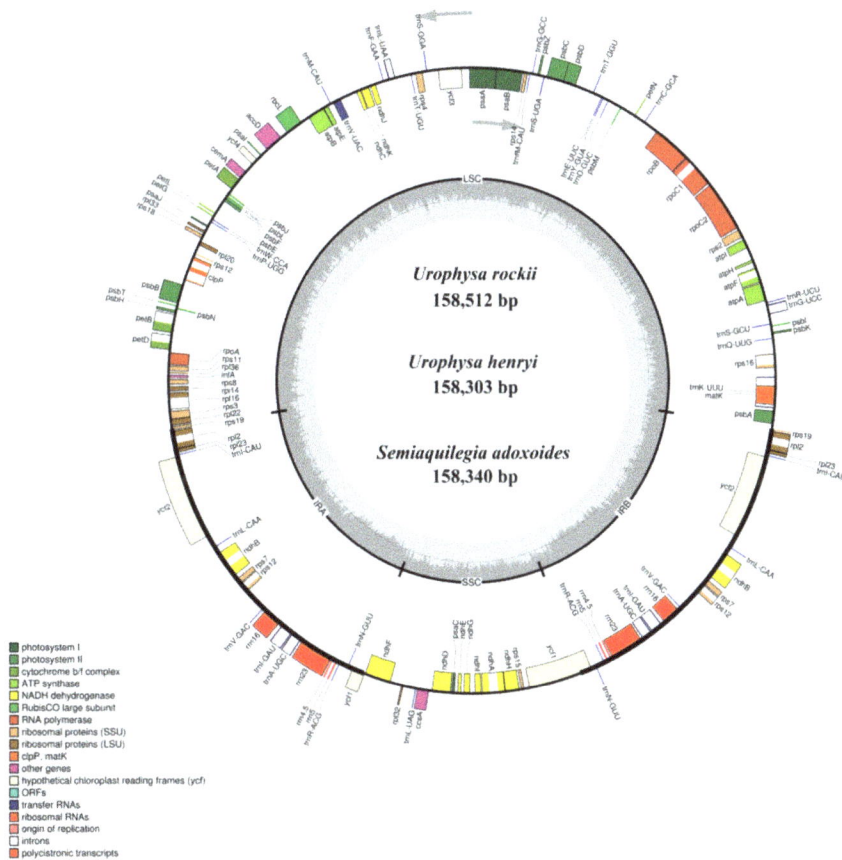

Figure 1. Gene maps of the *Urophysa rockii*, *Urophysa henryi* and *Semiquilegia adoxoides* chloroplast (cp) genomes. Genes shown inside the circle are transcribed clockwise, and those outside are transcribed counterclockwise. Genes belonging to different functional groups are color-coded. The darker gray color in the inner circle corresponds to the GC content, and the lighter gray color corresponds to the AT content. SSU: small subunit; LSU: large subunit; ORF: open reading frame.

Table 1. Summary of complete chloroplast genomes. LSC, large single-copy; SSC, small single-copy; IR, inverted repeat

Species	LSC			SSC			IR			Total	
	Length (bp)	GC%	Length (%)	Length (bp)	GC%	Length (%)	Length (bp)	GC%	Length (%)	Length (bp)	GC%
U. rockii	87,128	37.2	55.0	18,216	32.5	11.5	26,584	43.7	16.8	158,512	38.8
U. henryi	87,031	37.2	55.0	18,260	32.6	11.5	26,506	43.6	16.7	158,303	38.8
S. adoxoides	87,202	37.2	55.1	18,192	32.5	11.5	26,473	43.7	16.7	158,340	38.9
Tsuga chinensis	88,522	36.3	55.3	18,405	32.0	11.5	26,632	43.1	16.6	160,191	38.1
Aconitum austrokoreense	86,362	36.2	55.4	16,948	32.7	10.9	26,291	43.0	16.9	155,892	38.1
A. kusnezoffii	86,335	36.2	55.4	16,945	32.7	10.9	26,291	43.0	16.9	155,862	38.1
A. volubile	86,348	36.2	55.4	16,944	32.6	10.9	26,290	43.0	16.9	155,872	38.1
Ranunculus macranthus	84,637	36.0	54.6	18,909	31.0	12.2	25,791	43.5	16.6	155,129	37.9
R. occidentalis	83,532	35.9	54.1	21,269	31.6	13.8	24,831	43.6	16.1	154,474	37.8
R. austro-oreganus	83,582	35.9	54.1	21,249	31.6	13.8	24,831	43.6	16.1	154,493	37.8
Clematis terniflora	79,328	36.3	49.7	18,110	31.4	11.4	31,045	42.0	19.5	159,528	38.0
Coptis chinensis	84,567	36.4	54.4	17,376	32.1	11.2	26,762	43.0	17.2	155,484	38.2

Table 2. Comparison of the sizes of coding and non-coding regions among species.

Species	Protein-Coding			tRNA			rRNA		
	Length (bp)	GC%	Length (%)	Length (bp)	GC%	Length (%)	Length (bp)	GC%	Length (%)
U. rockii	78,867	39.2	49.8	2687	53.2	1.7	8602	55.8	5.4
U. henryi	78,769	39.2	49.8	2695	53.3	1.7	8602	55.8	5.4
S. adoxoides	78,498	39.3	49.6	2706	53.6	1.7	8602	55.8	5.4
T. chinensis	78,903	38.4	49.3	2716	53.1	1.7	9050	55.4	5.6
A. austrokoreense	79,575	38.3	51.0	2810	53.0	1.8	9050	55.4	5.8
A. kusnezoffii	78,294	38.4	50.2	2813	52.9	1.8	9046	55.3	5.8
A. volubile	79,560	38.3	51.0	2810	53.0	1.8	9050	55.5	5.8
R. macranthus	78,615	38.2	50.7	2738	53.1	1.8	7559	55.2	4.9
R. occidentalis	69,294	38.6	44.9	2717	53.1	1.8	9050	55.4	5.9
R. austro-oreganus	74,355	38.1	48.1	2796	52.9	1.8	9050	55.4	5.9
C. terniflora	81,819	38.3	51.3	2718	53.4	1.7	9050	55.4	5.7
C. chinensis	71,637	39.0	46.1	2716	53.2	1.7	9050	55.5	5.8

Table 3. List of genes encoded in two *Urophysa* species and *S. adoxoides*.

Category for Genes	Group of Genes	Name of Genes
Self-replication	transfer RNAs	trnA-UGC *, trnC-GCA, trnD-GUC, trnE-UUC, trnF-GAA, trnfM-CAU, trnG-GCC, trnG-UCC, trnI-CAU *, trnI-GAU *, trnK-UUU, trnL-CAA *, trnL-UAA, trnL-UAG, trnM-CAU, trnN-GUU *, trnP-UGG, trnQ-UUG, trnR-ACG *, trnR-UCU, trnS-GCU, trnS-GGA, trnS-UGA, trnT-GGU, trnT-UGU, trnV-GAC *, trnV-UAC, trnW-CCA, trnY-GUA
	ribosomal RNAs	rrn4.5 *, rrn5 *, rrn16 *, rrn23 *
	RNA polymerase	rpoA, rpoB, rpoC1, rpoC2
	Small subunit of ribosomal proteins (SSU)	rps2, rps3, rps4, rps7 *, rps8, rps11, rps12 *, rps14, rps15, rps16, rps18, rps19 *
	Large subunit of ribosomal proteins (LSU)	rpl2 *, rpl14, rpl16, rpl20, rpl22, rpl23 *, rpl32, rpl33, rpl36
Genes for photosynthesis	Subunits of NADH-dehydrogenase	ndhA, ndhB *, ndhC, ndhD, ndhE, ndhF, ndhG, ndhH, ndhI, ndhJ, ndhK
	Subunits of photosystem I	psaA, psaB, psaC, psaI, psaJ
	Subunits of photosystem II	psbA, psbB, psbC, psbD, psbE, psbF, psbH, psbI, psbJ, psbK, psbL, psbM, psbN, psbT, psbZ
	Subunits of cytochrome b/f complex	petA, petB, petD, petG, petL, petN
	Subunits of ATP synthase	atpA, atpB, atpE, atpF, atpH, atpI
	Large subunit of rubisco	rbcL
Other genes	Tanslational initiation factor	infA
	Protease	clpP
	Maturase	matK
	Subunit of Acetyl-CoA-carboxylase	accD
	Envelope membrane protein	cemA
	C-type cytochrome synthesis gene	ccsA
Genes of unknown function	hypothetical chloroplast reading frames (ycf)	ycf1 *, ycf2 *, ycf3, ycf4

* Gene with two copies.

2.2. Repeat Analysis

Chloroplast repeats are potentially useful genetic resources to investigate population genetics and biogeography of allied taxa [26]. Analyses of various cp genomes revealed that repeat sequences are essential to induce indels and substitutions [27]. Repeat analysis of the *U. rockii* cp genome revealed 22 palindromic repeats, 23 forward repeats, 5 reverse, and 1 complement repeats. Among them, 16 palindromic, 18 forward, and 5 reverse repeats are 20–40 bp in length. Six palindromic and five forward repeats are 41–60 in length (Figure 2). Similarly, 23 and 25 palindromic repeats, 21 and 22 forward repeats, 5 and 2 reverse repeats, and 1 complement repeats were detected, and the detailed repeats length distributions are shown in Figure 2. The number and length of the repeats indicate that *U. rockii* is more similar to *U. henryi* than to *S. aquilegia*. Previous studies suggested that the slipped-strand mispairing and improper recombination of repeat sequences can result in sequence variation and genome rearrangement [28–30]. These repeats are informative sources for developing genetic markers for phylogenetic and population studies [31].

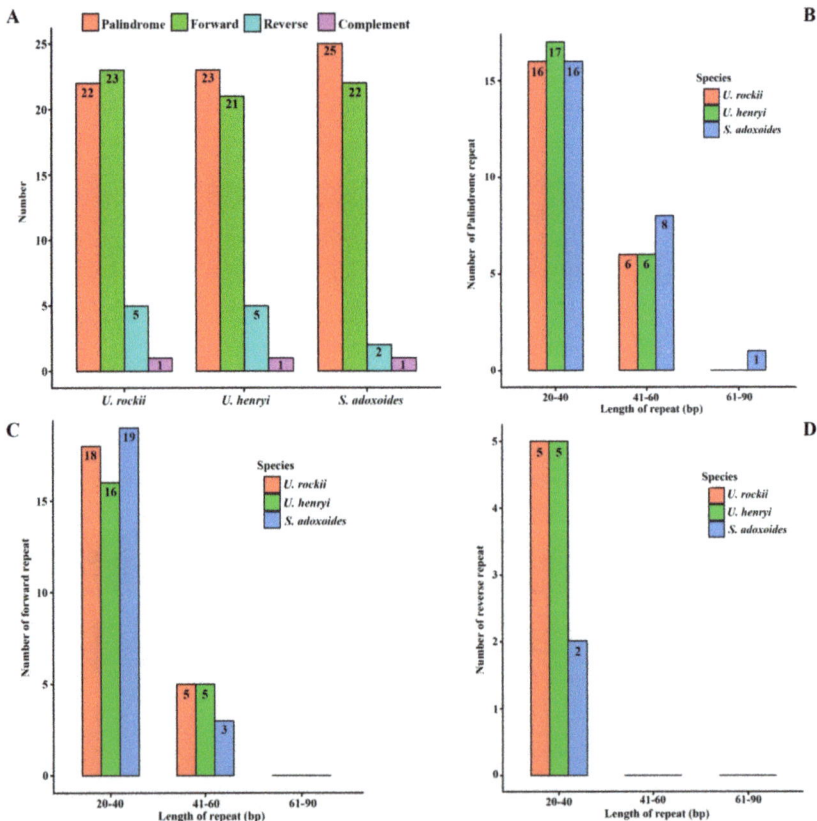

Figure 2. Analysis of repeated sequences in *U. rockii*, *U. henryi*, and *S. adoxoides* chloroplast genomes. (**A**) Total of four repeat types; (**B**) Frequency of the palindromic repeat by length; (**C**) Frequency of the forward repeat by length; (**D**) Frequency of the reverse repeat by length.

Simple sequence repeats (SSRs) in the cp genome can be highly variable at the intra-specific level and are therefore often used as genetic markers in population genetic and evolutionary studies [12,32–34]. Because of a high polymorphism rate at the species level, SSRs have been recognized as one of the main sources of molecular markers and have been extensively researched in phylogenetic and biogeographic studies of populations [35–37]. In this study, we analyzed the SSRs in the cp genomes. Five categories of perfect SSRs (mono-, di-, tri-, tetra-, and penta-nucleotide repeats) were detected in the cp genome of these three species, with an overall length ranging from 10 to 26 bp (Figure 3, Table S2). Certain parameters were set, because SSRs of 10 bp or longer are prone to slipped-strand mispairing, which is believed to be the main mutational mechanism for polymorphism [38–40].

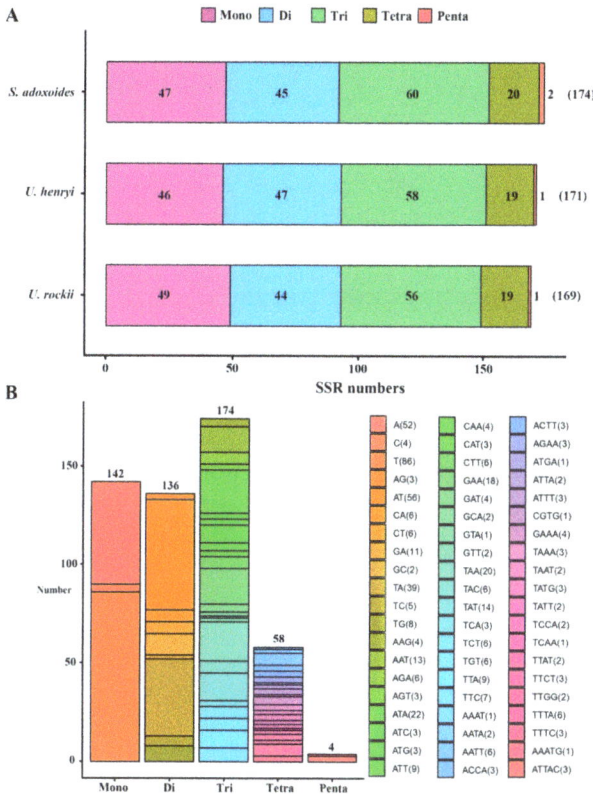

Figure 3. Analysis of simple sequence repeats (SSRs) in chloroplast genomes of the three species. (**A**) Number of different SSR types detected in each species; (**B**) type and frequency of each identified SSR.

A total of 169 microsatellites were detected in the *U. rockii* cp genome on the basis of the SSR analysis. Similarly, 171 and 174 SSRs were detected in *U. henryi* and *S. adoxoides*, respectively (Figure 3A). The most abundant were tri-nucleotide repeats, which accounted for about 33.85% of the total SSRs, and whose number varies from 56 in *U. rockii* to 60 in *S. adoxoides*, followed by mono-nucleotide repeats (27.63%), di-nucleotide repeats (26.46%), and tetra-nucleotides repeats (11.28%). Penta-nucleotide repeats were the least abundant (0.78%; Figure 3, Table S2). Most previous studies revealed that the richness of SSR types varies between species. In *Quercus* species, mono-nucleotide repeats are the most abundant, accounting for about 80% of the total SSRs [34]. In the cp genome of *Forthysia*, the number of di-nucleotide repeat is the highest [41]. Tri-nucleotide SSRs are most abundant in *Nicotiana species*, accounting for approximately 43.03% [42]. These results suggest that different repeats may contribute to the genetic variations differently among species. Thus, the SSR information will be important for understanding the genetic diversity status of *Urophysa* and its relatives.

In *U. rockii*, more than 96.2% mono-nucleotides are composed of A/T, and a majority of di-nucleotides (84.9%) is composed of A/T (Figure 3B, Table S2), which is consistent with *U. henryi* (97.8% mono-nucleotides and 83.0% di-nucleotides) and *S. aquilegia* (97.9% mono-nucleotides and 85.6% di-nucleotides). Our findings are comparable to previously reported observations that SSRs found in the chloroplast genome are generally composed of poly-thymine (polyT) or poly-adenine (polyA) repeats and infrequently contain tandem cytosine (C) and guanine (G) repeats [43]. Therefore, these SSRs contribute to the AT richness of the three species cp genome, as previously reported for different species [43,44]. SSRs were also detected in CDS regions of the *U. rockii* cp genome. The CDS regions account for approximately 49% of the total length. About 68.6% of SSRs (68.4% for *U. henryi* and 67.2% for *S. adoxoides*) were detected in non-coding regions, whereas only 28.9%of SSRs (29.2% for *U. henryi* and 30.5% for *S. adoxoides*) are present in the protein-coding region of *U. rockii*. Furthermore, about 62.1% of SSRs are present in the LSC region of *U. rockii* (66.1% for *U. henryi* and 68.9% for *S. adoxoides*), and a minority of SSRs exist in IR regions (17.8% in IRa and IRb in total). It was observed that 49 SSRs (28.9%) were located in 19 genes (CDS) regions (*atpF, rpoC1, rpoC2, rps14, rps15, rps19, psaB, psaA, rbcL, rpl33, rpl22, ndhB, ndhD, ndhF, ndhH, ccsA, ycf1, ycf2, ycf3*) in *U. rockii*. The detailed SSR location information is listed in Table S2. These results suggest an uneven distribution of SSRs in the *U. rockii*, *U. henryi*, and *S. adoxoides* cp genomes, as was also reported in different angiosperm cp genomes [44]. Moreover, the cp SSRs of the three species presented abundant variation and are useful for detecting genetic polymorphisms at population, intraspecific, and cultivar levels, as well as for comparing more distant phylogenetic relationships among species.

2.3. Genomes Sequence Divergence among the Three Species

In order to calculate the sequence divergence level, the nucleotide diversity values in the LSC, SSC, and IR regions of the chloroplast genomes were calculated (Figure 4, Table S3). In the LSC regions, these values varied from 0 to 0.05496, with a mean of 0.00705, in the IR regions they varied from 0 to 0.01265, with a mean of 0.00363, and only the SSC region had >0.010 average sequence nucleotide diversity, and its values varied from 0 to 0.02369, with a mean of 0.01048. All these results indicated that the differences among these genome regions were small. However, some highly variable loci, including *trnK-UUU, trnG-UCC, trnD-GUC, atpF, rps4, trnL-UAA, accD, cemA, rpl36, rpl22, rps19, ndhF, trnL-UAG, ccsA, ndhA*, and *ycf3* were more precisely located (Figure 4, Table S3). All these regions displayed higher nucleotide diversity values than other regions (value > 0.015). Twelve of these loci were found to be located in the LSC region, and four in the SSC region, but the nucleotide diversity in the IR regions appeared small, less than 0.015. Among these loci, *atpF, accD, ndhF, rpl22, ccsA*, and *ycf3* have been detected as highly variable regions in different plants [19,23,45,46]. On the basis of these results, we believe that *accD, rps4, ccsA, rpl36*, and *ndhF*, which have comparatively high sequence deviation, are good sources for interspecies phylogenetic analysis, as shown in previous studies [42,44].

Expansion and contraction at the borders of IR regions is the main reason for size variations in the cp genome and plays a vital role in its evolution [39,47,48]. The IR/LSC and IR/SSC junction regions were compared to identify IR expansion or contraction. The *rps19, ndhF, ycf1*, and *psbA* genes were located in the junctions of the LSC/IRa, IRa/SSC, SSC/IRb, and IRb/LSC regions, respectively (Figure 5). Despite the similar length of these three species IR regions, from 26,473 to 26,584 bp, some IR expansion and contraction were observed. The *rps19* gene traverses the LSC and IRb regions (LR line), with 104 bp located in the IR region. The RS line (the junction line between IRb and SSC) is located between *ycf1* and *ndhF*, and the variation in distances between the RS line and *ndhF* ranges from 33 to 36 bp across the three species. The SR line (the junction line between SSC and IRa) intersects the ycf1 gene, the SSC and IRa regions are the same in *U. rockii* and *U. henryi* (4259 bp in SSC and 1081 bp in IRb), while different in *S. adoxoides* (4229 bp in SSC and 1084 bp in IRb) (Figure 5). The distance between the *psbA* and RL line varies from 386 to 403 bp.

Compared to species of other genera, the IRb/SSC and SSC/IRa regions of *Urophysa* showed an expansion in *ycf1*, but a contraction in *rps19* (Figure 5). The expansion and contraction detected in the IR regions may act as a primary mechanism in creating the length variation of the cp genomes in *U. rockii*, *U. henryi*, and *S. adoxoides*, as previous studies suggested [32,34,42,49].

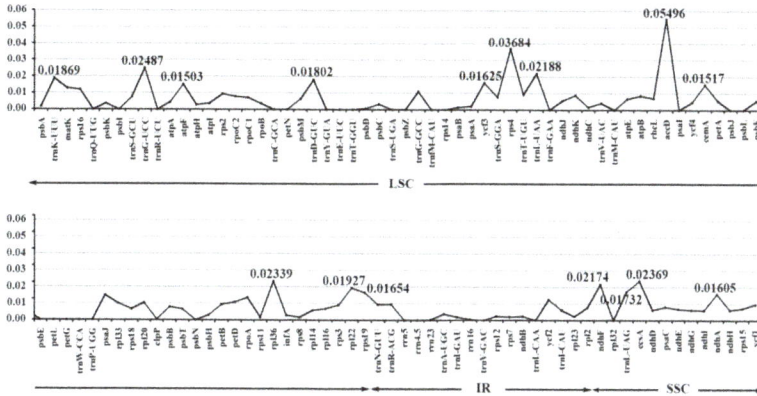

Figure 4. The nucleotide diversity of the whole chloroplast genomes of the three species. LSC: large single-copy region; IRs: inverted repeats region; SSC: small single-copy region.

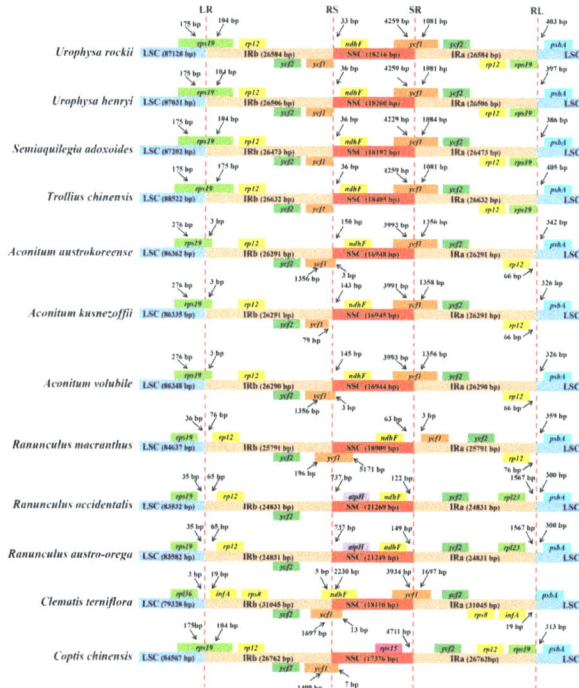

Figure 5. Comparison of the borders of the LSC, SSC, and IR regions of the chloroplast genomes of the three species. LR: junction line between LSC and IRb; RS: junction line between IRb and SSC; SR: junction line between SSC and IRa; RL: junction line between IRa and LSC.

2.4. Phylogenetic Analysis

To study the phylogenetic position of *U. rockii and U. henryi* within the Ranunculaceae family, we used 79 single-copy genes shared by the cp genomes of 12 Ranunculaceae members, representing seven genera (Figure 6). For Bayesian inference (BI) and maximum parsimony (MP), the posterior probabilities and bootstrap values were very high for each lineage, with all values ≥98%. Both the maximum likelihood (ML), BI, and MP phylogenetic results strongly supported that *U. rockii* is closely clustered with *U. henryi* within the genus *Urophysa*, with *S. adoxoides* as their closest relative with 100% bootstrap value (Figure 6), which is consistent with the results of previous molecular studies [50–52]. Furthermore, the species in each genus formed a single clade. The first clade is formed by species of the genera *Urophysa*, *Semiaquilegia*, and *Trollius*, the second clade was divided into two clades: one clade includes the *Ranunculus* and *Clematis* species, and the other clade consists of just the *Aconitum* species. Additionally, the topological structures from the whole complete chloroplast genome sequences and the CDS sequences are similar to that from single-copy genes (Figure S1), and all lineages possess high bootstrap values. These results suggest that there is no conflict among the entire genome data set, CDS sequences, and 79 shared single-copy genes of these cp genomes. Furthermore, these results are in accord with previous phylogeny research [53]. All these phylogenetic analyses are substantially increasing our understanding of the evolutionary relationship among species in Ranunculaceae.

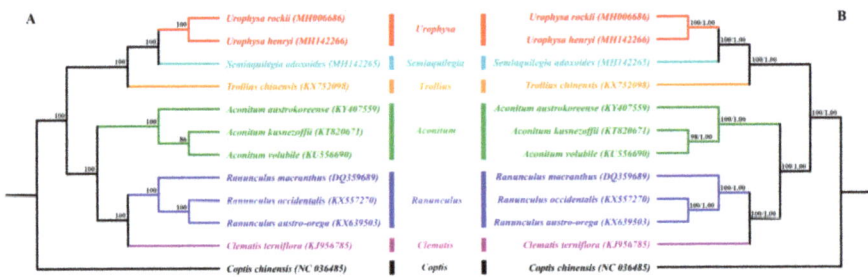

Figure 6. Phylogenetic relationship of *Urophysa* with related species based on 79 single-copy genes shared by all cp genomes. Tree constructed by (**A**) maximum likelihood (ML) with the bootstrap values of ML above the branches; (**B**) maximum parsimony (MP) and Bayesian inference (BI) with bootstrap values of MP and posterior probabilities of BI above the branches, respectively.

2.5. Positive Selected Analysis

Of 57 single-copy CDS genes initially considered for the positive selection analysis (Table S4), 47 were eventually selected (Table 4). No significant positive selection was detected for all genes (*p*-value > 0.05), but six genes that possess high posterior probabilities for codon sites were found in the Bayesian Empirical Bayes (BEB) test (*atpA*, *rpl20*, *psaA*, *atpB*, *ndhI*, and *rbcL*) (Figure 7, Figure S2 and Table 4). Previous studies suggested that codon sites with a high posterior probability should be regarded as positively selected sites [54], which means that these six genes may be under positive selection pressure [55]. After Jalview visualization, the results of the amino acid properties across each column of all species revealed that many amino acids vary between different genera, such as the 88th amino acid (G in *U. rockii* and *U. henryi*, R in other species) of the *rpl20* gene (Figure 7A) and other amino acids (marked with red blocks in Figure 7A). In the *ndhI* gene, two amino acids (the A in 168th and the P in 174th) were specific for *U. rockii* and *U. henryi*, and three amino acids (the 9th, 148th, and 165th, marked with red blocks in Figure 7B) were only possessed by *U. rockii*, *U. henryi*, and *S. adoxoides*. The amino acid properties of the other four genes (*atpA*, *atpB*, *rbcL*, and *psaA*) are shown in Figure S2. As we know, most amino acids may be under strong structural and functional constraints and not free to change [55]. We detected six genes with high posterior probability in codon site and many different amino acids among species, which may play an important role in *Urophysa* species evolution and environment adaptation. Populations of *U. rockii* and *U. henryi* are distributed only in karst regions of southern China, and the karst environments are characterized by low soil water content, insufficient light, and poor nutrient availability, which might have exerted strong selective forces on plant evolution [56].

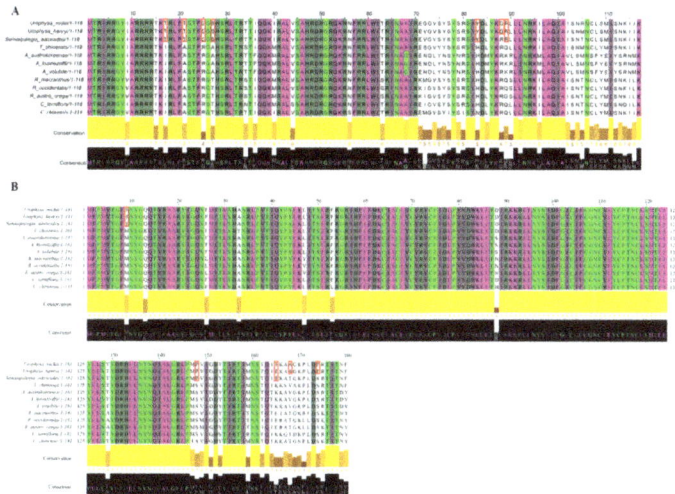

Figure 7. Two of the amino acids sequences that showed positive selection in the branch-site model test. (**A**) Amino acids sequences of the *rpl20* gene; (**B**) amino acids sequences of the *ndhI* gene. The red blocks represent the different amino acids.

Table 4. The potential positive selection test based on the branch-site model.

Gene Name	Null Hypothesis			Alternative Hypothesis			Significance Test		
	lnL	df	Omega ($\omega = 1$)	lnL	df	Omega ($\omega > 1$)	BEB	NEB	p-Value
psbI	−188.6475	26	1	−188.6475	27	3.40383	NA	NA	1
psbL	−164.11693	26	1	−164.1169	27	3.40719	NA	NA	1
rps14	−621.64162	26	1	−621.6416	27	3.40833	NA	NA	1
psaI	−214.67663	26	1	−214.6766	27	3.38764	NA	NA	1
atpH	−434.45059	26	1	−434.4506	27	3.35869	NA	NA	1
psaJ	−318.52192	26	1	−318.5219	27	3.4089	NA	NA	1
atpE	−868.20243	26	1	−868.2024	27	3.40891	NA	NA	1
atpA	−3297.629	26	1	−3297.41	27	69.43581	220, E, 0.794	NA	5.04×10^{-1}
petN	−126.25816	26	1	−126.2582	27	3.40693	NA	NA	1
rps11	−920.92455	26	1	−920.9246	27	1	NA	NA	1
psbT	−216.52331	26	1	−216.5233	27	1	NA	NA	1
ndhG	−1238.1161	26	1	−1238.116	27	3.33667	NA	NA	9.99×10^{-1}
ycf4	−1275.4093	26	1	−1275.409	27	3.40886	NA	NA	1
rps18	−567.98294	26	1	−567.9829	27	3.39414	NA	NA	1
petB	−1274.0507	26	1	−1274.051	27	3.403	NA	NA	1
rpl20	−1000.285	26	1	−999.941	27	112.30316	88, R, 0.683	NA	4.07×10^{-1}
psbN	−223.7602	26	1	−223.7602	27	3.40292	NA	NA	1
psbF	−198.46733	26	1	−198.4673	27	3.38407	NA	NA	1
petG	−206.74878	26	1	−206.7488	27	3.42095	NA	NA	1
psbK	−375.13705	26	1	−375.1371	27	3.4063	NA	NA	1
rpl36	−267.8099	26	1	−267.8099	27	1	NA	NA	1
rps2	−1620.734	26	1	−1620.734	27	3.40891	NA	NA	1
psbM	−179.71897	26	1	−179.719	27	3.4064	NA	NA	1
rpoB	−6830.0894	26	1	−6830.089	27	3.40847	NA	NA	9.99×10^{-1}
psaA	−4245.754	26	1	−4245.49	27	63.47379	28, R, 0.778	NA	4.66×10^{-1}
psbH	−540.92362	26	1	−540.9236	27	3.40123	NA	NA	1
ndhE	−616.75534	26	1	−616.7553	27	3.40218	NA	NA	1
atpB	−3133.747	26	1	−3133.75	27	1	115, N, 0.828	NA	1
ndhI	−1307.986	26	1	−1307.68	27	575.22179	174, S, 0.696	NA	4.35×10^{-1}
cemA	−1787.561	26	1	−1787.561	27	3.40891	NA	NA	1
ndhJ	−1001.4075	26	1	−1001.407	27	1	NA	NA	1
psbJ	−209.10513	26	1	−209.1051	27	3.38566	NA	NA	1

Table 4. *Cont.*

Gene Name	Null Hypothesis			Alternative Hypothesis			Significance Test		
	lnL	df	Omega ($\omega = 1$)	lnL	df	Omega ($\omega > 1$)	BEB	NEB	*p*-Value
petA	−1331.3789	26	1	−1331.379	27	3.4089	NA	NA	1
psbC	−2760.6743	26	1	−2760.674	27	1	NA	NA	1
ndhH	−2643.2896	26	1	−2643.29	27	1	NA	NA	9.98×10^{-1}
rbcL	−2937.477	26	1	−2937.41	27	5.22178	440, E, 0.736	NA	7.20×10^{-1}
clpP	−1301.1173	26	1	−1301.117	27	3.40876	NA	NA	1
ndhC	−731.03212	26	1	−731.0321	27	3.33544	NA	NA	1
ycf3	−935.76375	26	1	−935.7638	27	3.40891	NA	NA	1
psbD	−1922.7755	26	1	−1922.775	27	3.38592	NA	NA	1
psbA	−1960.3785	26	1	−1960.379	27	3.39639	NA	NA	1
petL	−172.24809	26	1	−172.2481	27	3.40087	NA	NA	1
rpl33	−413.59385	26	1	−413.5939	27	3.4089	NA	NA	1
psbE	−435.90511	26	1	−435.9051	27	3.40785	NA	NA	1
psaC	−498.98549	26	1	−498.9855	27	3.408	NA	NA	1
atpI	−1445.5558	26	1	−1445.556	27	3.39588	NA	NA	1
psaB	−4069.2947	26	1	−4069.295	27	3.41513	NA	NA	1

Bold types are positively selected sites. BEB: Bayesian Empirical Bayes; NEB: Naïve Empirical Bayes; Amino acid: (E: Glu; R: Arg; N: Asn; S: Ser).

However, five of the abovementioned six genes are involved in photosynthesis (*atpA, psaA, atpB, ndhI*, and *rbcL*) (Table 3). The gene *rpl20* is involved in translation, which is an important part of protein synthesis [57]. The genes *atpA* and *atpB* participate in ATP synthesis, which is the main source of energy for the functioning of living cells and all multicellular organisms [58]. Additionally, *rbcL* is the gene for the Rubisco large subunit protein, which is an important component of photosynthetic electron transport [59,60]. Most previous research has revealed that positive selection of the *rbcL* gene in land plants may be a common phenomenon [61]. All these genes might play important roles when founder effects occur in populations; both changes in selection pressures and genetic drift result in the rapid shift of these genes to a new, coadapted combination. Therefore, all these genes under positive selection give an indication of why *U. rockii* and *U. henryi* could adapt to the harsh environment of karst (characterized by low soil water content, periodic water deficiency, and poor nutrient availability). Moreover, the results of the gene effectiveness test (*rbcL* and *rpl20*) (Figure S3) suggested that these genes can distinguish the species of *Urophysa* and its relatives and can be used for future phylogenetic analyses. The six genes will not only provide insights into chloroplast genome evolution of species of *Urophysa*, but also offer valuable genetic markers for population phylogenomic studies of *Urophysa* and its close lineages.

3. Materials and Methods

3.1. Plant Materials and DNA Extraction

Fresh leaves of *U. rockii*, *U. henryi*, and *S. aquilegia* were collected from Jiangyou (Sichuan, China; coordinates: 31°59′ N, 104°51′ E), Yichang (Hubei, China; coordinates: 30°42′ N, 111°17′ E), and Nanchuan (Chongqing, China; coordinates: 30°04′ N, 90°33′ E), respectively. The fresh leaves from each site were immediately dried with silica gel for further DNA extraction. The total genomic DNA was extracted from leaf tissues with a modified Cetyl Trimethyl Ammonium (CTAB) method [62].

3.2. Chloroplast Genome Sequencing and Assembling

All cp genomes were sequenced using an Illumina Hiseq 2500 platform by Biomarker Technologies, Inc. (Beijing, China) In order to eliminate the interference from mitochondrial or nuclear DNAs, all the cp genome reads were extracted by mapping all raw reads to the reference cp genome of *Trollius chinensis* (KX752098) with Burrows Wheeler Alignment (BWA) [63]. High-quality reads were obtained using the CLC Genomics Workbench v7.5 (CLC Bio, Aarhus, Denmark) with the default parameters set. A few gaps in the assembled cp genomes were corrected by Sanger sequencing. The primers were designed using Lasergene 7.1 (DNASTAR, Madison, WI, USA). Primer synthesis and the sequencing of the polymerase chain reaction products were conducted by Sangon Biotech (Shanghai, China). The primers and amplifications are shown in Supplementary Table S5.

3.3. Genome Annotation and Analysis

The complete cp genomes were annotated using the online program DOGMA [64]. The annotation results were checked manually, and the codon positions were adjusted by comparing to a previously homologous gene from various chloroplast genomes present in the database using Geneious R11 (Biomatters, Ltd., Auckland, New Zealand). Furthermore, the OGDRAW1 program [65] was used to draw the circular plastid genome maps. GC content and codon usage were analyzed by the MEGA 6 software [66]. The complete cp genomes of *U. rockii*, *U. henryi*, and *S. adoxoides* are deposited in the GenBank under the accession numbers MH006686, MH142266, and MH142265, respectively.

3.4. Repeat Sequence Characterization and SSRs

Perl script MISA [67] was used to search for microsatellites (mono-, di-, tri-, tetra-, penta-, and hexa-nucleotides) loci in the cp genomes. The minimum numbers (thresholds) of the SSRs were 10, 5, 4, 3, 3, and 3 for mono-, di-, tri-, tetra-, penta-, and hexa-nucleotides, respectively. All the repeats were manually verified, and redundant results were removed. REPuter was employed to identify repeat sequences, including palindromic, forward, reverse, and complement, within the cp genome [68]. The following conditions for repeat identification were used: (1) Hamming distance of 3; (2) 90% or greater sequence identity; (3) a minimum repeat size of 30 bp.

3.5. Phylogenetic Analysis

Phylogenetic analysis was conducted using the single-copy genes of the three taxa, together with nine species downloaded from the NCBI GenBank (Tables S6 and S7). The sequences were aligned using MAFFT v5 [69] in GENEIOUS R11 (Biomatters, Ltd.) with the default parameters set and were manually adjusted in MEGA 6.0 [66]. Maximum parsimony (MP) analyses were conducted using PAUP [70]. All characters were equally weighted, gaps were treated as missing, and character states were treated as unordered. Heuristic search was performed with MULPARS option, tree bisection-reconnection (TBR) branch swapping, and random stepwise addition with 1000 replications. The maximum likelihood (ML) analyses were performed using RAxML 8.0 [71]. For ML analyses, the best-fit model, general time reversible (GTR) + G was used with 1000 bootstrap replicates. Bayesian inference (BI) was performed with Mrbayes v3.2 [72]. The Markov chain Monte Carlo (MCMC) analysis was run for 1×10^8 generations. The trees were sampled at every 1000 generations with the first 20% discarded as burn-in. The remaining trees were used to build a 50% majority-rule consensus tree. The stationarity was considered to be reached when the average standard deviation of split frequencies remained below 0.001. Additionally, in order to test the utility of different cp regions, phylogenetic analyses were performed for the complete chloroplast genome sequences and the CDS sequences, respectively.

3.6. Chloroplast Genome Nucleotide Diversity and Positive Selected Analysis

The cp genome sequences were aligned using MAFFT v5 [69] and adjusted manually. Furthermore, a sliding window analysis was conducted for nucleotide diversity in LSC, SSC, and IR regions of the cp genomes using the DnaSP version 5.1 [73]. In addition, to identify the genes under positive selection in *U. rockii* and *U. henryi*, endemic to special karst environment, an optimized branch-site model [74] combined with Bayesian Empirical Bayes (BEB) methods [55] were used by comparison with their relatives. We firstly extracted all CDS sequences from *U. rockii*, *U. henryi*, *S. adoxoides*, and nine closely related species downloaded from GenBank (Table S6). The single-copy CDS sequences between these twelve species were obtained (see the Table S4). Each single-copy CDS sequence of these twelve species was aligned according to their amino acid sequence alignment generated by MUSCLE [75], and the "number of gaps" in the alignments was further checked. Then, the alignments of the corresponding DNA codon sequences were further trimmed by TRIMAL [76], and the bona fide alignments were used to support the subsequent positive selection analysis. The optimized branch-site model in the CODEML program implemented in the PAML 4 package [77] was used to assess potential positive selection affecting individual codons along a specifically designated lineage, which was set as *U. rockii* and *U. henryi*. Selective pressure is measured by the ratio (ω) of the nonsynonymous substitution rate (dN) to the synonymous substitutions rate (dS). A ratio $\omega > 1$ indicates positive selection, $\omega = 1$ implies neutral selection, and $\omega < 1$ suggests negative selection [78]. Log-likelihood values were calculated in an alternative branch-site model (Model = 2; NSsites = 2; and Fix = 0) that allowed ω to vary among different codons along particular lineages and a neutral branch-site model (Model = 2; NSsites = 2; Fix = 1; Fix $\omega = 1$) that confined the codon sites under neutral selection ($\omega = 1$) on the basis of the likelihood ratio tests (LRT). The right-tailed chi-square test was performed to calculate the p values based on the difference in log-likelihood values between the alternative model and the neutral model with one degree of freedom to assess the model fit. Then, the p values were further adjusted according to multiple statistical tests [79]. A gene with an adjusted p value smaller than 0.05 and with positively selected sites was considered a positively selected gene (PSG). Moreover, in order to identify specific amino acid sites that are potentially under positive selection, a BEB method was implemented to calculate the posterior probabilities for sites classes. Codon sites with a high posterior probability were regarded as positively selected sites [54]. Jalview [80] was used to view the amino acid sequences of positively selected genes. In the end, in order to test the effectiveness of genes under positive selection, we randomly chose two genes to conduct the phylogenetic analyses.

Supplementary Materials: Supplementary Materials are available online at http://www.mdpi.com/1422-0067/19/7/1847/s1.

Author Contributions: D.-F.X., Y.Y., S.-D.Z., and X.-J.H. conceived and designed the experiment; D.-F.X., J.L., and S.-D.Z. collected the materials; D.-F.X., Y.-Q.D., Y.Y., and H.-Y.L. participated in data analysis and manuscript drafting; D.-F.X., Y.-Q.D., X.-J.H., and S.-D.Z. revised the manuscript; all authors read and approved the final manuscript.

Funding: This work was supported by the National Natural Science Foundation of China (Grant Nos. 31470009, 31570198, 31500188), the Specimen Platform of China, Teaching Specimen's sub-platform (Available website: http://mnh.scu.edu.cn/), the Science and Technology Basic Work (Grant No. 2013FY112100).

Acknowledgments: We acknowledge Fang-Yu Jin, Hao Li, Fu-Min Xie, and Xin Yang for their help in materials collection.

Conflicts of Interest: The authors declare no conflict of interest.

References

1. Fu, D.Z.; Orbelia, R.R. *Flora of China*; Science Press: Beijing, China, 2001; Volume 6, pp. 277–278.
2. Xie, D.F.; Li, M.J.; Tan, J.B.; Price, M.; Xiao, Q.Y.; Zhou, S.D.; He, X.J. Phylogeography and genetic effects of habitat fragmentation on endemic *Urophysa* (Ranunculaceae) in Yungui Plateau and adjacent regions. *PLoS ONE* **2017**, *12*, e0186378. [CrossRef] [PubMed]
3. Du, B.G.; Zhu, D.Y.; Yang, Y.J.; Shen, J.; Yang, F.L.; Su, Z.Y. Living situation and protection strategies of endangered *Urophysa rockii. Jiangsu J. Agri. Sci.* **2010**, *1*, 324–325.
4. Wang, J.X.; He, X.J.; Xu, W.; Meng, W.K.; Su, Z.Y. Preliminary study on *Urophysa rockii*. II. Biological characteristics, ecological characteristics and community analysis. *J. Sichuan For. Sci. Technol.* **2011**, *32*, 28–39.
5. Zhang, Y.X.; Hu, H.Y.; He, X.J. Genetic diversity of *Urophysa rockii* Ulbrich, an endangered and rare species, detected by ISSR. *Acta Bot. Boreal.-Occident. Sin.* **2013**, *33*, 1098–1105.
6. Zhang, Y.X.; Hu, H.Y.; Yang, L.J.; Wang, C.B.; He, X.J. Seed dispersal and germination of an endangered and rare species *Urophysa rockii* (Ranunculaceae). *Acta Bot. Boreal.-Occident. Sin.* **2013**, *35*, 303–309.
7. Park, M.; Park, H.; Lee, H.; Lee, B.H.; Lee, J. The complete plastome sequence of an antarctic bryophyte *Sanionia uncinata* (hedw.) loeske. *Int. J. Mol. Sci.* **2018**, *19*, 709. [CrossRef] [PubMed]
8. Dong, W.P.; Liu, H.; Xu, C.; Zuo, Y.J.; Chen, Z.J.; Zhou, S.L. A chloroplast genomic strategy for designing taxon specific DNA mini-barcodes: A case study on ginsengs. *BMC Genet.* **2014**, *15*, 138. [CrossRef] [PubMed]
9. Curci, P.L.; de Paola, D.; Danzi, D.; Vendramin, G.G.; Sonnante, G. Complete chloroplast genome of the multifunctional crop Globe artichoke and comparison with other Asteraceae. *PLoS ONE* **2015**, *10*, e0120589. [CrossRef] [PubMed]
10. Downie, S.R.; Jansen, R.K. A comparative analysis of whole plastid genomes from the Apiales: Expansion and contraction of the inverted repeat, mitochondrial to plastid transfer of DNA, and identification of highly divergent noncoding regions. *Syst. Bot.* **2015**, *40*, 336–351. [CrossRef]
11. Nadachowska-Brzyska, K.; Li, C.; Smeds, L.; Zhang, G.J.; Ellegren, H. Temporal dynamics of avian populations during pleistocene revealed by whole-genome sequences. *Curr. Biol.* **2015**, *25*, 1375–1380. [CrossRef] [PubMed]
12. Suo, Z.L.; Li, W.Y.; Jin, X.B.; Zhang, H.J. A new nuclear DNA marker revealing both microsatellite variations and single nucleotide polymorphic loci: A case study on classification of cultivars in *Lagerstroemia indica* L. *J. Microb. Biochem. Technol.* **2016**, *8*, 266–271. [CrossRef]
13. Saina, J.K.; Li, Z.Z.; Gichira, A.W.; Liao, Y.Y. The complete chloroplast genome sequence of tree of heaven (*Ailanthus altissima* (mill.) (Sapindales: Simaroubaceae), an important pantropical tree. *Int. J. Mol. Sci.* **2018**, *19*, 929. [CrossRef] [PubMed]
14. Yurina, N.P.; Odintsova, M.S. Comparative structural organization of plant chloroplast and mitochondrial genomes. *Genetika* **1998**, *34*, 5–22.

15. Jansen, R.K.; Raubeson, L.A.; Boore, J.L.; DePamphilis, C.W.; Chumley, T.W.; Haberle, R.C.; Wyman, S.K.; Alverson, A.; Peery, R.; Herman, S.J.; et al. Methods for obtaining and analyzing whole chloroplast genome sequences. *Method Enzymol.* **2005**, *395*, 348–384.

16. Jansen, R.K.; Ruhlman, T.A. Plastid Genomes of Seed Plants. In *Genomics of Chloroplasts and Mitochondria*; Bock, R., Knoop, V., Eds.; Springer: Dordrecht, The Netherlands, 2012; pp. 103–126.

17. Choi, K.S.; Chung, M.G.; Park, S. The complete chloroplast genome sequences of three *Veroniceae* species (Plantaginaceae): Comparative analysis and highly divergent regions. *Front. Plant Sci.* **2016**, *7*, 355. [CrossRef] [PubMed]

18. Dong, W.L.; Wang, R.N.; Zhang, N.Y.; Fan, W.B.; Fang, M.F.; Li, Z.H. Molecular evolution of chloroplast genomes of orchid species: Insights into phylogenetic relationship and adaptive evolution. *Int. J. Mol. Sci.* **2018**, *19*, 716. [CrossRef] [PubMed]

19. Dong, W.; Liu, J.; Yu, J.; Wang, L.; Zhou, S. Highly variable chloroplast markers for evaluating plant phylogeny at low taxonomic levels and for DNA barcoding. *PLoS ONE* **2012**, *7*, e35071. [CrossRef] [PubMed]

20. Krak, K.; Vít, P.; Belyayev, A.; Douda, J.; Hreusová, L.; Mandák, B. Allopolyploid origin of *Chenopodium album* s. str. (Chenopodiaceae): A molecular and cytogenetic insight. *PLoS ONE* **2016**, *11*, e0161063. [CrossRef] [PubMed]

21. Smith, D.R. Mutation rates in plastid genomes: They are lower than you might think. *Genome Biol. Evol.* **2015**, *7*, 1227–1234. [CrossRef] [PubMed]

22. Jansen, R.K.; Cai, Z.; Raubeson, L.A.; Daniell, H.; Depamphilis, C.W.; Leebensmack, J.; Müller, K.F.; Guisinger-Bellian, M.; Haberle, R.C.; Chumley, T.W.; et al. Analysis of 81 genes from 64 plastid genomes resolves relationships in angiosperms and identifies genome-scale evolutionary patterns. *Proc. Natl. Acad. Sci. USA* **2007**, *104*, 19369–19374. [CrossRef] [PubMed]

23. Qian, J.; Song, J.; Gao, H.; Zhu, Y.; Xu, J.; Pang, X. The complete chloroplast genome sequence of the medicinal plant *Salvia miltiorrhiza*. *PLoS ONE* **2013**, *8*, e57607. [CrossRef] [PubMed]

24. Asaf, S.; Waqas, M.; Khan, A.L.; Khan, M.A.; Kang, S.M.; Imran, Q.M.; Shahzad, R.; Bilal, S.; Yun, B.W.; Lee, I.J.; et al. The complete chloroplast genome of wild rice (*Oryza minuta*) and its comparison to related species. *Front. Plant Sci.* **2017**, *8*, 304. [CrossRef] [PubMed]

25. Gu, C.; Tembrock, L.R.; Zheng, S.; Wu, Z. The complete chloroplast genome of *Catha edulis*: A comparative analysis of genome features with related species. *Int. J. Mol. Sci.* **2018**, *19*, 525. [CrossRef] [PubMed]

26. Huang, J.; Chen, R.; Li, X. Comparative analysis of the complete chloroplast genome of four known *Ziziphus* species. *Genes* **2017**, *8*, 340. [CrossRef] [PubMed]

27. Yi, X.; Gao, L.; Wang, B.; Su, Y.J.; Wang, T. The complete chloroplast genome sequence of *Cephalotaxus oliveri* (Cephalotaxaceae): Evolutionary comparison of *Cephalotaxus* chloroplast DNAs and insights into the loss of inverted repeat copies in gymnosperms. *Genome Biol. Evol.* **2013**, *5*, 688–698. [CrossRef] [PubMed]

28. Cavalier-Smith, T. Chloroplast evolution: Secondary symbiogenesis and multiple losses. *Curr. Biol.* **2002**, *12*, 62–64. [CrossRef]

29. Asano, T.; Tsudzuki, T.; Takahashi, S.; Shimada, H.; Kadowaki, K. Complete nucleotide sequence of the sugarcane (*Saccharum officinarum*) chloroplast genome: A comparative analysis of four monocot chloroplast genomes. *DNA Res.* **2004**, *11*, 93–99. [CrossRef] [PubMed]

30. Timme, R.E.; Kuehl, J.V.; Boore, J.L.; Jansen, R.K. A comparative analysis of the *Lactuca* and *Helianthus* (Asteraceae) plastid genomes: Identification of divergent regions and categorization of shared repeats. *Am. J. Bot.* **2007**, *94*, 302–312. [CrossRef] [PubMed]

31. Nie, X.J.; Lv, S.Z.; Zhang, Y.X.; Du, X.H.; Wang, L.; Biradar, S.S.; Tan, X.F.; Wan, F.H.; Weining, S. Complete chloroplast genome sequence of a major invasive species, crofton weed (*Ageratina adenophora*). *PLoS ONE* **2012**, *7*, e36869. [CrossRef] [PubMed]

32. Dong, W.P.; Xu, C.; Li, D.L.; Jin, X.B.; Lu, Q.; Suo, Z.L. Comparative analysis of the complete chloroplast genome sequences in psammophytic *Haloxylon* species (Amaranthaceae). *Peer J.* **2016**, *4*, e2699. [CrossRef] [PubMed]

33. Kaur, S.; Panesar, P.S.; Bera, M.B.; Kaur, V. Simple sequence repeat markers in genetic divergence and marker-assisted selection of rice cultivars: A review. *Crit. Rev. Food Sci. Nutr.* **2015**, *55*, 41–49. [CrossRef] [PubMed]

34. Yang, Y.; Zhou, T.; Duan, D.; Yang, J.; Feng, L.; Zhao, G. Comparative analysis of the complete chloroplast genomes of five *Quercus* species. *Front. Plant Sci.* **2016**, *7*, 959. [CrossRef] [PubMed]

35. Powell, W.; Morgante, M.; McDevitt, R.; Vendramin, G.G.; Rafalski, J.A. Polymorphic simple sequence repeat regions in chloroplast genomes-applications to the population genetics of pines. *Proc. Natl. Acad. Sci. USA* **1995**, *92*, 7759–7763. [CrossRef] [PubMed]

36. Provan, J.; Corbett, G.; McNicol, J.W.; Powell, W. Chloroplast DNA variability in wild and cultivated rice (*Oryza* spp.) revealed by polymorphic chloroplast simple sequence repeats. *Genome* **1997**, *40*, 104–110. [CrossRef] [PubMed]

37. Pauwels, M.; Vekemans, X.; Gode, C.; Frerot, H.; Castric, V.; Saumitou-Laprade, P. Nuclear and chloroplast DNA phylogeography reveals vicariance among European populations of the model species for the study of metal tolerance, *Arabidopsis halleri* (Brassicaceae). *New Phytol.* **2012**, *193*, 916–928. [CrossRef] [PubMed]

38. Rose, O.; Falush, D. A threshold size for microsatellite expansion. *Mol. Biol. Evol.* **1998**, *15*, 613–615. [CrossRef] [PubMed]

39. Raubeson, L.A.; Peery, R.; Chumley, T.W.; Dziubek, C.; Fourcade, H.M.; Boore, J.L.; Jansen, R.K. Comparative chloroplast genomics: Analyses including new sequences from the angiosperms *Nuphar advena* and *Ranunculus macranthus*. *BMC Genom.* **2007**, *8*, 174. [CrossRef] [PubMed]

40. Huotari, T.; Korpelainen, H. Complete chloroplast genome sequence of *Elodea Canadensis* and comparative analyses with other monocot plastid genomes. *Gene* **2012**, *508*, 96–105. [CrossRef] [PubMed]

41. Wang, W.B.; Yu, H.; Wang, J.H.; Lei, W.J.; Gao, J.H.; Qiu, X.P.; Wang, J.S. The complete chloroplast genome sequences of the medicinal plant *Forsythia suspensa* (Oleaceae). *Int. J. Mol. Sci.* **2017**, *18*, 2288. [CrossRef] [PubMed]

42. Asaf, S.; Khan, A.L.; Khan, A.R.; Waqas, M.; Kang, S.M.; Khan, M.A.; Lee, S.M.; Lee, I.J. Complete chloroplast genome of *Nicotiana otophora* and its comparison with related species. *Front. Plant Sci.* **2016**, *7*, 447. [CrossRef] [PubMed]

43. Kuang, D.Y.; Wu, H.; Wang, Y.L.; Gao, L.M.; Zhang, S.Z.; Lu, L. Complete chloroplast genome sequence of *Magnolia kwangsiensis* (Magnoliaceae): Implication for DNA barcoding and population genetics. *Genome* **2011**, *54*, 663–673. [CrossRef] [PubMed]

44. Chen, J.; Hao, Z.; Xu, H.; Yang, L.; Liu, G.; Sheng, Y. The complete chloroplast genome sequence of the relict woody plant *Metasequoia glyptostroboides* Hu et Cheng. *Front. Plant Sci.* **2015**, *6*, 447. [CrossRef] [PubMed]

45. Kim, K.J.; Lee, H.L. Complete chloroplast genome sequences from Korean ginseng (*Panax schinseng* Nees) and comparative analysis of sequence evolution among 17 vascular plants. *DNA Res.* **2004**, *11*, 247–261. [CrossRef] [PubMed]

46. Hu, Y.; Woeste, K.E.; Zhao, P. Completion of the chloroplast genomes of five Chinese *Juglans* and their contribution to chloroplast phylogeny. *Front. Plant Sci.* **2017**, *7*, 1955. [CrossRef] [PubMed]

47. Wang, R.J.; Cheng, C.L.; Chang, C.C.; Wu, C.L.; Su, T.M.; Chaw, S.M. Dynamics and evolution of the inverted repeat-large single copy junctions in the chloroplast genomes of monocots. *BMC Evol. Biol.* **2008**, *8*, 36. [CrossRef] [PubMed]

48. Yang, M.; Zhang, X.; Liu, G.; Yin, Y.; Chen, K.; Yun, Q. The complete chloroplast genome sequence of date palm (*Phoenix dactylifera* L.). *PLoS ONE* **2010**, *5*, e12762. [CrossRef] [PubMed]

49. Li, Z.Z.; Saina, J.K.; Gichira, A.W.; Kyalo, C.M.; Wang, Q.F.; Chen, J.M. Comparative genomics of the balsaminaceae sister genera *Hydrocera triflora* and *Impatiens pinfanensis*. *Int. J. Mol. Sci.* **2018**, *19*, 319. [CrossRef] [PubMed]

50. Li, C.Y. *Classification and Systematics of the Aquilegiinae Tamura*; The Chinese Academy of Science: Beijing, China, 2006.

51. Bastida, J.M.; Alcántara, J.M.; Rey, P.J.; Vargas, P.; Herrera, C.M. Extended phylogeny of *Aquilegia*: The biogeographical and ecological patterns of two simultaneous but contrasting radiations. *Plant Syst. Evol.* **2010**, *284*, 171–185. [CrossRef]

52. Fior, S.; Li, M.; Oxelman, B.; Viola, R.; Hodges, S.A.; Ometto, L.; Varotto, C. Spatiotemporal reconstruction of the *Aquilegia* rapid radiation through next-generation sequencing of rapidly evolving cpDNA regions. *New Phytol.* **2013**, *198*, 579–592. [CrossRef] [PubMed]

53. Wei, W.; Lu, A.M.; Yi, R.; Endress, M.E.; Chen, Z.D. Phytogeny and classification of Ranunculales: Evidence from four molecular loci and morphological data. *Perspect. Plant Ecol. Evol. Syst.* **2009**, *11*, 81–110.

54. Lan, Y.; Sun, J.; Tian, R.M.; Bartlett, D.H.; Li, R.S.; Wong, Y.H.; Zhang, W.P.; Qiu, J.W.; Xu, T.; He, L.S.; et al. Molecular adaptation in the world's deepest-living animal: Insights from transcriptome sequencing of the hadal amphipod *Hirondellea gigas*. *Mol. Ecol.* **2017**, *26*, 3732–3743. [CrossRef] [PubMed]

55. Yang, Z.; Wong, W.S.; Nielsen, R. Bayes empirical Bayes inference of amino acid sites under positive selection. *Mol. Biol. Evol.* **2005**, *22*, 1107–1118. [CrossRef] [PubMed]

56. Ai, B.; Gao, Y.; Zhang, X.; Tao, J.; Kang, M.; Huang, H. Comparative transcriptome resources of eleven *Primulina* species, a group of 'stone plants' from a biodiversity hot spot. *Mol. Ecol. Resour.* **2015**, *15*, 619–632. [CrossRef] [PubMed]

57. Muto, A.; Ushida, C. Transcription and translation. *Methods Cell Biol.* **1995**, *48*, 483.

58. Romanovsky, Y.M.; Tikhonov, A.N. Molecular energy transducers of the living cell. Proton ATP synthase: A rotating molecular motor. *Physics-Uspekhi* **2010**, *53*, 931–956. [CrossRef]

59. Allahverdiyeva, Y.; Mamedov, F.; Mäenpää, P.; Vass, I.; Aro, E.M. Modulation of photosynthetic electron transport in the absence of terminal electron acceptors: Characterization of the rbcL deletion mutant of tobacco. *Biochim. Biophys. Acta Bioenerg.* **2005**, *1709*, 69–83. [CrossRef] [PubMed]

60. Piot, A.; Hackel, J.; Christin, P.A.; Besnard, G. One-third of the plastid genes evolved under positive selection in PACMAD grasses. *Planta* **2018**, *247*, 255–266. [CrossRef] [PubMed]

61. Kapralov, M.V.; Filatov, D.A. Widespread positive selection in the photosynthetic Rubisco enzyme. *BMC Evol. Biol.* **2007**, *7*, 73–82. [CrossRef] [PubMed]

62. Doyle, J.J.; Doyle, J.L. A rapid DNA isolation procedure for small quantities of fresh leaf tissue. *Phytochem Bull.* **1987**, *19*, 11–15.

63. Li, H.; Durbin, R. Fast and accurate short read alignment with Burrows-Wheeler Transform. *Bioinformatics* **2009**, *25*, 1754–1760. [CrossRef] [PubMed]

64. Wyman, S.K.; Jansen, R.K.; Boore, J.L. Automatic annotation of organellar genomes with DOGMA. *Bioinformatics* **2004**, *20*, 3252–3255. [CrossRef] [PubMed]

65. Lohse, M.; Drechsel, O.; Kahlau, S.; Bock, R. Organellar genome draw—A suite of tools for generating physical maps of plastid and mitochondrial genomes and visualizing expression data sets. *Nucleic Acids Res.* **2013**, *41*, 575. [CrossRef] [PubMed]

66. Kumar, S.; Nei, M.; Dudley, J.; Tamura, K. MEGA: A biologist centric software for evolutionary analysis of DNA and protein sequences. *Brief. Bioinform.* **2008**, *9*, 299–306. [CrossRef] [PubMed]

67. Thiel, T.; Michalek, W.; Varshney, R.; Graner, A. Exploiting EST databases for the development and characterization of gene derived SSR-markers in barley (*Hordeum vulgare* L.). *Theor. Appl. Genet.* **2003**, *106*, 411–422. [CrossRef] [PubMed]

68. Kurtz, S.; Choudhuri, J.V.; Ohlebusch, E.; Schleiermacher, C.; Stoye, J.; Giegerich, R. REPuter: The manifold applications of repeat analysis on a genomic scale. *Nucleic Acids Res.* **2001**, *29*, 4633–4642. [CrossRef] [PubMed]

69. Katoh, K.; Standley, D.M. MAFFT multiple sequence alignment software version 7: Improvements in performance and usability. *Mol. Biol. Evol.* **2013**, *30*, 772–780. [CrossRef] [PubMed]

70. Swofford, D.L. *PAUP*. Phylogenetic Analysis Using Parsimony (*and Other Methods)*; Version 4b10; Sinauer: Sunderland, MA, USA, 2003.

71. Stamatakis, A. RAxML-VI-HPC: Maximum likelihood-based phylogenetic analyses with thousands of taxa and mixed models. *Bioinformatics* **2006**, *22*, 2688–2690. [CrossRef] [PubMed]

72. Ronquist, F.; Teslenko, M.; van der Mark, P.; Ayres, D.L.; Darling, A.; Hohna, S.; Larget, B.; Liu, L.; Suchard, M.A.; Huelsenbeck, J. MrBayes 3.2: Efficient Bayesian phylogenetic inference and model choice across a large model space. *Syst. Biol.* **2012**, *61*, 539–542. [CrossRef] [PubMed]

73. Librado, P.; Rozas, J. DnaSP v5: A software for comprehensive analysis of DNA polymorphism data. *Bioinformatics* **2009**, *25*, 1451–1452. [CrossRef] [PubMed]

74. Yang, Z.; dos Reis, M. Statistical properties of the branch-site test of positive selection. *Mol. Biol. Evol.* **2011**, *28*, 1217–1228. [CrossRef] [PubMed]

75. Edgar, R.C. MUSCLE: Multiple sequence alignment with high accuracy and high throughput. *Nucleic Acids Res.* **2004**, *32*, 1792–1797. [CrossRef] [PubMed]

76. Capella-Gutierrez, S.; Silla-Martínez, J.M.; Gabaldon, T. TrimAl: A tool for automated alignment trimming in large-scale phylogenetic analyses. *Bioinformatics* **2009**, *25*, 1972–1973. [CrossRef] [PubMed]

77. Yang, Z. PAML 4: Phylogenetic analysis by maximum likelihood. *Mol. Biol. Evol.* **2007**, *24*, 1586–1591. [CrossRef] [PubMed]

78. Yang, Z.; Nielsen, R. Codon-substitution models for detecting molecular adaptation at individual sites along specific lineages. *Mol. Biol. Evol.* **2002**, *19*, 908–917. [CrossRef] [PubMed]

79. Benjamini, Y.; Hochberg, Y. Controlling the false discovery rate: A practical and powerful approach to multiple testing. *J. R. Stat. Soc. B* **1995**, *57*, 289–300.

80. Clamp, M.; Cuff, J.; Searle, S.M.; Barton, G.J. The Jalview java alignment editor. *Bioinformatics* **2004**, *20*, 426–427. [CrossRef] [PubMed]

International Journal of
Molecular Sciences

MDPI

Article

Exploring the History of Chloroplast Capture in *Arabis* Using Whole Chloroplast Genome Sequencing

Akira Kawabe *, Hiroaki Nukii and Hazuka Y. Furihata

Faculty of Life Sciences, Kyoto Sangyo University, Kyoto, Kyoto 603-8555, Japan;
g1447799@cc.kyoto-su.ac.jp (H.N.); hazuka.furi@cc.kyoto-su.ac.jp (H.Y.F.)
* Correspondence: akiraka@cc.kyoto-su.ac.jp; Tel.: +81-75-705-3126

Received: 26 January 2018; Accepted: 16 February 2018; Published: 18 February 2018

Abstract: Chloroplast capture occurs when the chloroplast of one plant species is introgressed into another plant species. The phylogenies of nuclear and chloroplast markers from East Asian *Arabis* species are incongruent, which indicates hybrid origin and shows chloroplast capture. In the present study, the complete chloroplast genomes of *A. hirsuta*, *A. nipponica*, and *A. flagellosa* were sequenced in order to analyze their divergence and their relationships. The chloroplast genomes of *A. nipponica* and *A. flagellosa* were similar, which indicates chloroplast replacement. If hybridization causing chloroplast capture occurred once, divergence between recipient species would be lower than between donor species. However, the chloroplast genomes of species with possible hybrid origins, *A. nipponica* and *A. stelleri*, differ at similar levels to possible maternal donor species *A. flagellosa*, which suggests that multiple hybridization events have occurred in their respective histories. The mitochondrial genomes exhibited similar patterns, while *A. nipponica* and *A. flagellosa* were more similar to each other than to *A. hirsuta*. This suggests that the two organellar genomes were co-transferred during the hybridization history of the East Asian *Arabis* species.

Keywords: *Arabis*; chloroplast capture; Brassicaceae

1. Introduction

The genus *Arabis* includes about 70 species that are distributed throughout the northern hemisphere. The genus previously included many more species, but a large number of these were reclassified into other genera, including *Arabidopsis*, *Turritis*, and *Boechera*, *Crucihimalaya*, *Scapiarabis*, and *Sinoarabis* [1–6]. Because of their highly variable morphology and life histories, *Arabis* species have been used for ecological and evolutionary studies of morphologic and phenotypic traits [7–11]. The whole genome of *Arabis alpina* has been sequenced, providing genomic information for evolutionary analyses [12,13].

Molecular phylogenetic studies of *Arabis* species have been conducted to determine species classification and also correlation to morphological evolution of *Arabis* species [10,14,15]. Despite having similar morphologies, *A. hirsuta* from Europe, North America, and East Asia have been placed in different phylogenetic positions and are now considered distinct species. For example, East Asian *A. hirsuta*, which was previously classified as *A. hirsuta* var. *nipponica,* is now designated as *A. nipponica* [16]. Meanwhile, nuclear ITS sequences indicated that *A. nipponica, A. stelleri,* and *A. takeshimana* were closely related to European *A. hirsuta*. However, chloroplast *trnLF* sequences indicated that the species were closely related to East Asian *Arabis* species [14,16]. Such incongruent nuclear and organellar phylogenies have been reported from in other plant species and this is generally known as "chloroplast capture" [17,18], which is a process that involves hybridization and many successive backcrosses [17]. When chloroplast capture happens, the chloroplast genome of a species is replaced by another species' chloroplast genome. *A. nipponica* may have originated from the hybridization of *A. hirsuta* or *A. sagittata* and East Asian *Arabis* species (similar to *A. serrata*, *A. paniculata*, and *A. flagellosa*), which act as paternal and maternal parents, respectively [14,16]. However, the evolutionary history and hybridization processes of *A. nipponica* and other East Asian *Arabis* species still need to be clarified. Because these conclusions for incongruence between nuclear and chloroplast phylogenies came from analyzing a small number of short sequences, hybridized species, the divergence level, and the classification of species are somewhat ambiguous. In the present study, the whole chloroplast genomes of three *Arabis* species were sequenced in order to analyze their divergence and evolutionary history. The whole chloroplast genome sequences also provide a basis for future marker development.

2. Results

2.1. Chloroplast Genome Structure of Arabis Species

The structures of the whole chloroplast genomes are summarized in Table 1, which also includes previously reported *Arabis* chloroplast genomes and the chloroplast genome of the closely related species *Draba nemorosa*. The chloroplast genome structure identified in the present study is shown as a circular map (see Figure 1). The complete chloroplast genomes of the *Arabis* species had total lengths of 152,866–153,758 base pairs, which included 82,338 to 82,811 base pair long single copy (LSC) regions and 17,938 to 18,156 base pair short single copy (SSC) regions, which were separated by a pair of 26,421 to 26,933 base pair inverted repeat (IR) regions. The structure and length are conserved, and are similar to other Brassicaceae species' chloroplast genome sequences [19–22]. The complete genomes contain 86 protein-coding genes, 37 tRNA genes, and eight rRNA genes. Of these, seven protein-coding genes, seven tRNA genes, and four rRNA genes were located in the IR regions, and were therefore duplicated. The *rps16* gene became a pseudogene in *A. flagellosa*, *A. hirsuta*, and *A. nipponica* strain Midori, which was previously reported as a related species [23]. In addition, the *rps16* sequences of *D. nemorosa*, *A. stelleri*, *A. flagellosa*, *A. hirsuta*, and *A. nipponica* shared a 10 base pair deletion in the first exon, while *A. stelleri*, *A. flagellosa*, *A. hirsuta*, and *A. nipponica* shared a 1 base pair deletion in the second exon and *D. nemorosa* lacked the second exon entirely. The *rps16* sequence of *A. alpina* also lacked part of the second exon and had mutations in the start and stop codons. Therefore, different patterns of *rps16* pseudogenization were observed in *A. alpina* and the other *Arabis* species, as was previously suggested [23]. The *A. alpina* lineage had acquired independent dysfunctional mutation(s). The patterns observed for the European *A. hirsuta* revealed that the pseudogenization of *rps16* in the other *Arabis* species might not have occurred independently but, instead, occurred before the divergence of *D. nemorosa* and other Arabis species after splitting from *A. alpina*.

Table 1. Summary of chloroplast genome structure in *Arabis* species.

Species	Strain	Nucleotide Length (bp)				GC Contents (%)				NCBI #	Reference
		Entire	LSC	SSC	IR	Entire	LSC	SSC	IR		
Draba nemorosa	JO21	153289	82457	18126	26353	36.47	34.27	29.3	42.39	AP009373 (NC009272)	
Arabis alpina		152866	82338	17938	26933	36.45	34.21	29.31	42.39	HF934132 (NC023367)	[25]
Arabis hirsuta	Brno	153758	82710	18156	26446	36.4	34.15	29.16	42.41	LC361350	this study
Arabis flagellosa	Kifune	153673	82775	18052	26423	36.4	34.13	29.22	42.41	LC361351	this study
Arabis stelleri		153683	82807	18030	26423	36.39	34.11	29.22	42.42	KY126841	[23]
Arabis nipponica	JO23	153689	82811	18036	26421	36.4	34.1	29.31	42.42	AP009369 (NC009268)	
Arabis nipponica	Midori	153668	82772	18052	26422	36.39	34.1	29.24	42.42	LC361349	this study

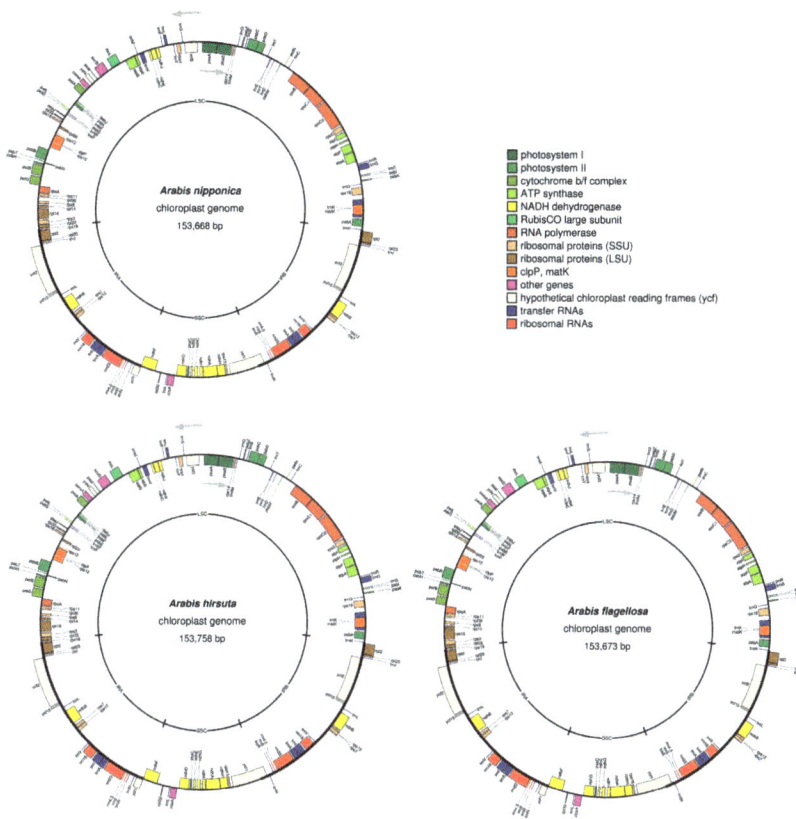

Figure 1. Chloroplast genome structure of *Arabis* species. Genes shown outside the map circles are transcribed clockwise, while those drawn inside are transcribed counterclockwise. Genes from different functional groups are color-coded according to the key at the top right. The positions of long single copy (LSC), short single copy (SSC), and two inverted repeat (IR: IRA and IRB) regions are shown in the inner circles.

2.2. Chloroplast Genome Divergence

Phylogenetic trees were generated by using whole chloroplast genome sequences and concatenated coding sequence (CDS) regions (see Figure 2). The inclusion of other Brassicaceae members revealed that *D. nemorosa* should be placed within *Arabis*, as previously reported [24]. In both trees, the two *A. nipponica* strains were grouped with *A. flagellosa* and *A. stelleri*. Although several nodes were supported by high bootstrap probabilities, the nearly identical sequences of the four East Asian *Arabis* species made them indistinguishable.

The divergence among the *Arabis* chloroplast genomes was shown using a VISTA plot (see Figure 3) and this was summarized in Table 2. The genome sequences of the two Japanese *A. nipponica* strains differed by only 55 nucleotide substitutions (0.036% per site), while those of *A. hirsuta* and *A. nipponica* differed by about 3500 sites (2.4% per site). The chloroplast genomes of *A. nipponica* and the other two East Asian *Arabis* species were also very similar (~100 nucleotide differences, <0.1% per site). Additionally, the 35 CDS regions, 29 tRNA genes, and four rRNA genes of the four East Asian *Arabis* species were identical, with three, 27, and four, respectively, also found to be identical in *A. hirsuta*. The levels of divergence between the East Asian *Arabis* species were similar to previously reported levels of variation within the local *A. alpina* population, in which 130 SNPs were identified among 24 individuals (Waterson's $\theta = 0.02\%$) [25]. If the hybridization event had facilitated chloroplast capture, the divergence between the *A. stelleri* and *A. nipponica* chloroplast genomes should have been less than their divergence from *A. flagellosa*. However, the divergence between the potential hybrid-origin species (*A. stelleri* and *A. nipponica*: 0.068 to 0.085) was similar to their divergence from *A. flagellosa* (0.056 to 0.086). Although the level of divergence was too low to make reliable comparisons, it is possible that *A. stelleri* and *A. nipponica* originated from independent hybridization events or the introgression process may still be ongoing.

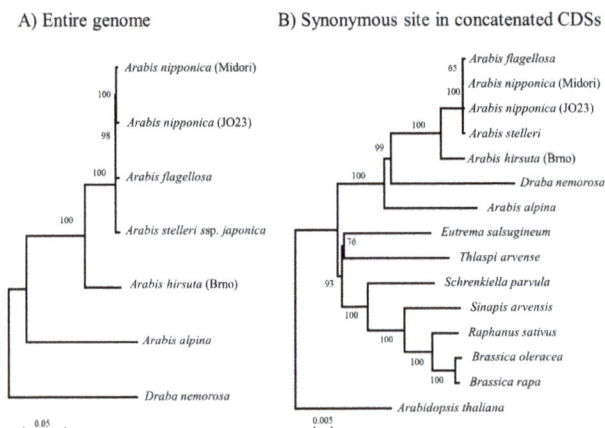

Figure 2. Chloroplast genome-based phylogenetic trees of *Arabis* species. The neighbor-joining trees were constructed using both (**A**) whole chloroplast genomes and (**B**) synonymous divergence from concatenated CDS. Numbers beside the nodes indicate bootstrap probabilities (%). Scale bars are shown at the bottom left of each tree.

Table 2. Divergence between species.

Compared Species			# of Differences	Divergence (%: Ks with JC Correction)
Draba nemorosa	vs.	*Arabis alpina*	4475	2.976
Draba nemorosa	vs.	*Arabis hirsuta*	4219	2.801
Draba nemorosa	vs.	*Arabis flagellosa*	4262	2.765
Draba nemorosa	vs.	*Arabis stelleri*	4171	2.771
Draba nemorosa	vs.	*Arabis nipponica* (JO23)	4150	2.757
Draba nemorosa	vs.	*Arabis nipponica* (Midori)	4131	2.745
Arabis alpina	vs.	*Arabis hirsuta*	3566	2.366
Arabis alpina	vs.	*Arabis flagellosa*	3571	2.371
Arabis alpina	vs.	*Arabis stelleri*	3565	2.366
Arabis alpina	vs.	*Arabis nipponica* (JO23)	3564	2.366
Arabis alpina	vs.	*Arabis nipponica* (Midori)	3547	2.355
Arabis hirsuta	vs.	*Arabis flagellosa*	1245	0.815
Arabis hirsuta	vs.	*Arabis stelleri*	1253	0.82
Arabis hirsuta	vs.	*Arabis nipponica* (JO23)	1234	0.808
Arabis hirsuta	vs.	*Arabis nipponica* (Midori)	1214	0.795
Arabis flagellosa	vs.	*Arabis stelleri*	132	0.086
Arabis flagellosa	vs.	*Arabis nipponica* (JO23)	111	0.072
Arabis flagellosa	vs.	*Arabis nipponica* (Midori)	86	0.056
Arabis stelleri	vs.	*Arabis nipponica* (JO23)	130	0.085
Arabis stelleri	vs.	*Arabis nipponica* (Midori)	104	0.068
Arabis nipponica (JO23)	vs.	*Arabis nipponica* (Midori)	55	0.036

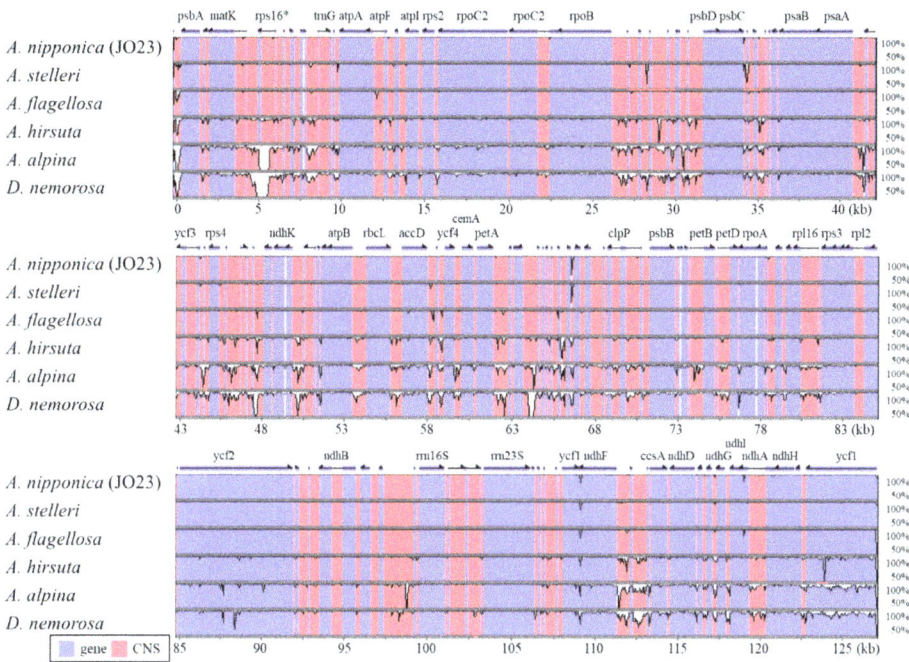

Figure 3. Alignment of the seven chloroplast genomes. VISTA-based identity plots of chloroplast genomes from six *Arabis* species and *Draba nemorosa* are compared to *A. nipponica* strain Midori. Arrows above the alignment indicate genes and their orientation. The names of genes ≥500 bp in length are also shown. A 70% identity cut-off was used for making the plots, and the Y-axis represents percent identity (50–100%), while the X-axis represents the location in the chloroplast genome. The blue and pink regions indicate genes and conserved noncoding sequences, respectively.

2.3. Distribution of Simple Sequence Repeats in the Chloroplast Genomes

Because the extremely low divergence among the East Asian *Arabis* species made it difficult to resolve their evolutionary relationships, other highly variable markers were needed. Therefore, simple sequence repeat (SSR) regions throughout the chloroplast genome were assessed for their ability to provide high-resolution species definition. A total of 74 mono-nucleotide, 22 di-nucleotide, and two tri-nucleotide repeat regions of ≥10 base pairs in length were identified (see Table 3). However, these repeat regions were still unable to completely resolve the relationships of the East Asian *Arabis* species. Fifty of the 98 SSRs exhibited no variation among the East Asian *Arabis* species, while only 29 SSRs exhibited species-specific variation, including nine in *A. flagellosa*, 15 in *A. stelleri*, four in *A. nipponica* strain JO23, and one in *A. nipponica* strain Midori. Five of the SSRs were shared by the two *A. nipponica* strains, which suggests that they were also species-specific. Although the two *A. nipponica* strains were similar to each other, *A. flagellosa*, *A. stelleri*, and *A. nipponica* differ to a similar degree in terms of of variable SSRs, which suggests that the occurrence of chloroplast capture would be independent or still ongoing. This was suggested by the patterns of nucleotide substitutions.

Table 3. Simple sequence repeats (SSRs) in *Arabis* chloroplast genome.

Position in A. nipponica (Midori) Genome		UNIT	A. nipponica		A. stelleri	A. flagellosa	A. hirsuta	A. alpina
from	to		Midori	JO23				
287	318	AT	16	15	15	12	13 with 2 mutations	29 bp with several mutations
1922	1932	A	11	11	9	12	11	9
3929	3938	T	10	9	10	9	7	7
4258	4270	T	13	18	18	17	13	13
7713	7727	T	15	15	15	15	12	11
7729	7738	A	10	10	10	9	10	10
8203	8216	TA	7	6	7	7	6	6
8273	8282	TA	5	5	5	5	5	6
8289	8302	AT	7	7	6	7	8	6
8321	8330	TA	5	5	4	5	5	deletion
9677	9690	T	14	14	T4GT10	15	14	14
9982	9991	TA	5	5	5	5	5	5
11,660	11,669	A	10	10	9	10	10	7
12,406	12,414	T	9	9	10	10	T3AT6	T3AT6
13,010	13,018	T	9	9	10	9	T7AT2	T10AT2
13,810	13,821	ATT	4	4	4	4	4	ATTATATTCTT
14,101	14,110	A	10	10	14	10	12	10
18,027	18,037	T	CDS	11	11	11	11	11	11
19,398	19,408	TA	5	5	5	5	5	5
22,549	22,558	T	10	10	11	10	9	15
25,777	25,786	T	CDS	10	10	10	10	10	10
27,601	27,611	G	11	11	11	15	12	9
28,808	28,817	T	10	10	9	10	10	T5CT3G2
30,293	30,310	A	18	17	17	18	12	A4CA5
30,737	30,751	T	15	15	14	15	15	5
30,830	30,839	A	10	9	11	10	8	6
30,918	30,929	TA	6	6	6	6	6	4
31,260	31,269	AT	5	5	5	5	5	3
35,309	35,316	G	8	11	10	11	7	10
35,516	35,525	AT	5	5	5	5	5	5
35,538	35,555	AT	9	9	9	9	3	deletion
41,508	41,522	T	15	13	11	13	18nt	101nt
41,768	41,778	A	11	12	12	11	A12GA4	11
43,656	43,665	A	10	10	10	11	A8TA2	A9TA2TA2
43,887	43,895	T	9	15	T4AT4AT4	T4AT4	4	4
45,038	45,046	T	9	9	9	10	8	7
45,771	45,788	T	18	18	18	18	13	9
46,034	46,057	A	24	24	24	24	16	11
46,116	46,133	AT	9	9	9	8	9	7
46,135	46,144	TA	5	5	5	5	5	3
46,782	46,791	T	10	11	10	11	14	10
47,368	47,378	A	11	11	12	12	10	13
47,586	47,595	T	10	10	10	10	13	TCT8
47,624	47,633	A	10	10	11	11	8	7
49,061	49,070	T	10	10	11	10	8	T3AT10
49,631	49,640	T	10	10	10	10	8	8
50,329	50,340	A	12	12	12	12	11	11
51,202	51,211	TA	5	5	5	5	19nt	deletion
51,215	51,230	T	16	17	17	16	13	13
53,088	53,097	T	CDS	10	10	10	10	10	12
53,592	53,601	C	10	11	9	12	9	9
55,477	55,490	T	14	14	14	14	complement A11	complement A13
55,891	55,906	T	16	16	16	16	13	15
56,476	56,485	T	10	10	10	10	10	A2T8
58,301	58,310	T	10	10	10	10	10	6
59,338	59,348	T	11	10	9	11	11	4
61,731	61,739	C	9	13	9	12	8	C3AC3
62,108	62,117	TA	5	5	5	5	6	4
62,161	62,182	T	22	22	22	31	2nt shorter	2nt shorter
62,202	62,210	A	9	10	9	9	A5TA3	A5TA3
63,523	63,538	T	16	15	15	T5GT10	16	7
64,629	64,639	T	11	11	11	11	11	T6GT3G
65,636	65,645	C	10	13	11	13	8	C2TCTGC7
66,253	66,262	AT	5	5	5	5	4	7
66,851	66,864	A	14	14	14	19	17	12
68,965	68,977	T	13	13	13	13	11	11
69,965	69,975	T	11	11	12	11	11	8
75,328	75,340	A	13	14	14	13	19	14

Table 3. *Cont.*

Position in A. nipponica (Midori) Genome		UNIT	A. nipponica		A. stelleri	A. flagellosa	A. hirsuta	A. alpina
76,614	76,626	T	13	13	13	13	13	13
78,154	78,162	TTG	3	5	3	3	4	2
80,484	80,493	A	10	11	10	10	10	9
81,019	81,035	T	17	17	17	17	17	17
81,178	81,191	T	14	14	14	14	18	8
82,568	82,578	A	11	10	9	10	9	10
83,489	83,498	TA	5	5	5	5	5	4
93,127	93,136	TA	5	5	5	5	5	4
97,975	97,984	A	10	10	10	10	12	9
98,781	98,791	T	11	11	11	11	10	14
107,287	107,295	AT	5	5	5	5	5	7
107,313	107,324	T	12	11	13	13	T2(AT)4T7	14
111,481	111,490	TA	5	5	5	5	TA2TGTA	4
111,589	111,598	AT	5	5	5	5	5	10
111,665	111,672	T	8	8	10	8	7	10
111,801	111,810	A	A7CA2	A7CA2	10	A7CA2	A7CA2	A7TAC
112,472	112,481	A	10	10	10	10	11	10
116,836	116,845	T	10	9	10	11	T7AT3	10
123,173	123,184	T	12	12	12	12	12	12
123,285	123,383	T	10	10	10	10	10	10
123,884	123,893	T	10	10	10	10	10	10
123,975	123,987	A	13	13	13	13	13	13
124,356	124,365	TA	5	5	5	5	5	5
124,874	124,886	T	13	13	13	13	13	13
125,029	125,041	A	13	13	13	13	13	13
126,052	125,385	T	15	15	15	15	15	17
126,087	126,097	T	11	11	11	11	11	11
126,117	126,128	A	12	12	12	12	12	12
126,952	126,962	T	11	11	11	11	T8CT2	T8CT2
127,241	127,252	A	12	12	12	12	6	6

2.4. Mitochondrial Genome Analysis

Chloroplast capture could have originated from hybridization events that also affected other cytoplasmic genomes. Due to this, variation in the mitochondrial genome sequences was analyzed. Mapping next-generation sequencing (NGS) reads to the *Eruca vesicaria* mitochondrial genome revealed that 29 sites with five or more mapped reads varied among the *A. nipponica* strain Midori, *A. flagellosa*, and *A. hirsuta* (see Table 4). Twenty-eight of the sites were conserved among *A. nipponica* and *A. flagellosa*. One site was specific to *A. nipponica* and provided 100% support for the relationship between *A. nipponica* and *A. flagellosa*. Even though reliability decreased, 123 of 125 sites with two or more reads (98.4%) also supported the similarity of the *A. nipponica* and *A. flagellosa* mitochondrial genomes. These findings suggest that the hybridization history of the species affects both the chloroplast and the mitochondrial genomes similarly.

Table 4. Nucleotide variation in the mitochondrial genome of *Arabis* species.

		Number of Mapped Reads			
		5 and More	4 and More	3 and More	2 and More
Number of variable sites	Total	29	46	74	129
Specific to	A. nipponica	1	1	4	12
	A. flagellosa	0	0	0	3
	A. hirsuta	14	25	35	62
Shared with	A. flagellosa and A. nipponica	14	19	31	46
	A. nipponica and A. hirsuta	0	0	1	1
	A. flagellosa and A. hirsuta	0	0	1	1
	other type	0	1	2	4

3. Discussion

Chloroplast capture results in the incongruence of chloroplast and nuclear phylogenies, which has been reported in many plant taxa and is considered common among plants [17,18,26–37]. Furthermore, it is possible that the introgression of chloroplast genomes occurs more frequently than that of nuclear genomes as a result of uniparental inheritance, lack of recombination, and low selective constraint [38–40]. Chloroplast capture could occur by using several factors including sampling error, convergence, evolutionary rate heterogeneity, wrong lineage sorting, and hybridization/introgression [17]. Introgression-induced chloroplast capture occurred through hybridization between distant but compatible species, which was followed by backcrossing with pollen donor species [41,42].

East Asian *Arabis* species have previously been reported to show evidence of chloroplast capture [14,16]. More specifically, detailed phylogenetic analyses of nuclear and chloroplast marker genes has suggested that *A. nipponica*, *A. stelleri*, and *A. takeshimana* originated from the hybridization of *A. hirsuta* (or *A. sugittata*) and East Asian *Arabis* species (close to *A. serrata*, *A. paniculata*, and *A. flagellosa*), which act as paternal and maternal parents, respectively [14,16]. In the present study, comparing the whole chloroplast genomes of four plants from three East Asian *Arabis* species (two *A. nipponica*, one each of *A. stelleri*, and *A. flagellosa*) revealed genome-wide similarities that indicated chloroplast capture by *A. nipponica* and *A. stelleri*. The study also compared the species' partial mitochondrial genomes, which indicated a closer relationship between *A. nipponica* and *A. flagellosa* than between the former and European *A. hirsuta*. This suggested that *A. nipponica* also has a history of mitochondrial capture. This is not surprising, because hybridization and backcrossing could have similar effects on both organellar genomes. Also, cyto-nuclear incompatibility caused by a mitochondrial genome could lead cytoplasmic replacement to exhibit chloroplast capture [17,41,42]. The pattern of variation in the mitochondrial genomes suggested that both the chloroplast and mitochondrial genomes were co-transmitted during the evolutionary history of East Asian *Arabis* species. Future research should focus on the process of chloroplast (organellar) capture. Simple backcrossing could show the mechanisms of cytoplasm replacement and could produce results in as few as a hundred generations under certain conditions [42]. In the present study, the divergence between the genomes of hybrid-origin species and putative pollen-donor species was similar to the divergence observed within species, which suggests that the hybridization event was relatively recent. Nuclear genome markers are needed to estimate the proportion of parental genome fragments in the current nuclear genome of *A. nipponica*.

4. Materials and Methods

4.1. Plant Materials

Arabis nipponica (*A. hirsuta* var. *nipponica*, sampled from Midori, Gifu Prefecture, Japan), *A. flagellosa* (sampled from Kifune, Kyoto Prefecture, Japan), and *A. hirsuta* (strain Brno from Ulm Botanical Garden, Germany) were used in the present study.

4.2. DNA Isolation, NGS Sequencing, and Genome Assembly

Chloroplasts were isolated from *A. hirsuta* and *A. nipponica* as described in Okegawa and Motohashi [43]. DNA was isolated from the chloroplasts using the DNeasy Plant Mini Kit (Qiagen, Valencia, CA, USA), while the total DNA was isolated from leaves of *A. flagellosa*. NGS libraries were constructed using the Nextera DNA Sample Preparation Kit (Illumina, San Diego, CA, USA) and sequenced as single-ended reads using the NextSeq500 platform (Illumina). About 2 Gb (1.4 Gb, 12 M clean reads) of sequences were obtained for *A. flagellosa* (43 Mb mapped reads, 282.69× coverage). Additionally, 400 Mb (300 Mb, 2.5 M clean reads) were obtained for both *A. hirsuta* (64 Mb mapped reads, 417.17× coverage) and *A. nipponica* (72 Mb mapped reads, 455.87× coverage). The generated reads were assembled using velvet 1.2.10 [44] and assembled into complete chloroplast genomes by mapping to previously published whole chloroplast genome sequences. Sequence gaps were resolved using Sanger sequencing. Genes were annotated using DOGMA [45] and BLAST. The newly constructed chloroplast genomes were deposited in the DDBJ database under the accession numbers LC361349-51. Finally, the circular chloroplast genome maps were drawn using OGDRAW [46].

4.3. Molecular Evolutionary Analyses

The whole chloroplast genome sequences of *A. nipponica* (strain JO23: AP009369), *A. stelleri* (KY126841) [23], *A. alpina* (HF934132) [25], and *D. nemorosa* (strain JO21: AP009373) in the GenBank were also used. Whole chloroplast sequences were aligned in order to construct neighbor-joining trees with Jukes and Cantor distances. The sequences of 77 known functional genes were linked in a series after excluding initiation and stop codons and were then used for phylogenetic analyses along with sequences from the related clade species *Brassica oleracea* (KR233156) [47], *B. rapa* (DQ231548), *Eutrema salsugineum* (KR584659) [48], *Raphanus sativus* (KJ716483) [49], *Scherenkiella parvula* (KT222186) [48], *Sinapis arvensis* (KU050690), and *Thlaspi arvense* (KX886351) [21] using *A. thaliana* (AP000423) [50] as an outgroup. The synonymous divergence of the concatenated CDS was estimated using the Nei and Gojobori method. All phylogenetic analyses were performed using MEGA 7.0 [51]. Levels of divergence throughout the chloroplast genome were visualized using mVISTA [52] with Shuffle-LAGAN alignment [53].

4.4. Mapping NGS Reads to Mitochondrial Genome Sequences

Because the chloroplast isolation method used in the present study did not completely exclude mitochondria, about 1% of the sequence reads were derived from mitochondrial genomes. Although this proportion is too low to be useful for assembling whole mitochondrial genomes, the reads were nevertheless mapped to the mitochondrial genome of *Eruca vesicaria* (KF442616) [54] in order to measure mitochondrial genome divergence. Regions with at least five mapped reads were used for the analysis.

Acknowledgments: This study was supported in part by JSPS KAKENHI Grant Number 17K19361 and Grants-in-Aid from MEXT-Supported Program for the Strategic Research Foundation at Private Universities (S1511023) to Akira Kawabe.

Author Contributions: Akira Kawabe designed the study. Hiroaki Nukii and Hazuka Y. Furihata performed the experiments. Akira Kawabe and Hazuka Y. Furihata analyzed the data. Akira Kawabe wrote the manuscript. All authors read and approved the final manuscript.

Conflicts of Interest: The authors declare no conflict of interest.

References

1. Al-Shehbaz, I.A. Transfer of most North American species of Arabis to Boechera (Brassicaceae). *Novon* **2003**, *13*, 381–391. [CrossRef]
2. O'Kane, S.L.; Al-Shehbaz, I.A. Phylogenetic Position and Generic Limits of Arabidopsis (Brassicaceae) Based on Sequences of Nuclear Ribosomal DNA. *Ann. Mo. Bot. Gard.* **2003**, *90*, 603–612. [CrossRef]
3. Al-Shehbaz, I.A.; Beilstein, M.A.; Kellogg, E.A. Systematics and phylogeny of the Brassicaceae (Cruciferae): An overview. *Plant Syst. Evol.* **2006**, *259*, 89–120. [CrossRef]
4. Al-Shehbaz, I.A.; German, D.A.; Karl, R.; Ingrid, J.T.; Koch, M.A. Nomenclatural adjustments in the tribe Arabideae (Brassicaceae). *Plant Div. Evol.* **2011**, *129*, 71–76. [CrossRef]
5. Koch, M.A.; Karl, R.; German, D.A.; Al-Shehbaz, I.A. Systematics, taxonomy and biogeography of three new Asian genera of Brassicaceae tribe Arabideae: An ancient distribution circle around the Asian high mountains. *Taxon* **2012**, *61*, 955–969.
6. Kiefer, M.; Schmickl, R.; German, D.A.; Mandáková, T.; Lysak, M.A.; Al-Shehbaz, I.A.; Franzke, A.; Mummenhoff, K.; Stamatakis, A.; Koch, M.A. BrassiBase: Introduction to a novel knowledge database on Brassicaceae evolution. *Plant Cell Physiol.* **2014**, *55*, e3. [CrossRef] [PubMed]

7.   Ansell, S.W.; Grundmann, M.; Russell, S.J.; Schneider, H.; Vogel, J.C. Genetic discontinuity, breeding-system change and population history of *Arabis alpina* in the Italian Peninsula and adjacent Alps. *Mol Ecol.* **2008**, *17*, 2245–2257. [CrossRef] [PubMed]

8.   Bergonzi, S.; Albani, M.C.; ver Loren van Themaat, E.; Nordström, K.J.; Wang, R.; Schneeberger, K.; Moerland, P.D.; Coupland, G. Mechanisms of age-dependent response to winter temperature in perennial flowering of *Arabis alpina*. *Science* **2013**, *340*, 1094–1097. [CrossRef] [PubMed]

9.   Karl, R.; Koch, M.A. A world-wide perspective on crucifer speciation and evolution: Phylogenetics, biogeography and trait evolution in tribe Arabideae. *Ann. Bot.* **2013**, *112*, 983–1001. [CrossRef] [PubMed]

10.   Toräng, P.; Vikström, L.; Wunder, J.; Wötzel, S.; Coupland, G.; Ågren, J. Evolution of the selfing syndrome: Anther orientation and herkogamy together determine reproductive assurance in a self-compatible plant. *Evolution* **2017**, *71*, 2206–2218. [CrossRef] [PubMed]

11.   Heidel, A.J.; Kiefer, C.; Coupland, G.; Rose, L.E. Pinpointing genes underlying annual/perennial transitions with comparative genomics. *BMC Genom.* **2016**, *17*, 921. [CrossRef] [PubMed]

12.   Willing, E.M.; Rawat, V.; Mandáková, T.; Maumus, F.; James, G.V.; Nordström, K.J.; Becker, C.; Warthmann, N.; Chica, C.; Szarzynska, B.; et al. Genome expansion of *Arabis alpina* linked with retrotransposition and reduced symmetric DNA methylation. *Nat. Plants* **2015**, *1*, 14023. [CrossRef] [PubMed]

13.   Jiao, W.B.; Accinelli, G.G.; Hartwig, B.; Kiefer, C.; Baker, D.; Severing, E.; Willing, E.M.; Piednoel, M.; Woetzel, S.; Madrid-Herrero, E.; et al. Improving and correcting the contiguity of long-read genome assemblies of three plant species using optical mapping and chromosome conformation capture data. *Genome Res.* **2017**, *27*, 778–786. [CrossRef] [PubMed]

14.   Koch, M.A.; Karl, R.; Kiefer, C.; Al-Shehbaz, I.A. Colonizing the American continent: Systematics of the genus Arabis in North America (Brassicaceae). *Am. J. Bot.* **2010**, *97*, 1040–1057. [CrossRef] [PubMed]

15.   Karl, R.; Kiefer, C.; Ansell, S.W.; Koch, M.A. Systematics and evolution of Arctic-Alpine *Arabis alpina* (Brassicaceae) and its closest relatives in the eastern Mediterranean. *Am. J. Bot.* **2012**, *99*, 778–794. [CrossRef] [PubMed]

16.   Karl, R.; Koch, M.A. Phylogenetic signatures of adaptation: The Arabis hirsuta species aggregate (Brassicaceae) revisited. *Perspect. Plant Ecol. Evol. Syst.* **2014**, *16*, 247–264. [CrossRef]

17.   Rieseberg, L.H.; Soltis, D.E. Phylogenetic consequences of cytoplasmic gene flow in plants. *Evolut. Trends Plants* **1991**, *5*, 65–84.

18.   Soltis, D.E.; Kuzoff, R.K. Discordance between nuclear and chloroplast phylogenies in the Heuchera group (Saxifragaceae). *Evolution* **1995**, *49*, 727–742. [CrossRef] [PubMed]

19.   Ruhfel, B.R.; Gitzendanner, M.A.; Soltis, P.S.; Soltis, D.E.; Burleigh, J.G. From algae to angiosperms-inferring the phylogeny of green plants (Viridiplantae) from 360 plastid genomes. *BMC Evol. Biol.* **2014**, *14*, 23. [CrossRef] [PubMed]

20.   Hohmann, N.; Wolf, E.M.; Lysak, M.A.; Koch, M.A. A Time-calibrated road map of brassicaceae species radiation and evolutionary history. *Plant Cell* **2015**, *27*, 2770–2784. [CrossRef] [PubMed]

21.   Guo, X.; Liu, J.; Hao, G.; Zhang, L.; Mao, K.; Wang, X.; Zhang, D.; Ma, T.; Hu, Q.; Al-Shehbaz, I.A.; Koch, M.A. Plastome phylogeny and early diversification of Brassicaceae. *BMC Genom.* **2017**, *18*, 176. [CrossRef] [PubMed]

22.   Mandáková, T.; Hloušková, P.; German, D.A.; Lysak, M.A. Monophyletic origin and evolution of the largest crucifer genomes. *Plant Physiol.* **2017**, *174*, 2062–2071. [CrossRef] [PubMed]

23.   Raman, G.; Park, V.; Kwak, M.; Lee, B.; Park, S. Characterization of the complete chloroplast genome of Arabis stellari and comparisons with related species. *PLoS ONE* **2017**, *12*, e0183197. [CrossRef] [PubMed]

24.   Jordon-Thaden, I.; Hase, I.; Al-Shehbaz, I.A.; Koch, M.A. Molecular phylogeny and systematics of the genus Draba (Brassicaceae) and identification of its most closely related genera. *Mol. Phylogenet. Evol.* **2010**, *55*, 524–540. [CrossRef] [PubMed]

25.   Melodelima, C.; Lobréaux, S. Complete *Arabis alpina* chloroplast genome sequence and insight into its polymorphism. *Meta Gene* **2013**, *1*, 65–75. [CrossRef] [PubMed]

26.   Acosta, M.C.; Premoli, A.C. Evidence of chloroplast capture in South American *Nothofagus* (subgenus *Nothofagus*, Nothofagaceae). *Mol. Phylogenet. Evol.* **2010**, *54*, 235–242. [CrossRef] [PubMed]

27.   Dorado, O.; Rieseberg, L.H.; Arias, D.M. Chloroplast DNA introgression in southern California sunflowers. *Evolution* **1992**, *46*, 566–572. [CrossRef] [PubMed]

28. Fehrer, J.; Gemeinholzer, B.; Chrtek, J.; Bräutigam, S. Incongruent plastid and nuclear DNA phylogenies reveal ancient intergeneric hybridization in *Pilosella* hawkweeds (*Hieracium*, Cichorieae, Asteraceae). *Mol. Phylogenet. Evol.* **2007**, *42*, 347–361. [CrossRef] [PubMed]

29. Gurushidze, M.; Fritsch, R.M.; Blattner, F.R. Species-level phylogeny of *Allium* subgenus *Melanocrommyum*: Incomplete lineage sorting, hybridization and *trnF* gene duplication. *Taxon* **2010**, *59*, 829–840.

30. Liston, A.; Kadereit, J.W. Chloroplast DNA evidence for introgression and long distance dispersal in the desert annual *Senecio flavus* (*Asteraceae*). *Plant Syst. Evol.* **1995**, *197*, 33–41. [CrossRef]

31. Mir, C.; Jarne, P.; Sarda, V.; Bonin, A.; Lumaret, R. Contrasting nuclear and cytoplasmic exchanges between phylogenetically distant oak species (*Quercus suber* L. and *Q. ilex* L.) in Southern France: Inferring crosses and dynamics. *Plant Biol.* **2009**, *11*, 213–226. [CrossRef] [PubMed]

32. Okuyama, Y.; Fujii, N.; Wakabayashi, M.; Kawakita, A.; Ito, M.; Watanabe, M.; Murakami, N.; Kato, M. Nonuniform concerted evolution and chloroplast capture: Heterogeneity of observed introgression patterns in three molecular data partition phylogenies of Asian *Mitella* (Saxifragaceae). *Mol. Biol. Evol.* **2005**, *22*, 285–296. [CrossRef] [PubMed]

33. Rieseberg, L.H.; Choi, H.C.; Ham, F. Differential cytoplasmic versus nuclear introgression in *Helianthus*. *J. Hered.* **1991**, *82*, 489–493. [CrossRef]

34. Schilling, E.E.; Panero, J.K. Phylogenetic reticulation in subtribe *Helianthinae*. *Am. J. Bot.* **1996**, *83*, 939–948. [CrossRef]

35. Wolfe, A.D.; Elisens, W.J. Evidence of chloroplast capture and pollen-mediated gene flow in Penstemon sect. Peltanthera (Scrophulariaceae). *Syst. Bot.* **1995**, *20*, 395–412. [CrossRef]

36. Yi, T.S.; Jin, G.H.; Wen, J. Chloroplast capture and intra-and inter-continental biogeographic diversification in the Asian–New World disjunct plant genus Osmorhiza (Apiaceae). *Mol. Phylogenet. Evol.* **2015**, *85*, 10–21. [CrossRef] [PubMed]

37. Yuan, Y.W.; Olmstead, R.G. A species-level phylogenetic study of the Verbena complex (Verbenaceae) indicates two independent intergeneric chloroplast transfers. *Mol. Phylogenet. Evol.* **2008**, *48*, 23–33. [CrossRef] [PubMed]

38. Avise, J.C. *Molecular Markers, Natural History and Evolution*, 2nd ed.; Sinauer: Sunderland, MA, USA, 2004.

39. Rieseberg, L.H.; Wendel, J. Introgression and its consequences in plants. In *Hybrid Zones and the Evolutionary Process*; Harrison, R., Ed.; Oxford University Press: New York, NY, USA, 1993; pp. 70–109.

40. Martinsen, G.D.; Whitham, T.G.; Turek, R.J.; Keim, P. Hybrid populations selectively filter gene introgression between species. *Evolution* **2001**, *55*, 1325–1335. [CrossRef] [PubMed]

41. Rieseberg, L.H. The role of hybridization in evolution: Old wine in new skins. *Am. J. Bot.* **1995**, *82*, 944–953. [CrossRef]

42. Tsitrone, A.; Kirkpatrick, M.; Levin, D.A. A model for chloroplast capture. *Evolution* **2003**, *57*, 1776–1782. [CrossRef] [PubMed]

43. Okegawa, Y.; Motohashi, K. Chloroplastic thioredoxin m functions as a major regulator of Calvin cycle enzymes during photosynthesis in vivo. *Plant J.* **2015**, *84*, 900–913. [CrossRef] [PubMed]

44. Zerbino, D.R.; Birney, E. Velvet: Algorithms for de novo short read assembly using de Bruijn graphs. *Genome Res.* **2008**, *18*, 821–829. [CrossRef] [PubMed]

45. Wyman, S.K.; Jansen, R.K.; Boore, J.L. Automatic annotation of organellar genomes with DOGMA. *Bioinformatics* **2004**, *20*, 3252–3255. [CrossRef] [PubMed]

46. Lohse, M.; Drechsel, O.; Bock, R. OrganellarGenomeDRAW (OGDRAW)—A tool for the easy generation of high-quality custom graphical maps of plastid and mitochondrial genomes. *Curr. Genet.* **2007**, *52*, 267–274. [CrossRef] [PubMed]

47. Seol, Y.J.; Kim, K.; Kang, S.H.; Perumal, S.; Lee, J.; Kim, C.K. The complete chloroplast genome of two *Brassica* species, *Brassica nigra* and *B. oleracea*. *Mitochondrial DNA Part A* **2017**, *28*, 167–168. [CrossRef] [PubMed]

48. He, Q.; Hao, G.; Wang, X.; Bi, H.; Li, Y.; Guo, X.; Ma, T. The complete chloroplast genome of *Schrenkiella parvula* (Brassicaceae). *Mitochondrial DNA Part A* **2016**, *27*, 3527–3528. [CrossRef] [PubMed]

49. Jeong, Y.M.; Chung, W.H.; Mun, J.H.; Kim, N.; Yu, H.J. De novo assembly and characterization of the complete chloroplast genome of radish (*Raphanus sativus* L.). *Gene* **2014**, *551*, 39–48. [CrossRef] [PubMed]

50. Sato, S.; Nakamura, Y.; Kaneko, T.; Asamizu, E.; Tabata, S. Complete structure of the chloroplast genome of *Arabidopsis thaliana*. *DNA Res.* **1999**, *6*, 283–290. [CrossRef] [PubMed]

51. Kumar, S.; Stecher, G.; Tamura, K. MEGA7: Molecular Evolutionary Genetics Analysis Version 7.0 for Bigger Datasets. *Mol. Biol. Evol.* **2016**, *33*, 1870–1874. [CrossRef] [PubMed]
52. Frazer, K.A.; Pachter, L.; Poliakov, A.; Rubin, E.M.; Dubchak, I. VISTA: Computational tools for comparative genomics. *Nucleic Acids Res.* **2004**, *32*, W273–W279. [CrossRef] [PubMed]
53. Brudno, M.; Malde, S.; Poliakov, A.; Do, C.B.; Couronne, O.; Dubchak, I.; Batzoglou, S. Glocal Alignment: Finding rearrangements during alignment. *Bioinformatics* **2003**, *19*, i54–i62. [CrossRef] [PubMed]
54. Wang, Y.; Chu, P.; Yang, Q.; Chang, S.; Chen, J.; Hu, M.; Guan, R. Complete mitochondrial genome of *Eruca sativa* Mill. (Garden rocket). *PLoS ONE* **2014**, *9*, e105748. [CrossRef] [PubMed]

International Journal of
Molecular Sciences

MDPI

Article

Mutational Biases and GC-Biased Gene Conversion Affect GC Content in the Plastomes of *Dendrobium* Genus

Zhitao Niu, Qingyun Xue, Hui Wang, Xuezhu Xie, Shuying Zhu, Wei Liu and Xiaoyu Ding *

College of Life Sciences, Nanjing Normal University, Nanjing 210023, China; niuzhitaonj@163.com (Z.N.);
qyxue1981@126.com (Q.X.); wanghui201711@163.com (H.W.); naive0312@126.com (X.X.);
zhushuy@126.com (S.Z.); liuwei4@njnu.edu.cn (W.L.)
* Correspondence: dingxynj@263.net

Received: 29 August 2017; Accepted: 20 October 2017; Published: 2 November 2017

Abstract: The variation of GC content is a key genome feature because it is associated with fundamental elements of genome organization. However, the reason for this variation is still an open question. Different kinds of hypotheses have been proposed to explain the variation of GC content during genome evolution. However, these hypotheses have not been explicitly investigated in whole plastome sequences. *Dendrobium* is one of the largest genera in the orchid species. Evolutionary studies of the plastomic organization and base composition are limited in this genus. In this study, we obtained the high-quality plastome sequences of *D. loddigesii* and *D. devonianum*. The comparison results showed a nearly identical organization in *Dendrobium* plastomes, indicating that the plastomic organization is highly conserved in *Dendrobium* genus. Furthermore, the impact of three evolutionary forces—selection, mutational biases, and GC-biased gene conversion (gBGC)—on the variation of GC content in *Dendrobium* plastomes was evaluated. Our results revealed: (1) consistent GC content evolution trends and mutational biases in single-copy (SC) and inverted repeats (IRs) regions; and (2) that gBGC has influenced the plastome-wide GC content evolution. These results suggest that both mutational biases and gBGC affect GC content in the plastomes of *Dendrobium* genus.

Keywords: *Dendrobium*; plastome assembly; selection; mutational biases; GC-biased gene conversion (gBGC); GC_{eq}

1. Introduction

Chloroplasts, responsible for photosynthesis and other biosynthesis processes in plants, have essential effects on plant growth and development. Their own genomes (plastomes) are usually uniparentally inherited and highly conserved in their quadripartite structure, which consists of a pair of inverted repeats (IRs) regions and two single-copy (SC) regions [1]. Comparative studies of the plastome sequence have revealed: (1) a relatively higher GC content in IR regions than that of SC regions and (2) varied GC content in different gene and non-coding loci e.g., [2,3]. The variation of GC content is a key genome feature because it is associated with fundamental elements of genome organization [4,5]. For instance, GC-rich regions exhibit higher gene density, more conserved mutation rates, and higher recombination rates, relative to GC-poor regions. Therefore, resolving the origin and causes of the variation in base composition has practical significance for a better understanding of the plastome organization.

Three major kinds of hypotheses have been proposed to explain the variation of GC content in genome evolution. The first hypothesis, "natural selection hypothesis", has suggested that high GC content can be selected for by their thermal stability [4,6]. Natural selection also affects the probability of fixation of a mutation based on the mutation fitness advantage or disadvantage of the organism [7,8].

The second hypothesis, the so-called "mutational biases hypothesis", is that the GC content is driven by the heterogeneous mutational biases along genomes [9]. The third hypothesis involves GC-biased gene conversion (gBGC), a process that takes place during meiotic recombination. The gBGC process prefers repairing DNA mismatches with GC bases and tends to increase the GC content of recombining DNA over evolutionary time [10,11]. This was recently confirmed by a broad range of genome comparison studies in eukaryotes and prokaryotes [12–14]. For example, in yeast genomes, gBGC occurred to repair the mismatches located at the extremities of the conversion regions [15]. The gBGC was also demonstrated to affect the GC content of third codon position and intron regions in grasses genomes [16]. Recently, Wu and Chaw, 2015 measured the gene conversion events of non-coding regions in cycads plastomes and reported the first case of plastome GC-biased gene conversion [17].

However, these hypotheses have not been explicitly investigated in whole plastome sequences. Orchids (Orchidaceae) are the largest family in the monocots, including about 25,000 species in 880 genera and five subfamilies [18]. Recent studies showed a diversified evolution of the plastome sequence among different orchid genera [19,20]. Moreover, orchids also present a peculiar plastomic structure: they exhibit variable IRs that caused the drastic reductions in small single-copy (SSC) regions (e.g., the expansion/contraction of IRs has led to the length of the SSC region in *Vanilla* being only about one-eighth of that in *Goodyera*) [20]. These features result in a variable GC content among different plastome sequences. However, there is still little known about the GC content evolution of the orchid plastome sequences.

Recently, plastome sequences have been made available for more than 20 orchid genera. However, in this study, we chose to focus on the *Dendrobium* genus because it is the only genus of orchids for which more than 30 plastome sequences have been sequenced [21,22]. *Dendrobium* is one of the largest genera in the orchid species. In China, there are about 80 *Dendrobium* species, some of which are well known for their high horticultural and medicinal value [23,24]. Although many plastome sequences have been published, the research of plastomic structure and base composition remains very limited in the *Dendrobium* species. In this study, we surveyed 10 plastomes of *Dendrobium* species, including two newly sequenced ones, and we addressed two key questions: (1) Is the plastome structure conserved among *Dendrobium* species or variable due to the expansion/contraction of IRs? (2) Which evolutionary forces—selection, mutational biases, or gBGC—have a significant impact on GC content in orchid plastomes?

To address the questions mentioned above, the complete plastome of two more *Dendrobium* species (*D. loddigesii* and *D. devonianum*) were sequenced and assembled by two different methods. The plastomic structures among *Dendrobium* plastomes were compared. Moreover, we evaluated the selection forces among the plastid protein-coding genes of 10 *Dendrobium* species. Meanwhile, biased mutations of protein-coding genes and non-coding regions were also measured on the basis of the estimated equilibrium GC content (GC_{eq}). Our results suggest that both mutational biases and gBGC affect GC content in the plastomes of the *Dendrobium* genus.

2. Results

2.1. Plastome Assembly of D. loddigesii and D. devonianum

A total of approximately 3.84 Gb of 150 bp pair-end reads each for *D. loddigesii* and *D. devonianum* was obtained from the Illumina paired-end sequencing. Reads with error probability >0.05 were discarded. After that, the plastomes of *D. loddigesii* and *D. devonianum* were assembled by two different ways: (1) de novo assembly by using SOAPdenovo version 1.12 and (2) using the reference-guided mapping method with CLC Genomics Workbench 6.0.1. For the de novo assembly analysis, 43,509 contigs of *D. loddigesii* and 32,667 contigs of *D. devonianum* were included in the initial assembly. After comparison with plant plastomes, 88 contigs and 67 contigs were obtained with E-values <10^{-10} and average coverage depth >30× for *D. loddigesii* and *D. devonianum*, respectively. Six of these contigs (length >17 kb and coverage depth >117×) resulted in a nearly complete draft genome for

D. loddigesii. Four contigs (length >25 kb and coverage depth >172×) were employed for the plastome assembly of *D. devonianum*. After assembly and gap closure, the plastome sequences of *D. loddigesii* and *D. devonianum* were 152,874 bp and 152,215 bp in length, respectively. For the reference-guided mapping analysis, the trimmed reads were mapped to the plastome sequence of *D. moniliforme* (AB893950), which served as a reference sequence. After gap closure, we obtained the plastome sequences of *D. loddigesii* and *D. devonianum* with 149,674 bp and 150,973 bp in length, respectively.

The assembled plastomes of *D. loddigesii* and *D. devonianum* were compared and plotted using mVISTA, with *D. moniliforme* as the reference (Figure S1). The comparison results showed that the differences—including single nucleotide polymorphism (SNP), insertion-deletion (InDel), and sequence repeat—between the plastomes were mainly distributed in the non-coding regions and *ndh* pseudo genes. Each type of these errors was corrected and validated by PCR amplification and Sanger sequencing. After that, the complete plastomes of *D. loddigesii* and *D. devonianum* were obtained with 152,498 bp and 151,715 bp respectively.

2.2. Highly Conserved Plastomic Structure and Organization in Dendrobium Genus

Genome maps of the newly sequenced two *Dendrobium* plastomes, including *D. loddigesii* and *D. devonianum*, were circular, as shown in Figure 1. The plastome contained a pair of inverted repeat (IR) regions, which separated the single-copy (SC) region into large SC (LSC) and small SC (SSC) regions. The LSC, SSC, and IR regions of *D. loddigesii* and *D. devonianum* ranged from 84,089 to 84,897 bp, 14,311 to 16,932 bp, and 25,736 to 25,800 bp, respectively. The overall GC content (37.38–37.56%) was similar with other *Dendrobium* plastomes [20,22], with 35.12–35.36% and 30.54–30.61% in LSC and SSC regions, respectively, and a higher content of 43.51–43.57% in the IR regions (Table 1). The two *Dendrobium* plastomes contained 102 unique functional coding genes, including 30 tRNA genes, four rRNA genes, and 68 protein-coding genes.

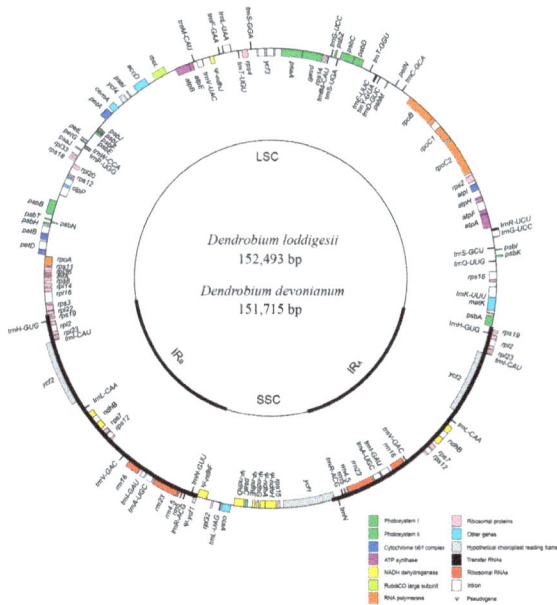

Figure 1. Genome map of two newly sequenced *Dendrobium* plastomes. Only the plastome of *D. loddigesii* is shown because it has an identical structure with *D. devonianum*. Genes outside and inside the circle are transcribed clockwise and counterclockwise, respectively. LSC: large single-copy; SSC: small single-copy; IR$_A$ and IR$_B$: two identical inverted repeats.

Table 1. Genome feature of the two newly sequenced *Dendrobium* plastomes.

Species	Plastome Length (bp)	LSC Length (bp)	SSC Length (bp)	IR Length (bp)	GC Content (%)	GC Content of LSC (%)	GC Content of SSC (%)	GC Content of IR (%)	Accession
Dendrobium loddigesii	152,493	84,089	16,932	25,736	37.38	35.63	30.54	43.57	LC317044
Dendrobium devonianum	151,715	84,897	14,311	25,800	37.56	35.12	30.61	43.51	LC317045

Abbreviations: LSC, large single-copy; SSC, small single-copy; IR, inverted repeats.

The plastome sequences of *D. loddigesii*, *D. devonianum*, *D. officinale*, and *D. moniliforme* were used for the plastome comparison (Figure 2). Comparative plastomes of these four *Dendrobium* species revealed distinct loss or retention of *ndh* genes, which indicated that the *ndh* genes have experienced independent loss during their evolution. Compared to the variable expansion/contraction of IRs in different orchid genera, e.g., [20,21], the IRs of plastomes in the *Dendrobium* genus were conserved. Overall, these *Dendrobium* plastomes appeared to have a nearly identical organization reflecting the highly conserved plastomic organization in the *Dendrobium* genus.

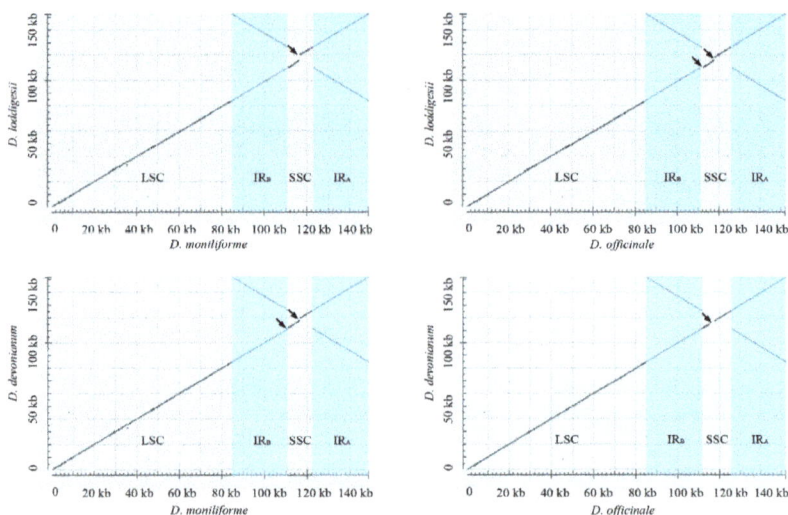

Figure 2. Dot-plot analysis of the four *Dendrobium* plastomes. The *Dendrobium* plastomes appear to have a nearly identical organization, which indicates that their plastomic organization is highly conserved. The black arrows indicate the different loss/retention of *ndh* genes.

2.3. Mutational Biases of Non-Coding Loci Are Associated with the Plastomic Structure

The phylogenetic relationship of these 10 *Dendrobium* species was inferred from the whole plastome sequence (Figure 3). The phylogenetic tree was highly resolved with the support value of all nodes = 100%. This tree was utilized to construct the ancestral sequences of non-coding loci, including intergenic and intronic loci for the 10 *Dendrobium* species. Then, the point mutations between ancestral and current sequences were calculated.

In all examined *Dendrobium* species, the nucleotide mutations were mainly distributed in the non-coding loci of SC regions. The frequencies of transversions are higher than that of transitions in both SC and IR regions. Figure 4 shows the relative rates of the six nucleotide pair mutations. The most common mutation is G/C to A/T mutations in SC regions. Therefore, we divided the nucleotide mutations into two groups: AT-rich (G/C to A/T) and GC-rich (A/T to G/C) mutations

(Table 2). The number of AT-rich mutations was larger than that of GC-rich mutations in the SC regions, while the two types of mutations did not display significant difference in the IR regions. These results indicated that the mutational biases in *Dendrobium* plastomes are associated with the plastomic structure. Moreover, the contrasting biased mutations between the SC and IR regions were also evident by the GC_{eq} values estimated from the non-coding loci of the SC and IR regions. In line with the counting results shown in Table 2, the GC_{eq} values for the SC regions were remarkably smaller than equilibrium (GC_{eq} = 50%) (Figure 5). However, in the IR regions, GC_{eq} values of all *Dendrobium* species were greater than 50%, except for *D. loddigesii*, *D. lohohense*, and *D. salaccense* (with GC_{eq} values 44.83%, 42.86% and 33.97%, respectively). This result disagreed with the data shown in Table 2, that the IR regions for *Dendrobium* species have no significant difference between GC-rich and AT-rich mutations. Because only few mutations were detected in the IR regions, the data shown in Table 1 might not precisely reflect the mutational biases of non-coding loci. Thus, based on the estimated GC_{eq} values, the mutational biases in the IR region were toward GC-richness. Considering the different mutational trends in *Dendrobium* plastomes: (1) toward AT-richness in SC regions and (2) toward GC-richness in IR regions, we proposed that the mutational biases of non-coding loci are associated with the plastomic structure in *Dendrobium* species.

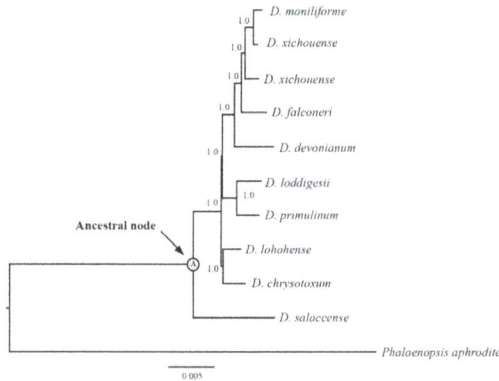

Figure 3. The BI tree inferred from whole plastome sequences of *Dendrobium* species. Tree node labeled with "A" denotes to the ancestor for each *Dendrobium* species. The values of posterior probabilities are showed for each node.

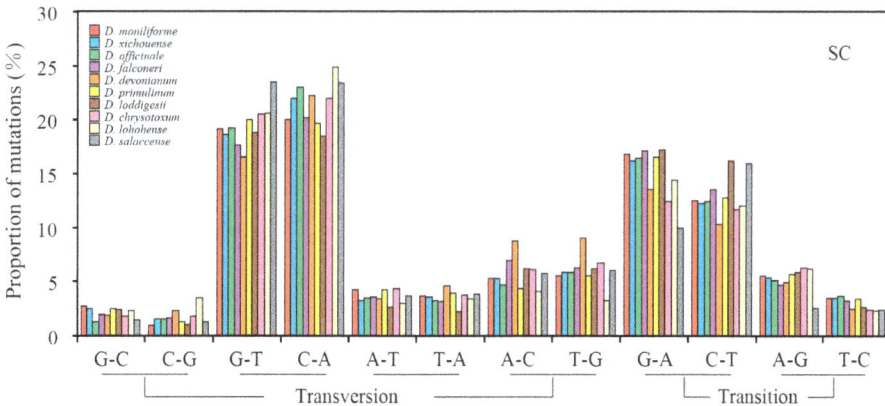

Figure 4. *Cont.*

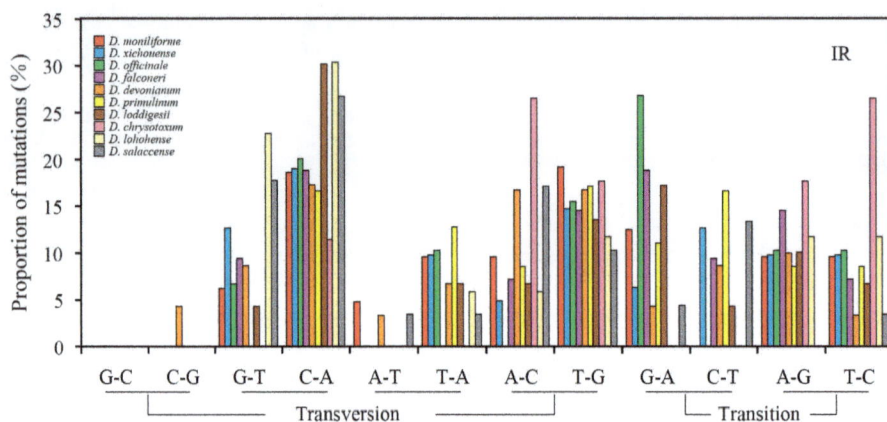

Figure 4. Proportion of the six nucleotide-pair mutations estimated from non-coding loci of the SC and IR regions of 10 *Dendrobium* species. The numbers of A to G mutations are normalized for the unequal nucleotide content of *Dendrobium* species. The frequencies of transversions are higher than that of transitions in both SC and IR regions.

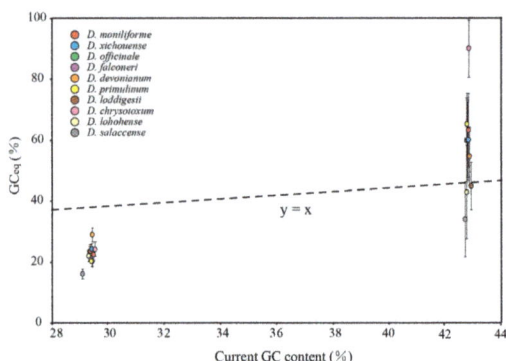

Figure 5. Comparison of equilibrium GC content (GC_{eq}) values between the SC and IR regions for the non-coding regions. Error bars depict 95% confidence intervals for GC_{eq}. Note that, the non-coding regions showed contrast GC_{eq} values (<50% in SC and >50% in IR) in SC and IR regions. Moreover, the estimated GC_{eq} values are lower than current GC content in SC regions, but higher than current GC content in IR regions.

Table 2. Summary of mutations in non-coding loci.

Species	SC		IR	
	Numbers of GC-AT Mutations	Numbers of Normalized AT-GC Mutations	Numbers of GC-AT Mutations	Numbers of Normalized AT-GC Mutations
D. moniliforme	229	66.65	6	7.74
D. xichouense	226	65.58	8	6.18
D. officinale	229	61.93	8	5.39
D. falconeri	248	76.54	6	4.63
D. devonianum	273	109.66	9	10.84
D. primulinum	221	60.78	8	7.71
D. loddigesii	210	62.25	13	8.62
D. chrysotoxum	188	61.10	1	7.74
D. lohohense	185	40.77	7	5.40
D. salaccense	452	104.40	14	6.92

2.4. Contrasting Mutational Biases of Protein-Coding Genes between SC and IR Regions

Three pairs of site models (M0 vs. M3, M1a vs. M2a, and M7 vs. M8) were used to test whether the evolution of plastid protein-coding genes was driven by positive selection. Among 68 plastid protein-coding genes, twelve genes (*accD, ccsA, matK, psaB, rbcL, rpl20, rpoC1, rpoC2, rps3, rps16, ycf1* and *ycf2*) were detected under positive selection (Table S1). The plastid protein-coding genes were classified into two categories, positive selected and non-positive selected data sets. The mutational biases were counted. However, the two data sets showed the same mutational trends: (1) the frequencies of transitions were higher than that of transversions (Figure S2); (2) counts of AT-rich mutations were larger than that of GC-rich mutations (Table S2); and (3) the GC$_{eq}$ values were smaller than equilibrium (GC$_{eq}$ = 50%) (Figure S3). Similar results were also observed in each codon position, which suggests that positive selection has no effect on gene conversions. Therefore, the plastid protein-coding genes were re-divided into SC and IR data sets based on their locations. The genes of *rpl22, rps12,* and *ycf1* were discarded because they were distributed in both SC and IR regions. Consistent with the counting results of non-coding loci, protein-coding genes also showed a contrast mutational bias between SC and IR regions (Table 3 and Figures 6 and 7).

Table 3. Summary of mutations in protein-coding loci.

Species	SC		IR	
	Numbers of GC-AT Mutations	Numbers of Normalized AT-GC Mutations	Numbers of GC-AT Mutations	Numbers of Normalized AT-GC Mutations
D. moniliforme	79	44.73	1	7.73
D. xichouense	115	45.95	8	5.78
D. officinale	100	51.09	8	10.29
D. falconeri	94	49.20	2	9.66
D. devonianum	89	53.71	5	5.79
D. primulinum	94	49.86	7	6.43
D. loddigesii	86	46.03	5	5.79
D. chrysotoxum	152	36.24	13	1.28
D. lohohense	59	48.04	3	5.79
D. salaccense	57	42.26	3	5.79

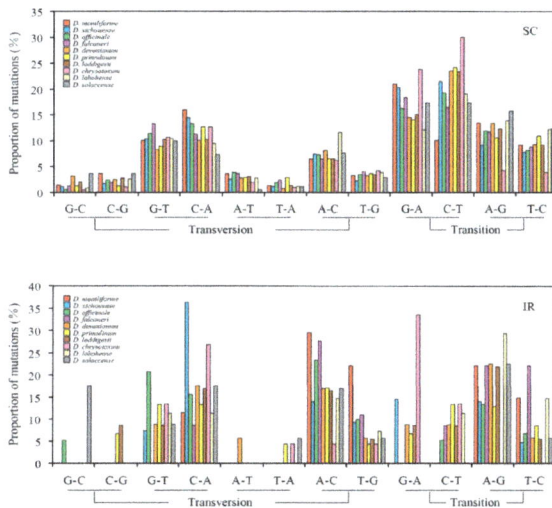

Figure 6. Proportion of the six nucleotide-pair mutations estimated from protein-coding genes. In contrast to the counting results of non-coding loci, the frequencies of transitions are higher than that of transversions in both SC and IR regions.

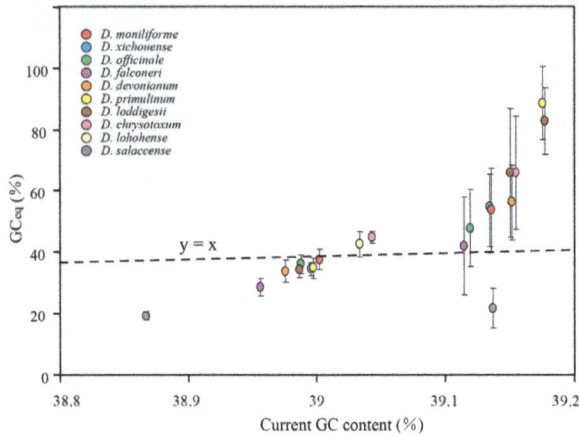

Figure 7. Comparison of GC_{eq} values between the SC and IR regions for the protein-coding genes. Error bars depict 95% confidence intervals for GC_{eq}. The protein-coding genes showed the same evolution trends with the non-coding loci.

2.5. Mutational Biases Have Directly Impact on the Evolution of GC Content in Dendrobium Plastomes

As mentioned above, our analysis revealed contrast mutational biases between SC and IR regions in both protein-coding genes and non-coding loci. To assess the effect of mutational biases on the evolution of GC content in *Dendrobium* plastomes, the GC_{eq} values were compared to current GC content. As shown in Figures 5 and 7, the GC_{eq} values of protein-coding genes and non coding loci in SC regions were estimated to be 19.25–44.88% and 16.05–28.92%, respectively. The estimated GC_{eq} values were lower than current GC content in SC regions indicating that the GC content of SC regions was evolved toward AT-richness. On the contrary, the results show that GC_{eq} values of IRs (42.86–90.02% for protein-coding genes and 21.72–88.54% for non-coding loci) were higher than current GC content, suggesting that the evolution trends of GC content in IRs were toward GC-richness. Considering the consistent evolution trends between mutational biases and the evolution of GC content in SC and IRs, we proposed that mutational biases have directly impact on the evolution of GC content in *Dendrobium* plastomes.

2.6. gBGC Has an Effect on Maintaining Higher GC Content in IR Regions

Figure S4 illustrated the non-synonymous (dn) and synonymous (ds) substitution rates of protein-coding genes and mutation rates estimated from the syntenic non-coding loci in the SC and IR regions. The two-sided Mann–Whitney test indicated that the substitution rates were higher in SC regions than that in IR regions of *Dendrobium* plastomes ($p < 0.01$). The reduced substitution rates of IR regions most likely resulted from gene conversion. The intraplastomic recombination via IR regions frequently occurred in the plastid because of the identical sequences in the two IR regions (Figure S5A). Therefore, the GC-biased mutation and reduced substitution rates in IRs could be caused by gBGC. To determine whether the gBGC has an effect on the evolution of GC content, we compared the GC content of IR regions between the two groups: (1) the one-IR group, including five plastomes from Pinaceae and cupressophytes, which contain only one copy of IR region; and (2) the two-IR group, including 10 *Dendrobium* plastomes, which contain two copies of IR regions. Our comparison revealed a significantly higher GC content in the IR regions of the two-IR group than that of the one-IR group (Mann–Whitney 2-sides, $p < 0.01$), which suggested that the gBGC has influenced the GC content in IR regions (Figure S5B). Furthermore, the significant correlation between the GC_{eq} values and current GC content in protein-coding genes and non-coding regions (Spearman's $r = 0.851, 0.841$, both $p < 0.01$) indicated that gBGC has a long-term effect on the GC content in *Dendrobium* plastomes.

3. Discussion

3.1. The High-Quality Complete Plastome Sequences That Could Represent for Dendrobium Species

Plastome sequences of seed plants, in general, have relatively small sizes, conserved gene contents, dense coding regions, and slower evolutionary rates as compared to nuclear and mitochondrial genomes [1]. These unique features have attracted intense attention from researchers, leading to numerous plastome sequences published, e.g., [20,22,25,26]. The first record of a *Dendrobium* plastome, *Dendrobium officinale*, was obtained by sequencing from the isolated chloroplast DNA [21]. However, it is extremely hard to extract the chloroplast from *Dendrobium* species due to their high percentage of polysaccharides, e.g., [27,28]. The plant whole genome sequencing data always contains high copy numbers of plastome sequence reads, which provide an opportunity for the complete plastome sequence assembly. Thus, a lot of methods were established to assemble the plastome sequence from the genome sequencing resources [29–31]. Among them, the reference-guided mapping method is a good approach for studies of related species with known reference sequences. Recently, more than 30 *Dendrobium* plastomes have been sequenced and assembled by using this approach [22]. However, tentative errors, such as SNP, InDel, and repeat sequences, may generate by using different assembly tools or mapping data to different references [31]. As shown in Figure S1, the sequences in the non-coding regions and *ndh* genes of *D. loddigesii* and *D. devonianum* were distinct from different assembly methods. Therefore, the high-quality plastome sequence that could be representing for *Dendrobium* species was urgently needed. In this study, two different methods were employed for the plastome assembly of *Dendrobium* species. According to the mVISTA analysis, the mis-assembled error sequences were detected, which indicated that at least two different methods should be used in the plastome assembly analysis to avoid assembling errors. After correcting and validating these errors by PCR amplification and Sanger sequencing, the complete plastomes of *D. loddigesii* and *D. devonianum* were obtained. Therefore, we proposed that the high-quality complete plastome sequences of *D. loddigesii* and *D. devonianum* were obtained in this study, which could represent for *Dendrobium* species for the future studies.

3.2. Both Mutational Biases and gBGC Affect GC Content in the Plastomes of Dendrobium Genus

The causes of GC content variation in plastome sequences were less clearly established than in nuclear genome sequences, and the role of gene conversion has only been investigated in non-coding regions more recently [17]. Previously, more than 30 plastomes of *Dendrobium* species have been published, which provides a huge resource for exploring the evolutionary mechanism of the variation of GC content in *Dendrobium* plastomes. Niu et al., (2017) proposed that evolutionary stasis should use at least 10 plastome sequences [22]. Here, we examined and discussed the three major hypotheses based on our newly sequenced data coupled with other eight *Dendrobium* plastomes.

3.2.1. Natural Selection

Natural selection has great impact on the variations in base composition. For example, in plastomes, the codon-usage of the *psbA* gene was proposed to be associated with positive selection for maintaining the translation efficiency [32]. Moreover, several studies have suggested a correlation between GC content and the function and expression of genes e.g., [5,33,34], which enhanced the hypothesis that natural selection has a direct impact on GC content. The "natural selection hypothesis" was also supported by the study of Shi et al., (2007), who found that the selection is the primary cause of the GC content variation in rice genes [35]. In this study, positive selection analysis was performed for the 68 plastid protein-coding genes. The higher GC content in the IRs than that in SC regions leads us to expect that genes located in IRs have a higher proportion under positive selection. However, only one protein-coding gene (*ycf2*) was identified under positive selection in the IR regions. Moreover, the same mutational trends were observed in both positive selected and non-positive selected data sets. These results suggested that natural selection does not appear to contribute significantly to the varied GC content of plastid protein-coding genes. Therefore, we ruled out the effect of natural selection, which leaves the mutational bias and gBGC as potential determinants for the evolution of GC content in the *Dendrobium* plastomes.

3.2.2. Mutational Bias

In this study, our analysis revealed a contrast mutational bias in *Dendrobium* plastomes that: (1) toward AT-richness in SC and (2) toward GC-richness in IR regions. The contrast mutational biases would lead to a lower GC content in SC regions. Indeed, the GC contents of LSC and SSC regions are lower than that of IRs. Moreover, the result estimated GC_{eq} values lower than current GC content in SC regions and higher than that of IRs, indicating that the GC content variation in SC regions is primarily determined by the mutational biases. Therefore, we proposed that mutational biases have direct impact on the variation of GC content in *Dendrobium* plastomes. Mutational biases toward AT-richness have been reported in gnetophytes [3]. Their plastomes were discovered to be compacting with enriched AT content, which was thought to have many advantages for resource consumption. For example, an A/T nucleotide contains seven atoms of nitrogen, one less than the G/C nucleotide. Moreover, AT-richness in base composition would benefit the rapid replication of plastomes [36]. Therefore, the sequence feature that mutated toward AT-richness would be favored in the plastome.

3.2.3. gBGC

The variation in base composition is proven to be affected by gene conversion, e.g., [37,38]. The gene conversion of the plastome sequences is typically modeled as an independent process, in which changes at any one site are independent of changes at any other site [39]. However, gene conversion is not an entirely stochastic process, as it acts according to certain deterministic biases that affect the GC content evolution. Over the past 10 years, GC-biased gene conversion (gBGC) has been clearly established as the main process affecting GC content evolution in the nuclear genome [40–43]. However, whether the gBGC affects the GC content in plastome sequences is poorly understood. In this study, although no direct evidence is detected, we cannot ignore the impact of gBGC to the GC content evolution in *Dendrobium* genus. Based on the comparison results of 10 *Dendrobium* plastomes, we proposed that the gBGC has influenced the plastome wide GC content evolution due to three reasons. Firstly, the higher GC content in IR regions could be explained by gBGC. In this study, the reduced substitution rates and GC-biased mutation indicated that gBGC occurred in IRs. The gBGC would affect GC/AT heterozygous sites yields more frequently to GC than to AT alleles, which lead to increasing the GC content over evolutionary time [13]. Secondly, the recombination of IRs provides an opportunity for gBGC to occur. IRs are the recombination hotspots in plastome sequence, which favors the gBGC [44–46]. Moreover, a gBGC model during recombination has been put forth to interpret the GC-biased mutations observed in the IR regions [17]. Through that model, GC to AT mutations were unfavored, whereas AT to GC ones were fixed after gBGC. After loss of one IR region, the GC content of IR regions in Pinaceae and cupressophytes species was evolved toward AT-richness. Thirdly, the significant correlation between the GC_{eq} values and GC content values in protein-coding genes and non-coding regions (Spearman's $r = 0.851$, 0.841, both $p < 0.01$) indicated that in *Dendrobium*, gBGC has shaped plastome-wide mutations and GC content. Therefore, we proposed that both mutational biases and gBGC affect GC content in the plastomes of the *Dendrobium* genus.

4. Materials and Methods

4.1. DNA Extraction and Plastome Sequencing

Two grams each of fresh leaves were harvested from individuals of *D. loddigesii* (voucher specimen: Niu14007) and *D. devonianum* (voucher specimen: Niu15008). They were cultivated in the greenhouse of Nanjing Normal University. The total DNA of each species was extracted using the Qiagen DNeasy Plant Mini Kit (Qiagen, Hilden, Germany), according to the manufacturer's instructions. DNA samples that met the quality of concentrations >300 ng/mL, A260/A280 ratio = 1.8–2.0, and A260/A230 ratio >1.7, were used for next-generation sequencing. The sequencing depth was 3.84 GB of 150 bp paired-end reads for each species.

4.2. Plastome Assembly, Annotation and Comparison

To obtain high-quality complete plastome sequences of *D. loddigesii* and *D. devonianum*, two different assembly methods: (1) de novo assembly methods and (2) reference-guided mapping method, were employed in this study. For the de novo assembly, raw reads were first trimmed with error probability <0.05, then de novo assembled using SOAPdenovo version 1.12 with the default parameters [47]. The de Bruijn graph approach was applied to assembly with an optimal K-mer size of 79. Contigs with length <200 bp were discarded. The remaining contigs were compared with the plastome sequences of *D. moniliforme* (AB893950) using BLAST searches. Matched contigs with E-values of <10^{-10} were designated as plastomic contigs. The gaps between plastomic contigs were closed by PCR amplification with specific primers and Sanger sequencing. For reference-guided assembly, the trimmed reads were mapped to the reference plastome sequence of *D. moniliforme* (AB893950) using CLC Genomics Workbench 6.0.1 (CLC Bio, Aarhus, Denmark). The four junctions between LSC/SSC and IRs and the regions with uncertain nucleotide "N" were validated with PCR amplification.

Protein-coding genes and ribosomal RNA genes were annotated using DOGMA [48]. The boundaries of each annotated gene were confirmed by their alignment with their orthologous genes from other *Dendrobium* plastomes. Genes of tRNA were predicted using tRNAscan-SE 1.21 [49]. The differences between plastome sequences of *D. loddigesii* and *D. devonianum* that assembled from different methods were compared by using the mVISTA software [50]. The plastome sequence of *D. moniliforme* (AB893950) was used as an outgroup.

4.3. Phylogenetic Analysis

The plastome sequences of 10 *Dendrobium* species and *Phalaenopsis aphrodite* were aligned based on the MAFFT method [51]. The gaps within alignment were excluded. The phylogenetic tree was constructed using MrBayes 3.2 [52]. For the Bayesian inference (BI) analysis, two simultaneous runs were conducted, each consisting of four chains. The parameters were set as "lset nst = 6 rates = γ". In total, chains were run for 2,000,000 generations, with topologies sampled every 100 generations. The first 25% of our sampled trees were discarded. The remaining trees were used to construct a majority-rule consensus tree and calculate posterior probabilities (PPs) for each node.

4.4. Ancestral Non-Coding Sequences Reconstruction and Calculation of the Relative Rates of the Six Nucleotide Pair Mutations

The non-coding loci, including intergenic and intronic loci, were manually retrieved from the 10 *Dendrobium* plastome sequences (Table S3). The loci were aligned by using MUSCLE [53] with the "Refining" option implemented in Mega 5.2 [54]. Gaps located at the 5′- and 3′-ends of alignments were excluded, and then concatenated using SequenceMatrix 1.8 [55]. The alignment of non-coding loci was divided into SC and IR data sets based on their locations. The ancestral non-coding sequences of 10 *Dendrobium* species were reconstructed using the maximum likelihood (ML) method with GTR + G model in Mega 5.2. The BI tree inferred from whole plastome sequences was designated as the "user tree". DAMBE 5 [56] was used to count the nucleotide changes between ancestral and current sequences. After that, we calculated the relative rates of the six nucleotide pair mutations. Firstly, we normalized the counts of the mutations from A/T to G/C, C/G, or T/A, and followed Hershberg and Petrov's method (2010) by multiplying them with current genome wide number of GC sites/AT sites [57]. For example, the normalized number of A to G mutations in *D. loddigesii* = the number of A to G changes between ancestral sequence and the current sequence of *D. loddigesii* × the current number of GC sites/AT sites. In this way, we could determine the expected number of mutations under equal GC and AT contents. Then, the relative rates of each nucleotide mutation were calculated with the formula = the number of mutations from G/C or the normalized number of mutations from A/T/ (the total number of mutations from G/C + the normalized total number of mutations from A/T) × 100.

4.5. Estimation of GC$_{eq}$

The GC$_{eq}$ values were estimated according to the method of Hershberg and Petrov (2010) [57] and Wu et al., 2015 [17]. The GC$_{eq}$ values were calculated as: rate of AT to GC/(rate of AT to GC + rate of GC to AT). The rate of AT to GC = the total number of AT to GC changes/the current number of AT sites, while the rate of GC to AT = the total number of GC to AT changes/the current number of GC sites. The SC and IR data sets including ancestral and current sequences were bootstrapped with 100 replications using PHYLIP 3.695 [58]. Then, the GC$_{eq}$ values were recalculated 100 times and the resulting values were used to estimate the 95% confidence intervals for GC$_{eq}$.

4.6. Natural Selection Analysis and Evolutionary Stasis of Protein-Coding Genes

Except for the 11 *ndh* genes, positive selection analysis was performed for the other 68 protein-coding genes using the codeml program in PAML vs. 4.9 [59]. The natural selection at the codon level was detected using the three pairs of site models (M0 vs. M3, M1a vs. M2a, and M7 vs. M8) as implemented in codeml. The likelihood ratio tests (LRTs) were used to compare the site models. The relative rates of the six nucleotide pair mutations and GC$_{eq}$ values of protein-coding genes were also counted.

4.7. Estimation of Substitution Rates

The synonymous (ds) and non-synonymous (dn) substitution rates were estimated with the codeml program of PAML [59]. The parameters were set to the following: seqtype = 1, runmodel = −2, and CodonFreq = 3. The protein-coding genes of *Phalaenopsis aphrodite* were used as the reference. The non-coding loci that flanked by the same genes/exons were identified as syntenic. The loci with sequence lengths less than 150 bp were discarded. The mutation rates between *P. aphrodite* and the 10 *Dendrobium* species were estimated with the baselml program of PAML using the REV model [59].

4.8. Statistical Analysis

Statistical analyses were performed using SPSS Statistics 20.0.

5. Conclusions

In conclusion, this study is the first to observe the impact of evolutionary forces, selection, mutational biases, and GC-biased gene conversion (gBGC) on the variation of GC content in the whole plastome sequences. The high-quality complete plastomes of *D. loddigesii* and *D. devonianum* were obtained and compared with eight other *Dendrobium* plastomes. The results indicate that the plastomes of *Dendrobium* species are highly conserved in plastomic organization. Because tentative errors generated by different assembly methods can lead to low quality of plastome sequence, we thus strongly suggest using two different assembly methods in the plastome assembly analysis. Furthermore, we examined three major hypotheses based on the plastome sequences of 10 *Dendrobium* species. Our results demonstrated that both mutational biases and gBGC affect GC content in the plastomes of the *Dendrobium* genus.

Supplementary Materials: Supplementary materials can be found at www.mdpi.com1422-0067/18/11/2307/s1.

Acknowledgments: This work was supported by the National Natural Science Foundation of China (Grant No. 31170300 and No. 31670330) and the Priority Academic Program Development of Jiangsu Higher Education Institutions to Xiaoyu Ding.

Author Contributions: Xiaoyu Ding and Zhitao Niu designed the study topic. Zhitao Niu, Hui Wang, Xuezhu Xie, and Shuying Zhu performed the experiments. Zhitao Niu, Qingyun Xue, and Wei Liu analyzed data. Zhitao Niu wrote the manuscript. Xiaoyu Ding, Zhitao Niu, and Qingyun Xue revised the final version of manuscript. All authors read and approved the final manuscript.

Conflicts of Interest: The authors declare no conflict of interest.

References

1. Raubeson, L.A.; Jansen, R.K. Chloroplast genomes of plants. In *Plant Diversity and Evolution: Genotypic and Phenotypic Variation in Higher Plants*; Henry, R.J., Ed.; CAB International: London, UK, 2005; pp. 45–68.
2. Wolfe, K.H.; Li, W.H.; Sharp, P.M. Rates of nucleotide substitution vary greatly among plant mitochondrial, chloroplast, and nuclear DNAs. *Proc. Natl. Acad. Sci. USA* **1987**, *84*, 9054–9058. [CrossRef] [PubMed]
3. Wu, C.S.; Lai, Y.T.; Lin, C.P.; Wang, Y.N.; Chaw, S.M. Evolution of reduced and compact chloroplast genomes (cpDNAs) in gnetophytes: Selection toward a lower-cost strategy. *Mol. Phylogenet. Evol.* **2009**, *52*, 115–124. [CrossRef] [PubMed]
4. Eyre-Walker, A.; Hurst, L.D. The evolution of isochores. *Nat. Rev. Genet.* **2001**, *2*, 549–555. [CrossRef] [PubMed]
5. Mukhopadhyay, P.; Basak, S.; Ghosh, T.C. Nature of selective constraints on synonymous codon usage of rice differs in GC-poor and GC-rich genes. *Gene* **2007**, *400*, 71–81. [CrossRef] [PubMed]
6. Bernardi, G. Isochores and the evolutionary genomics of vertebrates. *Gene* **2000**, *241*, 3–17. [CrossRef]
7. Wang, H.C.; Singer, G.A.C.; Hickey, D.A. Mutational bias affects protein evolution in flowering plants. *Mol. Biol. Evol.* **2004**, *21*, 90–96. [CrossRef] [PubMed]
8. Günther, T.; Lampei, C.; Schmid, K.J. Mutational bias and gene conversion affect the intraspecific nitrogen stoichiometry of the *Arabidopsis thaliana* transcriptome. *Mol. Biol. Evol.* **2013**, *30*, 561–568. [CrossRef] [PubMed]
9. Fryxell, K.J.; Zuckerkandl, E. Cytosine deamination plays a primary role in the evolution of mammalian isochores. *Mol. Biol. Evol.* **2000**, *17*, 1371–1383. [CrossRef] [PubMed]
10. Marais, G. Biased gene conversion: Implications for genome and sex evolution. *Trends Genet.* **2003**, *19*, 330–338. [CrossRef]
11. Duret, L.; Galtier, N. Biased gene conversion and the evolution of mammalian genomic landscapes. *Annu. Rev. Genom. Hum. Genet.* **2009**, *10*, 285–311. [CrossRef] [PubMed]
12. Smith, D.R.; Lee, R.W. Mitochondrial genome of the colorless green alga *Polytomella capuana*: A linear molecule with an unprecedented GC content. *Mol. Biol. Evol.* **2008**, *25*, 487–496. [CrossRef] [PubMed]
13. Pessia, E.; Popa, A.; Mousset, S.; Rezvoy, C.; Duret, L.; Marais, G.A.B. Evidence for widespread GC-biased gene conversion in eukaryotes. *Genome Biol. Evol.* **2012**, *4*, 675–682. [CrossRef] [PubMed]
14. Lassalle, F.; Périan, S.; Bataillon, T.; Nesme, X.; Duret, L.; Daubin, V. GC-content evolution in bacterial genomes: The biased gene conversion hypothesis expands. *PLoS Genet.* **2015**, *11*, e1004941. [CrossRef] [PubMed]
15. Lesecque, Y.; Mouchiroud, D.; Duret, L. GC-biased gene conversion in yeast is specifically associated with crossovers: Molecular mechanisms and evolutionary significance. *Mol. Biol. Evol.* **2013**, *30*, 1409–1419. [CrossRef] [PubMed]
16. Muyle, A.; Serres-Giardi, L.; Ressayre, A.; Escobar, J.; Glémin, S. GC-biased gene conversion and selection affect GC content in the *oryza* genus (rice). *Mol. Biol. Evol.* **2011**, *28*, 2695. [CrossRef] [PubMed]
17. Wu, C.S.; Chaw, S.M. Evolutionary stasis in cycad plastomes and the first case of plastome GC-biased gene conversion. *Genome Biol. Evol.* **2015**, *7*, 2000–2009. [CrossRef] [PubMed]
18. Givnish, T.J.; Spalink, D.; Ames, M.; Lyon, S.P.; Hunter, S.J.; Zuluaga, A.; Iles, W.J.D.; Clements, M.A.; Arroyo, M.T.K.; Leebens-Mack, J.; et al. Orchid phylogenomics and multiple drivers of their extraordinary diversification. *Proc. Biol. Sci. B* **2015**, *282*, 2108–2111. [CrossRef] [PubMed]

19. Shaw, J.; Shafer, H.L.; Leonard, O.R.; Kovach, M.J.; Schorr, M.; Morris, A.B. Chloroplast DNA sequence utility for the lowest phylogenetic and phylogeographic inferences in angiosperms: The tortoise and the hare IV. *Am. J. Bot.* **2014**, *101*, 1987–2004. [CrossRef] [PubMed]

20. Niu, Z.; Xue, Q.; Zhu, S.; Sun, J.; Liu, W.; Ding, X. The complete plastome sequences of four orchid species: Insights into the evolution of the Orchidaceae and the utility of plastomic mutational hotspots. *Front. Plant. Sci.* **2017**, *8*, 715. [CrossRef] [PubMed]

21. Luo, J.; Hou, B.W.; Niu, Z.T.; Liu, W.; Xue, Q.Y.; Ding, X.Y. Comparative chloroplast genomes of photosynthetic orchids: Insights into evolution of the Orchidaceae and development of molecular markers for phylogenetic applications. *PLoS ONE* **2014**, *9*, e99016. [CrossRef] [PubMed]

22. Niu, Z.; Zhu, S.; Pan, J.; Li, L.; Sun, J.; Ding, X. Comparative analysis of *Dendrobium* plastomes and utility of plastomic mutational hotspots. *Sci. Rep.* **2017**, *7*, 2073.

23. Wood, H.P. *The Dendrobiums*; Timber Press: Portland, OR, USA, 2006.

24. Feng, S.; Jiang, Y.; Wang, S.; Jiang, M.; Chen, Z.; Ying, Q.; Wang, H. Molecular identification of *Dendrobium* species (Orchidaceae) based on the DNA barcode ITS2 region and its application for phylogenetic study. *Int. J. Mol. Sci.* **2014**, *16*, 21975–21988. [CrossRef] [PubMed]

25. Niu, Z.; Pan, J.; Zhu, S.; Li, L.; Xue, Q.; Liu, W.; Ding, X. Comparative analysis of the complete plastomes of *Apostasia wallichii* and *Neuwiedia singapureana* (Apostasioideae) reveals different evolutionary dynamics of IR/SSC boundary among photosynthetic orchids. *Front. Plant. Sci.* **2017**, *8*, 1713. [CrossRef] [PubMed]

26. Wu, C.S.; Chaw, S.M. Large-scale comparative analysis reveals the mechanisms driving plastomic compaction, reduction, and inversions in conifers II (cupressophytes). *Genome Biol. Evol.* **2016**, *8*, 3740–3750. [CrossRef] [PubMed]

27. Xu, H.; Hou, B.; Zhang, J.; Min, T.; Yuan, Y.; Niu, Z.; Ding, X. Detecting adulteration of *Dendrobium officinale* by real-time PCR coupled with ARMS. *Int. J. Food Sci. Technol.* **2012**, *47*, 1695–1700. [CrossRef]

28. Yan, W.J.; Zhang, J.Z.; Zheng, R.; Sun, Y.L.; Ren, J.; Ding, X.Y. Combination of SYBR Green II and TaqMan Probe in the adulteration detection of *Dendrobium devonianum* by fluorescent quantitative PCR. *Int. J. Food Sci. Technol.* **2016**, *50*, 2572–2578. [CrossRef]

29. Straub, S.C.; Fishbein, M.; Livshultz, T.; Foster, Z.; Parks, M.; Weitemier, K.; Cronn, R.C.; Liston, A. Building a model: Developing genomic resources for common milkweed (*Asclepias syriaca*) with low coverage genome sequencing. *BMC Genom.* **2011**, *12*, 211. [CrossRef] [PubMed]

30. Wysocki, W.P.; Clark, L.G.; Kelchner, S.A.; Burke, S.V.; Pires, J.C.; Edger, P.P.; Mayfield, D.R.; Triplett, J.K.; Columbus, J.T.; Ingram, A.L.; et al. A multi-step comparison of short-read full plastome sequence assembly methods in grasses. *Taxon* **2014**, *63*, 899–910.

31. Kim, K.; Lee, S.C.; Lee, J.; Yu, Y.; Yang, T.J.; Choi, B.S.; Koh, H.; Waminal, N.E.; Choi, H.; Kim, N.; et al. Complete chloroplast and ribosomal sequences for 30 accessions elucidate evolution of *Oryza* AA genome species. *Sci. Rep.* **2015**, *5*, 15655. [CrossRef] [PubMed]

32. Morton, B.R. Chloroplast DNA codon use: Evidence for selection at the *psbA* locus based on tRNA availability. *J. Mol. Evol.* **1993**, *3*, 273–280. [CrossRef]

33. Tatarinova, T.V.; Alexandrov, N.N.; Bouck, J.B.; Feldmann, K.A. GC3 biology in corn, rice, sorghum and other grasses. *BMC Genom.* **2010**, *11*, 308. [CrossRef] [PubMed]

34. Tatarinova, T.; Elhaik, E.; Pellegrini, M. Cross-species analysis of genic GC3 content and DNA methylation patterns. *Genome Biol. Evol.* **2013**, *5*, 1443–1456. [CrossRef] [PubMed]

35. Shi, X.; Wang, X.; Li, Z.; Zhu, Q.; Yang, J.; Ge, S. Evidence that natural selection is the primary cause of the guanine-cytosine content variation in rice genes. *J. Integr. Plant Biol.* **2007**, *49*, 1393–1399. [CrossRef]

36. McCoy, S.R.; Kuehl, J.V.; Boore, J.L.; Raubeson, L.A. The complete plastid genome sequence of *Welwitschia mirabilis*: An unusually compact plastome with accelerated divergence rates. *BMC Evol. Biol.* **2008**, *8*, 130. [CrossRef] [PubMed]

37. Glémin, S.; Clément, Y.; David, J.; Ressayre, A. GC content evolution in coding regions of angiosperm genomes: A unifying hypothesis. *Trends Genet.* **2014**, *30*, 263. [CrossRef] [PubMed]

38. Zhu, A.; Guo, W.; Gupta, S.; Fan, W.; Mower, J.P. Evolutionary dynamics of the plastid inverted repeat: The effects of expansion, contraction, and loss on substitution rates. *New Phytol.* **2016**, *209*, 1747–1756. [CrossRef] [PubMed]

39. Drouin, G.; Daoud, H.; Xia, J. Relative rates of synonymous substitutions in the mitochondrial, chloroplast, and nuclear genomes of seed plants. *Mol. Phylogenet. Evol.* **2008**, *49*, 827–831. [CrossRef] [PubMed]

40. Galtier, N.; Duret, L.; Glémin, S.; Ranwez, V. GC-biased gene conversion promotes the fixation of deleterious amino acid changes in primates. *Trends Genet.* **2009**, *25*, 1–5. [CrossRef] [PubMed]
41. Escobar, J.S.; Glémin, S.; Galtier, N. GC-biased gene conversion impacts ribosomal DNA evolution in vertebrates, angiosperms, and other eukaryotes. *Mol. Biol. Evol.* **2011**, *28*, 2561–2575. [CrossRef] [PubMed]
42. Gotea, V.; Elnitski, L. Ascertaining regions affected by GC-biased gene conversion through weak-to-strong mutational hotspots. *Genomics* **2014**, *103*, 349–356. [CrossRef] [PubMed]
43. Goubert, C.; Modolo, L.; Vieira, C.; Valientemoro, C.; Mavingui, P.; Boulesteix, M. De novo assembly and annotation of the Asian tiger mosquito (*Aedes albopictus*) repeatome with dnaPipeTE from raw genomic reads and comparative analysis with the yellow fever mosquito (*Aedes aegypti*). *Genome Biol. Evol.* **2015**, *7*, 1192–1205. [CrossRef] [PubMed]
44. Glémin, S.; Arndt, P.F.; Messer, P.W.; Petrov, D.; Galtier, N.; Duret, L. Quantification of GC-biased gene conversion in the human genome. *Genome Res.* **2015**, *25*, 1215–1228. [CrossRef] [PubMed]
45. Palmer, J.D. Chloroplast DNA exists in two orientations. *Nature* **1983**, *301*, 92–93. [CrossRef]
46. Khakhlova, O.; Bock, R. Elimination of deleterious mutations in plastid genomes by gene conversion. *Plant J.* **2006**, *46*, 85–94. [CrossRef] [PubMed]
47. Li, R.; Fan, W.; Tian, G.; Zhu, H.; He, L. The sequence and de novo assembly of the giant panda genome. *Nature* **2010**, *463*, 311–317. [CrossRef] [PubMed]
48. Wyman, S.K.; Jansen, R.K.; Boore, J.L. Automatic annotation of organellar genomes with DOGMA. *Bioinformatics* **2004**, *20*, 3252–3255. [CrossRef] [PubMed]
49. Schattner, P.; Brooks, A.N.; Lowe, T.M. The tRNAscan-SE, snoscan and snoGPS web servers for the detection of tRNAs and snoRNAs. *Nucleic Acids Res.* **2005**, *33*, W686–W689. [CrossRef] [PubMed]
50. Frazer, K.A.; Pachter, L.; Poliakov, A.; Rubin, E.M.; Dubchak, I. VISTA: Computational tools for comparative genomics. *Nucleic Acids Res.* **2004**, *32*, W273–W279. [CrossRef] [PubMed]
51. Katoh, K.; Kuma, K.; Toh, H.; Miyata, T. MAFFT version 5: Improvement in accuracy of multiple sequence alignment. *Nucleic Acids Res.* **2005**, *33*, 511–518. [CrossRef] [PubMed]
52. Ronquist, F.; Teslenko, M.; Mark, P.V.D.; Ayres, D.L.; Darling, A.; Höhna, S.; Larget, B.; Liu, L.; Suchard, M.A.; Huelsenbeck, J.P. MrBayes 3.2: Efficient Bayesian phylogenetic inference and model choice across a large model space. *Syst. Biol.* **2012**, *61*, 539–542. [CrossRef] [PubMed]
53. Edgar, R.C. MUSCLE: Multiple sequence alignment with high accuracy and high throughput. *Nucleic Acids Res.* **2004**, *32*, 1792–1797. [CrossRef] [PubMed]
54. Tamura, K.; Peterson, D.; Peterson, N.; Stecher, G.; Nei, M.; Kumar, S. MEGA5: Molecular evolutionary genetics analysis using maximum likelihood, evolutionary distance, and maximum parsimony methods. *Mol. Biol. Evol.* **2011**, *28*, 2731–2739. [CrossRef] [PubMed]
55. Vaidya, G.; Lohman, D.J.; Meier, R. SequenceMatrix: Concatenation software for the fast assembly of multi-gene datasets with character set and codon information. *Cladistics* **2011**, *27*, 171–180. [CrossRef]
56. Xia, X. DAMBE5: A comprehensive software package for data analysis in molecular biology and evolution. *Mol. Biol. Evol.* **2013**, *30*, 1720–1728. [CrossRef] [PubMed]
57. Hershberg, R.; Petrov, D.A. Evidence that mutation is universally biased towards AT in bacteria. *PLoS Genet.* **2010**, *6*, e1001115. [CrossRef] [PubMed]
58. Felsenstein, J. *PHYLIP (Phylogeny Inference Package) Version 3.6*; Department of Genome Sciences, University of Washington: Seattle, DC, USA, 2005.
59. Yang, Z. PAML 4: Phylogenetic analysis by maximum likelihood. *Mol. Biol. Evol.* **2007**, *24*, 1586–1591. [CrossRef] [PubMed]

International Journal of
Molecular Sciences

MDPI

Brief Report

A Simple Method to Decode the Complete 18-5.8-28S rRNA Repeated Units of Green Algae by Genome Skimming

Geng-Ming Lin [1], Yu-Heng Lai [2], Gilbert Audira [3] and Chung-Der Hsiao [3,4,5,*]

[1] Laboratory of Marine Biology and Ecology, Third Institute of Oceanography, State Oceanic Administration, Xiamen 361005, China; lingengming@tio.org.cn
[2] Department of Chemistry, Chinese Culture University, Taipei 11114, Taiwan; lyh21@ulive.pccu.edu.tw
[3] Department of Bioscience Technology, Chung Yuan Christian University, Chung-Li 32023, Taiwan; gilbertaudira@yahoo.com
[4] Center for Biomedical Technology, Chung Yuan Christian University, Chung-Li 32023, Taiwan
[5] Center for Nanotechnology, Chung Yuan Christian University, Chung-Li 32023, Taiwan
* Correspondence: cdhsiao@cycu.edu.tw; Tel.: +886-3-265-3545

Received: 16 October 2017; Accepted: 3 November 2017; Published: 6 November 2017

Abstract: Green algae, *Chlorella ellipsoidea*, *Haematococcus pluvialis* and *Aegagropila linnaei* (Phylum Chlorophyta) were simultaneously decoded by a genomic skimming approach within 18-5.8-28S rRNA region. Whole genomic DNAs were isolated from green algae and directly subjected to low coverage genome skimming sequencing. After de novo assembly and mapping, the size of complete 18-5.8-28S rRNA repeated units for three green algae were ranged from 5785 to 6028 bp, which showed high nucleotide diversity (π is around 0.5–0.6) within ITS1 and ITS2 (Internal Transcribed Spacer) regions. Previously, the evolutional diversity of algae has been difficult to decode due to the inability design universal primers that amplify specific marker genes across diverse algal species. In this study, our method provided a rapid and universal approach to decode the 18-5.8-28S rRNA repeat unit in three green algal species. In addition, the completely sequenced 18-5.8-28S rRNA repeated units provided a solid nuclear marker for phylogenetic and evolutionary analysis for green algae for the first time.

Keywords: green algae; rRNA repeated unit; phylogeny; genome skimming

Green algae, a big group which contains at least 7000 species, has been found in wide range habitats from freshwater to sea water [1]. Similar to land plants, green algae contain chlorophyll *a* and chlorophyll *b* and store food as starch in plastids [2]. In the ecosystem, green algae play a role as primary photosynthetic eukaryotic organisms. The green algae have become powerful producers and providers of various natural substances, which may constitute the primary natural source, such as minerals, vitamins, nutrients, and fatty acids, as well as carotenoid pigments that include carotenes, xanthophylls, zeaxanthin, and lutein [3–5]. Currently, it is feasible to produce some carotenoids commercially through aquaculture [6]. In addition, because of their rapid growth and high oil content, some green algae also have been considered as a promising alternative feedstock for biodiesel production [7,8]. However, the challenge of developing green algae as a permanent fuel source will be to operate industries sustainably and compete with existing energy options with its costly investment [9].

In spite of their many unifying features, green algae exhibit remarkable in morphology and ecology reflecting their evolutionary diversification. Recently, according to cladistic classification and molecular analysis, the monophylogenic of green algae origin is still arguable [10–13], which referred that more approaches are required to validate phylogeny of green algae at molecular level by using different markers. The rapid increase in genomic data from a wide range of green algae has high potential to resolve large-scale green algal relationships. Furthermore, green algal genomes are important sources of information for the evolutionary origins of plant traits due to their evolutionary relationship to land plants [14,15]. Therefore, in this study, we demonstrated a genome skimming method to deduce the complete 18-5.8-28S rRNA repeated sequence (as a nuclear marker), which used as a molecular tool to reveal the relationships between *Haemotococcus pluvalis*, *Chlorella ellipsoidea* and *Aegagropila linnaei* to discuss the relationship between the freshwater and marine algae.

Genome skimming is an approach which reconstructs whole genome shotgun libraries faster and easily. This technique involves filtering millions of shotgun next generation sequencing (NGS) reads to find the few reads associated with particular DNA regions of interest [16], which is 18-5.8-28S rRNA repeated sequence for this case. Internal transcribed spacers (ITS) are sequences located in eukaryotic rRNA genes between the 18S and 5.8S rRNA coding regions (ITS1) and between the 5.8S and 28S rRNA coding regions (ITS2). The ITS is a non-coding region with high interspecific variability allowing differentiation of species within a genus, but low intra-specific variability preventing the separation of individuals or strains within the same species. These spacer sequences are present in all known nuclear rRNA genes of eukaryotes and have a high evolution rate. Previous restriction site variation studies in the ribosomal DNA (rDNA) have shown that the spacer regions are variable while coding regions are conserved. ITS are useful for phylogenetic analysis among related species and among populations within a species [17]. Combined with genome skimming, we implemented a rapid and cost-effective strategy for generating phylogenetically informative genomic data [18,19].

Haematococcus pluvialis, an unicellular green algae, is a promising microorganism that now showed potential to be a nutraceutical for human health because of the ability to produce astaxanthin, which is used as a coloring agent for aquaculture [20]. Moreover, astaxanthin not only helped to protect the skin against UV-induced damage, but was also used for tumor therapies and prevented neural damage associated with age-related degeneration [21,22]. *H. pluvalis* has developed into an organism that can be cultivated on an industrial scale [23]. In addition, *H. pluvalis* can also generate chlorophylls *a* and *b*, and primary carotenoid compounds namely neoxanthin, violaxanthin, zeaxanthin, lutein, and β-carotene, which suggested their great development and commercialization [21,24]. On the other hand, *Chlorella ellipsoidea* has been shown to possess bioactive substances that have various functional properties such as immunomodulatory, anti-inflammatory, and antioxidant effects [25,26]. Violaxanthin, the major component that was isolated from *C. ellipsoidae*, showed anti-inflammatory effect through inhibiting the NF-kB pathway [26]. Recent studies showed that *C. ellipsoidea* extract had significant apoptosis effect in human colon cancer cell line and suggested to have potential to prevent human cancer progression [6]. Moreover, *C. ellipsoidea* has been frequently used as a model organism in the field of genetics and the molecular biology in photosynthesis [6,27]. These observations indicated that microalgae have drawn more attention in scientific research. *Aegagropila linnaei* is a freshwater macroalga that is generally regarded as a rare and endangered species and belongs to *Cladophorales* order [28]. Velvety in appearance, these species can be excellent houseplants and are good to use in clear hanging vessels [29]. So far, the classification within the *Cladophorales* is still uncertain due to the polyphyletic nature of the large genus *Cladophora*, which results from a simple morphology with few specificity, extensive phenotypic plasticity, and both parallel and convergent evolutional character. Therefore, molecular data have contributed greatly a better understanding of *Cladophorales* evolution in recent years [30]. Unfortunately, *A. linnaei* has been used only in few molecular phylogenetic studies until now [31].

Three types of green algae were collected from a local commercial company (Available online: http://www.leadingtec.cn/). High quality genomic DNA was isolated by a modified CTAB method [32] to establish a genomic library with TrueSeq PCR-free kit and later we performed paired-end sequencing by Illumina HiSeq X Ten within 150 base pair (bp)-length unit. Paired-end deep sequencing reads were assembled by FLASH software [33] and then de novo assembled by CLCbio software (Available online: http://www.clcbio.com/) with default parameter settings (Kmer = 24, bubble size = 50). BLAST tool was applied to explore potential contigs that matched 18-5.8-28S rRNA repeats. Additional rRNA-related reads were obtained by repeatedly mapping with Geneious R9 software (Available online: http://www.geneious.com/) with 25–100 iterations. Finally, the complete 18-5.8-28S rRNA repeated unit consensus sequences were generated from the mapped reads, and were deposited to NCBI GenBank (detail information are listed in Table 1). In the three green algae tested, we found only 0.02 to 0.24% total reads were matched to 18-5.8-28S rRNA repeats. The average assembly coverage for 18-5.8-28S rRNA loci ranged from to 26 to 2736 X among the three green algal species (Table 1). These results suggested that the 18-5.8-28S rRNA copy numbers have great variation between different green algal species.

Table 1. Summary of three algae species tested in this study.

Species	*Chlorella ellipsoidea* (CE)	*Haematococcus pluvialis* (HP)	*Aegagropila linnaei* (AL)
Total reads	67,016,212	68,697,604	7,502,824
18-5.8-28S reads	159,445	74,748	1648
18-5.8-28S reads % *	0.24	0.11	0.02
Coverage (fold)	2736	825	26
Sequence (bp)	5785	5817	6028
NCBI accession number	KY364701	KY364700	KY364699

* This is defined as (18-5.8-28S reads/Total reads) × 100.

Next, we used rRNA prediction tool (Available online: http://weizhong-lab.ucsd.edu/metagenomic-analysis/server/hmm_rRNA/) and BLAST [34] to compare the gene annotation in other algal species and confirmed each identity and sequences of rRNA and ITS manually. The complete 18-5.8-28S repeats of three green algae ranged from 5785 to 6028 bps. The sequence identities of three species 18-5.8-28S rRNA repeats were confirmed by BLAST showing high identities (>99%) with previous published partial 18S rRNA sequences from the NCBI database (Figure S1). Sequence alignment of green algal rRNA repeats was generated by MAFFT [35] with default settings. High variation was detected in ITS1 and ITS2 regions (sequence identities ranged from 28.8% to 34.8% and 26.1% to 38.8%, respectively), while other regions of 18S, 5.8S and 28S (78.3% to 92.7%) were highly conserved among the three green algal species. We also used DnaSP V5 [36] to calculate the nucleotide diversity of 18-5.8-28S repeated units among three green algae. First, the nucleotide sequences were aligned by MAFFT and output as .meg file. Next, the nucleotide diversity of the aligned sequences was calculated in a sliding window with length and step size of 100 and 5 sites, respectively. Compared to rRNA regions, high nucleotide diversity (π is around 0.5–0.6) in ITS1 and ITS2 regions are detected (Figure 1).

To validate the phylogeny of three green algae, we used MEGA6 software [37] to construct a Maximum likelihood tree (with 500 bootstrap replicates and Kimura 2-parameter model), which contains 25 species derived from Phylum Chlorophyta. *Alexandrium tamarense* [38] derived from Phylum Dinoflagellata was used as outgroup for tree rooting. The result showed that, *Chlorella ellipsoidea* is closely related to *Micractinium reisseri*; *Haematococcus pluvialis* is closely related to *Chlamydomonas* sp. (Figure 2). The phylogenetic relationship obtained from complete 18-5.8-28S rDNA is consistent with previous research, which used short or partial sequences from 18S or other markers [39,40]. The freshwater macroalgae, *Aegagropila linnaei*, in contrast, was shown to be phylogenetically distinct from *Chlorella ellipsoidea*, *Haematococcus pluvialis* and other green algal species tested in this study (Figure 2). In conclusion, the 18-5.8-28S rRNA repeats deduced in this study provides an important DNA data for further phylogenetic and evolutionary analysis in green algae.

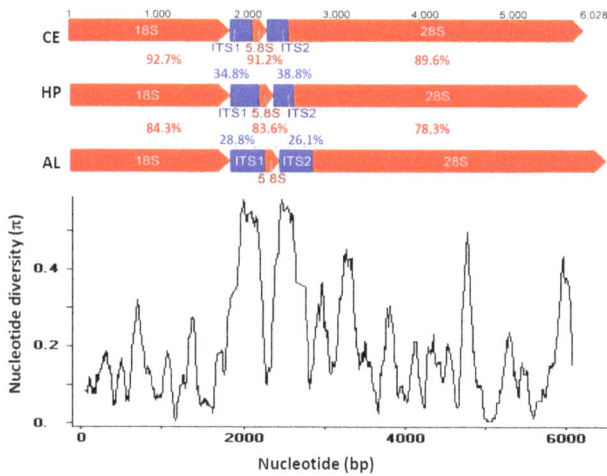

Figure 1. The complete 18-5.8-28S rRNA repeat unit of three green algae, *Chlorella ellipsoidea*, *Haematococcus pluvialis* and *Aegagropila linnaei* (Phylum Chlorophyta). The 18S, 5.8S and 28S rRNA genes are labeled in red, ITS1 and ITS2 sequences are labeled in blue. The nucleotide sequence identities are also highlighted for comparison. Lower panel shows the sliding window to compare the nucleotide diversity of 18-5.8-28S rDNA repeat unit assembly among three green algal species.

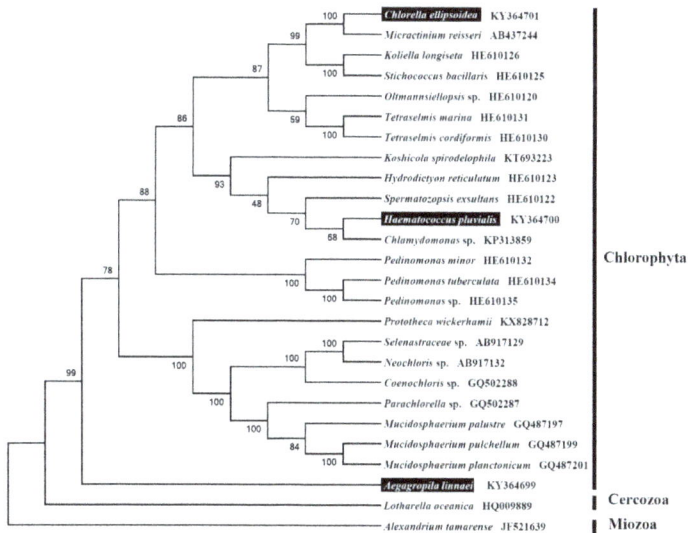

Figure 2. Molecular phylogeny of three green algae and related species in Phylum Chlorophyta based on complete 18-5.8-28S rRNA repeat units. The complete or partial 18-5.8-28S rRNA repeat sequences were downloaded from GenBank to construct a phylogenic tree by the Maximum likelihood method with 500 bootstrap replicates. Three targeted green algal *Chlorella ellipsoidea*, *Haematococcus pluvialis* and *Aegagropila linnaei*, are highlighted in black.

Molecular-based analysis on phylogeny has started on a new page of evolutionary relationship and substituted the previous morphological-based classification. Several markers, for example, small subunit (SSU) rDNA [31], large subunit (LSU) rDNA [41], have been established to study algal phylogeny and evolution. In addition, the detail inheritance may be accomplished by developing specific protein coding genes, such as *rbcL* and *matK* in chloroplast [42,43], internal transcribed spacer (ITS) [35], and nrDNA [44]. Generally, short and ease amplified sequence, such as ITS in fungi, *rbcL* and *matK* plastid loci in plants, and mitochondrial cytochrome c oxidase subunit 1 (*cox1*) in animals, have provided a convenient tool to perform DNA barcoding [45]. Moreover, NGS technology has advanced a genomics approach to differentiate more precisely among orthologous and paralogous regions at different loci within different species. The obstacle that scientists faced was that the short sequences may not support to all branches in a phylogeny [41]. Two-locus barcoding with *rbcL* and *matK* instead of single marker analyses has improved the accuracy and resolution of phylogenetic reconstruction [46]. However, the loss of deeper branches with short markers may also reduce its resolution in the case of two closely related species with nearly identical sequence. Although LSU was able to resolve ancient eukaryotic lineage [47], the SSU was more robust when decoding deeper divergence within LSU rRNA trees [48]. While *cox1* was a suitable marker for most of animals, the slow rate of *cox1* evolution in plants may impede its scientific application [49]. The same situation is reported in some coral species which show nearly identical *cox1* sequences between closely related species [50]. In contrast, high nucleotide diversity was reported in closely species or cryptic species [51]. This situation is very common in insects who display great genomic diversity and making it difficult to design universal primers to successfully amplify target marker genes [52,53]. The advantage of NGS is that hundreds to thousands markers can be easily discovered by high coverage whole genome or even reduced genome sequencing [54]. The 18-5.8-28S rRNA repeat is a highly copied genomic unit that plays an important function during protein translation [55]. The tandemly repeated head-to-tail organization has been considered the standard for eukaryotes, which has developed into a promising approach for phylogenetic reconstruction [55]. This high-copy-number trait makes it efficient to assemble with high coverage even in low coverage whole genome sequencing. Our approach is able to bypass the restricted and time-consuming works needed to design PCR primers to amplify the complete 18-5.8-28S rRNA repeat.

Intragenomic diversity is generally low due to concerted evolution [56]. This mechanism will lead individual repeats in the multigene family to evolve in concert, resulting in the homogenization of all the repeats in an array. However, recent findings in some species, including *Haematococcus*, pointed out the intragenomic variation of rRNA may be present, which indicated that more attention to be paid since this variation might affect the species delamination [57–60]. In our study, we obtained the complete 18-5.8-28S rRNA cluster by de novo assembly, mapping and consensus sequence generation approach. The parameter setting that we used to do de novo assembly and consensus sequence generation was able to remove low quality sequencing reads/errors, sequence variation, such as SNP or potential indels. The intragenomic variation reported in other species is around the level of 0–10% [61–63] and we believe that this tiny variation would be removed in our current analysis pipeline. Further studies using SNP/indel calling programs, such as SAMtools/BCFtools [64] or VarScan [65] allowed us to detect more possible intragenomic variation.

In conclusion, we provided a rapid and universal approach to deduce the complete 18-5.8-28S rRNA repeat sequences from evolutionarily diverse green algal species where the design of universal primers to amplify this locus is not possible. In addition, the complete 18-5.8-28S rRNA repeat unit sequence provides a good nuclear marker for phylogenetic and evolutionary analysis for green algae.

Supplementary Materials: The following are available online at www.mdpi.com/1422-0067/18/11/2341/s1.

Acknowledgments: This work was supported by the Scientific Research Foundation of Third Institute of Oceanography, SOA under contract No 20132017009 and The Public Science and Technology Research Funds Projects of Ocean under contract No 201305027 to Geng-Ming Lin.

Author Contributions: Geng-Ming Lin and Chung-Der Hsiao conceived and designed the experiments; Geng-Ming Lin and Yu-Heng Lai performed the experiments; Geng-Ming Lin, Yu-Heng Lai, Gilbert Audira and Chung-Der Hsiao analyzed the data; Geng-Ming Lin contributed reagents/materials/analysis tools; Geng-Ming Lin and Chung-Der Hsiao wrote the paper; all authors read and approved the final manuscript.

Conflicts of Interest: The authors declare no conflict of interest.

References

1. Guiry, M.D. How many species of algae are there? *J. Phycol.* **2012**, *48*, 1057–1063. [CrossRef] [PubMed]
2. Fondriest Environmental, I. "Algae, Phytoplankton and Chlorophyll." Fundamentals of Environmental Measurements. Available online: http://www.fondriest.com/environmental-measurements/parameters/water-quality/algae-phytoplankton-chlorophyll/ (accessed on 2 November 2017).
3. Cho, H.U.; Kim, H.G.; Kim, Y.M.; Park, J.M. Volatile fatty acid recovery by anaerobic fermentation from blue-green algae: Effect of pretreatment. *Bioresour. Technol.* **2017**, *244*, 1433–1438. [CrossRef] [PubMed]
4. Oheocha, C.; Haxo, F.T. Some atypical algal chromoproteins. *Biochim. Biophys. Acta* **1960**, *41*, 516–520. [CrossRef]
5. Wells, M.L.; Potin, P.; Craigie, J.S.; Raven, J.A.; Merchant, S.S.; Helliwell, K.E.; Smith, A.G.; Camire, M.E.; Brawley, S.H. Algae as nutritional and functional food sources: Revisiting our understanding. *J. Appl. Phycol.* **2017**, *29*, 949–982. [CrossRef] [PubMed]
6. Cha, K.H.; Koo, S.Y.; Lee, D.U. Antiproliferative effects of carotenoids extracted from chlorella ellipsoidea and chlorella vulgaris on human colon cancer cells. *J. Agric. Food Chem.* **2008**, *56*, 10521–10526. [CrossRef] [PubMed]
7. Posten, C.; Chen, S.F. *Microalgae Biotechnology*; Springer International Publishing: Basel, Switzerland, 2015; Vol. 153, p. 188.
8. Markou, G.; Nerantzis, E. Microalgae for high-value compounds and biofuels production: A review with focus on cultivation under stress conditions. *Biotechnol. Adv.* **2013**, *31*, 1532–1542. [CrossRef] [PubMed]
9. Hannon, M.; Gimpel, J.; Tran, M.; Rasala, B.; Mayfield, S. Biofuels from algae: Challenges and potential. *Biofuels* **2010**, *1*, 763–784. [CrossRef] [PubMed]
10. Fučíková, K.; Lewis, P.O.; Lewis, L.A. Chloroplast phylogenomic data from the green algal order sphaeropleales (chlorophyceae, chlorophyta) reveal complex patterns of sequence evolution. *Mol. Phylogenet. Evol.* **2016**, *98*, 176–183. [CrossRef] [PubMed]
11. Leliaert, F.; Verbruggen, H.; Vanormelingen, P.; Steen, F.; López-Bautista, J.M.; Zuccarello, G.C.; De Clerck, O. DNA-based species delimitation in algae. *Eur. J. Phycol.* **2014**, *49*, 179–196. [CrossRef]
12. Seckbach, J. *Life as We Know It*; Springer Netherlands: Amsterdam, The Netherlands, 2006; p. 765.
13. Smith, A.M.C.G.; Dolan, L.; Harberd, N.; Jones, J.; Martin, C.; Sablowski, R.; Amey, A. *Plant Biology*; Garland Science: New York, NY, USA, 2009; p. 679.
14. Boedeker, C.; Leliaert, F.; Zuccarello, G.C. Molecular phylogeny of the cladophoraceae (cladophorales, ulvophyceae), with the resurrection of acrocladus nageli and willeella borgesen, and the description of lurbica gen. nov. And pseudorhizoclonium gen. nov. *J. Phycol.* **2016**, *52*, 905–928. [CrossRef] [PubMed]
15. Leliaert, F.; Smith, D.R.; Moreau, H.; Herron, M.D.; Verbruggen, H.; Delwiche, C.F.; De Clerck, O. Phylogeny and molecular evolution of the green algae. *Crit. Rev. Plant Sci.* **2012**, *31*, 1–46. [CrossRef]
16. Ripma, L.A.; Simpson, M.G.; Hasenstab-Lehman, K. Geneious! Simplified genome skimming methods for phylogenetic systematic studies: A case study in oreocarya (boraginaceae). *Appl. Plant Sci.* **2014**, *2*, 1400062. [CrossRef] [PubMed]
17. Sridhar, K.R. *Frontiers in Fungal Ecology, Diversity and Metabolites*; IK International Publishing House: Delhi, India, 2008; p. 352.
18. Bleidorn, C. *Phylogenomics: An Introduction*; Springer: Basel, Switzerland, 2017; p. XIII, 222.
19. Webster, M.S. *The Extended Specimen: Emerging Frontiers in Collections-Based Ornithological Research*; CRC Press: Boca Raton, FL, USA, 2017; p. 240.
20. Lorenz, R.T.; Cysewski, G.R. Commercial potential for haematococcus microalgae as a natural source of astaxanthin. *Trends Biotechnol.* **2000**, *18*, 160–167. [CrossRef]
21. Cardozo, K.H.; Guaratini, T.; Barros, M.P.; Falcao, V.R.; Tonon, A.P.; Lopes, N.P.; Campos, S.; Torres, M.A.; Souza, A.O.; Colepicolo, P.; et al. Metabolites from algae with economical impact. *Comp. Biochem. Physiol. C Toxicol. Pharmacol.* **2007**, *146*, 60–78. [CrossRef] [PubMed]

22. Zhang, X.; Pan, L.; Wei, X.; Gao, H.; Liu, J. Impact of astaxanthin-enriched algal powder of haematococcus pluvialis on memory improvement in balb/c mice. *Environ. Geochem. Health* **2007**, *29*, 483–489. [CrossRef] [PubMed]

23. Boussiba, S. Carotenogenesis in the green alga haematococcus pluvialis: Cellular physiology and stress response. *Physiol. Plant.* **2000**, *108*, 111–117. [CrossRef]

24. Alasalvar, C.; Miyashita, K.; Shahidi, F.; Wanasundara, U. *Handbook of Seafood Quality, Safety and Health Applications*; Wiley: Hoboken, NJ, USA, 2011.

25. Herrero, M.; Ibanez, E.; Fanali, S.; Cifuentes, A. Quantitation of chiral amino acids from microalgae by mekc and lif detection. *Electrophoresis* **2007**, *28*, 2701–2709. [CrossRef] [PubMed]

26. Soontornchaiboon, W.; Joo, S.S.; Kim, S.M. Anti-inflammatory effects of violaxanthin isolated from microalga chlorella ellipsoidea in raw 264.7 macrophages. *Biol. Pharm. Bull.* **2012**, *35*, 1137–1144. [CrossRef] [PubMed]

27. Ko, S.C.; Kang, N.; Kim, E.A.; Kang, M.C.; Lee, S.H.; Kang, S.M.; Lee, J.B.; Jeon, B.T.; Kim, S.K.; Park, S.J. A novel angiotensin i-converting enzyme (ace) inhibitory peptide from a marine chlorella ellipsoidea and its antihypertensive effect in spontaneously hypertensive rats. *Process Biochem.* **2012**, *47*, 2005–2011. [CrossRef]

28. Acton, Q.A. *Issues in Earth Sciences, Geology, and Geophysics: 2011 Edition*; ScholarlyEditions: Atlanta, GA, USA, 2012.

29. Heibel, T.; De Give, T. *Rooted in Design: Sprout Home's Guide to Creative Indoor Planting*; Ten Speed Press: Emeryville, CA, USA, 2015; p. 224.

30. Boedeker, C.; Sviridenko, B.F. Cladophora koktschetavensis from kazakhstan is a synonym of aegagropila linnaei (cladophorales, chlorophyta) and fills the gap in the disjunct distribution of a widespread genotype. *Aquat. Bot.* **2012**, *101*, 64–68. [CrossRef]

31. Hanyuda, T.; Wakana, I.; Arai, S.; Miyaji, K.; Watano, Y.; Ueda, K. Phylogenetic relationships within cladophorales (ulvophyceae, chlorophyta) inferred from 18 s rrna gene sequences, with special reference to aegagropila linnaei 1. *J. Phycol.* **2002**, *38*, 564–571. [CrossRef]

32. Tiwari, K.; Jadhav, S.; Gupta, S. Modified ctab technique for isolation of DNA from some medicinal plants. *Res. J. Med. Plant* **2011**, *201*, 1.

33. Magoč, T.; Salzberg, S.L. Flash: Fast length adjustment of short reads to improve genome assemblies. *Bioinformatics* **2011**, *27*, 2957–2963. [CrossRef] [PubMed]

34. McGinnis, S.; Madden, T.L. Blast: At the core of a powerful and diverse set of sequence analysis tools. *Nucleic Acids Res.* **2004**, *32*, W20–W25. [CrossRef] [PubMed]

35. Katoh, K.; Misawa, K.; Kuma, K.I.; Miyata, T. Mafft: A novel method for rapid multiple sequence alignment based on fast fourier transform. *Nucleic Acids Res.* **2002**, *30*, 3059–3066. [CrossRef] [PubMed]

36. Librado, P.; Rozas, J. Dnasp v5: A software for comprehensive analysis of DNA polymorphism data. *Bioinformatics* **2009**, *25*, 1451–1452. [CrossRef] [PubMed]

37. Tamura, K.; Stecher, G.; Peterson, D.; Filipski, A.; Kumar, S. Mega6: Molecular evolutionary genetics analysis version 6.0. *Mol. Biol. Evol.* **2013**, *30*, 2725–2729. [CrossRef] [PubMed]

38. Orr, R.J.; Stüken, A.; Rundberget, T.; Eikrem, W.; Jakobsen, K.S. Improved phylogenetic resolution of toxic and non-toxic alexandrium strains using a concatenated rdna approach. *Harmful Algae* **2011**, *10*, 676–688. [CrossRef]

39. Heeg, J.S.; Wolf, M. Its2 and 18s rdna sequence-structure phylogeny of chlorella and allies (chlorophyta, trebouxiophyceae, chlorellaceae). *Plant Gene* **2015**, *4*, 20–28. [CrossRef]

40. Buchheim, M.A.; Sutherland, D.M.; Buchheim, J.A.; Wolf, M. The blood alga: Phylogeny of haematococcus (chlorophyceae) inferred from ribosomal rna gene sequence data. *Eur. J. Phycol.* **2013**, *48*, 318–329. [CrossRef]

41. Ali, A.B.; De Baere, R.; Van der Auwera, G.; De Wachter, R.; Van de Peer, Y. Phylogenetic relationships among algae based on complete large-subunit rrna sequences. *Int. J. Syst. Evol. Microbiol.* **2001**, *51*, 737–749. [PubMed]

42. Daugbjerg, N.; Andersen, R.A. A molecular phylogeny of the heterokont algae based on analyses of chloroplast-encoded rbcl sequence data. *J. Phycol.* **1997**, *33*, 1031–1041. [CrossRef]

43. Daugbjerg, N.; Andersen, R.A. Phylogenetic analyses of the rbcl sequences from haptophytes and heterokont algae suggest their chloroplasts are unrelated. *Mol. Biol. Evol.* **1997**, *14*, 1242–1251. [CrossRef] [PubMed]

44. An, S.; Friedl, T.; Hegewald, E. Phylogenetic relationships of scenedesmus and scenedesmus-like coccoid green algae as inferred from its-2 rdna sequence comparisons. *Plant Biol.* **1999**, *1*, 418–428. [CrossRef]

45. Ratnasingham, S.; Hebert, P.D. Bold: The barcode of life data system (http://www.Barcodinglife.Org). *Mol. Ecol. Resour.* **2007**, *7*, 355–364. [CrossRef] [PubMed]

46. Li, F.W.; Kuo, L.Y.; Rothfels, C.J.; Ebihara, A.; Chiou, W.L.; Windham, M.D.; Pryer, K.M. Rbcl and matk earn two thumbs up as the core DNA barcode for ferns. *PLoS ONE* **2011**, *6*, e26597. [CrossRef] [PubMed]

47. Van de Peer, Y.; Ben Ali, A.; Meyer, A. Microsporidia: Accumulating molecular evidence that a group of amitochondriate and suspectedly primitive eukaryotes are just curious fungi. *Gene* **2000**, *246*, 1–8. [CrossRef]

48. Van de Peer, Y.; De Wachter, R. Evolutionary relationships among the eukaryotic crown taxa taking into account site-to-site rate variation in 18s rrna. *J. Mol. Evol.* **1997**, *45*, 619–630. [CrossRef] [PubMed]

49. Robba, L.; Russell, S.J.; Barker, G.L.; Brodie, J. Assessing the use of the mitochondrial cox1 marker for use in DNA barcoding of red algae (rhodophyta). *Am. J. Bot.* **2006**, *93*, 1101–1108. [CrossRef] [PubMed]

50. Shearer, T.L.; Van Oppen, M.J.; Romano, S.L.; Worheide, G. Slow mitochondrial DNA sequence evolution in the anthozoa (cnidaria). *Mol. Ecol.* **2002**, *11*, 2475–2487. [CrossRef] [PubMed]

51. Vences, M.; Thomas, M.; Bonett, R.M.; Vieites, D.R. Deciphering amphibian diversity through DNA barcoding: Chances and challenges. *Philos. Trans. R. Soc. Lond. B Biol. Sci.* **2005**, *360*, 1859–1868. [CrossRef] [PubMed]

52. Virgilio, M.; Backeljau, T.; Nevado, B.; De Meyer, M. Comparative performances of DNA barcoding across insect orders. *BMC Bioinform.* **2010**, *11*, 206. [CrossRef] [PubMed]

53. Meusnier, I.; Singer, G.A.; Landry, J.F.; Hickey, D.A.; Hebert, P.D.; Hajibabaei, M. A universal DNA mini-barcode for biodiversity analysis. *BMC Genom.* **2008**, *9*, 214. [CrossRef] [PubMed]

54. Elshire, R.J.; Glaubitz, J.C.; Sun, Q.; Poland, J.A.; Kawamoto, K.; Buckler, E.S.; Mitchell, S.E. A robust, simple genotyping-by-sequencing (gbs) approach for high diversity species. *PLoS ONE* **2011**, *6*, e19379. [CrossRef] [PubMed]

55. Torres-Machorro, A.L.; Hernandez, R.; Alderete, J.F.; Lopez-Villasenor, I. Comparative analyses among the trichomonas vaginalis, trichomonas tenax, and tritrichomonas foetus 5s ribosomal rna genes. *Curr. Genet.* **2009**, *55*, 199–210. [CrossRef] [PubMed]

56. Eickbush, T.H.; Eickbush, D.G. Finely orchestrated movements: Evolution of the ribosomal rna genes. *Genetics* **2007**, *175*, 477–485. [CrossRef] [PubMed]

57. Ganley, A.R.; Kobayashi, T. Highly efficient concerted evolution in the ribosomal DNA repeats: Total rdna repeat variation revealed by whole-genome shotgun sequence data. *Genome Res.* **2007**, *17*, 184–191. [CrossRef] [PubMed]

58. Thornhill, D.J.; Lajeunesse, T.C.; Santos, S.R. Measuring rdna diversity in eukaryotic microbial systems: How intragenomic variation, pseudogenes, and pcr artifacts confound biodiversity estimates. *Mol. Ecol.* **2007**, *16*, 5326–5340. [CrossRef] [PubMed]

59. Leo, N.; Barker, S. Intragenomic variation in its2 rdna in the louse of humans, pediculus humanus: Its2 is not a suitable marker for population studies in this species. *Insect Mol. Biol.* **2002**, *11*, 651–657. [CrossRef] [PubMed]

60. Alanagreh, L.; Pegg, C.; Harikumar, A.; Buchheim, M. Assessing intragenomic variation of the internal transcribed spacer two: Adapting the illumina metagenomics protocol. *PLoS ONE* **2017**, *12*, e0181491. [CrossRef] [PubMed]

61. Simon, U.K.; Weiß, M. Intragenomic variation of fungal ribosomal genes is higher than previously thought. *Mol. Biol. Evol.* **2008**, *25*, 2251–2254. [CrossRef] [PubMed]

62. Pereira, T.J.; Baldwin, J.G. Contrasting evolutionary patterns of 28 s and its rrna genes reveal high intragenomic variation in cephalenchus (nematoda): Implications for species delimitation. *Mol. Phylogenet. Evol.* **2016**, *98*, 244–260. [CrossRef] [PubMed]

63. Shapoval, N.A.; Lukhtanov, V.A. Intragenomic variations of multicopy its2 marker in agrodiaetus blue butterflies (lepidoptera, lycaenidae). *Comp. Cytogenet.* **2015**, *9*, 483. [CrossRef] [PubMed]

Int. J. Mol. Sci. **2017**, *18*, 2341

64. Li, H.; Handsaker, B.; Wysoker, A.; Fennell, T.; Ruan, J.; Homer, N.; Marth, G.; Abecasis, G.; Durbin, R. The sequence alignment/map format and samtools. *Bioinformatics* **2009**, *25*, 2078–2079. [CrossRef] [PubMed]

65. Koboldt, D.C.; Chen, K.; Wylie, T.; Larson, D.E.; McLellan, M.D.; Mardis, E.R.; Weinstock, G.M.; Wilson, R.K.; Ding, L. Varscan: Variant detection in massively parallel sequencing of individual and pooled samples. *Bioinformatics* **2009**, *25*, 2283–2285. [CrossRef] [PubMed]

International Journal of
Molecular Sciences

MDPI

Article

Different Natural Selection Pressures on the *atpF* Gene in Evergreen Sclerophyllous and Deciduous Oak Species: Evidence from Comparative Analysis of the Complete Chloroplast Genome of *Quercus aquifolioides* with Other Oak Species

Kangquan Yin, Yue Zhang, Yuejuan Li and Fang K. Du *

College of Forestry, Beijing Forestry University, Beijing 100083, China; yinkq@im.ac.cn (K.Y.);
zhangyue2016@bjfu.edu.cn (Y.Z.); liyuejuan@bjfu.edu.cn (Y.L.)
* Correspondence: dufang325@bjfu.edu.cn; Tel.: +86-10-6233-8191

Received: 23 February 2018; Accepted: 27 March 2018; Published: 30 March 2018

Abstract: *Quercus* is an economically important and phylogenetically complex genus in the family Fagaceae. Due to extensive hybridization and introgression, it is considered to be one of the most challenging plant taxa, both taxonomically and phylogenetically. *Quercus aquifolioides* is an evergreen sclerophyllous oak species that is endemic to, but widely distributed across, the Hengduanshan Biodiversity Hotspot in the Eastern Himalayas. Here, we compared the fully assembled chloroplast (cp) genome of *Q. aquifolioides* with those of three closely related species. The analysis revealed a cp genome ranging in size from 160,415 to 161,304 bp and with a typical quadripartite structure, composed of two inverted repeats (IRs) separated by a small single copy (SSC) and a large single copy (LSC) region. The genome organization, gene number, gene order, and GC content of these four *Quercus* cp genomes are similar to those of many angiosperm cp genomes. We also analyzed the *Q. aquifolioides* repeats and microsatellites. Investigating the effects of selection events on shared protein-coding genes using the Ka/Ks ratio showed that significant positive selection had acted on the *atpF* gene of *Q. aquifolioides* compared to two deciduous oak species, and that there had been significant purifying selection on the *atpF* gene in the chloroplast of evergreen sclerophyllous oak trees. In addition, site-specific selection analysis identified positively selected sites in 12 genes. Phylogenetic analysis based on shared protein-coding genes from 14 species defined *Q. aquifolioides* as belonging to sect. *Heterobalanus* and being closely related to *Q. rubra* and *Q. aliena*. Our findings provide valuable genetic information for use in accurately identifying species, resolving taxonomy, and reconstructing the phylogeny of the genus *Quercus*.

Keywords: cp genome; repeat analysis; sequence divergence; non-synonymous substitution; electron transport chain; phylogeny

---

1. Introduction

The chloroplast (cp) is an organelle which plays an important role in photosynthesis and carbon fixation in plant cells. In angiosperms, the cp is a uniparentally inherited organelle, and it has its own circular, haploid, evolutionarily conserved genome. The cp genome is therefore considered to be a useful and informative genetic resource for studies on evolutionary relationships in the plant kingdom at various taxonomic levels [1]. In most cases, the cp genome is between 120 and 160 kb in size and has a structure composed of two copies of a large inverted repeat (IR) region, a large single copy (LSC) region, and a small single copy (SSC) region [2].

Oaks (*Quercus* L.), which comprise approximately 500 shrub and tree species, form a phylogenetically complex and economically important genus of the beech family, Fagaceae [3]. Distributed throughout much of the Northern Hemisphere, oaks are located in the northern temperate region, and they also occur in the Andes of South America and subtropical and tropical Asia [4]. Oaks are dominant in various habitats, such as temperate deciduous forest, oak-pine forest and temperate and subtropical evergreen forest [5]. They are intimately associated with many other organisms, including fungi, ferns, birds, mammals, and insects [4]. For this reason, their interactions have been the subject of a large number of ecological studies. Human beings have a close connection with oak, as throughout history it has been a common symbol of strength and courage and has been chosen as the national tree in many countries. Moreover, oaks are of great economic value, being used in, for example, the construction of fine furniture and the wine industry.

Oak species are notoriously difficult to classify taxonomically, due to morphological variation caused in part by hybridization [6–14]. Some studies stated that *Quercus* contained two subgenera, *Cyclobalanopsis* and *Quercus*, the latter including three sections: *Quercus* (white oaks), *Lobatae* (red oaks), and *Protobalanus* (golden cup or intermediate oaks) [3,15]. Because previous classifications of oaks have been based solely on morphological characters which are often homoplastic in oaks, these classifications have always been subject to debate [3,15]. With advances in molecular phylogenetics and techniques based on pollen morphology, views on oak classification are changing [15–19]. Recently, Denk et al. proposed an updated classification for *Quercus* with two subgenera: subgenus *Quercus*, the 'New World clade' or 'high-latitude clade', and subgenus *Cerris*, the exclusively Eurasian 'Old World clade' or 'mid-latitude clade' [19]. There are five sections (*Protobalanus*, *Ponticae*, *Virentes*, *Quercus*, and *Lobatae*) in subgenus *Quercus* and three sections (*Cyclobalanopsis*, *Ilex* and *Cerris*) in subgenus *Cerris*.

China, which is a center of *Quercus* diversity, has 35–51 species [20]. Based on morphological characters, including 25 qualitative and 18 quantitative characters, oaks in China were divided into five sections, namely *Aegilops*, *Quercus*, *Brachylepides*, *Engleriana*, and *Echinolepides*. Recently, we studied the phylogeography of *Quercus aquifolioides*, which is endemic to the Hengduanshan Biodiversity Hotspot, based on 58 populations distributed throughout the species range, using four chloroplast DNA fragments and 11 nuclear microsatellite loci [21]. Up till now, to our knowledge, very few studies have focused on the phylogenetic relationships and population genetics of oaks in China [22], in part due to the challenges arising from introgressive hybridization, lineage sorting, and molecular markers failing to give sufficient phylogenetic signals.

In this study, we produced the first cp genome sequence for *Q. aquifolioides* using next-generation sequencing technology. This complete cp genome, combined with previously reported cp genome sequences for other members of the genus, will enhance our understanding of the systematic evolution of *Quercus*. We analyzed the completely assembled cp genome of *Q. aquifolioides* and compared it to those of three other oak species to investigate common structural patterns and hotspot regions of sequence divergence in these four *Quercus* cp genomes, examined whether selection pressure had acted on protein coding genes, and reconstructed the phylogenetic relationships of the four *Quercus* species. Our findings will not only enrich the complete cp genome resources available for the genus *Quercus* but also provide abundant genetic information for use in subsequent taxonomic and phylogenetic identification of members of the genus, and assist geneticists and breeders in improving commercially-grown oak trees.

2. Results and Discussion

2.1. Chloroplast Genome Organization in Q. aquifolioides

The *Q. aquifolioides* cp genome is a typical circular double-stranded DNA molecule with a length of 160,415 bp, which falls within the normal angiosperm length range [23,24]. The cp genome has the usual quadripartite structure, featuring a LSC region (large single copy region, 89,493 bp), a SSC region (small single copy region, 16,594 bp), and a pair of IRs (inverted repeats, 25,857 bp) (Figure 1; GenBank accession No. KP340971). The GC contents of the LSC, SSC, and IR regions individually, and of the cp genome as a whole, are 34.8%, 31.2%, 42.7%, and 36.9%, respectively. These GC contents are within the range previously reported for other plant species. Approximately 48.0% of the cp genome encodes proteins, 5.6% encodes rRNAs and 1.3% encodes tRNAs. Noncoding regions (intergenic regions, introns and pseudogenes) constitute the remaining 45.1% of the genome. The *Q. aquifolioides* cp genome encodes 127 genes: 80 protein-coding genes, eight ribosomal RNA genes, and 39 tRNA genes. *ycf2* is the largest gene, having a length of 6834 bp. We found that 18 genes have one intron (10 protein coding genes and 8 tRNA genes) and two genes (*clpP* and *ycf3*) have two introns each. Two identical rRNA gene clusters (16S-23S-4.5S-5S) were found in the IR regions. There are two tRNA genes, *trnI* and *trnA*, in the 16S~23S spacer region of each cluster. The sequence of the rRNA coding region is highly conserved: sequence identities of four rRNA genes with those of *Arabidopsis thaliana* (L.) Heynh were over 98%.

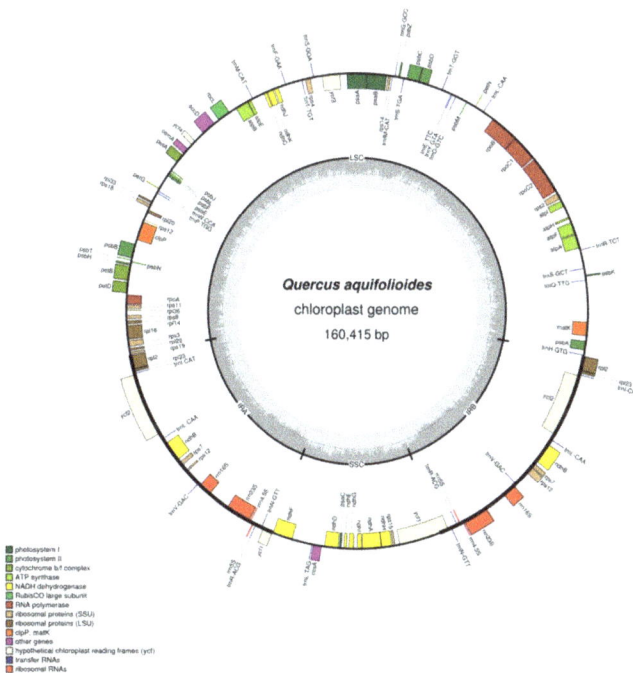

Figure 1. Gene map of the *Q. aquifolioides* chloroplast genome. The annotated chloroplast (cp) genome of *Q. aquifolioides* is represented as concentric circles. Genes shown outside the outer circle are transcribed counter-clockwise and genes indicated inside the outer circle are transcribed clockwise. Two inverted repeats (IRs), the large single copy (LSC) and the small single copy (SSC) are shown in the inner circle.

2.2. Repeat Sequence Analysis and Simple Sequence Repeats (SSR)

Repeat sequences have been used extensively for phylogeny, population genetics, genetic mapping, and forensic studies [25]. In the cp genome of *Q. aquifolioides*, 38 pairs of repeats longer than 30 bp were detected; they consisted of 24 palindromic repeats and 14 forward repeats (Figure 2). Among these repeats, 36 are 30–40 bp long, one is 44 bp long, and one is 64 bp long (Figure 2). A large proportion of the repeats (73.7%) are present in non-coding regions, but some repeats are embedded in coding regions, such as the *trnS-GCU*, *trnS-GGA*, *psaB*, *psaA*, *ycf1*, *ycf2*, and *accD* genes

(Table S1). As previous studies reported, many repeats were found in the *ycf2* gene [26–29]. Apart from the IR region, the longest repeats, which were 64 bp in length, were present in the *ndhD/psaC* intergenic region.

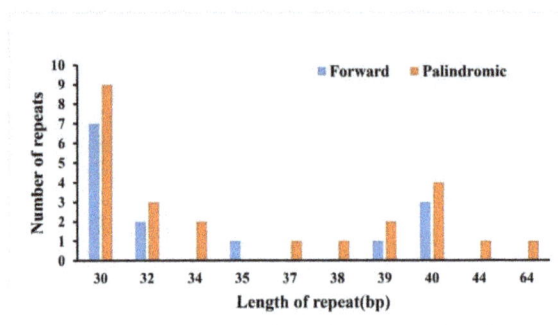

Figure 2. Analysis of repeated sequences in *Q. aquifolioides*.

SSR, also known as microsatellites, are highly polymorphic and thus widely used as molecular markers. A total of 78 perfect microsatellites were identified in the *Q. aquifolioides* cp genome. Among them, 70.51% were present in the LSC regions, whereas 10.26% and 19.23% were identified in the IR and SSC regions respectively (Figure 3A). This result is consistent with previous reports that SSRs are not evenly distributed in cp genomes [30]. Twelve of the SSRs were present in protein-coding regions, six were in introns, and 60 were located in intergenic spacers of the *Q. aquifolioides* cp genome (Figure 3B). Of the motifs forming these SSRs, 58 are mononucleotides, six are dinucleotides, five are trinucleotides, six are tetranucleotides, and three are pentanucleotides (Figure 3C). Most of the mononucleotides (98.28%) and dinucleotides (100%) are composed of A and T. (Figure 3C). These results are consistent with previous reports that SSRs in cp genomes generally consist of short polyA or polyT repeats [31]. The high AT content of cp SSRs contributes to the AT richness of the *Q. aquifolioides* cp genome, which is similar in this respect to other cp genomes [31].

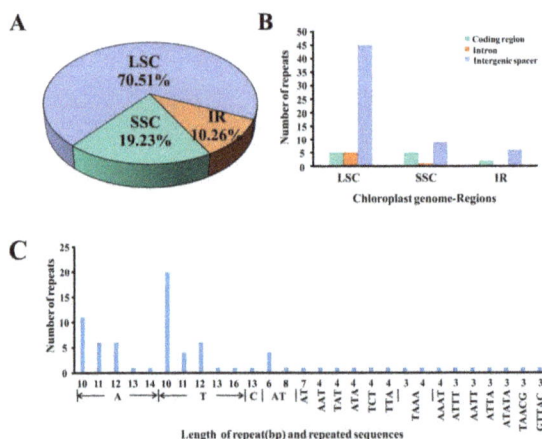

Figure 3. Analysis of simple sequence repeats (SSRs) in the *Q. aquifolioides* cp genome. (**A**) Frequency of SSRs identified in the LSC, SSC, and IR regions; (**B**) Frequency of SSRs identified in the coding regions, intergenic spacers and introns of the LSC, SSC and IR regions; (**C**) Frequency distribution of different classes of polymer in the cp genome of *Q. aquifolioides*.

2.3. Comparison of the cp Genomes of Q. aquifolioides and Three Related Quercus Species

Three complete cp genomes, those of *Q. aliena* (GenBank accession number: KP301144), *Q. rubra* (GenBank accession number: JX970937), and *Q. spinosa* (GenBank accession number: KM841421), belonging to three different sections within the *Quercus* genus, were selected for comparison with *Q. aquifolioides* (Table 1). *Q. rubra* has the largest cp genome; this is mostly attributable to variations in the lengths of the LSC and SSC regions. The GC content of these four cp genomes is very similar, at ~37%. *Q. aquifolioides* has the same number of protein coding genes and rRNA genes as the other three *Quercus* species. Although *Q. spinosa* has one tRNA fewer than the other three *Quercus* species, the total length of its tRNA genes is greater than that in any of the other three species. We found that *Q. aquifolioides* shared 80 protein-coding genes with the cp genomes of all three of the other *Quercus* species.

Table 1. Summary of the features of four complete *Quercus* plastomes.

Genome Features	Sect. *Heterobalanus*		Sect. *Lobatae*	Sect. *Quercus*
	Q. aquifolioides	*Q. spinosa*	*Q. rubra*	*Q. aliena*
Genome size/GC content	160,415/37.0	160,825/36.9	161,304/36.8	160,921/36.9
Coding genes: number/size	80(7)/80,270	80(7)/80,812	80(7)/80,946	80(7)/80,052
tRNA: number/size	39/10,625	38/11,402	39/10,756	39/10,753
rRNA: number/size	8/9048	8/9050	8/9050	8/9048
LSC: size/percent/GC content	89,807/56/34.8	90,371/56.2/34.7	90,541/56.1/34.7	90,258/56.1/34.7
SSC: size/percent/GC content	18,894/11.8/31.2	18,732/11.6/31.2	19,023/56.1/30.9	18,972/11.8/31.0
IR: size/percent/GC content	51,754/32.2/42.7	51,722/32.2/47.2	52,740/32.7/42.7	51,682/32.1/42.7
Introns: size/percent	20,473/12.8	19,757/12.3	20,217/12.5	20,014/12.4
Intergenic spacer: size/percent	49,548/31.0	50,207/31.2	47,473/29.4	47,304/29.3

Numbers in brackets denote the numbers of genes duplicated in the IR regions.

We compared the other three complete cp genomes with that of *Q. aquifolioides* (Figure 4). The sequence identity between these four *Quercus* cp genomes was analyzed. Our results revealed perfect conservation of gene order along the cp genomes of the four species and very high similarity between them.

Although the overall quadripartite structure, including the gene number and order, is usually well conserved, the IR region often undergoes expansion or contraction, a phenomenon called ebb and flow in cp genomes [32]. Generally, the expansion or contraction involves no more than a few hundred nucleotides. Kim and Lee proposed that length variation in angiosperm cp genomes was primarily caused by expansion and contraction of the IR region and the single-copy (SC) boundary regions [33]. The IR/SC boundary regions of these four complete *Quercus* cp genomes were compared, and found to exhibit clear differences in junction positions (Figure 5). The inverted repeat b (IRb)/SSC borders are located in the coding region of the *ycf1* gene with a region of 4590–4611 bp located in the SSC regions. The shortened *ycf1* gene crossed the inverted repeat a (IRa)/SSC borders, with 25–28 bp falling within the SSC regions, and the *ndhF* gene was located in the SSC region with its distance to the IRa/SSC borders ranging from 8 to 22 bp. At the LSC/IRa junction, the distances between *rps19* and the border ranged from 12 to 35 bp, while the distances between *rpl2* and the border were from 39 to 63 bp. At the LSC/IRb junction, the distances between *rpl2* and the border ranged from 54 to 226 bp and the distances between *trnH* and the border were the same, at 16 bp. Thus, variations at the IR/SC borders in these four cp genomes contribute to the differences in length of the cp genome sequence as a whole.

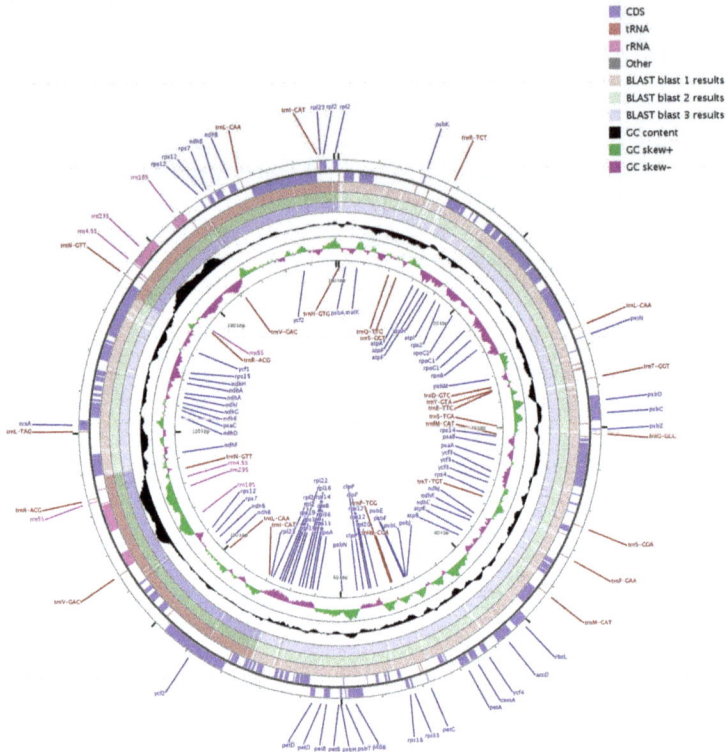

Figure 4. Comparison of four *Quercus* cp genome sequences. The outer four rings show the coding sequences, tRNA genes, rRNA genes, and other genes in the forward and reverse strands. The next three rings show the blast results between the cp genomes of *Q. aquifolioides* and three other *Quercus* species based on BlastN (blast 1–3 results: *Q. aquifolioides* Vs *Q. aliena*, *Q. rubra*, and *Q. spinosa*, respectively). The following black ring is the GC content curve for the *Q. aquifolioides* cp genome. The innermost ring is a GC skew curve for the *Q. aquifolioides* cp genome. GC skew+ (green) indicates G > C, GC skew− (purple) indicates G < C.

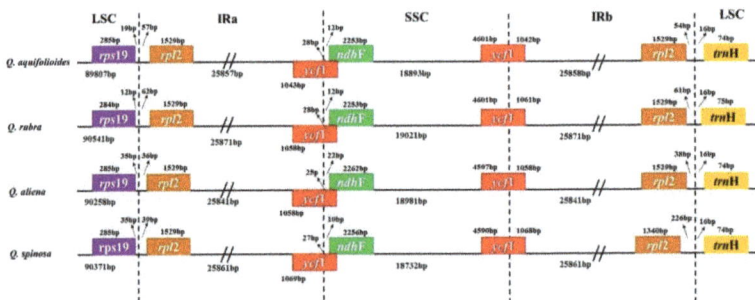

Figure 5. Comparisons of borders between neighboring genes and junctions of the LSC, SSC, and IR regions among the four *Quercus* cp genomes. Boxes above or below the main line indicate genes adjacent to borders. The figure is not to scale with regard to sequence length and shows only relative changes at or near (inverted repeats/single copy) IR/SC borders.

The whole-genome alignment revealed high sequence similarity across these four cp genomes, suggesting that *Quercus* cp genomes are well conserved (Figure 6). As observed in other angiosperms [34–36], we also found that among these four cp genomes the SC regions are more divergent than the IR regions, possibly due to error correction occurring via gene conversion between IRs [37]. Our results also showed that coding regions are more conserved than non-coding regions, as seen in other plants [38,39]. The most divergent coding region in these four *Quercus* cp genomes was *rpl22*. Non-coding regions showed various degrees of sequence divergence among these four *Quercus* cp genomes, with the *trnH-GUG/psbA* regions having the highest level of divergence. These hotspot regions furnish valuable information as a basis for developing molecular markers for phylogenetic studies and identification of *Quercus* species.

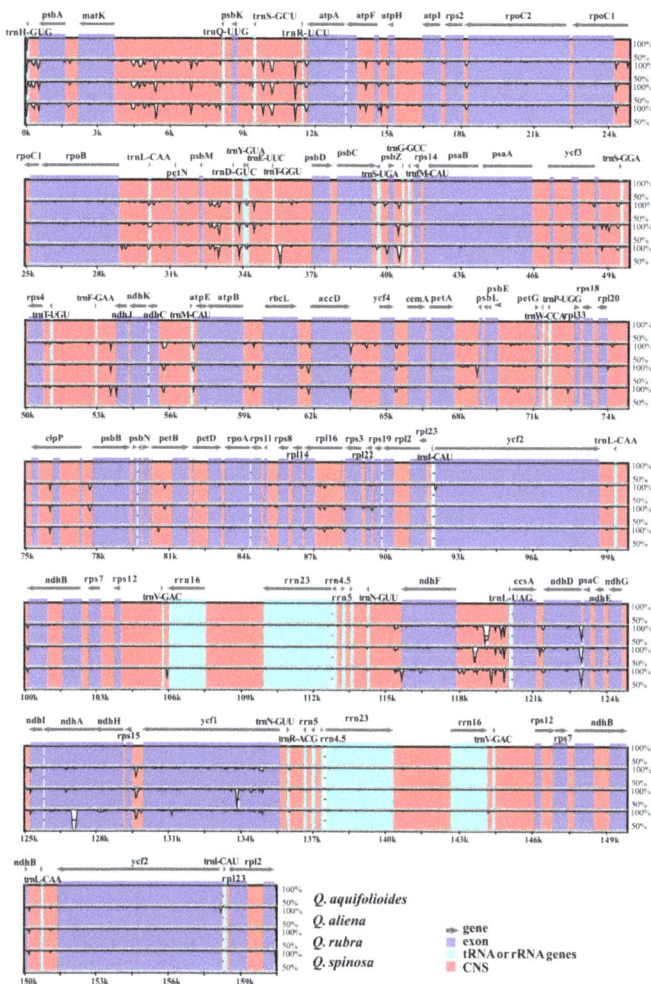

Figure 6. Alignment of four *Quercus* cp genome sequences. Sequence identity plot for four *Quercus* species, with *Q. aquifolioides* as a reference. The *X*-axis corresponds to coordinates within the cp genome. The *Y*-axis shows the percentage identity in the range 50% to 100%.

2.4. Genome Sequence Divergence among Quercus Species

To investigate the extent of sequence divergence among these four *Quercus* cp genomes, the nucleotide variability (*Pi*) values within 600 bp windows (200 bp stepwise moving) in the LSC, SSC, and IR regions of the genomes were calculated (Figure 7). In the LSC region, the values varied from 0 to 0.02389 with a mean of 0.00603, while the SSC regions were from 0 to 0.02 with a mean of 0.00863, and the IR regions were from 0 to 0.00417, with a mean of 0.00098. These results suggest that the differences between these genomic regions are very small. However, we also found certain highly variable regions in the LSC, SSC, and IRs. In the LSC, the highly variable regions were *trnH/psbA* and *petA/psbJ*, with *Pi* > 0.02. In the SSC, highly variable regions included *ndhF/rpl32*, *ndhA/ndhH* and *ycf1* (*Pi* > 0.015). In the IRs, two regions, *trnR/trnN* and *ndhB*, with *Pi* > 0.004 were identified (Figure 8). Four of these regions, *trnH/psbA*, *petA/psbJ*, *ndhF/rpl32*, and *ycf1*, have also been identified as highly variable in other plants [33,40–42]. On the basis of our results, five of these variable regions (*trnH/psbA*, *petA/psbJ*, *ndhF/rpl32*, *ndhA/ndhH*, and *ycf1*) show great potential as sources of useful phylogenetic markers for *Quercus*.

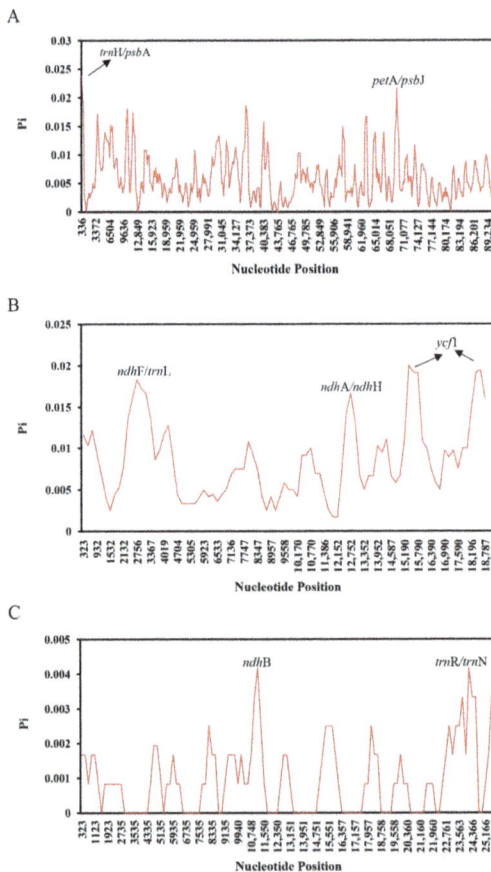

Figure 7. Comparative analysis of nucleotide variability (*Pi*) values among the four *Quercus* cp genome sequences. (**A**) Analysis of the LSC regions; (**B**) Analysis of the SSC regions; (**C**) Analysis of the IR regions. (Window length: 600 bp, step size: 200 bp). *X*-axis: position of the midpoint of a window, *Y*-axis: nucleotide diversity of each window.

2.5. Selection Events in Protein Coding Genes

The non-synonymous (Ka) to synonymous (Ks) nucleotide substitution rate ratio (denoted by Ka/Ks) is a very important tool used in studying in protein coding gene evolution. The Ka/Ks ratio is used to evaluate the rate of gene divergence and determine whether positive, purifying or neutral selection has been in operation. A Ka/Ks ratio of >1 indicates positive selection, while Ka/Ks < 1 (especially if it is less than 0.5) indicates purifying selection. A value of close to 1 indicates neutral selection [43].

In this study, we compared Ka/Ks ratios for 73 shared unique protein coding genes in the *Q. aquifolioides* cp genome and the cp genomes of three other related *Quercus* species: *Q. rubra*, *Q. spinosa*, and *Q. aliena* (Table S2). The results are shown in Figure 8. Interestingly, we found that the Ka/Ks ratios were both region specific and gene specific. The average Ka/Ks ratio of the 73 protein coding genes analyzed across the four cp genomes was 0.1653. The most conserved genes with average Ka/Ks values of 0, suggesting very strong purifying selective pressure, were *rpl23, rps7, psaC, rps12, psbA, psbK, psbI, atpH, atpI, rps2, petN, psbM, psbD, rps14, ycf3, ndhJ, ndhK, ndhC, psbJ, psbL, psbF, psbE, petL, rpl33, clpP, psbT, psbN, psbH, rpl36, rps8, rpl14, rpl16,* and *rps19* (Table S2). The averaging Ka/Ks method showed no gene with Ka/Ks > 1, which suggests that no gene had been under positive selection in the *Q. aquifolioides* cp genome. Average Ka/Ks value within the range 0.5 to 1, indicating relaxed selection, were observed for *accD, petA, rpl20, rpl22, ycf2, ndhF, ccsA, matK, atpF,* and *rpoC2*. The remaining genes showed average Ka/Ks values of between 0 and 0.49, which suggested that most genes in the *Q. aquifolioides* cp genome were under purifying selection.

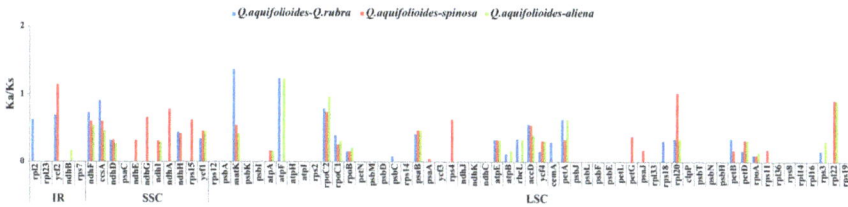

Figure 8. Ka/Ks ratios for protein-coding genes from *Q. rubra*, *Q. spinosa*, and *Q. aliena* chloroplast genome in comparison with *Q. aquifolioides*.

Although no genes were observed with average Ka/Ks > 1, there were four genes (*ycf2, matK, atpF,* and *rpl20*) with Ka/Ks > 1 in at least one pairwise comparison (Figure 8). Of these, only *atpF* had Ka/Ks > 1 in two pairwise comparisons. More interestingly, the Ka/Ks ratio for the *atpF* gene was more than 1 in the comparisons with the two deciduous *Quercus* species. In contrast, the Ka/Ks ratio for this gene was 0 in the comparison between *Q. aquifolioides* and *Q. spinosa*, which are both evergreen sclerophyllous oak trees. Deciduous oak trees completely lose their foliage during the winter or the dry season, usually as an adaptation to cold and/or drought, whereas evergreen sclerophyllous oak trees retain their leaves throughout of the year. For leaves to tolerate cold and drought stresses requires energy. The *atpF* gene encodes one of the subunits of the H^+-ATP synthase, which is essential for electron transport and photophosphorylation during photosynthesis [44]. This finding suggests that differential selection acting on the *atpF* gene may indicate that it has played a role in deciduous-evergreen oak tree divergence.

Alongside the Ka/Ks analysis, we also investigated site-specific selection events. We found a total of 12 genes exhibiting site-specific selection (Table 2). Of these, *rpoC2* was found to have 26 sites under positive selection. This gene encodes one of the four subunits of RNA polymerase type I (plastid-encoded polymerase, PEP), which is a key enzyme required for transcription of photosynthesis-related genes in the chloroplast [45]. Our identification of the positively-selected sites in this analysis could lead to a better understanding of the evolution of *Quercus* species.

Table 2. Positive selection sites identified by Selecton.

Gene	NULL (M8a)	POSITIVE (M8)	Putative Sites under Positive Selection
rpl2	−1177.05	−1174.54	1 (131 S)
ycf2	−9154.8	−9154.87	5 (96 K, 932 W, 1174 P, 1291 W, 2007 R)
rps7	−615.83	−615.616	1 (130 E)
ndhD	−2145.31	−2141.84	8 (170 T, 188 G, 200 L, 206 A, 362 R, 375 P, 413 Q, 504 F)
ycf1	−8057.78	−8051.12	8 (426 F, 529 L, 757 L, 761 L, 1007 I, 1490 Q, 1491 G, 1492 F)
rpoC2	−5814.68	−5814.8	26 (33 H, 131 P, 280 L, 364 I, 505 H, 542 E, 587 E, 595 P, 598 V, 626 N, 643 K, 691 G, 697 T, 815 Y, 849 G, 856 H, 898 D, 947 S, 1013 K, 1074 I, 1081 A, 1132 E, 1176 I, 1273 C, 1374 D, 1394 N)
rpoC1	−2857.37	−2855.15	1 (145 Y)
psaB	−2999.97	−2998.62	3 (145 L, 238 E, 239 K)
ndhJ	−633.793	−633.569	1 (107 A)
ndhC	−828.789	−828.761	2 (68 V, 86 F)
rpl36	−145.376	−145.135	1 (20 R)

Likelihood ratio test (LRT) analysis of models comparison M8 vs. M8a. M8 represents a model with positive selection; M8a represents null model without positive selection. Degree of freedom (df) = 1.

2.6. Phylogenetic Analysis of the cp Genomes of Q. aquifolioides and Related Quercus Species

The phylogeny of oak trees is complex due to extensive introgression, hybridization, incomplete lineage sorting, and convergent evolution [46]. However, phylogenetic issues in many angiosperms have been addressed successfully with the help of cp genome sequences [47–49]. Maximum parsimony (MP) analysis with 73 protein-coding genes from 12 Fragaceae with two tobacco species as outgroup revealed 10 out of 11 nodes with bootstrap values ≥ 95%, which is very high for an MP tree (Figure 9). The MP phylogenetic tree was even more strongly supported by eight 100% bootstrap values, showing that *Q. aquifolioides* was grouped with *Q. spinosa* within *Quercus*. Both of these are members of sect. *Heterobalanus*. The MP tree also revealed that *Q. rubra* and *Q. aliena* were the closest relatives of *Q. aquifolioides* and *Q. spinosa* (Figure 9). However, this phylogenetic tree is solely based on cp DNA. To fully understand their phylogenetic relationships, nuclear DNA is required to be investigated to assess the effect of introgression and hybridization on phylogeny.

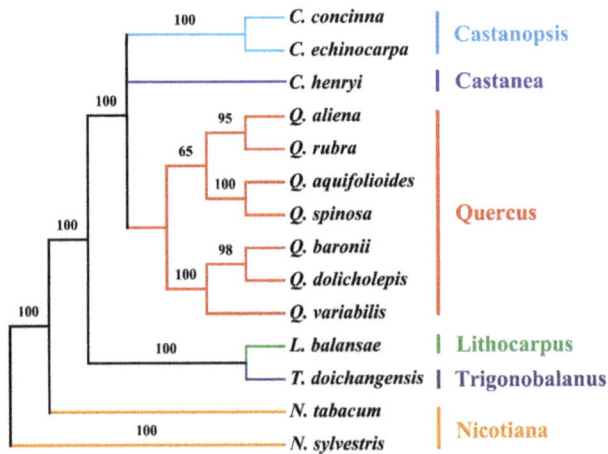

Figure 9. Phylogenetic relationship between *Q. aquifolioides* and related species, inferred from 73 protein-coding genes shared by all cp genomes. The phylogenetic tree was constructed by the maximum parsimony method, with two *Nicotiana* species as outgroups.

3. Materials and Methods

3.1. Plant Material

We collected a *Q. aquifolioides* tree less than 3 years old from Lijiang Alpine Botanic Garden, China and transplanted it to Beijing Forestry University. *Q. aquifolioides* is a common, non-endangered tree species in China. No specific protective policy was implemented in this area. The plants were grown in a growth chamber under 150 mmol·m^{-2}·s^{-1} light, with 16 h light/8 h dark cycles, at 24 °C with a constant humidity of 65%. Voucher specimens were deposited in the herbarium of Beijing Forestry University, Beijing, China.

3.2. Chloroplast Isolation, DNA Extraction, and Sequencing

A 0.3–0.5 g sample of the fresh young leaves was collected after the plant had grown in the dark for 24–36 h to promote starch degradation in chloroplasts (Nobel 1974). The chloroplast DNA (cpDNA) extraction and enrichment method followed the protocol developed by our group [50]. After amplifying the cpDNA by rolling circle amplification (RCA), we purified the RCA product and used 5 µg for library preparation. A 101-bp paired-end run was performed on an Illumina-HiSeq 1500 (Illumina, San Diego, CA, USA) at the gene sequencing platform of the School of Life Sciences, Tsinghua University, China. Briefly, library preparation was carried out following the manufacturer's instructions with an insert size of up to 500 bp. Base calling was performed with RTA v.1.6 (Illumina, San Diego, CA, USA).

3.3. Chloroplast Genome Assembly

We assembled the *Q. aquifolioides* chloroplast genome using a pipeline developed in our lab [21]. Briefly, we used an in-house Perl script to eliminate low quality (probability of error > 1%) nucleotides in each read. SOAPdenovo 2 [51] was used for de novo assembly with default parameters, except that an insert size of 500 bp was set. Next, the primary contigs were assembled using the *Quercus rubra* chloroplast genome (GenBank accession number: JX970937) as the reference sequence. Gaps between two neighbor contigs were filled with N. These gaps were resolved as previously described [50].

3.4. Genome Annotation

We used CpGAVAS [52] for chloroplast genome annotation then manually corrected the output. This program uses a chloroplast genome sequence in FASTA format to identify protein-coding genes by performing BLASTX searches against a custom database of known chloroplast genomes. The program also produces a circular map of the chloroplast genome, displaying the protein-coding genes, transfer RNAs (tRNAs), and ribosomal RNAs (rRNAs) based on the annotations.

3.5. Repeat Analysis

Simple sequence repeats (SSRs) in the cp genomes were detected using the Perl script MISA [53]. The thresholds set for the SSRs were 10, 6, 4, 3, 3, and 3 for mono-, di-, tri-, tetra-, penta-, and hexa-nucleotides, respectively. Tandem repeat sequences (>10 bp in length) were detected using the online program Tandem Repeats Finder [54]. The minimum alignment score and maximum period size were 90 and 500 respectively. The online REPuter software tool (Available online: https://bibiserv.cebitec.uni-bielefeld.de/reputer/) was used to identify forward, palindrome, reverse, and complement sequences with a minimum repeat size of 30 bp, and sequence identity greater than 90% (Hamming distance equal to 3) [55].

3.6. CCT Map

Comparative genome maps of *Q. aquifolioides* and the other three *Quercus* cp genomes were constructed by BLAST using CCT software [56] and the results were displayed as a circular map.

Additional features such as the Clusters of Orthologous Groups of proteins (COG) and GC Skew in the reference genome were also included.

3.7. Sequence Divergence Analysis

The alignments of the cp genomes of *Q. aquifolioides* and the other three *Quercus* cp genome were visualized using mVISTA [57] (Available online: http://genome.lbl.gov/vista/mvista/submit.shtml) in Shuffle-LAGAN mode [34] in order to show interspecific variation. The sequence divergences of four *Quercus* protein coding genes were evaluated using MEGA 7 [58]. A sliding window analysis was conducted to generate nucleotide diversity (*Pi*) values for the three data sets (the aligned LSC, SSC, and IR regions of the four complete *Quercus* cp genomes) using DnaSP 5 [59]. The step size was set to 200 bp, with a 600 bp window length. The Tamura 3-parameter (T92) model was selected to calculate pairwise sequence divergences [60].

3.8. Selection Pressure Analysis

To estimate selection pressures, non-synonymous (Ka) and synonymous (Ks) substitution rates of 73 protein coding genes between the cp genomes of *Q. aquifolioides* and the other three *Quercus* species were calculated using DnaSP 5. For identification of site-specific selection, protein coding gene alignments were analyzed using Selecton [61], with *Q. aquifolioides* as a reference sequence. Two models, M8 (allows for positive selection operating on the protein) and M8a (does not allow for positive selection), were used and likelihood scores estimated by models were evaluated using a log-likelihood ratio test (LRT) with degree of freedom (df) = 1. Only sites with posterior probabilities > 0.8 were selected.

3.9. Phylogenetic Analysis

The sequences were aligned using MAFFT 7 [62]. Maximum parsimony (MP) analysis was executed using PAUP 4 [63]. A total of 73 protein-coding genes shared by all cp genomes were used for this phylogenetic analysis, which included 12 Fagaceae species (*Q. aquifolioides* KP340971; *Q. aliena* KP301144; *Q. rubra* JX970937; *Q. spinosa* KM841421; *Q. variabilis* KU240009; *Q. dolicholepis* KU240010; *Q. baronii* KT963087; *Castanopsis concinna* KT793041; *C. echinocarpa* KJ001129; *Castanea henryi* KX954615; *Lithocarpus balansae* KP299291; *Trigonobalanus doichangensis* KF990556), with two *Nicotiana* species (*N. sylvestris* AB237912; *N. tabacum* Z00044) as outgroups.

Supplementary Materials: Supplementary materials can be found at http://www.mdpi.com/1422-0067/19/4/1042/s1.

Acknowledgments: This research was supported by the National Key Research and Development Plan "Research on protection and restoration of typical small populations of wild plants" (Grant No. 2016YFC0503106), Fundamental Research Funds for the Central Universities (No. 2015ZCQ-LX-03), the National Science Foundation of China (grant 41671039) and the Beijing Nova Program (grant Z151100000315056) to FKD.

Author Contributions: Fang K. Du and Kangquan Yin designed the research; Fang K. Du and Kangquan Yin collected the samples; Yue Zhang, Yuejuan Li and Kangquan Yin performed the experiments and analysis; Fang K. Du and Kangquan Yin wrote the manuscript; all authors revised the manuscript.

Conflicts of Interest: The authors declare no conflict of interest.

References

1. McCauley, D.E.; Stevens, J.E.; Peroni, P.A.; Raveill, J.A. The spatial distribution of chloroplast DNA and allozyme polymorphisms within a population of *Silene alba* (Caryophyllaceae). *Am. J. Bot.* **1996**, *83*, 727–731. [CrossRef]

2. Yurina, N.P.; Odintsova, M.S. Comparative structural organization of plant chloroplast and mitochondrial genomes. *Genetika* **1998**, *34*, 5–22.

3. Nixon, K.C. The genus *Quercus* in Mexico. In *Biological Diversity of Mexico: Origins and Distribution*; Oxford University Press: New York, NY, USA, 1993; pp. 447–458, ISBN 019506674X.

4. Keator, G.; Bazel, S. *The Life of an Oak: An Intimate Portrait*; Heyday Books: Berkeley, CA, USA; California Oak Foundation: Oakland, CA, USA, 1998; p. 256, ISBN 9780930588984.

5. Nixon, K.C. Global and neotropical distribution and diversity of oak (genus *Quercus*) and oak forests. In *Ecology and Conservation of Neotropical Montane Oak Forests*; Springer: Berlin/Heidelberg, Germany, 2006; pp. 3–13, ISBN 364206695X.

6. Trelease, W. The American oaks. *Mem. Natl. Acad. Sci.* **1924**, *20*, 1–255.

7. Palmer, E.J. Hybrid oaks of North America. *J. Arnold Arbor.* **1948**, *29*, 1–48.

8. Muller, C.H. Ecological control of hybridization in *Quercus*: A factor in the mechanism of evolution. *Evolution* **1952**, *6*, 147–161.

9. Tucker, J.M. Studies in the *Quercus undulata* complex. I. A preliminary statement. *Am. J. Bot.* **1961**, *48*, 202–208. [CrossRef]

10. Hardin, J.W. Hybridization and introgression in *Quercus alba*. *J. Arnold Arbor.* **1975**, *56*, 336–363.

11. Rushton, B.S. Natural hybridization within the genus *Quercus*. *Ann. For. Sci.* **1993**, *50*, 73s–90s. [CrossRef]

12. Spellenberg, R. On the hybrid nature of *Quercus basaseachicensis* (Fagaceae, sect. Quercus). *SIDA Contrib. Bot.* **1995**, *16*, 427–437.

13. Bacilieri, R.; Ducousso, A.; Petit, R.J.; Kremer, A. Mating system and asymmetric hybridization in a mixed stand of European oaks. *Evolution* **1996**, *50*, 900–908. [CrossRef] [PubMed]

14. Howard, D.J.; Preszler, R.W.; Williams, J.; Fenchel, S.; Boecklen, W.J. How discrete are oak species? Insights from a hybrid zone between *Quercus grisea* and *Quercus gambelii*. *Evolution* **1997**, *51*, 747–755. [CrossRef] [PubMed]

15. Manos, P.S.; Doyle, J.J.; Nixon, K.C. Phylogeny, biogeography, and processes of molecular differentiation in *Quercus* subgenus *Quercus* (Fagaceae). *Mol. Phylogenet. Evol.* **1999**, *12*, 333–349. [CrossRef] [PubMed]

16. Manos, P.S.; Stanford, A.M. The historical biogeography of Fagaceae: Tracking the tertiary history of temperate and subtropical forests of the Northern Hemisphere. *Int. J. Plant Sci.* **2001**, *162*, S77–S93. [CrossRef]

17. Grímsson, F.; Zetter, R.; Grimm, G.W.; Pedersen, G.K.; Pedersen, A.K.; Denk, T. Fagaceae pollen from the early Cenozoic of West Greenland: Revisiting Engler's and Chaney's Arcto-Tertiary hypotheses. *Plant Syst. Evol.* **2015**, *301*, 809–832. [CrossRef] [PubMed]

18. Simeone, M.C.; Grimm, G.W.; Papini, A.; Vessella, F.; Cardoni, S.; Tordoni, E.; Piredda, R.; Franc, A.; Denk, T. Plastome data reveal multiple geographic origins of *Quercus* Group Ilex. *PeerJ* **2016**, *4*, e1897. [CrossRef] [PubMed]

19. Denk, T.; Grimm, G.W.; Manos, P.S.; Deng, M.; Hipp, A.L. An updated infrageneric classification of the oaks: Review of previous taxonomic schemes and synthesis of evolutionary patterns. In *Oaks Physiological Ecology. Exploring the Functional Diversity of Genus Quercus L.*; Gil-Pelegrín, E., Peguero-Pina, J., Sancho-Knapik, D., Eds.; Springer: Cham, Switzerland, 2017; pp. 13–38. ISBN 978-3-319-69098-8.

20. Huang, C.J.; Zhang, Y.T.; Bartholomew, B. Fagaeeae. In *Flora of China*; Wu, Z.Y., Raven, P.H., Eds.; Science Press: Beijing, China, 1999; pp. 370–380, ISBN 0915279703.

21. Du, F.K.; Hou, M.; Wang, W.; Mao, K.S.; Hampe, A. Phylogeography of *Quercus aquifolioides* provides novel insights into the Neogene history of a major global hotspot of plant diversity in southwest China. *J. Biogeogr.* **2017**, *44*, 294–307. [CrossRef]

22. Yang, Y.; Zhou, T.; Duan, D.; Yang, J.; Feng, L.; Zhao, G. Comparative analysis of the complete chloroplast genomes of five *Quercus* species. *Front. Plant Sci.* **2016**, *7*, 959. [CrossRef] [PubMed]

23. Jansen, R.K.; Raubeson, L.A.; Boore, J.L.; dePamphilis, C.W.; Chumley, T.W.; Haberle, R.C.; Wyman, S.K.; Alverson, A.J.; Peery, R.; Herman, S.J.; et al. Methods for obtaining and analyzing whole chloroplast genome sequences. *Method Enzymol.* **2005**, *395*, 348–384.

24. Chumley, T.W.; Palmer, J.D.; Mower, J.P.; Fourcade, H.M.; Calie, P.J.; Boore, J.L.; Jansen, R.K. The complete chloroplast genome sequence of *Pelargonium* × *hortorum*: Organization and evolution of the largest and most highly rearranged chloroplast genome of land plants. *Mol. Biol. Evol.* **2006**, *23*, 2175–2190. [CrossRef] [PubMed]

25. Bull, L.N.; Pabón-Peña, C.R.; Freimer, N.B. Compound microsatellite repeats: Practical and theoretical features. *Genome Res.* **1999**, *9*, 830–838. [CrossRef] [PubMed]

26. Jansen, R.K.; Kaittanis, C.; Saski, C.; Lee, S.B.; Tomkins, J.; Alverson, A.J.; Daniell, H. Phylogenetic analyses of *Vitis* (Vitaceae) based on complete chloroplast genome sequences: Effects of taxon sampling and phylogenetic methods on resolving relationships among rosids. *BMC Evol. Biol.* **2006**, *6*, 32. [CrossRef] [PubMed]

27. Ruhlman, T.; Lee, S.B.; Jansen, R.K.; Hostetler, J.B.; Tallon, L.J.; Town, C.D.; Daniell, H. Complete plastid genome sequence of *Daucus carota*: Implications for biotechnology and phylogeny of angiosperms. *BMC Genom.* **2006**, *7*, 222. [CrossRef] [PubMed]

28. Silva, S.R.; Diaz, Y.C.; Penha, H.A.; Pinheiro, D.G.; Fernandes, C.C.; Miranda, V.F.; Todd, P.; Michael, T.P.; Varani, A.M. The Chloroplast Genome of *Utricularia reniformis* Sheds Light on the Evolution of the *ndh* Gene Complex of Terrestrial Carnivorous Plants from the Lentibulariaceae Family. *PLoS ONE* **2016**, *11*, e0165176. [CrossRef] [PubMed]

29. Liu, L.X.; Li, R.; Worth, J.R.; Li, X.; Li, P.; Cameron, K.M.; Fu, C.X. The Complete Chloroplast Genome of Chinese Bayberry (*Morella rubra*, Myricaceae): Implications for Understanding the Evolution of Fagales. *Front. Plant Sci.* **2017**, *8*, 968. [CrossRef] [PubMed]

30. Qian, J.; Song, J.; Gao, H.; Zhu, Y.; Xu, J.; Pang, X.; Yao, H.; Sun, C.; Li, X.; Li, C.; et al. The complete chloroplast genome sequence of the medicinal plant *Salvia miltiorrhiza*. *PLoS ONE* **2013**, *8*, e57607. [CrossRef] [PubMed]

31. Kuang, D.Y.; Wu, H.; Wang, Y.L.; Gao, L.M.; Zhang, S.Z.; Lu, L. Complete chloroplast genome sequence of *Magnolia kwangsiensis* (Magnoliaceae): Implication for DNA barcoding and population genetics. *Genome* **2011**, *54*, 663–673. [CrossRef] [PubMed]

32. Goulding, S.E.; Wolfe, K.H.; Olmstead, R.G.; Morden, C.W. Ebb and flow of the chloroplast inverted repeat. *Mol. Gen. Genet.* **1996**, *252*, 195–206. [CrossRef] [PubMed]

33. Kim, K.J.; Lee, H.L. Complete chloroplast genome sequences from Korean ginseng (*Panax schinseng* Nees) and comparative analysis of sequence evolution among 17 vascular plants. *DNA Res.* **2004**, *11*, 247–261. [CrossRef] [PubMed]

34. Dong, W.; Xu, C.; Cheng, T.; Zhou, S. Complete chloroplast genome of *Sedum sarmentosum* and chloroplast genome evolution in Saxifragales. *PLoS ONE* **2013**, *8*, e77965. [CrossRef] [PubMed]

35. Lu, R.; Li, P.; Qiu, Y. The complete chloroplast genomes of three *Cardiocrinum* (Liliaceae) species: Comparative genomic and phylogenetic analyses. *Front. Plant Sci.* **2016**, *7*, 2054. [CrossRef] [PubMed]

36. Zhang, Y.; Du, L.; Liu, A.; Chen, J.; Wu, L.; Hu, W.; Zhang, W.; Kim, K.; Lee, S.C.; Tae-Jin Yang, T.J.; et al. The Complete Chloroplast Genome Sequences of Five *Epimedium* Species: Lights into Phylogenetic and Taxonomic Analyses. *Front. Plant Sci.* **2016**, *7*, 306. [CrossRef] [PubMed]

37. Khakhlova, O.; Bock, R. Elimination of deleterious mutations in plastid genomes by gene conversion. *Plant J.* **2006**, *46*, 85–94. [CrossRef] [PubMed]

38. Liu, Y.; Huo, N.; Dong, L.; Wang, Y.; Zhang, S.; Young, H.A.; Feng, X.; Gu, Y.Q. Complete Chloroplast Genome Sequences of Mongolia Medicine *Artemisia frigida* and Phylogenetic Relationships with Other Plants. *PLoS ONE* **2013**, *8*, e57533. [CrossRef] [PubMed]

39. Nazareno, A.G.; Carlsen, M.; Lohmann, L.G. Complete Chloroplast Genome of *Tanaecium tetragonolobum*: The First Bignoniaceae Plastome. *PLoS ONE* **2015**, *10*, e0129930. [CrossRef] [PubMed]

40. Dong, W.; Liu, J.; Yu, J.; Wang, L.; Zhou, S. Highly variable chloroplast markers for evaluating plant phylogeny at low taxonomic levels and for DNA barcoding. *PLoS ONE* **2012**, *7*, e35071. [CrossRef] [PubMed]

41. Awad, M.; Fahmy, R.M.; Mosa, K.A.; Helmy, M.; El-Feky, F.A. Identification of effective DNA barcodes for *Triticum* plants through chloroplast genome-wide analysis. *Comput. Biol. Chem.* **2017**, *71*, 20–31. [CrossRef] [PubMed]

42. Shaw, J.; Lickey, E.B.; Schilling, E.E.; Small, R.L. Comparison of whole chloroplast genome sequences to choose noncoding regions for phylogenetic studies in angiosperms: The tortoise and the hare III. *Am. J. Bot.* **2007**, *94*, 275–288. [CrossRef] [PubMed]

43. Kimura, M. The neutral theory of molecular evolution and the world view of the neutralists. *Genome* **1983**, *31*, 24–31. [CrossRef]

44. Hudson, G.S.; Mason, J.G. The chloroplast genes encoding subunits of the H$^+$-ATP synthase. In *Molecular Biology of Photosynthesis*; Govindjee, Ed.; Springer: Dordrecht, The Netherlands, 1988; pp. 565–582, ISBN 978-94-010-7517-6.

45. Cummings, M.P.; King, L.M.; Kellogg, E.A. Slipped-strand mispairing in a plastid gene: *RpoC2* in grasses (*Poaceae*). *Mol. Biol. Evol.* **1994**, *11*, 1–8. [PubMed]

46. Aldrich, P.R.; Cavender-Bares, J. Quercus. In *Wild Crop Relatives: Genomic and Breeding Resources*; Kole, C., Ed.; Springer: Berlin/Heidelberg, Germany, 2011; pp. 89–129, ISBN 978-3-642-21249-9.

47. Jansen, R.K.; Cai, Z.; Raubeson, L.A.; Daniell, H.; dePamphilis, C.W.; Leebens-Mack, J.; Müller, K.F.; Guisinger-Bellian, M.; Haberle, R.C.; Hansen, A.K.; et al. Analysis of 81 genes from 64 plastid genomes resolves relationships in angiosperms and identifies genome-scale evolutionary patterns. *Proc. Natl. Acad. Sci. USA* **2007**, *104*, 19369–19374. [CrossRef] [PubMed]

48. Moore, M.J.; Bell, C.D.; Soltis, P.S.; Soltis, D.E. Using plastid genome-scale data to resolve enigmatic relationships among basal angiosperms. *Proc. Natl. Acad. Sci. USA* **2007**, *104*, 19363–19368. [CrossRef] [PubMed]

49. Yang, J.B.; Li, D.Z.; Li, H.T. Highly effective sequencing whole chloroplast genomes of angiosperms by nine novel universal primer pairs. *Mol. Ecol. Res.* **2014**, *14*, 1024–1031. [CrossRef] [PubMed]

50. Du, F.K.; Lang, T.; Lu, S.; Wang, Y.; Li, J.; Yin, K. An improved method for chloroplast genome sequencing in non-model forest tree species. *Tree Genet. Genomes* **2015**, *11*, 114. [CrossRef]

51. Luo, R.; Liu, B.; Xie, Y.; Li, Z.; Huang, W.; Yuan, J.; He, G.; Chen, Y.; Pan, Q.; Liu, Y.; et al. SOAPdenovo2: An empirically improved memory-efficient short-read de novo assembler. *Gigascience* **2012**, *1*, 18. [CrossRef] [PubMed]

52. Liu, C.; Shi, L.; Zhu, Y.; Chen, H.; Zhang, J.; Lin, X.; Guan, X. CpGAVAS, an integrated web server for the annotation, visualization, analysis, and GenBank submission of completely sequenced chloroplast genome sequences. *BMC Genom.* **2012**, *13*, 715. [CrossRef] [PubMed]

53. Thiel, T.; Michalek, W.; Varshney, R.; Graner, A. Exploiting EST databases for the development and characterization of gene-derived SSR-markers in barley (*Hordeum vulgare* L.). *Theor. Appl. Genet.* **2003**, *106*, 411–422. [CrossRef] [PubMed]

54. Benson, G. Tandem repeats finder: A program to analyze DNA sequences. *Nucleic Acids Res.* **1999**, *27*, 573. [CrossRef] [PubMed]

55. Kurtz, S.; Schleiermacher, C. REPuter: Fast computation of maximal repeats in complete genomes. *Bioinformatics* **1999**, *15*, 426–427. [CrossRef] [PubMed]

56. Grant, J.R.; Stothard, P. The CGView Server: A comparative genomics tool for circular genomes. *Nucleic Acids Res.* **2008**, *36*, W181–W184. [CrossRef] [PubMed]

57. Frazer, K.A.; Pachter, L.; Poliakov, A.; Rubin, E.M.; Dubchak, I. VISTA: Computational tools for comparative genomics. *Nucleic Acids Res.* **2004**, *32*, W273–W279. [CrossRef] [PubMed]

58. Kumar, S.; Stecher, G.; Tamura, K. MEGA7: Molecular evolutionary genetics analysis version 7.0 for bigger datasets. *Mol. Biol. Evol.* **2016**, *33*, 1870–1874. [CrossRef] [PubMed]

59. Librado, P.; Rozas, J. DnaSP v5: A software for comprehensive analysis of DNA polymorphism data. *Bioinformatics* **2009**, *25*, 1451–1452. [CrossRef] [PubMed]

60. Tamura, K. Estimation of the number of nucleotide substitutions when there are strong transition-transversion and G+C-content biases. *Mol. Biol. Evol.* **1992**, *9*, 678–687. [PubMed]

61. Stern, A.; Doron-Faigenboim, A.; Erez, E.; Martz, E.; Bacharach, E.; Pupko, T. Selecton 2007: Advanced models for detecting positive and purifying selection using a Bayesian inference approach. *Nucleic Acids Res.* **2007**, *35*, W506–W511. [CrossRef] [PubMed]

62. Katoh, K.; Standley, D.M. MAFFT multiple sequence alignment software version 7: Improvements in performance and usability. *Mol. Biol. Evol.* **2013**, *30*, 772–780. [CrossRef] [PubMed]

63. Swofford, D.L. *PAUP*. *Phylogenetic Analysis Using Parsimony (and Other Methods)*. *Version 4*; Sinauer Associates: Sunderland, MA, USA, 2003.

International Journal of
Molecular Sciences

MDPI

Article

Candidate Genes for Yellow Leaf Color in Common Wheat (*Triticum aestivum* L.) and Major Related Metabolic Pathways according to Transcriptome Profiling

Huiyu Wu [1], Narong Shi [1], Xuyao An [1], Cong Liu [1], Hongfei Fu [2], Li Cao [1], Yi Feng [1], Daojie Sun [1] and Lingli Zhang [1,*]

[1] College of Agronomy, Northwest A&F University, Yangling 712100, China; huiyuwu@nwafu.edu.cn (H.W.); narongshi@nwafu.edu.cn (N.S.); 16631407278@163.com (X.A.); congliu@nwafu.edu.cn (C.L.); caolinwafu@126.com (L.C.); fengyi92377504@126.com (Y.F.); daojie49124098@126.com (D.S.)

[2] College of Food Science and Engineering, Northwest A&F University, Yangling 712100, China; fuhongfei@nwafu.edu.cn

* Correspondence: zhanglingli@nwafu.edu.cn; Tel.: +86-29-8708-2845

Received: 30 April 2018; Accepted: 25 May 2018; Published: 29 May 2018

Abstract: The photosynthetic capacity and efficiency of a crop depends on the biosynthesis of photosynthetic pigments and chloroplast development. However, little is known about the molecular mechanisms of chloroplast development and chlorophyll (Chl) biosynthesis in common wheat because of its huge and complex genome. *Ygm*, a spontaneous yellow-green leaf color mutant of winter wheat, exhibits reduced Chl contents and abnormal chloroplast development. Thus, we searched for candidate genes associated with this phenotype. Comparative transcriptome profiling was performed using leaves from the yellow leaf color type (Y) and normal green color type (G) of the *Ygm* mutant progeny. We identified 1227 differentially expressed genes (DEGs) in Y compared with G (i.e., 689 upregulated genes and 538 downregulated genes). Gene ontology and pathway enrichment analyses indicated that the DEGs were involved in Chl biosynthesis (i.e., magnesium chelatase subunit H (CHLH) and protochlorophyllide oxidoreductase (POR) genes), carotenoid biosynthesis (i.e., β-carotene hydroxylase (BCH) genes), photosynthesis, and carbon fixation in photosynthetic organisms. We also identified heat shock protein (HSP) genes (*sHSP*, *HSP70*, *HSP90*, and *DnaJ*) and heat shock transcription factor genes that might have vital roles in chloroplast development. Quantitative RT-PCR analysis of the relevant DEGs confirmed the RNA-Seq results. Moreover, measurements of seven intermediate products involved in Chl biosynthesis and five carotenoid compounds involved in carotenoid-xanthophyll biosynthesis confirmed that CHLH and BCH are vital enzymes for the unusual leaf color phenotype in Y type. These results provide insights into leaf color variation in wheat at the transcriptional level.

Keywords: RNA-Seq; transcription factor; chlorophyll biosynthesis precursor; carotenoid composition; wheat; yellow-green leaf color mutant

---

1. Introduction

Photosynthesis is the basis of plant production, and at least 90% of grain yield is determined by photosynthesis [1]. Under irradiation by light, photosynthetic pigments fix light energy and convert it into chemical energy to synthesize carbohydrates. Chlorophyll (Chl) is the primary photosynthetic pigment in higher plants, where it is responsible for light harvesting in the antenna systems and driving electron transport in the reaction centers [2]. The entire Chl biosynthetic pathway from glutamyl-tRNA to Chl *a* and Chl *b* comprises about 20 different enzymatic steps [3]. Mutations in any of these genes may lead to variations in the Chl contents [4], abnormal chloroplast development [5], and decreased photosynthetic efficiency [6], thereby yielding leaf color mutants. Mutants deficient in Chl biosynthesis have been identified in many higher plants, such as rice [7,8], *Brassica napus* [9], *Arabidopsis thaliana* [10], barley [11], and *Camellia sinensis* [12]. Many of the reported chlorotic mutants exhibit reduced Chl biosynthesis due to the lower activity of magnesium chelatase (Mg-chelatase) [11,13–15]. Mg-chelatase (EC 6.6.1.1) is a key regulatory enzyme that catalyzes the insertion of Mg^{2+} into protoporphyrin IX (Proto IX) in an ATP-dependent manner as the first committed step in Chl biosynthesis, where this protein complex comprises magnesium chelatase subunit I (CHLI), D (CHLD), and H (CHLH) in higher plants [16], which are all required for its activity [17]. CHLI and CHLD belong to the large family of AAA^+ (ATPases associated with various cellular activities) proteins, but only the I-subunit has an ATPase activity [18]. The H-subunit binds the porphyrin substrate, and it is regarded as a catalytic subunit without ATPase activity [19]. The *GUN5* gene encodes the CHLH, and it has a specific function in the plastid signaling pathway where its activity is controlled by *GUN4* [20,21]. The *GUN4* gene encodes a protein that regulates Chl biosynthesis in plastids, and it has been implicated in plastid retrograde signaling via the regulation of Mg-protoporphyrin (Mg-Proto) synthesis or transport [21]. The activity of Mg-chelatase has essential regulatory roles in Chl biosynthesis and chloroplast development in higher plants. For example, in peas, the virus-induced gene silencing of *CHLI* and *CHLD* yields plants with yellow leaf phenotypes and reduced Mg-chelatase activities, as well as lower Chl accumulation correlated with undeveloped thylakoid membranes [22]. In addition, *CHLD* and *CHLI* silencing greatly reduces the levels of photosynthetic proteins, as well as being correlated with reactive oxygen species homeostasis [22]. A T-DNA insertion mutant *OsCHLH* in rice also exhibits underdeveloped chloroplasts with a low Chl content [7]. Previous studies have explored the semi-dominant leaf color mutants caused by Mg-chelatase. In barley, the mutants *chlorina-125*, *-157*, and *-161* have the same phenotypic ratio model (i.e., one green wild-type leaf, two light-green chlorina leaves, and one lethal yellow leaf at the seedling stage). The Mg-chelatase activity of the heterozygous chlorina seedlings is 25–50% of that in wild type seedlings [23,24]. In tobacco, the *Sulfur* mutant, a *CHLI* gene mutation due to the formation of inactive Mg-chelatase, is a semi-dominant aurea mutation, where homozygotes of the mutant are yellow seedling lethals, whereas the heterozygotes have reduced Chl contents and a yellow-green phenotype [25]. Moreover, a semi-dominant *CHLI* allele designated as *Oil Yellow1* (*Oy1*) has also been characterized in maize [26]. In rice, *chl1* and *chl9* mutants exhibit a yellowish-green leaf color phenotype where the abnormal leaf color is controlled by a single recessive gene. The *chl1* and *chl9* genes encode the CHLD and CHLI subunits of Mg-chelatase, and their mutation leads to underdeveloped chloroplasts and low Chl contents [13]. However, to the best of our knowledge, only a few *CHLD*, *CHLH*, and *CHLI* genes encoding the Mg-chelatase D, H, and I subunits have been reported in wheat leaf color mutants [27–29].

Common wheat (*Triticum aestivum* L.) is one of the most important food crops in the world. Two main types of Chl-deficient wheat mutant have been identified (i.e., albinism [30,31] and chlorina [32,33]), which have great research value for understanding the mechanisms of Chl biosynthesis and photosynthesis in wheat. However, only a few studies have reported the molecular mechanisms related to the changes in leaf color in wheat because of its large genome and high proportion of repetitive sequences (>80%) [30,34]. Most of these studies have focused on agronomic traits, photosynthetic characteristics, physiological and biochemical characteristics, and genetic mapping [31–33]. For example, controlled by cytogene, the wheat stage albinism mutant *FA85* exhibits albinism in cold early spring and returns to normal green gradually with the increase of temperature [35]. Chlorophyll precursors and key enzyme activities measurement of *FA85* have revealed that the accumulation of Proto IX, Mg-Proto IX, and Pchlide derives from the downregulated transcription level of Pchlide oxidoreductase and Chl synthase [36]. In addition, the five homeologous allelic chlorina mutants Driscoll's chlorina, chlorina-1, CD3, chlorina-214, and CDd-1 exhibit the reduction in Mg-chelatase activity which leads to the accumulation of Proto IX to different extents [28]. The above five homeologous allelic chlorina mutants have been mapped on the long arm of homoeologous group 7 chromosomes [37]. In recent years, due to the development of high-throughput sequencing technology, transcriptome sequencing (RNA-Seq) has emerged as a powerful tool for studying complex biological processes at the molecular level and for identifying candidate genes involved in specific biological functions [38,39]. RNA-Seq has been employed to investigate leaf color mechanisms in various plants, such as tea (*Camellia sinensis* L.) [12], *Anthurium andraeanum* [40], *Lagerstroemia indica* [41], and wheat [34]. These studies showed that the candidate genes related to leaf color mutation were involved in chloroplast development, Chl biosynthesis, photosynthesis, and transcription factors (TFs) such as phytochrome-interacting factor (PIF1 and PIF3), Golden2-like (GLK), and v-myb avian myeloblastosis viral oncogene homolog (MYB), which might participate in the pathways identified. Therefore, RNA-Seq can increase our understanding and provide new insights into wheat leaf color mutation at the genomics level.

Previously, we reported a spontaneous yellow-green leaf color mutant (*Ygm*) derived from a common wheat cultivar Xinong1718, with yellow (Y), yellow-green (Yg), and normal green (G) types under field temperature conditions from the jointing stage to the adult stage (Figure 1). Genetic analysis indicated that the yellow leaf color trait in the *Ygm* mutant was controlled by an incompletely dominant gene *Y1718* [42]. The homozygous dominant genotype (*Y1718Y1718*) of Y is extremely yellow, stunted, and sterile. The homozygous recessive genotype (*y1718y1718*) of G is normal green, whereas the heterozygous genotype (*Y1718y1718*) of Yg is yellow-green. Types Yg and G have similar agronomic traits to the wild type Xinong1718. The mutant is maintained steadily as the heterozygote genotype Yg. The *Y1718* gene was mapped on the short arm of chromosome 2B (2BS) between the simple sequence repeat marker *Xwmc25* and the expressed sequence tag-sequence tagged site marker *BE498358*, with genetic distances of 1.7 cM and 4.0 cM, respectively [42]. Thus, in the present study, the availability of a set of germplasm for the Y, Yg, and G genotypes allowed us to obtain insights into complex metabolic networks and certain biochemical traits, especially leaf color in common wheat.

Figure 1. Phenotypes of the Y, Yg, and G plants among the progeny of the *Ygm* mutant and wild type (WT, Xinong1718). (**A**) Jointing stage (9 April 2016); (**B**) Adult stage (28 April 2016); (**C**) Enlarged views of the leaves in different development states in G type at the jointing stage; (**D**) Enlarged views of the leaves in different development states in Y type at the jointing stage (WT, Xinong1718). G, normal green leaf color plant in the progeny of *Ygm*; Yg, yellow-green leaf color plant in the progeny of *Ygm*; Y, yellow leaf color plant in the progeny of *Ygm*. F_G and F_Y, fully-developed leaves in G and Y plants, respectively. H_G and H_Y, half-developed leaves in G and Y plants, respectively. L_G and L_Y, small leaf buds in G and Y plants, respectively. Bar = 5 cm.

In this study, comparative transcriptome profiles of the Y and G types in the progeny of the *Ygm* mutant were analyzed by RNA-Seq. Based on a combination of biochemical analysis and bioinformatics, we identified the major metabolic pathways related to leaf color and candidate genes for the loss of pigmentation in these plants. The concentrations of seven intermediate products involved in Chl biosynthesis and five carotenoid compounds involved in carotenoid-xanthophyll biosynthesis were measured to further understand the molecular mechanisms related to pigment biosynthesis in the *Ygm* mutant.

2. Results

2.1. Sequence Analysis Using RNA-Seq

To understand the molecular basis of leaf color polymorphism in the progeny of the *Ygm* mutant, cDNA libraries were constructed from the half-developed leaves of the Y and G types based on three biological replicates, which were then sequenced using the Illumina HiSeq^{TM} 2500 platform (Illumina, San Diego, CA, USA). The correlation coefficient values ranged from 0.947 to 0.989 (Table S1), thereby indicating strong correlations between the replicates. After cleaning and checking the read quality, 387,431,412 clean paired-end reads were generated, with 203,880,864 reads from the Y type and 183,550,548 from the G type, where the clean data GC content ranged from 57.29% to 59.25%, and the Q20 percentage exceeded 96.72%. The high-quality clean reads were then aligned with wheat genome sequences in the Unité de Recherche Génomique Info (URGI) database (http://wheat-urgi.versailles.inra.fr/Seq-Repository), where the alignment efficiency ranged from 69.67% to 70.71% (Table 1). Thus, the throughput and sequencing quality were sufficiently high to warrant further analysis.

Table 1. Summary of transcriptome sequencing data obtained using yellow leaves of Y plants and green leaves of G plants.

Groups	Total Reads	Clean Reads	GC (%)	Q20 (%)	Total Mapped Reads	Ratio (%)
G-1	59,480,874	58,804,014	58.37	96.76	41,285,997	70.21
G-2	61,434,418	60,676,246	58.68	96.54	42,662,324	70.31
G-3	64,966,228	64,070,288	58.68	96.65	44935,906	70.14
Y-1	67,083,278	66,278,074	59.25	96.87	46,615,892	70.33
Y-2	58,156,566	57,490,204	57.87	96.95	40,650,136	70.71
Y-3	81,071,740	80,112,586	57.29	96.61	55,817,063	69.67
Total	392,193,104	387,431,412	58.32	96.72	271,967,318	69.35

2.2. Identification and Functional Annotation of Differentially Expressed Genes (DEGs) in G and Y

The FPKM (fragments per kilobase of transcript per million mapped reads) method was used to analyze the gene expression. As a result, 74,937 and 75,211 genes were identified respectively in the cDNA library from G and Y leaves, of which 3879 and 4153 genes were expressed specifically in the leaves of G and Y, respectively (Figure 2). In total, 1227 DEGs (false discovery rate (FDR) <0.05 and |log$_2$Fold Change| (|log$_2$FC|)>1) were detected, where 689 were upregulated and 538 were downregulated in Y. In order to understand the functions of the DEGs and the biological processes involved with leaf color variation, all of the DEGs were searched for in the GenBank non-redundant (Nr) protein database as well as the Gene Ontology (GO), Clusters of Orthologous Groups (COG), and Kyoto Encyclopedia of Genes and Genomes (KEGG) databases. In total, 882 DEGs had BLASTx matches with known proteins and these 882 DEGs were assigned to one or more GO terms in the biological process (685 genes), molecular function (797 genes), and cellular component (247 genes) categories. Among these three categories, the metabolic process sub-category in the biological process category accounted for the majority of the GO annotations, followed by binding and catalytic activity in the molecular function category (Figure 3). Lipid metabolic process (50 genes) and phosphorus metabolic process (20 genes), which are closely related to cell membrane functions, were significantly enriched in the biological process GO term (Table S2). The nucleic acid binding transcription factor (TF) activity and sequence-specific DNA binding TF activity were the most significantly enriched GO terms in the molecular function category (Table S2). In the cellular component category, most of the DEGs were involved with cell, cell part, and membrane components (Figure 3), but the most significantly enriched component was membrane part (Table S2).

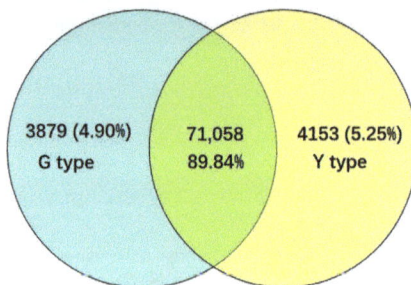

Figure 2. The numbers of specific genes and shared genes between G and Y.

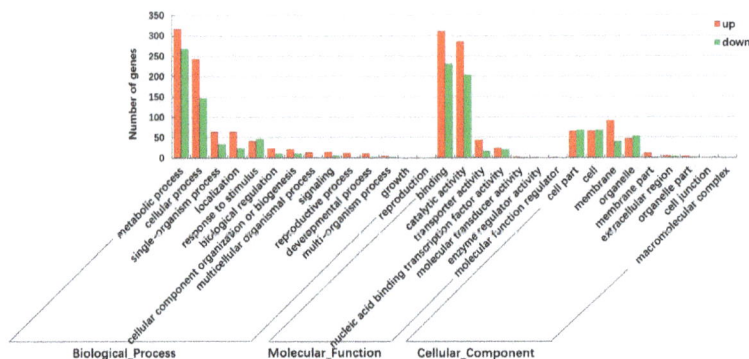

Figure 3. GO classifications of the DEGs in groups G and Y.

All of the 1227 DEGs were further annotated based on the COG database to obtain functional predictions and classifications (Figure 4). In total, 434 (35.37%) of the DEGs were finally mapped onto 22 different COG categories, where the "general function prediction only" cluster represented the largest group (108, 24.88%), followed by "posttranslational modification, protein turnover, chaperones" (97, 22.35%), "signal transduction mechanisms" (66, 15.20%), and "secondary metabolites biosynthesis, transport and catabolism" (59, 13.59%). In addition, "carbohydrate transport and metabolism" (42, 9.67%) and "lipid transport and metabolism" (35, 8.1%) were also annotated, which are closely related to energy metabolism and cell membrane function.

To identify biological pathways, the DEGs were annotated with the corresponding enzyme commission (EC) numbers from BLASTx alignments against the KEGG pathway databases. Among the 1227 DEGs, 323 (26.32%) were assigned to 82 KEGG pathways (Table S3), where the top 11 pathways were considered significant at a cut-off FDR corrected p-value (q-value) < 0.05 (Figure 5A). The 11 enriched pathways comprised protein processing in the endoplasmic reticulum, alpha-linolic acid metabolism, spliceosome, circadian rhythm-plant, linoleic acid metabolism, endocytosis, monoterpenoid biosynthesis, porphyrin and chlorophyll metabolism, brassinosteroid biosynthesis, glutathione metabolism, and thiamine metabolism. Most of the genes mapped in the first eight significantly enriched pathways had downregulated expression trends, except for the circadian rhythm-plant pathway. Chlorophyll and carotenoid biosynthesis are crucial for the leaf color and photosynthesis. We focused our analysis on the major genes related to pigment metabolism and photosynthesis. The results indicated that 33 genes related to either photosynthesis (five genes), Chl metabolism (nine genes), carotenoid biosynthesis (two genes), carbon fixation in photosynthetic organisms (five genes), or carbon metabolism (12 genes) pathways were differentially expressed, and the enrichment of each pathway is shown in Figure 5B. These pathways were investigated in greater detail and they may be important for the unusual leaf color phenotype in Y.

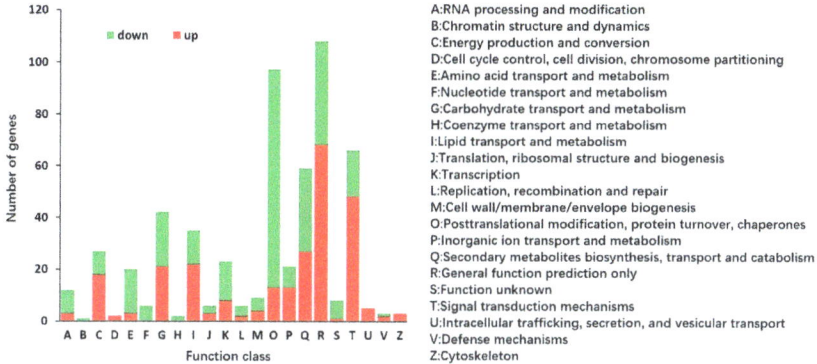

Figure 4. Clusters of Orthologous Groups (COG) classifications of the annotated 434 DEGs. The capital letters on the horizontal axis indicate the COG categories that are listed on the right of the histogram, and those on the vertical axis indicate the number of DEGs.

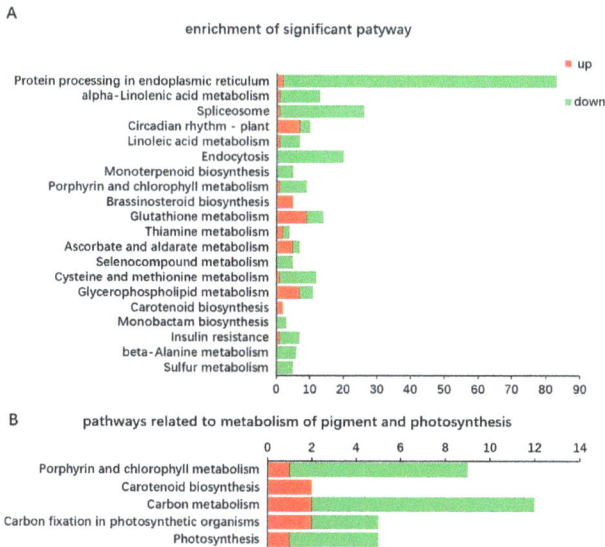

Figure 5. Kyoto Encyclopedia of Genes and Genomes (KEGG) classifications of DEGs. (**A**) Enrichment of the top 20 most significant pathways (p-value < 0.05). The vertical axis shows the annotations of the KEGG metabolic pathways. The horizontal axis represents the DEG numbers annotated in each pathway; (**B**) KEGG-based pathway assignments of the 33 DEGs (Y versus G) related to photosynthesis and pigment metabolism: photosynthesis (five genes), porphyrin and chlorophyll metabolism (nine genes), carotenoid biosynthesis (two genes), carbon fixation in photosynthetic organisms (five genes), and carbon metabolism (12 genes).

2.3. Identification of DEGs Related to Chl Metabolism and Carotenoid Biosynthesis in Y and G Plants

To further identify the key transcripts related to yellow leaf color formation in the mutant Y type, the DEGs in two pathways for Chl metabolism and carotenoid biosynthesis were compared in detail between the Y and G transcriptomes. The results demonstrated that eight genes in the porphyrin and Chl metabolism pathway had downregulated expression levels ($q < 0.05$, fold change > 2) (Figure 6A,C), including five genes encoding CHLH (i.e., *Traes_2DS_BB50DEEF8, Traes_2DS_DBD06E18F, Traes_2AS_BFBD75AB4, Traes_2AS_B6BA92570,* and *Traes_2BS_E67494A11*), and three genes encoding protochlorophyllide oxidoreductase (POR), which catalyzes the conversion of proto-chlorophyllide into chlorophyllide during Chl biosynthesis (i.e., *Traes_2BL_F22336B90, Traes_2AL_E0AC9DBC7,* and *Traes_2DL_3C229EB92*). Only one gene (i.e., *Traes_3AS_433192E29*) encoding chlorophyllase (Chlase) was upregulated in Y. Moreover, our comparison of the DEGs involved in carotenoid biosynthesis showed that two genes encoding β-carotene hydroxylase (BCH) (substrate: β-carotene; product: zeaxanthin) in the carotenoid biosynthesis pathway were significantly upregulated in Y (Figure 6B,C). The results indicate that photosynthetic pigment biosynthesis is important for the unusual phenotype of Y.

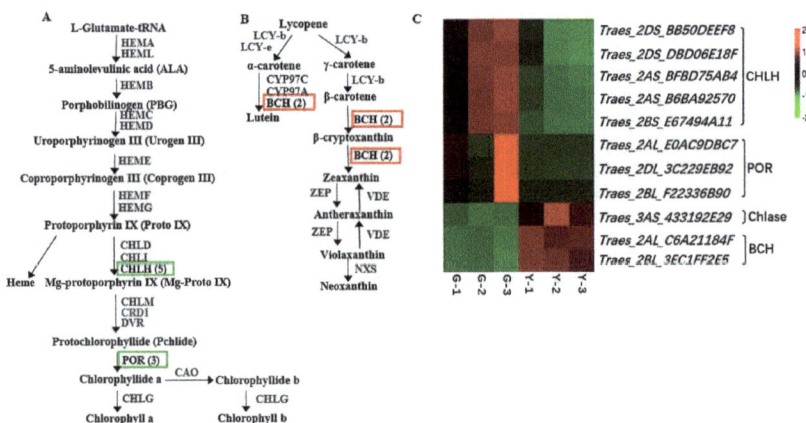

Figure 6. DEGs at the transcript level involved in chlorophyll and carotenoid biosynthesis pathways. (**A**) Chlorophyll biosynthesis pathway; (**B**) Carotenoid–xanthophyll biosynthesis pathway. In (**A,B**), upregulated genes are marked by red-line borders and downregulated genes by green-line borders. The numbers following each gene name indicate the number of corresponding DEGs identified in our database; (**C**) Expression profile clustering for chlorophyll and carotenoid biosynthesis pathways. Expression ratios are based on log$_2$ FPKM values (fragments per kilobase of transcript per million mapped reads), where each vertical column represents a sample (G-1, G-2, and G-3; Y-1, Y-2, and Y-3), and each horizontal row represents a single gene. CHLH, Mg-chelatase H subunit; POR, protochlorophyllide oxidoreductase; Chlase, chlorophyllase; BCH, β-carotene hydroxylase.

2.4. Identification of DEGs Related to Photosynthesis

The Chl content was closely related to photosynthesis. In this study, we identified five DEGs involved in photosynthesis, including two genes that encode photosystem II 47 kDa protein (PsbB), one gene that encodes photosystem II protein D2 (PsbD), one gene encode PsaC in photosystem I, and F-type ATPase β subunit, most of which were significantly downregulated in the Y type (Figure 7A,B). Moreover, most of the genes that mapped to the carbon metabolism and carbon fixation pathways in photosynthetic organisms exhibited decreased expression in Y compared with G, where the genes encoding ribulose-1,5-bisphosphate carboxylase/oxygenase (Rubisco), 6-phosphogluconolactonase (6PGL), glucose-6-phosphate 1-dehydrogenase (G6PDH), and 6-phosphogluconate dehydrogenase

(6PGD) were downregulated. However, two fructose-1,6-bisphosphatase genes were upregulated (Figure 7B). In addition, 18 genes related to early light-inducible proteins (ELIPs) were significantly upregulated in Y (Figure 7B). These results indicate that changes in the expression levels of these genes might have blocked photosynthesis, thereby influencing chlorophyll biosynthesis in the mutant Y type.

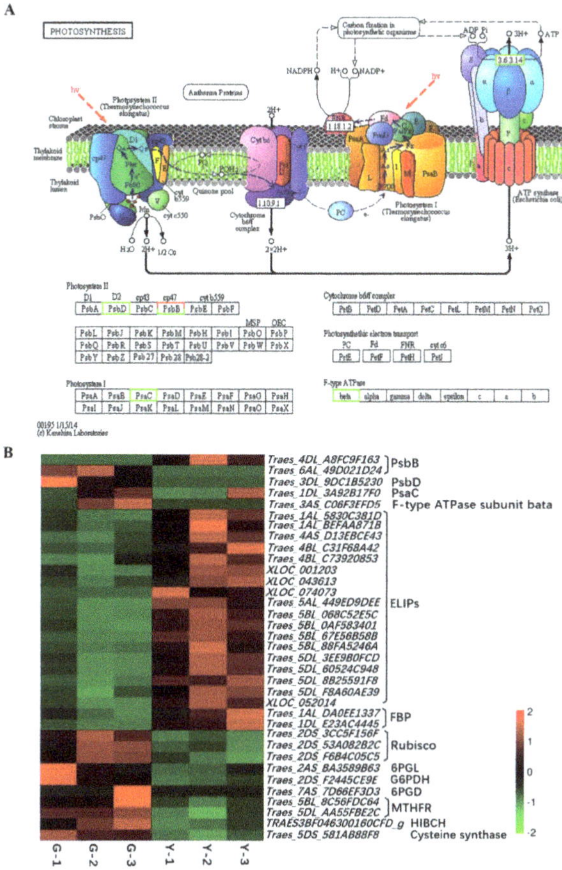

Figure 7. DEGs mapped onto the photosynthesis pathway. (**A**) Photosynthesis pathway. The image of the known photosynthesis pathway was obtained from the freely available KEGG database (http://www.kegg.jp/kegg-bin/show_pathway?ko00195). The green border denotes lower expression in Y compared with G, red color denotes higher expression, and half red/half green donates both up- and downregulated genes in Y compared to G. The blue dashed lines denote photosynthetic electron transport in the thylakoid membrane, red dashed lines denote light irradiation. The black dashed lines denote energy conversion of carbon fixation in photosynthetic organisms and the solid arrows denote molecular interaction or relation; (**B**) Expression profile clustering for genes involved in the photosynthesis and carbon metabolism pathway. Expression ratios are based on $\log_2$ FPKM values, where each vertical column represents a sample (G-1, G-2, and G-3; Y-1, Y-2, and Y-3), and each horizontal row represents a single gene. PsbB, photosystem II 47 kDa protein; PsbD, photosystem II protein D2; PsaC, photosystem I subunit VII; ELIPs, early light-inducible proteins; FBP, fructose-1,6-bisphosphatase; Rubisco, ribulose-1,5-bisphosphate carboxylase/oxygenase; 6PGL, 6-phosphogluconolactonase; G6PDH, glucose-6-phosphate 1-dehydrogenase; 6PGD, 6-phosphogluconate dehydrogenase; MTHFR, methylenetetrahydrofolate reductase; HIBCH, 3-hydroxyisobutyryl-CoA hydrolase-like protein.

2.5. Analysis of TFs and Heat Shock Proteins (HSPs) in Y and G Plants

Transcription factors are important regulators that can activate or repress the expression of genes in a sequence-specific manner, thereby affecting or controlling many biological processes. We found that the GO term "sequence-specific DNA binding transcription factor activity" was most significantly enriched among the entries in the molecular function category (Table S2). Therefore, we identified all of the TFs among the DEGs in Y and G using the plant transcription factor database (PlantTFDB) v4.0 (http://planttfdb.cbi.pku.edu.cn/). In total, 44 DEGs encoding putative TFs were identified and categorized into eight different common families (Table S4). The heat shock transcription factor (HSF) protein family was the most abundant (14 TFs, 32%), where all of exhibited downregulated expression in Y (Table S4), followed by the bZIP (12 TFs, 27%), WRKY (three TFs, 7%), ERF (three TFs, 7%), and GATA (three TFs, 7%) families (Figure 8).

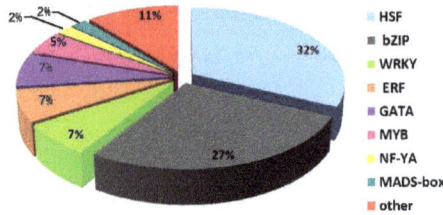

Figure 8. Percentages of different transcription factors involved in the "sequence-specific DNA binding transcription factor activity" GO term.

In order to obtain a more intuitive understanding of the roles of those downregulated HSFs, totally, seven HSF binding motifs were projected in the TFDB database (Figure S1). The binding motifs of these seven HSFs (Figure S1) were then searched to identify the cis-acting elements present in the promoters of the DEGs using FIMO software (version 4.9.0) [43]. The results showed that many genes interacted with these seven HSF genes (Table S5). Thus, we selected 37 genes related to leaf color variations as target nodes and genes annotated as HSFs as source nodes to generate an interaction network diagram using Cytoscape software (version 3.3.0) [44]. We found that genes encoding 30 HSPs (i.e., small HSPs (sHSPs) and HSP70) comprised the main group that interacted with the seven HSFs (Figure 9A). Interestingly, most of the genes annotated as heat shock proteins (HSPs) (e.g., sHSPs, HSP70, and HSP90) in the "protein processing in endoplasmic reticulum" pathway had significantly reduced mRNA levels in the Y type (Figure 9B,C and Table S6). In addition, we found that genes encoding ELIPs, CHLH, and BCH also interacted with the seven HSFs (Figure 9A). Thus, we suggest that downregulation of these seven HSFs may affect the regulation of leaf color formation.

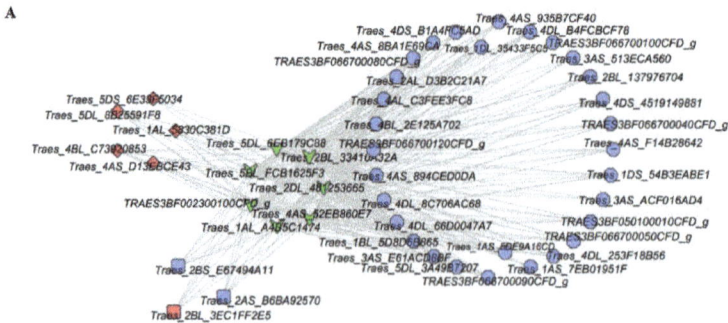

Figure 9. *Cont.*

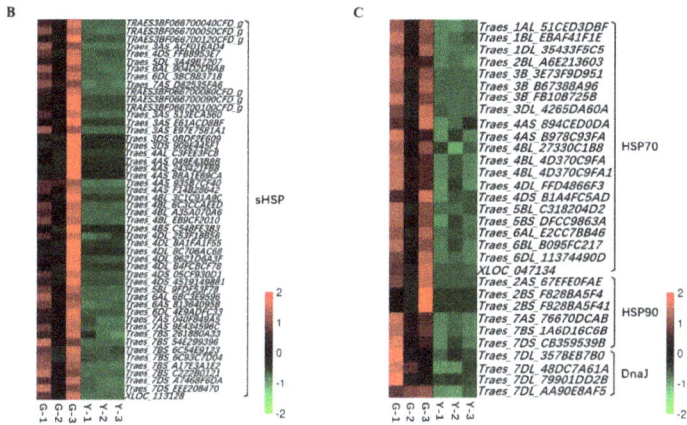

Figure 9. Gene interaction network diagrams and expression profile clustering for genes encoding heat shock proteins (HSPs). (**A**) Interactions between heat shock transcription factors (HSFs) and other genes. Green inverted triangles represent HSFs, HSFA6B, and HSFB2B. Pink rhombuses represent early light-inducible proteins (ELIPs). Blue circles represent HSPs. Blue squares represent CHLH. The pink square represents β-carotene hydroxylase (BCH); (**B,C**) Expression profile clustering for HSP encoding genes. Expression ratios are based on log$_2$ FPKM values, where each vertical column represents a sample (G-1, G-2, and G-3; Y-1, Y-2, and Y-3), and each horizontal row represents a single gene. sHSP, small heat shock protein; HSP70, heat shock cognate 70 kDa protein; HSP90, heat shock 90 kDa protein; DnaJ, chaperone protein DnaJ.

2.6. Quantitative Real-Time PCR (qRT-PCR) Analysis of Candidate Genes

To validate the reliability of the RNA-Seq data, 15 DEGs that we considered to have strong involvement with leaf color variation in Y were subjected to qRT-PCR analysis. The results showed that the expression patterns of the 14 genes detected by qRT-PCR were consistent with those in the transcriptome data, where only one gene (i.e., *Traes_4DL_A8FC9F163*) encoding PsbB had a different expression pattern (Figure 10). This discrepancy with the PsbB transcripts detected by RNA-Seq and qRT-PCR may have been caused by differences in the sensitivity of the method employed to analyze this gene. qRT-PCR is based on the specific amplification of single gene primers, and it has higher accuracy than RNA-Seq. Moreover, it is possible that this discrepancy is caused by low-expression levels of PsbB. Transcriptome results showed that *Traes_4DL_A8FC9F163* in G was almost no expression, and the expression level in Y was also low. Overall, most of the qRT-PCR data were consistent with the RNA-Seq results, thereby demonstrating the reliability of the RNA-Seq data.

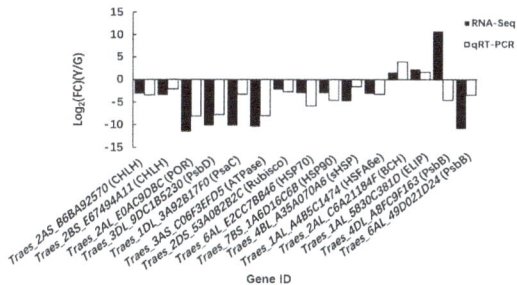

Figure 10. qRT-PCR validation of the RNA-Seq results for the candidate DEGs related to yellow leaf color formation in the Y type. Log$_2$(FC) represents the fold change in Y relative to that in G.

2.7. Comparison of Chl Precursors in Y and G Plants

Mg-chelatase is encoded by the *CHLD*, *CHLI*, and *CHLH* genes in higher plants, which catalyze the insertion of Mg^{2+} into Proto IX to form Mg-Proto IX in chlorophyll biosynthesis [45] (Figure 6A). To further investigate the effects of the downregulated expression of the *CHLH* gene, which is involved in chlorophyll biosynthesis in Y plants, seven intermediate products related to the metabolic process of Chl biosynthesis were compared in half-developed leaves from the Y and G types (i.e., H_Y and H_G, respectively, at the jointing stage). The 5-aminolevulinic acid (ALA), porphobilinogen (PBG), uroporphyrinogen III (Urogen III), coproporphyrinogen III (Coprogen III), and Proto IX contents of the half-developed leaves from Y plants were significantly higher than those from the G plants (Figure 11). In particular, the Proto IX contents (the substrate of Mg-chelatase) were two times higher in Y than G plants. However, the Mg-Proto IX (the product of Mg-chelatase), Pchlide, Chl *a*, and Chl *b* contents, decreased significantly in Y plants compared with G plants (Figure 11). These results indicate that the downregulated expression of the *CHLH* gene decreased the Mg-chelatase activity and reduced the production of Chl *a* and Chl *b* in Y plants.

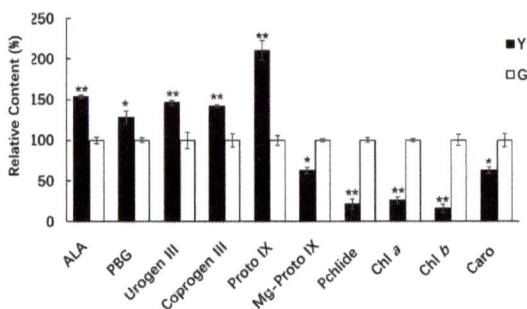

Figure 11. Comparison of the relative contents of chlorophyll precursors, chlorophyll, and carotenoids in H_Y and H_G leaves at the jointing stage. Three individuals were measured for each chlorophyll and chlorophyll precursor. Each plant was extracted once, and the chlorophyll contents were measured three times. Error bars indicate means ± SD based on three independent experiments. Significant differences were determined using the Student's *t*-test in Y compared with G plants (* $p < 0.05$, ** $p < 0.01$). ALA, 5-aminolevulinic acid; PBG, porphobilinogen; Urogen III, uroporphyrinogen III; Coprogen III, coproporphyrinogen III; Proto IX, protoporphyrin IX; Mg-Proto IX, Mg-protoporphyrin IX; Pchlide, protochlorophyllide; Chl *a*, chlorophyll *a*; Chl *b*, chlorophyll *b*; Caro, carotenoid.

2.8. Analysis of the Carotenoid Composition in Y and G Plants by HPLC

Our results showed that the total carotenoid level was significantly reduced in Y (Figure 11), but upregulated expression of the *BCH* gene related to carotenoid biosynthesis was also identified (Figure 6B,C). To further confirm the involvement of *BCH* in carotenoid-xanthophyll biosynthesis and the yellow leaf color phenotype in Y plants, we quantitatively analyzed five carotenoid compounds (i.e., lutein, zeaxanthin, β-cryptoxanthin, α-carotene, and β-carotene) in the half-developed leaves from the Y and G types using high performance liquid chromatography (HPLC), and the results are shown in Figure 12 and Table 2. The standard curve for carotenoids is shown in Table S7. β-carotene and lutein were the major carotenoids found in G plants, and a very small amount of zeaxanthin was also detected (Figure 12). Yellow plants differed in terms of the composition and levels of carotenoids. Consistent with the increased expression of *BCH* genes, the zeaxanthin level was significantly higher (more than 40 times) in Y, and it was the most abundant carotenoid, whereas the β-carotene and lutein levels were significantly lower than those in G plants (Figure 12 and Table 2). The β-cryptoxanthin level did not differ significantly and α-carotene was below the detection limit in G and Y (Figure 12 and Table 2). In addition, violaxanthin and neoxanthin were identified in Y and G according to individual peaks in the absorption spectra, but their levels were significantly lower in Y (Figure 12). Overall, our results showed that differences in the carotenoid composition could contribute to the yellow leaf phenotype in Y.

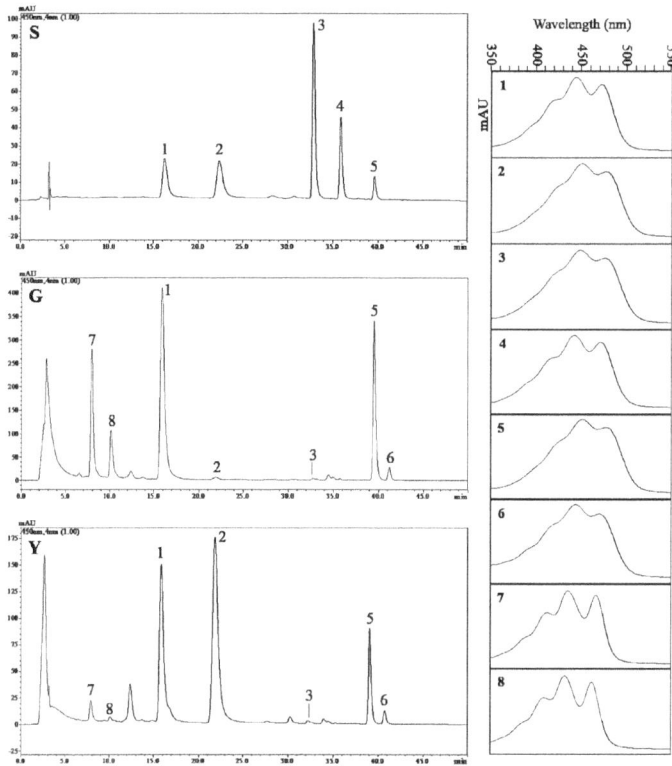

Figure 12. High performance liquid chromatography (HPLC) elution profiles for carotenoids accumulated in G and Y leaves at 450 nm. (**S**) HPLC elution profiles for five carotenoid standards. (**G**) HPLC elution profiles for carotenoids accumulated in G leaves. (**Y**) HPLC elution profiles for carotenoids accumulated in Y leaves. The vertical axis shows the absorbance (mAU) at 450 nm, and the horizontal axis represents the retention time for carotenoids. The right panel is the absorption spectra from peak 1 through 8 at 450 nm in Y and G types. The vertical axis shows the mAU, and the horizontal axis represents absorption wavelength. Peak 1, lutein (absorption peak λmax nm: 444, 472); peak 2, zeaxanthin (450, 476); peak 3, β-cryptoxanthin (451, 476); peak 4, α-carotene (445, 473); peak 5, β-carotene (452, 477); peak 6, 9-cis-β-carotene (446, 472); peak 7, violaxanthin (416, 438, 468); peak 8, neoxanthin (413, 435, 463).

Table 2. Comparison of the leaf carotenoids contents (μg/g Fresh Weight) in G and Y at the jointing stage according to HPLC.

Phenotype	Lutein	Zeaxanthin	β-Cryptoxanthin	α-Carotene	β-Carotene
G	366.60 ± 23.26	3.89 ± 0.66	0.85 ± 0.24	ND	494.94 ± 45.00
Y	140.18 ± 9.43 *	157.62 ± 15.63 **	0.56 ± 0.08	ND	105.73 ± 29.75 *

The carotenoids contents are expressed as the mean ± SD based on three independent replications. Asterisks indicate significant differences in the compound measured in the Y type compared with the G type (*t*-test, *n* = 3, *: $p < 0.05$; **: $p < 0.01$). ND denotes not detected or below the detection limit for HPLC.

3. Discussion

Leaf color formation is closely related to chloroplast development and photosynthetic pigments, and it is important for photosynthesis. Recently, many leaf color mutants have been identified in higher plants and they are valuable materials for investigating the biosynthesis of photosynthetic pigments and selective breeding for high photosynthetic efficiency. *Ygm* is a spontaneous yellow-green leaf color mutant of the cultivar Xinong1718 in common wheat, where it is an incomplete dominant semidominant mutant [42]. The dominant homozygotes (Y type) of *Ygm* have yellow leaves, whereas the heterozygotes (Yg type) have yellow-green leaves. The recessive homozygotes (G type) of *Ygm* are normal green. In this study, the G and Y types among the progeny of the *Ygm* mutant were selected as the materials, which we subjected to integrated biochemical analysis and transcriptome profiling to obtain insights into the differences in gene regulation and complex biological processes in G and Y.

3.1. Yellow Leaf Phenotype is Closely Associated with Chl and Carotenoid Pigment Metabolism

Leaf color variations are determined by complex biological processes. The yellow leaf color mainly depends on the Chl and carotenoid contents. Chlorophylls are essential molecules for harvesting solar energy in photosynthetic antenna systems, as well as for charge separation and electron transport within the reaction centers, where they are present in the thylakoid membrane in the form of a pigmented protein complex [2]. Comparative transcriptome profiling data for the Y type and G type among the *Ygm* mutant progeny showed that the expression of *CHLH* in Y plants was significantly downregulated compared with that in G plants (Figure 6A,C). Mg-chelatase comprises three subunits (i.e., CHLH, CHLI, and CHLD) which catalyze the insertion of Mg^{2+} into Proto IX to form Mg-Proto IX in chlorophyll biosynthesis [45]. CHLH is crucial for the Mg-chelatase activity as a catalytic subunit [19]. It has been reported that mutation of the CHLH gene leads to defective Chl and the chlorina or yellow phenotype in rice [8] and *Arabidopsis thaliana* [20]. Our analysis of seven intermediate products involved in Chl biosynthesis showed that the level of the substrate for Mg-chelatase increased significantly in Y plants whereas the level of the product from Mg-chelatase decreased significantly (Figure 11). Thus, our results indicate that the downregulated expression of the *CHLH* gene in Y plants may have decreased the Mg-chelatase activity and further reduced the production of Chl *a* and Chl *b* in Y plants.

Chl metabolism is a highly coordinated process, which is catalyzed by numerous enzymes. In addition to CHLH mentioned above, POR catalyzes the photoreduction of protochlorophyllide to chlorophyllide in Chl biosynthesis, which is a light-dependent enzyme that is present in all oxygen-producing photosynthetic organisms [46]. In our study, three genes encoding POR were significantly downregulated in Y compared with G (Figure 6A,C). Previous studies of *Arabidopsis* and rice have suggested that the Chl *a* content is directly related to the expression levels of *PORB and PORC*, and when *PORB* and *PORC* are absent in plants, the Chl *a* content is decreased and the thylakoid is not stacked [47,48]. Moreover, the *pgl10* mutant in rice has phenotypically pale-green leaves with significantly decreased Chl (Chl *a* and Chl *b*) and carotenoid contents, and less grana lamellae. Bioinformatics analysis indicates that *PGL10* encodes PORB [49]. Those results suggest that POR was inhibited in the Y type, which might have reduced the Chl content and led to the yellow leaf phenotype. Moreover, we found that one gene encoding Chlase in the Chl biosynthesis pathway was significantly upregulated in Y plants (Figure 6C). Chlase is considered to be a key enzyme in chlorophyll degradation [50,51]. However, some evidence does not support the involvement of Chlase in chlorophyll breakdown during leaf senescence [52–54]. For example, in *Arabidopsis*, *AtCHL1* and *AtCHL2* encode Chlase, but *CHL1* and *CHL2* single and double knockout mutants are still able to degrade chlorophyll during senescence [52]. Similarly, overexpression of the Chlase-encoding gene *ATHCOR1* in *Arabidopsis* leads to the increased breakdown of Chl *a*, but there are no substantial changes in the total amount of Chl [55]. In fact, Chlase has been shown to participate in chlorophyll breakdown in ethylene-treated citrus fruit, as well as in the tissue damage responses to fungi and bacteria [56–58]. The discovery of the involvement of pheophytin pheophorbide hydrolase (PPH) in early chlorophyll breakdown during leaf senescence [59] has made the biological role of Chlase controversial. Thus, the function and involvement of Chlase in Chl catabolic processes needs to be investigated further.

Carotenoids are essential components of the photosynthetic apparatus and photoprotection system, and their biosynthesis is coordinated with that of Chls in the chloroplasts [60]. In this study, HPLC analysis showed that the β-carotene and lutein levels were significantly lower in Y plants, whereas the zeaxanthin levels were more than 40 times higher than those in G plants (Figure 12 and Table 2). The changes in the abundances of carotenoids were accompanied by the altered expression of carotenoid biosynthesis genes [61]. β-carotene hydroxylase (BCH) is mainly responsible for the β-ring hydroxylation of β-carotene to produce zeaxanthin, and its activity overlaps slightly with the hydroxylation of the β-ring of α-carotene. [62]. In potato tubers, silencing the β-carotene hydroxylase genes *CHY1* and *CHY2* increases the levels of β-carotene and total carotenoids by up to 38 and 4.5 times, respectively, but reduces that of zeaxanthin [63]. By contrast, downregulating genes that encode BCH increases the β-carotene contents of sweet potato and the maize endosperm [64,65]. In our study, the expression levels of two genes that encode BCH were upregulated in Y plants (by ~1.5 and 1.8 time, respectively) (Figure 6B,C), which presumably enhanced the conversion of β-carotene into zeaxanthin, thereby increasing the zeaxanthin content and decreasing the β-carotene content. However, the increased zeaxanthin content was not converted into vioxanthin and neoxanthin, and their contents actually decreased significantly in Y plants (Figure 12). Zeaxanthin and violaxanthin are involved in the xanthophyll cycle, which is the main mechanism for photoprotection [66]. When plants are exposed to light stress, the photosynthetic organs are damaged and the conversion of violaxanthin into zeaxanthin is increased to protect against further light damage [67]. We suggest that the increase in zeaxanthin and the decrease in violaxanthin were related to damage to the photosynthetic system in Y plants, which enhanced the conversion of violaxanthin into zeaxanthin. We also found that the accumulation of lutein did not accompany the increased expression of *BCH*, and the lutein content actually decreased greatly in Y. This phenomenon has also been reported in transgenic tobacco, and it may be due to the limited effect of BCH on lutein [68]. In addition, significant reductions in the lutein content were also observed in previous studies in yellow-green tea mutants ZH1 and temperature-sensitive mutant Anji Baicha in the yellow-green stage [69,70].

3.2. Yellow Leaf Phenotype Affected by the Expression of Genes Related to Photosynthesis

Yellow leaf color formation is related to chloroplast development, where chloroplasts comprise the chloroplast membrane, thylakoid, and matrix, which are the main sites for photosynthesis. In higher plants, the multi-subunit pigment–protein complexes (i.e., photosystem (PS)I, PSII, light harvesting complexes, cytochrome b6/f, and ATP synthase) are embedded in the highly folded thylakoid membrane where they are responsible for light absorption and energy transfer [71,72]. The photosynthetic light reaction occurs in the thylakoid in chloroplasts, and thus we suggest that changes in gene expression might have been related to thylakoid development and photosynthesis in the two progeny of *Ygm*. Similar to the photosynthesis pathway, genes encoding PsbB (the transcriptome data indicated one upregulated gene and one downregulated gene, whereas both were downregulated according to qRT-PCR), PsbD, photosystem I subunit VII, and F-type ATPase β subunit were significantly repressed in Y plants (Figure 7). These results were consistent with our previous chloroplast ultrastructure analysis, which demonstrated that yellow leaf color formation is greatly affected by abnormal chloroplast development [42]. In addition, PSII is distributed mainly over the overlapping regions of the grana lamella [73], and the poorly stacked grana in Y might have been associated with the dramatic downregulation of the PSII protein complex [10,74,75]. In higher plants, ELIPs are light stress-induced, Chl *a*/*b*-binding proteins that accumulate in the thylakoid membranes, where their proposed function is in photoprotection [76–78]. The increased accumulation of ELIP transcripts and proteins is correlated with photodamage in the PSII reaction centers [79]. Overexpression of the *ELIP2* gene in *Arabidopsis* downregulates the level and activity of glutamyl tRNA reductase, CHLH, and CHLI, thereby reducing the accumulation of Chl and photosystems assembled in the thylakoid membranes [80]. In yellow leaves, we found that the accumulation of *ELIP* mRNAs was 2–4 times higher compared with that in green leaves (Figure 7B), which suggests that the yellow phenotype may be related to the upregulation of ELIP transcripts. Rubisco is a rate-limiting enzyme that participates in photosynthetic carbon fixation and it is a potential target for genetic manipulation to increase crop yields [81,82]. Albino or pale green phenotypes are observed in *Arabidopsis* transgenic lines due to the co-suppression of *Arabidopsis* Rubisco small subunit gene *RBCS3B*, where among these lines, *RBCS3B-7* exhibits abnormal thylakoid stacking and it is light sensitive under normal light [83]. Furthermore, the abundance of Rubisco is lower in *Brassica napus* and a wheat anther culture Chl-deficient mutant [30,74]. Similarly, we found that genes encoding the Rubisco small chain were significantly downregulated in Y in our study (Figure 7B). The pentose phosphate metabolic (PPP) pathway is part of the carbon metabolism pathway and some of the PPP intermediates such as ribulose-5-phosphate (Ru5P) are required for the synthesis of nucleotides, and they are shared in the carbon fixation pathways of photosynthetic organisms [84,85]. In the carbon metabolism pathway, three genes encoding glucose-6-phosphate 1-dehydrogenase (G6PDH), 6PGL, and 6PGD in the oxidative phase of PPP (product: Ru5P and DADPH) were downregulated in Y (Figure 7B). Thus, the altered expression of genes related to carbon metabolism and photosynthesis in Y may have caused abnormal chloroplast development and decreased the Chl content.

3.3. HSFs, HSPs, and Chloroplasts

In this study, all of the HSFs and HSPs had decreased expression levels in Y (Table S4 and Figure 9B,C). HSPs are among the most abundant protective proteins in plants where there are five conserved protein families: HSP100s, HSP90s, HSP70s, HSP60s, and sHSPs [86]. These HSPs act as molecular chaperones that participate in protein folding, assembly, and the prevention of irreversible protein aggregation to maintain cell homeostasis [87–89]. Therefore, HSPs play key roles in plant development and stress resistance processes. HSP-encoding genes are regulated by HSFs by specifically binding highly conserved characteristic palindromic sequence (5′-AGAAnnTTCT-3′) in the promoters of many HSP genes, thereby causing the accumulation of HSPs [90–92]. Seven HSFs target gene prediction analyses demonstrated that a large number of HSP genes (sHSPs and HSP70) as well as some photosynthesis and Chl biosynthesis related genes were target genes regulated by those seven HSFs (Figure 9A).

Some sHSP family members play roles in chloroplast development and photosynthesis under heat stress in many species. For example, after silencing *sHSP26* in maize, the abundances of four chloroplast proteins comprising ATP synthase subunit β, Chl *a/b* binding protein, oxygen-evolving enhancer protein 1, and photosystem I reaction center subunit IV declined greatly under heat stress [93]. The overexpression of *Oshsp26* in tall fescue also enhanced the photochemical efficiency of PSII (*Fv/Fm*) during heat stress [94]. In *Arabidopsis*, the cooperation between HSP21 and pTAC5 is required for chloroplast development under heat stress [95]. These results suggest that the inhibition of sHSPs in Y plants may be caused defective chloroplasts and led to the yellow leaf phenotype. In addition, the involvement of HSP70s with chloroplast development has been observed in *Arabidopsis* and rice. For example, suppressing the HSP70s homologues *cpHsc70-1/cpHsc70-2* double genes in *Arabidopsis* resulted in a white and stunted phenotype, and the chloroplasts in these plants had an unusual morphology with few or no thylakoid membranes [96]. The phenotype of the T-DNA inserted heat-sensitive rice mutant *OsHsp70CP1* varies with temperature, where a severe chlorotic phenotype and lower Chl contents are found in the leaves under a constant high temperature (40 °C), whereas plants grown at a constant temperature of 27 °C have a normal phenotype [97]. Furthermore, HSP70 has been implicated in photoprotection and the repair of PSII during and after photoinhibition [98]. DnaJ proteins are chaperones in the HSP40 family, where the J domain is generally used as a DnaJ co-chaperone to activate the HSP70 ATPase domain to allow stable substrate binding and the release of HSP70 [99,100]. In addition, DnaJ proteins are essential for normal chloroplast development. Silencing of DnaJ encoded gene *OsDjA7/8* in rice resulted in an albino lethal phenotype in the seedling stage due to disordered development of chloroplast [101]. The detection of HSPs in our transcriptome data has further proven that the expression levels of *HSPs* are closely related to chloroplast development and Chl biosynthesis in plant species.

Chloroplasts are semi-autonomous organelles and a complex network of regulatory signals exists between the nucleus and plastids. Plastid retrograde signaling is mediated by the tetrapyrrole intermediate Mg-Proto IX, and its methylester (Mg-Proto IX-ME) was identified in an *Arabidopsis* genome-uncoupled mutant (*GUN5*) that encodes the plastid-localized CHLH [20,102,103]. Mg-Proto IX can replace the action of light by inducing two nuclear HSP genes (*HSP70A* and *HSP70B*) [104]. In the present study, the Mg-Proto IX content and the expression of *HSP70* were significantly decreased in Y (Figures 9C and 11), thereby indicating the possible involvement of Mg-Proto IX in the expression of HSP genes. Furthermore, in the photosynthetic system, the responses of HSPs to plastid retrograde signaling have important roles in regulating the expression of nuclear genes involved in photosynthesis. Kindgren et al. [105] found that HSP90 proteins respond to the GUN5-mediated plastid signal to control the expression of photosynthesis-associated nuclear genes (*PhANG*) during the response to oxidative stress. Recently, it was shown that *Arabidopsis* chloroplast HSP21 is activated by the GUN5-dependent retrograde signaling pathway to maintain the stability of the PSII complex and thylakoid membranes under high temperature stress by directly binding to its core subunits, such as D1 and D2 proteins [106]. These results suggest that the responses of HSPs and Mg-Proto IX to plastid signaling might have important roles in photosynthesis.

4. Materials and Methods

4.1. Plant Materials

Ygm was produced from a spontaneous leaf color mutant in the winter wheat cultivar Xinong1718 following 14 generations of self-pollination and direct selection for the yellow-green phenotype in each generation. The progeny of the *Ygm* mutant exhibited three leaf color phenotypes (i.e., yellow leaf plants (Y), yellow-green leaf plants (Yg), and normal green leaf plants (G)) (Figure 1). Yellow leaf plants die after the flowering stage and do not produce seeds, so this genotype was maintained by sowing seeds from the Yg plants each year. Yg plants are similar to the wild type Xinong1718 in terms of their growth period and plant height, but the yield capacity is significantly lower than that of Xinong1718. Green leaf plants are similar to the wild type Xinong1718 in terms of their growth period, plant height, and yield capacity [42]. In total, 66 plant line populations (20 and 46 derived from G and Yg plants, respectively) were sown on 2 October 2015, and 125 seeds from each plant line population were used in a single plot with a spacing of 30 cm between the rows and 8 cm between the plants, where each row measured 2 m. All of the experimental materials were grown in an experimental field at Northwest A&F University, Yangling, China, according to the standard practices employed in the local area.

The colors of the young leaves and leaf developmental states in Y and G were observed visually in the jointing stage in the field (Figure 1A). Three types of young leaves were present in the Y and G seedlings (i.e., fully-developed leaves (F_Y and F_G, respectively), half-developed leaves (H_Y and H_G, respectively), and small leaf buds (L_Y and L_G, respectively)) (Figure 1C,D). We only collected H_Y and H_G in the jointing stage from the Y and G seedlings, respectively, to measure the concentrations of Chl precursors and carotenoids compounds, as well as for transcriptome sequencing analysis with the Illumina HiSeq™ 2500 sequencing system (Illumina, San Diego, CA, USA).

4.2. RNA Extraction, Library Construction, and RNA Sequencing

For transcriptome analysis, H_Y and H_G leaves were collected from Y and G seedlings at the jointing stage (31 March 2017) from 08:00 am to 10:00 am (Figure 1C,D). Six samples from three biological replicates of G and Y were used to construct cDNA libraries designated as G-1, G-2, G-3, Y-1, Y-2, and Y-3, respectively. The samples were frozen immediately in liquid nitrogen and stored at −80 °C until RNA extraction. Total RNA was extracted using Trizol Reagent (Invitrogen Life Technologies, Shanghai, China) and treated with RNase-free DNase I (TaKaRa, Dalian, China) according to the manufacturer's instructions. The quality of the total RNA was confirmed with a NanoDrop ND1000 spectrophotometer (Thermo Scientific, Wilmington, DE, USA) coupled with 1% agarose gel electrophoresis. The RNA integrity value (>8.0) was also verified using an Agilent 2100 Bioanalyzer (Agilent Technologies, Santa Clara, CA, USA). The cDNA library construction and sequencing of six RNA samples were completed by Guangzhou GENE DENOVO Biotechnology Co., Ltd. (Guangzhou, China).

4.3. Sequence Alignment and Functional Annotation

In order to obtain clean data, it is essential to remove adaptor sequences, more than 10% of the unknown nucleotides, and low quality reads with more than 50% of low quality (q-value ≤ 20) bases. After removing rRNA using the short reads alignment tool Bowtie2 (version 2.2.9) [107], the high-quality clean reads were then mapped to the wheat reference genome sequences in the URGI database (http://wheat-urgi.versailles.inra.fr/Seq-Repository) by TopHat2 (version 2.0.3.12) [108]. In order to identify new genes and new splice variants of known genes, the transcripts were reconstructed using Cufflinks (version 2.2.1) [109] based on reference annotation based transcripts (RABT). Cufflinks constructed faux reads according to references to compensate for the influence of low coverage sequencing. During the last assembly step, all of the reassembled fragments were aligned with reference genes and similar fragments were then removed. Cuffmerge was then employed to combine the assembly results for three samples of each biological duplicate for Y and G for further downstream differential expression analysis. For functional annotation, all of the transcripts including new gene transcripts (≥ 200 bp and exon number >2) were annotated using the BLASTx function with protein databases, including the NCBI Nr protein database (https://ftp.ncbi.nlm.nih.gov/blast/db/FASTA/), GO (http://www.geneontology.org/), and KEGG (http://www.genome.jp/kegg/kegg2.html) databases with a significance threshold of E value $< 10^{-5}$. GO annotation was conducted using Blast2GO (vision 2.5) software [110] and WEGO software (version 2.0) was then applied for gain GO function classification [111]. KEGG pathway analyses were conducted using the KEGG Automatic Annotation Server (KAAS) (http://www.genome.jp/tools/kaas/).

4.4. DEG Analysis

The FPKM values were used as a measure of normalized gene expression, and significance tests of differences in gene expression in Y and G (each with three biological replicates) were performed using the edgeR package (http://www.bioconductor.org/packages/release/bioc/html/edgeR.html). FDR < 0.05 and $|\log_2$ (fold change)$| \geq 1$ were used to assess the significance of differences in gene expression. Hierarchical clustering of the DEGs was performed using the OmicShare tools, which is a free online platform for data analysis (http://www.omicshare.com/tools).

The DEGs were then annotated with the COG database to predict and classify possible functions using BLASTx (E value $< 10^{-5}$). GO enrichment analysis of DEGs was implemented using the GOseq R package (Bioconductor version: release (3.7)) [112]. The enrichment of the DEGs in KEGG pathways was tested using the KOBAS software (version 2.0) [113]. GO terms and KEGG pathways with corrected q-values < 0.05 were considered significantly enriched for DEGs. The correlations between biological repeats in Y and G were expressed as Pearson's correlation coefficients.

4.5. qRT-PCR Analysis

In order to validate the results of the RNA-Seq and DEG analyses, the color-related DEGs were selected for qRT-PCR with specific primers designed using Primer Premier 5.0 software (Table S8). cDNA was synthesized according to the manufacturer's instructions using a PrimeScript™ RT reagent kit with gDNA Eraser (TaKaRa, Dalian, China). RNA (1 µg) extracted from the half-developed leaves was used as the template. qRT-PCR was performed using a SYBR Premix Ex Taq™ II Kit (TaKaRa,

Dalian, China) according to the manufacturer's instructions with a QuantStudio® 7 Flex Real-Time PCR system (Applied Biosystems, Shanghai, China). Wheat 18S rRNA was used as an internal control for normalization [114]. The amplification efficiencies of the primers were checked based on standard curve analysis. Each of the reactions was performed in triplicate. Dissociation curve analysis was performed after each assay to determine the target specificity. Relative gene expression levels were calculated according to the $2^{-\Delta\Delta Ct}$ comparative C_T method [115].

4.6. Determination of Photosynthetic Pigments and Chl Precursors

The Chl *a*, Chl *b*, and total carotenoid contents were determined using an UV-1800 spectrophotometer (Shanghai Mapada Instruments Co. Ltd., Shanghai, China) at 645, 663, and 470 nm according to the method described by Lichtenthaler [116]. The chlorophyll biosynthesis pathway is shown in Figure 6A. Seven precursors of chlorophyll biosynthesis were examined. The ALA contents were extracted and determined as described by Dei [117]. PBG, Urogen III, and Coprogen III were quantified as described by Bogorad [118]. Proto IX, Mg-Proto IX, and Pchlide were extracted according to the methods described by Rebeiz et al. [119]. The Proto IX, Mg-Proto IX, and protochlorophyllide contents were measured with a Hitachi F-4500 fluorescence spectrometer (Hitachi Instrument (Shanghai) Co., Ltd., Shanghai, China). The wavelengths used for detecting each porphyrin were as follows: Proto IX = excitation (Ex) 400 nm and emission (Em) 633 nm; Mg-Proto IX = Ex 440 nm and Em 595 nm; and protochlorophyllide = Ex 440 nm and Em 640 nm. Three individual plants were measured from the Y and G seedlings. The H_Y and H_G leaves from each plant in the jointing stage were extracted once and each sample was measured three times. Statistical analyses were performed using Microsoft Excel 2016 (Microsoft China, Beijing, China) with the one-way ANOVA test. The concentrations of pigments and Chl precursors in the G seedlings were set to 1, and the relative values for pigments and Chl precursors in the Y samples were expressed as fold changes relative to those in the G type samples.

4.7. Isolation and HPLC Analysis of Carotenoid Compounds

Carotenoids were extracted from half-developed wheat leaves from Y and G plants at the jointing stage according to the method described by Norris et al. [120] with appropriate modifications. All of the extraction procedures were conducted on ice with shielding from strong light. Briefly, 2 g of the fresh leaves were ground into a powder with liquid nitrogen and saponification was performed by adding 8 mL 20% *w/v* KOH and methanol. The homogenates were then transferred into 50 mL centrifuge tubes and heated at 60 °C for 30 min in darkness. After cooling to room temperature, each sample was ultrasonicated with 20 mL acetone:ethyl acetate (*v:v* = 2:1) for 40 min at 35 °C. The extract was then centrifuged at 8000× *g* at 4 °C for 5 min. The supernatant was then transferred to a fresh tube and concentrated using a nitrogen blowing instrument. The dried extract was re-suspended in 10 mL of acetone and ethyl acetate (*v:v* = 2:1) and then filtered through a 0.22 μm organic membrane for HPLC analysis.

Carotenoids were separated by reverse-phase HPLC analysis on a YMC C_{30} carotenoid column (150 mm × 4.6 mm, 3 μm) (YMC Co. Ltd., Shanghai, China) using a Shimadzu LC-20A HPLC system (Shimadzu, Tokyo, Japan). HPLC separation employed (A) methanol:acetonitrile (*v:v* = 3:1) and (B) methyl *tert*-butyl ether with a gradient of: 0–5 min, 0% B; 5–30 min, 0–35% B; 30–40 min, 35–45% B; 40–50 min, 45–0% B. The flow rate was 1 mL/min and the column temperature was 25 °C. The detection wavelength was 450 nm. β-carotene and zeaxanthin analytical standards were purchased from Sigma–Aldrich (Shanghai, China). β-cryptoxanthin and lutein analytical standards were purchased from Extrasynthese (Lyon, France). The α-carotene analytical standard was purchased from CaroteNature (Lupsingen, Switzerland). Each standard (1 mg) was dissolved in 1 mL of dimethyl sulfoxide. In order to establish a standard curve, mixed standard solutions of 4 μg/mL, 2 μg/mL, 1 μg/mL, 0.5 μg/mL, and 0.25 μg/mL were then prepared by diluting the stock solutions. Carotenoids were identified based on their retention time relative to known standards and absorption spectra for individual peaks compared with published spectra. Individual carotenoids were quantified based on the individual peak areas by using the standard curves and expressed as µg/g by fresh weight. Three biological replicates were analyzed in Y and G plants, and the carotenoid content was expressed as the mean ± SD based on three independent determinations.

5. Conclusions

In this study, transcriptome sequence analysis and physiological characterization were performed to identify the major molecular mechanisms related to leaf color variation in the mutant progeny of wheat *Ygm*. The transcriptome profiles differed considerably between Y and G, where various genes and pathways associated with yellow leaf formation were identified, including photosynthetic pigment synthesis, photosynthesis, and carbon fixation pathways. In addition, HSPs were shown to have important functions in response to plastid retrograde signaling and chloroplast development. Genes that interact with HSFs were shown to be associated with Chl biosynthesis. The measurements of Chl precursors indicated that the Y phenotype probably exhibited inhibited Chl biosynthesis due to the reduced activity of Mg-chelatase that is caused by the downregulation of *CHLH*. Moreover, the changes in the abundances of carotenoid composition may be associated with the yellow leaf phenotype. Overall, we speculated that the possible formation mechanism of yellow leaf phenotype in Y, which was shown in Figure 13. Our results provide new insights into the molecular mechanisms of yellow leaf formation in common wheat and they may facilitate selective breeding for high photosynthetic efficiency.

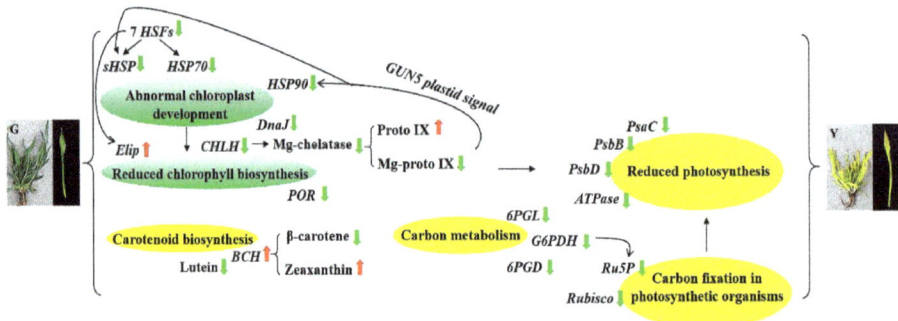

Figure 13. Possible formation pathway of yellow leaf phenotype of Y mutant. The red arrow indicates upregulated expression and the green arrow indicates downregulated expression. The green ovals indicate chlorophyll biosynthesis and chloroplast development. The yellow ovals indicate carotenoid biosynthesis, photosynthesis and energy metabolism.

Supplementary Materials: Supplementary materials can be found at www.mdpi.com/1422-0067/19/6/1594/s1

Author Contributions: L.Z. and H.W. conceived the original screening and research plans; L.Z. supervised the experiments; H.W. performed most of the experiments, analyzed the data, and prepared the figures and tables; L.C., N.S., C.L., and X.A. provided technical assistance to H.W.; H.F. provided help and guidance in carotenoid measurement experiment. Y.F. and D.S. advised on the analysis and interpretation of the results. H.W. wrote the article with contributions from all the authors. All authors approved the final manuscript.

Acknowledgments: This work was sponsored by the National Sci-Tech Support Foundation of China (Grant No. 2013BAD01B02) and the Zhonging Tang Crop Breeding Foundation of Northwest A&F University. We thank Guangzhou GENE DENOVO Biotechnology Co., Ltd. (Guangzhou, China) for their assistance with data processing and bioinformatics analysis.

Conflicts of Interest: The authors declare no conflict of interest.

Int. J. Mol. Sci. **2018**, *19*, 1594

References

1. Makino, A. Photosynthesis, grain yield, and nitrogen utilization in rice and wheat. *Plant Physiol.* **2011**, *155*, 125–129. [CrossRef] [PubMed]

2. Tanaka, A.; Tanaka, R. Chlorophyll metabolism. *Curr. Opin. Plant Biol.* **2006**, *9*, 248–255. [CrossRef] [PubMed]

3. Nagata, N.; Tanaka, R.; Satoh, S.; Tanaka, A. Identification of a vinyl reductase gene for chlorophyll synthesis in *Arabidopsis thaliana* and implications for the evolution of Prochlorococcus species. *Plant Cell* **2005**, *17*, 233–240. [CrossRef] [PubMed]

4. Li, W.; Tang, S.; Zhang, S.; Shan, J.; Tang, C.; Chen, Q.; Jia, G.; Han, Y.; Zhi, H.; Diao, X. Gene mapping and functional analysis of the novel leaf color gene SiYGL1 in foxtail millet [*Setaria italica* (L.) P. Beauv]. *Physiol. Plant* **2015**, *157*, 24–37. [CrossRef] [PubMed]

5. Wu, Z.; Zhang, X.; He, B.; Diao, L.; Sheng, S.; Wang, J.; Guo, X.; Su, N.; Wang, L.; Jiang, L.; et al. A chlorophyll-deficient rice mutant with impaired chlorophyllide esterification in chlorophyll biosynthesis. *Plant Physiol.* **2007**, *145*, 29–40. [CrossRef] [PubMed]

6. Zhu, X.; Shuang, G.; Zhongwei, W.; Qing, D.; Yadi, X.; Tianquan, Z.; Wenqiang, S.; Xianchun, S.; Yinghua, L.; Guanghua, H. Map-based cloning and functional analysis of YGL8, which controls leaf colour in rice (*Oryza sativa* L.). *BMC Plant Biol.* **2016**, *16*, 134. [CrossRef] [PubMed]

7. Jung, K.-H.; Hur, J.; Ryu, C.-H.; Choi, Y.; Chung, Y.-Y.; Miyao, A.; Hirochika, H.; An, G. Characterization of a Rice Chlorophyll-Deficient Mutant Using the T-DNA Gene-Trap System. *Plant Cell Physiol.* **2003**, *44*, 463–472. [CrossRef] [PubMed]

8. Zhao, S.; Long, W.; Wang, Y.; Liu, L.; Wang, Y.; Niu, M.; Zheng, M.; Wang, D.; Wan, J. A rice White-stripe leaf3 (wsl3) mutant lacking an HD domain-containing protein affects chlorophyll biosynthesis and chloroplast development. *J. Plant Biol.* **2016**, *59*, 282–292. [CrossRef]

9. Wang, Y.; He, Y.; Yang, M.; He, J.; Xu, P.; Shao, M.; Chu, P.; Guan, R. Fine mapping of a dominant gene conferring chlorophyll-deficiency in *Brassica napus*. *Sci. Rep.* **2016**, *6*, 31419. [CrossRef] [PubMed]

10. Kim, E.H.; Li, X.P.; Razeghifard, R.; Anderson, J.M.; Niyogi, K.K.; Pogson, B.J.; Chow, W.S. The multiple roles of light-harvesting chlorophyll a/b-protein complexes define structure and optimize function of *Arabidopsis* chloroplasts: A study using two chlorophyll b-less mutants. *Biochim. Biophys. Acta* **2009**, *1787*, 973–984. [CrossRef] [PubMed]

11. Braumann, I.; Stein, N.; Hansson, M. Reduced chlorophyll biosynthesis in heterozygous barley magnesium chelatase mutants. *Plant Physiol. Biochem.* **2014**, *78*, 10–14. [CrossRef] [PubMed]

12. Wu, Q.; Chen, Z.; Sun, W.; Deng, T.; Chen, M. De novo sequencing of the Leaf transcriptome reveals complex light-responsive regulatory networks in *Camellia sinensis* cv. *Baijiguan*. *Front. Plant Sci.* **2016**, *7*, 332. [CrossRef] [PubMed]

13. Zhang, H.; Li, J.; Yoo, J.H.; Yoo, S.C.; Cho, S.H.; Koh, H.J.; Seo, H.S.; Paek, N.C. Rice Chlorina-1 and Chlorina-9 encode ChlD and ChlI subunits of Mg-chelatase, a key enzyme for chlorophyll synthesis and chloroplast development. *Plant Mol. Biol.* **2006**, *62*, 325–337. [CrossRef] [PubMed]

14. Campoli, C.; Caffarri, S.; Svensson, J.T.; Bassi, R.; Stanca, A.M.; Cattivelli, L.; Crosatti, C. Parallel pigment and transcriptomic analysis of four barley Albina and Xantha mutants reveals the complex network of the chloroplast-dependent metabolism. *Plant Mol. Biol.* **2009**, *71*, 173–191. [CrossRef] [PubMed]

15. Campbell, B.W.; Mani, D.; Curtin, S.J.; Slattery, R.A.; Michno, J.-M.; Ort, D.R.; Schaus, P.J.; Palmer, R.G.; Orf, J.H.; Stupar, R.M. Identical Substitutions in Magnesium Chelatase Paralogs Result in Chlorophyll-Deficient Soybean Mutants. *G3 Genes Genomes Genet.* **2015**, *5*, 123–131. [CrossRef] [PubMed]

16. Papenbrock, J.; GrMe, S.; Kruse, E.; Hanel, F.; Grimm, B. Mg-chelatase of tobacco: Identification of a Chl D cDNA sequence encoding a third subunit, analysis of the interaction of the three subunits with the yeast two-hybrid system, and reconstitution of the enzyme activity by co-expression of recombinant CHL D, CHL H and CHL I. *Plant J.* **1997**, *12*, 981–990. [CrossRef] [PubMed]

17. Gibson, L.C.; Willows, R.D.; Kannangara, C.G.; Wettstein, D.V.; Hunter, C.N. Magnesium-protoporphyrin chelatase of Rhodobacter sphaeroides: Reconstitution of activity by combining the products of the bchH, -I, and -D genes expressed in *Escherichia coli*. *Proc. Natl. Acad. Sci. USA* **1995**, *92*, 1941–1944. [CrossRef] [PubMed]

18. Fodje, M.N.; Hansson, A.; Hansson, M.; Olsen, J.G.; Gough, S.; Willows, R.D.; Al-Karadaghi, S. Interplay Between an AAA Module and an Integrin I Domain May Regulate the Function of Magnesium Chelatase. *J. Mol. Biol.* **2001**, *311*, 111–122. [CrossRef] [PubMed]

19. Sirijovski, N.; Olsson, U.; Lundqvist, J.; Al-Karadaghi, S.; Willows, R.D.; Hansson, M. ATPase activity associated with the magnesium chelatase H-subunit of the chlorophyll biosynthetic pathway is an artefact. *Biochem. J.* **2006**, *400*, 477–484. [CrossRef] [PubMed]

20. Mochizuki, N.; Brusslan, J.A.; Larkin, R.; Nagatani, A.; Chory, J. *Arabidopsis* genomes uncoupled 5 (GUN5) mutant reveals the involvement of Mg-chelatase H subunit in plastid-to-nucleus signal transduction. *Proc. Natl. Acad. Sci. USA* **2000**, *98*, 2053–2058. [CrossRef] [PubMed]

21. Larkin, R.M.; Alonso, J.M.; Ecker, J.R.; Chory, J. GUN4, a Regulator of Chlorophyll Synthesis and Intracellular Signaling. *Science* **2003**, *299*, 902–906. [CrossRef] [PubMed]

22. Luo, T.; Luo, S.; Araujo, W.L.; Schlicke, H.; Rothbart, M.; Yu, J.; Fan, T.; Fernie, A.R.; Grimm, B.; Luo, M. Virus-induced gene silencing of pea *CHLI* and *CHLD* affects tetrapyrrole biosynthesis, chloroplast development and the primary metabolic network. *Plant Physiol. Biochem.* **2013**, *65*, 17–26. [CrossRef] [PubMed]

23. Hansson, A.; Willows, R.D.; Roberts, T.H.; Hansson, M. Three semidominant barley mutants with single amino acid substitutions in the smallest magnesium chelatase subunit form defective AAA$^+$ hexamers. *Proc. Natl. Acad. Sci. USA* **2002**, *99*, 13944–13949. [CrossRef] [PubMed]

24. Hansson, A.; Kannangara, C.G.; Wettstein, D.V.; Hansson, M. Molecular basis for semidominance of missense mutations in the XANTHA-H (42-kDa) subunit of magnesium chelatase. *Proc. Natl. Acad. Sci. USA* **1999**, *96*, 1744–1749. [CrossRef] [PubMed]

25. Fitzmaurice, W.P.; Nguyen, L.V.; Wernsman, E.A.; Thompson, W.F.; Conkling, M.A. Transposon tagging of the sulfur gene of tobacco using engineered maize *Ac/Ds* elements. *Genetics* **1999**, *153*, 1919–1928. [PubMed]

26. Sawers, R.J.; Viney, J.; Farmer, P.R.; Bussey, R.R.; Olsefski, G.; Anufrikova, K.; Hunter, C.N.; Brutnell, T.P. The maize *Oil yellow1 (Oy1)* gene encodes the I subunit of magnesium chelatase. *Plant Mol. Biol.* **2006**, *60*, 95–106. [CrossRef] [PubMed]

27. Williams, N.D.; Joppa, L.R.; Duysen, M.E.; Freeman, T.P. Inheritance of Three Chlorophyll-Deficient Mutants of Common Wheat. *Crop Sci.* **1985**, *25*, 1023–1025. [CrossRef]

28. Watanabe, N.; Koval, S.F. Mapping of chlorina mutant genes on the long arm of homoeologous group 7 chromosomes in common wheat with partial deletion lines. *Euphytica* **2003**, *129*, 259–265. [CrossRef]

29. Ansari, M.J.; Al-Ghamdi, A.; Kumar, R.; Usmani, S.; Al-Attal, Y.; Nuru, A.; Mohamed, A.A.; Singh, K.; Dhaliwal, H.S. Characterization and gene mapping of a chlorophyll-deficient mutant clm1 of *Triticum monococcum* L. *Biol. Plant.* **2013**, *57*, 442–448. [CrossRef]

30. Zhao, P.; Wang, K.; Zhang, W.; Liu, H.Y.; Du, L.P.; Hu, H.R.; Ye, X.G. Comprehensive analysis of differently expressed genes and proteins in albino and green plantlets from a wheat anther culture. *Biol. Plant.* **2017**, *61*, 255–265. [CrossRef]

31. Zhao, H.B.; Guo, H.J.; Zhao, L.S.; Gu, J.Y.; Zhao, S.R.; Li, J.H.; Liu, L.X. Agronomic Traits and Photosynthetic Characteristics of Chlorophyll-Deficient Wheat Mutant Induced by Spaceflight Environment. *Acta Agron. Sin.* **2011**, *37*, 119–126. [CrossRef]

32. Kosuge, K.; Watanabe, N.; Kuboyama, T. Comparative genetic mapping of homoeologous genes for the chlorina phenotype in the genus Triticum. *Euphytica* **2011**, *179*, 257–263. [CrossRef]

33. Brestic, M.; Zivcak, M.; Kunderlikova, K.; Sytar, O.; Shao, H.; Kalaji, H.M.; Allakhverdiev, S.I. Low PSI content limits the photoprotection of PSI and PSII in early growth stages of chlorophyll b-deficient wheat mutant lines. *Photosynth. Res.* **2015**, *125*, 151–166. [CrossRef] [PubMed]

34. Shi, K.; Gu, J.; Guo, H.; Zhao, L.; Xie, Y.; Xiong, H.; Li, J.; Zhao, S.; Song, X.; Liu, L. Transcriptome and proteomic analyses reveal multiple differences associated with chloroplast development in the spaceflight-induced wheat albino mutant mta. *PLoS ONE* **2017**, *12*, e0177992. [CrossRef] [PubMed]

35. Hou, D.Y.; Xu, H.; Du, G.Y.; Lin, J.T.; Duan, M.; Guo, A.G. Proteome analysis of chloroplast proteins in stage albinism line of winter wheat (*Triticum aestivum*) FA85. *BMB Rep.* **2009**, *42*, 450–455. [CrossRef] [PubMed]

36. Liu, X.G.; Xu, H.; Zhang, J.Y.; Liang, G.W.; Liu, Y.T.; Guo, A.G. Effect of low temperature on chlorophyll biosynthesis in albinism line of wheat (*Triticum aestivum*) FA85. *Physiol. Plant.* **2012**, *145*, 384–394. [CrossRef] [PubMed]

37. Falbel, T.G.; Staehelin, L.A. Characterization of a family of chlorophyll-deficient wheat (*Triticum*) and barley (*Hordeum vulgare*) mutants with defects in the magnesium-lnsertion step of chlorophyll biosynthesis. *Plant Physiol.* **1994**, *104*, 639–648. [CrossRef] [PubMed]

38. Zhang, N.; Zhang, H.J.; Zhao, B.; Sun, Q.Q.; Cao, Y.Y.; Li, R.; Wu, X.X.; Weeda, S.; Li, L.; Ren, S.; et al. The RNA-seq approach to discriminate gene expression profiles in response to melatonin on cucumber lateral root formation. *J. Pineal Res.* **2014**, *56*, 39–50. [CrossRef] [PubMed]

39. Bellieny-Rabelo, D.; De Oliveira, E.A.; Ribeiro, E.S.; Costa, E.P.; Oliveira, A.E.; Venancio, T.M. Transcriptome analysis uncovers key regulatory and metabolic aspects of soybean embryonic axes during germination. *Sci. Rep.* **2016**, *6*, 36009. [CrossRef] [PubMed]

40. Yang, Y.; Chen, X.; Xu, B.; Li, Y.; Ma, Y.; Wang, G. Phenotype and transcriptome analysis reveals chloroplast development and pigment biosynthesis together influenced the leaf color formation in mutants of *Anthurium andraeanum* 'Sonate'. *Front. Plant Sci.* **2015**, *6*, 139. [CrossRef] [PubMed]

41. Li, Y.; Zhang, Z.; Wang, P.; Wang, S.A.; Ma, L.; Li, L.; Yang, R.; Ma, Y.; Wang, Q. Comprehensive transcriptome analysis discovers novel candidate genes related to leaf color in a *Lagerstroemia indica* yellow leaf mutant. *Genes Genom.* **2017**, *37*, 851–863. [CrossRef]

42. Zhang, L.L.; Liu, C.; An, X.Y.; Wu, H.Y.; Feng, Y.; Wang, H.; Sun, D.J. Identification and genetic mapping of a novel incompletely dominant yellow leaf color gene, *Y1718*, on chromosome 2BS in wheat. *Euphytica* **2017**, *213*, 141. [CrossRef]

43. Grant, C.E.; Bailey, T.L.; Noble, W.S. Fimo: Scanning for occurrences of a given motif. *Bioinformatics* **2011**, *27*, 1017–1018. [CrossRef] [PubMed]

44. Shannon, P.; Markiel, A.; Ozier, O.; Baliga, N.S.; Wang, J.T.; Ramage, D.; Amin, N.; Schwikowski, B.; Ideker, T. Cytoscape: A software environment for integrated models of biomolecular interaction networks. *Genome Res.* **2003**, *13*, 2498–2504. [CrossRef] [PubMed]

45. Willows, R.D.; Gibson, L.C.; Kanangara, C.G.; Hunter, C.N.; Wettstein, D. Three separate proteins constitute the magnesium chelatase of *Rhodobacter Sphaeroides. Eur. J. Biochem.* **1996**, *235*, 438–443. [CrossRef] [PubMed]

46. Masuda, T.; Takamiya, K. Novel insights into the enzymology, regulation and physiological functions of light-dependent protochlorophyllide oxidoreductase in angiosperms. *Photosynth. Res.* **2004**, *81*, 1–29. [CrossRef] [PubMed]

47. Su, Q.; Frick, G.; Armstrong, G.; Apel, K. POR C of *Arabidopsis thaliana*: A third light- and NADPH-dependent protochlorophyllide oxidoreductase that is differentially regulated by light. *Plant Mol. Biol.* **2001**, *47*, 805–813. [CrossRef] [PubMed]

48. Sakuraba, Y.; Rahman, M.L.; Cho, S.H.; Kim, Y.S.; Koh, H.J.; Yoo, S.C.; Paek, N.C. The rice faded green leaf locus encodes protochlorophyllide oxidoreductase B and is essential for chlorophyll synthesis under high light conditions. *Plant J.* **2013**, *74*, 122–133. [CrossRef] [PubMed]

49. Yang, Y.L.; Xu, J.; Rao, Y.C.; Zeng, Y.J.; Liu, H.J.; Zheng, T.T.; Zhang, G.-H.; Hu, J.; Guo, L.B.; Qian, Q.; et al. Cloning and functional analysis of pale-green leaf (PGL10) in rice (*Oryza sativa* L.). *Plant Growth Regul.* **2016**, *78*, 69–77. [CrossRef]

50. Takamiya, K.-I.; Tsuchiya, T.; Ohta, H. Degradation pathway(s) of chlorophyll: What has gene cloning revealed? *Trends Plant Sci.* **2000**, *5*, 426–431. [CrossRef]

51. Harpaz-Saad, S.; Azoulay, T.; Arazi, T.; Ben-Yaakov, E.; Mett, A.; Shiboleth, Y.M.; Hortensteiner, S.; Gidoni, D.; Gal-On, A.; Goldschmidt, E.E.; et al. Chlorophyllase is a rate-limiting enzyme in chlorophyll catabolism and is posttranslationally regulated. *Plant Cell* **2007**, *19*, 1007–1022. [CrossRef] [PubMed]

52. Schenk, N.; Schelbert, S.; Kanwischer, M.; Goldschmidt, E.E.; Dormann, P.; Hortensteiner, S. The chlorophyllases ATCLH1 and ATCLH2 are not essential for senescence-related chlorophyll breakdown in *Arabidopsis thaliana. FEBS Lett.* **2007**, *581*, 5517–5525. [CrossRef] [PubMed]

53. Hu, X.; Makita, S.; Schelbert, S.; Sano, S.; Ochiai, M.; Tsuchiya, T.; Hasegawa, S.F.; Hortensteiner, S.; Tanaka, A.; Tanaka, R. Reexamination of chlorophyllase function implies its involvement in defense against chewing herbivores. *Plant Physiol.* **2015**, *167*, 660–670. [CrossRef] [PubMed]

54. Hortensteiner, S. Chlorophyll degradation during senescence. *Annu. Rev. Plant Biol.* **2006**, *57*, 55–77. [CrossRef] [PubMed]

55. Benedetti, C.E.; Arruda, P. Altering the expression of the chlorophyllase gene ATHCOR1 in transgenic *Arabidopsis* caused changes in the chlorophyll-to-chlorophyllide ratio. *Plant Physiol.* **2002**, *128*, 1255–1263. [CrossRef] [PubMed]

56. Kariola, T.; Brader, G.; Li, J.; Palva, E.T. Chlorophyllase 1, a damage control enzyme, affects the balance between defense pathways in plants. *Plant Cell* **2005**, *17*, 282–294. [CrossRef] [PubMed]

57. Jacob-Wilk, D.; Goldschmidt, D.H.E.E.; Riov, J.; Eyal, Y. Chlorophyll breakdown by chlorophyllase: Isolation and functional expression of the Chlase1 gene from ethylene-treated Citrus fruit and its regulation during development. *Plant J.* **1999**, *20*, 653–661. [CrossRef] [PubMed]

58. Azoulay Shemer, T.; Harpaz-Saad, S.; Belausov, E.; Lovat, N.; Krokhin, O.; Spicer, V.; Standing, K.G.; Goldschmidt, E.E.; Eyal, Y. Citrus chlorophyllase dynamics at ethylene-induced fruit color-break: A study of chlorophyllase expression, posttranslational processing kinetics, and in situ intracellular localization. *Plant Physiol.* **2008**, *148*, 108–118. [CrossRef] [PubMed]

59. Schelbert, S.; Aubry, S.; Burla, B.; Agne, B.; Kessler, F.; Krupinska, K.; Hortensteiner, S. Pheophytin pheophorbide hydrolase (pheophytinase) is involved in chlorophyll breakdown during leaf senescence in *Arabidopsis. Plant Cell* **2009**, *21*, 767–785. [CrossRef] [PubMed]

60. Demmig-Adams, B.; Adams, W.W., III. The role of xanthophyll cycle carotenoids in the protection of photosynthesis. *Trends Plant Sci.* **1996**, *1*, 21–26. [CrossRef]

61. Zhang, L.; Ma, G.; Kato, M.; Yamawaki, K.; Takagi, T.; Kiriiwa, Y.; Ikoma, Y.; Matsumoto, H.; Yoshioka, T.; Nesumi, H. Regulation of carotenoid accumulation and the expression of carotenoid metabolic genes in citrus juice sacs in vitro. *J. Exp. Bot.* **2012**, *63*, 871–886. [CrossRef] [PubMed]

62. Kim, J.; Smith, J.J.; Tian, L.; Dellapenna, D. The evolution and function of carotenoid hydroxylases in *Arabidopsis. Plant Cell Physiol.* **2009**, *50*, 463–479. [CrossRef] [PubMed]

63. Diretto, G.; Welsch, R.; Tavazza, R.; Mourgues, F.; Pizzichini, D.; Beyer, P.; Giuliano, G. Silencing of beta-carotene hydroxylase increases total carotenoid and beta-carotene levels in potato tubers. *BMC Plant Biol.* **2007**, *7*, 11. [CrossRef] [PubMed]

64. Kim, S.H.; Ahn, Y.O.; Ahn, M.J.; Lee, H.S.; Kwak, S.S. Down-regulation of beta-carotene hydroxylase increases beta-carotene and total carotenoids enhancing salt stress tolerance in transgenic cultured cells of sweetpotato. *Phytochemistry* **2012**, *74*, 69–78. [CrossRef] [PubMed]

65. Berman, J.; Zorrilla-Lopez, U.; Sandmann, G.; Capell, T.; Christou, P.; Zhu, C. The silencing of carotenoid beta-hydroxylases by RNA interference in different maize genetic backgrounds increases the beta-carotene content of the endosperm. *Int. J. Mol. Sci.* **2017**, *18*. [CrossRef] [PubMed]

66. Latowski, D.; Grzyb, J.; Strzałka, K. The xanthophyll cycle-molecular mechanism and physiological significance. *Acta Physiol. Plant.* **2004**, *26*, 197. [CrossRef]

67. Demmig-Adams, B.; Adams, W.W. Photoprotection and other responses of plants to high light stress. *Annu. Rev. Plant Physiol. Plant Mol. Biol.* **1992**, *43*, 599–626. [CrossRef]

68. Zhao, Q.; Wang, G.; Ji, J.; Jin, C.; Wu, W.; Zhao, J. Over-expression of *Arabidopsis thaliana* β-carotene hydroxylase (*chyB*) gene enhances drought tolerance in transgenic tobacco. *J. Plant Biochem. Biotechnol.* **2014**, *23*, 190–198. [CrossRef]

69. Wang, L.; Cao, H.; Chen, C.; Yue, C.; Hao, X.; Yang, Y.; Wang, X. Complementary transcriptomic and proteomic analyses of a chlorophyll-deficient tea plant cultivar reveal multiple metabolic pathway changes. *J. Proteom.* **2016**, *130*, 160–169. [CrossRef] [PubMed]

70. Li, C.F.; Xu, Y.X.; Ma, J.Q.; Jin, J.Q.; Huang, D.J.; Yao, M.Z.; Ma, C.L.; Chen, L. Biochemical and transcriptomic analyses reveal different metabolite biosynthesis profiles among three color and developmental stages in 'Anji Baicha' (*Camellia sinensis*). *BMC Plant Biol.* **2016**, *16*, 195. [CrossRef] [PubMed]

71. Bashir, H.; Qureshi, M.I.; Ibrahim, M.M.; Iqbal, M. Chloroplast and photosystems: Impact of cadmium and iron deficiency. *Photosynthetica* **2015**, *53*, 321–335. [CrossRef]

72. Allen, J.F.; de Paula, W.B.; Puthiyaveetil, S.; Nield, J. A structural phylogenetic map for chloroplast photosynthesis. *Trends Plant Sci.* **2011**, *16*, 645–655. [CrossRef] [PubMed]

73. Albertsson, P. A quantitative model of the domain structure of the photosynthetic membrane. *Trends Plant Sci.* **2001**, *6*, 349–354. [CrossRef]

74. Chu, P.; Yan, G.X.; Yang, Q.; Zhai, L.N.; Zhang, C.; Zhang, F.Q.; Guan, R.Z. iTRAQ-based quantitative proteomics analysis of *Brassica napus* leaves reveals pathways associated with chlorophyll deficiency. *J. Proteom.* **2015**, *113*, 244–259. [CrossRef] [PubMed]

75. Ma, C.; Cao, J.; Li, J.; Zhou, B.; Tang, J.; Miao, A. Phenotypic, histological and proteomic analyses reveal multiple differences associated with chloroplast development in yellow and variegated variants from *Camellia sinensis. Sci. Rep.* **2016**, *6*, 33369. [CrossRef] [PubMed]

76. Hutin, C.; Nussaume, L.; Moise, N.; Moya, I.; Kloppstech, K.; Havaux, M. Early light-induced proteins protect *Arabidopsis* from photooxidative stress. *Proc. Natl. Acad. Sci. USA* **2003**, *100*, 4921–4926. [CrossRef] [PubMed]

77. Beck, J.; Lohscheider, J.N.; Albert, S.; Andersson, U.; Mendgen, K.W.; Rojas-Stutz, M.C.; Adamska, I.; Funck, D. Small One-Helix Proteins Are Essential for Photosynthesis in *Arabidopsis*. *Front. Plant Sci.* **2017**, *8*, 7. [CrossRef] [PubMed]

78. Adamska, I.; Kruse, E.; Kloppstech, K. Stable insertion of the early light-induced proteins into etioplast membranes requires chlorophyll A. *J. Biol. Chem.* **2001**, *276*, 8582–8587. [CrossRef] [PubMed]

79. Heddad, M.; Noren, H.; Reiser, V.; Dunaeva, M.; Andersson, B.; Adamska, I. Differential expression and localization of early light-induced proteins in *Arabidopsis*. *Plant Physiol.* **2006**, *142*, 75–87. [CrossRef] [PubMed]

80. Tzvetkova-Chevolleau, T.; Franck, F.; Alawady, A.E.; Dall'Osto, L.; Carrière, F.; Bassi, R.; Grimm, B.; Nussaume, L.; Havaux, M. The light stress-induced protein ELIP2 is a regulator of chlorophyll synthesis in *Arabidopsis thaliana*. *Plant J.* **2007**, *50*, 795–809. [CrossRef] [PubMed]

81. Parry, M.; Andralojc, P.J.; Scales, J.C.; Salvucci, M.E.; Carmo-Silva, A.E.; Alonso, H.; Whitney, S.M. Rubisco activity and regulation as targets for crop improvement. *J. Exp. Bot.* **2013**, *64*, 717–730. [CrossRef] [PubMed]

82. Andersson, I.; Backlund, A. Structure and function of Rubisco. *Plant Physiol. Biochem.* **2008**, *46*, 275–291. [CrossRef] [PubMed]

83. Zhan, G.M.; Li, R.J.; Hu, Z.Y.; Liu, J.; Deng, L.B.; Lu, S.Y.; Hua, W. Cosuppression of RBCS3B in *Arabidopsis* leads to severe photoinhibition caused by ROS accumulation. *Plant Cell Rep.* **2014**, *33*, 1091–1108. [CrossRef] [PubMed]

84. Esposito, S. Nitrogen assimilation, Abiotic stress and glucose 6-phosphate dehydrogenase: The full circle of reductants. *Plants* **2016**, *5*, 24. [CrossRef] [PubMed]

85. Berg, J.; Tymoczko, J.; Stryer, L. 20.3 the pentose phosphate pathway generates NADPH and synthesizes five-carbon sugars. In *Biochemistry*, 5th ed.; W H Freeman: New York, NY, USA, 2002; ISBN-10 0-7167-3051-0.

86. Waters, E.R.; Lee, G.J.; Vierling, E. Evolution, structure and function of the small heat shock proteins in plants. *J. Exp. Bot.* **1996**, *47*, 325–338. [CrossRef]

87. Waters, E.R. The evolution, function, structure, and expression of the plant sHSPs. *J. Exp. Bot.* **2013**, *64*, 391–403. [CrossRef] [PubMed]

88. Wang, W.; Vinocur, B.; Shoseyov, O.; Altman, A. Role of plant heat-shock proteins and molecular chaperones in the abiotic stress response. *Trends Plant Sci.* **2004**, *9*, 244–252. [CrossRef] [PubMed]

89. Timperio, A.M.; Egidi, M.G.; Zolla, L. Proteomics applied on plant abiotic stresses: Role of heat shock proteins (HSP). *J. Proteom.* **2008**, *71*, 391–411. [CrossRef] [PubMed]

90. Xue, G.P.; Drenth, J.; McIntyre, C.L. TaHsfA6f is a transcriptional activator that regulates a suite of heat stress protection genes in wheat (*Triticum aestivum* L.) including previously unknown Hsf targets. *J. Exp. Bot.* **2015**, *66*, 1025–1039. [CrossRef] [PubMed]

91. Westerheidea, S.D.; Raynesa, R.; Powella, C.; Xueb, B.; Uversky, V.N. HSF transcription factor family, heat shock response, and protein intrinsic disorder. *Curr. Protein Pept. Sci.* **2012**, *13*, 86–103. [CrossRef]

92. Nover, L.; Scharf, K.D. Heat stress proteins and transcription factors. *Cell. Mol. Life Sci.* **1997**, *53*, 80–103. [CrossRef] [PubMed]

93. Hu, X.; Yang, Y.; Gong, F.; Zhang, D.; Zhang, L.; Wu, L.; Li, C.; Wang, W. Protein sHSP26 improves chloroplast performance under heat stress by interacting with specific chloroplast proteins in maize (*Zea mays*). *J. Proteom.* **2015**, *115*, 81–92. [CrossRef] [PubMed]

94. Kim, K.H.; Alam, I.; Kim, Y.G.; Sharmin, S.A.; Lee, K.W.; Lee, S.H.; Lee, B.H. Overexpression of a chloroplast-localized small heat shock protein OsHSP26 confers enhanced tolerance against oxidative and heat stresses in tall fescue. *Biotechnol. Lett.* **2012**, *34*, 371–377. [CrossRef] [PubMed]

95. Zhong, L.; Zhou, W.; Wang, H.; Ding, S.; Lu, Q.; Wen, X.; Peng, L.; Zhang, L.; Lu, C. Chloroplast small heat shock protein HSP21 interacts with plastid nucleoid protein pTAC5 and is essential for chloroplast development in *Arabidopsis* under heat stress. *Plant Cell* **2013**, *25*, 2925–2943. [CrossRef] [PubMed]

96. Latijnhouwers, M.; Xu, X.M.; Møller, S.G. *Arabidopsis* stromal 70-kDa heat shock proteins are essential for chloroplast development. *Planta* **2010**, *232*, 567–578. [CrossRef] [PubMed]

97. Kim, S.R.; An, G. Rice chloroplast-localized heat shock protein 70, OsHsp70CP1, is essential for chloroplast development under high-temperature conditions. *J. Plant Physiol.* **2013**, *170*, 854–863. [CrossRef] [PubMed]

98. Schroda, M.; Vallon, O.; Wollman, F.A.; Beck, C.F. A chloroplast-targeted heat shock protein 70 (HSP70) contributes to the photoprotection and repair of photosystem II during and after photoinhibition. *Plant Cell* **1999**, *11*, 1165–1178. [CrossRef] [PubMed]

99. Greene, M.K.; Maskos, K.; Landry, S.J. Role of the J-domain in the cooperation of HSP40 with HSP70. *Proc. Natl. Acad. Sci. USA.* **1998**, *95*, 6108–6113. [CrossRef] [PubMed]

100. Cheetham, M.E.; Caplan, A.J. Structure, function and evolution of DnaJ: Conservation and adaptation of chaperone function. *Cell Stress Chaperones* **1998**, *3*, 28–36. [CrossRef]

101. Zhu, X.; Liang, S.; Yin, J.; Yuan, C.; Wang, J.; Li, W.; He, M.; Wang, J.; Chen, W.; Ma, B.; et al. The DnaJ OsDjA7/8 is essential for chloroplast development in rice (*Oryza sativa*). *Gene* **2015**, *574*, 11–19. [CrossRef] [PubMed]

102. Strand, A.; Asami, T.; Alonso, J.; Chory, J. Chloroplast to nucleus communication triggered by accumulation of Mg-ProtoporphyrinIX. *Nature* **2003**, *421*, 79–83. [CrossRef] [PubMed]

103. Chi, W.; Sun, X.; Zhang, L. Intracellular signaling from plastid to nucleus. *Annu. Rev. Plant Biol.* **2013**, *64*, 559–582. [CrossRef] [PubMed]

104. Kropat, J.; Oster, U.; Rüdiger, W.; Beck, C.F. Chlorophyll precursors are signals of chloroplast origin involved in light induction of nuclear heat-shock genes. *Proc. Natl. Acad. Sci. USA* **1997**, *94*, 14168–14172. [CrossRef] [PubMed]

105. Kindgren, P.; Noren, L.; Lopez Jde, D.; Shaikhali, J.; Strand, A. Interplay between heat shock protein 90 and HY5 controls *phANG* expression in response to the GUN5 plastid signal. *Mol. Plant* **2012**, *5*, 901–913. [CrossRef] [PubMed]

106. Chen, S.T.; He, N.Y.; Chen, J.H.; Guo, F.Q. Identification of core subunits of photosystem II as action sites of HSP21, which is activated by the GUN5-mediated retrograde pathway in *Arabidopsis*. *Plant J.* **2017**, *89*, 1106–1118. [CrossRef] [PubMed]

107. Langmead, B.; Salzberg, S.L. Fast gapped-read alignment with Bowtie 2. *Nat. Methods* **2012**, *9*, 357–359. [CrossRef] [PubMed]

108. Kim, D.; Pertea, G.; Trapnell, C.; Pimentel, H.; Kelley, R.; Salzberg, S.L. TopHat2: Accurate alignment of transcriptomes in the presence of insertions, deletions and gene fusions. *Genome Biol.* **2013**, *14*, R36. [CrossRef] [PubMed]

109. Trapnell, C.; Williams, B.A.; Pertea, G.; Mortazavi, A.; Kwan, G.; van Baren, M.J.; Salzberg, S.L.; Wold, B.J.; Pachter, L. Transcript assembly and quantification by RNA-Seq reveals unannotated transcripts and isoform switching during cell differentiation. *Nat. Biotechnol.* **2010**, *28*, 511–515. [CrossRef] [PubMed]

110. Conesa, A.; Gotz, S. Blast2GO: A comprehensive suite for functional analysis in plant genomics. *Int. J. Plant Genom.* **2008**, *2008*. [CrossRef] [PubMed]

111. Ye, J.; Fang, L.; Zheng, H.; Zhang, Y.; Chen, J.; Zhang, Z.; Wang, J.; Li, S.; Li, R.; Bolund, L.; et al. WEGO: A web tool for plotting GO annotations. *Nucleic Acids Res.* **2006**, *34*, W293–W297. [CrossRef] [PubMed]

112. Young, M.D.; Wakefield, M.J.; Smyth, G.K.; Oshlack, A. Gene ontology analysis for RNA-seq: Accounting for selection bias. *Genome Biol.* **2010**, *11*, R14. [CrossRef] [PubMed]

113. Wu, J.; Mao, X.; Cai, T.; Luo, J.; Wei, L. Kobas server: A web-based platform for automated annotation and pathway identification. *Nucleic Acids Res.* **2006**, *34*, W720–W724. [CrossRef] [PubMed]

114. Wang, G.P.; Hou, W.Q.; Zhang, L.; Wu, H.Y.; Zhao, L.F.; Du, X.Y.; Ma, X.; Li, A.F.; Wang, H.W.; Kong, L.R. Functional analysis of a wheat pleiotropic drug resistance gene involved in Fusarium head blight resistance. *J. Integr. Agric.* **2016**, *15*, 2215–2227. [CrossRef]

115. Schmittgen, T.D.; Livak, K.J. Analyzing real-time PCR data by the comparative C_T method. *Nat. Protoc.* **2008**, *3*, 1101–1108. [CrossRef] [PubMed]

116. Lichtenthaler, H.K. Chlorophylls and carotenoids: Pigments of photosynthetic biomembranes. *Methods Enzymol.* **1987**, *148*, 350–382. [CrossRef]

117. Dei, M. Benzyladenine-induced stimulation of 5-aminolevulinic acid accumulation under various light intensities in levulinic acid-treated cotyledons of etiolated cucumber. *Physiol. Plant.* **1985**, *64*, 153–160. [CrossRef]

118. Bogorad, L. Porphyrin synthesis. In *Methods in Enzymology*; Colowick, S.P., Kaplan, N.O., Eds.; Academic Press: New York, NY, USA, 1962; Volume 5, pp. 885–895, ISBN 0076-6879.

119. Rebeiz, C.A.; Mattheis, J.R.; Smith, B.B.; Rebeiz, C.C.; Dayton, D.F. Chloroplast biogenesis: Biosynthesis and accumulation of protochlorophyll by isolated etioplasts and developing chloroplasts. *Arch. Biochem. Biophys.* **1975**, *171*, 549–567. [CrossRef]

120. Norris, S.R.; Barrette, T.R.; DellaPenna, D. Genetic dissection of camtenoid synthesis in *Arabidopsis* defines plastoquinone as an essential component of phytoene desaturation. *Plant Cell* **1995**, *7*, 2139–2149. [CrossRef] [PubMed]

International Journal of
Molecular Sciences

MDPI

Article

Stable Membrane-Association of mRNAs in Etiolated, Greening and Mature Plastids

Julia Legen and Christian Schmitz-Linneweber *

Institut of Biology, Department of the Life Sciences, Humboldt-Universität Berlin, Philippstraße 11–13,
Grüne Amöbe, 10115 Berlin, Germany; legenjul@staff.hu-berlin.de
* Correspondence: smitzlic@rz.hu-berlin.de; Tel.: +49-30-2093-49700

Received: 26 July 2017; Accepted: 28 August 2017; Published: 31 August 2017

Abstract: Chloroplast genes are transcribed as polycistronic precursor RNAs that give rise to a multitude of processing products down to monocistronic forms. Translation of these mRNAs is realized by bacterial type 70S ribosomes. A larger fraction of these ribosomes is attached to chloroplast membranes. This study analyzed transcriptome-wide distribution of plastid mRNAs between soluble and membrane fractions of purified plastids using microarray analyses and validating RNA gel blot hybridizations. To determine the impact of light on mRNA localization, we used etioplasts, greening plastids and mature chloroplasts from *Zea mays* as a source for membrane and soluble extracts. The results show that the three plastid types display an almost identical distribution of RNAs between the two organellar fractions, which is confirmed by quantitative RNA gel blot analyses. Furthermore, they reveal that different RNAs processed from polycistronic precursors show transcript-autonomous distribution between stroma and membrane fractions. Disruption of ribosomes leads to release of mRNAs from membranes, demonstrating that attachment is likely a direct consequence of translation. We conclude that plastid mRNA distribution is a stable feature of different plastid types, setting up rapid chloroplast translation in any plastid type.

Keywords: chloroplast; etioplast; membrane; organelle; ribosome; RNA processing; translation; *Zea mays*

1. Introduction

Subcellular RNA localization is an important means of gene regulation in eukaryotic organisms [1], but also in bacteria [2]. In plants, insights into intracellular RNA localization are limited and have been mostly studied for the localization of mRNAs for endosperm storage proteins [3]. Plant cells do not only contain RNAs in the nucleo-cytosolic compartments, but also within the two DNA-containing organelles, the mitochondria and chloroplasts. Chloroplast genomes of angiosperms code for about 80 proteins. In addition, there are genes for tRNAs and rRNAs; rRNAs are 70S ribosomes that translate all chloroplast mRNAs.

The sub-organellar localization of chloroplast RNAs has rarely been addressed, but there are a few notable exceptions. In the green algae *Chlamydomonas reinhardtii*, several chloroplast mRNAs (e.g., *psbC*, *psbD*, *rbcL*) have been demonstrated to be enriched around the chloroplast pyrenoid [4,5] by fluorescence in situ hybridization (FISH). The pyrenoid seems to be a dominant site of mRNA translation and thylakoid membrane complex assembly in this algae [6]. A large fraction of all Chlamydomonas chloroplast 70S ribosomes is attached to thylakoid membranes, as evidenced both by electron microscopy as well as cell fractionation [7–9]. This suggested early on that chloroplast mRNAs are translated in close association with chloroplast membranes. Since many chloroplast mRNAs encode integral membrane proteins, it was assumed that translating ribosomes get trapped on membranes because the nascent protein chain is translated directly into the thylakoid membrane.

This is supported by the finding that membrane attachment of RNAs depends on active translation [10] and is proportional to overall protein synthesis within thylakoid membranes [11]. These findings imply that the membrane-bound ribosomes are associated with mRNAs. This is supported by in vitro experiments, where washed thylakoids are capable of synthesizing proteins when supplemented with soluble factors [12,13]. Recently, ribosome-protected mRNA fragments in the soluble fraction of chloroplasts were compared with ribosome-protected mRNA fragments in the membrane fraction. This demonstrated transcriptome-wide association of mRNAs coding for membraneous proteins to chloroplast membranes [14]. Noteworthy, mRNA association via nascent-chain bound ribosomes with membranes is astonishingly stable and occurs even in the absence of chlorophyll production [15].

Chloroplast mRNAs are produced as polycistronic units and are subsequently processed [16]. RNA splicing, endonucleolytic cleavage and exonucleolytic trimming and decay gives rise to complex transcript patterns. For most transcript units, the biological importance of this complexity is unclear. Some processing events have been shown to be required for producing translatable messages (e.g., [17]). In maize chloroplast development, a correlation of the accumulation of monocistronic transcript isoforms with their translational efficiency has been noted [18]. In general, however, both unprocessed polycistronic as well as processed monocistronic forms are supposed to be translated [14,19]. How prevalent differential translation of longer or shorter transcript isoforms is in the chloroplast transcriptome is largely unknown.

We set out to determine which mRNAs are associated with chloroplast membranes versus free mRNAs in the chloroplast stroma fraction. Our results suggests that individual transcripts from a single operon show preference for either stromal or membrane association. In rare cases such associations are subject to developmental change, but in general, membrane-enrichment of plastid RNAs is constant during chloroplast development.

2. Results and Discussion

2.1. Preparation of Sub-Organellar Fractions Highly Enriched in Membrane and Stroma Marker Proteins

We prepared membranes and stroma extracts from isolated plastids from 9-day-old *Zea mays* seedlings that were either grown in the dark (i.e., are etiolated), from greening seedlings (grown in the dark and then irradiated for 16 h, then kept in the dark for an 8 h night and harvested next morning after an additional 75 min of light exposure), or from plants grown under standard long-day conditions. The three tissue types represent etioplasts, greening plastids, and mature chloroplasts, respectively. The fractions were analyzed immunologically for the presence of the marker proteins PetD and RbcL (Figure 1). PetD is a membrane protein of the cytochrome b_6f complex; RbcL is a stromal protein. There is a slight signal for PetD in stroma from etiolated plants, while the protein is beyond the detection limit in the two other stroma preparations. A small fraction of RbcL can be found within each membrane fraction, indicating a slight contamination of our membrane preparation with stromal proteins. Nevertheless, this immunological analysis shows successful enrichment for both, membrane and stroma proteins, in the two fractions.

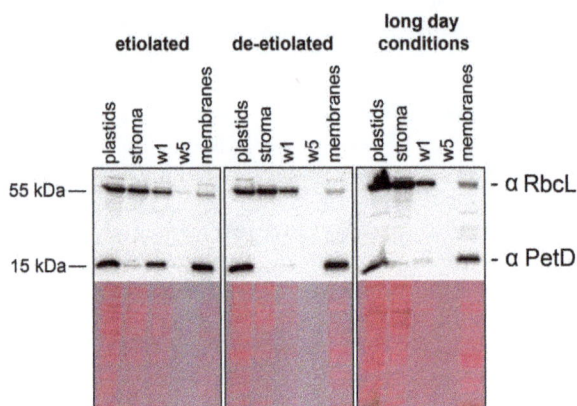

Figure 1. Enrichment of marker proteins in chloroplast membrane and stroma fractions. Western blot analysis was performed from each type of tissue using RbcL and PetD antisera as markers for stroma and membrane, respectively. RNAs from these fractions were used for microarray analysis. A seven hundredth of each sample was analyzed by sodium dodecyl sulfate polyacrylamide gel electrophoresis (SDS-PAGE). The membrane fraction was washed five times prior to RNA extraction. Aliquots from the supernatants of the first and the last wash were analyzed here as well (W1 and W5).

2.2. Global Association of Chloroplast mRNAs for Membrane Proteins with Chloroplast and Etioplast Membranes

To investigate the association of RNA with membranes on a global level, we labelled RNA from stroma and membrane fractions with fluorescent dyes and competitively hybridized them to a whole-genome tiling array of the maize chloroplast genome [20]. We opted for stroma preparations versus total chloroplast RNA as a control sample, since at least some RNA degradation is expected to occur during membrane/stroma preparation, but is largely absent from total chloroplast RNA preparations. The ratio of membrane-signal over stroma signal (the membrane enrichment value (MEV)) was calculated for all probes. MEVs from four biological replicates were normalized, combined and plotted against the genome position of the probes (Figure 2a). A gene ontology (GO) analysis was carried out for the top 10% probes with the highest MEVs in comparison to all probes on the array (Figure 2b; full data set can be found in Table S1). This revealed that probes representing components of photosystem I, II and the cytochrome b_6f complex are overrepresented in the top 10% MEV probes. Several operons containing genes for these complexes are represented by multiple probes of similar MEVs, for example the *psbB–petB* and the *psbD/C* operons, which is further support for the validity of the assay. Membrane-based translation of mRNAs coding for photosynthetic proteins has been demonstrated before by ribosome profiling [14]. By contrast, probes for ribosomal proteins, the NADH dehydrogenase (NDH) complex and the ATP synthase are underrepresented in the top 10% MEV probes. Probes for *rbcL*, tRNAs, the plastid RNA polymerase and uncharacterized open reading frames (ORFs) are not found at all among the top 10% MEV probes. This distribution of MEVs was found for each of the three tissues analyzed, i.e., noteworthy also for etiolated tissue (Figure 2 and Table S2).

While most RNAs within the top 10% membrane-enriched fraction are encoding proteins with photosynthetic functions, there are also interesting exceptions, namely *rps14* and *cemA*. Both are represented by multiple probes within the top 10% MEV probes. *rps14* is the only mRNA coding for a ribosomal protein that shows such high enrichment in membranes (Table S2). It is suggested that the *cemA* gene product is required for the import of inorganic carbon across the envelope membrane, but it was not found to be attached to membranes in recent ribosome profiling analysis [14]. Both *rps14* and *cemA* are part of main peaks in this analysis, represented by multiple probes corresponding to the *psaA/B* and the *cemA–petA* operon (Figure 2a). *rps14* is accumulating as part of a tricistronic transcript together with *psaA* and *psaB* and *cemA* is located within an operon adjacent to *petA*. *petA* and *psaA/B* encode core membrane proteins of photosystem I and the *cytochrome b_6f* complex. Thus, *rps14* and *cemA* are likely drawn to the membrane as part of a larger transcript tethered to the thylakoid membrane via translated *psaA/B* and *petA*, respectively.

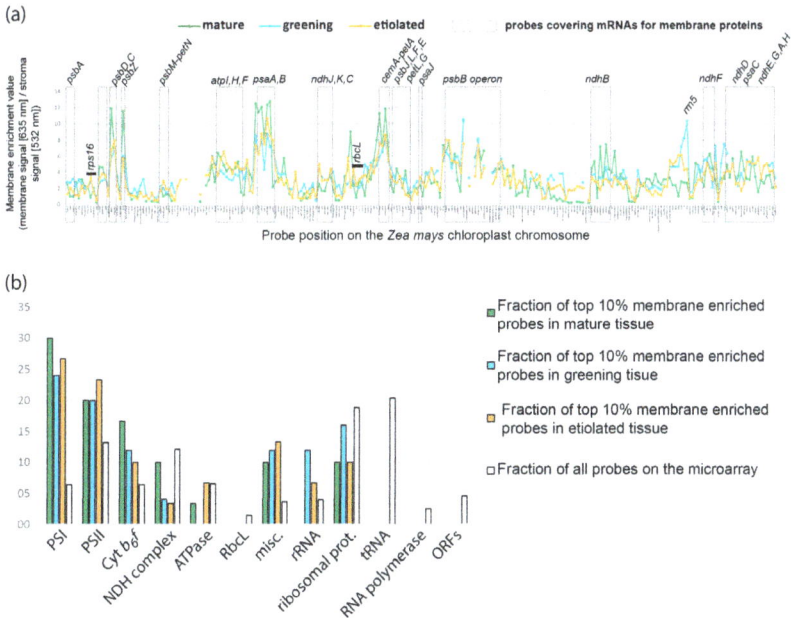

Figure 2. Microarray analysis of RNAs enriched at chloroplast membranes. (**a**) Membrane-enrichment of RNA along the chloroplast chromosome. Isolated chloroplast from maize seedlings grown at three different light regimes (leading to mature, greening, and etiolated plastids, respectively) were processed into stroma and membrane fractions. Two µg of RNA isolated from each fraction were differentially labeled and hybridized to a microarray representing the maize chloroplast genome in a tiling fashion. The ratio of membrane versus stroma signal was plotted against the genomic position on the *Zea mays* chloroplast genome (acc. NC_001666). The graphs shown represent four biological replicates. Normalization between conditions is based on the sum of membrane enrichment values (MEVs) of all probes for each condition. Probes corresponding to mRNAs coding for known membrane proteins are highlighted by dashed boxes. Selected soluble RNAs are labelled as well. The data underlying this chart are deposited in Table S1; (**b**) Gene ontology (GO) term analysis of membrane-enriched RNAs in the chloroplast. The probes showing the top 10 percent MEVs binned into 12 functional categories (PSI = photosystem I; PSII = photosystem II; Prot. = proteins; ORFs = open reading frames of unknown function; misc. = miscellaneous). Probes covering more than one category were counted in each relevant bin. The numbers of probes within each bin are expressed as the fraction of the total number of probes for each condition (in %). This is compared to the distribution of all probes on the array across the bins (open bars). Probes for photosystems I and II are over-represented in the top 10% MEV probes of all three tissue types.

2.3. Membrane Association of Chloroplast RNAs via Ribosomes

The suggested mechanism of membrane association of chloroplast mRNAs is via actively translating ribosomes [7,9,14]. To test directly for the importance of ribosomes for mRNA-membrane association, we treated membranes with ethylenediaminetetraacetic acid (EDTA) to separate the small and large subunit of the ribosome and thus disrupt membrane and mRNA association. RNA retrieved from supernatants and membranes after EDTA treatment was analyzed by RNA gel blot hybridization using a probe specific for the *psbD* gene (Figure 3a). We observed that in mock-treated samples, the signal for *psbD* mRNAs is almost completely retained in the membrane fractions. Only shorter transcripts of 1.0 to 1.5 kb are found in the stroma of etiolated and greening tissue. These transcripts likely represent monocistronic *psbD* (the *psbD* coding region is 1.0 kb in length) and are in excess of longer precursors. Possibly, such abundant shorter transcripts are not all engaged with ribosomes and can thus partially remain in the stroma. Larger transcripts encompassing *psbD*, *psbC* and further genes are only found in membrane fractions prior to EDTA treatment, which could be explained by their more efficient translation relative to the smaller monocistronic RNAs. After EDTA treatment, no *psbD* remains detectable in membranes and the RNA is instead found in the supernatant. This is further evidence for ribosomes being required for membrane attachment of chloroplast mRNAs. Still, it cannot be ruled out that larger ribonucleoprotein (RNP) granules that would co-enrich with membranes despite a physical interaction, dissolve due to EDTA treatment and are thus released in the supernatant. Such granules have been described in *Chlamydomonas reinhardtii* [21].

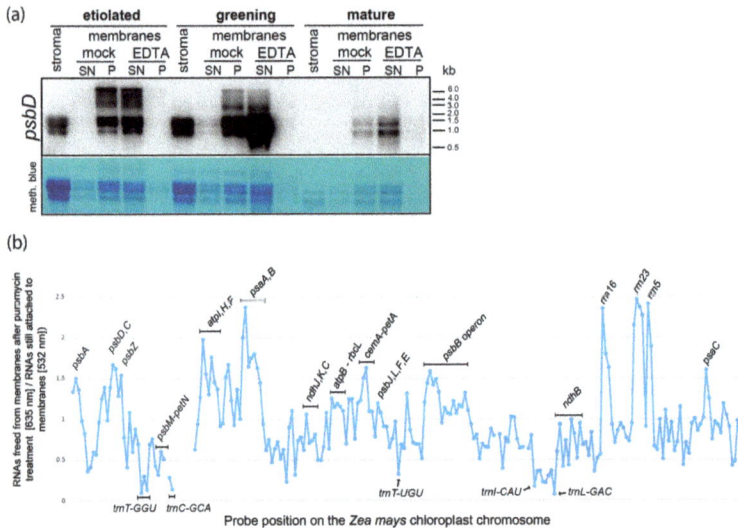

Figure 3. RNAs are associated to chloroplast membranes via translating ribosomes. (**a**) Disruption of ribosomes by ethylenediaminetetraacetic acid (EDTA) releases *psbD* mRNAs into the soluble fraction. Membrane fractions from the same number of plastids from three different tissue types were either treated with EDTA in order to separate the large and small subunit of the ribosome or were treated with buffer (mock). Membranes were spun down after treatment (P) and the supernatant was collected as well (S). RNA was extracted from all fractions and fractionated on 1.2% agarose gel and analyzed by hybridization to a radiolabeled *psbD* probe (see Table S4 for primer sequences). As a quality control, the rRNAs were stained with methylene blue; (**b**) Release of mRNAs from chloroplast membranes after puromycin treatment. Membranes were isolated from mature chloroplasts that were purified from four-week-old plants grown under long-day conditions. Membranes were treated with puromycin to release nascent chains and ribosomes from the mRNAs and thus secondarily free mRNAs from the membrane. The membrane was washed multiple times. RNA from the supernatant after puromycin treatment and from the washed membranes was analyzed by a whole genome tiling array of the maize chloroplast genome. The ratio of freed RNA over membrane-bound RNA was calculated for each probe and plotted against genome position. Selected probes are labelled by the gene-names they represent. The isoaceptor tRNAs are named by their respective codon (GGU, GCA, UGU, CAU and GAC, respectively).

Therefore, to validate the hypothesis that nascent chains emerging from ribosomes tether polysomes to chloroplast membranes, we treated mature membranes with puromycin. Puromycin leads to premature peptide chain termination and thus releases ribosomes from mRNAs. We analyzed RNAs released into the supernatant versus membrane-bound RNAs on our microarray after puromycin treatment. As expected, puromycin led to a massive release of ribosomal RNAs from membranes, demonstrating that the antibiotic treatment was successful (Figure 3b). This rRNA release was accompanied by a loss of mRNAs from membranes. mRNAs that showed high MEV values in our previous analysis were particularly prone to detachment from membranes after puromycin treatment. This was for example seen for the *psbD/C* and *psaA/B* operons (Figures 2a and 3b). A similar set of RNAs was found when we treated membranes with EDTA and analyzed the RNAs on microarrays (Figure S1). By contrast, tRNAs were released to a far lesser extent (Figure 3b). tRNAs and also the known stromal mRNAs *rbcL* were found in membranes even after extensive washing. This could be caused by the protection of RNAs within vesicles that are forming from membranes during preparation. In sum, association of mRNAs with membranes was mediated globally by translating ribosomes tethered to membranes via nascent peptide chains.

2.4. tRNAs, mRNAs for Stromal Proteins and Short mRNAs for Membrane Proteins Are Enriched in the Soluble Fraction of Plastids

As expected, many tRNAs show low MEVs, in fact the lowest of the entire analysis (Table S2). 79% of all genes represented in the 10% least enriched RNAs across all three tissue types are tRNA genes. An additional 16% of the low MEVs correspond to ORFs of unknown function (Table S2). In order to visualize, which mRNAs show low MEVs, we removed probes for tRNAs and rRNAs from the analysis (Table S3). Expectedly, mRNAs like *rbcL*, *clpP*, *infA* and *matK* that code for soluble proteins do show low MEVs. Similarly, many mRNAs for ribosomal proteins display low MEVs. In addition, few probes for photosynthetic mRNAs coding for membrane proteins have low MEVs as well. Of the latter, almost all represent short reading frames, including *petN* (29 codons), *petL* (31), *petG* (37), *psaI* (36), *psaJ* (42), *psbM* (34) and *psbJ* (40). As has recently been discussed for ribosome profiling, such short mRNAs are fully translated before a sufficiently long peptide chain emerges from the large subunit's exit channel that would enable contact with the membrane [14]. An exception within this group is *psbA*, which encodes the 40 kDa membrane protein D1 and is 354 codons long. Several probes for *psbA* appear at the bottom of the MEV list for all three tissue types. mRNA for *psbA* is accumulating in large amounts in chloroplasts, mostly untranslated [22,23], and is believed to be required as a reservoir during stress-induced demand for D1 production [24]. Thus, the untranslated excess of *psbA* mRNA likely leads to the observed low membrane enrichment values. In sum, our analysis demonstrates on a transcriptome-wide scale that mRNAs coding for soluble proteins localize to the stroma. In addition, short mRNAs for membrane proteins as well as the *psbA* mRNA are predominantly stroma-localized as well. Paralleling what we observed for membrane-associated RNAs, the list of soluble RNAs is similar between all three tissue types examined.

2.5. Etioplasts Display Membrane Enrichment of RNAs Similar to Chloroplasts

Etioplasts differ dramatically in their internal structure from structures known in developing and mature chloroplasts, showing a para-crystalline prolammelar body (PLB) instead of thylakoid membranes. Thus, it came as a surprise that the membrane enrichment curves of plastid RNAs are similar between the three plastid types analyzed (Figure 2a). To uncover potential differences, we did a pairwise comparison of the MEV datasets (Figure S2). This corroborated the large similarities found in the operon (Figure 2a) and GO term analysis (Figure 2b). Intriguingly, greatest similarities are not found between the two green plastid types, but between etioplasts and plastids from greening tissue (Pearson correlation $r_{E/DE} = 0.83$; Figure S2). By contrast, transcript enrichment at membranes in mature chloroplasts are less similar to the two other plastid preparations ($r_{DE/G} = 0.66$; $r_{E/G} = 0.72$, Figure S2). Possibly, the short time span of irradiation has not sufficed to erase the etioplast transcript pattern in greening plastids, thus making these two plastid types more similar on the RNA level and thus also membrane enrichment level than either is to mature chloroplasts. By contrast, the translational activity will have much increased in the greening plastids, but has not led to stronger membrane attachment in our dataset nor in a previous analysis of selected mRNAs [25].

Overall, the most pronounced difference was found for the aforementioned ORFs. The region encompassing ORF46 to ORF137 shows a higher membrane enrichment in etiolated and greening versus mature tissue (Figure S2). The ORFs in question are remnants of a long reading frame of unknown function called *ycf2* [26]. Their low spot count, indicative of a low expression level, suggests that further, more sensitive assays will be necessary to ascertain the differential membrane association seen here.

Why should there be mRNA association with membranes in etioplasts at all? 95% of all prolammelar body (PLB) proteins are protochlorophyllid oxidoreductases (PORA and PORB), which are encoded in the nuclear genome [27–31]. In addition, a number of proteins involved in the light reactions of photosynthesis has been identified. This includes the soluble chloroplast-encoded proteins AtpA, AtpB, RbcL, as well as membranous, chloroplast-encoded proteins like the subunits of the cytochrome b6f complex, and the PSII subunit PsbE [28,30,32–34]. Thus, etioplast do have an active translation system. We conclude, that the observed association of many mRNAs coding for membrane proteins with etioplast membranes is reflecting this translational activity (Figures 2 and 3; [14,35]). It remains however unclear, to which membrane type these RNAs are bound, since we did not differentiate between etioplast envelope membranes, the PLB and lamellar prothylakoid membranes present in various cell types of etiolated maize leafs [36]. Also, we cannot exclude that the green light we utilize during etioplast preparation affects translational activity and thus mRNA localization.

In contrast to the examples listed above, many other of the chloroplast-encoded proteins of the thylakoid membrane complexes have not been detected in etioplasts. These appear to be only accumulating upon illumination. For example, the photosynthetic membrane proteins D1, PsbB, PsbC or the PSI reaction core proteins are not or only barely detectable in etioplasts, but have been shown to exhibit a massive increase in expression during photomorphogenesis [25,33,37]. We show here on a genome-wide scale that this increased expression is not reflected by a comparable increase in association of the corresponding mRNAs to chloroplast membranes in greening versus etiolated tissue. Rather, membrane association of plastid mRNAs appears stable across the three plastid types analyzed here. Even *psbA* and *psbB*, which show induction of mRNA levels in greening tissue, exhibit only a mild change in the ratio of membrane-bound versus stroma-bound RNAs. This is in line with previous analyses that show only a slight increase in *psbA* membrane attachment during de-etiolation in barley [38].

2.6. Quantitative Analysis of Membrane-Bound and Stromal mRNAs on a "Per-Chloroplast" Base Uncovers Transcript-Autonomous Localization Patterns

Our array analysis indicates association of RNAs with chloroplast membranes, but the resolution of this approach is limited by the number and length of probes on the microarray. Thus, the microarray analysis misses the great variety and complexity of transcripts representing individual genes within each operon. To understand which RNA species are bound to membranes, we performed quantitative RNA gel blot hybridization assays. We again extracted chloroplasts from the three maize cultures and followed the same procedure for RNA preparation from membrane and stroma fractions as for the microarray approach, with the exception that we counted chloroplasts prior to lysis and cell fractionation. This way, we could normalize the RNA amount from stroma and membrane preparations to chloroplast numbers and thus represent the quantitative differences in RNA abundance between the two fractions more accurately than when total RNA amounts are used for normalization. We utilized probes for eight different chloroplast genes that can be divided into two groups: a group of genes coding for membrane-proteins; and a second group including mRNAs coding for soluble proteins and RNAs that are not translated. *atpH, petA, psbB, psbA* represent the former group, while *psaC, rps16, rbcL*, *rrn5* and *rrn16* represent the latter. The RNAs we recover from the chloroplast fractions are intact: there are no bands found that would not also be seen in total RNA preparations (Figure 4).

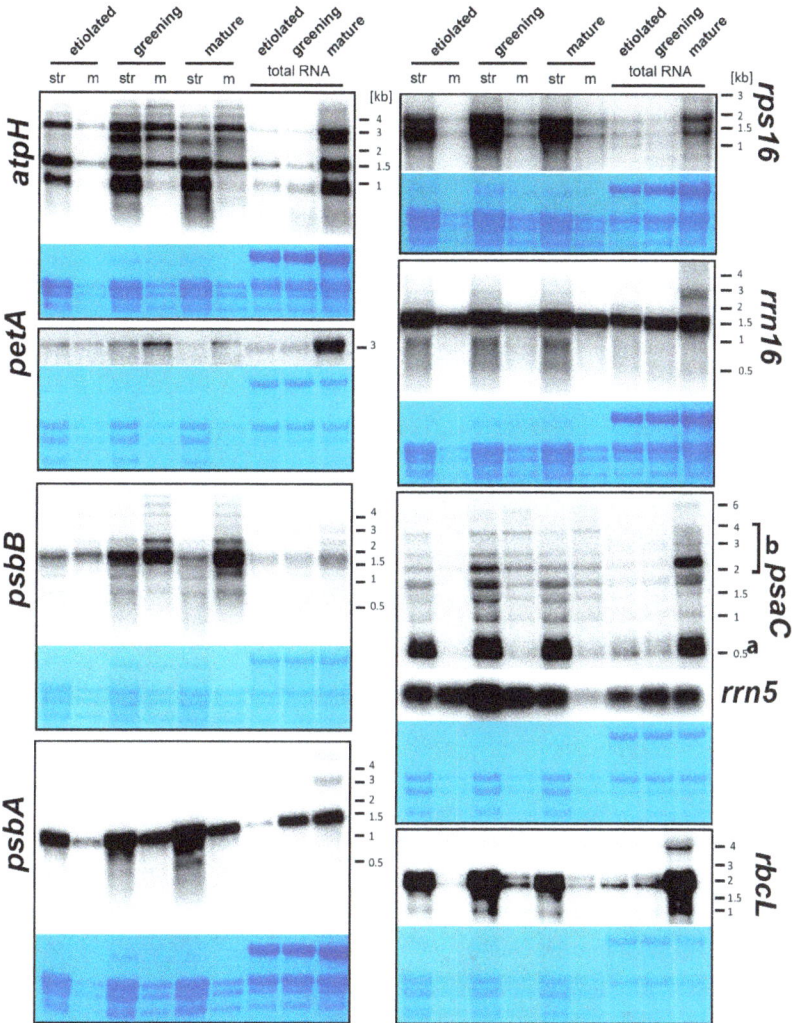

Figure 4. Analysis of transcript accumulation in membrane and stroma preparations on a per-chloroplast base. Equal aliquots of RNA from membrane and stroma fractions of purified plastids from three different tissues were extracted from the same number of chloroplasts. In addition, five micrograms of total leaf RNA was analyzed as well. The RNAs were fractionated on 1.2% agarose gel and analyzed by hybridization to radiolabeled probes for the plastid RNAs indicated (see Table S4 for primer sequences). As a quality control, the rRNAs were stained with methylene blue. Str = RNA from stroma fractions; m = RNA from membrane fractions.

2.6.1. The Analysis of Equal Amounts of Total Cellular RNA from Different Tissues Does Not Accurately Reflect per Organelle RNA Levels for Most Plastid Genes

All RNAs analyzed are found to be expressed in etiolated, greening and mature plastids. In etiolated tissue, there is overall less RNA per plastid than in the two photosynthetic tissues examined

(Figure 4). This is most pronounced for RNAs related to the light reactions of photosynthesis (*psaC, petA, psbB, atpH, psbA*) and to a lesser extent also for *rbcL* and genes of the genetic apparatus (*rps16, rrn5*). 16S rRNA accumulation remains constant across all three conditions. For a number of RNAs, peak accumulation levels are found for greening tissue rather than for mature tissue (e.g., *rrn5, rbcL, atpH*). Hence, in line with previous analyses, there is a global dependency of chloroplast RNA accumulation on light signals and chloroplast stage. Whether the observed smaller differences in the extent and timing of induction are biologically relevant, is unclear at present.

When considering total RNA preparations (lanes loaded with equal total cellular RNA amounts), there is dramatically less chloroplast RNAs in etiolated and greening tissue than in mature tissue for most genes analyzed. Apparently, the amount of RNA per chloroplast remains constant during initial chloroplast development while, later, the ratio of plastid RNA to cytoplasmic ribosomal RNAs increases towards mature chloroplasts. Again, only the 16S ribosomal RNAs displays an approximately equal accumulation in total RNA across the three conditions. In addition, *psbA* and *rrn5* display induction of RNA levels already in total RNA preparations of greening tissue. Apparently, RNA accumulation is reacting faster to irradiation for these two genes than for the other genes assessed here. A rapid response of *psbA* on all levels of gene expression has indeed been described in various plant species [33,39]. In general, our analysis demonstrates that changes observed for chloroplast-encoded transcripts in different tissues or conditions are pronouncedly dependent on whether normalization is made according to total cellular or total chloroplast RNAs.

2.6.2. mRNAs Are Enriched at Chloroplast Membranes in a Transcript-Autonomous Fashion

The RNA gel blot hybridizations corroborate microarray findings. For example, RNAs for *psbB* and *petA* that display the most pronounced membrane-bias in the northern analysis do also display high MEVs in microarrays. RNAs found enriched in the stroma in RNA gel blots like *rbcL* and *rps16* have comparatively low MEVs (compare these four mRNAs in Figures 2a and 4). In line with this, mRNAs for *atpH, psbA*, and the extrinsic membrane proteins *psaC*, display low membrane enrichment values in the microarray analysis and are also found predominantly in stroma fractions in RNA gel blots. This demonstrates a qualitative congruence of the two assays.

For all probes, we noted that the qualitative transcript patterns are similar between stroma and membrane fractions: All transcripts found in the stroma can also be detected in the membrane fraction for any of the probes used (Figure 4). Intriguingly, different transcripts detected with the same probe display distinct distributions between membrane and stroma. A striking example for this is the *psaC* probing, which shows several transcripts highly enriched in the stroma fraction, and in addition larger transcripts that are approximately equally distributed. Most prominent is a band of less than 0.5 kb (labelled "a" in Figure 4), which corresponds to the monocistronic form of *psaC* [40] and is found almost exclusively in the stroma fraction. This RNA species has been described as the major translatable RNA species in tobacco [40,41]. Given that PsaC is translated as a soluble protein and only later assembled into PSI [42,43], it is not surprising that its translated mRNA is found in the stroma. Longer transcripts are however found in membrane fractions (summarized as "b" in Figure 4). These are co-transcripts including cistrons encoding membraneous subunits of the NDH complex (e.g., NdhD). These likely draw the *psaC*-cistron to membranes in the process of their translation. Similarly, monocistronic *atpH* transcripts are found almost exclusively in the stroma, while the long, polycistronic precursors are co-fractioning also considerably with membranes. An additional cistrons in the longer transcripts is represented by *atpF*, which is translated on membranes [14]. Thus, the membrane-attachment of polycistronic *atpH–atpF* transcripts can be explained by the affinity of the *atpF* mRNA for membranes. As a general trend, different transcripts from complex operons can show differential association with membranes in chloroplasts.

2.6.3. Differential Membrane-Association of rRNA Species

If a number of chloroplast mRNAs display membrane-association and if this association is mediated by ribosomes, then we should find rRNA attached to chloroplast membranes as well. Indeed, we do observe that 16S rRNA and 5S rRNA co-fractionate with membranes (Figure 4). The unexpected finding is, that different rRNA species show different MEVs. While the ratios of membrane versus stroma signal of 16S and 23S rRNAs remain constant across the three conditions analyzed, 5S rRNA displays a strong decline of membrane association in tissue with mature chloroplasts, which is also reflected by our microarray analysis (see *rrn5* in Figure 2a). 5S rRNA and 23S rRNA are part of the large ribosomal subunit. Thus, the finding that a subpopulation of 5S rRNA localizes to the stroma in mature chloroplasts suggests that it does so independently of the ribosome. Alternatively, a recently discovered antisense RNA to 5S may cause this discrepancy [44]. We cannot distinguish between sense and antisense transcripts with our microarray nor with the double-stranded probe used to detect 5S rRNA on RNA gel blots. Whether the tissue-dependent localization of 5S rRNA is of functional significance remains to be determined.

2.6.4. Constant Association of Plastid RNAs with Chloroplast Membranes Suggests Altered Translational Rates during Chloroplast Greening rather than Increased Accumulation or Improved Recruitment of mRNAs to Membranes

Consistent with the microarray analyses, the RNA gel blot hybridizations also shows a striking similarity in RNA distribution between membranes and stroma in the three tissue types (Figure 4). mRNAs for the soluble proteins *rbcL*, *rps16* and *psaC* display almost identical preferential localization to the stroma fractions with only minor amounts within the membrane preparation in all three tissues. A noticeable, albeit small shift of RNAs towards membrane pools in mature chloroplasts was observed for *psbB* and *petA*, and a minor shift also for *psbA*. *psbA* translation is massively induced after irradiation [33,39], far exceeding the minor increase in membrane attachment seen here. Given that in general, translational induction in the light is known to be massive for many mRNAs [33], we have to assume that RNA tethering to membranes does not directly mirror translational activity. This is reflecting previous analyses in *Chlamydomonas* that found EF-Tu and *psaA/B* mRNAs are strongly and consistently attached to chloroplast membranes throughout dark-light cycles, while translation occurs only in the light [45]. Similarly, the barley *psaA/B* mRNA associates with chloroplast membranes in etioplasts without noticeable *psaA/B* translation [25,38]. It was suggested that there is active repression of *psaA/B* translation in the dark since translation off of isolated etioplast membranes could be initiated in vitro after adding a translation-competent extract [25]. This is supported by studies in pea that report that membrane-bound, non-polysomal RNAs are moving within 8 min into polysomes after illumination, suggesting that the initial association with membranes is either mediated by one or few ribosomes or by other means [10]. Increasing translation from low levels to high levels would not necessarily lead to an increased detection of mRNAs in membranes, since at least in theory, it would not matter for our approach if few or many ribosomes tether an mRNA to thylakoid membranes. Quantitative ribosome profiling experiments in etioplasts versus chloroplasts could solve this problem.

In sum, our analyses demonstrate that etioplasts are poised for a rapid translational answer to light signals on the level of global mRNA attachment to chloroplast membranes. This likely goes hand in hand with parallel processes that stage the necessary production of mRNAs and the corresponding expression factors in etioplasts [30,32]. It remains to be shown, how important constant attachment of mRNAs to membranes is for rapid photomorphogenesis versus other processes, like the provision of RNA binding proteins for translational initiation.

3. Materials and Methods

3.1. Plant Material

Wild type maize seedlings were grown on soil for 8 days with 16 h light/8 h dark cycle at 25 °C. Etiolated tissue was grown without light for 8 days and harvested directly. Greening tissue was generated by growing plants for 6 days without light and the light-exposure on the seventh day for 16 h and subsequently return into darkness for 8 h. On the 8th day the tissue was exposed for a further hour and 15 min prior to harvest, which was carried out in safety green light. Tissue with mature chloroplasts was generated by growing plants for 8 days under a standard light regime (16 h light/8 h dark). On the 8th day, mature seedlings were exposed for one additional hour to light after which the tissue was harvested. During plastid isolation from etiolated and greening tissue, safety green light was used, so additional exposure to photomorphogenic light was eliminated. All plants were grown at 26 °C.

3.2. Extraction of Stroma and Membrane Fractions

Intact plastids from three different tissue types were isolated from 8-day-old seedlings [46] with the following modifications. After Percoll gradient purification, washed plastids were resuspended in resuspension buffer (50 mM HEPES/KOH pH 8, 330 mM sorbitol) by gentle agitation at 4 °C. Dilution of extracted plastids was used for estimation of the number of plastids per microliter microscopically. Plastids from etiolated, deetiolated and green tissue was adjusted to the same number of plastids per microliter. The plastid pellet was resuspended in hypotonic polysomal buffer [19] without detergents, containing heparin, chloramphenicol, cycloheximide, β-mercatoethanol and protease inhibitor cocktail without EDTA. Chloramphenicol stabilizes ribosomes on mRNAs. Lysis of plastids was performed by passing the extract for 40 to 50 times through a 0.5 mm × 25 mm syringe in a hypotonic buffer designed to keep polysomes intact [19,47]. Membranes and stroma were separated by centrifugation at 40,000× g at 4 °C for 30 min. The membrane pellets were washed five to six times in polysomal buffer, and finally resuspended in the same volume as the stroma volume obtained. From equal volume proportions, RNA was isolated. For RNA gel blot analyses, RNAs were loaded on volume basis.

3.3. EDTA and Puromycin Treatments of Purified Plastid Membranes

Plastid membrane fractions from all three tissue types were obtained from 8-day-old seedlings. Aliquots (300 µL) were incubated at room temperature with 100 mM EDTA (final concentration) for 15 min with gentle shaking. Control reactions were performed without addition of EDTA. After incubation, RNAs released from membrane particles, were separated by centrifugation for 10 min at 21,000× g at 4 °C. Membrane pellets were resuspended in lysis buffer in same volume as the one of the corresponding supernatant. From each of the fractions RNA was extracted by phenol/chloroform/isoamylalcohol (24:24:1).

For puromycin treatment, membrane fractions were incubated in 500 ng/µL Puromycin in polysome buffer [19] for 15 min at room temperature. Afterwards, membranes were pelleted by centrifugation at 21,000× g for 10 min at 4 °C. The membrane pellet was washed five times in polysome buffer. RNA was extracted from supernatant (freed RNA) and pellet (membrane-bound RNA) fractions. Two µg of RNA from each fraction were labelled and subjected to microarray analysis see below.

3.4. RNA Gel Blot Analyses

Total RNA was extracted from etiolated, deetiolated and green tissue with TRIzol reagent (Invitrogen, Carlsbad, CA, USA) according to the manufacturer's instructions. For each individual sample a pool of collected leaves was used. Five micrograms of total RNAs from tissue or RNAs extracted from stroma and membrane fractions were separated on an agarose gel containing 1.2%

formaldehyde and transferred to uncharged nylon membranes (Hybond N, GE Healthcare, Little Chalfont, UK). After transfer, the blots were UV cross-linked (150 mJ/cm^2) and then stained with methylene blue to check RNA integrity and loading.

Polymerase chain reaction (PCR) based probes were used for in vitro transcription via T7 polymerase (Thermo Fisher, Waltham, MA, USA) in the presence of 32P-UTP according to the manufacturer's instructions. Primers used are listed in Table S4. For all northern blots, except *rrn16* and *rrn5*, single stranded in vitro transcripts were used for hybridization. The *rrn16* and *rrn5* RNA gel blots were hybridized with a double-stranded PCR probe labelled in presence of 32P-CTP via a Klenow exo- fragment (Thermo Fisher, Waltham, MA, USA) according to the manufacturer's instructions. After pre-incubation in Church buffer (0.5 M sodium phosphate buffer, pH 7.2 and 7% sodium dodecyl sulfate (SDS) for 1 h at 68 °C, hybridization of radiolabelled probe was performed overnight at 68 °C in the same buffer, followed by at least three 15-min washes in 1× SSC; 0.1% SDS, 0.5× SSC/0.1% SDS and 0.2× SSC/0.1% SDS, respectively. Ribosomal RNA related probes, *psbA*, and *rbcL* probes were washed additionally for 15 min at 68 °C with 0.1× SSC/0.1% SDS and 0.05× SSC/0.1% SDS. Signals were detected by autoradiography with the Personal Molecular Imager system (Bio-Rad, Hercules, CA, USA).

3.5. Immunoblot Analyses

Proteins from stroma and membrane fractions were loaded on a volume-per-volume basis. Proteins were separated by sodium dodecyl sulfate polyacrylamide gel electrophoresis (SDS-PAGE) and transferred to Hybond-C Extra Nitrocellulose membranes (GE healthcare). Integrity and loading of protein samples was detected by Ponceau S stain of the membrane prior incubation with antibodies. Antibody hybridization was performed for 1 h in 2% skim milk powder in TBST buffer for primary antibody and secondary horseradish peroxidase coupled antibody in TBST.

3.6. Tiling Microarray Design

Overlapping PCR products covering the whole maize chloroplast genome were generated by using self-made Taq-polymerase and were purified via the QIAquick PCR purification kit (Qiagen, Hilden, Germany). A total of 500 ng of each PCR product was transferred into a 384-well plate, dried and resuspended in 5 μL of 1M betaine in 3× SSC buffer. DNA fragments were transferred on to silanated glass slides (Vantage Silanated Amine Slides; CEL Associates, Los Angeles, CA, USA) using an OmniGrid Accent microarrayer (GeneMachines, San Francisco, CA, USA). Array design as described [20].

3.7. Microarray Hybridisation

Microarrays were cross-linked at 250 mJ/cm^2 in a UV cross linker (GS Gene Linker, Bio-Rad), and blocked with BSA buffer (1% BSA, 0.1% SDS and 5× SSC) for 1 h at 42 °C. The labelled RNAs from stroma and membrane fractions (approximately 15 μL) were mixed, loaded on to array and covered with a cover slip. Hybridisation was performed overnight in a 42 °C water bath in Corning microarray hybridization chambers. Unspecifically bound RNA was washed off the slide by incubation in 1× SSC, 0.2× SSC, and 0.05× SSC for 8 min each on a horizontal shaker at 180 rpm. Slides were dried by centrifugation at 1500 rpm in plate centrifuge and scanned using the ScanArrayGx Plus microarray scanner (Perkin Elmer, Shelton, CT, USA).

3.8. Microarray Analysis

Data from four replicate experiments (this totals 48 spots per probe since we were using 12 replicate spots per probe per array) were filtered against elements with low signal-to-noise ratios, and local background was calculated according to default parameters in Genepix Pro 7.0 software (Molecular Devices, Sunnyvale, CA, USA). Only spots with a signal-to-background ratio >4 and for which 60% of pixels have a F532 fluorescent signal >2 SD above background were chosen for further analysis. Fragments for which fewer than half the number of all spots (i.e., less than 24 spots) per array passed these cutoffs were not used for subsequent analyses and appear as gaps when enrichment ratios are plotted according to chromosomal position. Background-subtracted data were used to calculate the median (median of ratios (membrane RNA = 635 nm: stroma RNA = 532 nm)) as described in [21]. This value is called the membrane enrichment value (MEV). Normalization for Figure 2a and Table S1 was done according to the median (median of ratios (F635/F532)) value for all probes with signal above background on each array.

Supplementary Materials: Supplementary materials can be found at www.mdpi.com/1422-0067/18/9/1881/s1.

Acknowledgments: Generous funding by the DFG to CSL (as part of the collaborative research centre TRR175 project A02) is gratefully acknowledged. The antisera against PetD were kindly provided by Alice Barkan (University of Oregon). We thank Alice Barkan for critical reading and comments on the manuscript. The authors have no conflict of interest to declare.

Author Contributions: Julia Legen and Christian Schmitz-Linneweber conceived and designed the experiments; Julia Legen performed the experiments; Julia Legen and Christian Schmitz-Linneweber analyzed the data; Christian Schmitz-Linneweber wrote the paper.

Conflicts of Interest: The authors declare no conflict of interest. The founding sponsors had no role in the design of the study; in the collection, analyses, or interpretation of data; in the writing of the manuscript, and in the decision to publish the results.

Abbreviations

MEV   Membrane Enrichment Value
NDH   NADH Dehydrogenase

References

1. Jung, H.; Gkogkas, C.G.; Sonenberg, N.; Holt, C.E. Remote control of gene function by local translation. *Cell* **2014**, *157*, 26–40. [CrossRef] [PubMed]
2. Keiler, K.C. RNA localization in bacteria. *Curr. Opin. Microbiol.* **2011**, *14*, 155–159. [CrossRef] [PubMed]
3. Doroshenk, K.A.; Crofts, A.J.; Morris, R.T.; Wyrick, J.J.; Okita, T.W. Ricerbp: A resource for experimentally identified RNA binding proteins in *Oryza sativa*. *Front. Plant Sci.* **2012**, *3*, 90. [CrossRef] [PubMed]
4. Uniacke, J.; Zerges, W. Chloroplast protein targeting involves localized translation in *Chlamydomonas*. *Proc. Natl. Acad. Sci. USA* **2009**, *106*, 1439–1444. [CrossRef] [PubMed]
5. Uniacke, J.; Zerges, W. Photosystem II assembly and repair are differentially localized in *Chlamydomonas*. *Plant Cell* **2007**, *19*, 3640–3654. [CrossRef] [PubMed]
6. Schottkowski, M.; Peters, M.; Zhan, Y.; Rifai, O.; Zhang, Y.; Zerges, W. Biogenic membranes of the chloroplast in *Chlamydomonas reinhardtii*. *Proc. Natl. Acad. Sci. USA* **2012**, *109*, 19286–19291. [CrossRef] [PubMed]
7. Chua, N.-H.; Blobel, G.; Siekevitz, P.; Palade, G.E. Attachment of chloroplast polysomes to thylakoid membranes in *Chlamydomonas reinhardtii*. *Proc. Natl. Acad. Sci. USA* **1973**, *70*, 1554–1558. [CrossRef] [PubMed]
8. Chua, N.H.; Blobel, G.; Siekevitz, P.; Palade, G.E. Periodic variations in the ratio of free to thylakoid-bound chloroplast ribosomes during the cell cycle of *Chlamydomonas reinhardtii*. *J. Cell Biol.* **1976**, *71*, 497–514. [CrossRef] [PubMed]
9. Margulies, M.M.; Michaels, A. Ribosomes bound to chloroplast membranes in *Chlamydomonas reinhardtii*. *J. Cell Biol.* **1974**, *60*, 65–77. [CrossRef] [PubMed]
10. Fish, L.E.; Jagendorf, A.T. Light-induced increase in the number and activity of ribosomes bound to pea chloroplast thylakoids in vivo. *Plant Physiol.* **1982**, *69*, 814–824. [CrossRef] [PubMed]
11. Hurewitz, J.; Jagendorf, A.T. Further characterization of ribosome binding to thylakoid membranes. *Plant Physiol.* **1987**, *84*, 31–34. [CrossRef] [PubMed]
12. Michaels, A.; Margulies, M.M. Amino acid incorporation into protein by ribosomes bound to chloroplast thylakoid membranes: Formation of discrete products. *Biochim. Biophys. Acta* **1975**, *390*, 352–362. [CrossRef]
13. Alscher-Herman, R.; Jagendorf, A.T.; Grumet, R. Ribosome-thylakoid association in peas: Influence of anoxia. *Plant Physiol.* **1979**, *64*, 232–235. [CrossRef] [PubMed]

14. Zoschke, R.; Barkan, A. Genome-wide analysis of thylakoid-bound ribosomes in maize reveals principles of cotranslational targeting to the thylakoid membrane. *Proc. Natl. Acad. Sci. USA* **2015**, *112*, E1678–E1687. [CrossRef] [PubMed]

15. Zoschke, R.; Chotewutmontri, P.; Barkan, A. Translation and co-translational membrane engagement of plastid-encoded chlorophyll-binding proteins are not influenced by chlorophyll availability in maize. *Front. Plant Sci.* **2017**, *8*, 385. [CrossRef] [PubMed]

16. Barkan, A. Expression of plastid genes: Organelle-specific elaborations on a prokaryotic scaffold. *Plant Physiol.* **2011**, *155*, 1520–1532. [CrossRef] [PubMed]

17. Adachi, Y.; Kuroda, H.; Yukawa, Y.; Sugiura, M. Translation of partially overlapping *psbD–psbC* mRNAs in chloroplasts: The role of 5′-processing and translational coupling. *Nucleic Acids Res.* **2012**, *40*, 3152–3158. [CrossRef] [PubMed]

18. Chotewutmontri, P.; Barkan, A. Dynamics of chloroplast translation during chloroplast differentiation in maize. *PLoS Genet.* **2016**, *12*, e1006106. [CrossRef] [PubMed]

19. Barkan, A. Proteins encoded by a complex chloroplast transcription unit are each translated from both monocistronic and polycistronic mRNAs. *EMBO J.* **1988**, *7*, 2637–2644. [PubMed]

20. Schmitz-Linneweber, C.; Williams-Carrier, R.; Barkan, A. RNA immunoprecipitation and microarray analysis show a chloroplast pentatricopeptide repeat protein to be associated with the 5′ region of mRNAs whose translation it activates. *Plant Cell* **2005**, *17*, 2791–2804. [CrossRef] [PubMed]

21. Zerges, W.; Rochaix, J.D. Low density membranes are associated with RNA-binding proteins and thylakoids in the chloroplast of *Chlamydomonas reinhardtii*. *J. Cell Biol.* **1998**, *140*, 101–110. [CrossRef] [PubMed]

22. Klaff, P.; Gruissem, W. A 43 kd light-regulated chloroplast RNA-binding protein interacts with the *psbA* 5′ non-translated leader RNA. *Photosyn. Res.* **1995**, *46*, 235–248. [CrossRef] [PubMed]

23. Hotto, A.M.; Huston, Z.E.; Stern, D.B. Overexpression of a natural chloroplast-encoded antisense RNA in tobacco destabilizes 5S rRNA and retards plant growth. *BMC Plant Biol.* **2010**, *10*, 213. [CrossRef] [PubMed]

24. Mulo, P.; Sakurai, I.; Aro, E.M. Strategies for *psbA* gene expression in cyanobacteria, green algae and higher plants: From transcription to psii repair. *Biochim. Biophys. Acta* **2012**, *1817*, 247–257. [CrossRef] [PubMed]

25. Klein, R.R.; Mason, H.S.; Mullet, J.E. Light-regulated translation of chloroplast proteins. I. Transcripts of *psaA–psaB*, *psbA*, and *rbcL* are associated with polysomes in dark-grown and illuminated barley seedlings. *J. Cell Biol.* **1988**, *106*, 289–301. [CrossRef] [PubMed]

26. Maier, R.M.; Neckermann, K.; Igloi, G.L.; Kossel, H. Complete sequence of the maize chloroplast genome: Gene content, hotspots of divergence and fine tuning of genetic information by transcript editing. *J. Mol. Biol.* **1995**, *251*, 614–628. [CrossRef] [PubMed]

27. Blomqvist, L.A.; Ryberg, M.; Sundqvist, C. Proteomic analysis of highly purified prolamellar bodies reveals their significance in chloroplast development. *Photosynth. Res.* **2008**, *96*, 37–50. [CrossRef] [PubMed]

28. Blomqvist, L.A.; Ryberg, M.; Sundqvist, C. Proteomic analysis of the etioplast inner membranes of wheat (*Triticum aestivum*) by two-dimensional electrophoresis and mass spectrometry. *Physiol. Plant.* **2006**, *128*, 368–381. [CrossRef]

29. Ikeuchi, M.; Murakami, S. Behaviour of the 36,000-dalton protein in the integral membranes of squash etioplasts during greening. *Plant Cell Physiol.* **1982**, *23*, 575–583.

30. Von Zychlinski, A.; Kleffmann, T.; Krishnamurthy, N.; Sjolander, K.; Baginsky, S.; Gruissem, W. Proteome analysis of the rice etioplast: Metabolic and regulatory networks and novel protein functions. *Mol. Cell. Proteom.* **2005**, *4*, 1072–1084. [CrossRef] [PubMed]

31. Selstam, E.; Sandelius, A.S. A comparison between prolamellar bodies and prothylakoid membranes of etioplasts of dark-grown wheat concerning lipid and polypeptide composition. *Plant Physiol.* **1984**, *76*, 1036–1040. [CrossRef] [PubMed]

32. Kleffmann, T.; von Zychlinski, A.; Russenberger, D.; Hirsch-Hoffmann, M.; Gehrig, P.; Gruissem, W.; Baginsky, S. Proteome dynamics during plastid differentiation in rice. *Plant Physiol.* **2007**, *143*, 912–923. [CrossRef] [PubMed]

33. Kanervo, E.; Singh, M.; Suorsa, M.; Paakkarinen, V.; Aro, E.; Battchikova, N.; Aro, E.M. Expression of protein complexes and individual proteins upon transition of etioplasts to chloroplasts in pea (*Pisum sativum*). *Plant Cell Physiol.* **2008**, *49*, 396–410. [CrossRef] [PubMed]

34. Lonosky, P.M.; Zhang, X.; Honavar, V.G.; Dobbs, D.L.; Fu, A.; Rodermel, S.R. A proteomic analysis of maize chloroplast biogenesis. *Plant Physiol.* **2004**, *134*, 560–574. [CrossRef] [PubMed]

35. Margulies, M.M.; Tiffany, H.L.; Hattori, T. Photosystem I reaction center polypeptides of spinach are synthesized on thylakoid-bound ribosomes. *Arch. Biochem. Biophys.* **1987**, *254*, 454–461. [CrossRef]

36. Mackender, R.O. Etioplast development in dark-grown leaves of *Zea mays* L. *Plant Physiol.* **1978**, *62*, 499–505. [CrossRef] [PubMed]

37. Klein, R.R.; Mullet, J.E. Control of gene expression during higher plant chloroplast biogenesis. Protein synthesis and transcript levels of *psbA*, *psaA–psaB*, and *rbcL* in dark-grown and illuminated barley seedlings. *J. Biol. Chem.* **1987**, *262*, 4341–4348. [PubMed]

38. Laing, W.; Kreuz, K.; Apel, K. Light-dependent, but phytochrome-independent, translational control of the accumulation of the p700 chlorophyll-A protein of photosystem I in barley (*Hordeum vulgare* L.). *Planta* **1988**, *176*, 269–276. [CrossRef] [PubMed]

39. Klein, R.R.; Mullet, J.E. Regulation of chloroplast-encoded chlorophyll-binding protein translation during higher plant chloroplast biogenesis. *J. Biol. Chem.* **1986**, *261*, 11138–11145. [PubMed]

40. Ruf, S.; Kossel, H.; Bock, R. Targeted inactivation of a tobacco intron-containing open reading frame reveals a novel chloroplast-encoded photosystem I-related gene. *J. Cell Biol.* **1997**, *139*, 95–102. [CrossRef] [PubMed]

41. Hirose, T.; Sugiura, M. Both RNA editing and RNA cleavage are required for translation of tobacco chloroplast *NdhD* mRNA: A possible regulatory mechanism for the expression of a chloroplast operon consisting of functionally unrelated genes. *EMBO J.* **1997**, *16*, 6804–6811. [CrossRef] [PubMed]

42. Antonkine, M.L.; Jordan, P.; Fromme, P.; Krauss, N.; Golbeck, J.H.; Stehlik, D. Assembly of protein subunits within the stromal ridge of photosystem I. Structural changes between unbound and sequentially PSI-bound polypeptides and correlated changes of the magnetic properties of the terminal iron sulfur clusters. *J. Mol. Biol.* **2003**, *327*, 671–697. [CrossRef]

43. Li, N.; Zhao, J.D.; Warren, P.V.; Warden, J.T.; Bryant, D.A.; Golbeck, J.H. PsaD is required for the stable binding of PsaC to the photosystem I core protein of synechococcus sp. Pcc 6301. *Biochemistry* **1991**, *30*, 7863–7872. [CrossRef] [PubMed]

44. Sharwood, R.E.; Hotto, A.M.; Bollenbach, T.J.; Stern, D.B. Overaccumulation of the chloroplast antisense RNA *as5* is correlated with decreased abundance of 5S rRNA in vivo and inefficient 5S rRNA maturation in vitro. *RNA* **2011**, *17*, 230–243. [CrossRef] [PubMed]

45. Breidenbach, E.; Leu, S.; Michaels, A.; Boschetti, A. Synthesis of EF-Tu and distribution of its mRNA between stroma and thylakoids during the cell cycle of *Chlamydomonas reinhardii*. *Biochim. Biophys. Acta* **1990**, *1048*, 209–216. [CrossRef]

46. Voelker, R.; Barkan, A. Two nuclear mutations disrupt distinct pathways for targeting proteins to the chloroplast thylakoid. *EMBO J.* **1995**, *14*, 3905–3914. [PubMed]

47. Finster, S.; Eggert, E.; Zoschke, R.; Weihe, A.; Schmitz-Linneweber, C. Light-dependent, plastome-wide association of the plastid-encoded RNA polymerase with chloroplast DNA. *Plant J.* **2013**, *76*, 849–860. [CrossRef] [PubMed]

International Journal of
Molecular Sciences

MDPI

Article

Effects of TROL Presequence Mutagenesis on Its Import and Dual Localization in Chloroplasts

Lea Vojta *, Andrea Čuletić † and Hrvoje Fulgosi

Laboratory for Plant Molecular Biology and Biotechnology, Division of Molecular Biology, Ruđer Bošković Institute, Bijenička cesta 54, 10000 Zagreb, Croatia; aculetic@bot.uni-kiel.de (A.Č.); fulgosi@irb.hr (H.F.)
* Correspondence: lvojta@irb.hr; Tel.: +385-1-468-0238; Fax: +385-1-456-1177
† Current address: Biologie der Pflanzenzelle, Botanisches Institut, Christian-Albrechts-Universität zu Kiel, Am Botanischen Garten 1-9, 24118 Kiel, Germany.

Received: 11 January 2018; Accepted: 11 February 2018; Published: 14 February 2018

Abstract: Thylakoid rhodanase-like protein (TROL) is involved in the final step of photosynthetic electron transport from ferredoxin to ferredoxin: NADP$^+$ oxidoreductase (FNR). TROL is located in two distinct chloroplast compartments—in the inner envelope of chloroplasts, in its precursor form; and in the thylakoid membranes, in its fully processed form. Its role in the inner envelope, as well as the determinants for its differential localization, have not been resolved yet. In this work we created six N-terminal amino acid substitutions surrounding the predicted processing site in the presequence of TROL in order to obtain a construct whose import is affected or localization limited to a single intrachloroplastic site. By using in vitro transcription and translation and subsequent protein import methods, we found that a single amino acid exchange in the presequence, Ala67 to Ile67 interferes with processing in the stroma and directs the whole pool of in vitro translated TROL to the inner envelope of chloroplasts. This result opens up the possibility of studying the role of TROL in the chloroplast inner envelope as well as possible consequence/s of its absence from the thylakoids.

Keywords: TROL; protein import; chloroplasts; dual localization; ATP; inner envelope membrane; thylakoids

1. Introduction

TROL (thylakoid rhodanase-like protein, At4g01050) is an integral membrane component of non-appressed thylakoid membranes, responsible for the anchoring of FNR (ferredoxin:NADP$^+$ oxidoreductase) [1]. This 66 kDa protein is firmly attached to the membrane by its two transmembrane helices, resisting extraction by high salt, urea, or high pH treatments [1]. The N-terminus of TROL consists of a presequence that directs the protein to chloroplasts. Two predicted transmembrane helices surround the centrally positioned inactive rhodanase-like domain, RHO, which is orientated towards the thylakoid lumen. The C-terminus of the protein resides in the cytosol and consists of a single hydrophobic FNR-binding region, ITEP (highly conserved module of TROL necessary for establishing high-affinity interaction with FNR), and a region upstream of the ITEP domain, designated PEPE (Pro-Val-Pro repeat-rich region), which is followed by a possible PVP hinge, proposed to introduce flexibility to the FNR-binding region. In our previous research we proposed that the TROL–FNR interaction is dynamic [2,3], in which binding and release of FNR from TROL can regulate the flow of photosynthetic electrons before the pseudo-cyclic electron transfer pathway becomes activated [4]. By studying Arabidopsis *trol* mutants we proposed that the TROL–FNR interaction is the bifurcating point between electron-conserving and electron-dissipating pathways [4].

In addition to being mainly located at the stroma thylakoids, in its fully processed form of 66 kDa, TROL can also be found embedded in the chloroplast inner envelope membrane (IM) in its non-processed form (70 kDa), which indicates a possible role in the electron transfer chain specific for this membrane [1]. This dual localization has been proven by the proteome analysis of chloroplast envelopes and thylakoid membranes [5] and was proposed to be dependent on the NADP$^+$/NADPH ratio in the chloroplasts, as shown for the shuttling of the Tic62 protein [6]. Although there were attempts to investigate the structure and the function of TROL as the FNR anchor [1,7–9] its role in the IM remains undefined.

TROL is a nuclear-encoded protein, synthesized with a cleavable N-terminal presequence that targets the protein to and across the chloroplast envelope. During chloroplast protein import, the targeting sequences are sequentially decoded resulting in localization of the polypeptide to the appropriate organelle subcompartment [10]. Since TROL has been found both in the IM and the thylakoids [1,5], the determinants for this dual localization were the subject of investigation in this research.

Most of the IM proteins are directed to the general import pathway (Toc and Tic complex) through their cleavable transit peptides on their way to the chloroplasts. Some preproteins are released from the translocon at the level of the IM, as instructed by a hydrophobic stop-transfer signal in their sequence [11]. Others are first targeted to the stroma by their stroma-targeting presequence, and their processed mature form is subsequently re-exported into the IM, by so-called conservative sorting [11]. Proteins targeted to the thylakoid membrane require dual targeting signals that direct their import across the chloroplast envelope membranes and subsequent transport to the thylakoids. Stroma-targeting domains of preproteins are recognized and removed by the stromal processing peptidase (SPP) and lumen-targeting domain by a second processing protease [10,12]. A single bipartite transit sequence can carry the information for targeting to the thylakoids, with the stromal targeting domain located at the N-terminal region and the thylakoid lumen targeting domain at the C-terminal region of the presequence [10,13]. Preplastocyanin (prePC), and the subunits of the oxygen evolving complex (preOE16, preOE23, and preOE33) are configured as described. Integral membrane proteins, such as the precursor to the light-harvesting chlorophyll a/b binding protein (preLHCP) and the precursor to the 20-kDa subunit of the CP24 complex [13], contain information only for envelope transport within the presequence. The signals for targeting of these proteins to the thylakoids seem to reside within the primary structure of the mature polypeptide [14]. Since a single targeting domain is predicted for the transit sequence of TROL, this protein might belong to the latter group.

Targeting sequences are of various sizes, from 30 to 120 amino acids, and contain a high number of hydroxylated residues, but lacking acidic ones. The N-terminal 10–15 residues are devoid of Gly, Pro, and charged residues, a variable middle region is rich in Ser, Thr, Lys, and Arg, lacking acidic residues; and a C-terminal proteolytic processing site [15]. It contains a loosely conserved Ile/Val-x-Ala/Cys-Ala motif, recognized by SPP.

It has been postulated that chloroplast-localized proteins have various energy requirements for their import, according to their localization. ATP in the intermembrane space, usually less than 50 µM, drives the transport of the precursor across the outer envelope membrane [16,17], and import through the IM into the stroma progresses if the ATP concentration is higher than 100 µM [18,19]. This ATP is needed for the action of molecular chaperones in the stroma, which provide the driving force to complete import into the organelle [20]. Proteins localized in the same organellar compartment might have different energy needs for their import, depending on the import pathway they use [17]. Upon entering the stroma, the transit sequence is removed by SPP [20]. Further import into thylakoids requires additional translocators and energy demand. Four different mechanisms are known to target proteins from the stroma to the thylakoids [20–22].

The aim of this work was to introduce mutations in the presequence of the TROL protein, around the processing site, in an attempt to obtain a construct that would direct TROL to a single sub-chloroplast compartment, namely to a single membrane: either the inner envelope or the thylakoids. In addition, we wanted to characterize TROL import properties and requirements in more detail.

2. Results

After amino acid comparison between TROL from *A. thaliana* and other vascular plants, the N-terminal conserved region was determined around the SPP cleavage site. In this conserved region, AKSLTYEEALQQ (aa 67–78), we have chosen six potentially significant amino acids that could influence the import and localization of protein TROL in chloroplasts. Changes made to the presequece were as following: e1: 67Ala→67Ile, e2: 71Thr→71Asn, e3: 72Tyr→72Val, e4: 73Glu74Glu→73Gln74Gln, e5: 76Leu→76Thr, e6: 78Gln→78Val (Table 1). Hydrophobicity was checked for each amino acid substitution, according to Kyte and Doolittle [23] (Figure 1). For substitutions 73Glu74Glu→73Gln74Gln, the exchange of polarity has been compared according to Zimmerman et al. [24] (Figure 1). After the mutations were introduced into the gene *At4g01050*, constructs in the pZL1 vector were checked by DNA sequencing.

Table 1. Mutations introduced to the TROL presequence by the QuikChange Multi Site-Directed Mutagenesis and corresponding primer sequences.

Change	Mutation/Primer Name	Primer Sequence
e1	g199a_c200t	aagcagtgcaacagctcctattaaatccctgacgtacgag
e2	c212a_g213t	gctcctgctaaatccctgaattacgaggaagctctgcaac
e3	t214g_a215t	ctgctaaatccctgacggtcgaggaagctctgcaac
e4	g217c_g220c	gctaaatccctgacgtaccagcaagctctgcaacaatcta
e5	c226a_t227c	ccctgacgtacgaggaagctacgcaacaatctatgacca
e6	c232g_a233t	cgtacgaggaagctctgcaagtatctatgaccacttcttca
	TROLfor	caccatggaagctctgaaaaccgca
	TROLrev	gggctgcgatggcatcg

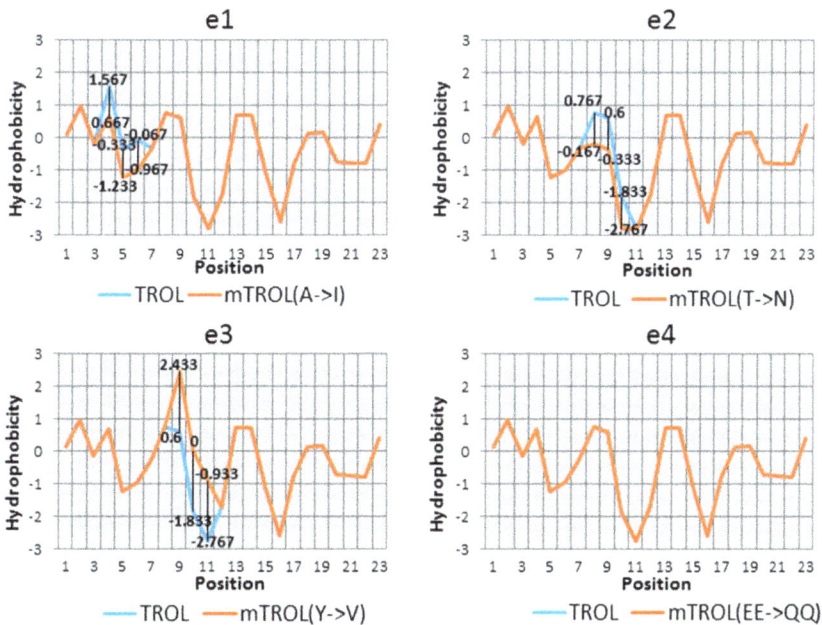

Figure 1. *Cont.*

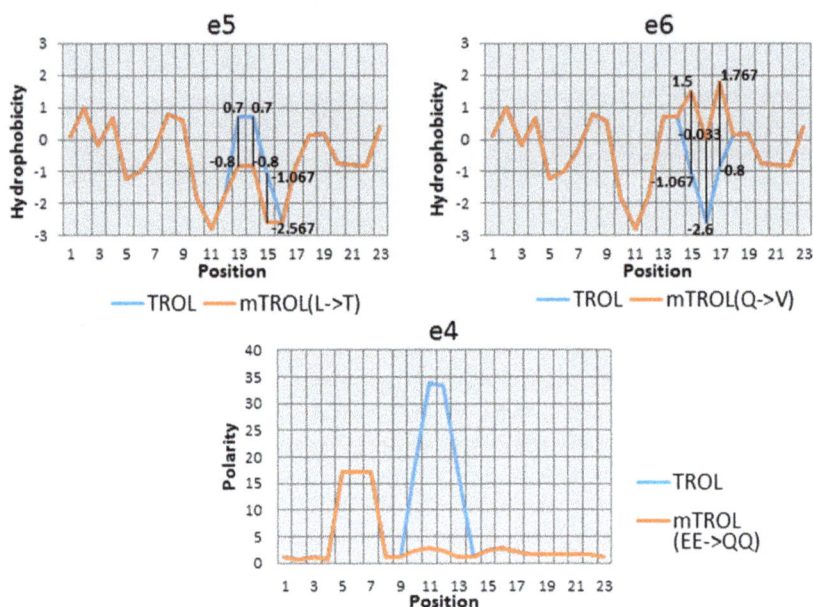

Figure 1. Kyte and Doolittle plots representing hydrophobicity change as a consequence of amino acid substitution/s in the TROL presequence [23]. Amino acids 67–78 of the partially conserved N-terminal part of the presequence around the predicted transit peptide cleavage site, AKSLTYEEALQQ, were substituted as follows: e1: 67Ala→67Ile, e2: 71Thr→71Asn, e3: 72Tyr→72Val, e4:73Glu74Glu→73Gln74Gln, e5: 76Leu→76Thr, e6:78Gln→78Val. For the e6 substitution, a polarity check according to Zimmerman was performed [24].

The wild type precursor of TROL and its e1–e6 presequence mutants were labeled with [35S]-methionine during in vitro translation and imported into isolated intact pea chloroplasts under various conditions. First, the standard import experiment was performed, in the absence or the presence of externally added 3 mM ATP. After a 20-min long import at 25 °C, chloroplasts were re-isolated on a Percoll cushion and treated with the protease thermolysin, to distinguish precursors loosely bound to the chloroplast envelope (early intermediates) from the firmly incorporated ones [17,18]. We observed that TROL requires ATP for successful processing and its import into the thylakoid membrane. After import with ATP, the majority of TROL was found in thylakoids, in its fully processed form, and just a small portion was located in the chloroplast envelope, in its precursor form, protected from thermolysin (Figure 2a, WT, lane 5). Presequence mutants e2, e3, e4, and e6 showed identical import behaviour as the wild type TROL (Figure 2a,b). In contrast, mutant e1 locates almost exclusively to the envelope membrane (Figure 2a, e1, lane 10, Figure 2b, lane 4). Only a very small portion was processed to the mature form and transported to the thylakoids. It seems that just a single amino acid substitution in the TROL presequence (67Ala→67Ile) results in almost exclusive envelope localization (Figure 2a,b). Presequence mutant e5, although having very similar import properties to the wild-type TROL, seems to have a slightly increased portion of envelope-bound form after import into chloroplasts (Figure 2a, e5, lanes 9 and 10, Figure 2b, lane 12).

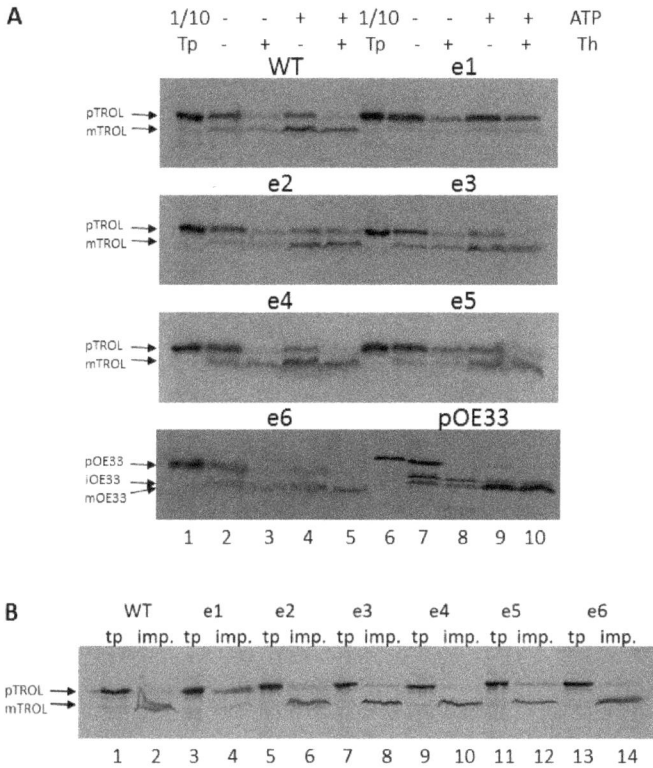

Figure 2. (**A**) Wt TROL and its presequence mutants e1–e6 are imported into pea chloroplasts. In vitro synthesized [^{35}S]-TROL and e1–e6, as well as the control protein pOE33 from thylakoid lumen were incubated with isolated intact chloroplasts at 25 °C for 20 min, in a standard import reaction containing 3 mM ATP (lanes 4, 5, 9, and 10) or without ATP (lanes 2, 3, 7, and 8). After import, samples were re-isolated on a Percoll cushion and treated with 0.5 µg thermolysin (Th) per µg chlorophyll (lanes 3, 5, 8, and 10). Untreated samples are shown in lanes 2, 4, 7, and 9. The results were analyzed by SDS-PAGE. Lanes 1 and 6 represent 10% of the translation product (Tp) used for the import reactions. The positions of pTROL, mTROL, pOE33, iOE33, and mOE33 are indicated by arrows. (**B**) Comparison of import of TROL and its presequence mutants e1–e6 into chloroplasts. In vitro synthesized [^{35}S]-TROL and e1–e6 were incubated with isolated intact chloroplasts at 25 °C for 20 min, in a standard import reaction containing 3 mM ATP. After import, samples were re-isolated on a Percoll cushion and treated with 0.5 µg thermolysin (Th) per µg chlorophyll (lanes 2, 4, 6, 8, 10, 12, and 14). The results were analyzed by SDS-PAGE. Lanes 1, 3, 5, 7, 9, 11, and 13 represent 10% of the respective translation product. The positions of pTROL and mTROL are indicated by arrows.

Further, we wanted to investigate the energy requirements for the import of TROL and its presequence mutants e1 and e5. After isolating chloroplasts from peas grown in the dark for at least 8 h, and incubating isolated intact chloroplasts in the dark and on ice to minimize internal ATP production, we added 0, 30, 300, or 3000 µM of external ATP to the import reaction. After 20 min. import at 25 °C and subsequent chloroplast re-isolation, we observed that even the smallest amount of ATP enabled the import of TROL into the thylakoids to some extent (Figure 3, lanes 3 and 5). Additional ATP resulted in a slight increase in the quantity of processed protein, while at the same time the portion of precursor form, located in the inner envelope, decreased (Figure 3, lanes 7 and 9). Mutant e5 seems to require higher amounts of external ATP to complete the import (Figure 3, e5). Also,

Int. J. Mol. Sci. **2018**, *19*, 569

there is more envelope form present in this mutant (Figure 3, lanes 5, 7, 9). Mutant e1 localizes almost exclusively to the chloroplast inner envelope, needing some external ATP (30–300 µM) for complete protection from thermolysin, as required for inner envelope incorporation (Figure 3, e1). As a control, oxygen evolving complex protein of 33 kDa (pOE33) was imported along with the TROL constructs. This is a well-characterized protein which is localized to the thylakoid lumen and forms a soluble translocation intermediate in the stroma. It uses the general import pathway into the chloroplasts and is therefore a suitable control for TROL import experiments. It seems that pOE33 requires more ATP to complete its import. With the lowest ATP concentration, stromal intermediate iOE33 is visible, and import seems to be complete after the addition of more than 300 µM ATP (Figure 3, pOE33). The higher energy need for pOE33 import results probably from its two processing events (in the stroma and in the thylakoid lumen) and its luminal localization.

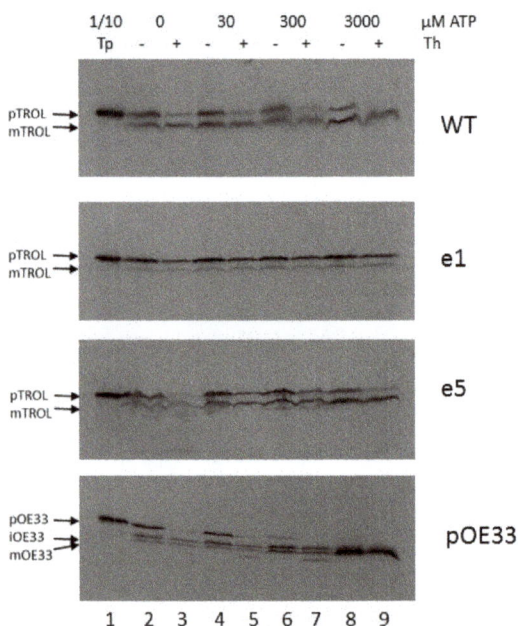

Figure 3. Energy requirement for import of wt TROL and its presequence mutants e1 and e5 into pea chloroplasts. Import into intact pea chloroplasts was performed under standard conditions, by incubating in vitro synthesized [^{35}S]-TROL, e1, e5, and control protein pOE33 from thylakoid lumen with chloroplasts corresponding to 20 µg chlorophyll at 25 °C. ATP-scale import into intact pea chloroplasts was performed using increasing concentrations of ATP from 0 to 3000 µM. After import, chloroplasts were re-isolated on a Percoll cushion and samples were treated with 0.5 µg thermolysin (Th) per µg chlorophyll (lanes 3, 5, 7 and 9). Untreated samples are shown in lanes 2, 4, 6, and 8. The results were analyzed by SDS/PAGE. The respective precursor, intermediate, and mature forms are indicated by arrow heads. Lane 1 represents 1/10 of the translation product (Tp) used for the import reaction.

We also wanted to explore how the import of TROL and e1 and e5 mutants proceeds on a temporal scale. Radioactively labeled precursors were added to the standard import reaction including 3 mM ATP and chloroplasts corresponding to 20 µg chlorophyll. Imports were performed for 0.5, 2, 5, 10, and 20 min at 25 °C. All samples were subsequently re-isolated on Percoll cushion, treated with thermolysin, lysed, and separated into membrane (P) and soluble fractions (S). After the analysis of radioactive signals on films, it was clearly visible that as time advances the amount of proteins in the pellet fractions increases for all tested precursors (Figure 4). Between 5 and 10 min of import are needed for around 50% of TROL import to be accomplished. For complete import, more than 10 min were required (Figure 4, lane 10). For the e1 mutant, the portion of imported protein that locates to the IM visibly increases with time, while the amount of mature, thylakoid-localized form starts to appear after 10 min. of import (Figure 4, lane 8). In this experiment, mutant e5 showed nearly identical properties to wt TROL.

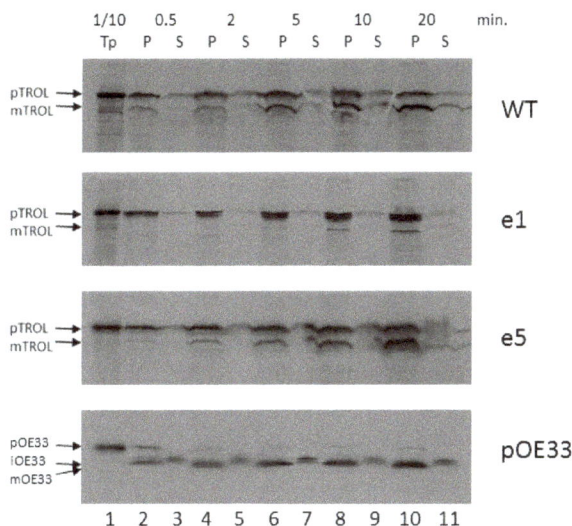

Figure 4. Time-scale import of wt TROL and e1 and e5 presequence mutants into intact pea chloroplasts. Radioactively labeled TROL, e1, e5, and control protein pOE33 were imported using increasing times in standard import reactions at 25 °C in the presence of 3 mM ATP. Import was performed for 0.5 (lanes 2 and 3), 2 (lanes 4 and 5), 5 (lanes 6 and 7), 10 (lanes 8 and 9), or 20 min (lanes 10 and 11). After import chloroplasts were separated into the pellet (P, lanes 2, 4, 6, 8, and 10) and soluble (S, lanes 3, 5, 7, 9, and 11) fractions. Lane 1 indicates 1/10 of the respective translation product (Tp) used for the import reaction. Precursor (p), intermediate (i), and mature (m) forms of TROL, e1, e5, and OE33 are indicated by arrows.

Finally, we investigated to what extent is TROL, imported both to the IM and the thylakoids, associated with the membranes. The strength and the nature of this association was tested by applying either 6 M Urea, 0.1 M Na_2CO_3 pH 11.5 for the separation of integral from peripheral membrane proteins, or 1 M NaCl, which decreases electrostatic interactions between proteins and charged lipids. 6 M Urea only partially extracted imported TROL from the membranes, mostly its envelope-located portion, while the mature forms from thylakoid membranes remained almost fully intact (Figure 5, lanes 2 and 3). Treatment of the envelopes with Na_2CO_3 at pH 11.5 converts membrane vesicles to sheets and disrupts protein–protein interactions, while protein–lipid interactions remain and the bilayer is otherwise intact. In this way we could determine if an integral membrane protein has achieved stable insertion into the bilayer. Carbonate treatment extracted only a very small portion of the tested proteins from the membrane, indicating that TROL and the tested mutants are strongly attached to the lipid bilayer (Figure 5, lanes 4 and 5). High salt did not extract TROL from the membranes (Figure 5, lanes 6 and 7), indicating strong ionic interactions between TROL and the membranes.

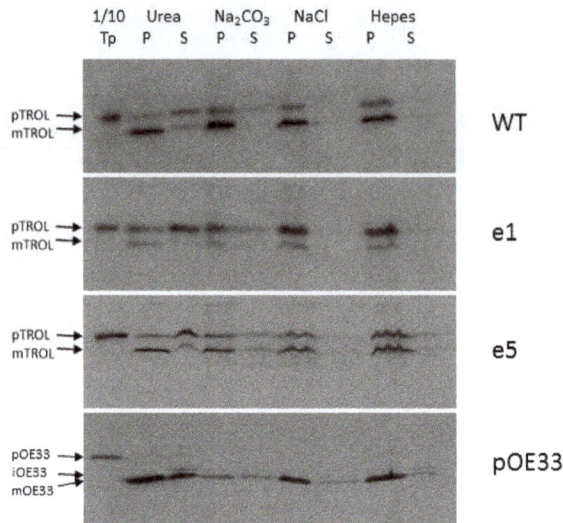

Figure 5. Extraction of wt TROL and its presequence mutants e1 and e5 from the membranes. In vitro synthesized [^{35}S]-TROL, e1, e5, and control protein pOE33 were incubated with isolated intact chloroplasts at 25 °C for 20 min, in a standard import reaction containing 3 mM ATP. Subsequent to import, chloroplasts were re-isolated, washed, and separated into membrane and soluble fractions. Isolated membranes, containing imported proteins, were treated with 6 M Urea in 10 mM HEPES/KOH pH 7.6 (lanes 2 and 3), 0.1 M Na$_2$CO$_3$ pH 11.5 (lanes 4 and 5), or 1 M NaCl (lanes 6 and 7). All incubations were performed for 20 min on RT. As a control, membranes were incubated solely in 10 mM HEPES/KOH pH 7.6 for 30 min on ice (lanes 8 and 9). Afterwards, samples were centrifuged at 265,000× g for 10 min at 4 °C, and both pellets (P, lanes 2, 4, 6, and 8) and the supernatants (S, lanes 3, 5, 7, and 9) were analyzed by SDS–PAGE and by exposure on X-ray films. Lane 1 represents 10% of the respective translation product (Tp) used for the import reactions. The positions of pTROL, mTROL, pOE33, iOE33, and mOE33 are indicated by arrows.

3. Discussion

In the N-terminal conserved region of TROL, around the SPP cleavage site, we have chosen six potentially significant amino acids for its import and localization in chloroplasts. Selection of the amino acids was largely based on their hydrophobic or hydrophilic character, which we tried to change by substitutions, in an attempt to interfere with stromal processing. The structure and influence of the neighboring amino acids have also been taken into consideration. We have chosen those amino acids whose neighboring hydrophobic/hydrophilic signals were as indiscernible as possible. In the e1 mutation, Ala67 was changed to Ile. Ala67 is a part of the AXS motif, predicted to represent a signal for cleavage of the presequence by the SPP. Ile is similar in nature to Ala, but contains three more methyl groups that make it more hydrophobic. In the e2 mutation, neutral and polar Thr71 was changed to hydrophobic Asn. The corresponding Kyte and Doolittle plot [23] resulted in a neutral to slightly hydrophilic character of the changed amino acid site. In e3, by Tyr72 to Val exchange, strong hydrophilicity was substituted by medium to strong hydrophobicity. In this example the structure has also been changed by removing the cyclic ring in Tyr, which could additionally influence the hydrophobicity. In e4, where Glu73 and Glu74 were changed to two Gln, hydrophobicity remained the same, but the loss of polarity (according to Zimmerman et al. [24]) could influence protein sorting. In e5 hydrophobic Leu76 was exchanged for neutral Thr, leading to a more hydrophilic character. In e6 Gln78 to Val exchange leads to a large hydrophilicity loss and change to hydrophobicity.

After successfully introduced mutations into TROL presequence, constructs e1–e6, as well as the wild type, were incorporated into the pZL1 vector and further utilized for investigation of import characteristics of this dually localized protein.

In organello import into isolated chloroplasts was performed to investigate chloroplast localization and integration of wt TROL and its presequence mutants into the chloroplast membranes. We expected that some of the amino acid substitutions made to the presequence might interfere with proper processing in the stromal compartment and/or result in alternative localization of TROL. The labeled precursor was imported into organelles and processed into a smaller mature protein of around 66 kDa (Figure 2, mTROL), previously shown to co-purify entirely with the thylakoid fraction [1]. A weaker signal of the size of the labeled precursor was also detected (Figure 2, pTROL). This signal was protected from protease digestion (Figure 2a, lane 5, Figure 2b, lane 2), indicating a portion of TROL located in the IM of chloroplasts, in its non-processed form of 70 kDa [1,5]. In contrast to the wt, the TROL e1 comprised almost entirely of the IM portion, indicating an influence of the 67Ala/67Ile presequence mutation on TROL localization. In this mutant the SPP recognition motif AKS has been changed to IXS, leading to inhibition of stromal processing and the increment of the IM portion of TROL. Compared to wt TROL, it is obvious that for the e1 mutant not only import into thylakoids is impaired, but the IM portion of the protein is highly increased. Since this portion is protected from thermolysin action, as well as from extraction with high salt and carbonate concentrations, we conclude that envelope-TROL is firmly incorporated into the membrane.

Experiments using increasing external ATP concentrations indicate that TROL requires more than 300 mM ATP for completion of its import into thylakoids, while only 30 mM is necessary for IM incorporation, as visible after thermolysin treatment (Figure 3). This result implies the stop-transfer mechanism of TROL import and its lateral insertion to the IM [11,22]. Time-scale experiments in the presence of 3 mM ATP indicate that the incorporation of TROL into the IM happens very early, and the smaller thylakoid form starts to appear after 2–5 min of import, reaching a maximum between 10–20 min (Figure 4). We observed that after addition of 3 mM ATP in wt, but also in e1, to some extent, there is much less of the IM form of TROL compared to the 300 mM experiment. The distribution of TROL between these two membranes might be influenced by the energy distribution between thylakoids and envelope compartments. TROL could be switching the intensity of its action between the IM (role unknown) and the thylakoids (regulation of photosynthetic electron transfer), influenced/directed by the energetic state of those compartments.

Jurić et al. [1] have shown that imported TROL could not be extracted from membranes by high salt, urea, or high pH treatments, indicating that At4g01050 is an integral thylakoid membrane protein. The same extraction procedures were used to investigate the membrane incorporation character of the e1–e6 TROL mutants (Figure 5). Only 6 M urea extracted a portion of protein, mainly the IM incorporated one, while TROL in the thylakoids remained fully protected. Mutation e1 causes visible changes in TROL localization, directing most of the protein to the IM of chloroplasts. Once there, the incorporated TROL resists the extraction procedures in the same way as the wild type protein, in which only urea solubilized a portion of the protein from the membranes. After import, the e5 mutation exhibits slightly more IM portion of TROL than the wt, but not significant enough to use it in further experiments.

TROL is not an isolated case in terms of dual localization in chloroplasts. Seventeen other proteins have been identified in both chloroplastic envelope and thylakoid membranes [22]. All of them are predicted to carry an N-terminal signal for chloroplast import. These proteins belong to one of the following groups according to their function: protein transport, tetrapyrrole biosynthesis, membrane dynamics, and transport of nucleotides and inorganic phosphate [22]. One of these proteins is Tic62, a redox sensor in the inner envelope and the thylakoids, proposed to anchor FNR, just like TROL [7].

The nature and the mechanism of dual localization of TROL remains unresolved. The inner envelope of chloroplasts and thylakoid membrane share a similar lipid composition, but perform very different functions. The way in which TROL incorporates into these membranes is at the moment just a speculation. Up to now, no pathway that would catalyse dual targeting has been found. All known nuclear-encoded thylakoid proteins and some IEM proteins are targeted to the respective membranes via a stromal intermediate. TROL, as a dually localized protein, could also use a similar pathway. This has been confirmed by the import properties of the e1 mutant. However, absence of stromal processing for IM located TROL favours the possibility of lateral insertion to the IM. Stromal processing seems to be the key moment for further sorting of TROL to thylakoids. Tha4 and Hcf106 components of the tat pathway are integral membrane proteins that insert into thylakoid membrane by unassisted insertion [25]. The same pathway use PSII subunits W, X, and Y [26], PSI subunit K [27], and the CFoII subunit of the ATP synthase [28], and these have been shown to insert into the membrane by the unassisted pathway. As already mentioned, some IM proteins insert laterally into the membrane by a stop transfer mechanism upon entering the chloroplast. Some of the dual-localized integral proteins, like TROL, may be similarly held at the envelope first, then sorted to the thylakoids. Low ATP demands for insertion of TROL into the IM points to this conclusion.

Future prospects for the e1 mutant would be the production of transgenic *A. thaliana* plants containing TROL located only/mostly to the IM of chloroplasts. In this way we could study the effect of the absence of TROL from thylakoids on photosynthesis and subsequent electron transfer/dissipation events, as well as the so far unknown role of this protein in the inner envelope membrane.

4. Materials and Methods

4.1. TROL Presequence Substitutions

Using the QuikChange Multi Site-Directed Mutagenesis method, various mutations were introduced into the TROL presequence. Amino acids 67–78 of the presequence, AKSLTYEEALQQ, represent a partially conserved N-terminal part of the sequence, around the predicted transit peptide cleavage site. In this sequence we have chosen six amino acids potentially significant for TROL processing that might influence its import into and localization inside the chloroplasts. Changes made to the presequence were as following:

e1: 67Ala→67Ile, e2: 71Thr→71Asn, e3: 72Tyr→72Val, e4:73Glu74Glu→73Gln74Gln, e5: 76Leu→76Thr, e6:78Gln→78Val. Hydrophobicity was checked for each amino acid substitution, according to Kyte and Doolittle [23] (Figure 1). For substitutions 73Glu74Glu→73Gln74Gln, exchange polarity has been compared according to Zimmerman et al. [24] (Figure 1).

Subsequent to the introduction of mutations into the TROL presequence by PCR, constructs were transformed into competent bacteria, multiplied, purified, and checked by restriction enzymes and DNA sequencing.

4.2. In Vitro Transcription and Translation

The coding region for TROL from *Arabidopsis thaliana* was cloned into the vector pZL1 under the control of the T7 promoter and pOE33 from *Pisum sativum* was cloned into the vector pGEM4Z under the control of the SP6 promoter. Transcription and translation were carried out using the TNT® Quick Coupled Transcription/Translation System (Promega, Madison, WI, USA) in the presence of [^{35}S]-methionine (185 MBq, PerkinElmer, Boston, MA, USA) for radioactive labelling. After translation, the reaction mixture was centrifuged at 50,000× *g* for 20 min at 4 °C and the post-ribosomal supernatant was used for import experiments.

4.3. Chloroplast Isolation and Protein Import

Chloroplasts were isolated from leaves of 8 days old pea seedlings (*P. sativum* var. Letin, Agricultural Institute Osijek, Osijek, Croatia) and purified through Percoll density gradients as described [11,17]. A standard import reaction contained chloroplasts equivalent to 20 µg chlorophyll in 100 µL import buffer (330 mM sorbitol, 50 mM HEPES/KOH pH 7.6, 3 mM MgSO$_4$, 10 mM Met, 10 mM Cys, 20 mM K-gluconate, 10 mM NaHCO$_3$, 2% BSA (*w/v*)), up to 3 mM ATP and maximal 10%

(*v/v*) [^{35}S]-labeled translation products. Import reactions were initiated by the addition of translation product and carried out for 20 min at 25 °C, unless indicated otherwise. Reactions were terminated by separation of chloroplasts from the reaction mixture by centrifugation through a 40 % (*v/v*) Percoll cushion. Chloroplasts were washed once in 330 mM sorbitol, 50 mM HEPES/KOH pH 7.6, and 0.5 mM CaCl$_2$, lysed in 10 mM HEPES/KOH pH 7.6 for 30 min on ice and separated into membrane and soluble fractions by centrifugation at 265,000× *g* for 10 min at 4 °C. Import products were separated by SDS–PAGE and radiolabeled proteins analysed by exposure on X-ray films.

For the purpose of investigating the energy requirement, prior to import, ATP was depleted from chloroplasts and the translation product. For chloroplast isolation, plants were taken from the dark and isolated intact chloroplasts were further incubated in the dark on ice for 30 min. to diminish internal ATP production. For the import experiment, 0, 30, 30, and 3000 µM ATP was used and chloroplasts corresponding to 20 µg chlorophyll, in a 20 min import reaction.

Some experiments included chloroplast protease posttreatment, by using thermolysin after import. Thermolysin in concentration of 0.5 µg per µg chlorophyll was applied for 20 min on ice. The reaction was stopped by adding 5 mM EDTA. Chloroplasts were pelleted and resuspended in Laemmli buffer [29].

4.4. Membrane Extraction of Imported Proteins

Subsequent to import of 20 min at 25 °C, chloroplasts were re-isolated, washed, and separated to the membrane and soluble fractions. Isolated membranes, containing imported proteins, were treated with 6 M Urea in 10 mM HEPES/KOH pH 7.6, 0.1 M Na$_2$CO$_3$ pH 11.5, or 1 M NaCl. All incubations were performed for 20 min on RT. As a control, membranes were incubated solely in 10 mM HEPES/KOH pH 7.6 for 30 min on ice. Afterwards, samples were centrifuged at 265,000× *g* for 10 min. at 4 °C, and both pellets and the supernatants were analysed by SDS–PAGE and by the exposure on X-ray films.

Acknowledgments: This work has been funded by a Grant IP-2014-09-1173 from the Croatian Science Foundation to Hrvoje Fulgosi. We thank Dr. Mary Sopta for the language editing and critical reading of the manuscript.

Author Contributions: Lea Vojta designed the import experiments, grew experimental plants, isolated chloroplasts, performed in vitro transcription and translation and all import experiments. Lea Vojta also analyzed results and wrote the manuscript. Andrea Čuletić evaluated the character of substituted amino acids and performed molecular cloning. Hrvoje Fulgosi conceived this research and designed presequence amino acid substitutions and analyzed and discussed the results.

Conflicts of Interest: The authors declare no conflict of interest. The founding sponsors had no role in the design of the study; in the collection, analyses, or interpretation of data; in the writing of the manuscript, and in the decision to publish the results.

Abbreviations

TROL	thylakoid rhodanase-like protein
FNR	ferredoxin:NADP$^+$ oxidoreductase
RHO	rhodanase-like domain
ITEP	highly conserved module of TROL necessary for establishing high-affinity interaction with FNR
PEPE	Pro-Val-Pro repeat-rich region
IM	inner envelope membrane
SPP	stromal processing peptidase

References

1. Jurić, S.; Hazler-Pilepić, K.; Tomašić, A.; Lepeduš, H.; Jeličić, B.; Puthiyaveetil, S.; Bionda, T.; Vojta, L.; Allen, J.F.; Schleiff, E.; et al. Tethering of ferredoxin:NADP$^+$ oxidoreductase to thylakoid membranes is mediated by novel chloroplast protein TROL. *Plant J.* **2009**, *60*, 783–794. [CrossRef] [PubMed]

2.   Vojta, L.; Fulgosi, H. Energy conductance from thylakoid complexes to stromal reducing equivalents. In *Advances in Photosynthesis—Fundamental Aspects*; Najafpour, M.M., Ed.; InTech: Rijeka, Croatia, 2012; pp. 175–190, ISBN 978-953-307-928-8.

3.   Vojta, L.; Horvat, L.; Fulgosi, H. Balancing chloroplast redox status—Regulation of FNR binding and release. *Period. Biol.* **2012**, *114*, 25–31.

4.   Vojta, L.; Carić, D.; Cesar, V.; Antunović Dunić, J.; Lepeduš, H.; Kveder, M.; Fulgosi, H. TROL-FNR interaction reveals alternative pathways of electron partitioning in photosynthesis. *Sci. Rep.* **2015**, *5*, 10085. [CrossRef] [PubMed]

5.   Peltier, J.B.; Ytterberg, A.J.; Sun, Q.; van Wijk, K.J. New functions of the thylakoid membrane proteome of *Arabidopsis thaliana* revealed by a simple, fast, and versatile fractionation strategy. *J. Biol. Chem.* **2004**, *279*, 49367–49383. [CrossRef] [PubMed]

6.   Stengel, A.; Benz, P.; Balsera, M.; Soll, J.; Bölter, B. TIC62 redox-regulated translocon composition and dynamics. *J. Biol. Chem.* **2008**, *283*, 6656–6667. [CrossRef] [PubMed]

7.   Benz, J.P.; Lintala, M.; Soll, J.; Mulo, P.; Bölter, B. A new concept for ferredoxin-NADPH oxidoreductase binding to plant thylakoids. *Trends Plant Sci.* **2010**, *15*, 608–613. [CrossRef] [PubMed]

8.   Twachtmann, M.; Altmann, B.; Muraki, N.; Voss, I.; Okutani, S.; Kurisu, G.; Hase, T.; Hanke, G.T. N-terminal structure of maize ferredoxin: NADP$^+$ reductase determines recruitment into different thylakoid membrane complexes. *Plant Cell* **2012**, *24*, 2979–2991. [CrossRef] [PubMed]

9.   Vojta, L.; Fulgosi, H. Data supporting the absence of FNR dynamic photosynthetic membrane recruitment in *trol* mutants. *Data Brief* **2016**, *7*, 393–396. [CrossRef] [PubMed]

10. Cline, K.; Henry, R. Import and routing of nucleus-encoded chloroplast proteins. *Annu. Rev. Cell Dev. Biol.* **1996**, *12*, 1–26. [CrossRef] [PubMed]

11. Vojta, L.; Soll, J.; Bölter, B. Requirements for a conservative protein translocation pathway in chloroplasts. *FEBS Lett.* **2007**, *581*, 2621–2624. [CrossRef] [PubMed]

12. Teixeira, P.F.; Glaser, E. Processing peptidases in mitochondria and chloroplasts. *BBA Mol. Cell Res.* **2013**, *1833*, 360–370. [CrossRef] [PubMed]

13. Schnell, D.J. Protein targeting to the thylakoid membrane. *Annu. Rev. Plant Physiol. Plant Mol. Biol.* **1998**, *49*, 97–126. [CrossRef] [PubMed]

14. Lamppa, G.K. The chlorophyll a/b-binding protein inserts into the thylakoids independent of its cognate transit peptide. *J. Biol. Chem.* **1988**, *263*, 14996–14999. [PubMed]

15. Von Heijne, G.; Steppuhn, J.; Herrmann, R.G. Domain structure of mitochondrial and chloroplast targeting peptides. *Eur. J. Biochem.* **1989**, *180*, 535–545. [CrossRef] [PubMed]

16. Scott, S.V.; Theg, S.M. A new chloroplast protein import intermediate reveals distinct translocation machineries in the two envelope membranes: Energetics and mechanistic implications. *J. Cell Biol.* **1996**, *132*, 63–75. [CrossRef] [PubMed]

17. Vojta, L.; Soll, J.; Bölter, B. Protein transport in chloroplasts—Targeting to the intermembrane space. *FEBS J.* **2007**, *274*, 5043–5054. [CrossRef] [PubMed]

18. Theg, S.M.; Scott, S.V. Protein import into chloroplasts. *Trends Cell Biol.* **1993**, *3*, 186–190. [CrossRef]

19. Schnell, D.J.; Blobel, G. Identification of intermediates in the pathway of protein import into chloroplasts and their localization to envelope contact sites. *J. Cell Biol.* **1993**, *120*, 103–115. [CrossRef] [PubMed]

20. Soll, J.; Robinson, C.; Heins, L. The import and sorting of protein into chloroplasts. In *Protein Targeting, Transport and Translocation*; Dalbey, R., von Heijne, G., Eds.; Elsevier: Amsterdam, The Netherlands, 2002; pp. 240–267, ISBN 978-0-12-200731-6.

21. Jarvis, P.; Robinson, C. Mechanisms of protein import and routing in chloroplasts. *Curr. Biol.* **2004**, *14*, R1064–R1077. [CrossRef] [PubMed]

22. Klasek, L.; Inoue, K. Dual protein localization to the envelope and thylakoid membranes within the chloroplast. *Int. Rev. Cell Mol. Biol.* **2016**, *323*, 231–263. [CrossRef] [PubMed]

23. Kyte, J.; Doolittle, R.F. A simple method for displaying the hydropathic character of a protein. *J. Mol. Biol.* **1982**, *157*, 105–132. [CrossRef]

24. Zimmerman, J.M.; Eliezer, N.; Simha, R. The characterization of amino acid sequences in proteins by statistical methods. *J. Theor. Biol.* **1968**, *21*, 170–201. [CrossRef]

25. Fincher, V.; Dabney-Smith, C.; Cline, K. Functional assembly of thylakoid deltapH-dependent/Tat protein transport pathway components in vitro. *Eur. J. Biochem.* **2003**, *270*, 4930–4941. [CrossRef] [PubMed]

26. Woolhead, C.A.; Thompson, S.J.; Moore, M.; Tissier, C.; Mant, A.; Rodger, A.; Henry, R.; Robinson, C. Distinct Albino3-dependent and -independent pathways for thylakoid membrane protein insertion. *J. Biol. Chem.* **2001**, *276*, 40841–40846. [CrossRef] [PubMed]

27. Mant, A.; Woolhead, C.A.; Moore, M.; Henry, R.; Robinson, C. Insertion of PsaK into the thylakoid membrane in a "Horseshoe" conformation occurs in the absence of signal recognition particle, nucleoside triphosphates, or functional albino3. *J. Biol. Chem.* **2001**, *276*, 36200–36206. [CrossRef] [PubMed]

28. Michl, D.; Robinson, C.; Shackleton, J.B.; Herrmann, R.G.; Klösgen, R.B. Targeting of proteins to the thylakoids by bipartite presequences: CFoII is imported by a novel, third pathway. *EMBO J.* **1994**, *13*, 1310–1317. [PubMed]

29. Laemmli, U.K. Cleavage of structural proteins during the assembly of the head of bacteriophage T4. *Nature* **1970**, *227*, 680–685. [CrossRef] [PubMed]

International Journal of
Molecular Sciences

MDPI

Article

Nitric Oxide Enhancing Resistance to PEG-Induced Water Deficiency is Associated with the Primary Photosynthesis Reaction in *Triticum aestivum* L.

Ruixin Shao *, Huifang Zheng, Shuangjie Jia, Yanping Jiang, Qinghua Yang * and Guozhang Kang *

Collaborative Innovation Center of Henan Grain Crops and State Key Laboratory of Wheat and Maize Crop Science/College of Agronomy, Henan Agricultural University, Zhengzhou 450001, China; zhenghuifang@gmail.com (H.Z.); jiashuangjie2018@gmail.com (S.J.); jiangyanping.up@gmail.com (Y.J.)

* Correspondence: shaoruixin@henau.edu.cn (R.S.); yangqinghua@henau.edu.cn (Q.Y.); kangguozhang@henau.edu.cn (G.K.); Tel.: +86-371-5699-0186 (R.S., Q.Y. & G.K.)

Received: 9 August 2018; Accepted: 12 September 2018; Published: 18 September 2018

Abstract: Photosynthesis is affected by water-deficiency (WD) stress, and nitric oxide (NO) is a free radical that participates in the photosynthesis process. Previous studies have suggested that NO regulates excitation-energy distribution of photosynthesis under WD stress. Here, quantitative phosphoproteomic profiling was conducted using iTRAQ. Differentially phosphorylated protein species (DEPs) were identified in leaves of NO- or polyethylene glycol (PEG)-treated wheat seedlings (D), and in control seedlings. From 1396 unique phosphoproteins, 2257 unique phosphorylated peptides and 2416 phosphorylation sites were identified. Of these, 96 DEPs displayed significant changes ($\geq$1.50-fold, $p < 0.01$). These DEPs are involved in photosynthesis, signal transduction, etc. Furthermore, phosphorylation of several DEPs was upregulated by both D and NO treatments, but downregulated only in NO treatment. These differences affected the chlorophyll A–B binding protein, chloroplast post-illumination chlorophyll-fluorescence-increase protein, and SNT7, implying that NO indirectly regulated the absorption and transport of light energy in photosynthesis in response to WD stress. The significant difference of chlorophyll (Chl) content, Chl a fluorescence-transient, photosynthesis index, and trapping and transport of light energy further indicated that exogenous NO under D stress enhanced the primary photosynthesis reaction compared to D treatment. A putative pathway is proposed to elucidate NO regulation of the primary reaction of photosynthesis under WD.

Keywords: nitric oxide; 20% PEG-induced water deficiency; phosphoproteomic; *Triticum aestivum* L.; primary reaction of photosynthesis

---

1. Introduction

Nitric oxide (NO) is a small gaseous signaling molecule that has attracted significant interest. As a bioactive molecule, NO is involved in numerous physiological processes in animals, and has also been reported as a mediator of both biotic and abiotic stress responses in plants [1]. Recently, various enzymatic and nonenzymatic pathways for its synthesis have been reported [2,3]. Important advancements have been achieved to elucidate the roles of NO in plant growth and development. The role of NO as a signal molecule has been established as an activator of reactive oxygen species (ROS), scavenging enzymes under abiotic stress [4–7]. Interestingly, NO can be rapidly induced by environmental stimuli, and also acts as a secondary messenger during environmental stress-related signal transduction [4,8]. Exogenous NO, supplied to plants via sodium nitroprusside (SNP) or potassium nitrite, can also improve plant tolerance for environmental stress.

Water deficiency (WD) is an environmental factor that significantly impairs both crop growth and productivity [9]. Recent trends in climate change have increased the frequency and severity of WD stress, indicating it as one of the main environmental issues of global concern [10]. Plants grown under WD conditions have evolved various mechanisms to resist WD stress. One vital adaptation is the regulation of photosynthesis as a countermeasure to oxidative stress arising in WD conditions. Some of these adaptations involve NO [11–13].

Understanding the complex effect of NO on plants requires a detailed analysis of both physiological and molecular changes. The adaptability of plants to WD at the cellular and physiological levels is implemented by either induction or repression of relevant genes. Recently, a large number of analyses of plant responses to NO have been conducted using different techniques: transcriptional analyses identified 510 NO-related genes in *Arabidopsis thaliana* [14]; comparative proteomics identified 92 NO-related proteins in *Oryza sativa*, indicating that exogenous NO alleviated Al^{3+} toxicity [15]; and 166 proteins were identified in *Anemone vitifolia Buch* using label-free quantitative proteomics after NO treatment without stress [16]. The above-identified NO-related genes or proteins were mostly involved in photosynthesis, stress, signaling, and secondary metabolism. NO has also been reported to be either directly or indirectly involved in the regulation of translation, and in post-translational modifications (PTMs). An established example is phosphorylation, which is one of the most prevalent and functionally important PTMs [17]. Phosphorylation is also a frequent post-translational modification because thousands of kinase genes occupy 3%–4% of functional genes in plants. Phosphorylation has therefore been widely employed to investigate the molecular mechanism of biotic/abiotic stress and the resulting hormone response.

Wheat (*Triticum aestivum* L.) is one of the most-produced cereal crops. Although previous studies indicated that NO can alleviate WD stress, no systematic profiling studies by "omics" have to date been published that dissect the effect of NO on wheat plants in response to WD stress. Here, comparative phosphoproteomics were performed to analyze the phosphorylation dynamics in response to exogenous NO donors (SNP) and 20% polyethylene glycol (PEG) 6000-induced WD (D) in wheat seedlings. Based on most protein species and in combination with the physiological level, NO-induced WD resistance was further verified, and a possible mechanism was suggested.

2. Results

2.1. Morphological Change in Response to NO and D Stress

Leaf relative water content (RWC) showed a gradual decline both in "D" and "S + D" treatments; however, "S + D" exhibited higher RWC than "D" plants (Figure 1) during 72 h of D stress. Biomass results showed that D without SNP treatment inhibited the growth of wheat, while the growth state in the "S + D" treatment exceeded that of the "D" treatment. Here, this was also reflected in fresh and dry biomass accumulation as well as in plant height; S + D increased by 28.8%, 6.0%, and 12.2%, respectively, compared to D treatment (Figure 2A,B).

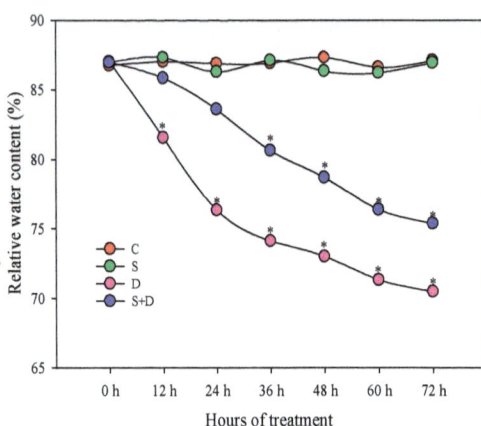

Figure 1. Change of relative water content (RWC) in leaves of wheat seedlings under normal water conditions and in response to three different treatments at different time points; a total of three independent biological replicates were conducted ($n = 10$). C, normal water conditions; S, pretreated with 150 µmol/L sodium nitroprusside; D, water deficiency induced by 20% polyethylene glycol (PEG)-6000; S + D, pretreated with 150 µmol/L sodium nitroprusside and water deficiency stress by 20% PEG-6000. Asterisks indicate significant differences among the four treatments at $p < 0.05$.

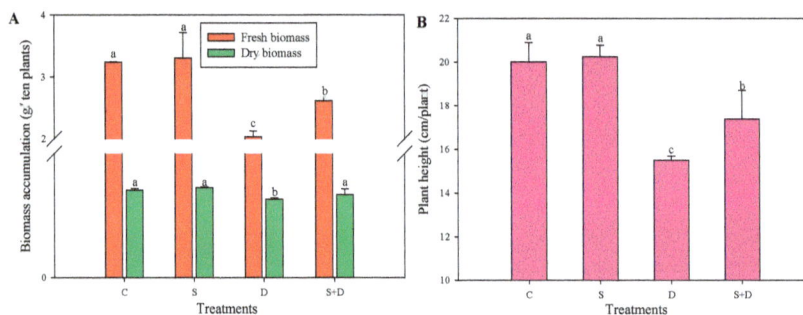

Figure 2. Phenotypic (**A**,**B**) changes in leaves of wheat seedlings under normal water conditions and in response to three different treatments; a total of three independent biological replicates were used ($n = 10$). For a detailed description of treatment conditions, please refer to the legend of Figure 1. Different lowercase letters indicate a statistically significant difference at $p < 0.05$.

2.2. Quantitative Identification of Phosphoproteins Using iTRAQ

A total of 2257 unique phosphopeptides were identified among all four experimental groups (Table S1); these peptides originated from 1396 unique phosphoproteins (Table S2) searched in the *pooideae* Uniprot database, and 2416 phosphorylation sites, 2110 (87%) of which contained serine (Ser) residues (Figure S1 and Table S3). A fold-change value >1 indicates that the examined phosphorylated peptide is more abundant (suggesting its upregulation), while a fold-change value <1 indicates a less abundant phosphorylated peptide (suggesting its downregulation). A total of 96 differentially expressed phosphorylated proteins (DEPs) were identified in D/C consisting of 37 increased and 61 decreased proteins; (S + D)/C had less DEPs than D/C, including four less-increased and four less-decreased proteins (Table 1). However, only 25 and 24 DEPs were found in S/C and (S + D)/D. In total, the four groups had 148 DEPs with 172 differentially expressed phosphorylated peptides (DEPPs) (>1.5-fold change and $p < 0.05$, see Figure 3). The peptide-sequence and phosphorylation-site (probabilities > 75%), protein description, peptide score, and fold changes of 198 spots associated with NO or WD are provided in Table S4. At least one phosphorylated S/T-Q site in most DEPPs (85%) was found to be localized to the nucleus.

Table 1. Number of differentially expressed significant phosphorylated proteins in the leaves of wheat seedlings.

Groups	Increase	Decrease	Each Total	Total Phosphorylated Proteins
S/C	12	13	25	
D/C	37	61	96	
(S + D)/C	33	57	88	148
(S + D)/D	7	17	24	

Notes: The rates of four groups exceeded +/−1.5, and $p < 0.05$. For a detailed description of treatment conditions, please refer to the legend of Figure 1.

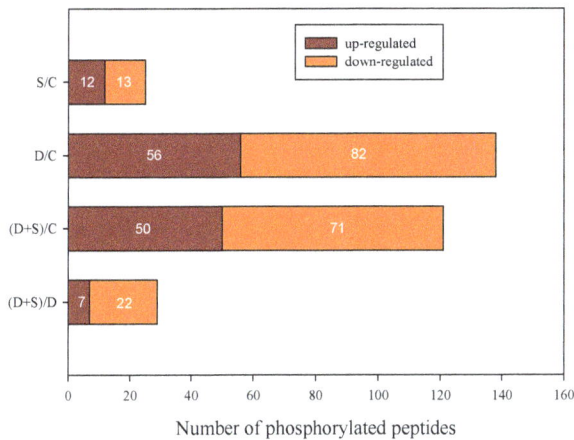

Figure 3. Upregulated and downregulated significant phosphorylated peptides among four groups. For a detailed description of treatment conditions, please refer to the legend of Figure 1.

2.3. Gene ontology and Principle Component Analysis (PCA) Analysis of Differentially Expressed Phosphorylated Protein Species in Response to D Stress and NO

Hierarchical clustering analysis of DEPs detected different expression patterns in all four comparison groups (Figure 4A). Gene ontology (GO) analysis of DEPs indicated that most DEPs could be classified into 17 different biological processes. Fewer DEPs were found to have molecular function than those associated with cellular component and biological processes, and most DEPs were involved in binding. In biological processes, most DEPs were related to cellular processes, single-organism processes, biological regulation, metabolic processes, and response to stimulus (Figure S2). D/C or (S + D)/C exhibited two-fold increased DEPs compared to S/C or (S + D)/D, respectively (Figure S3). The function of the phosphorylated proteins listed in Table S4 was mainly divided into photosynthesis metabolism, signaling, stress defense, protein modification, translation, and DNA binding (Figure 4B).

PCA was performed to analyze the reason for the observed DEPs within the four groups. The variation explained by the first principle component (PC1, *x* axis) can be largely attributed to D; PC1 explains 50.6% of the variation. PC2 (*y* axis) explains an additional 10.8% of the variation, much of which is attributable to NO treatment (Figure S4).

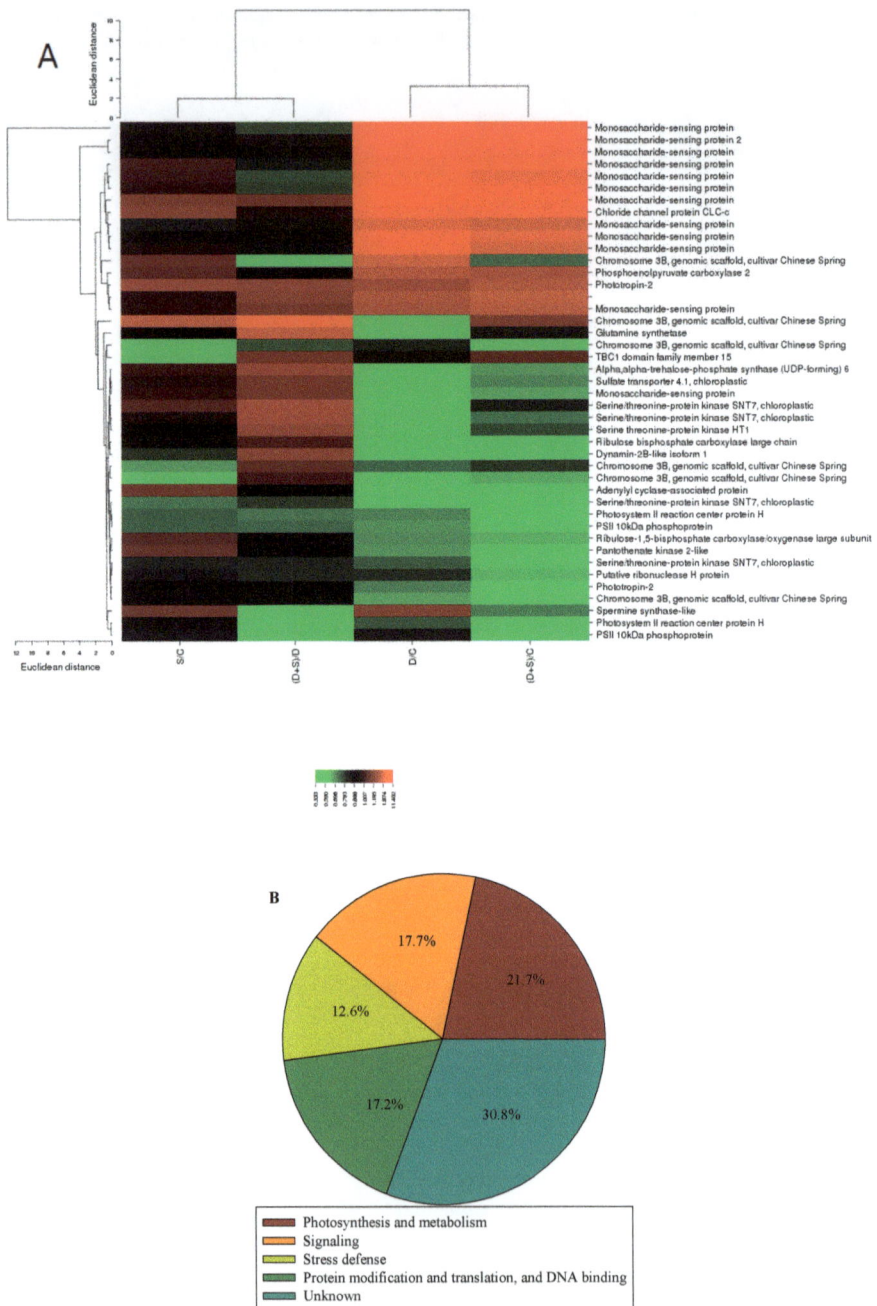

Figure 4. (**A**) Cluster analysis and (**B**) functional classification of significant differentially phosphorylated peptides in leaves of winter wheat among four different treatments. The color scale bar at the left of the hierarchical cluster analysis indicates the increased (red) and the decreased (green) peptides. For a detailed description of treatment conditions, please refer to the legend of Figure 1.

2.4. Metabolism-Related Differentially Expressed Phosphorylated Protein Species Accumulation in Response to NO and D Stress

Based on their classification, most DEPs (photosynthesis metabolism-related, 43 species) correlated with the photosynthesis metabolism (Table S4). No change in S/C was detected. Not surprisingly, most of the DEPs were downregulated when wheat seedlings suffered from D (D/C). However, some of these proteins were significantly upregulated (>1.5-fold change and $p < 0.05$) in (S + D)/D (Figure 4A), such as the chlorophyll a-b binding protein (Lhcb), chloroplast postillumination chlorophyll-fluorescence-increase protein (PIFI), and SNT7. The chosen DEPs are listed in Table 2, including their accession, coverage, proteins, unique peptides, Molecular weight, calc, and fold ratio. Only one peptide was phosphorylated with class II, the others were phosphorylated with class I. In these phosphorylation sites of class I, 11 DEPs have one or two phosphosites, located in either serine (Ser) or threonine (Thr) residue.

Int. J. Mol. Sci. **2018**, *19*, 2819

Table 2. Phosphorylated peptides related to the primary photosynthesis metabolism that were differentially and abundantly accumulated in response to nitric oxide (NO) and D stress.

Description	Phosphosite [a]	Accession	Coverage [b] (%)	Proteins [c]	Unique Peptide [d]	MW [kDa] [e]	Calc. pI [f]	Fold Ratio [g] (S/C, D/C, (S + D)/C, (S + D)/D)
Lhcb	S(11): 94.0	F2CRC1	12.24	4	2	26.72	6.11	0.73/0.22/0.27/1.23 ↑
SNT7	T(1): 100.0; S(5): 100.0 T(4): 100.0; S(8): 100.0 S(5): 100.0 S(5): 100.0	M8CIW2	2.59	1	2	68.90	9.01	1.04/0.62/0.82/1.32 ↑ 0.95/0.59/0.75/1.26 ↑ 0.82/0.76/0.62/0.82 0.74/0.63/0.49/0.78
PIFI	Class II T(2): 100.0	W5I170 B3TN78	14.98 19.18	3 1	1 1	24.69 7.83	4.77 8.53	0.79/0.60/0.72/1.20 ↑ 0.75/0.69/0.52/0.75

Notes: Significantly upregulated phosphorylated peptides ($p < 0.5$) are marked with ↑. If they were downregulated in D/C, they have been restored after sodium nitroprusside (SNP) pretreatment ((S + D)/C) to a certain extent. [a] Potential phosphorylation sites of 0.75 or above that belong to class I are highly reliable; $0.75 > p \geq 0.5$ belongs to class II. Each line corresponds to the last column. S refers to serine, T refers to thrine. [b] Percentage of peptides assigned to the predicted protein. [c] Number of proteins in each proteome species. [d] Number of unique peptides assigned to each protein. [e] Molecular weight of proteins. [f] Isoelectric point of proteins. [g] Ratio of NO or drought-treated protein levels to control protein levels (S/C, D/C, (S + D)/C), NO plus drought-treated protein levels to drought-treated protein levels ((S + D)/D). Ratio corresponds to the group in the same column. For a detailed description of treatment conditions, please refer to the legend of Figure 1.

2.5. Physiological Changes of Photosynthesis-Related Parameters in Response to NO and D Stress

Figure 5 shows the physiological changes associated with photosynthesis performance in response to both NO and D. No significant differences were found between C and S; however, D decreased by 48.2%, 45.1%, 37.87%, 34.2%, 20.3%, and 18.5% in Chl a, Chl b, Pn, RC/ABS, PI_{ABS}, and P_{ET}. When pretreated with SNP and then stressed by D, their values increased significantly; however, they could not completely recover to the control level.

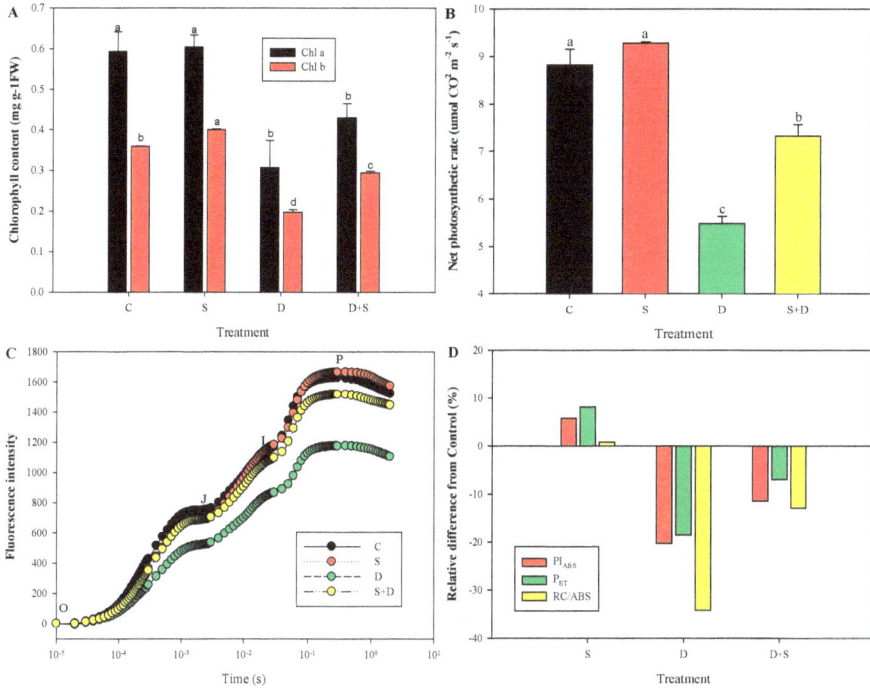

Figure 5. (**A**) Chlorophyll (Chl) content (Chl a, Chl b), (**B**) net photosynthetic rate (Pn), (**C**) fluorescence intensity, PI_{ABS}, P_{ET}, and (**D**) RC/ABS in leaves of wheat seedlings under normal water conditions and in response to NO or PEG-induced water deficiency treatments. C, normal water conditions; S, pretreated with 150 μmol/L sodium nitroprusside; D, water deficiency induced by treatment with 20% polyethylene glycol (PEG)-6000; S + D, pretreated with 150 μmol/L sodium nitroprusside and then water deficiency stressed by 20% PEG-6000. O, J, I and P mean that the analysis of the transient-considered fluorescence values at 50 ms (Fo, step O), 2 ms (F2 ms, step J), 30 ms (F30 ms, step I), and maximal level (FM, step P), respectively. For a detailed description of treatment conditions, please refer to the legend of Figure 1. Different lowercase letters in (**A,B**) indicate a statistically significant difference at $p < 0.05$.

3. Discussion

In abiotic stress studies, morphological, comparative physiological, and phosphoproteomic analyses are the most prevalent and effective strategies to investigate the mechanisms of abiotic stress resistance [18]. The results of the present study suggest that morphological and RWC changes after exogenous NO pretreatment contribute to D resistance. Furthermore, a comparative phosphoproteomic analysis among NO or D treatments was successfully performed. The results indicate that the phosphorylated peptides participated in key biological processes, which might work cooperatively to establish a new cellular homeostasis in response to NO.

3.1. Phosphoprotomics Elucidates Molecular Mechanism of NO-Induced D Tolerance in Wheat Seedlings

A number of studies indicated that phosphorylation played an important role in drought and chilling tolerance [17,18]. Here, 148 significant DEPs were identified in response to D or NO treatments. D was the single largest contributor to the variation among all four phosphoproteomic datasets obtained via PCA, followed by NO. This indicates that D treatment potentially impacts the phosphorylation of these proteins more significantly than NO treatment.

To date, quantitative phosphoproteomic studies in plants have shown that a majority of phosphoproteins only had one or two phosphorylation sites. Here, similar findings were obtained for wheat, and 76.9% of the phosphoproteins in four groups were found to possess fewer than two phosphorylation sites. However, only few DEPs were shown to have multiple phosphorylation sites. The distribution of phosphorylation Ser was maximal (87.3%) compared to the distributions of Thr and Lys, suggesting that these are preferentially highly sensitive and specific for Ser sites under D and NO treatment, followed by Thr and Lys residues. Identification of these phosphoproteins and phosphorylation sites provides a basis for the understanding of the molecular mechanisms underlying NO-regulated D resistance.

3.2. Comparison between Present Phosphoproteomic Data and Previous Studies

Only one NO-induced phosphoproteomic profile has been reported in cotton plants [19]; furthermore, no study on NO with a D stress-related phosphoproteomic profile has been published for higher plants. Two studies reported the D-induced phosphoproteomic profile in wheat [18,20]; however, one of these studies sampled wheat grains. Consequently, the results reported in these three References [18–20] were compared to the findings of this study. Thirty-four of the NO-induced phosphoprotein species found here were identical to those found by Fan et al. [19]; furthermore, nine and seven D–induced phosphoprotein species identical to those found here were identified in the remaining two studies [18,20], respectively. Differences and commonalities are listed in Table S5, most of which are related to photosynthesis metabolism and signaling; however, differential mechanisms could well be induced by NO with or without D stress in cotton or wheat. This difference is caused by different WD durations and extent, as well as wheat variety and sampling organization. The data reported herein aid the exploration of the molecular mechanism of NO-induced phosphorylation in response to severe D exposure.

3.3. Phosphorylation is Involved in the Photosynthesis Metabolism against D

Previous studies indicated that NO enhanced the stress resistance in other crops by regulating genes related to photosynthesis, metabolism, and stress signals [15,21]. Here, most of the phosphoproteins also participated in the photosynthesis metabolism, most of which in turn were downregulated after D treatment (Table 1, Figure S4). However, others were upregulated after NO plus D treatment, suggesting that the phosphorylation of these upregulated proteins was regulated by the interaction between NO and D; NO without D stress did not activate protein phosphorylation, which has also been reported before [19], suggesting that the effect of NO treatment is indirect.

Light-harvesting chlorophyll a/b protein (CAB) located in light-harvesting complex II (LHCII) can switch between light-harvesting antenna systems for photosystem (PS)I or PSII to maintain optimal excitation balance. This switch (termed state transition), involves the phosphorylation of CAB by specific thylakoid-bound Ser/Thr kinases (Stt7, SNT7, and TAKs) [22]. Polverari et al. [23] reported that NO modulated the CAB gene, SNT7 is required for the plastoquinone (PQ) redox state and provides a link between short- and long-term photosynthetic acclimation [24]. The novel nucleus-encoded chloroplast protein PIFI indicates the nonphotochemical reduction of PQ and has been reported to be involved in the NAD(P)H dehydrogenase complex-mediated chlororespiratory electron transport [25]. Here, the phosphorylation level at Ser or Thr sites showed significant decreases due to D stress, and no significant difference was found in response to NO exposure. However, phosphorylation of several peptides was significantly activated by the interaction of NO and D, indicating that, if these peptides are phosphorylated, they play an important role in the adaptation to D stress, which may be linked to the induction of NO.

3.4. NO-Enhanced D Tolerance Correlated with Primary Reaction of Photosynthesis

The central processes of photosynthesis are the absorption of light energy, electron transfer, and conversion of energy. Chl is an important pigment for photosynthesis and includes Chl a and Chl b. Chl a plays an important role as primary electron donor in the electron-transport chain and transfers resonance energy to the antenna complex, ending at the reaction center. Chl b aids the photosynthesis by absorbing light energy [26]. Chl a and Chl b concentrations influence the ability to absorb light energy. Numerous previous studies indicated that NO increased biomass accumulation and photosynthetic-performance driving force (PI_{ABS}) due to both the trapping of excitation energy (RC/ABS) and electron transport (P_{ET}) [2,5,12].

Here, changes of Chl a and Chl b content, Chl fluorescence intensity, PIABS, RC/ABS, and PET in response to NO and D treatment showed that absorption, transfer, and conversion of light energy were regulated by NO under D stress. The energy derived from the primary reaction of the photosynthesis is used in particular pathways to achieve the final result of photosynthesis [27].

3.5. Putative Mechanism of NO Regulates Primary Reaction of Photosynthesis in Response to D Stress

Combining the results reported here with those of previous studies [2,7,28–31] suggests that the primary reaction of photosynthesis may be regulated by NO involved in LHCII phosphorylation under D stress. The specific putative mechanism in wheat seedlings is shown in Figure 6.

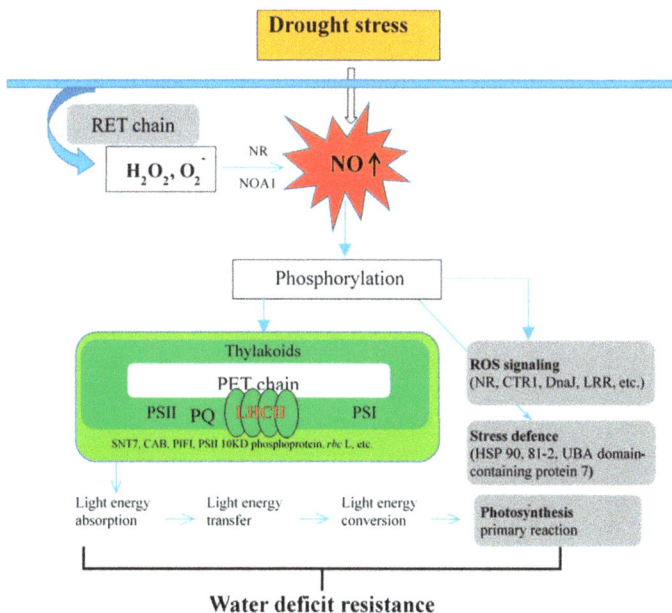

Figure 6. Schematic representation of NO involved in the regulation of photosynthesis to improve water-deficiency stress resistance. The contribution of NO to photosynthesis primary reaction under water deficit stress flows in the direction indicated by the arrow.

In this study, in response to D stress, the ROS that was generated via oxidative burst (intersystem electron transport (PET) from chloroplast and respiratory chain (RET) of mitochondria) is suggested to trigger NO synthesis. Under D stress, NO activates the phosphorylation of photosynthesis, signaling, and stress defense-related peptides, thus promoting stress tolerance. Additionally, when seedlings suffered from D stimuli, PSII momentarily ran faster than PSI; thus, this led to the uneven excitation of both the PS-generated redox and ROS signals that are then transmitted as retrograde signals to the cytosol, mitochondrion, and nucleus. During PET, including the PQ pool, the LHCII kinase SNT7 activates the phosphorylated LHCII. For example, the CAB protein moved from PSII to PSI, compensates the imbalance, and upregulates the PIFI phosphorylation level. This improved absorption, transfer, and conversion of light energy. These changes therefore enhance the primary reaction of photosynthesis, which carries implications for the mechanisms with which NO improves WD resistance.

4. Materials and Methods

4.1. Plant Materials and D and NO Treatments

Seeds of common wheat cv. Zhoumai 18 were germinated for 14 h in glass petri dishes (15 cm) in an illuminated incubator (Laifu Technology, Ningbo, China). Standard growth conditions for wheat were used (14 h light: 10 h dark photoperiod; 25 °C day and 15 °C night; relative humidity of 60% during the day and 75% at night). Uniform seedlings were transferred to full-strength Hoagland's liquid growth medium [32], and each petri dish contained approximately 60 seedlings. After 15 further days of growth, 3 dishes of each sample group were treated: for the C treatment, seedlings were incubated in Hoagland medium without SNP and PEG; for the D treatment, seedlings were incubated in Hoagland medium supplemented with 20% PEG (6000) (−0.8 MPa osmotic potential) (Sigma Chemical Co., St. Louis, MO, USA); for the S treatment, seedlings were incubated in Hoagland solution supplemented with 150 μmol/L SNP (Sigma Chemical Co., St. Louis, MO, USA); for the S + D treatment, seedlings were incubated in Hoagland medium supplemented with 150 μmol/L SNP for 3 days before also adding 20% PEG solution. At all stages of the experiment and for all sample groups, liquid media were changed daily. After 3 days of 20% PEG treatment, the uppermost fully developed leaves of all 4 treatments were collected, immediately frozen in liquid nitrogen, and stored at −80 °C until further physiological and phosphorylated proteomic analysis.

4.2. Phenotypic Analysis

Ten individual wheat seedlings were randomly harvested from each dish. The RWC in the leaves was determined with the method described by Wu and Xia [33]. RWC percentage was calculated using the following formula: RWC (%) = 100 × (FW − DW)/(SW − DW), where FW represents fresh weight, DW represents dry weight, and SW represents saturated weight. SW was determined after letting the leaflet float on distilled water for 24 h at room temperature.

Plant height and plant biomass (fresh and dry weights) were measured. Samples were then placed in an oven at 105 °C for 15 min followed by drying to a constant weight at 75 °C.

4.3. Physiological Analysis

Chl was extracted using the nonmaceration method developed by Hiscox and Israelstam [34]. Leaf samples (0.05 g) were incubated in 5 mL dimethyl sulfoxide at 65 °C for 4 h. Absorbance of the chloroplast pigments was measured at both 645 and 663 nm, and the Chl a and Chl b contents were calculated accordingly [35].

Fast Chl fluorescence-induction kinetics were measured using a Handy PEA Chl fluorimeter (Hansatech Instruments Ltd., Norflok, UK). Induced by red saturating light, the OJIP rising transients were measured with leaves that had been dark-adapted for 20 min. The analysis of the transient-considered fluorescence values at 50 ms (Fo, step O), 2 ms (F2 ms, step J), 30 ms (F30 ms, step I), and maximal level (FM, step P). On a logarithmic time scale, the increasing transient from Fo (F measured at 50 ms) to FP (where FP = FM under saturating excitation light, of which the excitation intensity was sufficiently high to ensure the closure of all RCs of PSII) showed polyphasic behavior. The JIP test represents a translation of original data into biophysical parameters that quantify the energy flow. The absorption of light energy (ABS), RC/ABS = ((F2 ms − Fo)/4(F300 ms − Fo))(Fv/FM) represents the active RC density on a Chl basis. A decrease in RC/ABS indicates an increase in the size of the Chl antenna serving each RC [2]. The contribution of dark reactions was calculated with the formula (ψo/(1 − ψo)) = (FM − F2 ms)/(F2 ms − F50 ms). The performance index of photosynthesis (PI$_{ABS}$) represents the driving force of the performance. One component is performance due to the conversion of excitation energy into electron transport (P$_{ET}$). The net leaf photosynthetic rate (Pn) was measured using a Li-Cor 6400 photosynthetic system (Li-Cor, Lincoln, NE, USA).

To dissect the impact of NO on the growth of wheat under WD, the second newly expanded leaves from 3 treated and untreated (Control) samples were harvested for protein extraction, iTRAQ labeling, and identification of phosphorylated peptides.

4.4. Protein Extraction

For each sample group, 5 g of tissue from each of the 3 biological replicates were ground with a mortar and pestle and transferred into chilled 50 mL conical tubes. Protein extraction was conducted according to Pi et al. [36].

4.5. Protein Digestion and iTRAQ Labeling

Protein digestion was performed using a method involving filter-aided sample preparation (FASP Digestion) [37]. After the resulting peptides were collected by centrifugation at 14,000× g for 10 min, the filters were then rinsed with 40 μL 106 DS buffer (50 mM triethylammonium bicarbonate at pH 8.5) and centrifuged a final time. Peptide concentration was estimated by UV light spectral density at 280 nm using an extinctions coefficient of 1.1 of 0.1% (g/L) solution that was calculated on the basis of the frequencies of tryptophan and tyrosine in vertebrate proteins. For iTRAQ labeling, 100 μg of sample peptides was separately labeled for each biological replicate using an iTRAQ Reagent4plex Multiplex Kit (AB Sciex, Framingham, USA). 114, 115, 116, and 117 labels were used for the C, S, D, and S + D sample groups, respectively. Labeling was conducted according to the manufacturer's instructions.

4.6. Enrichment of Phosphorylated Peptides Using TiO$_2$ Beads

The labeled peptide samples were concentrated using a vacuum concentrator and resuspended in 500 μL buffer (2% glutamic acid, 65% acetonitrile, and 2% trifluoroacetic acid). Then, 500 μg TiO$_2$ beads (GL Sciences, Tokyo, Japan) was added, and samples were agitated for 40 min before centrifugation for 1 min at 5000× g. The supernatant was then discarded. The beads were washed 3 times with 50 μL 30% acetonitrile and 3% TFA and 3 times with 50 μL of 80% acetonitrile and 0.3% trifluoroacetic acid to remove the remaining nonadsorbed material. Finally, the phosphopeptides were eluted with 50 μL of elution buffer (40% acetonitrile, 15% NH$_4$OH) and lyophilized [38]. Each iTRAQ sample (5 μL) was mixed with 15 μL of 0.1% trifluoroacetic acid (*v*/*v*) for mass spectrometry analysis.

4.7. Liquid Chromatograph -Mass Spectrometry Analysis

A 5 μL mixture from each sample was added to a Thermo Scientific EASY precolumn (2 cm × 100 μm × 5 μm C18) in buffer A (0.1% formic acid) and separated with a linear gradient of buffer B (80% acetonitrile and 0.1% formic acid) at a flow rate of 250 nL/min on a Thermo Scientific | | instrument| | nLC (ProxeonBiosystems, now Thermo Fisher Scientific, Waltham, MA, USA) EASY column (75 μm × 250 mm × 3 μm C18). The peptides were eluted with a gradient of 0–55% buffer B from 0 to 220 min, 55–100% buffer B from 220 to 228 min, and 100% buffer B from 228 to 240 min.

Following nano-Liquid Chromatograph (LC) separation, Mass Spectrometry (MS)/MS analysis was performed on a Q exactive mass spectrometer (Thermo Fisher Scientific). The mass spectrometer was operated in positive-ion mode. MS data were acquired using a data-dependent (top-10) method that dynamically selects the most abundant precursor ions from a survey scan (350–1800 m/z) for HCD fragmentation and uses the automatic gain control function as target value. The dynamic exclusion duration was 30.0 s. Survey scans were acquired at a resolution of 70,000 at m/z 200, and the resolution for the HCD spectra was set to 17,500 at m/z 200. Isolation width was set to 2 m/z. Normalized collision energy was 29 eV, and the underfill ratio (which specifies the minimum percentage of the target value likely to be reached by the maximum fill time), was set to 0.1%. The instrument was operated with enabled peptide recognition mode. Each iTRAQ experiment was analyzed 3 times.

4.8. Data Workflow

MS/MS spectra were searched using Mascot 2.2 (Matrix Science, Boston, MA, USA) and Proteome Discoverer 1.4 software (Thermo Fisher Scientific,) against both the *pooideae* Uniprot database (294, e962 sequences, downloaded 9 November 2015) and a decoy database. For protein identification, the following parameters were used: fragment mass tolerance, 0.1Da; peptide mass tolerance, 20 ppm; enzyme, trypsin; max missed cleavages, 2; fixed modifications, carbamidomethyl (C), iTRAQ 4plex (N-term), iTRAQ 4plex (K); variable modifications, oxidation (M), phosphorylation (ST), phosphorylation (Y). The score threshold for peptide identification was set at a false-discovery rate (FDR) of ≤0.01. Phosphorylated sites on the identified peptides were assigned using the PhosphoRS algorithm; for each phosphorylation site on all phosphopeptides, PhosphoRS probabilities above 75% were interpreted to indicate that a site is truly phosphorylated, and PhosphoRS scores above 50% were considered to indicate good peptide spectral match [39,40]. Please note that the expression change of phosphorylated peptides can be attributed to either the change in abundance of the corresponding protein, the phosphorylation level of the protein, or both; in this study, we did not attempt to dissect the relative contributions of these two factors. All datasets were processed using inhouse Perlscripts.

Phosphopeptide quantification was evaluated based on the intensity of the reporter ions. We normalized the phosphopeptide ratios by dividing the value for a given peptide by the median ratio of all identified peptides. The log2 fold-change values (S/C; D/C; (S + D)/C; (S + D)/D) for each treatment were calculated for each phosphopeptide. Only phosphopeptides detected in at least 2 out of the 3 biological replicates were used for the assessment of significant changes.

4.9. Identification of Differentially Phosphorylated Peptides

Differentially phosphorylated peptides were identified on a per-peptide basis and in a pairwise manner. Briefly, the abundance values for a particular phosphorylated peptide from 3 biological replicates of each experimental condition were compared to the abundances of the same peptide from a different experimental condition using a 2-sample Student's t-test with default parameters. The peptide was considered to be differentially phosphorylated if the reported 2-sided p-value was below 0.05. The fold-change was then calculated for the peptide as the ratio between the mean value of the biological replicates of one experiment and the mean value of the other.

Multiple distinct phosphorylated peptides can be mapped to the same protein. Our analyses showed that these multiple peptides exhibit similar trends, i.e., they are either all significantly upregulated or all significantly downregulated between two experimental conditions, highlighting the accuracy and high quality of our data. To gain indepth insight of their phosphorylation patterns, we conducted hierarchical cluster analysis of the different phosphorylated peptides based on their phosphorylation intensities.

Additionally, PCA was performed using the built-in R function 'prcomp'. PCA is often used to investigate the internal structure of the data in a way that best explains the variance in the data. In our case, PCA was used to illustrate the similarities among sample groups.

Accession numbers of sequence data can be accessed here: ProteomeXchange PXD010724.

Int. J. Mol. Sci. **2018**, *19*, 2819

4.10. Statistical Analysis

All experiments were repeated independently 3 times. Growth and physiological parameters were statistically analyzed using one-way analysis of variance (ANOVA) and Duncan's multiple-range test to determine significant differences among group means. Significant differences from control values were determined at the $p < 0.05$ level.

5. Conclusions

In this study, 20% PEG greatly inhibited both growth and leaf RWC of wheat seedlings; however, NO could alleviate this PEG-induced inhibition. Phosphorylation has been widely employed to understand the molecular mechanism of biotic/abiotic stress and hormone responses. Here, quantitative phosphoproteomic analysis was performed using iTRAQ in seedlings treated with S, D, and S + D compared with control. In total, 2257 unique phosphopeptides, 1396 unique phosphoproteins, and 2416 phosphorylation sites were identified among four experimental groups. In total, the four groups contained 148 DEPs with 172 different expressed DEPPs (>1.5-fold change and $p < 0.05$, Figure 3). Most species were involved in both photosynthetic metabolism and signaling. Most DEPPs decreased in response to D stress. However, several were increased after NO plus D treatment, such as part of the peptides of CAB, PIFI, and SNT7, which may be closely related to the absorption of light energy in the LHCII antenna and electron transport, as a primary reaction of photosynthesis. We suggest that the NO signal could play an important role in the phosphorylation of these peptides. The corresponding physiological change further confirmed the effect of NO on the primary reaction of photosynthesis in wheat seedlings suffering from D stress. This study provides insight into how the molecular mechanisms of NO improving D resistance in higher plants.

Supplementary Materials: Supplementary materials can be found at http://www.mdpi.com/1422-0067/19/9/2819/ s1.

Author Contributions: R.S. and G.K. conceived the study, analyzed the data, and wrote the manuscript; H.Z., S.J., and Y.J. performed most of the experiments and consulted references; Q.Y. provided the financial support. All authors have read and approved the final manuscript.

Funding: This study was financially supported by the National Natural Science Foundation of China (31401304).

Conflicts of Interest: The authors declare no conflict of interest.

References

1. Khan, M.N.; Mobin, M.; Mohammad, F.; Corpas, F.J. *Nitric Oxide Action in Abiotic Stress Responses in Plants*; Springer International Publishing: Basel, Switzerland, 2015; pp. 51–52.
2. Shao, R.X.; Wang, K.B.; Shangguan, Z.P. Cytokinin-induced photosynthetic adaptability of *Zea mays* L. to drought stress associated with nitric oxide signal: Probed by ESR spectroscopy and fast OJIP fluorescence rise. *J. Plant Physiol.* **2010**, *167*, 472–479. [CrossRef] [PubMed]
3. Zweier, J.L.; Samouilov, A.; Kuppusamy, P. Non-enzymatic nitric oxide synthesis in biological systems. *Biochim. Biophys. Acta Bioenerg.* **1999**, 250–262. [CrossRef]
4. Sahay, S.; Gupta, M. An update on nitric oxide and its benign role in plant responses under metal stress. *Nitric Oxide-Biol. Chem.* **2017**, *67*, 39–52. [CrossRef] [PubMed]
5. Tan, J.F.; Zhao, H.J.; Hong, J.P.; Han, Y.L.; Li, H.; Zhao, W.C. Effects of exogenous nitric oxide on photosynthesis, antioxidant capacity and proline accumulation in wheat seedlings subjected to osmotic stress. *World J. Agric. Sci.* **2008**, *4*, 307–313.
6. Liao, W.B.; Huang, G.B.; Yu, J.H.; Zhang, M.L. Nitric oxide and hydrogen peroxide alleviate drought stress in marigold explants and promote its adventitious root development. *Plant Physiol. Biochem.* **2012**, *58*, 6–15. [CrossRef] [PubMed]

7. Cantrel, C.; Vazquez, T.; Puyaubert, J.; Rezé, N.; Lesch, M.; Kaiser, W.M.; Dutilleul, C.; Guillas, I.; Zachowski, A.; Baudouin, E. Nitric oxide participates in cold-responsive phosphosphingolipid formation and gene expression in *Arabidopsis thaliana*. *New Phytol.* **2011**, *189*, 415–427. [CrossRef] [PubMed]

8. Nawaz, F.; Shabbir, N.; Shahbaz, M.; Majeed, S.; Raheel, M.; Hassan, W.; Amir Sohail, M. Cross talk between nitric oxide and phytohormones regulate plant development during abiotic stresses. In *Phytohormones-Signaling Mechanisms and Crosstalk in Plant Development and Stress Responses*; InTech: Kowloon, Hong Kong, 2017. [CrossRef]

9. Bian, Y.; Deng, X.; Yan, X.; Zhou, J.; Yuan, L.; Yan, Y. Integrated proteomic analysis of *Brachypodium distachyon* roots and leaves reveals a synergistic network in the response to drought stress and recovery. *Sci. Rep.* **2017**, *7*, 46183. [CrossRef] [PubMed]

10. Lee, J.H.; Kim, C.J. A multimodel assessment of the climate change effect on the drought severity-duration-frequency relationship. *Hydrol. Process.* **2013**, *27*, 2800–2813. [CrossRef]

11. Shao, R.X.; Yang, Q.H.; Xin, L.F.; Shangguan, Z.P. Effects of exogenous nitric oxide and cytokinin on the growth and photosynthesis of wheat seedlings under water deficiency. *J. Food Agric. Environ.* **2012**, *10*, 1451–1456.

12. Procházková, D.; Haisel, D.; Wilhelmová, N.; Pavlíková, D.; Száková, J. Effects of exogenous nitric oxide on photosynthesis. *Photosynthetica* **2013**, *51*, 483–489. [CrossRef]

13. Shao, R.X.; Li, L.L.; Zheng, H.F.; Zhang, J.Y.; Yang, S.J.; Ma, Y.; Xin, L.F.; Su, X.Y.; Ran, W.L.; Mao, J.; et al. Effects of exogenous nitric oxide on photosynthesis of maize seedlings under drought stress. *Sci. Agric. Sin.* **2016**, *2*, 251–259.

14. Shi, H.T.; Ye, T.T.; Zhu, J.K.; Chan, Z.L. Constitutive production of nitric oxide leads to enhanced drought stress resistance and extensive transcriptional reprogramming in Arabidopsis. *J. Exp. Bot.* **2014**, *65*, 4119–4131. [CrossRef] [PubMed]

15. Yang, L.M.; Tian, D.G.; Todd, C.D.; Luo, Y.; Hu, X. Comparative proteome analyses reveal that nitric oxide is an important signal molecule in the response of rice to aluminum toxicity. *J. Proteome Res.* **2013**, *12*, 1316–1330. [CrossRef] [PubMed]

16. Meng, Y.Y.; Liu, F.; Pang, C.Y.; Fan, S.L.; Song, M.Z.; Wang, D.; Li, W.H.; Yu, S.X. Label-free quantitative proteomics analysis of cotton leaf response to nitric oxide. *J. Proteome Res.* **2011**, *10*, 5416–5432. [CrossRef] [PubMed]

17. Bonhomme, L.; Valot, B.; Tardieu, F.; Zivy, M. Phosphoproteome dynamics upon changes in plant water status reveal early events associated with rapid growth adjustment in maize leaves. *Mol. Cell. Proteom.* **2012**, *11*, 957–972. [CrossRef] [PubMed]

18. Zhang, M.; Lv, D.W.; Ge, P.; Bian, Y.W.; Chen, G.X.; Zhu, G.R.; Li, X.H.; Yan, Y.M. Phosphoproteome analysis reveals new drought response and defense mechanisms of seedling leaves in bread wheat (*Triticum aestivum* L.). *J. Proteome* **2014**, *109*, 290–308. [CrossRef] [PubMed]

19. Fan, S.L.; Meng, Y.Y.; Song, M.Z.; Pang, C.Y.; Wei, H.L.; Liu, J.; Zhan, X.J.; Lan, J.Y.; Feng, C.H.; Zhang, S.X.; et al. Quantitative phosphoproteomics analysis of nitric oxide-responsive phosphoproteins in cotton leaf. *PLoS ONE* **2014**, *9*, e94261. [CrossRef] [PubMed]

20. Zhang, M.; Ma, C.Y.; Lv, D.W.; Zhen, S.M.; Li, X.H.; Yan, Y.M. Comparative phosphoproteome analysis of the developing grains in bread wheat (*Triticum aestivum* L.) under well-watered and water-deficiency conditions. *J. Proteome Res.* **2014**, *13*, 4281–4297. [CrossRef] [PubMed]

21. Parankusam, S.; Adimulam, S.S.; Bhatnagarmathur, P.; Sharma, K.K. Nitric oxide (NO) in plant heat stress tolerance: Current knowledge and perspectives. *Front. Plant Sci.* **2017**, *8*, 1–18. [CrossRef] [PubMed]

22. Tikkanen, M.; Grieco, M.; Kangasjärvi, S.; Aro, E.M. Thylakoid protein phosphorylation in higher plant chloroplasts optimizes electron transfer under fluctuating light. *Plant Physiol.* **2010**, *152*, 723–735. [CrossRef] [PubMed]

23. Polverari, A.; Molesini, B.; Pezzotti, M.; Buonaurio, R.; Marte, M.; Delledonne, M. Nitric oxide-mediated transcriptional changes in Arabidopsis thaliana. *Mol. Plant Microbe Interact.* **2003**, *16*, 1094–1105. [CrossRef] [PubMed]

24. Pesaresi, P.; Hertle, A.; Pribil, M.; Schneider, A.; Kleine, T.; Leister, D. Optimizing photosynthesis under fluctuating light: The role of the Arabidopsis SNT7 kinase. *Plant Signal. Behav.* **2010**, *5*, 21–25. [CrossRef] [PubMed]

Int. J. Mol. Sci. **2018**, *19*, 2819

25. Wang, D.; Portis, A.R. A novel nucleus-encoded chloroplast protein, PIFI, is involved in NAD(P)H dehydrogenase complex-mediated chlororespiratory electron transport in Arabidopsis. *Plant Physiol.* **2007**, *144*, 1742–1752. [CrossRef] [PubMed]

26. Scheibe, R.; Dietz, K.J. Reduction–oxidation network for flexible adjustment of cellular metabolism in photoautotrophic cells. *Plant Cell Environ.* **2012**, *35*, 202–216. [CrossRef] [PubMed]

27. Noctor, G.; Reichheld, J.P.; Foyer, C.H. ROS-related redox regulation and signaling in plants. *Semin. Cell Dev. Biol.* **2018**, *80*, 3–12. [CrossRef] [PubMed]

28. Wientjes, E.; van Amerongen, H.; Croce, R. LHCII is an antenna of both photosystems after long-term acclimation. *Biochim. Biophys. Acta Bioenerg.* **1827**, 420–426. [CrossRef] [PubMed]

29. Bos, I.; Bland, K.M.; Tian, L.J.; Croce, R.; Frankel, L.K.; van Amerongen, H.; Bricker, T.M.; Wientjes, E. Multiple LHCII antennae can transfer energy efficiently to a single Photosystem I. *Biochim. Biophys. Acta Bioenerg.* **1858**, 371–378. [CrossRef] [PubMed]

30. Koziol, A.G.; Borza, T.; Ishida, K.I.; Keeling, P.; Lee, R.W.; Durnford, D.G. Tracing the evolution of the light-harvesting antennae in chlorophyll a/b-containing organisms. *Plant Physiol.* **2007**, *143*, 1802–1816. [CrossRef] [PubMed]

31. Bolaños, J.P.; HerreroMendez, A.; FernandezFernandez, S.; Almeida, A. Linking glycolysis with oxidative stress in neural cells: A regulatory role for nitric oxide. *Biochem. Soc. Trans.* **2007**, *35*, 1224–1227. [CrossRef] [PubMed]

32. Kang, G.Z.; Li, G.Z.; Xu, W.; Peng, X.Q.; Han, Q.X.; Zhu, Y.J.; Guo, T.C. Proteomics reveals the effects of salicylic acid on growth and tolerance to subsequent drought stress in wheat. *J. Proteome Res.* **2012**, *11*, 6066–6079. [CrossRef] [PubMed]

33. Wu, Q.S.; Xia, R.X. Arbuscular mycorrhizal fungi influence growth, osmotic adjustment and photosynthesis of citrus under well-watered and water stress conditions. *J. Plant Physiol.* **2006**, *163*, 417–425. [CrossRef] [PubMed]

34. Hiscox, J.D.; Israelstam, G.F. Erratum: A method for the extraction of chlorophyll from leaf tissue without maceration. *Can. J. Bot.* **1980**, *58*, 403. [CrossRef]

35. Arnon, D.I. Copperenzymesin isolated chloroplasts. Polyphenoloxidase in beta vulgaris. *Plant Physiol.* **1949**, *24*, 1–15. [PubMed]

36. Pi, E.; Qu, L.; Hu, J.; Huang, Y.; Qiu, L.; Lu, H.; Jiang, B.; Liu, C.; Peng, T.; Zhao, Y.; et al. Mechanisms of soybean roots' tolerances to salinity revealed by proteomic and phosphoproteomic comparisons between two cultivars. *Mol. Cell. Proteom.* **2016**, *15*, 266–288. [CrossRef] [PubMed]

37. Wisniewski, J.; Zougman, A.; Nagaraj, N.; Mann, M. Universal sample preparation method for proteome analysis. *Nat. Methods* **2009**, *6*, 359–362. [CrossRef] [PubMed]

38. Larsen, M.R.; Thingholm, T.E.; Jensen, O.N.; Roepstorff, P.; Jørgensen, T.J. Highly selective enrichment of phosphorylated peptides from peptide mixturesusing titanium dioxide microcolumns. *Mol. Cell. Proteom.* **2005**, *4*, 873–886. [CrossRef] [PubMed]

39. Beausoleil, S.A.; Villén, J.; Gerber, S.A.; Rush, J.; Gygi, S.P. A probability-based approach for high-throughput protein phosphorylation analysis and site localization. *Nat. Biotechnol.* **2006**, *24*, 1285–1292. [CrossRef] [PubMed]

40. Snyder, G.L.; Stacey, G.; Hendrick, J.P.; Hemmings, H.C. General anesthetics selectively modulate glutamatergic and dopaminergic signaling site-specific phosphorylation in vivo. *Neuropharmacology* **2007**, *53*, 619–630. [CrossRef] [PubMed]

International Journal of
Molecular Sciences

MDPI

Review

Metabolic Reprogramming in Chloroplasts under Heat Stress in Plants

Qing-Long Wang [1], Juan-Hua Chen [1], Ning-Yu He [1] and Fang-Qing Guo [1,2,*]

[1] The National Key Laboratory of Plant Molecular Genetics, Institute of Plant Physiology & Ecology, Chinese Academy of Sciences, 300 Fenglin Road, Shanghai 200032, China; qlwang@sibs.ac.cn (Q.-L.W.); jhchen02@sibs.ac.cn (J.-H.C.); nyhe@sibs.ac.cn (N.-Y.H.)
[2] CAS Center for Excellence in Molecular Plant Sciences, Institute of Plant Physiology & Ecology, Chinese Academy of Sciences, 300 Fenglin Road, Shanghai 200032, China
*   Correspondence: fqguo@sibs.ac.cn; Tel.: +86-021-5492-4098

Received: 29 January 2018; Accepted: 5 March 2018; Published: 14 March 2018

Abstract: Increases in ambient temperatures have been a severe threat to crop production in many countries around the world under climate change. Chloroplasts serve as metabolic centers and play a key role in physiological adaptive processes to heat stress. In addition to expressing heat shock proteins that protect proteins from heat-induced damage, metabolic reprogramming occurs during adaptive physiological processes in chloroplasts. Heat stress leads to inhibition of plant photosynthetic activity by damaging key components functioning in a variety of metabolic processes, with concomitant reductions in biomass production and crop yield. In this review article, we will focus on events through extensive and transient metabolic reprogramming in response to heat stress, which included chlorophyll breakdown, generation of reactive oxygen species (ROS), antioxidant defense, protein turnover, and metabolic alterations with carbon assimilation. Such diverse metabolic reprogramming in chloroplasts is required for systemic acquired acclimation to heat stress in plants.

Keywords: heat stress; metabolic reprogramming; chloroplasts; chlorophyll breakdown; reactive oxygen species (ROS); antioxidant defense; protein turnover; Photosystem II (PSII) core subunit; PSII repair cycle; carbon assimilation

---

1. Introduction

Photosynthesis is thought to be the most important photo-chemical reaction, during which sunlight is trapped and converted into biological energy in plants. In general, leaves function as a highly specialized organ that are basically appointed in the photosynthetic process in higher plants. Leaf photosynthesis is substantially affected, often lethally, by high temperature stress, usually 10–15 °C above an optimum temperature for plant growth as plants are not capable of moving to more favorable environments [1–3]. Housed in chloroplasts, the photosynthetic apparatus is susceptible to be damaged by heat stress and the chloroplasts have been demonstrated to play an essential role in activation of cellular heat stress signaling [4–6]. Given that Photosystem II (PSII) is the most susceptible target within the chloroplast thylakoid membrane protein complexes for heat stress, heat stress commonly causes severe thermal damages to PSII, dramatically affecting photosynthetic electron transfer and ATP synthesis [1–3,7,8]. Under heat stress, heat stress-induced damages lead to alterations of photochemical reactions in thylakoid lamellae of chloroplast, reflected as a significant reduction in the ratio of variable fluorescence to maximum fluorescence (Fv/Fm) [1,2,9–11]. Exposure to high temperature stress causes oxidative stress in plants, particularly leading to the dissociation of oxygen evolving complex (OEC) in PSII, which further results in inhibition of the electron transportation from OEC to the acceptor side of PSII [1,2,12,13]. Heat stress causes cleavage of the reaction center-binding protein D1 of PSII in spinach thylakoids and induces dissociation of a manganese (Mn)-stabilizing 33-kDa proteins from PSII reaction center complex [14]. As a photosynthetic carbon fixation cycle, the Calvin–Benson cycle is responsible for the fixation of CO_2 into carbohydrates, as well as assimilation, transport, and utilization of photoassimilates as the organic products of photosynthesis. In addition to the disruption of OEC in PSII, heat stress also results in dysfunction in the system of carbon assimilation metabolism in the stroma of chloroplast [2,8]. It has been observed that the disruption of electron transport and inactivation of the oxygen evolving enzymes of PSII dramatically inhibit the rate of ribulose-1,5-bisphosphate (RuBP) regeneration [10,15]. Heat stress-induced inhibition in the activity of ribulose-1,5-bisphosphate carboxylase/oxygenase (Rubisco) mainly results from inactivation of Rubisco activase that is extremely sensitive to heat stress because the enzyme Rubisco of higher plants is heat-stable [8,15]. In addition to the early effects on photochemical reactions and carbon assimilation, alterations in the microscopic ultrastructures of chloroplast and the integrity of thylakoid membranes were reported to be severely disrupted, including membrane destacking and reorganization when subjected to heat stress [2,16–19].

In photosynthetic organisms, chloroplasts respond to a variety of environmental stresses, including heat stress, with adjustments for major metabolic processes to optimize carbon fixation and growth requirements [2,3,6,8,20]. As one of the subcellular energy organelles, chloroplast in plant cells conducts major metabolic reprogramming processes including chlorophyll breakdown, generation and scavenging of reactive oxygen species (ROS), protein turnover, and metabolic alterations of carbon assimilation in response to heat stress (Figure 1). Here we review studies that have investigated the effects of heat stress on photosynthesis and its associated metabolic adaptations for optimizing plant growth and development under stress conditions. We will assess the role of chloroplast from an organellar perspective to begin building insights into better understanding the importance and significance of metabolic reprogramming within this organelle during high temperature stresses.

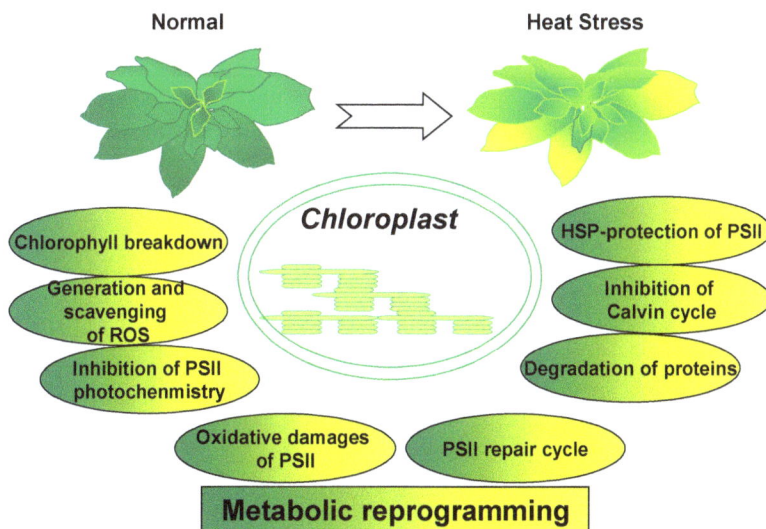

Figure 1. Extensive and transient metabolic reprogramming in chloroplasts under heat stress. Major events of metabolic reprogramming in response to heat stress include chlorophyll breakdown, generation of reactive oxygen species (ROS), antioxidant defense, protein turnover, and metabolic alterations with carbon assimilation. With respect to the systemic acquired acclimation to heat stress in plants, diverse metabolic reprogramming in chloroplasts is required for optimizing plant growth and development during high temperature stresses.

2. Chlorophyll Breakdown under Heat Stress

Chlorophyll (Chl) functions in harvesting light energy and driving electron transfer during the initial and indispensable processes of photosynthesis, and is a pigment consisting of two moieties: a chlorin ring containing a magnesium ion at its center and a long hydrophobic phytol chain, joined by an ester bond. It should be noted that when the photosynthetic apparatus is overexcited, and oxygen receives the absorbed energy from Chl, Chl acts as a harmful molecule, negatively affecting plant cells including most porphyrins [21,22]. Importantly, the breakdown of Chl is a physiological process as a prerequisite for protecting plant cells from hazardous effects of phototoxic pigments in association with recycling nitrogen sources from Chl-binding proteins in chloroplasts when leaves are senescing [22–25].

Recent studies have provided advances in better understanding of the pathway of Chl catabolism in higher plants. In brief, the removal of the phytol residue and the central Mg by chlorophyllase occurs at the initial reaction in the Chl breakdown pathway. Chlorophyllase genes, termed *CLHs*, were reported to be involved in Chl breakdown [26], but recent studies questioned the involvement of CLHs in Chl degradation in vivo during leaf senescence [23,27]. In 2009, functional analysis of pheophytinase (PPH) supports its role in porphyrin-phytol hydrolysis involved in senescence-related chlorophyll breakdown in vivo [28]. Several reports have revealed that the resulting pheophorbide (pheide) a is converted into a primary fluorescent chlorophyll catabolite (pFCC) in the next-step reactions, and two enzymes are involved in the reactions, including pheide a oxygenase (PAO) and red chl catabolite reductase (RCCR) [29–32]. *PAO* encodes a Rieske-type iron-sulfur oxygenase that is bound to chloroplast envelop, also named as *ACCELERATED CELL DEATH* (*ACD1*) [30]. PAO functions as a key enzyme that catalyzing the cleavage of the porphyrin ring in Chl breakdown pathway, and the red chlorophyll catabolite resulted from reactions can be further catalyzed into pFCC by RCCR. Next, a primary active transport system is responsible for exporting the resulting primary fluorescent catabolite pFCCs from the plastid and importing into the vacuole [33,34]. Based on the genetic analysis of the stay-green mutants, *NON-YELLOW COLORING1* (*NYC1*) and *NYC1-LIKE* (*NOL*) were cloned and found to selectively retain photosystem II (PSII) light-harvesting complex subunits [35,36]. The first half of chlorophyll *b* can be catalyzed into chlorophyll *a* by chlorophyll *b* reductase that is composed of two subunits NYC1 and NOL [36]. In addition, leaves of the loss-of-function mutant *pao* showed a stay-green phenotype under dark-induction conditions and a light-dependent lesion mimic phenotype was also observed because of the increase in levels of phototoxic pheide *a* [31,37,38]. Interestingly, the *Bf993* mutant of *Festuca pratensis* is classified into the third type of stay-green mutants in which the stay-green gene named as *senescence-induced degradation* (*SID*) is defective [39–41]. *SGR* (*STAY-GREEN*), designated as the orthologous gene of *SID*, has been cloned and characterized in a lot of plant species, including *Arabidopsis* [42], rice [43–45], pea [45,46], bell pepper [47,48], and tomato [47]. Based on the accumulated evidence, the direct interacting relationship has been characterized to exist between SGR and a subset of the protein components of the light harvesting chlorophyll *a/b*-protein complex II (LHCPII), suggesting that SGR is likely to play a role in making pigment-protein complexes unstable as a prerequisite for the enzymes in chlorophyll breakdown pathway to reach their substrate when leaves are in senescing processes [22,44,49].

A visible sign of leaf senescence and fruit ripening, loss of green color is resulted from massive Chl breakdown into nonphototoxic breakdown products in combination with carotenoid retention or anthocyanin accumulation [22,23,50]. Under normal growth conditions, chlorophyll maintains at a steady level due to the balance of its biosynthesis and degradation without a visible change in chlorophyll content [23,25]. In contrast, chlorophylls undergo turnover or breakdown when the partial or complete dismantling of the photosynthetic machinery occurs in response to environmental stresses, including heat stress. Heat stress symptoms in plants are typically characterized by leaf senescence or chlorosis due to a decline in chlorophyll content [2,51,52]. Heat-induced leaf chlorosis has been observed in a variety of plant species, including *Arabidopsis* [4,5,53], Soybean (*Glycine max* L. Merr.) [54], sorghum (*Sorghum bicolor*) [55], wheat (*Triticum aestivum*) [56,57] and creeping bentgrass (*Agrostis stolonifera*) [58]. However, the underlying mechanisms by which leaf senescence is regulated during heat stress remain elusive. Further studies are needed for answering the question of whether heat-induced chlorophyll loss is caused by heat-induced inhibition of chlorophyll synthesis and/or heat-enhanced chlorophyll degradation in plant leaves.

In *Arabidopsis*, chlorophyll *a* (Chl *a*) content was gradually reduced and chlorophyllase (Chlase) activity substantially increased during the high temperature treatment [53]. Upon heat treatment, no significant difference was detected in the enzyme activity of a key chlorophyll-synthesizing enzyme, porphobilinogen deaminase across all the lines of bentgrass (*Agrostis* spp.). However, the activities of chlorophyll-degrading enzymes, including chlorophyllase and chlorophyll-degrading peroxidase, increased significantly after heat stress whereas pheophytinase activity was unchanged [52]. Interestingly, lower activities of chlorophyll-degrading enzymes were detected in heat-tolerant transgenic lines in which the expression of isopentenyl transferase (*ipt*) gene is driven by a senescence-activated promoter (*SAG12*) or heat shock promoter (*HSP18.2*) in modulation of cytokinin biosynthesis when compared with the WT under heat stress [52]. The authors suggested that the enhanced degradation of chlorophyll under heat stress could result in a severe loss of the chlorophyll content in heat-challenged bentgrass. Studies on genetic variations contrasting in heat tolerance in the strength of leaf senescence under heat stress in hybrids of colonial (*Agrostis capillaris*) x creeping bentgrass (*Agrostis stolonifera*) supported that heat-induced loss of chlorophyll in leaves was caused by the rapid breakdown of chlorophyll, as manifested by the high-level activation of genes encoding chlorophyllase and pheophytinase, and the activity of pheophytinase (PPH) [59]. Recently, the map-based cloning of a semidominant, heat-sensitive, missense allele (*cld1-1*) led to identify a putative hydrolase, named as CHLOROPHYLL DEPHYTYLASE1 (CLD1) that is capable of dephytylating chlorophyll [60]. Their findings suggest that CLD1 is conserved in oxygenic photosynthetic organisms and plays a key role as the long-sought enzyme in making the phytol chain removed from chlorophyll in its degradation process at the steady-state level in chloroplasts.

Unlike studies on hormonal regulation of leaf senescence, less attention has been paid to effects of hormones on heat-induced chlorophyll loss. Exogenous application of a synthetic form of cytokinin, zeatin riboside (ZR) helped maintain higher leaf chlorophyll content creeping bentgrass exposed to heat stress by slowing down the action of protease and by induction or upregulation of heat-shock proteins [61]. The ethylene-inhibiting compound 1-methylcyclopropene (1-MCP) treatment can delay leaf senescence in cotton plants under high temperature by reducing lipid peroxidation, membrane leakage, soluble sugar content, and increasing chlorophyll content [62]. It was reported that the novel ethylene antagonist, 3-cyclopropyl-1-enyl-propanoic acid sodium salt (CPAS), increases grain yield in wheat by delaying leaf senescence under extreme weather conditions [63].

Plant chlorophyll retention-staygreen is considered a valuable trait under heat stress. Accumulating data indicate that understanding the physiological and molecular mechanisms of "STAY-GREEN" trait or delayed leaf senescence is required for regulating photosynthetic capability and may be also a key to break the plateau of productivity associated with adaptation to high temperature [64]. Staygreen traits are associated with heat tolerance in bread wheat and quantitative trait loci (QTL) for staygreen and related traits were identified across the genome co-located with agronomic and physiological traits associated to plant performance under heat stress, confirming that the staygreen phenotype is a useful trait for productivity enhancement in hot-irrigated environments [65]. In soybean, high temperatures and drought stress can lead to chlorophyll retention in mature seeds, termed as "green seed problem", which is usually related to lower oil and bad seed quality, thus inducing a yield loss of soybean seeds. A "mild" stay-green phenotype was observed in a susceptible soybean cultivar when subjected to combined abiotic stresses of heat and drought stress and also the transcript levels of the *STAY-GREEN 1* and *STAY-GREEN 2* (*D1, D2*), *PHEOPHORBIDASE 2* (*PPH2*) and *NON-YELLOW COLORING 1* (*NYC1*) genes were downregulated in soybean seeds, indicating that the high-level transcriptional activation of these genes mentioned above in fully mature seeds is critical for a tolerant cultivar to cope with stresses and conduct a rapid and complete turnover of chlorophyll [66].

It is well-established that the major responses of crop plants to heat stress are classified into several aspects including the enhancement of leaf senescence, reduction of photosynthesis, deactivation of photosynthetic enzymes, and generation of oxidative damages to the chloroplasts. With respect to crop yield, heat stress also reduces grain number and size by affecting grain setting, assimilate translocation and duration and growth rate of grains [57]. In wheat, delayed senescence, or stay-green, contributes to a long grain-filling period and stable yield under heat stress [67]. Waxy maize (*Zea mays* L. sinensis Kulesh) is frequently exposed to high temperatures during grain filling in southern China. The heat-sensitive waxy maize variety exhibited a significant increase in the translocation amount and rate of assimilates pre-pollination and the accelerated leaf senescence phenotype under heat stress [68]. Short heat waves during grain filling can reduce grain size and consequently yield in wheat (*Triticum aestivum* L.). The four susceptible varieties of wheat showed greater heat-triggered reductions in final grain weight, grain filling duration and chlorophyll contents in flag leaves under heat stress, suggesting that grain size effects of heat may be driven by premature senescence [69,70].

3. Generation and Homeostasis of ROS in Chloroplasts under Heat Stress

As a significant source of reactive oxygen species (ROS) in plant cells, the chloroplast produces a variety of ROS such as hydrogen peroxide (H_2O_2), superoxide, hydroxyl radicals ($^{\bullet}OH$), and $^1O_2$ during photosynthesis [71]. The transfer of excitation energy in the PSII antenna complex and the electron transport in the PSII reaction center can be inhibited by a variety of abiotic stresses, resulting in the formation of ROS in algae and higher plants [72]. Given that ROS generation is an unavoidable consequence of aerobic metabolism, plants have evolved a large array of ROS-scavenging mechanisms [71]. ROS such as $^1O_2$ is formed by the excitation energy transfer, whereas superoxide anion radical ($O_2^{\bullet-}$), H_2O_2 and $^{\bullet}OH$ are formed by the electron transport [73]. In chloroplasts, ROS are mainly generated in the reaction centers of PSI and PSII in the chloroplast thylakoid membranes [71]. Photoreduction of oxygen to superoxide occurs in PSI and in PSII, oxygen of the ground (triplet) of oxygen ($^3O_2$) is excited to the excited singlet state of oxygen ($^1O_2$) by the P680 reaction center chlorophyll [71]. Under extreme conditions, these ROS synthesis rates can increase, leading to an oxidative stress in both organelle and whole-cell functions [72,74]. It is well established that the chloroplast is extremely sensitive to high temperature stress during photosynthesis [1,2,9,10]. Accumulated data indicate that oxidative bursts of superoxide and/or hydrogen peroxide can be induced rapidly in response to a variety of environmental stresses, including also heat stress in plants [75–77]. It was reported that ROS, produced in PSI, PSII as well as in the Calvin-Benson cycle, can cause irreversible oxidative damage to cells when plants were subjected to heat stress [71,72,78]. Under high temperature conditions, large amounts of ROS were generated in tobacco cells for initiating signaling events involved in PCD, which is consistent with the role of applying the antioxidants ascorbate or superoxide dismutase (SOD) to the cultures in supporting the survival of cells [79]. In the leaves of tobacco (*Nicotiana tabacum*) defective in ndhC-ndhK-ndhJ (Delta ndhCKJ), hydrogen peroxide was rapidly produced in response to heat treatment, implying a role of the NAD(P)H dehydrogenase-dependent pathway in repressing generation of reactive oxygen species in chloroplasts [80]. In exposure to moderate heat treatment conditions, the oxidative damages of the reaction center-binding D1 protein of photosystem II increased and showed a tight positive relationship with the accumulated levels of $^1O_2$ and $^{\bullet}OH$ in spinach PSII membranes, implying that inhibition of a water-oxidizing manganese complex led to a rapid production of ROS through lipid peroxidation under heat stress [81]. In *Arabidopsis*, large amounts of chlorophyllide *a* caused a surge of phototoxic singlet oxygen in the chlorophyll synthase mutant (*chlg-1*) under heat stress, suggesting that chlorophyll synthase acts in maintenance of ROS homeostasis in response to heat stress [82].

It is well known that ROS burst can trigger the oxidative damage to pigments, proteins, and lipids in the thylakoid membrane [71,73,83]. ROS are primarily agents of damage, but this view is questioned by data proving their beneficial role and particularly their signaling function. The chloroplast harbors ROS-producing centers (triplet chlorophyll, ETC in PSI and PSII) and a diversified ROS-scavenging network (antioxidants, SOD, APX-glutathione cycle, and a thioredoxin system) to keep the equilibrium between ROS production and scavenging [71]. The non-enzymatic and enzymatic ROS scavenging systems are engaged in preventing harmful effects of ROS on the thylakoid membrane components to keep ROS level in chloroplasts under control [71,73,74,83–87]. The efficient enzymatic scavenging systems are composed of several key enzymes, including superoxide dismutase (SOD), catalase (CAT), ascorbate peroxidase (APX), glutathione reductase (GR), monodehydroascorbate reductase (MDHAR), dehydroascorbate reductase (DHAR), glutathione peroxidase (GPX) and glutathione-*S*-transferase (GST) and non-enzymatic systems contain antioxidants such as ascorbic acid (ASH), glutathione (GSH), phenolic compounds, alkaloids, non-protein amino acids and alpha-tocopherols [21,71,73,83,87–90]. These antioxidant defense systems work in concert to maintain homeostasis of ROS, protecting plant cells from oxidative damage by scavenging of ROS.

In addition to triggering ROS burst, heat stress also affects the scavenging systems in plants. Heat induced degradation of chloroplast Cu/Zn superoxide dismutase as shown by reduced protein levels and isozyme-specific SOD activity [91]. Loss of Cu/Zn SOD and induction of catalase activity would explain the altered balance between hydrogen peroxide and superoxide under stress. The authors proposed that degradation of PSII could thus be caused by the loss of components of chloroplast antioxidant defense systems and subsequent decreased function of PSII [91]. In a recent study aiming at identifying heat protective mechanisms promoted by CO_2 in coffee crop, the results likely favored that the maintenance of reactive oxygen species (ROS) at controlled levels contributed to mitigate of PSII photoinhibition under the high temperature stress [92]. Exogenous application of spermidine protected rice seedlings from heat-induced damage as marked by lower levels of malondialdehyde (MDA), H_2O_2, and proline content coupled with increased levels of Ascorbate (AsA), GSH, AsA and GSH redox status [93]. The authors conclude that heat exposure provoked an oxidative burden while enhancement of the antioxidative and glyoxalase systems by spermidine rendered rice seedlings more tolerant to heat stress. Under the later high temperature stress, heat priming contributed to a better redox homeostasis, as exemplified by the higher activities of superoxide dismutase (SOD) in chloroplasts and glutathione reductase (GR), and of peroxidase (POD) in mitochondria, which led to the lower superoxide radical production rate and malondialdehyde concentration in both chloroplasts and mitochondria [94]. These results suggested that heat priming effectively improved thermo-tolerance of wheat seedlings subjected to a later high temperature stress, which could be largely ascribed to the enhanced anti-oxidation at the subcellular level. In Kentucky bluegrass (*Poa pratensis*), higher activities of superoxide dismutase (SOD), catalase (CAT), peroxidase (POD), ascorbate peroxidase (APX), and glutathione reductase were detected in plants of heat-tolerant "Midnight" after long-term heat stress (21 and 28 d) in comparison with plants of heat-sensitive "Brilliant" [95]. Meanwhile, transcript levels of chloroplastic Cu/Zn SOD, Fe SOD, CAT, POD and cytosolic (cyt) APX were significantly higher in "Midnight" than in "Brilliant" under long-term heat stress. These results supported the hypothesis that enzymatic ROS scavenging systems could play predominant roles in antioxidant protection against oxidative damages from long-term heat stress. In tomato, studies on the role of a tomato (*Lycopersicon esculentum*) chloroplast-targeted DnaJ protein (LeCDJ1) showed that the sense transgenic tomato plants were more tolerant to heat stress due to the higher activities of ascorbate peroxidase (APX) and superoxide dismutase (SOD) [96]. The high temperature stress in southern China is one of the major factors leading to loss of the yield and quality of wucai (*Brassica campestris* L.). Comparative investigations on two cultivars of wucai (heat-sensitive and heat-tolerant) showed that greater severity of damage to the photosynthetic apparatus and membrane system was observed in heat-sensitive cultivar probably due to a high-level accumulation of ROS and malondialdehyde (MDA) [97]. In line with the studies mentioned above, plants of heat-tolerant Cucurbit species exhibited comparatively little oxidative damage, with the lowest hydrogen peroxide (H_2O_2), superoxide (O^{2-}) and malondialdehyde (MDA) compared with the thermolabile and moderately heat-tolerant interspecific inbred line [98]. The enzyme activities of superoxide dismutase (SOD), ascorbate peroxidase (APX), catalase (CAT) and peroxidase (POD) were found to be increased with heat stress in tolerant genotypes and the significant inductions of FeSOD, MnSOD, APX2, CAT1 and CAT3 isoforms in tolerant genotypes suggested their participation in heat tolerance [98].

Generally, the chloroplast possesses a variety of constitutively expressed antioxidant defense mechanisms to scavenge the various types of ROS generated during heat stress in preventing and minimizing oxidative damage to biological macromolecules (Figure 2). Therefore, the capacity of antioxidant defense is critical for heat stress adaptation and its strength is correlated with acquisition of thermotolerance with respect to buffering the effect of heat stress on the metabolic system [3,74,99,100]. On the other hand, ROS are produced in chloroplasts can function as plastid signals to inform the nucleus to activate the expression of genes encoding antioxidant enzyme and to adjust the stress-responsive machinery for more efficient adaptation to environmental stresses [6,47,72,74,100,101].

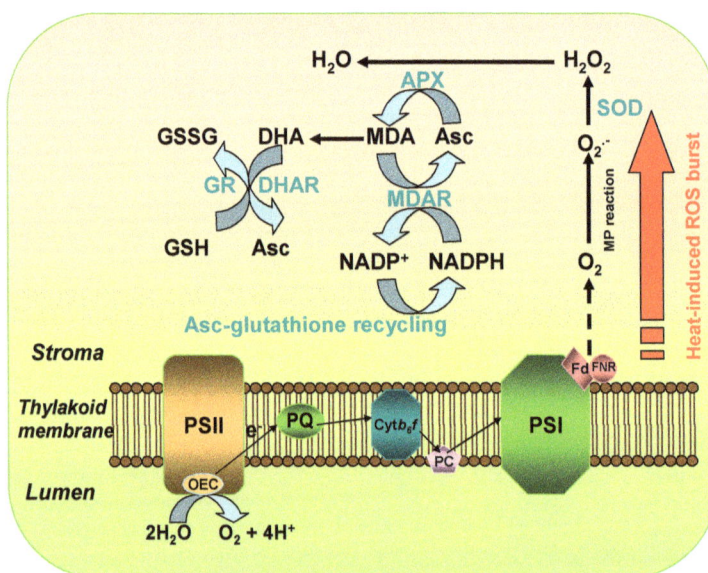

Figure 2. A representative scheme of reactive oxygen species (ROS) generation and scavenging in chloroplasts under heat stress. High temperature stress triggers oxidative bursts of superoxide and/or hydrogen peroxide in plants. The transfer of excitation energy in the photosystem II (PSII) antenna complex and the electron transport in the PSII reaction center can be inhibited by heat stress. It has been established that ROS are generated both on the electron acceptor and the electron donor side of PSII under heat stress during which electron transport from the manganese complex to plastoquinone (PQ) is limited. The leakage of electrons to molecular oxygen on the electron acceptor side of PSII forms $O_2^{\bullet-}$, inducing initiation of a cascade reaction leading to the formation of H_2O_2. A diversified ROS-scavenging network functions in concert in chloroplasts, mainly including antioxidants and APX-glutathione cycle, to keep the equilibrium between ROS production and scavenging. The efficient enzymatic scavenging systems are composed of several key enzymes, including superoxide dismutase (SOD), ascorbate peroxidase (APX), glutathione reductase (GR), monodehydroascorbate reductase (MDHAR), dehydroascorbate reductase (DHAR), glutathione peroxidase (GPX) and glutathione-*S*-transferase (GST) and non-enzymatic systems contain antioxidants such as ascorbic acid (Asc) and glutathione (GSH).

4. Turnover of PSII Core Subunits and PSII Protection under Heat Stress

The deleterious effects of high temperatures on proteins in chloroplasts include protein denaturation and aggregation [18,102,103]. ROS generation is one of the earliest responses of plant cells in response to heat stress and chloroplasts are the main targets of ROS-linked damage [71,102]. Given that chloroplasts are a major site of protein degradation, turnover of the damaged proteins is critical for plants to adapt to heat stress through the process of acclimation. As a multisubunit thylakoid membrane pigment-protein complex, PSII is vulnerable to light and heat damages that inhibit light-driven oxidation of water and reduction of plastoquinone [2,71]. PSII produces ROS, responsible for the frequent damage and turnover of this megacomplex that occur under physiological and stress conditions [71]. Accumulating data suggest that although more than 40 proteins are known to associate with PSII, the damage mainly targeted to one of its core proteins, the D1 protein under light and heat stress conditions [18,102,104,105]. Importantly, D1 protein has been demonstrated as a very susceptible target of $^1O_2$; on the other hand, it appears to function as a major scavenger of $^1O_2$ due to its close localization to the site of $^1O_2$ formation in the reaction center of PSII [106]. In addition to the D1 protein, β-carotene, plastoquinol, and a-tocopherol have also shown to play a role in scavenging $^1O_2$ and protect PSII against photo-oxidative damage [107,108]. Under heat stress conditions, two processes have been characterized in the thylakoids: dephosphorylation of the D1 protein in the stroma thylakoids, and aggregation of the phosphorylated D1 protein in the grana [102]. Heat stress also induced the release of the extrinsic Photosystem II Subunit O (PsbO), P and Q proteins from Photosystem II, which affected D1 degradation and aggregation significantly [102]. In spinach thylakoids, cleavage of the D1 protein was detected in response to moderate heat stress (40 °C, 30 min), producing an N-terminal 23-kDa fragment, a C-terminal 9-kDa fragment, and aggregation of the D1 protein [102,103]. ROS are known to specifically modify PSII proteins. Using high-resolution tandem mass spectrometry, oxidative modifications were identified on 36 amino acid residues on the lumenal side of PSII, in the core PSII proteins D1, D2, and CP43 of the cyanobacterium *Synechocystis* sp. PCC 6803, providing the compelling evidence to date of physiologically relevant oxidized residues in PSII [109]. Taken together, the oxidative damage of the D1 protein is caused by reactive oxygen species, mostly singlet oxygen, and also by endogenous cationic radicals generated by the photochemical reactions of photosystem II [71,73]. Under heat stress, the damage to the D1 protein by moderate heat stress is due to reactive oxygen species produced by lipid peroxidation near photosystem II [73,110]. Moreover, the damage of the D1 protein has been shown to be directly proportional to light intensity [111,112] or strength of heat stress [102]. D1 protein is well characterized as a protein of high turnover rate due to its rapid degradation when oxidized after its interaction with 1O2 and its replacement by newly synthesized D1 polypeptides [105]. It was reported that the D1 protein was shown to have the half-life of 2.4 h, the fourth fastest turn-over rate of barley (*Hordeum vulgare*) proteins when plants were growing under normal growth light intensity (500 μmol · m^{-2} · s^{-2}) [113]. Meanwhile, higher degradation rates of the D2, CP43 and PsbH subunits were also detected when compared with the other PSII subunits [113–115]. The maintenance of PSII activity is critical, but also a challenge for oxygenic photosynthetic organisms to survive under normal growth or stress conditions.

Thus, replacing the photo- or heat-damaged D1 with a newly synthesized copy is essential for maintaining PSII activity [105,116]. In chloroplasts, the damaged D1 is degraded by a concerted action of particular filamentation temperature sensitive H (FtsH) protease and Deg (for degradation of periplasmic proteins) isoforms during its rapid and specific turnover and replaced with a de novo synthesized one in a system which is termed the PSII repair cycle [105]. It has been a subject of extensive studies about the basic concept for replacement of the damaged D1 protein by a newly-synthetized copy in the PSII repair cycle [105,116–119]. Repair of damaged D1 protein in PSII includes five steps: (1) migration of damaged PSII core complex to the stroma thylakoid, (2) partial PSII disassembly of the PSII core monomer, (3) access of protease degrading damaged D1, (4) concomitant D1 synthesis, and (5) reassembly of PSII into grana thylakoid [105,116,120,121]. Deg /HtrA (for high temperature requirement A) proteases, a family of serine-type ATP-independent proteases, have been shown in higher plants to be involved in the degradation of the Photosystem II reaction center protein D1 [105,116,122]. In *Arabidopsis*, five DEGs (Deg1, Deg2, Deg5, Deg7, and Deg8) have shown to be peripherally attached to the thylakoid membrane of chloroplasts: Deg1, Deg5, and Deg8 are localized on the lumenal side, and Deg2 and Deg7 are localized on the stromal side [123–125]. DEG5 and DEG8 may have synergistic function in degradation of D1 protein in the repair cycle of PSII under heat stress based on functional analysis of *deg5*, *deg8* and the double mutant *deg5 deg8* of *Arabidopsis thaliana* [126]. Under photoinhibitory conditions, cooperative degradation of D1 by Deg and FtsH has been demonstrated in vivo, in which Deg cleavage assists FtsH processive degradation [127]. In Arabidopsis, FtsH11-encoded protease have shown to play a direct role in thermotolerance, a function previously reported for bacterial and yeast FtsH proteases [128]. It should be noted that photosynthetic capability and PSII quantum yield are greatly reduced in the leaves of FtsH11 mutants when exposed to the moderately high temperature whereas under high light conditions, FtsH11 mutants and wild-type plants showed no significant difference in photosynthesis capacity [128]. In general, several possible mechanisms have been proposed for activation of these proteases, which depend on oligomerization of the monomer subunits [129]. In line with the hypothesis that hexamers of the FtsH proteases are probably localized near the Photosystem II complexes at the grana, degradation of the D1 protein could take place in the grana rather than in the stroma thylakoids to circumvent long-distance migration of both the Photosystem II complexes containing the photodamaged D1 protein and the proteases [129]. Under high light conditions, the lumen-exposed loops of the D1 protein at specific sites were cleaved by all of three lumenal serine proteases, Deg1, Deg5 and Deg8, during PSII repair cycle [130–133]. It was reported that Deg5 and Deg8 interact to form an active protease complex under high light [133]. Interestingly, Deg1 is activated when the Deg1 monomers are transformed into a proteolytically active hexamer at acidic pH upon protonation of a histidine amino acid residue [134]. In addition to functioning as a protease, Deg1 also plays a novel role as a chaperone/assembly factor of PSII [135]. In addition, Deg1 has been shown to be responsible for the proteolytic activity against the PsbO protein in vitro [136]. During the PSII repair cycle, only the damaged reaction center protein D1 and occasionally also the D2, CP43 and PsbH subunits are replaced while the other protein components of the complex are recycled, indicating that many aspects of PSII repair cycle and de novo biogenesis are partially overlapping [113–115,137]. Originally, the vulnerability of PSII to high light or heat stress was taken as an inherent fault of the photosynthetic machinery. However, recent studies strongly support that the constant, yet highly regulated, photodamage and repair cycle of PSII are of a strong physiological basis. Collectively, the photodamage of PSII is likely to act as a PSI protection mechanism instead of being considered solely as an undesired consequence of the highly oxidizing chemistry of the water splitting PSII [105].

The synthesis of heat shock proteins (HSPs) is characterized as a major response of all organisms responding to heat stress. The HSPs act as chaperones by assisting in protein folding and preventing irreversible protein aggregation [138,139]. A chloroplast-localized sHSP, HSP21, has been identified in diverse higher plant species, including both dicots and monocots [139,140] and its precursor polypeptide is ~5 kD larger than the mature protein [141,142]. HSP21 is thought to protect photosynthetic electron transport, specifically that of Photosystem II, during heat stress [143–147] and oxidative stress [145,148]. In addition, HSP21 has been demonstrated to play a dual role in protecting PSII from oxidative stress and promoting color changes during fruit maturation whereas no protective effects for the transgene were detected on PSII thermotolerance [149]. Importantly, studies using the transgenic tomato plants overexpressing HSP21 have shown that this protein associates with proteins of Photosystem II and does not reactivate heat-denatured Photosystem II, but instead protects this complex from damage during heat stress [150]. Interestingly, around two thirds of chloroplast HSP21 proteins are translocated into the thylakoid membranes in response to heat treatment in plants, suggesting that the association with membranes should be considered to fully understand the role of sHsps in physiological adaptation processes under stress conditions [151]. Despite extensive studies on HSP21, the specific roles of HSP21 in protecting PSII from heat stress remain elusive.

Recently, HSP21 has been demonstrated to protect PSII from heat stress-induced damages by directly binding to D1 and D2 proteins, the core subunits of PSII. Importantly, heat-responsive transcriptional activation of *HSP21* is regulated by the chloroplast retrograde signaling pathway in which GUN5 acts as a determinant upstream signaling component in *Arabidopsis* [5]. Based on these findings, an auto-adaptation loop working module has emerged in which the GUN5-dependent plastid signal(s) is triggered in response to heat stress and in turn communicated into the nucleus to activate the heat-responsive expression of *HSP21* for optimizing particular demands of chloroplasts in making photosynthetic complexes stable during adaptation to heat stress in plants [5,6].

5. Effects of Heat Stress on Metabolic Flux through the Calvin-Benson-Bassham Cycle

In the Calvin-Benson-Bassham Cycle, Ribulose-1,5-bisphosphate carboxylase/oxygenase (Rubisco) plays a critical role in catalyzing the carboxylation of the 5-carbon sugar ribulose-1,5-bisphosphate (RuBP) when atmospheric CO_2 during is fixed during photosynthesis. Rubisco activase (RCA) regulates the activity of Rubisco by facilitating the dissociation of inhibitory sugar phosphates from the active site of Rubisco in an ATP-dependent manner [152]. Extensive evidence supports the conclusion that reduction of plant photosynthesis arises primarily from thermal inactivation of Rubisco activity due to the inhibition of RCA under moderately elevated temperatures [8,152–155]. In addition to Rubisco activation, electron transport activity, ATP synthesis, and RuBP regeneration are also inhibited by moderately heat stress [156–158]. As the temperature increases further above the thermal optimum, the physical integrity of electron transport components of the photosynthetic apparatus can be severely damaged, leading to the increased limitation in photosynthesis [15]. It has been the subject of extensive investigations of elucidating the biochemical basis for the decrease in Rubisco activation state under heat stress [8,15,153,154,159]. Initially, studies on thermal stability of purified RCA showed that heat treatment only slightly inhibited the activities of this enzyme [160,161] and later experiments confirmed that heat stress caused thermal denaturation of activase in wheat and cotton leaves [162]. In line with the thermal denaturation of RCA under heat stress, Feller et al. (1998) suggested that RCA exhibited the exceptional thermal lability in vivo and the thermal stabilities of activases were different in plants from contrasting thermal environments [15,159]. Thus, loss of activase activity during heat stress is caused by an exceptional sensitivity of the protein to thermal denaturation and is responsible, in part, for deactivation of Rubisco [163]. It has been assumed that the stability of RCA could be influenced by heat-induced changes either in redox state [11,164,165] or the concentrations of ions, nucleotides, or other chloroplast constituents in plants [154]. On the other hand, the thermotolerance of Rubisco activase has been proposed to be responsible for restricting the distribution of certain plant species [154] as demonstrated by the response of Rubisco activase activity to temperature for cotton, a warm-season species, and *Camelina sativa*, a cool-season species. With respect to the effects of high growth temperature on the relative contribution of diffusive and biochemical limitations to photosynthesis, our knowledge is limited although there is abundant evidence that photosynthesis can acclimate to temperature [166,167]. Accumulating data suggest that the biochemical mechanisms about the decrease in Rubisco activation can be attributed to: (1) more rapid de-activation of Rubisco caused by a faster rate of dead-end product formation; and (2) slower re-activation of Rubisco by activase [168]. In a word, the resulting consequence is that RCA becomes less effective in keeping Rubisco catalytically competent as temperature increases.

Inhibition of net photosynthesis by heat stress has been attributed to an inability of Rubisco activase to maintain Rubisco in an active form because of the low thermal stability of Rubisco's chaperone, activase. These results support a role for RCA in limiting photosynthesis at high temperature when the temperature exceeds the optimum range for plants. In cotton (*Gossypium hirsutum* L.), activase gene expression is influenced by post-transcriptional mechanisms that may contribute to acclimation of photosynthesis during extended periods of heat stress [169]. In wheat, northern blot analysis showed maximum accumulation of *TaRCA1* transcript in thermotolerant cv. during mealy-ripe stage, as compared to thermosusceptible ones [170]. To test the hypothesis that thermostable RCA can improve photosynthesis under elevated temperatures, gene shuffling technology was used to generate several *Arabidopsis thaliana* RCA1 (short isoform) variants exhibiting improved thermostability [171]. In line with the findings mentioned above, transgenic Arabidopsis lines expressing a thermostable chimeric activase showed higher rates of photosynthesis than the wild type after a short exposure to higher temperatures and they also recovered better, when they were returned to the normal temperature [172]. The results showed that photosynthesis and growth were improved under moderate heat stress in transgenic Arabidopsis expressing these thermotolerant RCA isoforms, providing evidence that manipulation of activase properties can improve C3 photosynthesis. In addition, the transcriptional level of wheat *RCA* (45–46 kDa) positively correlated with the yield of plants under heat-stress conditions in a very significant and linear manner [173]. At present, accumulating data indicate that RCA could affect plant productivity in relation to its endogenous levels under temperature stress conditions. Critically, RCA as the molecular chaperone plays a key role in constant engagement and remodeling of Rubisco to maintain metabolic flux through the Calvin-Benson-Bassham cycle as Rubisco is characterized as a dead-end inhibited complex in higher plants. In plants of the crassulacean acid metabolism (CAM), it has been assumed that possessing thermostable RCA is necessary for these plants to support the metabolic flux of Calvin-Benson-Bassham cycle when closure of stomata is a limitation factor during the day [174]. It is interesting that the CAM Rca isoforms (*Agave tequilana*) were found to be approximately 10 °C more thermostable when compared with the C3 isoforms of Rca isolated from rice (*Oryza sativa*) [174]. Interestingly, sequence analysis and immuno-blotting identified the beta-subunit of chaperonin-60 (cpn60 beta), the chloroplast GroEL homologue, as a protein that was bound to Rubisco activase from leaf extracts prepared from heat-stressed, but not control plants [175]. Rubisco requires RCA, an AAA+ ATPase that reactivates Rubisco by remodeling the conformation of inhibitor-bound sites. RCA is regulated by the ratio of ADP:ATP, with the precise response potentiated by redox regulation of the alpha-isoform [176]. Given that RCA uses the energy from ATP hydrolysis to restore catalytic competence to Rubisco, manipulation of RCA by redox regulation of the a-isoform might provide a strategy for enhancing photosynthetic performance in *Arabidopsis* [177]. In rice, heat stress significantly induced the expression of *RCA* large isoform (*RCAL*) as determined by both mRNA and protein levels and correlative analysis indicated that and RCA small isoform (RCAS) protein content was very tightly correlated to Rubisco initial activity and net photosynthetic rate under both heat stress and normal conditions [178]. In two *Populus* species adapted to contrasting thermal environments, the difference in the primary sequence of Rubisco activases between the species is more significant in the regions conferring ATPase activity and Rubisco recognition, suggesting that the genotypic distinctive characterizations in Rubisco activase are likely to underlie the specificities with respect to the heat-sensitive strength of Rubisco activase and photosynthesis under moderate high temperature conditions [179]. Recent studies on the effects of heat and drought on three major cereal crops, including rice, wheat, and maize, indicate that reductions in Rubisco activation might be not dependent on the amount of Rubisco and RCA, but could be resulted from the inhibition of RCA activity, as evidenced by the mutual reduction and positive relationship existed between the activation state of Rubisco and the rate of electron transport [153]. Critically, Rubisco activase acts as a key player in photosynthesis under heat stress conditions (non-stomatal limitation) [180]. When exposed to a moderate heat stress, Rca can be inhibited reversibly, but is irreversibly inhibited under a higher temperature and/or longer exposure due to heat stress-induced insolubilization and degradation of the Rca protein [180].

6. Conclusions and Perspectives

In many regions of the world, high temperature stress is one of the most important constraints to plant growth and productivity, especially for crop plants. The mechanism underlying the development of heat-tolerance for important agricultural crops as well as plant responses and adaptation to elevated temperatures needs to be better understood. Extensive studies have shown that metabolic regulation of adaptation processes during heat stress is not only an important developmental process, but also allows for flexibility of physiological responses to heat stress. In photosynthetic organisms, heat stress can affect photosynthesis through altered carbon assimilation metabolism in chloroplasts with remobilizing their starch reserve to release energy, sugars and derived metabolites in order to help mitigate the stress. This is thought to be an essential process for plant fitness with important implications for plant productivity under high temperature stress. One future challenges is to dissect the complex interaction networks between heat stress sensing, signal transduction and activations of key genes involved in metabolic reprogramming in coordination with developmental programmes. Accumulation and modification of metabolites in chloroplasts under heat stress may play a key role in the regulation of adaptation processes at cellular levels in plants, allowing plants to interact with their environment and to activate cellular heat stress responses at the optimal time in order to maintain photosynthesis. This kind of metabolic reprogramming is critical for plants to survive stress periods, and to prevent further damage to the whole plant.

The role of chloroplast in the metabolic regulation of heat stress responses has attracted increasing attention and extensive investigations from an organellar perspective have provided insights into better understanding the hypothesis stated that the heat stress-induced reprogramming, including decline in photosynthesis and alterations in photosynthetic metabolites which, in turn, could act as signal(s) or trigger the initial signal cascades to activate cellular heat stress responses. The present knowledge concerning the interplay between the chloroplast and nucleus in heat stress signal perception and activation of cellular heat stress responses is emerging, but more efforts are needed to reach a detailed overview. It can be predicted that uncovering the molecular mechanisms of heat sensing will pave the way to engineering plants capable of tolerating heat stress. It is well known that the ability of plants adapting to different climate regimes vary dramatically across and within species. Identification and functional analysis of the valuable heat-tolerant genetic resources will bring about a further significant improvement in manipulation of photosynthesis to increase crop yield based on a direct comparative analysis between the different manipulations with all the transgenic and wild-type plants grown and assessed in parallel under filed growth conditions. Thus, in-depth analyses of the interactions between the chloroplast and nucleus in heat stress responses are likely to be in focus during forthcoming years. On the other hand, Rubisco activase and enzymes functioning in the detoxification of reactive oxygen species are thought to be critical targets for breeding heat-tolerant crop plants with high yields under high temperature stress.

Acknowledgments: This study was supported by Chinese Academy of Sciences (Strategic Priority Research Program XDPB0404), the Ministry of Science and Technology of China (National Key R&D Program of China, 2016YFD0100405), and the National Natural Science Foundation of China (31770314 and 31570260).

Author Contributions: Fang-Qing Guo and Qing-Long Wang conceived and designed the outline and contents of this review article; Fang-Qing Guo and Qing-Long Wang wrote the paper with inputs from Juan-Hua Chen and Ning-Yu He.

Conflicts of Interest: The authors declare no conflict of interest. The founding sponsors had no role in the design of the study; in the collection, analyses, or interpretation of data; in the writing of the manuscript, and in the decision to publish the results.

Abbreviations

1-MCP	1-methylcyclopropene
$^1O_2$	singlet state of oxygen
$^3O_2$	triplet state of oxygen
ACD1	Accelerated cell death
APX	ascorbate peroxidase
AsA	Ascorbate
ASH	ascorbic acid
ATP	adenosine 5′-triphosphate
CAM	crassulacean acid metabolism
CAT	catalase

Int. J. Mol. Sci. **2018**, *19*, 849

Chl	Chlorophyll
Chl *a*	Chlorophyll *a*
Chlase	chlorophyllase
CLD1	CHLOROPHYLL DEPHYTYLASE1
CLHs	Chlorophyllase genes
CPAS	3-cyclopropyl-1-enyl-propanoic acid sodium salt
cpn60 beta	beta-subunit of chaperonin-60
Deg/HtrA	high temperature requirement A
DHAR	dehydroascorbate reductase
FtsH	filamentation temperature sensitive H
Fv/Fm	variable fluorescence to maximum fluorescence
GPX	glutathione peroxidase
GR	glutathione reductase
GSH	glutathione
GST	glutathione-*S*-transferase
GUN5	genomes uncoupled 5
H_2O_2	hydrogen peroxide
HSPs	heat shock proteins
ipt	isopentenyl transferase
LeCDJ1	tomato (*Lycopersicon esculentum*) chloroplast-targeted DnaJ protein
LHCPII	light harvesting chlorophyll *a/b*-protein complex II
MDA	malondialdehyde
MDHAR	monodehydroascorbate reductase
NAD(P)H	nicotinamide adenine dinucleotide phosphate (NADP)
NOL	NYC1-LIKE
NYC1	NON-YELLOW COLORING1
$O_2^{\bullet-}$	superoxide anion radical
OEC	oxygen evolving complex
OH	hydroxyl radicals
PAO	pheide a oxygenase
pFCC	primary fluorescent chlorophyll catabolite
pheide	pheophorbide
POD	peroxidase
PPH	pheophytinase
PPH2	PHEOPHORBIDASE 2
PsbO	PHOTOSYSTEM II SUBUNIT O
PSII	Photosystem II
QTL	quantitative trait loci
RCA	Rubisco activase
RCAL	RCA large isoform
RCAS	RCA small isoform
RCCR	Red chl catabolite reductase
ROS	Reactive oxygen species
Rubisco	Ribulose-1,5-bisphosphate carboxylase/oxygenase
RuBP	Ribulose-1,5-bisphosphate
SGR	STAY-GREEN
SID	senescence-induced degradation
SOD	superoxide dismutase
ZR	zeatin riboside

References

1. Wahid, A.; Gelani, S.; Ashraf, M.; Foolad, M.R. Heat tolerance in plants: An overview. *Environ. Exp. Bot.* **2007**, *61*, 199–223. [CrossRef]

2. Allakhverdiev, S.I.; Kreslavski, V.D.; Klimov, V.V.; Los, D.A.; Carpentier, R.; Mohanty, P. Heat stress: An overview of molecular responses in photosynthesis. *Photosynth. Res.* **2008**, *98*, 541–550. [CrossRef] [PubMed]

3. Berry, J.; Bjorkman, O. Photosynthetic Response and Adaptation to Temperature in Higher-Plants. *Annu. Rev. Plant Physiol. Plant Mol. Biol.* **1980**, *31*, 491–543. [CrossRef]

4. Yu, H.-D.; Yang, X.-F.; Chen, S.-T.; Wang, Y.-T.; Li, J.-K.; Shen, Q.; Liu, X.-L.; Guo, F.-Q. Downregulation of Chloroplast RPS1 Negatively Modulates Nuclear Heat-Responsive Expression of HsfA2 and Its Target Genes in *Arabidopsis. PLoS Genet.* **2012**, *8*, e1002669. [CrossRef] [PubMed]

5. Chen, S.-T.; He, N.-Y.; Chen, J.-H.; Guo, F.-Q. Identification of core subunits of photosystem II as action sites of HSP21, which is activated by the GUN5-mediated retrograde pathway in *Arabidopsis. Plant J.* **2017**, *89*, 1106–1118. [CrossRef] [PubMed]

6. Sun, A.-Z.; Guo, F.-Q. Chloroplast Retrograde Regulation of Heat Stress Responses in Plants. *Front. Plant Sci.* **2016**, *7*. [CrossRef] [PubMed]

7. Havaux, M. Characterization of Thermal-Damage to the Photosynthetic Electron-Transport System in Potato Leaves. *Plant Sci.* **1993**, *94*, 19–33. [CrossRef]

8. Sharkey, T.D. Effects of moderate heat stress on photosynthesis: Importance of thylakoid reactions, rubisco deactivation, reactive oxygen species, and thermotolerance provided by isoprene. *Plant Cell Environ.* **2005**, *28*, 269–277. [CrossRef]

9. Yamada, M.; Hidaka, T.; Fukamachi, H. Heat tolerance in leaves of tropical fruit crops as measured by chlorophyll fluorescence. *Sci. Hortic.* **1996**, *67*, 39–48. [CrossRef]

10. Wise, R.R.; Olson, A.J.; Schrader, S.M.; Sharkey, T.D. Electron transport is the functional limitation of photosynthesis in field-grown *Pima cotton* plants at high temperature. *Plant Cell Environ.* **2004**, *27*, 717–724. [CrossRef]

11. Sharkey, T.D.; Schrader, S.M. High temperature stress. In *Physiology and Molecular Biology of Stress Tolerance in Plants*; Springer: New York, NY, USA, 2006; pp. 101–129.

12. Havaux, M.; Tardy, F. Temperature-dependent adjustment of the thermal stability of photosystem II in vivo: Possible involvement of xanthophyll-cycle pigments. *Planta* **1996**, *198*, 324–333. [CrossRef]

13. Klimov, V.V.; Baranov, S.V.; Allakhverdiev, S.I. Bicarbonate protects the donor side of photosystem II against photoinhibition and thermoinactivation. *FEBS Lett.* **1997**, *418*, 243–246. [CrossRef]

14. Yamane, Y.; Kashino, Y.; Koike, H.; Satoh, K. Effects of high temperatures on the photosynthetic systems in spinach: Oxygen-evolving activities, fluorescence characteristics and the denaturation process. *Photosynth. Res.* **1998**, *57*, 51–59. [CrossRef]

15. Salvucci, M.E.; Crafts-Brandner, S.J. Relationship between the heat tolerance of photosynthesis and the thermal stability of Rubisco activase in plants from contrasting thermal environments. *Plant Physiol.* **2004**, *134*, 1460–1470. [CrossRef] [PubMed]

16. Gounaris, K.; Brain, A.R.R.; Quinn, P.J.; Williams, W.P. Structural Reorganization of Chloroplast Thylakoid Membranes in Response to Heat-Stress. *Biochim. Biophys. Acta* **1984**, *766*, 198–208. [CrossRef]

17. Semenova, G.A. Structural reorganization of thylakoid systems in response to heat treatment. *Photosynthetica* **2004**, *42*, 521–527. [CrossRef]

18. Yamamoto, Y.; Aminaka, R.; Yoshioka, M.; Khatoon, M.; Komayama, K.; Takenaka, D.; Yamashita, A.; Nijo, N.; Inagawa, K.; Morita, N.; et al. Quality control of photosystem II: Impact of light and heat stresses. *Photosynth. Res.* **2008**, *98*, 589–608. [CrossRef] [PubMed]

19. Vani, B.; Saradhi, P.P.; Mohanty, P. Alteration in chloroplast structure and thylakoid membrane composition due to in vivo heat treatment of rice seedlings: Correlation with the functional changes. *J. Plant Physiol.* **2001**, *158*, 583–592. [CrossRef]

20. Kmiecik, P.; Leonardelli, M.; Teige, M. Novel connections in plant organellar signalling link different stress responses and signalling pathways. *J. Exp. Bot.* **2016**, *67*, 3793–3807. [CrossRef] [PubMed]

21. Apel, K.; Hirt, H. Reactive oxygen species: Metabolism, oxidative stress, and signal transduction. *Annu. Rev. Plant Biol.* **2004**, *55*, 373–399. [CrossRef] [PubMed]

22. Hortensteiner, S. Stay-green regulates chlorophyll and chlorophyll-binding protein degradation during senescence. *Trends Plant Sci.* **2009**, *14*, 155–162. [CrossRef] [PubMed]

23. Hortensteiner, S. Chlorophyll degradation during senescence. *Annu. Rev. Plant Biol.* **2006**, *57*, 55–77. [CrossRef] [PubMed]

24. Hortensteiner, S.; Feller, U. Nitrogen metabolism and remobilization during senescence. *J. Exp. Bot.* **2002**, *53*, 927–937. [CrossRef] [PubMed]

25. Ginsburg, S.; Schellenberg, M.; Matile, P. Cleavage of Chlorophyll-Porphyrin—Requirement for Reduced Ferredoxin and Oxygen. *Plant Physiol.* **1994**, *105*, 545–554. [CrossRef] [PubMed]

26. Tsuchiya, T.; Ohta, H.; Okawa, K.; Iwamatsu, A.; Shimada, H.; Masuda, T.; Takamiya, K. Cloning of chlorophyllase, the key enzyme in chlorophyll degradation: Finding of a lipase motif and the induction by methyl jasmonate. *Proc. Natl. Acad. Sci. USA* **1999**, *96*, 15362–15367. [CrossRef] [PubMed]

27. Schenk, N.; Schelbert, S.; Kanwischer, M.; Goldschmidt, E.E.; Doermann, P.; Hoertensteiner, S. The chlorophyllases AtCLH1 and AtCLH2 are not essential for senescence-related chlorophyll breakdown in *Arabidopsis thaliana*. *FEBS Lett.* **2007**, *581*, 5517–5525. [CrossRef] [PubMed]

28. Schelbert, S.; Aubry, S.; Burla, B.; Agne, B.; Kessler, F.; Krupinska, K.; Hoertensteiner, S. Pheophytin Pheophorbide Hydrolase (Pheophytinase) Is Involved in Chlorophyll Breakdown during Leaf Senescence in *Arabidopsis*. *Plant Cell* **2009**, *21*, 767–785. [CrossRef] [PubMed]

29. Rodoni, S.; Muhlecker, W.; Anderl, M.; Krautler, B.; Moser, D.; Thomas, H.; Matile, P.; Hortensteiner, S. Chlorophyll breakdown in senescent chloroplasts—Cleavage of pheophorbide a in two enzymic steps. *Plant Physiol.* **1997**, *115*, 669–676. [CrossRef] [PubMed]

30. Wuthrich, K.L.; Bovet, L.; Hunziker, P.E.; Donnison, I.S.; Hortensteiner, S. Molecular cloning, functional expression and characterisation of RCC reductase involved in chlorophyll catabolism. *Plant J.* **2000**, *21*, 189–198. [CrossRef] [PubMed]

31. Pruzinska, A.; Tanner, G.; Anders, I.; Roca, M.; Hortensteiner, S. Chlorophyll breakdown: Pheophorbide a oxygenase is a Rieske-type iron-sulfur protein, encoded by the accelerated cell death 1 gene. *Proc. Natl. Acad. Sci. USA* **2003**, *100*, 15259–15264. [CrossRef] [PubMed]

32. Pruzinska, A.; Anders, I.; Aubry, S.; Schenk, N.; Tapernoux-Luthi, E.; Muller, T.; Krautler, B.; Hortensteiner, S. In vivo participation of red chlorophyll catabolite reductase in chlorophyll breakdown. *Plant Cell* **2007**, *19*, 369–387. [CrossRef] [PubMed]

33. Hinder, B.; Schellenberg, M.; Rodon, S.; Ginsburg, S.; Vogt, E.; Martinoia, E.; Matile, P.; Hortensteiner, S. How plants dispose of chlorophyll catabolites—Directly energized uptake of tetrapyrrolic breakdown products into isolated vacuoles. *J. Biol. Chem.* **1996**, *271*, 27233–27236. [CrossRef] [PubMed]

34. Tommasini, R.; Vogt, E.; Fromenteau, M.; Hortensteiner, S.; Matile, P.; Amrhein, N.; Martinoia, E. An ABC-transporter of *Arabidopsis thaliana* has both glutathione-conjugate and chlorophyll catabolite transport activity. *Plant J.* **1998**, *13*, 773–780. [CrossRef] [PubMed]

35. Kusaba, M.; Ito, H.; Morita, R.; Iida, S.; Sato, Y.; Fujimoto, M.; Kawasaki, S.; Tanaka, R.; Hirochika, H.; Nishimura, M.; et al. Rice NON-YELLOW COLORING1 is involved in light-harvesting complex II and grana degradation during leaf senescence. *Plant Cell* **2007**, *19*, 1362–1375. [CrossRef] [PubMed]

36. Sato, Y.; Morita, R.; Katsuma, S.; Nishimura, M.; Tanaka, A.; Kusaba, M. Two short-chain dehydrogenase/reductases, NON-YELLOW COLORING 1 and NYC1-LIKE, are required for chlorophyll b and light-harvesting complex II degradation during senescence in rice. *Plant J.* **2009**, *57*, 120–131. [CrossRef] [PubMed]

37. Pruzinska, A.; Tanner, G.; Aubry, S.; Anders, I.; Moser, S.; Muller, T.; Ongania, K.H.; Krautler, B.; Youn, J.Y.; Liljegren, S.J.; et al. Chlorophyll breakdown in senescent *Arabidopsis* leaves. Characterization of chlorophyll catabolites and of chlorophyll catabolic enzymes involved in the degreening reaction. *Plant Physiol.* **2005**, *139*, 52–63. [CrossRef] [PubMed]

38. Tanaka, R.; Hirashima, M.; Satoh, S.; Tanaka, A. The Arabidopsis-accelerated cell death gene ACD1 is involved in oxygenation of pheophorbide a: Inhibition of the pheophorbide a oxygenase activity does not lead to the "Stay-Green" phenotype in *Arabidopsis*. *Plant Cell Physiol.* **2003**, *44*, 1266–1274. [CrossRef] [PubMed]

39. Thomas, H. Sid—A Mendelian Locus Controlling Thylakoid Membrane Disassembly in Senescing Leaves of Festuca-Pratensis. *Theor. Appl. Genet.* **1987**, *73*, 551–555. [CrossRef] [PubMed]

40. Armstead, I.; Donnison, I.; Aubry, S.; Harper, J.; Hoertensteiner, S.; James, C.; Mani, J.; Moffet, M.; Ougham, H.; Roberts, L.; et al. From crop to model to crop: Identifying the genetic basis of the staygreen mutation in the Lolium/Festuca forage and amenity grasses. *New Phytol.* **2006**, *172*, 592–597. [CrossRef] [PubMed]

41. Armstead, I.; Donnison, I.; Aubry, S.; Harper, J.; Hortensteiner, S.; James, C.; Mani, J.; Moffet, M.; Ougham, H.; Roberts, L.; et al. Cross-species identification of Mendel's/locus. *Science* **2007**, *315*, 73. [CrossRef] [PubMed]

42. Ren, G.; An, K.; Liao, Y.; Zhou, X.; Cao, Y.; Zhao, H.; Ge, X.; Kuai, B. Identification of a novel chloroplast protein AtNYE1 regulating chlorophyll degradation during leaf senescence in *Arabidopsis*. *Plant Physiol.* **2007**, *144*, 1429–1441. [CrossRef] [PubMed]

43. Jiang, H.; Li, M.; Liang, N.; Yan, H.; Wei, Y.; Xu, X.; Liu, J.; Xu, Z.; Chen, F.; Wu, G. Molecular cloning and function analysis of the stay green gene in rice. *Plant J.* **2007**, *52*, 197–209. [CrossRef] [PubMed]

44. Park, S.-Y.; Yu, J.-W.; Park, J.-S.; Li, J.; Yoo, S.-C.; Lee, N.-Y.; Lee, S.-K.; Jeong, S.-W.; Seo, H.S.; Koh, H.-J.; et al. The senescence-induced staygreen protein regulates chlorophyll degradation. *Plant Cell* **2007**, *19*, 1649–1664. [CrossRef] [PubMed]

45. Sato, Y.; Morita, R.; Nishimura, M.; Yamaguchi, H.; Kusaba, M. Mendel's green cotyledon gene encodes a positive regulator of the chlorophyll-degrading pathway. *Proc. Natl. Acad. Sci. USA* **2007**, *104*, 14169–14174. [CrossRef] [PubMed]

46. Aubry, S.; Mani, J.; Hortensteiner, S. Stay-green protein, defective in Mendel's green cotyledon mutant, acts independent and upstream of pheophorbide a oxygenase in the chlorophyll catabolic pathway. *Plant Mol. Biol.* **2008**, *67*, 243–256. [CrossRef] [PubMed]

47. Pogson, B.J.; Woo, N.S.; Foerster, B.; Small, I.D. Plastid signalling to the nucleus and beyond. *Trends Plant Sci.* **2008**, *13*, 602–609. [CrossRef] [PubMed]

48. Borovsky, Y.; Paran, I. Chlorophyll breakdown during pepper fruit ripening in the chlorophyll retainer mutation is impaired at the homolog of the senescence-inducible stay-green gene. *Theor. Appl. Genet.* **2008**, *117*, 235–240. [CrossRef] [PubMed]

49. Barry, C.S. The stay-green revolution: Recent progress in deciphering the mechanisms of chlorophyll degradation in higher plants. *Plant Sci.* **2009**, *176*, 325–333. [CrossRef]

50. Hoertensteiner, S. Update on the biochemistry of chlorophyll breakdown. *Plant Mol. Biol.* **2013**, *82*, 505–517. [CrossRef] [PubMed]

51. Lim, P.O.; Kim, H.J.; Nam, H.G. Leaf senescence. *Annu. Rev. Plant Biol.* **2007**, *58*, 115–136. [CrossRef] [PubMed]

52. Rossi, S.; Burgess, P.; Jespersen, D.; Huang, B. Heat-Induced Leaf Senescence Associated with Chlorophyll Metabolism in Bentgrass Lines Differing in Heat Tolerance. *Crop Sci.* **2017**, *57*, S169–S178. [CrossRef]

53. Todorov, D.T.; Karanov, E.N.; Smith, A.R.; Hall, M.A. Chlorophyllase activity and chlorophyll content in wild type and eti 5 mutant of *Arabidopsis thaliana* subjected to low and high temperatures. *Biol. Plant.* **2003**, *46*, 633–636. [CrossRef]

54. Djanaguiraman, M.; Prasad, P.V.V.; Boyle, D.L.; Schapaugh, W.T. High-Temperature Stress and Soybean Leaves: Leaf Anatomy and Photosynthesis. *Crop Sci.* **2011**, *51*, 2125–2131. [CrossRef]

55. Djanaguiraman, M.; Prasad, P.V.V.; Murugan, M.; Perumal, R.; Reddy, U.K. Physiological differences among sorghum (*Sorghum bicolor* L. Moench) genotypes under high temperature stress. *Environ. Exp. Bot.* **2014**, *100*, 43–54. [CrossRef]

56. Ristic, Z.; Momcilovic, I.; Fu, J.M.; Callegaric, E.; DeRidder, B.P. Chloroplast protein synthesis elongation factor, EF-Tu, reduces thermal aggregation of Rubisco activase. *J. Plant Physiol.* **2007**, *164*, 1564–1571. [CrossRef] [PubMed]

57. Akter, N.; Islam, M.R. Heat stress effects and management in wheat. A review. *Agron. Sustain. Dev.* **2017**, *37*, 37. [CrossRef]

58. Liu, X.Z.; Huang, B.R. Heat stress injury in relation to membrane lipid peroxidation in creeping bentgrass. *Crop Sci.* **2000**, *40*, 503–510. [CrossRef]

59. Jespersen, D.; Zhang, J.; Huang, B. Chlorophyll loss associated with heat-induced senescence in bentgrass. *Plant Sci.* **2016**, *249*, 1–12. [CrossRef] [PubMed]

60. Lin, Y.-P.; Wu, M.-C.; Charng, Y.-Y. Identification of a Chlorophyll Dephytylase Involved in Chlorophyll Turnover in *Arabidopsis*. *Plant Cell* **2016**, *28*, 2974–2990. [CrossRef] [PubMed]

61. Veerasamy, M.; He, Y.; Huang, B. Leaf senescence and protein metabolism in creeping bentgrass exposed to heat stress and treated with cytokinins. *J. Am. Soc. Hortic. Sci.* **2007**, *132*, 467–472.

62. Chen, Y.; Cothren, J.T.; Chen, D.-H.; Ibrahim, A.M.H.; Lombardini, L. Ethylene-inhibiting compound 1-MCP delays leaf senescence in cotton plants under abiotic stress conditions. *J. Integr. Agric.* **2015**, *14*, 1321–1331. [CrossRef]

63. Huberman, M.; Riov, J.; Goldschmidt, E.E.; Apelbaum, A.; Goren, R. The novel ethylene antagonist, 3-cyclopropyl-1-enyl-propanoic acid sodium salt (CPAS), increases grain yield in wheat by delaying leaf senescence. *Plant Growth Regul.* **2014**, *73*, 249–255. [CrossRef]

64. Abdelrahman, M.; El-Sayed, M.; Jogaiah, S.; Burritt, D.J.; Lam-Son Phan, T. The "STAY-GREEN" trait and phytohormone signaling networks in plants under heat stress. *Plant Cell Rep.* **2017**, *36*, 1009–1025. [CrossRef] [PubMed]

65. Suzuky Pinto, R.; Lopes, M.S.; Collins, N.C.; Reynolds, M.P. Modelling and genetic dissection of staygreen under heat stress. *Theor. Appl. Genet.* **2016**, *129*, 2055–2074. [CrossRef] [PubMed]

66. Teixeira, R.N.; Ligterink, W.; Franca-Neto, J.D.B.; Hilhorst, H.W.M.; da Silva, E.A.A. Gene expression profiling of the green seed problem in Soybean. *BMC Plant Biol.* **2016**, *16*, 37. [CrossRef] [PubMed]

67. Vijayalakshmi, K.; Fritz, A.K.; Paulsen, G.M.; Bai, G.; Pandravada, S.; Gill, B.S. Modeling and mapping QTL for senescence-related traits in winter wheat under high temperature. *Mol. Breed.* **2010**, *26*, 163–175. [CrossRef]

68. Chen, Y.-E.; Zhang, C.-M.; Su, Y.-Q.; Ma, J.; Zhang, Z.-W.; Yuan, M.; Zhang, H.-Y.; Yuan, S. Responses of photosystem II and antioxidative systems to high light and high temperature co-stress in wheat. *Environ. Exp. Bot.* **2017**, *135*, 45–55. [CrossRef]

69. Shirdelmoghanloo, H.; Cozzolino, D.; Lohraseb, I.; Collins, N.C. Truncation of grain filling in wheat (*Triticum aestivum*) triggered by brief heat stress during early grain filling: Association with senescence responses and reductions in stem reserves. *Funct. Plant Biol.* **2016**, *43*, 919–930. [CrossRef]

70. Shirdelmoghanloo, H.; Lohraseb, I.; Rabie, H.S.; Brien, C.; Parent, B.; Collins, N.C. Heat susceptibility of grain filling in wheat (*Triticum aestivum* L.) linked with rapid chlorophyll loss during a 3-day heat treatment. *Acta Physiol. Plant.* **2016**, *38*, 208. [CrossRef]

71. Asada, K. Production and scavenging of reactive oxygen species in chloroplasts and their functions. *Plant Physiol.* **2006**, *141*, 391–396. [CrossRef] [PubMed]

72. Suzuki, N.; Koussevitzky, S.; Mittler, R.; Miller, G. ROS and redox signalling in the response of plants to abiotic stress. *Plant Cell Environ.* **2012**, *35*, 259–270. [CrossRef] [PubMed]

73. Pospisil, P.; Prasad, A. Formation of singlet oxygen and protection against its oxidative damage in Photosystem II under abiotic stress. *J. Photochem. Photobiol. B Biol.* **2014**, *137*, 39–48. [CrossRef] [PubMed]

74. Mittler, R.; Vanderauwera, S.; Gollery, M.; Van Breusegem, F. Reactive oxygen gene network of plants. *Trends Plant Sci.* **2004**, *9*, 490–498. [CrossRef] [PubMed]

75. Foyer, C.H.; LopezDelgado, H.; Dat, J.F.; Scott, I.M. Hydrogen peroxide- and glutathione-associated mechanisms of acclimatory stress tolerance and signalling. *Physiol. Plant.* **1997**, *100*, 241–254. [CrossRef]

76. Dat, J.F.; Lopez-Delgado, H.; Foyer, C.H.; Scott, I.M. Parallel changes in H_2O_2 and catalase during thermotolerance induced by salicylic acid or heat acclimation in mustard seedlings. *Plant Physiol.* **1998**, *116*, 1351–1357. [CrossRef] [PubMed]

77. Vallelian-Bindschedler, L.; Schweizer, P.; Mosinger, E.; Metraux, J.P. Heat-induced resistance in barley to powdery mildew (*Blumeria graminis* f.sp. hordei) is associated with a burst of active oxygen species. *Physiol. Mol. Plant Pathol.* **1998**, *52*, 185–199. [CrossRef]

78. Foyer, C.H.; Noctor, G. Redox Signaling in Plants. *Antioxid. Redox Signal.* **2013**, *18*, 2087–2090. [CrossRef] [PubMed]

79. Vacca, R.A.; Valenti, D.; Bobba, A.; Merafina, R.S.; Passarella, S.; Marra, E. Cytochrome c is released in a reactive oxygen species-dependent manner and is degraded via caspase-like proteases in tobacco bright-yellow 2 cells en route to heat shock-induced cell death. *Plant Physiol.* **2006**, *141*, 208–219. [CrossRef] [PubMed]

80. Wang, P.; Duan, W.; Takabayashi, A.; Endo, T.; Shikanai, T.; Ye, J.Y.; Mi, H.L. Chloroplastic NAD(P)H dehydrogenase in tobacco leaves functions in alleviation of oxidative damage caused by temperature stress. *Plant Physiol.* **2006**, *141*, 465–474. [CrossRef] [PubMed]

81. Yamashita, A.; Nijo, N.; Pospisil, P.; Morita, N.; Takenaka, D.; Aminaka, R.; Yamamoto, Y.; Yamamoto, Y. Quality control of photosystem II—Reactive oxygen species are responsible for the damage to photosystem II under moderate heat stress. *J. Biol. Chem.* **2008**, *283*, 28380–28391. [CrossRef] [PubMed]

82. Lin, Y.-P.; Lee, T.-Y.; Tanaka, A.; Charng, Y.-Y. Analysis of an *Arabidopsis* heat-sensitive mutant reveals that chlorophyll synthase is involved in reutilization of chlorophyllide during chlorophyll turnover. *Plant J.* **2014**, *80*, 14–26. [CrossRef] [PubMed]

83. Edreva, A. Generation and scavenging of reactive oxygen species in chloroplasts: A submolecular approach. *Agric. Ecosyst. Environ.* **2005**, *106*, 119–133. [CrossRef]

84. Foyer, C.H.; Neukermans, J.; Queval, G.; Noctor, G.; Harbinson, J. Photosynthetic control of electron transport and the regulation of gene expression. *J. Exp. Bot.* **2012**, *63*, 1637–1661. [CrossRef] [PubMed]

85. Foyer, C.H.; Noctor, G. Oxidant and antioxidant signalling in plants: A re-evaluation of the concept of oxidative stress in a physiological context. *Plant Cell Environ.* **2005**, *28*, 1056–1071. [CrossRef]

86. Foyer, C.H.; Noctor, G. Redox homeostasis and antioxidant signaling: A metabolic interface between stress perception and physiological responses. *Plant Cell* **2005**, *17*, 1866–1875. [CrossRef] [PubMed]

87. Foyer, C.H.; Noctor, G. Redox Regulation in Photosynthetic Organisms: Signaling, Acclimation, and Practical Implications. *Antioxid. Redox Signal.* **2009**, *11*, 861–905. [CrossRef] [PubMed]

88. Foyer, C.H.; Noctor, G. Ascorbate and Glutathione: The Heart of the Redox Hub. *Plant Physiol.* **2011**, *155*, 2–18. [CrossRef] [PubMed]

89. Foyer, C.H.; Noctor, G. Managing the cellular redox hub in photosynthetic organisms. *Plant Cell Environ.* **2012**, *35*, 199–201. [CrossRef] [PubMed]

90. Gill, S.S.; Tuteja, N. Reactive oxygen species and antioxidant machinery in abiotic stress tolerance in crop plants. *Plant Physiol. Biochem.* **2010**, *48*, 909–930. [CrossRef] [PubMed]

91. Sainz, M.; Diaz, P.; Monza, J.; Borsani, O. Heat stress results in loss of chloroplast Cu/Zn superoxide dismutase and increased damage to Photosystem II in combined drought-heat stressed *Lotus japonicus*. *Physiol. Plant.* **2010**, *140*, 46–56. [CrossRef] [PubMed]

92. Martins, M.Q.; Rodrigues, W.P.; Fortunato, A.S.; Leitao, A.E.; Rodrigues, A.P.; Pais, I.P.; Martins, L.D.; Silva, M.J.; Reboredo, F.H.; Partelli, F.L.; et al. Protective Response Mechanisms to Heat Stress in Interaction with High [CO$_2$] Conditions in *Coffea* spp. *Front. Plant Sci.* **2016**, *7*. [CrossRef] [PubMed]

93. Mostofa, M.G.; Yoshida, N.; Fujita, M. Spermidine pretreatment enhances heat tolerance in rice seedlings through modulating antioxidative and glyoxalase systems. *Plant Growth Regul.* **2014**, *73*, 31–44. [CrossRef]

94. Wang, X.; Cai, J.; Liu, F.; Dai, T.; Cao, W.; Wollenweber, B.; Jiang, D. Multiple heat priming enhances thermo-tolerance to a later high temperature stress via improving subcellular antioxidant activities in wheat seedlings. *Plant Physiol. Biochem.* **2014**, *74*, 185–192. [CrossRef] [PubMed]

95. Du, H.; Zhou, P.; Huang, B. Antioxidant enzymatic activities and gene expression associated with heat tolerance in a cool-season perennial grass species. *Environ. Exp. Bot.* **2013**, *87*, 159–166. [CrossRef]

96. Kong, F.; Deng, Y.; Wang, G.; Wang, J.; Liang, X.; Meng, Q. LeCDJ1, a chloroplast DnaJ protein, facilitates heat tolerance in transgenic tomatoes. *J. Integr. Plant Biol.* **2014**, *56*, 63–74. [CrossRef] [PubMed]

97. Zou, M.; Yuan, L.; Zhu, S.; Liu, S.; Ge, J.; Wang, C. Effects of heat stress on photosynthetic characteristics and chloroplast ultrastructure of a heat-sensitive and heat-tolerant cultivar of wucai (*Brassica campestris* L.). *Acta Physiol. Plant.* **2017**, *39*. [CrossRef]

98. Ara, N.; Nakkanong, K.; Lv, W.; Yang, J.; Hu, Z.; Zhang, M. Antioxidant Enzymatic Activities and Gene Expression Associated with Heat Tolerance in the Stems and Roots of Two Cucurbit Species ("Cucurbita maxima" and "Cucurbita moschata") and Their Interspecific Inbred Line "Maxchata". *Int. J. Mol. Sci.* **2013**, *14*, 24008–24028. [CrossRef] [PubMed]

99. Miller, G.; Shulaev, V.; Mittler, R. Reactive oxygen signaling and abiotic stress. *Physiol. Plant.* **2008**, *133*, 481–489. [CrossRef] [PubMed]

100. Mittler, R.; Vanderauwera, S.; Suzuki, N.; Miller, G.; Tognetti, V.B.; Vandepoele, K.; Gollery, M.; Shulaev, V.; Van Breusegem, F. ROS signaling: The new wave? *Trends Plant Sci.* **2011**, *16*, 300–309. [CrossRef] [PubMed]

101. Xiao, Y.; Wang, J.; Dehesh, K. Review of stress specific organelles-to-nucleus metabolic signal molecules in plants. *Plant Sci.* **2013**, *212*, 102–107. [CrossRef] [PubMed]

102. Komayama, K.; Khatoon, M.; Takenaka, D.; Horie, J.; Yamashita, A.; Yoshioka, M.; Nakayama, Y.; Yoshida, M.; Ohira, S.; Morita, N.; et al. Quality control of photosystem II: Cleavage and aggregation of heat-damaged D1 protein in spinach thylakoids. *Biochim. Biophys. Acta Bioenerg.* **2007**, *1767*, 838–846. [CrossRef] [PubMed]

103. Yoshioka, M.; Uchida, S.; Mori, H.; Komayama, K.; Ohira, S.; Morita, N.; Nakanishi, T.; Yamamoto, Y. Quality control of photosystem II—Cleavage of reaction center D1 protein in spinach thylakoids by FtsH protease under moderate heat stress. *J. Biol. Chem.* **2006**, *281*, 21660–21669. [CrossRef] [PubMed]

104. Khatoon, M.; Inagawa, K.; Pospisil, P.; Yamashita, A.; Yoshioka, M.; Lundin, B.; Horie, J.; Morita, N.; Jajoo, A.; Yamamoto, Y.; et al. Quality Control of Photosystem II thylakoid unstacking is necessary to avoid further damage to the D1 protein and to facilitate D1 degradation under light stress in spinach thylakoids. *J. Biol. Chem.* **2009**, *284*, 25343–25352. [CrossRef] [PubMed]

105. Jarvi, S.; Suorsa, M.; Aro, E.M. Photosystem II repair in plant chloroplasts—Regulation, assisting proteins and shared components with photosystem II biogenesis. *Biochim. Biophys. Acta Bioenerg.* **2015**, *1847*, 900–909. [CrossRef] [PubMed]

106. Vass, I.; Cser, K. Janus-faced charge recombinations in photosystem II photoinhibition. *Trends Plant Sci.* **2009**, *14*, 200–205. [CrossRef] [PubMed]

107. Telfer, A.; Dhami, S.; Bishop, S.M.; Phillips, D.; Barber, J. Beta-Carotene Quenches Singlet Oxygen Formed by Isolated Photosystem-II Reaction Centers. *Biochemistry* **1994**, *33*, 14469–14474. [CrossRef] [PubMed]

108. Kruk, J.; Trebst, A. Plastoquinol as a singlet oxygen scavenger in photosystem II. *BBA-Bioenergetics* **2008**, *1777*, 154–162.

109. Weisz, D.A.; Gross, M.L.; Pakrasi, H.B. Reactive oxygen species leave a damage trail that reveals water channels in Photosystem II. *Sci. Adv.* **2017**, *3*. [CrossRef] [PubMed]

110. Pospisil, P.; Yamamoto, Y. Damage to photosystem II by lipid peroxidation products. *Biochim. Biophys. Acta Gen. Subj.* **2017**, *1861*, 457–466. [CrossRef] [PubMed]

111. Park, Y.I.; Chow, W.S.; Anderson, J.M. Light Inactivation of Functional Photosystem-Ii in Leaves of Peas Grown in Moderate Light Depends on Photon Exposure. *Planta* **1995**, *196*, 401–411. [CrossRef]

112. Tyystjarvi, E.; Aro, E.M. The rate constant of photoinhibition, measured in lincomycin-treated leaves, is directly proportional to light intensity. *Proc. Natl. Acad. Sci. USA* **1996**, *93*, 2213–2218. [CrossRef] [PubMed]

113. Nelson, C.J.; Alexova, R.; Jacoby, R.P.; Millar, A.H. Proteins with High Turnover Rate in Barley Leaves Estimated by Proteome Analysis Combined with in Planta Isotope Labeling. *Plant Physiol.* **2014**, *166*, 91–108. [CrossRef] [PubMed]

114. Bergantino, E.; Brunetta, A.; Touloupakis, E.; Segalla, A.; Szabo, I.; Giacometti, G.M. Role of the PSII-H subunit in photoprotection—Novel aspects of D1 turnover in Synechocystis 6803. *J. Biol. Chem.* **2003**, *278*, 41820–41829. [CrossRef] [PubMed]

115. Rokka, A.; Suorsa, M.; Saleem, A.; Battchikova, N.; Aro, E.M. Synthesis and assembly of thylakoid protein complexes: Multiple assembly steps of photosystem II. *Biochem. J.* **2005**, *388*, 159–168. [CrossRef] [PubMed]

116. Kato, Y.; Sakamoto, W. Protein Quality Control in Chloroplasts: A Current Model of D1 Protein Degradation in the Photosystem II Repair Cycle. *J. Biochem.* **2009**, *146*, 463–469. [CrossRef] [PubMed]

117. Barber, J.; Andersson, B. Too Much of a Good Thing—Light Can Be Bad for Photosynthesis. *Trends Biochem. Sci.* **1992**, *17*, 61–66. [CrossRef]

118. Aro, E.M.; Virgin, I.; Andersson, B. Photoinhibition of Photosystem-2—Inactivation, Protein Damage and Turnover. *Biochim. Biophys. Acta* **1993**, *1143*, 113–134. [CrossRef]

119. Adir, N.; Zer, H.; Shochat, S.; Ohad, I. Photoinhibition—A historical perspective. *Photosynth. Res.* **2003**, *76*, 343–370. [CrossRef] [PubMed]

120. Aro, E.M.; Suorsa, M.; Rokka, A.; Allahverdiyeva, Y.; Paakkarinen, V.; Saleem, A.; Battchikova, N.; Rintamaki, E. Dynamics of photosystem II: A proteomic approach to thylakoid protein complexes. *J. Exp. Bot.* **2005**, *56*, 347–356. [CrossRef] [PubMed]

121. Baena-Gonzalez, E.; Aro, E.M. Biogenesis, assembly and turnover of photosystem II units. *Philos. Trans. R. Soc. Lond. Ser. B Biol. Sci.* **2002**, *357*, 1451–1459. [CrossRef] [PubMed]

122. Cheregi, O.; Wagner, R.; Funk, C. Insights into the Cyanobacterial Deg/HtrA Proteases. *Front. Plant Sci.* **2016**, *7*. [CrossRef] [PubMed]

123. Huesgen, P.F.; Schuhmann, H.; Adamska, I. Deg/HtrA proteases as components of a network for photosystem II quality control in chloroplasts and cyanobacteria. *Res. Microbiol.* **2009**, *160*, 726–732. [CrossRef] [PubMed]

124. Schuhmann, H.; Huesgen, P.F.; Adamska, I. The family of Deg/HtrA proteases in plants. *BMC Plant Biol.* **2012**, *12*. [CrossRef] [PubMed]

125. Schuhmann, H.; Adamska, I. Deg proteases and their role in protein quality control and processing in different subcellular compartments of the plant cell. *Physiol. Plant.* **2012**, *145*, 224–234. [CrossRef] [PubMed]

126. Sun, X.; Wang, L.; Zhang, L. Involvement of DEG5 and DEG8 proteases in the turnover of the photosystem II reaction center D1 protein under heat stress in *Arabidopsis thaliana*. *Chin. Sci. Bull.* **2007**, *52*, 1742–1745. [CrossRef]

127. Kato, Y.; Sun, X.; Zhang, L.; Sakamoto, W. Cooperative D1 Degradation in the Photosystem II Repair Mediated by Chloroplastic Proteases in *Arabidopsis*. *Plant Physiol.* **2012**, *159*, 1428–1439. [CrossRef] [PubMed]

128. Chen, J.P.; Burke, J.J.; Velten, J.; Xin, Z.U. FtsH11 protease plays a critical role in *Arabidopsis thermotolerance*. *Plant J.* **2006**, *48*, 73–84. [CrossRef] [PubMed]

129. Yoshioka, M.; Yamamoto, Y. Quality control of Photosystem II: Where and how does the degradation of the D1 protein by FtsH proteases start under light stress?—Facts and hypotheses. *J. Photochem. Photobiol. B Biol.* **2011**, *104*, 229–235. [CrossRef] [PubMed]

130. Peltier, J.B.; Emanuelsson, O.; Kalume, D.E.; Ytterberg, J.; Friso, G.; Rudella, A.; Liberles, D.A.; Soderberg, L.; Roepstorff, P.; von Heijne, G.; et al. Central functions of the lumenal and peripheral thylakoid proteome of *Arabidopsis* determined by experimentation and genome-wide prediction. *Plant Cell* **2002**, *14*, 211–236. [CrossRef] [PubMed]

131. Schubert, M.; Petersson, U.A.; Haas, B.J.; Funk, C.; Schroder, W.P.; Kieselbach, T. Proteome map of the chloroplast lumen of *Arabidopsis thaliana*. *J. Biol. Chem.* **2002**, *277*, 8354–8365. [CrossRef] [PubMed]

132. Kapri-Pardes, E.; Naveh, L.; Adam, Z. The thylakoid lumen protease Deg1 is involved in the repair of photosystem II from photoinhibition in *Arabidopsis*. *Plant Cell* **2007**, *19*, 1039–1047. [CrossRef] [PubMed]

133. Sun, X.; Peng, L.; Guo, J.; Chi, W.; Ma, J.; Lu, C.; Zhang, L. Formation of DEG5 and DEG8 complexes and their involvement in the degradation of photodamaged photosystem II reaction center D1 protein in *Arabidopsis*. *Plant Cell* **2007**, *19*, 1347–1361. [CrossRef] [PubMed]

134. Kley, J.; Schmidt, B.; Boyanov, B.; Stolt-Bergner, P.C.; Kirk, R.; Ehrmann, M.; Knopf, R.R.; Naveh, L.; Adam, Z.; Clausen, T. Structural adaptation of the plant protease Deg1 to repair photosystem II during light exposure. *Nat. Struct. Mol. Biol.* **2011**, *18*, 728–731. [CrossRef] [PubMed]

135. Sun, X.; Ouyang, M.; Guo, J.; Ma, J.; Lu, C.; Adam, Z.; Zhang, L. The thylakoid protease Deg1 is involved in photosystem-II assembly in *Arabidopsis thaliana*. *Plant J.* **2010**, *62*, 240–249. [CrossRef] [PubMed]

136. Chassin, Y.; Kapri-Pardes, E.; Sinvany, G.; Arad, T.; Adam, Z. Expression and characterization of the thylakoid lumen protease DegP1 from *Arabidopsis*. *Plant Physiol.* **2002**, *130*, 857–864. [CrossRef] [PubMed]

137. Aro, E.M.; McCaffery, S.; Anderson, J.M. Photoinhibition and D1 Protein-Degradation in Peas Acclimated to Different Growth Irradiances. *Plant Physiol.* **1993**, *103*, 835–843. [CrossRef] [PubMed]

138. Tyedmers, J.; Mogk, A.; Bukau, B. Cellular strategies for controlling protein aggregation. *Nat. Rev. Mol. Cell Biol.* **2010**, *11*, 777–788. [CrossRef] [PubMed]

139. Vierling, E. The Roles of Heat-Shock Proteins in Plants. *Annu. Rev. Plant Physiol. Plant Mol. Biol.* **1991**, *42*, 579–620. [CrossRef]

140. Chen, Q.; Vierling, E. Analysis of Conserved Domains Identifies a Unique Structural Feature of a Chloroplast Heat-Shock Protein. *Mol. Gen. Genet.* **1991**, *226*, 425–431. [CrossRef] [PubMed]

141. Chen, Q.; Lauzon, L.M.; Derocher, A.E.; Vierling, E. Accumulation, Stability, and Localization of a Major Chloroplast Heat-Shock Protein. *J. Cell Biol.* **1990**, *110*, 1873–1883. [CrossRef] [PubMed]

142. Vierling, E.; Harris, L.M.; Chen, Q. The Major Low-Molecular-Weight Heat-Shock Protein in Chloroplasts Shows Antigenic Conservation among Diverse Higher-Plant Species. *Mol. Cell. Biol.* **1989**, *9*, 461–468. [CrossRef] [PubMed]

143. Heckathorn, S.A.; Downs, C.A.; Sharkey, T.D.; Coleman, J.S. The small, methionine-rich chloroplast heat-shock protein protects photosystem II electron transport during heat stress. *Plant Physiol.* **1998**, *116*, 439–444. [CrossRef] [PubMed]

144. Heckathorn, S.A.; Ryan, S.L.; Baylis, J.A.; Wang, D.F.; Hamilton, E.W.; Cundiff, L.; Luthe, D.S. In vivo evidence from an *Agrostis stolonifera* selection genotype that chloroplast small heat-shock proteins can protect photosystem II during heat stress. *Funct. Plant Biol.* **2002**, *29*, 933–944. [CrossRef]

145. Kim, K.-H.; Alam, I.; Kim, Y.-G.; Sharmin, S.A.; Lee, K.-W.; Lee, S.-H.; Lee, B.-H. Overexpression of a chloroplast-localized small heat shock protein OsHSP26 confers enhanced tolerance against oxidative and heat stresses in tall fescue. *Biotechnol. Lett.* **2012**, *34*, 371–377. [CrossRef] [PubMed]

146. Shakeel, S.; Ul Haq, N.; Heckathorn, S.A.; Hamilton, E.W.; Luthe, D.S. Ecotypic variation in chloroplast small heat-shock proteins and related thermotolerance in *Chenopodium album*. *Plant Physiol. Biochem.* **2011**, *49*, 898–908. [CrossRef] [PubMed]

147. Wang, D.F.; Luthe, D.S. Heat sensitivity in a bentgrass variant. Failure to accumulate a chloroplast heat shock protein isoform implicated in heat tolerance. *Plant Physiol.* **2003**, *133*, 319–327. [CrossRef] [PubMed]

148. Harndahl, U.; Hall, R.B.; Osteryoung, K.W.; Vierling, E.; Bornman, J.F.; Sundby, C. The chloroplast small heat shock protein undergoes oxidation-dependent conformational changes and may protect plants from oxidative stress. *Cell Stress Chaperones* **1999**, *4*, 129–138. [CrossRef]

149. Neta-Sharir, I.; Isaacson, T.; Lurie, S.; Weiss, D. Dual role for tomato heat shock protein 21: Protecting photosystem II from oxidative stress and promoting color changes during fruit maturation. *Plant Cell* **2005**, *17*, 1829–1838. [CrossRef] [PubMed]

150. Downs, C.A.; Coleman, J.S.; Heckathorn, S.A. The chloroplast 22-Ku heat-shock protein: A lumenal protein that associates with the oxygen evolving complex and protects photosystem II during heat stress. *J. Plant Physiol.* **1999**, *155*, 477–487. [CrossRef]

151. Bernfur, K.; Rutsdottir, G.; Emanuelsson, C. The chloroplast-localized small heat shock protein Hsp21 associates with the thylakoid membranes in heat-stressed plants. *Protein Sci. Publ. Protein Soc.* **2017**, *26*, 1773–1784. [CrossRef] [PubMed]

152. Spreitzer, R.J.; Salvucci, M.E. Rubisco: Structure, regulatory interactions, and possibilities for a better enzyme. *Annu. Rev. Plant Biol.* **2002**, *53*, 449–475. [CrossRef] [PubMed]

153. Perdomo, J.A.; Capo-Bauca, S.; Carmo-Silva, E.; Galmes, J. Rubisco and Rubisco Activase Play an Important Role in the Biochemical Limitations of Photosynthesis in Rice, Wheat, and Maize under High Temperature and Water Deficit. *Front. Plant Sci.* **2017**, *8*. [CrossRef] [PubMed]

154. Sage, R.F.; Way, D.A.; Kubien, D.S. Rubisco, Rubisco activase, and global climate change. *J. Exp. Bot.* **2008**, *59*, 1581–1595. [CrossRef] [PubMed]

155. Portis, A.R. Rubisco activase—Rubisco's catalytic chaperone. *Photosynth. Res.* **2003**, *75*, 11–27. [CrossRef] [PubMed]

156. Schrader, S.M.; Wise, R.R.; Wacholtz, W.F.; Ort, D.R.; Sharkey, T.D. Thylakoid membrane responses to moderately high leaf temperature in *Pima cotton*. *Plant Cell Environ.* **2004**, *27*, 725–735. [CrossRef]

157. Yamori, W.; Noguchi, K.; Kashino, Y.; Terashima, I. The role of electron transport in determining the temperature dependence of the photosynthetic rate in spinach leaves grown at contrasting temperatures. *Plant Cell Physiol.* **2008**, *49*, 583–591. [CrossRef] [PubMed]

158. Carmo-Silva, A.E.; Salvucci, M.E. The activity of Rubisco's molecular chaperone, Rubisco activase, in leaf extracts. *Photosynth. Res.* **2011**, *108*, 143–155. [CrossRef] [PubMed]

159. Salvucci, M.E.; Crafts-Brandner, S.J. Mechanism for deactivation of Rubisco under moderate heat stress. *Physiol. Plant.* **2004**, *122*, 513–519. [CrossRef]

160. Robinson, S.P.; Streusand, V.J.; Chatfield, J.M.; Portis, A.R. Purification and Assay of Rubisco activase from Leaves. *Plant Physiol.* **1988**, *88*, 1008–1014. [CrossRef] [PubMed]

161. Holbrook, G.P.; Galasinski, S.C.; Salvucci, M.E. Regulation of 2-Carboxyarabinitol 1-Phosphatase. *Plant Physiol.* **1991**, *97*, 894–899. [CrossRef] [PubMed]

162. Feller, U.; Crafts-Brandner, S.J.; Salvucci, M.E. Moderately high temperatures inhibit ribulose-1,5-bisphosphate carboxylase/oxygenase (Rubisco) activase-mediated activation of Rubisco. *Plant Physiol.* **1998**, *116*, 539–546. [CrossRef] [PubMed]

163. Salvucci, M.E.; Osteryoung, K.W.; Crafts-Brandner, S.J.; Vierling, E. Exceptional sensitivity of Rubisco activase to thermal denaturation in vitro and in vivo. *Plant Physiol.* **2001**, *127*, 1053–1064. [CrossRef] [PubMed]

164. Schrader, S.M.; Kleinbeck, K.R.; Sharkey, T.D. Rapid heating of intact leaves reveals initial effects of stromal oxidation on photosynthesis. *Plant Cell Environ.* **2007**, *30*, 671–678. [CrossRef] [PubMed]

165. Zhang, R.; Sharkey, T.D. Photosynthetic electron transport and proton flux under moderate heat stress. *Photosynth. Res.* **2009**, *100*, 29–43. [CrossRef] [PubMed]

166. Way, D.A.; Yamori, W. Thermal acclimation of photosynthesis: On the importance of adjusting our definitions and accounting for thermal acclimation of respiration. *Photosynth. Res.* **2014**, *119*, 89–100. [CrossRef] [PubMed]

167. Yamori, W.; Hikosaka, K.; Way, D.A. Temperature response of photosynthesis in C-3, C-4, and CAM plants: Temperature acclimation and temperature adaptation. *Photosynth. Res.* **2014**, *119*, 101–117. [CrossRef] [PubMed]

168. Salvucci, M.E.; Crafts-Brandner, S.J. Inhibition of photosynthesis by heat stress: The activation state of Rubisco as a limiting factor in photosynthesis. *Physiol. Plant.* **2004**, *120*, 179–186. [CrossRef] [PubMed]

169. DeRidder, B.P.; Salvucci, M.E. Modulation of Rubisco activase gene expression during heat stress in cotton (*Gossypium hirsutum* L.) involves post-transcriptional mechanisms. *Plant Sci.* **2007**, *172*, 246–254. [CrossRef]

170. Kumar, R.R.; Goswami, S.; Singh, K.; Dubey, K.; Singh, S.; Sharma, R.; Verma, N.; Kala, Y.K.; Rai, G.K.; Grover, M.; et al. Identification of Putative RuBisCo Activase (TaRca1)-The Catalytic Chaperone Regulating Carbon Assimilatory Pathway in Wheat (*Triticum aestivum*) under the Heat Stress. *Front. Plant Sci.* **2016**, *7*, 986. [CrossRef] [PubMed]

171. Kurek, I.; Chang, T.K.; Bertain, S.M.; Madrigal, A.; Liu, L.; Lassner, M.W.; Zhu, G. Enhanced thermostability of *Arabidopsis* Rubisco activase improves photosynthesis and growth rates under moderate heat stress. *Plant Cell* **2007**, *19*, 3230–3241. [CrossRef] [PubMed]

172. Kumar, A.; Li, C.; Portis, A.R., Jr. *Arabidopsis thaliana* expressing a thermostable chimeric Rubisco activase exhibits enhanced growth and higher rates of photosynthesis at moderately high temperatures. *Photosynth. Res.* **2009**, *100*, 143–153. [CrossRef] [PubMed]

173. Ristic, Z.; Momcilovic, I.; Bukovnik, U.; Prasad, P.V.V.; Fu, J.; DeRidder, B.P.; Elthon, T.E.; Mladenov, N. Rubisco activase and wheat productivity under heat-stress conditions. *J. Exp. Bot.* **2009**, *60*, 4003–4014. [CrossRef] [PubMed]

174. Shivhare, D.; Mueller-Cajar, O. In Vitro Characterization of Thermostable CAM Rubisco Activase Reveals a Rubisco Interacting Surface Loop. *Plant Physiol.* **2017**, *174*, 1505–1516. [CrossRef] [PubMed]

175. Salvucci, M.E. Association of Rubisco activase with chaperonin-60 beta: A possible mechanism for protecting photosynthesis during heat stress. *J. Exp. Bot.* **2008**, *59*, 1923–1933. [CrossRef] [PubMed]

176. Scales, J.C.; Parry, M.A.J.; Salvucci, M.E. A non-radioactive method for measuring Rubisco activase activity in the presence of variable ATP: ADP ratios, including modifications for measuring the activity and activation state of Rubisco. *Photosynth. Res.* **2014**, *119*, 355–365. [CrossRef] [PubMed]

177. Carmo-Silva, A.E.; Salvucci, M.E. The Regulatory Properties of Rubisco activase Differ among Species and Affect Photosynthetic Induction during Light Transitions. *Plant Physiol.* **2013**, *161*, 1645–1655. [CrossRef] [PubMed]

178. Wang, D.; Li, X.-F.; Zhou, Z.-J.; Feng, X.-P.; Yang, W.-J.; Jiang, D.-A. Two Rubisco activase isoforms may play different roles in photosynthetic heat acclimation in the rice plant. *Physiol. Plant.* **2010**, *139*, 55–67. [CrossRef] [PubMed]

179. Hozain, M.D.I.; Salvucci, M.E.; Fokar, M.; Holaday, A.S. The differential response of photosynthesis to high temperature for a boreal and temperate *Populus* species relates to differences in Rubisco activation and Rubisco activase properties. *Tree Physiol.* **2010**, *30*, 32–44. [CrossRef] [PubMed]

180. Feller, U. Drought stress and carbon assimilation in a warming climate: Reversible and irreversible impacts. *J. Plant Physiol.* **2016**, *203*, 69–79. [CrossRef] [PubMed]

International Journal of
Molecular Sciences

MDPI

Review

Chloroplast Protein Turnover: The Influence of Extraplastidic Processes, Including Autophagy

Masanori Izumi [1,2,3,*] and Sakuya Nakamura [2]

1   Frontier Research Institute for Interdisciplinary Sciences, Tohoku University, Sendai 980-8578, Japan
2   Department of Environmental Life Sciences, Graduate School of Life Sciences, Tohoku University, Sendai 980-8577, Japan
3   Precursory Research for Embryonic Science and Technology (PRESTO), Japan Science and Technology Agency, Kawaguchi 332-0012, Japan
*   Correspondence: m-izumi@ige.tohoku.ac.jp

Received: 2 February 2018; Accepted: 6 March 2018; Published: 12 March 2018

Abstract: Most assimilated nutrients in the leaves of land plants are stored in chloroplasts as photosynthetic proteins, where they mediate CO_2 assimilation during growth. During senescence or under suboptimal conditions, chloroplast proteins are degraded, and the amino acids released during this process are used to produce young tissues, seeds, or respiratory energy. Protein degradation machineries contribute to the quality control of chloroplasts by removing damaged proteins caused by excess energy from sunlight. Whereas previous studies revealed that chloroplasts contain several types of intraplastidic proteases that likely derived from an endosymbiosed prokaryotic ancestor of chloroplasts, recent reports have demonstrated that multiple extraplastidic pathways also contribute to chloroplast protein turnover in response to specific cues. One such pathway is autophagy, an evolutionarily conserved process that leads to the vacuolar or lysosomal degradation of cytoplasmic components in eukaryotic cells. Here, we describe and contrast the extraplastidic pathways that degrade chloroplasts. This review shows that diverse pathways participate in chloroplast turnover during sugar starvation, senescence, and oxidative stress. Elucidating the mechanisms that regulate these pathways will help decipher the relationship among the diverse pathways mediating chloroplast protein turnover.

Keywords: autophagy; chlorophagy; chloroplasts; Rubisco-containing bodies; photooxidative damage; plants; senescence; sugar starvation; ubiquitin proteasome system; vacuole

1. Introduction

Chloroplasts are a type of plastid in plants and algae. In land plants, chloroplasts are present in green tissues, such as leaves, that are required for photosynthetic energy production. Within chloroplasts, thylakoid membranes contain pigments and proteins that form the light harvesting complex and electron transport chain, and the stroma contains soluble proteins that mediate the assimilation of carbon dioxide (CO_2) via the Calvin cycle. In mature plant leaves, assimilated nutrients are largely stored in chloroplasts as photosynthetic proteins. For instance, the nitrogen in chloroplasts accounts for around 75% of the total leaf nitrogen in C3 species [1]. The CO_2-fixing enzyme ribulose-1,5-bisphosphate carboxylase/oxygenase (Rubisco) in the stroma is especially abundant, accounting for 10–30% of the total leaf nitrogen, and constituting around half of the total soluble proteins in leaves [2,3]. Chloroplast proteins are degraded, and the amino acids and other molecules released during this process are reutilized in growth. Leaf senescence is a well-established developmental process during which chloroplast proteins are degraded en masse; the released amino acids are remobilized to generate juvenile tissues and produce seeds [4,5].

A portion of the photoassimilate is accumulated in chloroplasts as starch during the day, and is degraded at night to produce sucrose as the major source for respiratory energy production within mitochondria [6]. Since the availability of solar energy fluctuates under the ever-changing environment, plants occasionally need alternatives to sugars for producing the energy required for continuous growth. Stress conditions can also interfere with photosynthetic energy production; for instance, stomatal closure due to drought stress inhibits CO_2 intake and thereby reduces photosynthetic activity in leaves [7,8]. Plants must metabolically produce alternative energy sources to survive under photosynthesis-limited conditions. Amino acids derived from chloroplast protein degradation via catabolic pathways can serve as alternative respiratory substrates [9].

Chloroplast protein degradation is also vital for maintaining chloroplast function, as chloroplast proteins constantly accumulate damage caused by sunlight during photosynthesis. Photoinhibition occurs when the photosynthetic apparatus is damaged by excess energy from strong visible light (with wavelengths of between 400 and 700 nm) [10–13]. Although chloroplasts cannot use ultraviolet-B (UVB; with wavelengths of between 280 and 315 nm) for photosynthesis, various macromolecules, such as proteins, lipids, and nucleotides, directly absorb UVB, which may result in cumulative damage [14]. To maintain photosynthetic activity and avoid the overproduction of reactive oxygen species (ROS) in response to sunlight irradiation, damaged components within chloroplasts must be removed.

Turnover of chloroplastic components is required for efficient nutrient recycling during plant senescence, respiratory energy production under photoassimilate-starved conditions, and quality control in individual chloroplasts under photooxidative damage. Chloroplasts contain various types of intraplastidic proteases, which are thought to have been derived from an endosymbiosed prokaryotic ancestor of chloroplasts [15,16]. Recent studies of chloroplast protein turnover have further demonstrated the contribution of extraplastidic protein degradation systems to nutrient or energy recycling and the removal of damaged proteins. In this review, we describe the extraplastidic pathways that facilitate the degradation of chloroplastic components, and compare the physiological roles of these pathways and the environmental and developmental stimuli that activate them.

2. Autophagic Degradation of Rubisco-Containing Bodies

Autophagy is an evolutionarily conserved process in eukaryotes whereby the cell sequesters a portion of cytoplasm, including organelles, for subsequent transport into lytic organelles [17–19]. During autophagy, a nascent double membrane-bound vesicle called an autophagosome encloses a portion of the cytoplasm. The outer membrane of autophagosomes then fuses with the vacuolar or lysosomal membrane to release the inner-membrane structures, referred to as autophagic bodies, into the vacuolar or lysosomal lumen for digestion. The basic mechanism of autophagosome formation was described in the budding yeast *Saccharomyces cerevisiae* through the identification of autophagy (*ATG*) genes [20].

The *ATG* genes required for the initiation or elongation of autophagosomal membranes are referred as core *ATGs* (*ATG1–10, 12–14, 16, 18*), and these genes are also required for all types of autophagy [17]. Many core ATGs function in two conjugation cascades that are required for ATG8 lipidation and autophagosomal membrane elongation. ATG7 and ATG10 conjugate ATG12 to ATG5, and the resulting ATG12-ATG5 conjugate then interacts with ATG16 to form the ATG12-ATG5-ATG16 complex. ATG8 is processed by the protease ATG4. The resulting mature ATG8 is activated by ATG7, transferred to ATG3, and is eventually conjugated with phosphatidylethanolamine with the aid of the ATG12-ATG5-ATG16 complex. Orthologues of the yeast core *ATGs* are conserved in plant species [21–23], and studies of autophagy-deficient *atg* mutants of *Arabidopsis thaliana* show that they have similar functions [24–32].

During leaf senescence, the amount of chloroplast stromal proteins, including Rubisco, decreases prior to the reduction in the number of chloroplasts [33–35]. Therefore, stromal proteins appear to be degraded either inside or outside the chloroplast without the breakdown of the entire chloroplast. An immuno-electron microscopy (EM) analysis of Rubisco degradation in senescing wheat (*Triticum aestivum*) leaves revealed the presence of cytosol-localized small vesicles that contained Rubisco, but not thylakoid proteins such as light-harvesting chlorophyll a/b protein of Photosystem II (LHC II), α, β-subunits of coupling factor 1 in ATPase, or cytochrome *f* [36]. These vesicles, which are around 1 μm in diameter and are frequently surrounded by autophagosome-like double membranes, were originally referred to as Rubisco-containing bodies (RCBs). The development of live-cell imaging techniques using fluorescent protein markers allowed for the visualization of RCBs in vivo in Arabidopsis and rice (*Oryza sativa*) leaves expressing stroma-targeted green fluorescent protein (GFP) or GFP-labeled Rubisco [37,38]. This technique further demonstrated that RCBs are not produced in the mutant *atg5* or *atg7* lines, and that RCBs labeled with stroma-targeted red fluorescent proteins (RFPs) are co-localized with an autophagosomal marker, GFP-ATG8. These observations revealed that RCBs are a type of autophagic body that delivers a portion of the stromal proteins into the vacuole. Thus, the RCB pathway was established as an autophagic process that mobilizes stromal proteins to the vacuole (Figure 1a).

Endosomal sorting complex required for transport (ESCRT) proteins are part of an evolutionarily conserved system that is responsible for the remodeling of endosomal membranes in eukaryotes [39]. A recent study in Arabidopsis indicated that the ESCRT-III paralogs charged multivesicular body protein 1A (CHMP1A) and CHMP1B are required for the delivery of RCBs to the vacuole [40]. In *chmp1a chmp1b* double mutant plants, RCBs were produced but accumulated in the cytoplasm; therefore, CHMP1 proteins are required for the vacuolar sorting of chloroplast-derived RCBs or the fusion of autophagosomes enclosing RCBs. How a portion of stroma is separated as RCBs, and how RCBs are then recruited for autophagic transport remain unclear.

The RCB pathway is particularly active in sugar-starved, excised Arabidopsis leaves in darkness or the presence of photosynthesis inhibitors [41]. Starch is the major carbohydrate form for energy storage. The starchless mutants, *phosphoglucomutase* (*pgm*) and *ADP-glucose pyrophosphorylase1* (*adg1*), which lack starch, exhibited enhanced production of RCBs [41,42]. Moreover, starchless and *atg* double mutants exhibited reduced growth and enhanced cell death during developmental senescence compared to the respective single mutants [42]. These results indicate that the RCB pathway plays a role in the response to sugar starvation. Recent studies found that in the sugar-starved leaves of Arabidopsis plants maintained in complete darkness for several days, autophagy deficiency compromises the release of free amino acids, especially free branched chain amino acids (BCAAs) like isoleucine, leucine, and valine [43,44]. Arabidopsis mutants with defects in the enzymes involved in BCAA catabolism have reduced tolerance to sugar starvation due to prolonged complete darkness [9,45–49]; thus, BCAAs are a particularly important energy source for mitochondrial respiration as alternatives to sugars. The RCB pathway might supply free amino acids, especially BCAAs, derived from vacuolar degradation of stromal proteins as an alternative energy source during periods of impaired photosynthesis (Figure 1a).

Photosynthetic energy production can be perturbed by various types of suboptimal conditions, including shading, flooding, or drought. The importance of core autophagy machinery during submergence-induced hypoxia or draught stress was reported in Arabidopsis plants [50,51]. The RCB pathway might alleviate the energy limitation that is caused by some types of abiotic stresses.

RCB production is also activated during accelerated leaf senescence induced in leaves that were individually covered to impair photosynthesis [52]. This activation of senescence corresponds to chloroplast shrinkage. In addition to direct observations of RCBs labeled with stroma-localized fluorescent proteins, the activity of the RCB pathway can be monitored by biochemical detection of free GFP or RFP derived from vacuolar degradation of Rubisco-GFP or -RFP fusion proteins, which are mobilized to the vacuole via RCBs [53]. This technique indicated that autophagy contributes substantially to the degradation of Rubisco in individually darkened leaves and in those shaded by the leaves of neighboring Arabidopsis plants [53]. Such biochemical methods of monitoring the RCB pathway have also been established in rice plants, and have shown that Rubisco is degraded via RCBs in individually darkened rice leaves [38]. In autophagy-deficient *atg* mutant rice plants, *osatg7*, Rubisco degradation was attenuated in senescing leaves, which is consistent with the partly compromised nitrogen remobilization from lower leaves to newly developing upper leaves [54]. These findings further indicate that the RCB pathway mediates nitrogen remobilization from older leaves that cannot acquire sufficient light due to shading of developing leaves by upper tissues.

Analyses of Arabidopsis and maize (*Zea mays*) plants harboring the *atg* mutation indicated that autophagy contributes to nitrogen remobilization from vegetative tissues to reproductive tissues, including seeds [55–57]. However, such a role for autophagy in rice plants was not evaluated, because autophagy-deficient rice plants exhibit male sterility due to impaired pollen maturation [58].

3. Chlorophagy: Degradation of Entire Chloroplasts

Whereas the amount of stromal proteins decreases during the earlier stages of leaf senescence in wheat or barley (*Hordeum vulgare*) plants, the number of chloroplasts per cell decreases during the later stages [33–35]. In individually darkened leaves of wild-type Arabidopsis plants, RCB production and subsequent shrinkage of chloroplasts occur during the earlier stages of senescence, and the chloroplast population decreases during the later stages of senescence [52]. This decrease in chloroplast number is suppressed in *atg4* mutants. Some isolated vacuoles from the darkened leaves of wild-type plants contained chloroplasts that exhibited chlorophyll autofluorescence signals. These findings suggest that shrunken chloroplasts, which are produced through the active separation of their components in the RCB pathway, become the targets of autophagic transport as entire organelles, a process known as chlorophagy [59] (Figure 1b).

In yeast and mammals, autophagy is also recognized as a major quality control system for organelles through the selective removal of dysfunctional organelles [19]. In Arabidopsis *atg* plants, oxidized peroxisomes containing aggregated catalase accumulate in the cytoplasm of senescing leaves [60–62]. During germination, enzymes in peroxisomes catalyze β-oxidation and the glyoxylate cycle, thereby allowing lipids stored in seeds to be used as energy before photosynthetic machinery within chloroplasts are developed. As photosynthetic growth is established several days after germination, peroxisomes are remodeled to carry out the glycolate pathway, which is required for photorespiration. This functional conversion of peroxisomes was partly compromised in Arabidopsis *atg* plants in which peroxisome aggregates accumulate in mesophyll cells containing mature chloroplasts [63,64]. Thus, plant peroxisomes are likely targets of a process of selective autophagy known as pexophagy during senescence or seedling development. Autophagic degradation of the endoplasmic reticulum (ER) during ER stress due to tunicamycin treatment was also observed in Arabidopsis roots [65,66]. Selective degradation of ER by autophagy termed ER-phagy may function in plants.

A recent study investigated the involvement of autophagy in the turnover of chloroplasts under photooxidative stress conditions and demonstrated that chlorophagy is induced in Arabidopsis leaves damaged by UVB exposure [67]. A subset of the chloroplasts in the cytoplasm of UVB-damaged *atg5* and *atg7* plants exhibited irregular shapes and disorganized thylakoid structures. Chlorophagy was also induced by chloroplast damage caused by exposure to strong visible light or natural sunlight. Therefore, chlorophagy may remove entire photo-damaged chloroplasts by transporting them into the vacuole [67,68] (Figure 1c).

The chloroplast-targeted RCB pathway and chlorophagy differ in individually darkened leaves and in leaves subjected to UVB damage [67,69]. During sugar starvation in individually darkened leaves, RCBs were observed after 1 d of treatment, whereas chlorophagy was rarely observed during 3 days of dark treatment. By contrast, in leaves subjected to UVB-mediated oxidative stress, chlorophagy was actively induced 2 days after treatment without prior RCB production. These observations suggest that the induction of these two types of autophagy is individually controlled by distinct upstream mechanisms in response to environmental or developmental conditions (Figure 1).

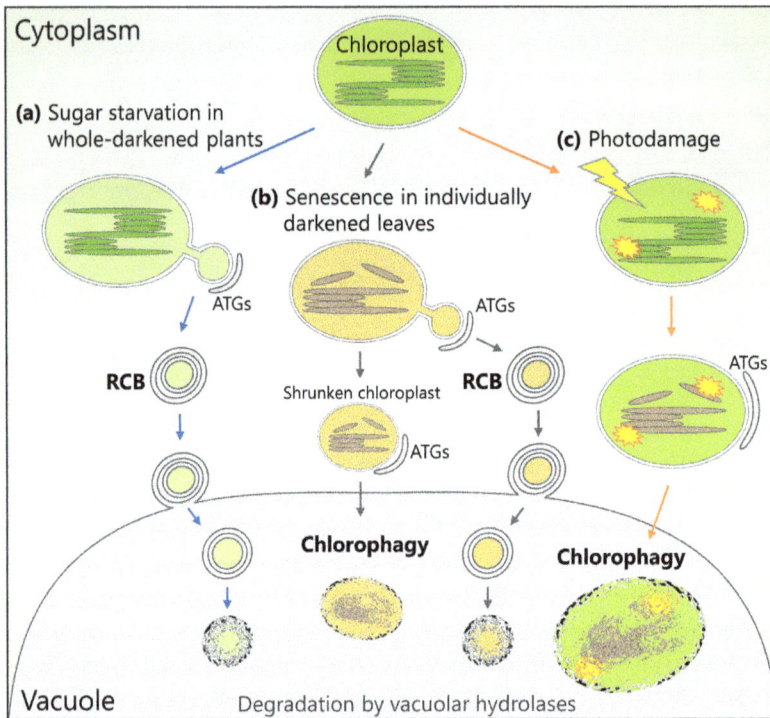

Figure 1. Schematic model for the Rubisco-containing body (RCB) pathway and chlorophagy forms of chloroplast-related autophagy. (**a**) When photosynthetic energy production of whole plants is impaired due to complete darkness, a portion of the chloroplast stroma is transported to the central vacuole via RCBs, which are a type of autophagic compartment that specifically contains stromal proteins. The RCB pathway can facilitate the recycling of amino acids as an energy source. (**b**) When senescence is accelerated in individually darkened leaves, the active production of RCBs leads to chloroplast shrinkage, thereby allowing the transport of entire chloroplasts to the vacuole via chlorophagy. (**c**) Photodamage from exposure to ultraviolet-B (UV-B), strong visible light, or natural sunlight causes chloroplasts to collapse. The collapsed chloroplasts are then transported to the vacuole without prior activation of RCBs. This process is suggested to serve as a quality control mechanism that removes damaged chloroplasts.

4. ATI Body-Mediated Chloroplast Degradation

ATG8 is a core ATG protein that builds up the autophagosomal membrane by conjugating with phosphatidylethanolamine [70]. In yeast, several types of organelle-targeted autophagy are controlled by ATG proteins containing an ATG8-interacting motif (AIM) [71]. ATG32 triggers the removal of dysfunctional or excess mitochondria by interacting with autophagosomal membrane-anchored ATG8 on the mitochondrial outer envelope [72,73]. ATG39 and ATG40 were also identified as ATG8-interacting proteins that control nucleus- or ER-targeted selective autophagy, respectively [74].

ATG8-interacting protein 1 (ATI1) and ATI2 were identified in a yeast two-hybrid screen for candidates that interact with the Arabidopsis ATG8 isoform, ATG8f [75]. These proteins were found to associate with plastids in addition to the ER as small vesicles of approximately 1 μm in diameter, which are referred to as ATI bodies [76]. A screen of potential ATI1-interacting proteins and microscopy observations of fluorescent marker proteins indicated that plastid-associated ATI bodies transport some thylakoid, stroma, and envelope proteins into the vacuole, especially under dark-induced energy limitation [76]. These delivery cargos differ from those of the RCBs that specifically contain a portion of stroma [36,37]; however, the vacuolar transport of plastid-associated ATI bodies is an autophagy-dependent process, as this body was not produced in the *atg5* mutants [76]. Therefore, ATI bodies represent a distinct form of autophagy vesicles that transport some stroma, thylakoid, and envelope components into the vacuole (Figure 2a). Plastid-associated ATI bodies are also observed inside the chloroplast, and ATI1 interacts with some thylakoid proteins in vivo [76]. It is thus proposed that plastid-associated ATI bodies form in chloroplasts and are then delivered into the vacuole via autophagosome-mediated transport (Figure 2a), although how such bodies are evacuated from chloroplasts remains unclear.

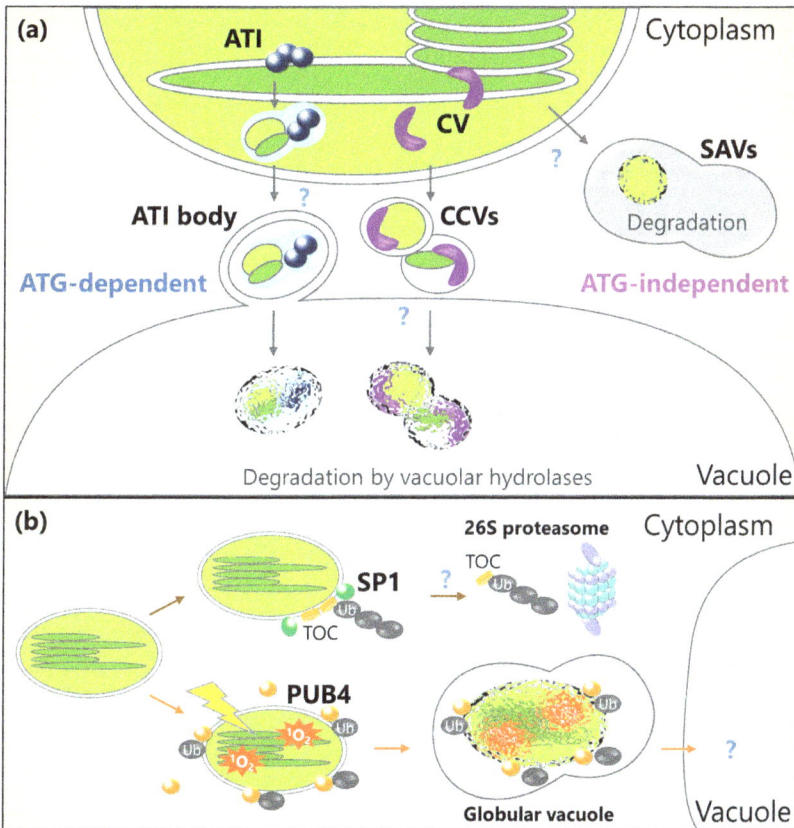

Figure 2. Schematic model for chloroplast protein turnover mediated by ATI bodies, CV-containing vesicles (CCVs), senescence-associated vacuoles (SAVs), or ubiquitination. (**a**) Plastid-associated ATI bodies are produced in chloroplasts and are then delivered into the central vacuole via an autophagy-dependent pathway. ATI bodies transport thylakoid, stroma, and envelope proteins. CV protein also interacts with thylakoid and stroma proteins, and then induces the production of CCVs that transport thylakoid, stroma, and envelope proteins into the central vacuole via an autophagy-independent pathway. SAVs are small lytic compartments that form in the cytoplasm. Stroma components are incorporated into the SAVs for digestion. (**b**) Chloroplast outer envelope-anchored E3 ligase, SP1, ubiquitinates TOC proteins and facilitates their degradation by 26S proteasome. Cytoplasmic E3 ligase PUB4 ubiquitinates oxidative chloroplasts accumulating $^1O_2$ for the digestion of such chloroplasts in their entirety.

Int. J. Mol. Sci. **2018**, *19*, 828

The appearance of plastid-associated ATI bodies in energy-starved seedlings or senescing leaves suggests that ATI bodies also contribute to amino acid recycling during starvation or senescence as part of the autophagy process, although the link between the induction level of the ATI bodies and changes in free amino acid content has not been evaluated. Additionally, plastid-associated ATI bodies are produced under salt stress, and ATI-knockdown plants have reduced salt tolerance [76]. The activation of autophagosome production and the reduced tolerance of *atg* mutants to salt stress were also observed in Arabidopsis plants [50,77]. These findings suggest that ATI bodies are involved in salt stress-induced chloroplast protein turnover.

5. Senescence-Associated Vacuoles

The formation of small, lytic senescence-associated vacuoles (SAVs) was reported when senescing leaves of Arabidopsis, soybean (*Glycine max*), and tobacco (*Nicotiana tabacum*) plants were stained with R-6502 dye, which emits strong fluorescence upon the hydrolytic activity of cysteine proteases [78,79]. Senescence-associated gene 12 (SAG12) is a senescence-induced cysteine protease localized within SAVs. SAVs are formed in the peripheral cytoplasmic region of mesophyll cells and are much smaller than the central vacuole, being approximately 0.7 μm in diameter. In addition, SAVs have greater lytic activity than the central vacuole and are strongly stained by lysotracker red or neutral red, fluorescent markers of acidic organelles.

SAV numbers increase as leaf senescence progresses [80]. Proteomic analysis of isolated SAVs in tobacco plants indicated that SAVs contain stromal proteins such as Rubisco and glutamine synthetase, but not thylakoid proteins such as LHCII and the reaction center D1 protein in photosystem II [79]. Treatment with a specific inhibitor of cysteine proteases, E-64, partially suppressed Rubisco degradation in the tobacco leaf discs [80]. These observations suggest that SAVs contribute to senescence-induced Rubisco degradation, similar to RCBs; however, *atg7* mutants produced SAVs [79]. Therefore, SAVs may be an autophagy-independent, extra-chloroplastic route for the degradation of stromal proteins in senescent leaves (Figure 2a). How stromal proteins are transported into the SAVs remains uncertain.

6. Autophagy-Independent Vesicles Derived from Chloroplasts

The *chloroplast vesiculation* (*CV*) gene encodes a plastid-targeted protein in rice plants that is strongly upregulated under abiotic stress and downregulated by cytokinin [81]. In Arabidopsis, expression of the *CV-GFP* construct under the control of the dexamethasone-inducible promoter caused the formation of a type of chloroplast-derived vesicle exhibiting strong CV-GFP signal referred to as CV-containing vesicles (CCVs) [81]. CCVs are around 1 μm in diameter and contain stroma, envelope, and thylakoid proteins, as demonstrated by immunoblot analysis of some chloroplast proteins, co-immunoprecipitation assays of potential CV-interacting protein, and confocal microscopy of fluorescent marker proteins of chloroplast stroma [81]. CCVs do not associate with the autophagosome marker GFP-ATG8a, and the *atg5* mutation does not affect the production of CCVs. Additionally, CCVs do not associate with SAVs stained with lysotracker red. Thus, CCVs are part of a vacuolar degradation process for chloroplasts that is independent of autophagy and SAVs (Figure 2a).

Immuno-EM analysis showed that CV-GFP was associated with the thylakoid or envelope membranes before CCV production [81]. The interaction of CV with PsbO protein, a subunit in the thylakoid-bound photosystem II complex, was confirmed by co-immunoprecipitation detection and a bimolecular fluorescence complementation (BiFC) assay. These results indicate that CV interacts with some proteins inside the chloroplast before CCVs form. The C-terminal domain of CV, which is largely conserved among CV orthologs of various plant species, is required for CCV production [81]; however, how chloroplast-targeted CV induces the formation of CCVs and chloroplast destabilization has not been evaluated.

In Arabidopsis plants, endogenous *CV* was upregulated in senescing leaves and leaves subjected to oxidative stress or salt stress [81]. Consistent with this, transient expression of *CV* caused accelerated leaf senescence, and the suppression of *CV* transcript by miRNA led to increased leaf longevity under salt stress. Similarly, in rice plants, the RNAi silencing of *CV* expression led to delayed leaf senescence under water deficit stress, and the transient overexpression of *CV-GFP* under the control of the β-estradiol-inducible promoter accelerated leaf senescence symptoms [82]. Elevated *CV* transcript levels were observed in UVB-damaged Arabidopsis leaves [67]. *CV* may activate the destabilization and degradation of chloroplasts through the formation of CCVs during senescence, especially under stress conditions in Arabidopsis and rice plants.

7. Ubiquitin E3 Ligase-Associated Chloroplast Degradation

The ubiquitin proteasome system (UPS) is an evolutionarily conserved major protein degradation system in eukaryotic cells [83–85]. During UPS-mediated proteolysis, the polypeptide ubiquitin acts as a sorting signal for the degradation of specific proteins by the 26S proteasome in the ubiquitination cascade. Ubiquitin is activated by E1 proteins and then transferred to E2 ubiquitin conjugating enzymes. The transfer of ubiquitin from E2s to target proteins requires E3 ubiquitin ligases. The resulting ubiquitinated proteins are selectively incorporated into the 26S proteasome complex for breakdown. Eukaryotic genomes generally encode a large family of E3s and Arabidopsis plants can theoretically express more than 1500 of these proteins [86–88]. The ubiquitination of specific proteins by individual E3s allows for highly controlled, selective protein degradation by the UPS.

The UPS was shown to contribute to the degradation of chloroplast proteins in an experiment using suppressor of ppi1 locus 1 (SP1) isolated from Arabidopsis plants [89]. SP1 is a chloroplast outer envelope-anchored E3 ligase that ubiquitinates some proteins of the translocon on the outer chloroplast membrane (TOC) complex (Figure 2b). Most nucleus-encoded chloroplast proteins are imported into chloroplasts through the TOC and translocon on the inner chloroplast membrane (TIC) complexes [90]. During the greening of etiolated seedlings, etioplasts, which are a type of plastid present in non-green tissues, are converted to mature chloroplasts; therefore, large amount of photosynthetic proteins encoded in the nuclear genome are expressed and imported into the plastid via TIC-TOC complexes. *sp1* mutant plants exhibit delayed maturation of chloroplasts during the greening of etiolated seedlings [89]. Thus, SP1 likely serves as a control for protein import into chloroplasts via the turnover of the TOC complex when etioplasts develop into functional chloroplasts. *sp1* mutants also showed delayed leaf yellowing during dark-induced accelerated senescence; conversely, SP1-overexpressing plants showed an enhanced decline of photosynthetic efficiency [89]. SP1-mediated TOC turnover may further regulate protein import into chloroplasts when functional chloroplasts are actively degraded during senescence.

SP1 induces the degradation of TOC during oxidative stress caused by salt or osmotic stress, thereby attenuating protein import into the chloroplasts [91]. Under these stress conditions, accumulation of hydrogen peroxide (H_2O_2), a type of ROS, was enhanced in *sp1* mutants and was alleviated in SP1-overexpressing plants. SP1-mediated degradation of the TOC complex by UPS suppresses photosynthetic activity and thereby limits ROS production [91], since ROS are produced during photosynthesis. SP1-mediated TOC turnover may therefore control the chloroplast proteome and photosynthetic capacity in response to stress. It is still unclear how ubiquitinated proteins on outer-envelope proteins are solubilized to allow degradation by UPS localized in the cytoplasm.

A recent study reported that a cytosol-localized E3 ligase functions in the degradation of entire chloroplasts [92]. When dark-germinated, etiolated seedlings of the Arabidopsis mutant of plastid *ferrochelatase 2* (*fc2*) are transferred from darkness to light, their chloroplasts over-accumulate singlet oxygen ($^1O_2$), thereby leading to the death of photosynthetic cells and compromised greening of plants. A suppressor mutant (referred to as *pub4–6* [92]) of this inhibited greening phenomenon had an amino acid substitution in *plant u-box 4* (*PUB4*), which encodes a cytosolic ubiquitin E3 ligase. Although EM analysis indicated that entire chloroplasts were digested in the cytoplasm during compromised greening in *fc2* plants, this degradation of chloroplasts was lower in *fc2 pub4–6* plants, even though $^1O_2$ accumulation was not suppressed. Therefore, PUB4-related ubiquitination triggers the digestion of entire chloroplasts that are accumulating $^1O_2$ (Figure 2b). However, unlike the *pub4–6* mutation, the T-DNA insertional knockout mutations of *PUB4* (referred to as *pub4–1* and *pub4–2*) did not suppress the phenotype of the *fc2* mutant during greening [92]. Therefore, it is unclear how PUB4 is involved in the ubiquitination of $^1O_2$ accumulating chloroplasts and their subsequent degradation.

In mammalian cells, ubiquitination largely acts as a trigger of autophagic removal of dysfunctional organelles [93]. During mitophagy, depolarized mitochondria are ubiquitinated by the E3 ligase Parkin, allowing for the autophagic removal of targeted mitochondria into the lysosome [94–97]. During the greening of *fc2* mutants, some chloroplasts appeared to be degraded in the cytoplasm, and the interaction between degrading chloroplasts and the vacuole via a globule-like structure was observed [92]. Such observations were distinct from the vacuolar chloroplasts that result from chlorophagy in leaves exposed to strong visible light (1200–2000 $\mu mol \cdot m^{-2} \cdot s^{-1}$), where entire chloroplasts exhibiting thylakoid membranes are localized in the central vacuole in EM imaging [67]. Furthermore, the *pub4–6* and *atg10* mutants are phenotypically distinct, as *atg10* plants showed accelerated senescence during dark treatment compared to wild-type plants, but *pub4–6* plants did not [92]. Therefore, PUB4-related ubiquitination is unlikely a simple trigger of autophagy.

8. Future Perspectives

Our understanding of the diverse extraplastidic pathways mediating chloroplast protein degradation has progressed in the past decades. Table 1 compares their relationships to core autophagy machinery, plant species, induction stimuli, and degradation targets. It is clear that multiple pathways are induced during diverse stress conditions, such as sugar starvation, senescence, and oxidative stress (Table 1). Thus, new questions about chloroplast turnover arise, including why plants have multiple processes for chloroplast protein turnover, and how these processes are differentially utilized. Future research should examine how extraplastidic systems are coordinated with intraplastidic proteolysis. The RCB pathway is activated during the earlier stages of dark treatment and chlorophagy is induced during the later stages [52,67], suggesting that several pathways are induced at distinct time points during leaf senescence and stress responses. An important role of intrachloroplastic proteases in chloroplast protein turnover during photodamage was largely demonstrated [15,16]. Therefore, extraplastidic pathways that are induced during photooxidative stress might be triggered when intraplastidic proteolysis is insufficient for maintaining chloroplast functions.

Table 1. List of extraplastidic degradation pathways described.

Pathway	Relationship to Core Autophagy Machinery	Analyzed Species	Degradation Targets	Stimuli [b]	References
RCBs (Rubisco-containing bodies)	dependent	Arabidopsis, rice, wheat	stroma, envelope	sugar starvation, senescence	[36–38,41,42,44,52–54]
Chlorophagy	dependent	Arabidopsis	entire chloroplasts	photodamage, senescence	[52,67]
ATI bodies	dependent	Arabidopsis	stroma, thylakoid, envelope	sugar starvation, salt stress, senescence	[76]
SAVs (Senescence-associated vacuoles)	independent	Arabidopsis, soybean, tobacco	stroma	senescence	[78–80]
CCVs (Chloroplast vesiculation-containing vesicles)	independent	Arabidopsis, rice	stroma, thylakoid, envelope	senescence, salt stress, oxidative stress	[81,82]
E3 ligase SP1	- [a]	Arabidopsis	TOC proteins on outer envelope	senescence, greening, oxidative stress	[89,91]
E3 ligase PUB4	- [a]	Arabidopsis	entire chloroplasts	Oxidative stress ($^1O_2$)	[92]

[a] The link of the E3 ubiquitin ligases to autophagy has not been directly examined. [b] Stimuli inducing the respective pathways.

Int. J. Mol. Sci. **2018**, *19*, 828

In Arabidopsis *atg* plants, both *CV* expression and proteasome activity are increased [43,98], suggesting a complementary relationship among some of the chloroplast-associated degradation systems. However, since senescence symptoms are largely accelerated in *atg* mutants due to the over-accumulation of salicylic acid [99], the increase in *CV* expression or proteasome activity in *atg* plants can also be interpreted as a result of accelerated senescence and cell death. To better understand the process of chloroplast protein turnover and to decipher the relationships among the diverse pathways mediating this process, the mechanisms regulating these pathways will need to be elucidated. It would be fascinating to determine whether the distinct pathways that mediate chloroplast degradation share a common upstream regulatory mechanism, or whether they are regulated independently. In addition, how small vesicles delivering portions of chloroplasts, including RCBs, ATI bodies, and CCVs, are derived from entire chloroplasts largely remains to be explained.

The extraplastidic routes for chloroplast protein turnover were largely identified using Arabidopsis plants (Table 1). This advance greatly expanded our understanding of chloroplast protein turnover in important cereals, such as rice and maize [38,54,56,82]. Chloroplast degradation is strongly linked to nitrogen remobilization and the changes of photosynthetic capacity that are important determinants of productivity in crop plants. Therefore, manipulating chloroplast protein turnover might be an effective strategy to improve the productivity of crops. In rice plants, RNAi-mediated silencing of *CV* led to an increase in grain yield under water deficit stress [74]. Studies showed that Arabidopsis plants overexpressing one of the core *ATGs* had an enhanced stress tolerance [100,101]. In addition, SP1-overexpressing Arabidopsis plants had improved tolerance to oxidative stress [91]. Therefore, elucidating the molecular basis of multiple processes for chloroplast protein turnover in Arabidopsis plants may suggest strategies to improve the productivity and stress tolerance of crop plants.

Acknowledgments: This work was supported, in part, by KAKENHI (Grant Numbers 17H05050, awarded to Masanori Izumi and 16J03408, awarded to Sakuya Nakamura), the JSPS Research Fellowship for Young Scientists (awarded to Sakuya Nakamura), Building of Consortia for the Development of Human Resources in Science and Technology (awarded to Masanori Izumi), JST PRESTO (Grant Number JPMJPR16Q1, awarded to Masanori Izumi), and the Program for Creation of Interdisciplinary Research at Frontier Research Institute for Interdisciplinary Sciences, Tohoku University, Japan (awarded to Masanori Izumi).

Author Contributions: Masanori Izumi conceived the topic of this review; Masanori Izumi and Sakuya Nakamura wrote the paper; Sakuya Nakamura designed the figures with the support of Masanori Izumi.

Conflicts of Interest: The authors declare no conflicts of interest.

References

1. Makino, A.; Osmond, B. Effects of nitrogen nutrition on nitrogen partitioning between chloroplasts and mitochondria in pea and wheat. *Plant Physiol.* **1991**, *96*, 355–362. [CrossRef] [PubMed]
2. Makino, A.; Sakuma, H.; Sudo, E.; Mae, T. Differences between maize and rice in N-use efficiency for photosynthesis and protein allocation. *Plant Cell Physiol.* **2003**, *44*, 952–956. [CrossRef] [PubMed]
3. Evans, J.R. Photosynthesis and nitrogen relationships in leaves of C_3 plants. *Oecologia* **1989**, *78*, 9–19. [CrossRef] [PubMed]
4. Mae, T.; Ohira, K. The remobilization of nitrogen related to leaf growth and senescence in rice plants (*Oryza sativa* L.). *Plant Cell Physiol.* **1981**, *22*, 1067–1074. [CrossRef]
5. Masclaux-Daubresse, C.; Daniel-Vedele, F.; Dechorgnat, J.; Chardon, F.; Gaufichon, L.; Suzuki, A. Nitrogen uptake, assimilation and remobilization in plants: Challenges for sustainable and productive agriculture. *Ann. Bot.* **2010**, *105*, 1141–1157. [CrossRef] [PubMed]
6. Stitt, M.; Zeeman, S.C. Starch turnover: Pathways, regulation and role in growth. *Curr. Opin. Plant Biol.* **2012**, *15*, 282–292. [CrossRef] [PubMed]
7. Baena-González, E.; Sheen, J. Convergent energy and stress signaling. *Trends Plant Sci.* **2008**, *13*, 474–482. [CrossRef] [PubMed]
8. Chaves, M.M.; Flexas, J.; Pinheiro, C. Photosynthesis under drought and salt stress: Regulation mechanisms from whole plant to cell. *Ann. Bot.* **2009**, *103*, 551–560. [CrossRef] [PubMed]

9. Araújo, W.L.; Tohge, T.; Ishizaki, K.; Leaver, C.J.; Fernie, A.R. Protein degradation—An alternative respiratory substrate for stressed plants. *Trends Plant Sci.* **2011**, *16*, 489–498. [CrossRef] [PubMed]
10. Sonoike, K. Various aspects of inhibition of photosynthesis under light/chilling stress: "Photoinhibition at chilling temperatures" versus "Chilling damage in the light". *J. Plant Res.* **1998**, *111*, 121–129. [CrossRef]
11. Li, Z.R.; Wakao, S.; Fischer, B.B.; Niyogi, K.K. Sensing and responding to excess light. *Annu. Rev. Plant Biol.* **2009**, *60*, 239–260. [CrossRef] [PubMed]
12. Tikkanen, M.; Mekala, N.R.; Aro, E.M. Photosystem II photoinhibition-repair cycle protects Photosystem I from irreversible damage. *Biochim. Biophys. Acta* **2014**, *1837*, 210–215. [CrossRef] [PubMed]
13. Takahashi, S.; Badger, M.R. Photoprotection in plants: A new light on photosystem II damage. *Trends Plant Sci.* **2011**, *16*, 53–60. [CrossRef] [PubMed]
14. Kataria, S.; Jajoo, A.; Guruprasad, K.N. Impact of increasing Ultraviolet-B (UV-B) radiation on photosynthetic processes. *J. Photochem. Photobiol B* **2014**, *137*, 55–66. [CrossRef] [PubMed]
15. Nishimura, K.; Kato, Y.; Sakamoto, W. Chloroplast proteases: Updates on proteolysis within and across suborganellar compartments. *Plant Physiol.* **2016**, *171*, 2280–2293. [CrossRef] [PubMed]
16. Van Wijk, K.J. Protein maturation and proteolysis in plant plastids, mitochondria, and peroxisomes. *Annu Rev. Plant Biol.* **2015**, *66*, 75–111. [CrossRef] [PubMed]
17. Nakatogawa, H.; Suzuki, K.; Kamada, Y.; Ohsumi, Y. Dynamics and diversity in autophagy mechanisms: Lessons from yeast. *Nat. Rev. Mol. Cell Biol.* **2009**, *10*, 458–467. [CrossRef] [PubMed]
18. Mizushima, N.; Komatsu, M. Autophagy: Renovation of cells and tissues. *Cell* **2011**, *147*, 728–741. [CrossRef] [PubMed]
19. Anding, A.L.; Baehrecke, E.H. Cleaning house: Selective autophagy of organelles. *Dev. Cell* **2017**, *41*, 10–22. [CrossRef] [PubMed]
20. Tsukada, M.; Ohsumi, Y. Isolation and characterization of autophagy-defective mutants of *Saccharomyces erevisiae*. *FEBS Lett.* **1993**, *333*, 169–174. [CrossRef]
21. Chung, T.; Suttangkakul, A.; Vierstra, R.D. The ATG autophagic conjugation system in maize: ATG transcripts and abundance of the ATG8-lipid adduct are regulated by development and nutrient availability. *Plant Physiol.* **2009**, *149*, 220–234. [CrossRef] [PubMed]
22. Xia, K.F.; Liu, T.; Ouyang, J.; Wang, R.; Fan, T.; Zhang, M.Y. Genome-wide identification, classification, and expression analysis of autophagy-associated gene homologues in rice (*Oryza sativa* L.). *DNA Res.* **2011**, *18*, 363–377. [CrossRef] [PubMed]
23. Meijer, W.H.; van der Klei, I.J.; Veenhuis, M.; Kiel, J.A.K.W. *ATG* genes involved in non-selective autophagy are conserved from yeast to man, but the selective Cvt and pexophagy pathways also require organism-specific genes. *Autophagy* **2007**, *3*, 106–116. [CrossRef] [PubMed]
24. Yoshimoto, K.; Hanaoka, H.; Sato, S.; Kato, T.; Tabata, S.; Noda, T.; Ohsumi, Y. Processing of ATG8s, ubiquitin-like proteins, and their deconjugation by ATG4s are essential for plant autophagy. *Plant Cell* **2004**, *16*, 2967–2983. [CrossRef] [PubMed]
25. Suzuki, N.N.; Yoshimoto, K.; Fujioka, Y.; Ohsumi, Y.; Inagaki, F. The crystal structure of plant ATG12 and its biological implication in autophagy. *Autophagy* **2005**, *1*, 119–126. [CrossRef] [PubMed]
26. Xiong, Y.; Contento, A.L.; Bassham, D.C. AtATG18a is required for the formation of autophagosomes during nutrient stress and senescence in *Arabidopsis thaliana*. *Plant J.* **2005**, *42*, 535–546. [CrossRef] [PubMed]
27. Doelling, J.H.; Walker, J.M.; Friedman, E.M.; Thompson, A.R.; Vierstra, R.D. The APG8/12-activating enzyme APG7 is required for proper nutrient recycling and senescence in *Arabidopsis thaliana*. *J. Biol. Chem.* **2002**, *277*, 33105–33114. [CrossRef] [PubMed]
28. Phillips, A.R.; Suttangkakul, A.; Vierstra, R.D. The ATG12-conjugating enzyme ATG10 is essential for autophagic vesicle formation in *Arabidopsis thaliana*. *Genetics* **2008**, *178*, 1339–1353. [CrossRef] [PubMed]
29. Suttangkakul, A.; Li, F.Q.; Chung, T.; Vierstra, R.D. The ATG1/ATG13 protein kinase complex Is both a regulator and a target of autophagic recycling in *Arabidopsis*. *Plant Cell* **2011**, *23*, 3761–3779. [CrossRef] [PubMed]
30. Chung, T.; Phillips, A.R.; Vierstra, R.D. ATG8 lipidation and ATG8-mediated autophagy in Arabidopsis require ATG12 expressed from the differentially controlled *ATG12A* and *ATG12B* loci. *Plant J.* **2010**, *62*, 483–493. [CrossRef] [PubMed]

31. Thompson, A.R.; Doelling, J.H.; Suttangkakul, A.; Vierstra, R.D. Autophagic nutrient recycling in Arabidopsis directed by the ATG8 and ATG12 conjugation pathways. *Plant Physiol.* **2005**, *138*, 2097–2110. [CrossRef] [PubMed]

32. Li, F.; Chung, T.; Vierstra, R.D. AUTOPHAGY-RELATED11 plays a critical role in general autophagy- and senescence-induced mitophagy in Arabidopsis. *Plant Cell* **2014**, *26*, 788–807. [CrossRef] [PubMed]

33. Mae, T.; Kai, N.; Makino, A.; Ohira, K. Relation between ribulose bisphosphate carboxylase content and chloroplast number in naturally senescing primary leaves of wheat. *Plant Cell Physiol.* **1984**, *25*, 333–336. [CrossRef]

34. Ono, K.; Hashimoto, H.; Katoh, S. Changes in the number and size of chloroplasts during senescence of primary leaves of wheat grown under different conditions. *Plant Cell Physiol.* **1995**, *36*, 9–17. [CrossRef]

35. Martinoia, E.; Heck, U.; Dalling, M.J.; Matile, P. Changes in chloroplast number and chloroplast constituents in senescing barley leaves. *Biochem. Physiol. Pflanz.* **1983**, *178*, 147–155. [CrossRef]

36. Chiba, A.; Ishida, H.; Nishizawa, N.K.; Makino, A.; Mae, T. Exclusion of ribulose-1,5-bisphosphate carboxylase/oxygenase from chloroplasts by specific bodies in naturally senescing leaves of wheat. *Plant Cell Physiol.* **2003**, *44*, 914–921. [CrossRef] [PubMed]

37. Ishida, H.; Yoshimoto, K.; Izumi, M.; Reisen, D.; Yano, Y.; Makino, A.; Ohsumi, Y.; Hanson, M.R.; Mae, T. Mobilization of rubisco and stroma-localized fluorescent proteins of chloroplasts to the vacuole by an *ATG* gene-dependent autophagic process. *Plant Physiol.* **2008**, *148*, 142–155. [CrossRef] [PubMed]

38. Izumi, M.; Hidema, J.; Wada, S.; Kondo, E.; Kurusu, T.; Kuchitsu, K.; Makino, A.; Ishida, H. Establishment of monitoring methods for autophagy in rice reveals autophagic recycling of chloroplasts and root plastids during energy limitation. *Plant Physiol.* **2015**, *167*, 1307–1320. [CrossRef] [PubMed]

39. Gao, C.J.; Zhuang, X.H.; Shen, J.B.; Jiang, L.W. Plant ESCRT complexes: Moving beyond endosomal sorting. *Trends Plant Sci.* **2017**, *22*, 986–998. [CrossRef] [PubMed]

40. Spitzer, C.; Li, F.Q.; Buono, R.; Roschzttardtz, H.; Chung, T.J.; Zhang, M.; Osteryoung, K.W.; Vierstra, R.D.; Otegui, M.S. The endosomal protein CHARGED MULTIVESICULAR BODY PROTEIN1 regulates the autophagic turnover of plastids in Arabidopsis. *Plant Cell* **2015**, *27*, 391–402. [CrossRef] [PubMed]

41. Izumi, M.; Wada, S.; Makino, A.; Ishida, H. The autophagic degradation of chloroplasts via rubisco-containing bodies is specifically linked to leaf carbon status but not nitrogen status in Arabidopsis. *Plant Physiol.* **2010**, *154*, 1196–1209. [CrossRef] [PubMed]

42. Izumi, M.; Hidema, J.; Makino, A.; Ishida, H. Autophagy contributes to nighttime energy availability for growth in Arabidopsis. *Plant Physiol.* **2013**, *161*, 1682–1693. [CrossRef] [PubMed]

43. Barros, J.A.S.; Cavalcanti, J.H.F.; Medeiros, D.B.; Nunes-Nesi, A.; Avin-Wittenberg, T.; Fernie, A.R.; Araujo, W.L. Autophagy deficiency compromises alternative pathways of respiration following energy deprivation in *Arabidopsis thaliana*. *Plant Physiol.* **2017**, *175*, 62–76. [CrossRef] [PubMed]

44. Hirota, T.; Izumi, M.; Wada, S.; Makino, A.; Ishida, H. Vacuolar protein degradation via autophagy provides substrates to amino acid catabolic pathways as an adaptive response to sugar starvation in *Arabidopsis thaliana*. *Plant Cell Physiol.* **2018**. [CrossRef] [PubMed]

45. Hildebrandt, T.M.; Nesi, A.N.; Araujo, W.L.; Braun, H.P. Amino acid catabolism in plants. *Mol. Plant* **2015**, *8*, 1563–1579. [CrossRef] [PubMed]

46. Araújo, W.L.; Ishizaki, K.; Nunes-Nesi, A.; Larson, T.R.; Tohge, T.; Krahnert, I.; Witt, S.; Obata, T.; Schauer, N.; Graham, I.A.; et al. Identification of the 2-hydroxyglutarate and Isovaleryl-CoA dehydrogenases as alternative electron donors linking lysine catabolism to the electron transport chain of *Arabidopsis* mitochondria. *Plant Cell* **2010**, *22*, 1549–1563. [CrossRef] [PubMed]

47. Ishizaki, K.; Larson, T.R.; Schauer, N.; Fernie, A.R.; Graham, I.A.; Leaver, C.J. The critical role of *Arabidopsis* electron-transfer flavoprotein: Ubiquinone oxidoreductase during dark-induced starvation. *Plant Cell* **2005**, *17*, 2587–2600. [CrossRef] [PubMed]

48. Ishizaki, K.; Schauer, N.; Larson, T.R.; Graham, I.A.; Fernie, A.R.; Leaver, C.J. The mitochondrial electron transfer flavoprotein complex is essential for survival of Arabidopsis in extended darkness. *Plant J.* **2006**, *47*, 751–760. [CrossRef] [PubMed]

49. Peng, C.; Uygun, S.; Shiu, S.H.; Last, R.L. The impact of the branched-chain ketoacid dehydrogenase complex on amino acid homeostasis in Arabidopsis. *Plant Physiol.* **2015**, *169*, 1807–1820. [CrossRef] [PubMed]

50. Liu, Y.; Xiong, Y.; Bassham, D.C. Autophagy is required for tolerance of drought and salt stress in plants. *Autophagy* **2009**, *5*, 954–963. [CrossRef] [PubMed]

51. Chen, L.; Liao, B.; Qi, H.; Xie, L.J.; Huang, L.; Tan, W.J.; Zhai, N.; Yuan, L.B.; Zhou, Y.; Yu, L.J.; et al. Autophagy contributes to regulation of the hypoxia response during submergence in *Arabidopsis thaliana*. *Autophagy* **2015**, *11*, 2233–2246. [CrossRef] [PubMed]

52. Wada, S.; Ishida, H.; Izumi, M.; Yoshimoto, K.; Ohsumi, Y.; Mae, T.; Makino, A. Autophagy plays a role in chloroplast degradation during senescence in individually darkened leaves. *Plant Physiol.* **2009**, *149*, 885–893. [CrossRef] [PubMed]

53. Ono, Y.; Wada, S.; Izumi, M.; Makino, A.; Ishida, H. Evidence for contribution of autophagy to rubisco degradation during leaf senescence in *Arabidopsis thaliana*. *Plant Cell Environ.* **2013**, *36*, 1147–1159. [CrossRef] [PubMed]

54. Wada, S.; Hayashida, Y.; Izumi, M.; Kurusu, T.; Hanamata, S.; Kanno, K.; Kojima, S.; Yamaya, T.; Kuchitsu, K.; Makino, A.; et al. Autophagy supports biomass production and nitrogen use efficiency at the vegetative stage in rice. *Plant Physiol.* **2015**, *168*, 60–73. [CrossRef] [PubMed]

55. Guiboileau, A.; Yoshimoto, K.; Soulay, F.; Bataillé, M.P.; Avice, J.C.; Masclaux-Daubresse, C. Autophagy machinery controls nitrogen remobilization at the whole-plant level under both limiting and ample nitrate conditions in Arabidopsis. *New Phytol.* **2012**, *194*, 732–740. [CrossRef] [PubMed]

56. Li, F.Q.; Chung, T.; Pennington, J.G.; Federico, M.L.; Kaeppler, H.F.; Kaeppler, S.M.; Otegui, M.S.; Vierstra, R.D. Autophagic recycling plays a central role in maize nitrogen remobilization. *Plant Cell* **2015**, *27*, 1389–1408. [CrossRef] [PubMed]

57. Guiboileau, A.; Avila-Ospina, L.; Yoshimoto, K.; Soulay, F.; Azzopardi, M.; Marmagne, A.; Lothier, J.; Masclaux-Daubresse, C. Physiological and metabolic consequences of autophagy deficiency for the management of nitrogen and protein resources in Arabidopsis leaves depending on nitrate availability. *New Phytol.* **2013**, *199*, 683–694. [CrossRef] [PubMed]

58. Kurusu, T.; Koyano, T.; Hanamata, S.; Kubo, T.; Noguchi, Y.; Yagi, C.; Nagata, N.; Yamamoto, T.; Ohnishi, T.; Okazaki, Y.; et al. OsATG7 is required for autophagy-dependent lipid metabolism in rice postmeiotic anther development. *Autophagy* **2014**, *10*, 878–888. [CrossRef] [PubMed]

59. Ishida, H.; Izumi, M.; Wada, S.; Makino, A. Roles of autophagy in chloroplast recycling. *Biochim. Biophys. Acta* **2014**, *1837*, 512–521. [CrossRef] [PubMed]

60. Kim, J.; Lee, H.; Lee, H.N.; Kim, S.H.; Shin, K.D.; Chung, T. Autophagy-related proteins are required for degradation of peroxisomes in Arabidopsis hypocotyls during seedling growth. *Plant Cell* **2013**, *25*, 4956–4966. [CrossRef] [PubMed]

61. Shibata, M.; Oikawa, K.; Yoshimoto, K.; Kondo, M.; Mano, S.; Yamada, K.; Hayashi, M.; Sakamoto, W.; Ohsumi, Y.; Nishimura, M. Highly oxidized peroxisomes are selectively degraded via autophagy in Arabidopsis. *Plant Cell* **2013**, *25*, 4967–4983. [CrossRef] [PubMed]

62. Yoshimoto, K.; Shibata, M.; Kondo, M.; Oikawa, K.; Sato, M.; Toyooka, K.; Shirasu, K.; Nishimura, M.; Ohsumi, Y. Organ-specific quality control of plant peroxisomes is mediated by autophagy. *J. Cell Sci.* **2014**, *127*, 1161–1168. [CrossRef] [PubMed]

63. Goto-Yamada, S.; Mano, S.; Nakamori, C.; Kondo, M.; Yamawaki, R.; Kato, A.; Nishimura, M. Chaperone and protease functions of LON protease 2 modulate the peroxisomal transition and degradation with autophagy. *Plant Cell Physiol.* **2014**, *55*, 482–496. [CrossRef] [PubMed]

64. Farmer, L.M.; Rinaldi, M.A.; Young, P.G.; Danan, C.H.; Burkhart, S.E.; Bartel, B. Disrupting autophagy restores peroxisome function to an *Arabidopsis lon2* mutant and reveals a role for the LON2 protease in peroxisomal matrix protein degradation. *Plant Cell* **2013**, *25*, 4085–4100. [CrossRef] [PubMed]

65. Liu, Y.; Burgos, J.S.; Deng, Y.; Srivastava, R.; Howell, S.H.; Bassham, D.C. Degradation of the endoplasmic reticulum by autophagy during endoplasmic reticulum stress in Arabidopsis. *Plant Cell* **2012**, *24*, 4635–4651. [CrossRef] [PubMed]

66. Yang, X.C.; Srivastava, R.; Howell, S.H.; Bassham, D.C. Activation of autophagy by unfolded proteins during endoplasmic reticulum stress. *Plant J.* **2016**, *85*, 83–95. [CrossRef] [PubMed]

67. Izumi, M.; Ishida, H.; Nakamura, S.; Hidema, J. Entire photodamaged chloroplasts are transported to the central vacuole by Autophagy. *Plant Cell* **2017**, *29*, 377–394. [CrossRef] [PubMed]

68. Izumi, M.; Nakamura, S. Vacuolar digestion of entire damaged chloroplasts in *Arabidopsis thaliana* is accomplished by chlorophagy. *Autophagy* **2017**, *13*, 1239–1240. [CrossRef] [PubMed]

69. Izumi, M.; Nakamura, S. Partial or entire: Distinct responses of two types of chloroplast autophagy. *Plant Signal. Behav.* **2017**, *12*, e1393137. [CrossRef] [PubMed]

70. Ichimura, Y.; Kirisako, T.; Takao, T.; Satomi, Y.; Shimonishi, Y.; Ishihara, N.; Mizushima, N.; Tanida, I.; Kominami, E.; Ohsumi, M.; et al. A ubiquitin-like system mediates protein lipidation. *Nature* **2000**, *408*, 488–492. [CrossRef] [PubMed]

71. Noda, N.N.; Ohsumi, Y.; Inagaki, F. Atg8-family interacting motif crucial for selective autophagy. *FEBS Lett.* **2010**, *584*, 1379–1385. [CrossRef] [PubMed]

72. Kanki, T.; Wang, K.; Cao, Y.; Baba, M.; Klionsky, D.J. Atg32 is a mitochondrial protein that confers selectivity during mitophagy. *Dev. Cell* **2009**, *17*, 98–109. [CrossRef] [PubMed]

73. Okamoto, K.; Kondo-Okamoto, N.; Ohsumi, Y. Mitochondria-anchored receptor Atg32 mediates degradation of mitochondria via selective autophagy. *Dev. Cell* **2009**, *17*, 87–97. [CrossRef] [PubMed]

74. Mochida, K.; Oikawa, Y.; Kimura, Y.; Kirisako, H.; Hirano, H.; Ohsumi, Y.; Nakatogawa, H. Receptor-mediated selective autophagy degrades the endoplasmic reticulum and the nucleus. *Nature* **2015**, *522*, 359–362. [CrossRef] [PubMed]

75. Honig, A.; Avin-Wittenberg, T.; Ufaz, S.; Galili, G. A new type of compartment, defined by plant-specific atg8-interacting proteins, is induced upon exposure of Arabidopsis plants to carbon starvation. *Plant Cell* **2012**, *24*, 288–303. [CrossRef] [PubMed]

76. Michaeli, S.; Honig, A.; Levanony, H.; Peled-Zehavi, H.; Galili, G. Arabidopsis ATG8-INTERACTING PROTEIN1 is involved in autophagy-dependent vesicular trafficking of plastid proteins to the vacuole. *Plant Cell* **2014**, *26*, 4084–4101. [CrossRef] [PubMed]

77. Luo, L.M.; Zhang, P.P.; Zhu, R.H.; Fu, J.; Su, J.; Zheng, J.; Wang, Z.Y.; Wang, D.; Gong, Q.Q. Autophagy is rapidly induced by salt stress and is required for salt tolerance in Arabidopsis. *Front. Plant Sci.* **2017**, *8*, 1459. [CrossRef] [PubMed]

78. Otegui, M.S.; Noh, Y.S.; Martinez, D.E.; Vila Petroff, M.G.; Andrew Staehelin, L.; Amasino, R.M.; Guiamet, J.J. Senescence-associated vacuoles with intense proteolytic activity develop in leaves of Arabidopsis and soybean. *Plant J.* **2005**, *41*, 831–844. [CrossRef] [PubMed]

79. Martinez, D.E.; Costa, M.L.; Gomez, F.M.; Otegui, M.S.; Guiamet, J.J. 'Senescence-associated vacuoles' are involved in the degradation of chloroplast proteins in tobacco leaves. *Plant J.* **2008**, *56*, 196–206. [CrossRef] [PubMed]

80. Carrion, C.A.; Costa, M.L.; Martinez, D.E.; Mohr, C.; Humbeck, K.; Guiamet, J.J. In vivo inhibition of cysteine proteases provides evidence for the involvement of 'senescence-associated vacuoles' in chloroplast protein degradation during dark-induced senescence of tobacco leaves. *J. Exp. Bot.* **2013**, *64*, 4967–4980. [CrossRef] [PubMed]

81. Wang, S.H.; Blumwald, E. Stress-induced chloroplast degradation in Arabidopsis is regulated via a process independent of autophagy and senescence-associated vacuoles. *Plant Cell* **2014**, *26*, 4875–4888. [CrossRef] [PubMed]

82. Sade, N.; Umnajkitikorn, K.; Rubio Wilhelmi, M.D.M.; Wright, M.; Wang, S.; Blumwald, E. Delaying chloroplast turnover increases water-deficit stress tolerance through the enhancement of nitrogen assimilation in rice. *J. Exp. Bot.* **2017**, *69*, 867–878. [CrossRef] [PubMed]

83. Komander, D.; Rape, M. The ubiquitin code. *Annu. Rev. Biochem.* **2012**, *81*, 203–229. [CrossRef] [PubMed]

84. Vierstra, R.D. The expanding universe of ubiquitin and ubiquitin-like modifiers. *Plant Physiol.* **2012**, *160*, 2–14. [CrossRef] [PubMed]

85. Shu, K.; Yang, W.Y. E3 ubiquitin ligases: Ubiquitous actors in plant development and abiotic stress responses. *Plant Cell Physiol.* **2017**, *58*, 1461–1476. [CrossRef] [PubMed]

86. Hua, Z.H.; Vierstra, R.D. The cullin-ring ubiquitin-protein ligases. *Annu. Rev. Plant Biol.* **2011**, *62*, 299–334. [CrossRef] [PubMed]

87. Kraft, E.; Stone, S.L.; Ma, L.G.; Su, N.; Gao, Y.; Lau, O.S.; Deng, X.W.; Callis, J. Genome analysis and functional characterization of the E2 and RING-type E3 ligase ubiquitination enzymes of Arabidopsis. *Plant Physiol.* **2005**, *139*, 1597–1611. [CrossRef] [PubMed]

88. Stone, S.L.; Hauksdottir, H.; Troy, A.; Herschleb, J.; Kraft, E.; Callis, J. Functional analysis of the RING-type ubiquitin ligase family of Arabidopsis. *Plant Physiol.* **2005**, *137*, 13–30. [CrossRef] [PubMed]

89. Ling, Q.H.; Huang, W.H.; Baldwin, A.; Jarvis, P. Chloroplast biogenesis is regulated by direct action of the ubiquitin-proteasome system. *Science* **2012**, *338*, 655–659. [CrossRef] [PubMed]

90. Jarvis, P.; López-Juez, E. Biogenesis and homeostasis of chloroplasts and other plastids. *Nat. Rev. Mol. Cell Biol.* **2013**, *14*, 787–802. [CrossRef] [PubMed]

91. Ling, Q.H.; Jarvis, P. Regulation of chloroplast protein import by the ubiquitin E3 Ligase SP1 is important for stress tolerance in plants. *Curr. Biol.* **2015**, *25*, 2527–2534. [CrossRef] [PubMed]

92. Woodson, J.D.; Joens, M.S.; Sinson, A.B.; Gilkerson, J.; Salom, P.A.; Weigel, D.; Fitzpatrick, J.A.; Chory, J. Ubiquitin facilitates a quality-control pathway that removes damaged chloroplasts. *Science* **2015**, *350*, 450–454. [CrossRef] [PubMed]

93. Kraft, C.; Peter, M.; Hofmann, K. Selective autophagy: Ubiquitin-mediated recognition and beyond. *Nat. Cell Biol.* **2010**, *12*, 836–841. [CrossRef] [PubMed]

94. Matsuda, N.; Sato, S.; Shiba, K.; Okatsu, K.; Saisho, K.; Gautier, C.A.; Sou, Y.S.; Saiki, S.; Kawajiri, S.; Sato, F.; et al. PINK1 stabilized by mitochondrial depolarization recruits Parkin to damaged mitochondria and activates latent Parkin for mitophagy. *J. Cell Biol.* **2010**, *189*, 211–221. [CrossRef] [PubMed]

95. Narendra, D.; Tanaka, A.; Suen, D.F.; Youle, R.J. Parkin is recruited selectively to impaired mitochondria and promotes their autophagy. *J Cell Biol.* **2008**, *183*, 795–803. [CrossRef] [PubMed]

96. Narendra, D.P.; Jin, S.M.; Tanaka, A.; Suen, D.F.; Gautier, C.A.; Shen, J.; Cookson, M.R.; Youle, R.J. PINK1 is selectively stabilized on impaired mitochondria to activate Parkin. *PLoS Biol.* **2010**, *8*, e1000298. [CrossRef] [PubMed]

97. Vives-Bauza, C.; Zhou, C.; Huang, Y.; Cui, M.; de Vries, R.L.; Kim, J.; May, J.; Tocilescu, M.A.; Liu, W.; Ko, H.S.; et al. PINK1-dependent recruitment of Parkin to mitochondria in mitophagy. *Proc. Natl. Acad. Sci. USA* **2010**, *107*, 378–383. [CrossRef] [PubMed]

98. Have, M.; Balliau, T.; Cottyn-Boitte, B.; Derond, E.; Cueff, G.; Soulay, F.; Lornac, A.; Reichman, P.; Dissmeyer, N.; Avice, J.C.; et al. Increase of proteasome and papain-like cysteine protease activities in autophagy mutants: Backup compensatory effect or pro cell-death effect? *J. Exp. Bot.* **2017**. [CrossRef]

99. Yoshimoto, K.; Jikumaru, Y.; Kamiya, Y.; Kusano, M.; Consonni, C.; Panstruga, R.; Ohsumi, Y.; Shirasu, K. Autophagy negatively regulates cell death by controlling NPR1-dependent salicylic acid signaling during senescence and the innate immune response in *Arabidopsis*. *Plant Cell* **2009**, *21*, 2914–2927. [CrossRef] [PubMed]

100. Wang, P.; Sun, X.; Jia, X.; Ma, F. Apple autophagy-related protein MdATG3s afford tolerance to multiple abiotic stresses. *Plant Sci.* **2017**, *256*, 53–64. [CrossRef] [PubMed]

101. Xia, T.M.; Xiao, D.; Liu, D.; Chai, W.T.; Gong, Q.Q.; Wang, N.N. Heterologous expression of ATG8c from soybean confers tolerance to nitrogen deficiency and increases yield in Arabidopsis. *PLoS ONE* **2012**, *7*, e37217. [CrossRef] [PubMed]

International Journal of
Molecular Sciences

MDPI

Review

Insights into the Mechanisms of Chloroplast Division

Yamato Yoshida

Department of Science, College of Science, Ibaraki University, Ibaraki 310-8512, Japan;
yamato.yoshida.sci@vc.ibaraki.ac.jp

Received: 28 January 2018; Accepted: 1 March 2018; Published: 4 March 2018

Abstract: The endosymbiosis of a free-living cyanobacterium into an ancestral eukaryote led to the evolution of the chloroplast (plastid) more than one billion years ago. Given their independent origins, plastid proliferation is restricted to the binary fission of pre-existing plastids within a cell. In the last 25 years, the structure of the supramolecular machinery regulating plastid division has been discovered, and some of its component proteins identified. More recently, isolated plastid-division machineries have been examined to elucidate their structural and mechanistic details. Furthermore, complex studies have revealed how the plastid-division machinery morphologically transforms during plastid division, and which of its component proteins play a critical role in generating the contractile force. Identifying the three-dimensional structures and putative functional domains of the component proteins has given us hints about the mechanisms driving the machinery. Surprisingly, the mechanisms driving plastid division resemble those of mitochondrial division, indicating that these division machineries likely developed from the same evolutionary origin, providing a key insight into how endosymbiotic organelles were established. These findings have opened new avenues of research into organelle proliferation mechanisms and the evolution of organelles.

Keywords: chloroplast division; mitochondrial division; endosymbiotic organelle; FtsZ; dynamin-related protein; glycosyltransferase protein

1. Introduction

Chloroplasts (plastids) produce organic molecules and oxygen via photosynthesis, directly and indirectly providing a diverse array of living organisms with the materials they need to grow and develop. The activity of plastids over the past billion years has resulted in the dramatic greening of the Earth. Due to their endosymbiotic origin, plastids contain their own genomes and multiply by the binary fission of pre-existing plastids [1–4]. Although the mechanisms driving this division were long unclear, the discovery of a specialized ring structure at the division site of plastids in primitive unicellular alga provided groundbreaking insights into this process [5], and the rise of genomics and proteomics has further accelerated studies in this field. Consequently, some components for plastid division have been identified in the last 25 years [1–4,6]. It is now known that plastid division is carried out by a ring-shaped supermolecule termed the plastid-division machinery, which contains three or more types of ring structures: the plastid-dividing (PD) ring, which forms the main framework of the division machinery and comprises a ring-shaped bundle of nanofilaments on the cytosolic side of the outer envelope membrane of the corresponding organelle [4,5,7]; the FtsZ ring, a single ring constructed from homologs of the bacterial fission protein FtsZ located beneath the inner envelope membrane at the division site [8]; and the dynamin ring, a disconnected ring-like structure formed of a dynamin superfamily member on the cytosolic side of the outer envelope membrane at the site of organelle division (Figure 1) [9,10]. The importance of each component for plastid division has been well studied and summarized in detail elsewhere [3,11]; however, the functional mechanical details of the plastid-division machinery remain unclear. Various types of functional domains have been identified in the component proteins, including GTPase and glycosyltransferase domains, some of which are conserved not only within the plant kingdom but also in bacteria and non-photosynthetic eukaryotes; therefore, considerations and comparisons of these component proteins in other species will support our understanding of their fundamental functions during plastid division. Furthermore, it is now known that the mitochondrial- and peroxisome-division machineries carry out similar division processes to those of the plastids, suggesting that the elucidation of the plastid-division system will also provide insights into the proliferation of other membranous organelles within eukaryotic cells. This review summarizes and considers the domain architectures of the major proteins involved in plastid division, enabling the further exploration of the proliferation mechanisms in plastids as well as the proliferation mechanisms in mitochondria.

2. Structure and Assembly of the Plastid-Division Machinery

As mentioned above, plastids divide under the regulation of their division machinery, supramolecular complexes comprising a dynamic trio of rings, the PD ring, FtsZ ring, and dynamin ring, which span the plastid double membrane. Although the molecular function of each ring in the division machinery has not been fully revealed, studies using the alga *Cyanidioschyzon merolae* and the model dicot *Arabidopsis thaliana* have begun to uncover the molecular mechanisms by which this machinery functions. The formation of the plastid-division machinery is executed in a specific order during plastid division (Figure 1). As in bacteria, the assembly of the FtsZ ring in the stromal region is the first known event to occur at the plastid-division site, followed by the appearance of the inner PD ring beneath the inner envelope membrane at the division site. After the formation of the inner rings, the interaction between FtsZ and certain membrane proteins might transfer positional information about the FtsZ ring to the outside of the plastid, resulting in the binding of the glycosyltransferase protein PLASTID-DIVIDING RING 1 (PDR1) to the outer membrane, where it assembles the outer PD ring [12]. Another possibility was suggested following a series of electron microscopy (EM) observations in *C. caldarium*; small electron-dense vesicles were visualized along the putative division site before the assembly of the outer PD ring. Interestingly, the boundary of the vesicles appeared to coincide with one end of the PD ring filament, suggesting that these filaments might be biosynthesized on the surface of these vesicles using their components [5].

These inner and outer rings appeared to be linked to each other through nano-scale holes that appear on the groove of the division site, as revealed by scanning EM [13]. Recent yeast two-hybrid studies using several plastid-division proteins in *A. thaliana* showed that the FtsZ ring interacts with the inner membrane proteins ACCUMULATION AND REPLICATION OF CHLOROPLAST6 (ARC6) and PALAROG OF ARC6 (PARC6) in the stromal region [14,15]. ARC6 and PARC6 then further interact with the outer membrane proteins PDV2 and PDV1 in the intermembrane space to form the ARC6-PDV2 and PARC6-PDV1 complexes [15–17]. The intermembrane structure enables the translation of positional information regarding the FtsZ ring from the stromal region to the outer envelope membrane. In the final assembly step of the plastid-division machinery, dynamin-related protein (Dnm2/DRP5B) molecules cross-link the outer PD ring filaments. The sequence of assembly was confirmed in a range of knockdown/knockout experiments involving the plastid-division genes. When the expression of *PDR1* was downregulated, the FtsZ ring was assembled, but the Dnm2 proteins were not recruited to the division site [12]. Meanwhile, expression of a gene encoding the Dnm2 K135A mutant protein, which corresponds to a GTP-binding deficient mutant (K44A) of human dynamin 1, disturbed the formation of the dynamin ring and inhibited plastid division in *C. merolae*, despite the normal assembly of the FtsZ and PD rings [18–20]. Thus, neither FtsZ ring formation nor PD ring formation depends on the subsequent assembly of the plastid-division machinery, while the localization of Dnm2 relies on formation of the outer PD ring.

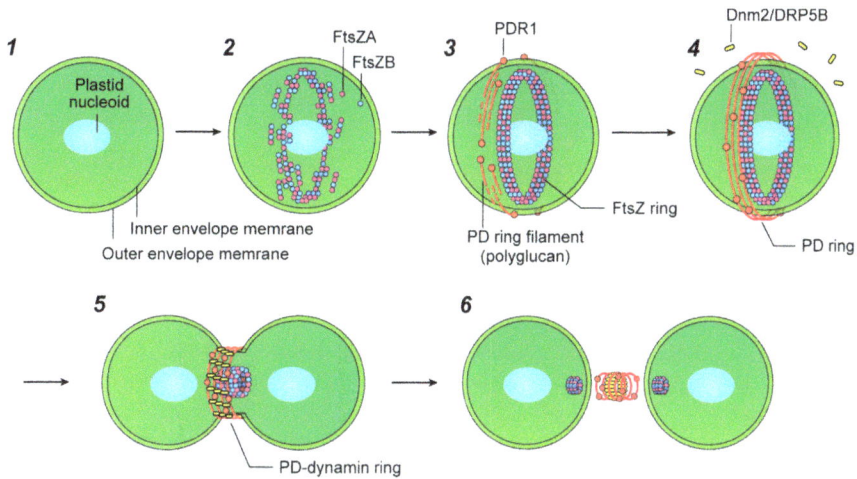

Figure 1. Representation of the plastid-division process. Plastid division occurs as follows: (**1,2**) Two types of FtsZ protein assemble in a heterodimer in the stromal region, then polymerize to form the FtsZ ring in the center of the plastid. To tether to the inner envelope membrane, FtsZ proteins bind to several membrane proteins. (**3,4**) PDR1 proteins attach to the outer envelope membrane above the site of the FtsZ ring, and it is hypothesized that PDR1 biosynthesizes polyglucan nanofilaments to form the PD ring from UDP-glucose molecules. (**5**) The GTPase protein Dnm2 (also known as DRP5B) binds to the PD ring filaments and is likely to generate the motive force for constriction. (**6**) Dnm2 proteins accumulate at the contracting bridge of two daughter plastids and pinch off the membranes. After the abscission of the plastids, the division machinery is disassembled. The inner PD ring and membrane proteins such as ARC6, and PDV2 are omitted from this representation. Modified from Yoshida et al. (2016) [21].

A series of ultrastructural studies provided structural insights into the contractile mechanism of the plastid-division machinery. Sequential EM observations of cells during plastid division revealed that the thickness of the inner PD ring does not change, but its volume decreases at a constant rate during contraction [22]. By contrast, the width and thickness of the outer PD ring monotonically increase during contraction in *C. merolae*, retaining its density and volume [22,23]. Similar EM observations of the outer PD ring changes have also been made in the green alga *Nannochloris bacillaris* [24], and in the land plants [25]. These morphological transitions of the outer PD ring during plastid division led to the idea that the outer PD ring filaments can slide and squeeze the plastid membranes. The establishment of a technique for isolating the intact plastid-division machinery from *C. merolae* cells provided a major breakthrough on this issue [13]; the isolated plastid-division machinery not only formed a circular structure, but also formed super-twisted and spiral structures featuring both clockwise and anticlockwise spirals. The existence of plastid-division machinery in these twisted states indicates the motive force involved in the contraction of plastid membranes; indeed, the plastid-division machinery autonomously contracted when plastid membranes were dissolved by detergent. In addition, reconstituted plastid FtsZ rings also displayed contractile ability. Although the detailed molecular mechanism involved is still unclear, the contraction process of the FtsZ ring accompanies the transition of the FtsZ protofilament from a less dynamic to a more dynamic state; therefore, this dynamic transition of protofilament states is assumed to induce a decrease in the average protofilament length in the FtsZ ring and thus cause contraction [21]. This led to the undertaking of complex studies to reveal how the plastid-division machinery is transformed morphologically during contraction, and which of its component proteins play a key role in generating this force.

3. The FtsZ Ring

Plastid FtsZ is a homolog of the bacterial fission protein FtsZ [8,26,27], and was the first factor found to be responsible for plastid division, assembling into a ring beneath the inner envelope membrane at the division site [22,28,29]. Both plastid and bacterial FtsZ proteins are composed of two functional domains, a GTP-binding domain at the N-terminal and a GTPase-activating domain at the C-terminal, which interact with the opposing terminal of other FtsZ proteins to form a polymer strand (Figure 2) [30,31]. Interestingly, whereas bacteria have one *FtsZ* gene in their genome, the plastid *FtsZ* gene underwent duplication and these duplicated loci, now present in the nuclear genomes, are widely conserved throughout photosynthetic eukaryotes [8,32,33]. Thus, the *C. merolae* genome contains two *FtsZ* genes, *FtsZ2-1* (*FtsZ2* in *A. thaliana*; FtsZA group) and *FtsZ2-2* (*FtsZ1* in *A. thaliana*; FtsZB group), for plastid division [27,32,34]. Phylogenetic studies have found that FtsZA is more ancestral [32], containing a conserved C-terminal core motif similar to that of bacterial FtsZ, which is assumed to interact with other plastid-division proteins to tether the FtsZ ring to the inner membrane [6,8]. The other FtsZ group, FtsZB, lacks the C-terminal core motif and probably evolved following the duplication of the original *FtsZ* gene or its precursor during the establishment of the plastids after the endosymbiotic event.

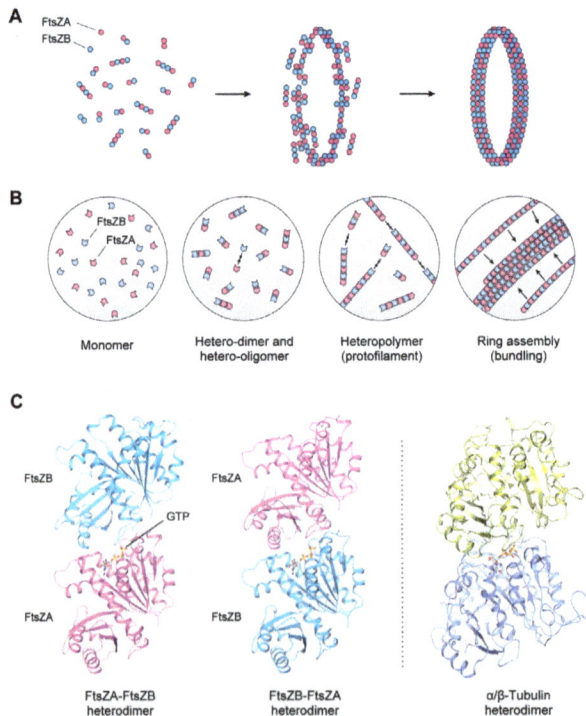

Figure 2. Assembly of the FtsZ ring. (**A,B**) FtsZ molecules assemble into hetero-oligomers, then these FtsZ protofilaments bundle and assemble into a ring structure in the stroma region. (**C**) The two types of FtsZs can assemble into heteropolymer structures via FtsZA-FtsZB and FtsZB-FtsZA hetero-interactions. The protein structure of the tubulin heterodimer (PDB: 1TUB) is also shown on the right. The protein structures of *A. thaliana* FtsZ2 (shown as FtsZA in the Figure) and FtsZ1 (FtsZB in the Figure) were obtained using homology modeling in the Modeller program [35], and structural data for each protein molecule were visualized using CueMol: Molecular Visualization Framework software (http://www.cuemol.org/). Reproduced and modified from Yoshida et al. (2016) [21].

Int. J. Mol. Sci. **2018**, *19*, 733

Although the mechanistic details of FtsZ ring assembly and dynamics remain unclear, recent studies using a heterologous yeast system revealed that the two types of plastid FtsZs spontaneously formed heteropolymers then assembled into a single ring that could generate contractile force in the absence of any other related proteins (Figure 2A,B) [21,36]. Interestingly, the plastid FtsZ heteropolymer had higher kinetic dynamics for protofilament assembly, mobility, and flexibility than either homopolymer, suggesting that the gene duplication of plastid *FtsZ* led to innovation in the kinetic functions of the FtsZ ring for plastid division [21]. The FtsZ heteropolymers are structurally similar to the microtubules comprised of α- and β-tubulin, suggesting the convergent evolution of functions in the plastid FtsZs and eukaryotic tubulins (Figure 2C). Consistent with molecular genetic studies using *A. thaliana*, the assembly of the plastid FtsZ ring in vivo is further promoted by several regulatory factors. ARC6 positively regulates the assembly of the FtsZ protofilament through interactions with the FtsZA molecules [6,8]. PARC6 also interacts with FtsZA, inducing the remodeling of the FtsZ protofilaments for ring formation with support from ARC3 [6,8,15]. Moreover, ARC3 and GIANT CHLOROPLAST1 (GC1) negatively regulate FtsZ ring assembly [37,38]; however, as these FtsZ-associated genes are not well conserved in the plant kingdom, another regulatory system with unknown factors might also coordinate the assembly of the FtsZ ring.

4. The Dynamin Ring

A member of the dynamin superfamily, Dnm2 (also known as DRP5B or ARC5 in *A. thaliana*), is also observed in a ring structure at the plastid-division site (Figure 1) [9,10]. Originally, the classical dynamin protein was identified as a 100-kDa mechanochemical GTPase required for the scission of clathrin-coated vesicles from the plasma membrane [39,40]. Dynamin-related proteins are now known to be involved in diverse membrane remodeling events; for example, Dnm1 (also known as DRP1 in animals) is involved in mitochondrial and peroxisome divisions [40,41]. The dynamin superfamily proteins possess several functional domain subunits, including a large GTPase domain (approximately 300 amino acids), a middle domain, a pleckstrin-homology (PH) domain, a GTPase effector domain (GED), and a carboxy-terminal proline-rich domain (PRD) (Figure 3A,B, upper) [40,42,43]. Although the molecular mass of Dnm2 is similar to that of classical dynamin, a conserved domain search identified only the GTPase domain at its amino-terminal region (Figure 3A). The three uncharacterized regions occupying the rest of the sequence are highly conserved between Dnm2 proteins in plant species, implying that these regions might be responsible for key roles during plastid division (Figure 3A,B, bottom).

The functional importance of Dnm2 proteins during plastid division has been demonstrated in *C. merolae*, the moss *Physcomitrella patens*, and *A. thaliana* [9,10,44]. Dnm2 proteins formed a discontinuous ring structure at the plastid-division site in the early phases of plastid division, after which they appeared to link and assemble into a single ring [9,10]. Surprisingly, the plastid-division dynamin-related proteins are phylogenetically related to a group of dynamin-related proteins involved in cytokinesis [45]. Considering that the primitive algal genome encodes only two dynamin-related proteins, Dnm2 for plastid division and Dnm1 for mitochondrial/peroxisome division [34,46,47], these findings raise the question of the original function of the ancestral dynamin protein.

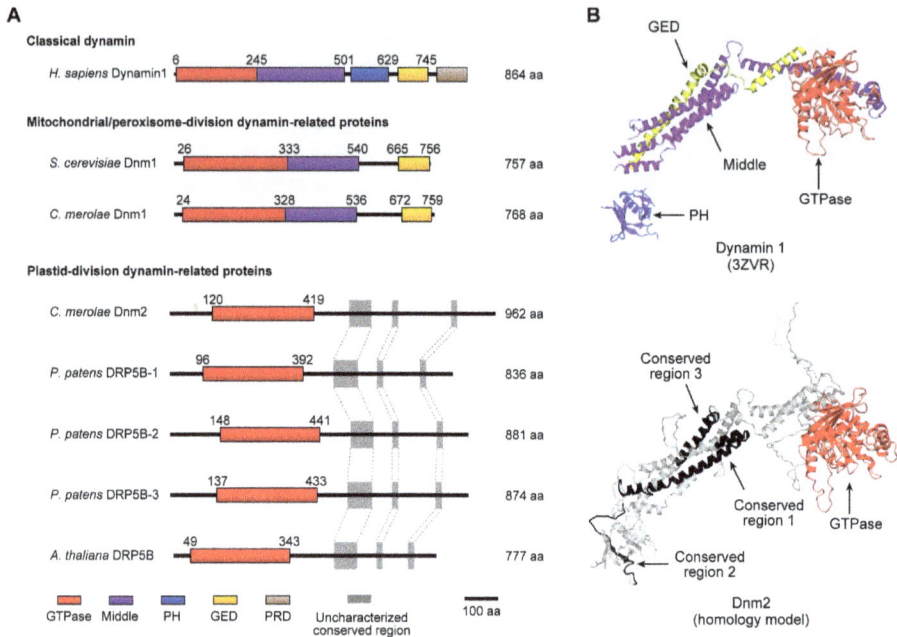

Figure 3. Structures of the dynamin superfamily. (**A**) Domain architectures of the dynamin superfamily. Dynamin 1 catalyzes clathrin-coated vesicle scission at the plasma membrane. Dnm1/DRP1 is involved in the division of mitochondria and peroxisomes. Dnm2 (also known as DRP5B/ARC5 in *A. thaliana* and moss) is involved in plastid division. GTPase domain (red); middle domain (purple); pleckstrin homology domain (PH, blue); GTPase effector domain (GED, yellow); and proline-rich domain (PRD, light brown). Domain architectures were identified using a conserved-domain search program [48]. (**B**) Protein structures of human dynamin 1 (classical dynamin) for vesicle scission and *C. merolae* Dnm2 for plastid division. The structure of dynamin 1 is represented with crystal structure data from an assembly-deficient dynamin 1 mutant, G397D ΔPRD (PDB: 3ZVR) [43], while the structure of Dnm2 is visualized using homology modeling based on the dynamin 1 structure. The functional domains in dynamin 1 are shown in red (GTPase domain), purple (middle domain), blue (PH domain), and yellow (GED domain); the proline-rich domain (PRD) is not shown. Uncharacterized conserved regions in Dnm2 are shown in black. The protein structure of Dnm2 was modeled as described in Figure 2.

5. The PD Ring

In the 1980s and 1990s, EM observations revealed the presence of the PD ring at the division site of plastids in numerous photosynthetic eukaryotes [5,49]. Two (or three) types of electron-dense specialized ring structures comprise the PD ring: the outer PD ring is the main skeletal structure of the plastid-division machinery, and is composed of a ring-shaped bundle of nanofilaments, 5 to 7 nm in width, on the cytosolic side of the outer envelope membrane (Figure 1) [13,23]. In addition, an inner 5-nm-thick belt-like PD ring forms on the stromal side of the inner envelope membrane [22,50]. The outer and inner PD rings have been widely observed in the plant kingdom [5,49], while a third, intermediate, PD ring has only been identified in the intermembrane space of *C. merolae* and the green alga *N. bacillaris* [24,51]. Many angiosperms also form double rings at the constricted isthmi of dividing plastids, including proplastids, amyloplasts, and chloroplasts [5,49]. The PD ring structure can be observed only during the late phases of plastid division in land plants [49,52], suggesting that the number and electron density of the PD rings in these organisms might be too low to detect during the early phases of plastid division using EM, making their morphological dynamics more difficult to study in these species. Interestingly, a phylogenetic study of the PD ring identified a clear trend, in which the PD rings of the primitive unicellular eukaryotes with smaller genomes are larger than those of the multiplastidic cells of land plants [5,49]; therefore, *C. merolae* is one of the most suitable organisms for studying the involvement of the PD ring in plastid division. Based on these previous studies, the PD rings, especially the outer rings, were concluded to be universal across the plant kingdom, where they play an important role in plastidokinesis.

Although the molecular components of the outer PD ring have not yet been elucidated, despite more than 25 years passing since its discovery, a chemical-staining screen shed light on this issue [12]. Periodic acid-horseradish peroxidase staining indicated that the outer PD ring is likely to contain saccharic components, which led us to perform a proteomic analysis of the isolated plastid-division machinery fraction and identify a novel glycosyltransferase protein, PDR1 [12]. Interestingly, the expression of *PDR1* was specifically detected in the plastid-division phase, and PDR1 proteins assembled a single-ring structure at the plastid-division site. Ultrastructural studies clearly showed that PDR1 proteins localized on the outer PD ring filaments. Furthermore, analyses of the components of the purified PD ring filaments revealed that glucan molecules were components of the outer PD ring. PDR1 has sequence similarity to glycogenin, which acts as a priming protein for glycogen biosynthesis (Figure 4A) [53]; therefore, it is now hypothesized that PDR1 can elongate the glucan chain to biosynthesize PD ring filaments from UDP-glucose molecules, analogous to the biosynthesis of glycogen (Figure 4B). Taking these findings together, although the biosynthesis mechanism is still unclear, PDR1 is probably involved in the biosynthesis of the PD ring polyglucan filaments. Potential orthologs of *PDR1* have also been identified in land plant genomes.

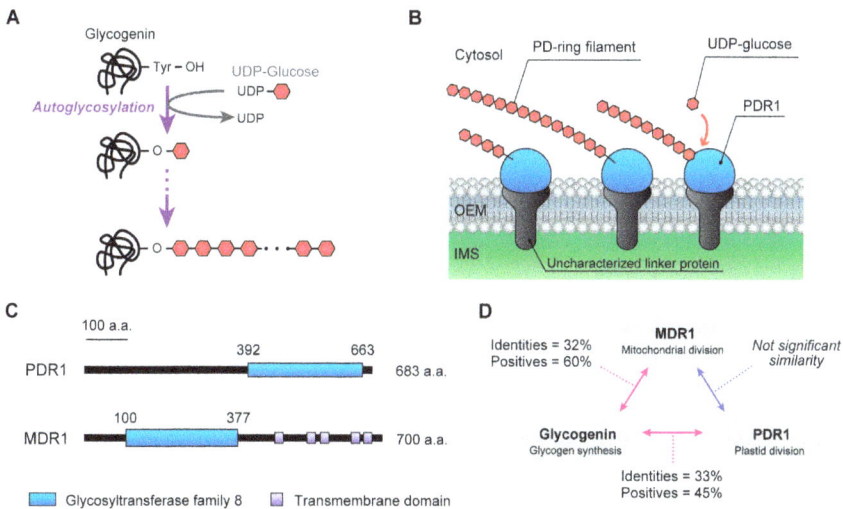

Figure 4. *Cont.*

E

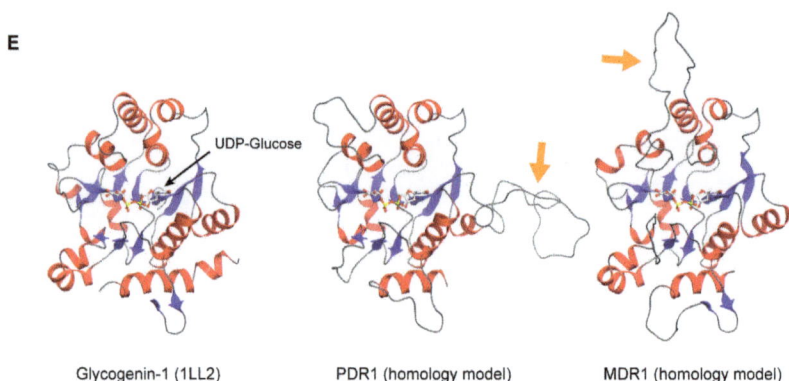

| Glycogenin-1 (1LL2) | PDR1 (homology model) | MDR1 (homology model) |

Figure 4. Working models of the glycosyltransferases glycogenin and PDR1. (**A**) Glycogenin is required for the initiation of glycogen biosynthesis, and can be autoglycosylated at a specific tyrosine residue to form a short oligosaccharide chain of glucose molecules to act as a priming chain for the subsequent biosynthesis of glycogen. (**B**) A schematic representation of PD ring filament biosynthesis by PDR1. A series of results suggested that PD ring filaments are composed of both PDR1 and glucose molecules. Considering the sequence similarity with glycogenin, the glycosyltransferase domain of PDR1 may biosynthesize the polyglucan nanofilaments from UDP-glucose residues to form the PD ring filaments. OEM, outer envelope membrane; IMS, intermembrane space. (**C**) Schematic of *C. merolae* PDR1 and MDR1 domain structures. The glycosyltransferase domains of PDR1 and MDR1 identified them as type-8 subgroup members of the glycosyltransferase family. (**D**) Protein sequence similarities between PDR1, MDR1 and glycogenin-1. (**E**) Comparisons of the protein structure of glycogenin-1 (PDB: 1LL2) and the putative structures of the glycosyltransferase domains of PDR1 and MDR1. Orange arrows indicate specific insertion regions in the glycosyltransferase domains of PDR1 and MDR1. The protein structures of the PDR1 and MDR1 glycosyltransferase domains were modeled as described in Figure 2.

6. The Homology between Plastid- and Mitochondrial-Division Machinery

Over the past 25 years, the mode of plastid division has been unveiled and several key components of the division machinery have been identified. Furthermore, studies have revealed that the other endosymbiotic organelles, mitochondria, also proliferate via the activity of a supramolecular complex, the mitochondrial-division machinery, which, like the plastid-division machinery, comprises an electron-dense specialized ring called the mitochondrion-dividing (MD) ring which is the counterpart of the PD ring in mitochondrial division, the FtsZ ring, and a dynamin ring (Figure 5) [1,54,55]. Interestingly, some of the mechanisms of mitochondrial division are also very similar to those of the plastid-division machinery. In addition, a recent multi-omics analysis of isolated plastid- and mitochondrial-division machineries showed that 185 proteins, including 54 uncharacterized proteins, were present in the fraction, indicating that many unknown components may be involved in these machineries [54]. Indeed, a glycosyltransferase homologous to PDR1, MITOCHONDRION-DIVIDING RING1 (MDR1) (Figure 4C), was identified from these candidates, and a series of analyses showed that MDR1 is required for the assembly of the MD ring, which also consists of polyglucan filaments and is required for mitochondrial division [54]. Despite the low sequence similarity between PDR1 and MDR1 (Figure 4D,E), these proteins both have a glycosyltransferase domain belongs to the type-8 subgroup of the glycosyltransferase family and they have homologous functions in plastid and mitochondrial division [54]. Given that both plastids and mitochondria evolved from free-living bacteria, the compelling structural and mechanical similarity between the plastid- and mitochondrial-division machineries indicates that they were established in host cells to dominate and control the proliferation of these endosymbiotic organelles during their early evolution.

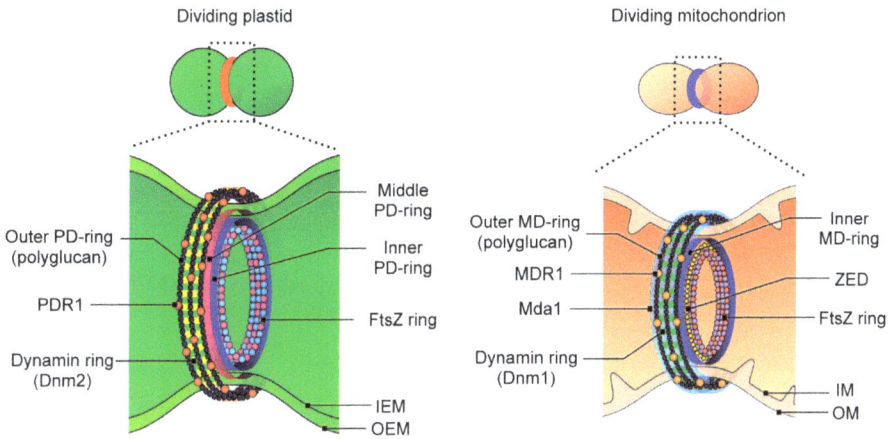

Figure 5. Schematic representations of the division machinery in plastids and mitochondria. IEM, inner envelope membrane; OEM, outer envelope membrane; IM, inner membrane; OM, outer membrane. For details on the mitochondrial-division machinery, see Refs. [1,54–58]. Modified from Yoshida et al. (2016) [21] and Yoshida et al. (2017) [54].

7. Conclusions and Perspectives

One of the current hot topics in this field is the elucidation of the system coordinating the cell-division cycle and the plastid/mitochondrial-division cycle [59–61]. Combinational analyses using genetic engineering and synchronized *C. merolae* cells has revealed the existence of a plastid-division checkpoint in the cell cycle, which is very likely to contribute to the permanent possession of plastids [19]; however, the fundamental mechanisms driving this process are still unclear. To identify the coordination system, the selection of suitable organisms for each analysis will be very important. The unicellular alga *C. merolae* and the land plant *A. thaliana* have enabled the molecular study of plastid division via their species-specific advantages. As a model organism, *A. thaliana* enabled many types of genetic studies to be conducted, leading to the identification of some of the genes responsible for plastid division (see for more details Chen et al. (2018) [6]). Studies in this area also benefit from using *C. merolae* cells, which can be synchronized to enable the isolation of intact plastid/mitochondrial-division machineries [13,54,62]. Furthermore, many genetic-engineering techniques have been recently established in *C. merolae* [18,63–66]. These innovations will enable dramatic advances in the investigation of the molecular mechanisms driving plastid division over the next decade; for example, an impressive recent study elucidated the crystal structure of the ARC6-PDV2 complex to reveal how protein-protein interactions translate information regarding plastid division across the double membrane, from the stromal region to the cytosol [17]. It is expected that the further identification of three-dimensional protein structures involved in the plastid-division machinery will open up a completely new avenue in the field.

Acknowledgments: This work was supported by Human Frontier Science Program Long-Term Fellowship LT000356/2011-L and a Japan Society for the Promotion of Science Postdoctoral Research Fellowship for Research Abroad.

Conflicts of Interest: The author declares no conflict of interest.

References

1. Osteryoung, K.W.; Nunnari, J. The division of endosymbiotic organelles. *Science* **2003**, *302*, 1698–1704. [CrossRef] [PubMed]

2. Kuroiwa, T.; Misumi, O.; Nishida, K.; Yagisawa, F.; Yoshida, Y.; Fujiwara, T.; Kuroiwa, H. Vesicle, mitochondrial, and plastid division machineries with emphasis on dynamin and electron-dense rings. *Int. Rev. Cell Mol. Biol.* **2008**, *271*, 97–152. [CrossRef] [PubMed]

3. Miyagishima, S.Y.; Nakanishi, H.; Kabeya, Y. Structure, regulation, and evolution of the plastid division machinery. *Int. Rev. Cell Mol. Biol.* **2011**, *291*, 115–153. [CrossRef] [PubMed]

4. Yoshida, Y.; Miyagishima, S.Y.; Kuroiwa, H.; Kuroiwa, T. The plastid-dividing machinery: Formation, constriction and fission. *Curr. Opin. Plant Biol.* **2012**, *15*, 714–721. [CrossRef] [PubMed]

5. Kuroiwa, T.; Kuroiwa, H.; Sakai, A.; Takahashi, H.; Toda, K.; Itoh, R. The division apparatus of plastids and mitochondria. *Int. Rev. Cytol.* **1998**, *181*, 1–41. [PubMed]

6. Chen, C.; MacCready, J.S.; Ducat, D.C.; Osteryoung, K.W. The molecular machinery of chloroplast division. *Plant Physiol.* **2018**, *176*, 138–151. [CrossRef] [PubMed]

7. Miyagishima, S.Y.; Nishida, K.; Kuroiwa, T. An evolutionary puzzle: Chloroplast and mitochondrial division rings. *Trends Plant Sci.* **2003**, *8*, 432–438. [CrossRef]

8. TerBush, A.D.; Yoshida, Y.; Osteryoung, K.W. FtsZ in chloroplast division: Structure, function and evolution. *Curr. Opin. Cell Biol.* **2013**, *25*, 461–470. [CrossRef] [PubMed]

9. Miyagishima, S.Y.; Nishida, K.; Mori, T.; Matsuzaki, M.; Higashiyama, T.; Kuroiwa, H.; Kuroiwa, T. A plant-specific dynamin-related protein forms a ring at the chloroplast division site. *Plant Cell Online* **2003**, *15*, 655–665. [CrossRef]

10. Gao, H.; Kadirjan-Kalbach, D.; Froehlich, J.E.; Osteryoung, K.W. ARC5, a cytosolic dynamin-like protein from plants, is part of the chloroplast division machinery. *Proc. Natl. Acad. Sci. USA* **2003**, *100*, 4328–4333. [CrossRef] [PubMed]

11. Osteryoung, K.W.; Pyke, K.A. Division and dynamic morphology of plastids. *Annu. Rev. Plant Biol.* **2014**, *65*, 443–472. [CrossRef] [PubMed]

12. Yoshida, Y.; Kuroiwa, H.; Misumi, O.; Yoshida, M.; Ohnuma, M.; Fujiwara, T.; Yagisawa, F.; Hirooka, S.; Imoto, Y.; Matsushita, K.; et al. Chloroplasts divide by contraction of a bundle of nanofilaments consisting of polyglucan. *Science* **2010**, *329*, 949–953. [CrossRef] [PubMed]

13. Yoshida, Y.; Kuroiwa, H.; Misumi, O.; Nishida, K.; Yagisawa, F.; Fujiwara, T.; Nanamiya, H.; Kawamura, F.; Kuroiwa, T. Isolated chloroplast division machinery can actively constrict after stretching. *Science* **2006**, *313*, 1435–1438. [CrossRef] [PubMed]

14. Glynn, J.M.; Yang, Y.; Vitha, S.; Schmitz, A.J.; Hemmes, M.; Miyagishima, S.-Y.; Osteryoung, K.W. PARC6, a novel chloroplast division factor, influences FtsZ assembly and is required for recruitment of PDV1 during chloroplast division in Arabidopsis. *Plant J.* **2009**, *59*, 700–711. [CrossRef] [PubMed]

15. Zhang, M.; Chen, C.; Froehlich, J.E.; Terbush, A.D.; Osteryoung, K.W. Roles of Arabidopsis PARC6 in coordination of the chloroplast division complex and negative regulation of FtsZ assembly. *Plant Physiol.* **2016**, *170*, 250–262. [CrossRef] [PubMed]

16. Glynn, J.M.; Froehlich, J.E.; Osteryoung, K.W. Arabidopsis ARC6 coordinates the division machineries of the inner and outer chloroplast membranes through interaction with PDV2 in the intermembrane space. *Plant Cell* **2008**, *20*, 2460–2470. [CrossRef] [PubMed]

17. Wang, W.; Li, J.; Sun, Q.; Yu, X.; Zhang, W.; Jia, N.; An, C.; Li, Y.; Dong, Y.; Han, F.; et al. Structural insights into the coordination of plastid division by the ARC6-PDV2 complex. *Nat. Plants* **2017**, *3*, 1–9. [CrossRef] [PubMed]

18. Sumiya, N.; Fujiwara, T.; Kobayashi, Y.; Misumi, O.; Miyagishima, S.Y. Development of a heat-shock inducible gene expression system in the red alga *Cyanidioschyzon merolae*. *PLoS ONE* **2014**, *9*, 1–11. [CrossRef] [PubMed]

19. Sumiya, N.; Fujiwara, T.; Era, A.; Miyagishima, S.Y. Chloroplast division checkpoint in eukaryotic algae. *Proc. Natl. Acad. Sci. USA* **2016**, *113*, E7629–E7638. [CrossRef] [PubMed]

20. Sumiya, N.; Miyagishim, S.Y. Hierarchal order in the formation of chloroplast division machinery in the red alga *Cyanidioschyzon merolae*. *Commun. Integr. Biol.* **2017**, *10*, 1–5. [CrossRef] [PubMed]

21. Yoshida, Y.; Mogi, Y.; TerBush, A.D.; Osteryoung, K.W. Chloroplast FtsZ assembles into a contractible ring via tubulin-like heteropolymerization. *Nat. Plants* **2016**, *2*, 16095. [CrossRef] [PubMed]

22. Miyagishima, S.Y.; Takahara, M.; Mori, T.; Kuroiwa, H.; Higashiyama, T.; Kuroiwa, T. Plastid division is driven by a complex mechanism that involves differential transition of the bacterial and eukaryotic division rings. *Plant Cell* **2001**, *13*, 2257–2268. [CrossRef]

23. Miyagishima, S.Y.; Takahara, M.; Kuroiwa, T. Novel filaments 5 nm in diameter constitute the cytosolic ring of the plastid division apparatus. *Plant Cell* **2001**, *13*, 707–721. [CrossRef] [PubMed]

24. Sumiya, N.; Hirata, A.; Kawano, S. Multiple FtsZ ring formation and reduplicated chloroplast DNA in *Nannochloris bacillaris* (Chlorophyta, Trebouxiophyceae) under phosphate-enriched culture. *J. Phycol.* **2008**, *44*, 1476–1489. [CrossRef] [PubMed]

25. Kuroiwa, H.; Mori, T.; Takahara, M.; Miyagishima, S.Y.; Kuroiwa, T. Chloroplast division machinery as revealed by immunofluorescence and electron microscopy. *Planta* **2002**, *215*, 185–190. [CrossRef] [PubMed]

26. Osteryoung, K.W.; Vierling, E. Conserved cell and organelle division. *Nature* **1995**, *376*, 473–474. [CrossRef] [PubMed]

27. Takahara, M.; Takahashi, H.; Matsunaga, S.Y.; Miyagishima, S.; Takano, H.; Sakai, A.; Kawano, S.; Kuroiwa, T. A putative mitochondrial FtsZ gene is present in the unicellular primitive red alga *Cyanidioschyzon merolae*. *Mol. Gen. Genet.* **2000**, *264*, 452–460. [CrossRef] [PubMed]

28. Mori, T.; Kuroiwa, H.; Takahara, M.; Miyagishima, S.Y.; Kuroiwa, T. Visualization of an FtsZ ring in chloroplasts of *Lilium longiflorum* leaves. *Plant Cell Physiol.* **2001**, *42*, 555–559. [CrossRef] [PubMed]

29. Vitha, S.; McAndrew, R.S.; Osteryoung, K.W. FtsZ ring formation at the chloroplast division site in plants. *J. Cell Biol.* **2001**, *153*, 111–120. [CrossRef] [PubMed]

30. Löwe, J.; Amos, L.A. Crystal structure of the bacterial cell-division protein FtsZ. *Nature* **1998**, *391*, 203–206. [CrossRef]

31. Oliva, M.A.; Cordell, S.C.; Löwe, J. Structural insights into FtsZ protofilament formation. *Nat. Struct. Mol. Biol.* **2004**, *11*, 1243–1250. [CrossRef] [PubMed]

32. Miyagishima, S.Y.; Nozaki, H.; Nishida, K.; Nishida, K.; Matsuzaki, M.; Kuroiwa, T. Two types of FtsZ proteins in mitochondria and red-lineage chloroplasts: The duplication of FtsZ is implicated in endosymbiosis. *J. Mol. Evol.* **2004**, *58*, 291–303. [CrossRef] [PubMed]

33. Schmitz, A.J.; Glynn, J.M.; Olson, B.J.S.C.; Stokes, K.D.; Osteryoung, K.W. Arabidopsis FtsZ2-1 and FtsZ2-2 are functionally redundant, but FtsZ-based plastid division is not essential for chloroplast partitioning or plant growth and development. *Mol. Plant* **2009**, *2*, 1211–1222. [CrossRef] [PubMed]

34. Matsuzaki, M.; Misumi, O.; Shin-I, T.; Maruyama, S.; Takahara, M.; Miyagishima, S.Y.; Mori, T.; Nishida, K.; Yagisawa, F.; Nishida, K.; et al. Genome sequence of the ultrasmall unicellular red alga *Cyanidioschyzon merolae* 10D. *Nature* **2004**, *428*, 653–657. [CrossRef] [PubMed]

35. Eswar, N.; John, B.; Mirkovic, N.; Fiser, A.; Ilyin, V.A.; Pieper, U.; Stuart, A.C.; Marti-Renom, M.A.; Madhusudhan, M.S.; Yerkovich, B.; et al. Tools for comparative protein structure modeling and analysis. *Nucleic Acids Res.* **2003**, *31*, 3375–3380. [CrossRef] [PubMed]

36. TerBush, A.D.; Osteryoung, K.W. Distinct functions of chloroplast FtsZ1 and FtsZ2 in Z-ring structure and remodeling. *J. Cell Biol.* **2012**, *199*, 623–637. [CrossRef] [PubMed]

37. Shimada, H.; Koizumi, M.; Kuroki, K.; Mochizuki, M.; Fujimoto, H.; Ohta, H.; Masuda, T.; Takamiya, K. ARC3, a chloroplast division factor, is a chimera of prokaryotic FtsZ and part of eukaryotic phosphatidylinositol-4-phosphate 5-kinase. *Plant Cell Physiol.* **2004**, *45*, 960–967. [CrossRef] [PubMed]

38. Maple, J.; Fujiwara, M.T.; Kitahata, N.; Lawson, T.; Baker, N.R.; Yoshida, S.; Møller, S.G. GIANT CHLOROPLAST 1 is essential for correct plastid division in *Arabidopsis*. *Curr. Biol.* **2004**, *14*, 776–781. [CrossRef] [PubMed]

39. Shpetner, H.S.; Vallee, R.B. Identification of dynamin, a novel mechanochemical enzyme that mediates interactions between microtubules. *Cell* **1989**, *59*, 421–432. [CrossRef]

40. Praefcke, G.J.K.; McMahon, H.T. The dynamin superfamily: Universal membrane tubulation and fission molecules? *Nat. Rev. Mol. Cell Biol.* **2004**, *5*, 133–147. [CrossRef] [PubMed]

41. Bleazard, W.; McCaffery, J.M.; King, E.J.; Bale, S.; Mozdy, A.; Tieu, Q.; Nunnari, J.; Shaw, J.M. The dynamin-related GTPase Dnm1 regulates mitochondrial fission in yeast. *Nat. Cell Biol.* **1999**, *1*, 298–304. [CrossRef] [PubMed]

42. Faelber, K.; Posor, Y.; Gao, S.; Held, M.; Roske, Y.; Schulze, D.; Haucke, V.; Noé, F.; Daumke, O. Crystal structure of nucleotide-free dynamin. *Nature* **2011**, *477*, 556–560. [CrossRef] [PubMed]

43. Ford, M.G.J.; Jenni, S.; Nunnari, J. The crystal structure of dynamin. *Nature* **2011**, *477*, 561–566. [CrossRef] [PubMed]

44. Sakaguchi, E.; Takechi, K.; Sato, H.; Yamada, T.; Takio, S.; Takano, H. Three dynamin-related protein 5B genes are related to plastid division in *Physcomitrella patens*. *Plant Sci.* **2011**, *180*, 789–795. [CrossRef] [PubMed]

45. Miyagishima, S.Y.; Kuwayama, H.; Urushihara, H.; Nakanishi, H. Evolutionary linkage between eukaryotic cytokinesis and chloroplast division by dynamin proteins. *Proc. Natl. Acad. Sci. USA* **2008**, *105*, 15202–15207. [CrossRef] [PubMed]

46. Nishida, K.; Takahara, M.; Miyagishima, S.Y.; Kuroiwa, H.; Matsuzaki, M.; Kuroiwa, T. Dynamic recruitment of dynamin for final mitochondrial severance in a primitive red alga. *Proc. Natl. Acad. Sci. USA* **2003**, *100*, 2146–2151. [CrossRef] [PubMed]

47. Imoto, Y.; Kuroiwa, H.; Yoshida, Y.; Ohnuma, M.; Fujiwara, T.; Yoshida, M.; Nishida, K.; Yagisawa, F.; Hirooka, S.; Miyagishima, S.; Misumi, O.; Kawano, S.; Kuroiwa, T. Single-membrane-bounded peroxisome division revealed by isolation of dynamin-based machinery. *Proc. Natl. Acad. Sci. USA* **2013**, *110*, 9583–9588. [CrossRef] [PubMed]

48. Marchler-Bauer, A.; Derbyshire, M.K.; Gonzales, N.R.; Lu, S.; Chitsaz, F.; Geer, L.Y.; Geer, R.C.; He, J.; Gwadz, M.; Hurwitz, D.I.; et al. CDD: NCBI's conserved domain database. *Nucleic Acids Res.* **2015**, *43*, D222–D226. [CrossRef] [PubMed]

49. Kuroiwa, T. The primitive red algae *Cyanidium caldarium* and *Cyanidioschyzon merolae* as model system for investigating the dividing apparatus of mitochondria and plastids. *BioEssays* **1998**, *20*, 344–354. [CrossRef]

50. Hashimoto, H. Double ring structure around the constricting neck of dividing plastids of *Avena sativa*. *Protoplasma* **1986**, *135*, 166–172. [CrossRef]

51. Miyagishima, S.Y.; Itoh, R.; Toda, K.; Takahashi, H.; Kuroiwa, H.; Kuroiwa, T. Identification of a triple ring structure involved in plastid division in the primitive red alga *Cyanidioschyzon merolae*. *J. Electron Microsc.* **1998**, *47*, 269–272. [CrossRef]

52. Kuroiwa, H.; Mori, T.; Takahara, M.; Miyagishima, S.Y.; Kuroiwa, T. Multiple FtsZ rings in a pleomorphic chloroplast in embryonic cap cells of *Pelargonium zonale*. *Cytologia* **2001**, *66*, 227–233. [CrossRef]

53. Lomako, J.; Lomako, W.M.; Whelan, W.J. Glycogenin: The primer for mammalian and yeast glycogen synthesis. *Biochim. Biophys. Acta* **2004**, *1673*, 45–55. [CrossRef] [PubMed]

54. Yoshida, Y.; Kuroiwa, H.; Shimada, T.; Yoshida, M.; Ohnuma, M.; Fujiwara, T.; Imoto, Y.; Yagisawa, F.; Nishida, K.; Hirooka, S.; et al. Glycosyltransferase MDR1 assembles a dividing ring for mitochondrial proliferation comprising polyglucan nanofilaments. *Proc. Natl. Acad. Sci. USA* **2017**, *114*, 13284–13289. [CrossRef] [PubMed]

55. Kuroiwa, T.; Nishida, K.; Yoshida, Y.; Fujiwara, T.; Mori, T.; Kuroiwa, H.; Misumi, O. Structure, function and evolution of the mitochondrial division apparatus. *Biochim. Biophys. Acta* **2006**, *1763*, 510–521. [CrossRef] [PubMed]

56. Van der Bliek, A.M.; Shen, Q.; Kawajiri, S. Mechanisms of mitochondrial fission and fusion. *Cold Spring Harb. Perspect. Biol.* **2013**, *5*, 1–16. [CrossRef] [PubMed]

57. Friedman, J.R.; Nunnari, J. Mitochondrial form and function. *Nature* **2014**, *505*, 335–343. [CrossRef] [PubMed]

58. Roy, M.; Reddy, P.H.; Iijima, M.; Sesaki, H. Mitochondrial division and fusion in metabolism. *Curr. Opin. Cell Biol.* **2015**, *33*, 111–118. [CrossRef] [PubMed]

59. Kobayashi, Y.; Kanesaki, Y.; Tanaka, A.; Kuroiwa, H.; Kuroiwa, T.; Tanaka, K. Tetrapyrrole signal as a cell-cycle coordinator from organelle to nuclear DNA replication in plant cells. *Proc. Natl. Acad. Sci. USA* **2009**, *106*, 803–807. [CrossRef] [PubMed]

60. Kobayashi, Y.; Imamura, S.; Hanaoka, M.; Tanaka, K. A tetrapyrrole-regulated ubiquitin ligase controls algal nuclear DNA replication. *Nat. Cell Biol.* **2011**, *13*, 483–487. [CrossRef] [PubMed]

61. Miyagishima, S.Y.; Fujiwara, T.; Sumiya, N.; Hirooka, S.; Nakano, A.; Kabeya, Y.; Nakamura, M. Translation-independent circadian control of the cell cycle in a unicellular photosynthetic eukaryote. *Nat. Commun.* **2014**, *5*, 3807. [CrossRef] [PubMed]

62. Suzuki, K.; Ehara, T.; Osafune, T.; Kuroiwa, H.; Kawano, S.; Kuroiwa, T. Behavior of mitochondria, chloroplasts and their nuclei during the mitotic cycle in the ultramicroalga *Cyanidioschyzon merolae*. *Eur. J. Cell Biol.* **1994**, *63*, 280–288. [PubMed]

63. Ohnuma, M.; Yokoyama, T.; Inouye, T.; Sekine, Y.; Tanaka, K. Polyethylene glycol (PEG)-mediated transient gene expression in a red alga, *Cyanidioschyzon merolae* 10D. *Plant Cell Physiol.* **2008**, *49*, 117–120. [CrossRef] [PubMed]

64. Ohnuma, M.; Misumi, O.; Fujiwara, T.; Watanabe, S.; Tanaka, K.; Kuroiwa, T. Transient gene suppression in a red alga, *Cyanidioschyzon merolae* 10D. *Protoplasma* **2009**, *236*, 107–112. [CrossRef] [PubMed]

65. Fujiwara, T.; Kanesaki, Y.; Hirooka, S.; Era, A.; Sumiya, N.; Yoshikawa, H.; Tanaka, K.; Miyagishima, S.Y. A nitrogen source-dependent inducible and repressible gene expression system in the red alga *Cyanidioschyzon merolae*. *Front. Plant Sci.* **2015**, *6*, 1–10. [CrossRef] [PubMed]

66. Fujiwara, T.; Ohnuma, M.; Kuroiwa, T.; Ohbayashi, R.; Hirooka, S.; Miyagishima, S.Y. Development of a double nuclear gene-targeting method by two-step transformation based on a newly established chloramphenicol-selection system in the red alga *Cyanidioschyzon merolae*. *Front. Plant Sci.* **2017**, *8*, 1–10. [CrossRef] [PubMed]

International Journal of
Molecular Sciences

MDPI

Review

Bacterial Heterologous Expression System for Reconstitution of Chloroplast Inner Division Ring and Evaluation of Its Contributors

Hiroki Irieda [1],* and Daisuke Shiomi [2],*

[1]   Academic Assembly, Institute of Agriculture, Shinshu University, Nagano 399-4598, Japan
[2]   Department of Life Science, College of Science, Rikkyo University, Tokyo 171-8501, Japan
*   Correspondence: irieda@shinshu-u.ac.jp (H.I.); dshiomi@rikkyo.ac.jp (D.S.);
     Tel.: +81-265-77-1428 (H.I.); +81-339-85-2401 (D.S.)

Received: 20 January 2018; Accepted: 8 February 2018; Published: 11 February 2018

Abstract: Plant chloroplasts originate from the symbiotic relationship between ancient free-living cyanobacteria and ancestral eukaryotic cells. Since the discovery of the bacterial derivative *FtsZ* gene—which encodes a tubulin homolog responsible for the formation of the chloroplast inner division ring (Z ring)—in the *Arabidopsis* genome in 1995, many components of the chloroplast division machinery were successively identified. The knowledge of these components continues to expand; however, the mode of action of the chloroplast dividing system remains unknown (compared to bacterial cell division), owing to the complexities faced in in planta analyses. To date, yeast and bacterial heterologous expression systems have been developed for the reconstitution of Z ring-like structures formed by chloroplast FtsZ. In this review, we especially focus on recent progress of our bacterial system using the model bacterium *Escherichia coli* to dissect and understand the chloroplast division machinery—an evolutionary hybrid structure composed of both bacterial (inner) and host-derived (outer) components.

Keywords: chloroplast division; Z ring; membrane-tethering; heterologous expression; *E. coli*; AtFtsZ1; AtFtsZ2; ARC6; ARC3

1. Introduction

Plant chloroplasts evolved from free-living cyanobacteria through primary endosymbiosis, which started with an engulfment of ancient cyanobacteria by ancestral eukaryotic cells approximately a billion years ago [1,2]. As in bacteria, the proliferation of chloroplasts is achieved by binary fission via a hybrid division machinery comprised of inner (stromal) bacterial and outer (cytosolic) host-derived elements [3–5]. This machinery mainly consists of four rings: two (the Z ring and the inner plastid-dividing (PD) ring) are inside, while the other two (the dynamin-related protein5B (DRP5B) ring and the outer PD ring) are outside the chloroplast [3,6,7]. Of those, the Z ring is broadly conserved from bacteria to chloroplasts [3,4,8–11]. Each Z ring is composed of FtsZ homolog tubulin-like GTPase, and is believed to generate a constrictive force for division both in bacteria and chloroplasts of *Arabidopsis thaliana* (although it has also been reported that the motive force was provided by the outer DRP5B ring, but not inner Z ring, in chloroplast of the red alga *Cyanidioschyzon merolae*) [8–10,12–15]. While the four rings and other division-related components might coordinately function to constrict the mid-chloroplast, the initial and critical event is the assembly of FtsZ into the Z ring just beneath the inner envelope membrane (IEM) at the future site of division. The Z ring then works as a scaffold, to which other components are recruited in a specific order to drive a division complex [4].

In contrast to the bacterial Z ring that is composed of a single FtsZ protein, the components of the chloroplast Z ring are two phylogenetically distinct FtsZ proteins, FtsZ1 and FtsZ2, which heteropolymerize into FtsZ filaments in vivo and in vitro [15–19]. FtsZ1 probably emerged from FtsZ2 through a gene duplication event, because the C-terminal amino acid sequence (which is critical for the membrane-tethering in bacterial FtsZ—see below) is conserved only in FtsZ2 [20,21]. FtsZ1 and FtsZ2 show high amino acid sequence identity and similarity regarding their GTPase core domain with bacterial FtsZ, but play distinct roles in the formation of the FtsZ polymers; FtsZ2 dominantly forms the backbone of the filament, while FtsZ1 assists in its remodeling [16,19,22]. In *A. thaliana*, FtsZ2 was additionally duplicated into two functionally redundant paralogs: AtFtsZ2-1 and AtFtsZ2-2. These two paralogs are functionally interchangeable with respect to in vivo chloroplast division activity, although the distinct contributions by these two AtFtsZ2 conforming to the shape of the chloroplast have also been reported [22,23].

The assembly and dynamics of the chloroplast Z ring is elaborately regulated by many components that negatively or positively affect the Z ring formation [3–5]. The stromal protein Accumulation and Replication of Chloroplasts 3 (ARC3) directly interacts with both AtFtsZ1 and AtFtsZ2 and inhibits the assembly of the Z ring at non-division sites [19,24–26], resembling the function of bacterial division inhibitor protein MinC for the positioning of the Z ring [27], whereas the IEM-spanning protein Accumulation and Replication of Chloroplasts 6 (ARC6) directly interacts only with AtFtsZ2 and promotes Z ring assembly in the stroma [22,26,28,29]. ARC3 and ARC6 were identified in the native AtFtsZ1-AtFtsZ2 complex isolated from *Arabidopsis* chloroplast [30]. Another IEM-spanning protein Paralog of ARC6 (PARC6) also directly interacts with AtFtsZ2, and in addition to the stromal proteins MinD and MinE, indirectly affects the Z ring formation through a direct interaction with the inhibitor protein ARC3 [25,26,31–34]. Briefly, in the working model of Z ring regulation in chloroplast division, ARC6 promotes the formation of a Z ring composed of AtFtsZ1-AtFtsZ2 heteropolymer, possibly by the tethering of AtFtsZ2 to the IEM. ARC3, MinD, and MinE act together as a Z ring positioning system and accurately confine the Z ring to the mid-chloroplast. During the remodeling and constriction of the Z ring, ARC3 may also function as an inhibitor of Z ring assembly after being recruited by PARC6 to the division site [3–5]. For an in-depth review of the many contributors to the chloroplast Z ring dynamics that include bacterial- and host-derivatives, we refer readers to previously published reviews [3–5].

To understand the chloroplast division comprehensively, in planta molecular analysis of Z ring assembly and dynamics is important. However, it can be challenging owing to the complexity of the plant cell, wherein many division-related components act together. Furthermore, plant breeding and genetic manipulation require more time compared to model microorganisms, even in the model plant *A. thaliana*. This situation has led to the development of some heterologous expression systems for *Arabidopsis* FtsZ proteins and other related components using single-celled model microorganisms, such as the yeasts *Schizosaccharomyces pombe* and *Pichia pastoris*, as well as the bacterium *Escherichia coli* [15,19,26,35]. The fission yeast *S. pombe* system was established as a cellular model for the functional analysis of bacterial actin-related protein MreB and FtsZ ahead of chloroplast FtsZ [36,37]. At present, together with recent methylotrophic yeast *P. pastoris* system, the yeast systems have shown the value of using heterologous expression systems for chloroplast division-related proteins—particularly filament and ring formation by FtsZs—to analyze their inherent functions [15,19,26,35]. On the other hand, based on the evolutionary background of the chloroplast and the fact that the Z ring-driven division system indeed functions in bacteria, as well as other practical advantages of a model bacterium, the *E. coli* system could be a good tool for the research of chloroplast FtsZ. However, the previous report showed that the chloroplast FtsZ produced in *E. coli* cells did not successfully form the Z ring or Z ring-like structure, but only formed long filaments and aberrant clusters; therefore, this system is lagging behind yeast expression systems [15,19,26,35].

Recently, we progressively developed the *E. coli* system to reconstitute Z ring or Z ring-like structures composed of the *A. thaliana* FtsZ protein AtFtsZ2-1 (hereafter called AtFtsZ2) [38]. Our system plausibly reflects the dynamic properties of AtFtsZ2, where the AtFtsZ2 assembles into long filaments or Z ring-like structures depending on the conditions. In Figure 1, we summarize the proposed molecular mechanism of action of AtFtsZ2 and its positive contributor ARC6, which has been demonstrated to anchor the chloroplast Z ring to the membrane in our system. In the following sections, we describe the development of this system and the important factors contributing to the filament morphology of AtFtsZ2, which includes N-terminal extended region of AtFtsZ2, membrane-tethering of the AtFtsZ2 filament, the negative regulator ARC3, and the positive regulators ARC6 and AtFtsZ1.

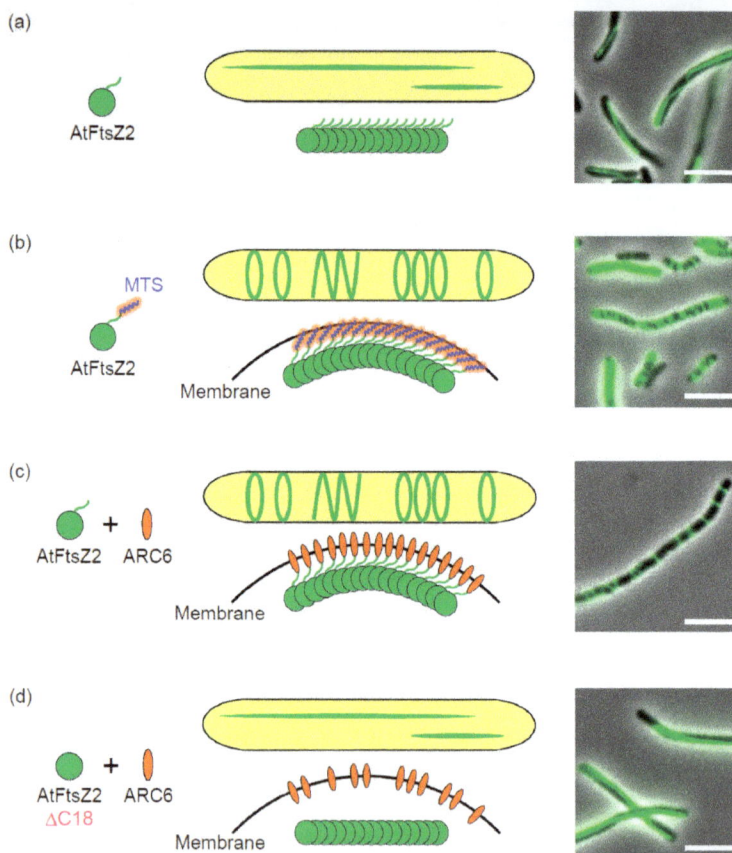

Figure 1. The formation of Z ring-like structures of *Arabidopsis* chloroplast FtsZ2 in the bacterial heterologous expression system. Schematic illustration of the proposed molecular behavior of chloroplast division-related components in *E. coli* cells and its merged microscopic image of phase-contrast and GFP are shown when expressing (a) super folder GFP (sfGFP)-AtFtsZ2, (b) sfGFP-AtFtsZ2-2MTS, (c) sfGFP-AtFtsZ2, and Accumulation and Replication of Chloroplasts 6 (ARC6) and (d) sfGFP-AtFtsZ2ΔC18 (C-terminal 18-residue truncated form of AtFtsZ2) and ARC6. *E. coli* cells were grown in L broth (1% bactotryptone, 0.5% yeast extract, 0.5% NaCl) to the stationary phase at 22 °C. Scale bars: 5 μm. MTS: membrane-targeting sequence. To reduce the complexity in the diagrams, bundling of the FtsZ2 filaments was omitted.

2. Optimization of *E. coli* System for Heterologous Expression of Chloroplast FtsZ

2.1. Fluorescent Tagging of AtFtsZ2 and Culture Condition

FtsZ proteins can polymerize, assemble into filament bundles, and eventually form the Z ring in the division system [3–5,8,9,39]. In each organism (bacterium or plant), visualizing the FtsZ protein using immunoelectron microscopy, immunofluorescence, or fluorescent protein (FP) labeling techniques clearly showed the Z ring formation at the division site [40–47]. In particular, FP labeling enables us to monitor the FtsZ protein dynamics in live cells. As for heterologous expression systems for chloroplast FtsZs tagged with FPs, in *S. pombe* yeast cells, AtFtsZ1 and/or AtFtsZ2 formed linear and ring-shaped filaments that were free-floating in the cytosol [19,35]. Furthermore, AtFtsZ1 and/or forcibly membrane-targeted AtFtsZ2 expressed in *P. pastoris* yeast cells assembled into ring structures, and the rings including membrane-tethered AtFtsZ2 showed contractible ability [15]. Thus, yeast systems can be available for the analysis of chloroplast FtsZs.

By contrast, as mentioned above, heterologous expression of AtFtsZ2 in *E. coli* cells only showed long filaments with aberrant clusters at 42 °C [19]. In bacterial cells, recombinant proteins frequently aggregate into inclusion-bodies because of high growth temperature and/or high-level expression of the heterologous protein [48]. In this context, we confirmed that the expression level of AtFtsZ2 protein introduced in *E. coli* was not too high, and successfully removed aberrant aggregation of AtFtsZ2 by decreasing the growth temperature to 22 °C—an optimal temperature for *A. thaliana* (Figure 1a) [38]. We also found that the growth phase of the bacteria strongly affected the AtFtsZ2 filamentation; sampling in the stationary phase is more suitable to form long filaments than in the logarithmic phase [38]. Some theories for this phenomenon include (i) reduced dynamics and a lower turnover rate of AtFtsZ2 compared with those of *E. coli* FtsZ (EcFtsZ) [19], and (ii) less competition with actively assembling EcFtsZ during the stationary phase. In the paper, we critically investigated the fusion terminus of FP [38]. In most papers to-date, irrespective of the derivatives from bacteria and plants, FPs were tagged at the C-terminus of FtsZ. On the other hand, as previously reported, we confirmed that N-terminal FP fusions of EcFtsZ were also precisely localized at the middle of the *E. coli* cell [38,49,50]. As for chloroplast FtsZ, we conducted heterologous expression of both N- and C-terminal FP-fused AtFtsZ2 and concluded that—at least in our *E. coli* system—filamentation of C-terminally FP-tagged AtFtsZ2 was repressed compared with that of an N-terminally fused one (Figure 1a) [38]. Since the C-terminal domain of FtsZ is important for its function, C-terminal FP-tagging might partially interfere with the filamentation ability of AtFtsZ2 [19,51]. Furthermore, the C-terminal FP fusions of AtFtsZ2 showed aberrant aggregations at higher temperature (37 °C) compared to N-terminal FP fusions [38]. As a consequence, we selected N-terminal FP-fused AtFtsZ2 for our expression system. Importantly, C-terminal FP fusions of FtsZ are generally assembly-competent, and the AtFtsZ2-FP fusion forms a Z ring at the middle of the chloroplast in *A. thaliana* [19,52]. Thus, this issue might need to be discussed. Many chloroplast proteins—including FtsZs—are encoded in the nuclear genome and are eventually transported into the chloroplast via its N-terminal transit peptide [12,45,46,53]. The transit peptide is cleaved upon its import into the chloroplast, and this is one reason why the C-terminal FP-fusion of AtFtsZ was preferred for in planta analysis [12,47]. Therefore, if the N-terminal FP-fused FtsZ is expressed in planta, it is necessary that the FP does not disturb the function of the transit peptide. In heterologous expression systems, the transit peptide-lacking chloroplast FtsZs are generally used, and this issue does not need to be considered.

2.2. N-Terminal Region of AtFtsZ2

Besides transit peptide, the chloroplast AtFtsZ2 protein harbors an extended N-terminal region compared with eubacterial FtsZ proteins [35,38]. AtFtsZ1 lacks this "extended" region, but many cyanobacterial and chloroplast FtsZ proteins exhibit N-terminal extension regardless of their amino acid sequence conservation, possibly implying the additional trait(s) in these FtsZs [38]. In the *S. pombe* yeast system, it has been reported that the N-terminal-extended region of AtFtsZ2 promotes its polymer bundling and turnover, suggesting that the N-terminus of AtFtsZ2—as with its C-terminus—is also important for its function [35]. Consistent with this, we confirmed the dependency of AtFtsZ2 filamentation on its N-terminus in *E. coli* cells, where the N-terminally-truncated AtFtsZ2 (here we deleted amino acid residues from 49 to 112 in addition to the transit peptide, AtFtsZ2ΔN) with

N-terminal FP showed considerably shorter filaments than did full-length FP-AtFtsZ2 fusion [38]. This suggests that the behavior of AtFtsZ2 produced in *E. coli* cytosol reflects its inherent properties.

2.3. Membrane-Tethering of AtFtsZ2

To establish the *E. coli* reconstitution system completely for the analysis of chloroplast division-related components, the formation of the Z ring composed of AtFtsZ2 in *E. coli* cells is critical. The FtsZ protein itself has no membrane spanning or anchoring domains and requires the cognate membrane associating and/or transmembrane proteins FtsA and ZipA in *E. coli* and Ftn2/ZipN in cyanobacteria, which interact with the C-terminus of FtsZ and target it to the membrane [42,54–62]. In green lineage chloroplasts, IEM protein ARC6—an Ftn2/ZipN ortholog—was believed to tether the Z ring, though no direct evidence has been reported for this so far [22,28,29,63,64]. *E. coli* has no ARC6 homolog, and FtsA and ZipA might not target AtFtsZ2 to the membrane, despite partial conservation of the FtsA-interacting sequence in the AtFtsZ2 C-terminus, which was supported by the fact that AtFtsZ2 did not form ring-like structures (Figure 1a) [38].

Previous reconstitution systems using liposomes revealed that the membrane-tethering of FtsZ is required for Z ring formation, in which reconstitutions of contractile EcFtsZ ring were achieved by artificial membrane-tethering with the C-terminal membrane-targeting sequence (MTS) or natural tethering through a co-introduced FtsA. This membrane anchored EcFtsZ could actually constrict a liposome, indicating that membrane-tethering is critical to form the *E. coli* Z ring and generate a constriction force [14,62]. Similarly, in our *E. coli* system, an artificial membrane-tethering of AtFtsZ2 by MTS gave AtFtsZ2 the ability to form multiple Z ring-like structures in both wild-type and *ftsZ*-depleted *E. coli*, indicating the intrinsic property of AtFtsZ2 to form Z ring-like structures in *E. coli* cells (Figure 1a,b). However, these Z ring-like structures did not constrict a cell (probably because AtFtsZ2 could not interact with FtsA and ZipA, which stabilize Z ring composed of EcFtsZ) [38]. Around the same time, it was independently shown that the MTS-tagged chloroplast FtsZ derived from *Galdieria sulphuraria* thermophilic red alga (GsFtsZ) formed multiple Z ring-like structures in *E. coli* cells [65]. Together with the recent success in the *P. pastoris* yeast system that showed reconstitution of the ring-like structure of AtFtsZ2 by MTS-tagging [15], these reports strongly demonstrated that membrane-tethering is a necessary and sufficient factor for bacterial and chloroplast FtsZ proteins to form Z ring or Z ring-like structures.

Interestingly, the diameter of the ring-like structure formed by membrane-tethered AtFstZ2 was much larger than that of EcFtsZ when reconstituted in *P. pastoris* cells, resembling the size of their corresponding rings in vivo [15]. This has suggested that the structure of each FtsZ protein determines the curvature, and consequently the size, of each ring [15]. However, we successfully reconstituted the Z ring-like structures of AtFtsZ2 in *E. coli* cells [38]. Furthermore, in *A. thaliana*, the chloroplast size increases during leaf development, and leaf epidermal cells contained small chloroplasts with smaller AtFtsZ1 and AtFtsZ2 rings compared to leaf mesophyll cells [52,66]. Thus, we presume that FtsZ proteins have the potential to form Z rings with various diameters according to the cell or chloroplast diameter.

3. The Function of Negative and Positive Contributors in Bacterial Reconstitution Systems

3.1. The Negative Regulator ARC3

In rod-shaped bacteria such as *E. coli* and *Bacillus subtilis*, mid-cell positioning of the Z ring is tightly regulated by the Min system, in which the negative regulator MinC inhibits FtsZ polymerization at cell poles [67,68]. Spatial regulation of the Z ring by MinC is also conserved in cyanobacteria [69]. In contrast, except for certain algal lineages and the moss *Physcomitrella patens*, the chloroplast has no MinC homolog, but instead acquired the plant-specific stromal protein ARC3 as a functional analog of bacterial MinC [19,24–26,52,70,71]. Indeed, *Arabidopsis arc3* mutants showed multiple Z rings and nonuniform chloroplast size and number, whereas ARC3-overexpressing mutants exhibited a small number of enlarged chloroplasts with fragmented AtFtsZ filaments [25,26,72,73]. ARC3 directly interacts with both AtFtsZ1 and AtFtsZ2, and these interactions were inhibited by the C-terminal membrane-occupation-and-recognition nexus (MORN) domain of ARC3 [25,26]. The MORN domain is a binding site of PARC6, which is believed to recruit and activate ARC3 at the chloroplast division site [32,34]. In a yeast heterologous expression system which does not contain the PARC6 homolog, recombinant ARC3 lacking the MORN domain was used to analyze the ARC3 inhibitory effects on AtFtsZ filaments [19,26]. We also co-expressed this mutant ARC3 and AtFtsZ2 with an FP in *E. coli* cells and evaluated its function in our bacterial system. Consistent with previous reports, we confirmed the inhibition of the AtFtsZ2 assembly by ARC3 regardless of the presence or absence of the MTS tag (Figure 2a,b) [38]. Since it has already been reported in the yeast system that ARC3 inhibited the assembly of cytosolic free-floating AtFtsZ filaments (linear and ring-shaped structures), our bacterial system presented the first example of ARC3 in inhibiting AtFtsZ filaments in membrane-tethered Z ring-like structures in a heterologous expression system (Figure 2b) [19,26,38]. It is worth noting that in our *E. coli* system— like the yeast systems—there might be no factors that affect ARC3 behavior. The consistency of the inhibitory effects of ARC3 on FtsZ filament assembly among yeast, *E. coli*, and in planta analyses strongly demonstrates the clear function of ARC3 in chloroplast Z ring regulation. This is further supported by a recent study in which in vitro assays showed that ARC3 promoted AtFtsZ2 debundling and disassembly by enhancing its GTPase activity and 3D reconstruction using single-particle analysis, suggesting that PARC6 mediated ARC3–AtFtsZ2 interaction [74]. Bacterial MinC also promotes the debundling and disassembly of FtsZ, but does not affect its GTPase activity [27]. In addition, chloroplast ARC3 binds to both MinD and MinE, but bacterial MinC only binds to MinD [25,75]. The analogous function of ARC3 to MinC is indisputable, but there might be differences between ARC3 and MinC in their mode of action in each division system.

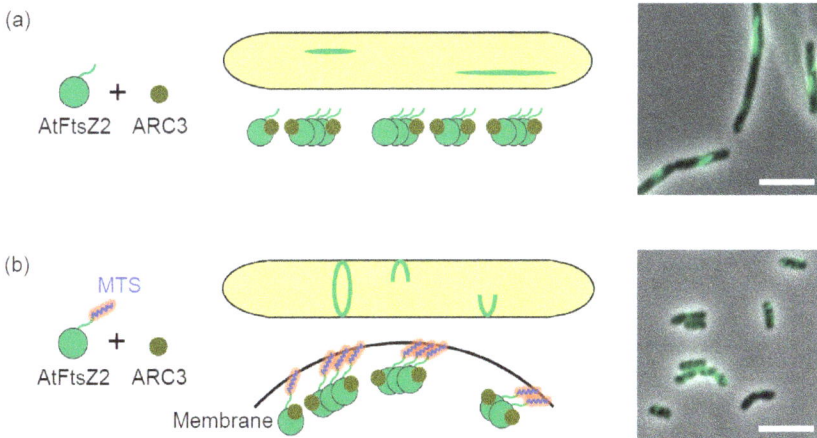

Figure 2. The effects of Accumulation and Replication of Chloroplasts 3 (ARC3) on the filaments of *Arabidopsis* chloroplast FtsZ2 in the bacterial heterologous expression system. Schematic illustration of the proposed molecular behavior of chloroplast division-related components in *E. coli* cells and its merged microscopic image of phase-contrast and GFP are shown when expressing (**a**) sfGFP-AtFtsZ2 and ARC3, and (**b**) sfGFP-AtFtsZ2-2MTS and ARC3. *E. coli* cells were grown in L broth (1% bactotryptone, 0.5% yeast extract, 0.5% NaCl) to the stationary phase at 22 °C. Scale bars: 5 μm. MTS: membrane-targeting sequence. To reduce the complexity in the diagrams, bundling of the FtsZ filaments was omitted.

3.2. The Positive Regulator ARC6

Using liposomes and purified EcFtsZ proteins, Osawa and Erickson (2013) demonstrated that the reconstitution of the Z ring (or Z ring-like structure) membrane-tethered by its natural partner is one goal for protein-free or heterologous expression systems in order to study the molecular mechanisms of the Z ring-centered division machinery [62]. In the case of chloroplast division machinery, a Z ring-anchoring factor has not yet been identified, but a great deal of indirect evidence implied the IEM protein ARC6 as a potential candidate [22,28,29,34,45,63,64]. In *Arabidopsis*, FP-labeled ARC6 concentrated at the chloroplast constriction sites in the shape of a ring, and ARC6 mutants exhibited a Z ring-defective phenotype, consequently leading to a small number of enlarged chloroplasts [28,45,63,64]. Using the yeast two-hybrid system, a direct interaction between ARC6 and the C-terminal conserved sequence of AtFtsZ2 was demonstrated [22,29,34]. FP-labeled ARC6 and AtFtsZ2 co-localized in the yeast cytosol, which mostly depended on the AtFtsZ2 C-terminus [34]. Collectively, all these reports revealed that ARC6 is a positive regulator of Z ring formation in chloroplasts. Therefore, the next challenge will test whether ARC6 truly anchors the chloroplast Z ring to the membrane.

We applied Osawa and Erickson's strategy to our *E. coli* heterologous expression system, where the MTS-untagged AtFtsZ2 (which itself can only form linear filaments) and ARC6 were co-expressed, and evaluated the effects of ARC6 on AtFtsZ2 filament morphology [38]. Fortunately, our challenge was a success—AtFtsZ2 polymer drastically altered its morphology from the linear filaments to Z ring-like or helical structures dependent on ARC6 (Figure 1a,c). FP-labeling of both AtFtsZ2 and ARC6 showed co-localization of these two proteins in the ring-like structures [38]. These Z ring-like structures completely depended on the ARC6-interacting sequence at the AtFtsZ2 C-terminus (here we truncated 18 amino acids, AtFtsZ2ΔC18), suggesting ARC6-mediated tethering of AtFtsZ2 filaments to the membrane (Figure 1c,d). Membrane-fractionation assays further supported the membrane attachment of AtFtsZ2 by ARC6 through their direct interaction [38]. The C-terminal region of ARC6 protrudes into the chloroplast IEM and directly interacts with the outer envelope membrane (OEM) protein Plastid Division 2 (PDV2), being able to transfer the Z ring positioning information from the stromal division machinery to the cytosolic one [76,77]. Together with previous results, our data obtained from the bacterial reconstitution system clarified that the other N-terminal side of ARC6 interacts with AtFtsZ2—a backbone protein of the Z ring through which ARC6 directly anchors the Z ring to IEM [22,28,29,34,38,45,63,64]. Bacteria such as *E. coli* and *B. subtilis* have no ARC6 ortholog, but the membrane-interacting protein FtsA interacts with FtsZ and anchors it to the membrane, hence stabilizing the Z ring [42,57,58,62,78,79]. By contrast, cyanobacteria uniquely evolved Ftn2/ZipN—an ancestor of chloroplast ARC6—as a functional analog of FtsA for Z ring-tethering [59–61]. The successful reconstitution of chloroplast FtsZ2 ring in bacterial cells by membrane-tethering through the chloroplast ARC6 indicates high stability and plasticity of the Z ring-centered division machinery that is conserved from bacteria to chloroplasts.

Additionally, in the *S. pombe* yeast expression system, it has been demonstrated that ARC6 stabilizes AtFtsZ2 filaments independent of its tethering ability [35]. On the other hand, a recent report revealed a new function of the *E. coli* FtsA in aligning FtsZ protofilaments in the unbundled state and stabilizing them, in addition to its membrane-tethering ability [80]. Thus, these functional analogs commonly work for Z ring-tethering but have additional function(s) as a positive regulator in each division system.

3.3. The Positive Regulator AtFtsZ1

In our paper, we described the unexpected function of AtFtsZ1 in positively contributing to the filament morphology of AtFtsZ2 in the bacterial expression system [38]. As mentioned above, the chloroplast Z ring is composed of AtFtsZ2 and AtFtsZ1, but the former dominantly determines the filament morphology, while the latter plays a regulatory role [15,19]. It is worth noting that the *Arabidopsis ftsZ1* mutant (like *ftsZ2* mutants) showed a small number of enlarged chloroplasts, implying the indispensable function of AtFtsZ1 in chloroplast division, although it has also been observed that chloroplasts in the *ftsZ1* mutant still exhibited a single mid-plastid constriction [22,26,30,81]. As expected, FP-labeled AtFtsZ1 and AtFtsZ2 co-localized in the *E. coli* expression system, consistent with previous observations that AtFtsZ1 and AtFtsZ2 can form a heteropolymer both in vitro and in yeast systems (Figure 3) [15,17–19,38]. However, the AtFtsZ1 surprisingly induced a morphological transformation of AtFtsZ2 filaments into ring-like and helical structures in *E. coli* cells, resembling Z ring-like structures tethered by ARC6 (Figure 3) [38]. This phenomenon was also observed in the case of AtFtsZ2ΔC18 [38]. AtFtsZ1 possesses no C-terminal sequences responsible for membrane-tethering, and independently expressed AtFtsZ1 showed only a dispersed pattern in the bacterial cytoplasm, which suggests that AtFtsZ1 itself is unlikely to interact with any of the *E. coli* endogenous components supporting membrane-tethering [38]. Thus, it still remains an open question as to how AtFtsZ2 filaments form Z ring-like structures depending on AtFtsZ1 in the bacterial expression system. In the *P. pastoris* yeast system, a ring-like structure of MTS-untagged AtFtsZ1 was observed in the absence of other related component(s), although we did not detect any AtFtsZ1 filaments or rings in *E. coli* cells [15,38]. As for membrane-tethering, these data led to the speculation that AtFtsZ1 itself may interact with the membrane. Nevertheless, it is clear that AtFtsZ1 works positively to form the chloroplast Z ring. The conservation of two distinct FtsZ proteins in green lineage indicates a unique mechanism to regulate chloroplast Z ring assembly and dynamics compared to bacterial cell division. We hope future studies will reveal unknown mechanisms of action of AtFtsZ1 apart from its ability to increase the FtsZ filament turnover rate [19].

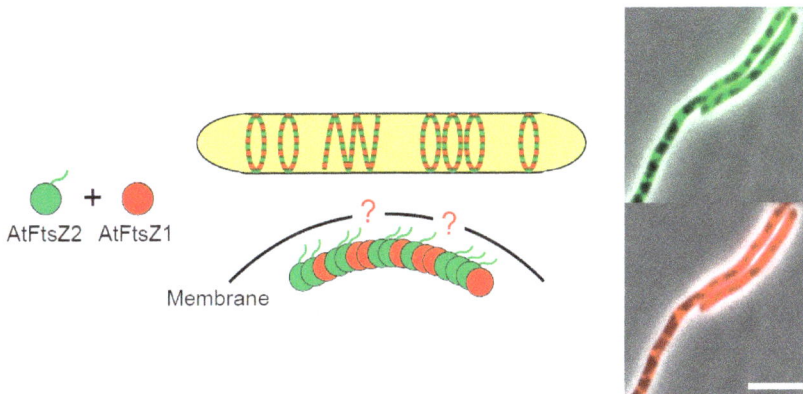

Figure 3. The effects of *Arabidopsis* chloroplast FtsZ1 on the formation of Z ring-like structures of *Arabidopsis* chloroplast FtsZ2 in the bacterial heterologous expression system. Schematic illustration of the Z ring-like structures composed of AtFtsZ1 and AtFtsZ2 heterooligomer in *E. coli* cells, although the mechanism by which the *Arabidopsis* FtsZ filaments are tethered to the membrane is unclear (indicated by red question marks in the illustration), and its merged microscopic images of phase-contrast and GFP (upper panel), and phase-contrast and mCherry (lower panel) are shown. *E. coli* cells expressing sfGFP-AtFtsZ2 and mCherry-AtFtsZ1 were grown in L broth (1% bactotryptone, 0.5% yeast extract, 0.5% NaCl) to the stationary phase at 22 °C. Scale bar: 5 μm. To reduce the complexity in the diagram, bundling of the FtsZ filaments was omitted.

4. Conclusions

The *E. coli* reconstitution system has now become one of the many useful heterologous expression systems for studying the FtsZ-centered chloroplast division machinery. Although in planta analysis is the best way to comprehensively examine the components employed in chloroplast division, heterologous expression systems lacking such native division-related contributors also provide important insights into inherent properties of the introduced component(s). In general, microbial manipulations have practical advantages, such as rapid growth, axenic culture, and genetic accessibility. Yeast systems allow the advantage of a eukaryotic environment completely lacking plant derivatives. The advantages of bacterial systems are as follows: (i) bacteria is an evolutionary progenitor of chloroplast and bacterial cytosol, wherein the Z ring-centered division system still works and is topologically equivalent to chloroplast stroma; and (ii) bacteria—especially *E. coli*—are distantly related to cyanobacteria, and partly lack the homologs of chloroplast division-related proteins. Indeed, using the *E. coli* system, we reconfirmed the negative regulation of the chloroplast FtsZ2 filaments by plant-specific ARC3 and found that this regulation was also effective for the membrane-tethered one (Z ring-like structures) [38]. We also directly demonstrated the Z ring-tethering ability of ARC6, which is unique to cyanobacteria and chloroplasts [38]. The components of the chloroplast division machinery in the stromal side (including IEM) have changed over the course of evolution [3–5]. Besides ARC3 and ARC6, the homologous proteins of PARC6 and MULTIPLE CHLOROPLAST DIVISION SITE 1 (MCD1) are absent in *E. coli* [32–34,82]; hence, they are capable of being applied to our bacterial system. Additionally, bacteria-derived chloroplast MinD and MinE can be analyzed by using the minCDE deletion mutant of *E. coli*, which creates minicells but also shows normal- and long-sized cells, as in the case of AtFtsZ2 expressed in the ftsZ-depleted *E. coli* strain [38,83]. By contrast, in the bacterial system (and in the yeast systems), the analyses of division-related components employed in the plant cytosolic side (including OEM), such as DRP5B, Plastid Division 1 (PDV1), and PDV2, appear more challenging [7,13,76,77,84,85]. Accordingly, the integration of data obtained from yeast and *E. coli* systems into in planta results is important, and will continue to accelerate the research on chloroplast division.

Acknowledgments: We thank all the members of Cellular Function Laboratory (Shiomi lab) for the helpful discussions. This work was supported by the Strategic Research Foundation Grant-aided Project for Private Universities (S1201003 to Daisuke Shiomi) from the Ministry of Education, Culture, Sports, Science and Technology, Japan.

Author Contributions: Hiroki Irieda and Daisuke Shiomi contributed to the writing of this review.

Conflicts of Interest: The authors declare no conflict of interest.

References

1. Gould, S.B.; Waller, R.F.; McFadden, G.I. Plastid evolution. *Annu. Rev. Plant Biol.* **2008**, *59*, 491–517. [CrossRef] [PubMed]
2. Keeling, P.J. The number, speed, and impact of plastid endosymbiosis in eukaryotic evolution. *Annu. Rev. Plant Biol.* **2013**, *64*, 583–607. [CrossRef] [PubMed]
3. Miyagishima, S.Y.; Nakanishi, H.; Kabeya, Y. Structure, regulation, and evolution of the plastid division machinery. *Int. Rev. Cell Mol. Biol.* **2011**, *291*, 115–153. [CrossRef] [PubMed]
4. Osteryoung, K.W.; Pyke, K.A. Division and dynamic morphology of plastids. *Annu. Rev. Plant Biol.* **2014**, *65*, 443–472. [CrossRef] [PubMed]
5. Chen, C.; MacCready, J.S.; Ducat, D.C.; Osteryoung, K.W. The molecular machinery of chloroplast division. *Plant Physiol.* **2017**. [CrossRef] [PubMed]
6. Kuroiwa, T.; Kuroiwa, H.; Sakai, A.; Takahashi, H.; Toda, K.; Itoh, R. The division apparatus of plastids and mitochondria. *Int. Rev. Cytol.* **1998**, *181*, 1–41. [CrossRef] [PubMed]
7. Yoshida, Y.; Kuroiwa, H.; Misumi, O.; Yoshida, M.; Ohnuma, M.; Fujiwara, T.; Yagisawa, F.; Hirooka, S.; Imoto, Y.; Matsushita, K.; et al. Chloroplasts divide by contraction of a bundle of nanofilaments consisting of polyglucan. *Science* **2010**, *329*, 949–953. [CrossRef] [PubMed]
8. Adams, D.W.; Errington, J. Bacterial cell division: Assembly, maintenance and disassembly of the Z ring. *Nat. Rev. Microbiol.* **2009**, *7*, 642–653. [CrossRef] [PubMed]
9. Erickson, H.P.; Anderson, D.E.; Osawa, M. FtsZ in bacterial cytokinesis: Cytoskeleton and force generator all in one. *Microbiol. Mol. Biol. Rev.* **2010**, *74*, 504–528. [CrossRef] [PubMed]

10. Mingorance, J.; Rivas, G.; Vélez, M.; Gómez-Puertas, P.; Vicente, M. Strong FtsZ is with the force: Mechanisms to constrict bacteria. *Trends Microbiol.* **2010**, *18*, 348–356. [CrossRef] [PubMed]

11. Terbush, A.D.; MacCready, J.S.; Chen, C.; Ducat, D.C.; Osteryoung, K.W. Conserved dynamics of chloroplast cytoskeletal FtsZ proteins across photosynthetic lineages. *Plant Physiol.* **2018**, *176*, 295–306. [CrossRef] [PubMed]

12. Osteryoung, K.W.; Vierling, E. Conserved cell and organelle division. *Nature* **1995**, *376*, 473–474. [CrossRef] [PubMed]

13. Yoshida, Y.; Kuroiwa, H.; Misumi, O.; Nishida, K.; Yagisawa, F.; Fujiwara, T.; Nanamiya, H.; Kawamura, F.; Kuroiwa, T. Isolated chloroplast division machinery can actively constrict after stretching. *Science* **2006**, *313*, 1435–1438. [CrossRef] [PubMed]

14. Osawa, M.; Anderson, D.E.; Erickson, H.P. Reconstitution of contractile FtsZ rings in liposomes. *Science* **2008**, *320*, 792–794. [CrossRef] [PubMed]

15. Yoshida, Y.; Mogi, Y.; TerBush, A.D.; Osteryoung, K.W. Chloroplast FtsZ assembles into a contractible ring via tubulin-like heteropolymerization. *Nat. Plants* **2016**, *2*, 16095. [CrossRef] [PubMed]

16. Osteryoung, K.W.; Stokes, K.D.; Rutherford, S.M.; Percival, A.L.; Lee, W.Y. Chloroplast division in higher plants requires members of two functionally divergent gene families with homology to bacterial FtsZ. *Plant Cell* **1998**, *10*, 1991–2004. [CrossRef] [PubMed]

17. Olson, B.J.; Wang, Q.; Osteryoung, K.W. GTP-dependent heteropolymer formation and bundling of chloroplast FtsZ1 and FtsZ2. *J. Biol. Chem.* **2010**, *285*, 20634–20643. [CrossRef] [PubMed]

18. Smith, A.G.; Johnson, C.B.; Vitha, S.; Holzenburg, A. Plant FtsZ1 and FtsZ2 expressed in a eukaryotic host: GTPase activity and self-assembly. *FEBS Lett.* **2010**, *584*, 166–172. [CrossRef] [PubMed]

19. TerBush, A.D.; Osteryoung, K.W. Distinct functions of chloroplast FtsZ1 and FtsZ2 in Z ring structure and remodeling. *J. Cell Biol.* **2012**, *199*, 623–637. [CrossRef] [PubMed]

20. Miyagishima, S.Y.; Nozaki, H.; Nishida, K.; Nishida, K.; Matsuzaki, M.; Kuroiwa, T. Two types of FtsZ proteins in mitochondria and red-lineage chloroplasts: The duplication of FtsZ is implicated in endosymbiosis. *J. Mol. Evol.* **2004**, *58*, 291–303. [CrossRef] [PubMed]

21. TerBush, A.D.; Yoshida, Y.; Osteryoung, K.W. FtsZ in chloroplast division: Structure, function and evolution. *Curr. Opin. Cell Biol.* **2013**, *25*, 461–470. [CrossRef] [PubMed]

22. Schmitz, A.J.; Glynn, J.M.; Olson, B.J.; Stokes, K.D.; Osteryoung, K.W. *Arabidopsis* FtsZ2-1 and FtsZ2-2 are functionally redundant, but FtsZ-based plastid division is not essential for chloroplast partitioning or plant growth and development. *Mol. Plant* **2009**, *2*, 1211–1222. [CrossRef] [PubMed]

23. Karamoko, M.; El-Kafafi, E.S.; Mandaron, P.; Lerbs-Mache, S.; Falconet, D. Multiple FtsZ2 isoforms involved in chloroplast division and biogenesis are developmentally associated with thylakoid membranes in *Arabidopsis*. *FEBS Lett.* **2011**, *585*, 1203–1208. [CrossRef] [PubMed]

24. Shimada, H.; Koizumi, M.; Kuroki, K.; Mochizuki, M.; Fujimoto, H.; Ohta, H.; Masuda, T.; Takamiya, K. ARC3, a chloroplast division factor, is a chimera of prokaryotic FtsZ and part of eukaryotic phosphatidylinositol-4-phosphate 5-kinase. *Plant Cell Physiol.* **2004**, *45*, 960–967. [CrossRef] [PubMed]

25. Maple, J.; Vojta, L.; Soll, J.; Møller, S.G. ARC3 is a stromal Z ring accessory protein essential for plastid division. *EMBO Rep.* **2007**, *8*, 293–299. [CrossRef] [PubMed]

26. Zhang, M.; Schmitz, A.J.; Kadirjan-Kalbach, D.K.; TerBush, A.D.; Osteryoung, K.W. Chloroplast division protein ARC3 regulates chloroplast FtsZ ring assembly and positioning in *Arabidopsis* through interaction with FtsZ2. *Plant Cell* **2013**, *25*, 1787–1802. [CrossRef] [PubMed]

27. Hu, Z.; Mukherjee, A.; Pichoff, S.; Lutkenhaus, J. The MinC component of the division site selection system in *Escherichia coli* interacts with FtsZ to prevent polymerization. *Proc. Natl. Acad. Sci. USA* **1999**, *96*, 14819–14824. [CrossRef] [PubMed]

28. Vitha, S.; Froehlich, J.E.; Koksharova, O.; Pyke, K.A.; van Erp, H.; Osteryoung, K.W. ARC6 is a J-domain plastid division protein and an evolutionary descendant of the cyanobacterial cell division protein Ftn2. *Plant Cell* **2003**, *15*, 1918–1933. [CrossRef] [PubMed]

29. Maple, J.; Aldridge, C.; Møller, S.G. Plastid division is mediated by combinatorial assembly of plastid division proteins. *Plant J.* **2005**, *43*, 811–823. [CrossRef] [PubMed]

30. McAndrew, R.S.; Olson, B.J.; Kadirjan-Kalbach, D.K.; Chi-Ham, C.L.; Vitha, S.; Froehlich, J.E.; Osteryoung, K.W. In vivo quantitative relationship between plastid division proteins FtsZ1 and FtsZ2 and identification of ARC6 and ARC3 in a native FtsZ complex. *Biochem. J.* **2008**, *412*, 367–378. [CrossRef] [PubMed]

31. Maple, J.; Chua, N.H.; Møller, S.G. The topological specificity factor AtMinE1 is essential for correct plastid division site placement in *Arabidopsis*. *Plant J.* **2002**, *31*, 269–277. [CrossRef] [PubMed]

32. Glynn, J.M.; Yang, Y.; Vitha, S.; Schmitz, A.J.; Hemmes, M.; Miyagishima, S.Y.; Osteryoung, K.W. PARC6, a novel chloroplast division factor, influences FtsZ assembly and is required for recruitment of PDV1 during chloroplast division in Arabidopsis. *Plant J.* **2009**, *59*, 700–711. [CrossRef] [PubMed]

33. Zhang, M.; Hu, Y.; Jia, J.; Li, D.; Zhang, R.; Gao, H.; He, Y. CDP1, a novel component of chloroplast division site positioning system in *Arabidopsis*. *Cell Res.* **2009**, *19*, 877–886. [CrossRef] [PubMed]

34. Zhang, M.; Chen, C.; Froehlich, J.E.; TerBush, A.D.; Osteryoung, K.W. Roles of Arabidopsis PARC6 in coordination of the chloroplast division complex and negative regulation of FtsZ assembly. *Plant Physiol.* **2016**, *170*, 250–262. [CrossRef] [PubMed]

35. TerBush, A.D.; Porzondek, C.A.; Osteryoung, K.W. Functional analysis of the chloroplast division complex using *Schizosaccharomyces pombe* as a heterologous expression system. *Microsc. Microanal.* **2016**, *22*, 275–289. [CrossRef] [PubMed]

36. Srinivasan, R.; Mishra, M.; Murata-Hori, M.; Balasubramanian, M.K. Filament formation of the Escherichia coli actin-related protein, MreB, in fission yeast. *Curr. Biol.* **2007**, *17*, 266–272. [CrossRef] [PubMed]

37. Srinivasan, R.; Mishra, M.; Wu, L.; Yin, Z.; Balasubramanian, M.K. The bacterial cell division protein FtsZ assembles into cytoplasmic rings in fission yeast. *Genes Dev.* **2008**, *22*, 1741–1746. [CrossRef] [PubMed]

38. Irieda, H.; Shiomi, D. ARC6-mediated Z ring-like structure formation of prokaryote-descended chloroplast FtsZ in *Escherichia coli*. *Sci. Rep.* **2017**, *7*, 3492. [CrossRef] [PubMed]

39. Haeusser, D.P.; Margolin, W. Splitsville: Structural and functional insights into the dynamic bacterial Z ring. *Nat. Rev. Microbiol.* **2016**, *14*, 305–319. [CrossRef] [PubMed]

40. Bi, E.F.; Lutkenhaus, J. FtsZ ring structure associated with division in *Escherichia coli*. *Nature* **1991**, *354*, 161–164. [CrossRef] [PubMed]

41. Levin, P.A.; Losick, R. Transcription factor Spo0A switches the localization of the cell division protein FtsZ from a medial to a bipolar pattern in *Bacillus subtilis*. *Genes Dev.* **1996**, *10*, 478–488. [CrossRef] [PubMed]

42. Ma, X.; Ehrhardt, D.W.; Margolin, W. Colocalization of cell division proteins FtsZ and FtsA to cytoskeletal structures in living *Escherichia coli* cells by using green fluorescent protein. *Proc. Natl. Acad. Sci. USA* **1996**, *93*, 12998–13003. [CrossRef] [PubMed]

43. Addinall, S.G.; Bi, E.; Lutkenhaus, J. FtsZ ring formation in *fts* mutants. *J. Bacteriol.* **1996**, *178*, 3877–3884. [CrossRef] [PubMed]

44. Sun, Q.; Margolin, W. FtsZ dynamics during the division cycle of live *Escherichia coli* cells. *J. Bacteriol.* **1998**, *180*, 2050–2056. [PubMed]

45. McAndrew, R.S.; Froehlich, J.E.; Vitha, S.; Stokes, K.D.; Osteryoung, K.W. Colocalization of plastid division proteins in the chloroplast stromal compartment establishes a new functional relationship between FtsZ1 and FtsZ2 in higher plants. *Plant Physiol.* **2001**, *127*, 1656–1666. [CrossRef] [PubMed]

46. Mori, T.; Kuroiwa, H.; Takahara, M.; Miyagishima, S.Y.; Kuroiwa, T. Visualization of an FtsZ ring in chloroplasts of *Lilium longiflorum* leaves. *Plant Cell Physiol.* **2001**, *42*, 555–559. [CrossRef] [PubMed]

47. Vitha, S.; McAndrew, R.S.; Osteryoung, K.W. FtsZ ring formation at the chloroplast division site in plants. *J. Cell Biol.* **2001**, *153*, 111–120. [CrossRef] [PubMed]

48. Baneyx, F.; Mujacic, M. Recombinant protein folding and misfolding in *Escherichia coli*. *Nat. Biotechnol.* **2004**, *22*, 1399–1408. [CrossRef] [PubMed]

49. Bernhardt, T.G.; de Boer, P.A. SlmA, a nucleoid-associated, FtsZ binding protein required for blocking septal ring assembly over chromosomes in *E. coli*. *Mol. Cell* **2005**, *18*, 555–564. [CrossRef] [PubMed]

50. Osawa, M.; Erickson, H.P. Probing the domain structure of FtsZ by random truncation and insertion of GFP. *Microbiology* **2005**, *151*, 4033–4043. [CrossRef] [PubMed]

51. Ma, X.; Margolin, W. Genetic and functional analyses of the conserved C-terminal core domain of *Escherichia coli* FtsZ. *J. Bacteriol.* **1999**, *181*, 7531–7544. [PubMed]

52. Johnson, C.B.; Shaik, R.; Abdallah, R.; Vitha, S.; Holzenburg, A. FtsZ1/FtsZ2 turnover in chloroplasts and the role of ARC3. *Microsc. Microanal.* **2015**, *21*, 313–323. [CrossRef] [PubMed]

53. Fujiwara, M.; Yoshida, S. Chloroplast targeting of chloroplast division FtsZ2 proteins in *Arabidopsis*. *Biochem. Biophys. Res. Commun.* **2001**, *287*, 462–467. [CrossRef] [PubMed]

54. Hale, C.A.; de Boer, P.A. Direct binding of FtsZ to ZipA, an essential component of the septal ring structure that mediates cell division in *E. coli*. *Cell* **1997**, *88*, 175–185. [CrossRef]

55. Liu, Z.; Mukherjee, A.; Lutkenhaus, J. Recruitment of ZipA to the division site by interaction with FtsZ. *Mol. Microbiol.* **1999**, *31*, 1853–1861. [CrossRef] [PubMed]

56. Mosyak, L.; Zhang, Y.; Glasfeld, E.; Haney, S.; Stahl, M.; Seehra, J.; Somers, W.S. The bacterial cell-division protein ZipA and its interaction with an FtsZ fragment revealed by X-ray crystallography. *EMBO J.* **2000**, *19*, 3179–3191. [CrossRef] [PubMed]

57. Pichoff, S.; Lutkenhaus, J. Unique and overlapping roles for ZipA and FtsA in septal ring assembly in *Escherichia coli*. *EMBO J.* **2002**, *21*, 685–693. [CrossRef] [PubMed]

58. Pichoff, S.; Lutkenhaus, J. Tethering the Z ring to the membrane through a conserved membrane targeting sequence in FtsA. *Mol. Microbiol.* **2005**, *55*, 1722–1734. [CrossRef] [PubMed]

59. Koksharova, O.A.; Wolk, C.P. A novel gene that bears a DnaJ motif influences cyanobacterial cell division. *J. Bacteriol.* **2002**, *184*, 5524–5528. [CrossRef] [PubMed]

60. Mazouni, K.; Domain, F.; Cassier-Chauvat, C.; Chauvat, F. Molecular analysis of the key cytokinetic components of cyanobacteria: FtsZ, ZipN and MinCDE. *Mol. Microbiol.* **2004**, *52*, 1145–1158. [CrossRef] [PubMed]

61. Marbouty, M.; Saguez, C.; Cassier-Chauvat, C.; Chauvat, F. ZipN, an FtsA-like orchestrator of divisome assembly in the model cyanobacterium *Synechocystis* PCC6803. *Mol. Microbiol.* **2009**, *74*, 409–420. [CrossRef] [PubMed]

62. Osawa, M.; Erickson, H.P. Liposome division by a simple bacterial division machinery. *Proc. Natl. Acad. Sci. USA* **2013**, *110*, 11000–11004. [CrossRef] [PubMed]

63. Pyke, K.A.; Rutherford, S.M.; Robertson, E.J.; Leech, R.M. *arc6*, a fertile *Arabidopsis* mutant with only two mesophyll cell chloroplasts. *Plant Physiol.* **1994**, *106*, 1169–1177. [CrossRef] [PubMed]

64. Johnson, C.B.; Tang, L.K.; Smith, A.G.; Ravichandran, A.; Luo, Z.; Vitha, S.; Holzenburg, A. Single particle tracking analysis of the chloroplast division protein FtsZ anchoring to the inner envelope membrane. *Microsc. Microanal.* **2013**, *19*, 507–512. [CrossRef] [PubMed]

65. Chen, Y.; Porter, K.; Osawa, M.; Augustus, A.M.; Milam, S.L.; Joshi, C.; Osteryoung, K.W.; Erickson, H.P. The chloroplast tubulin homologs FtsZA and FtsZB from the red alga *Galdieria sulphuraria* co-assemble into dynamic filaments. *J. Biol. Chem.* **2017**, *292*, 5207–5215. [CrossRef] [PubMed]

66. Okazaki, K.; Kabeya, Y.; Suzuki, K.; Mori, T.; Ichikawa, T.; Matsui, M.; Nakanishi, H.; Miyagishima, S. The PLASTID DIVISION1 and 2 Components of the Chloroplast Division Machinery Determine the Rate of Chloroplast Division in Land Plant Cell Differentiation. *Plant Cell* **2009**, *21*, 1769–1780. [CrossRef] [PubMed]

67. Lutkenhaus, J. Assembly dynamics of the bacterial MinCDE system and spatial regulation of the Z ring. *Annu. Rev. Biochem.* **2007**, *76*, 539–562. [CrossRef] [PubMed]

68. Rowlett, V.W.; Margolin, W. The bacterial Min system. *Curr. Biol.* **2013**, *23*, R553–R556. [CrossRef] [PubMed]

69. MacCready, J.S.; Schossau, J.; Osteryoung, K.W.; Ducat, D.C. Robust Min-system oscillation in the presence of internal photosynthetic membranes in cyanobacteria. *Mol. Microbiol.* **2017**, *103*, 483–503. [CrossRef] [PubMed]

70. Yang, Y.; Glynn, J.M.; Olson, B.J.; Schmitz, A.J.; Osteryoung, K.W. Plastid division: Across time and space. *Curr. Opin. Plant Biol.* **2008**, *11*, 577–584. [CrossRef] [PubMed]

71. Miyagishima, S.Y.; Kabeya, Y. Chloroplast division: Squeezing the photosynthetic captive. *Curr. Opin. Microbiol.* **2010**, *13*, 738–746. [CrossRef] [PubMed]

72. Pyke, K.A.; Leech, R.M. Chloroplast division and expansion is radically altered by nuclear mutations in *Arabidopsis thaliana*. *Plant Physiol.* **1992**, *99*, 1005–1008. [CrossRef] [PubMed]

73. Glynn, J.M.; Miyagishima, S.Y.; Yoder, D.W.; Osteryoung, K.W.; Vitha, S. Chloroplast division. *Traffic* **2007**, *8*, 451–461. [CrossRef] [PubMed]

74. Shaik, R.S.; Sung, M.W.; Vitha, S.; Holzenburg, A. Chloroplast division protein ARC3 acts on FtsZ2 by preventing filament bundling and enhancing GTPase activity. *Biochem. J.* **2018**, *475*, 99–115. [CrossRef] [PubMed]

75. Huang, J.; Cao, C.; Lutkenhaus, J. Interaction between FtsZ and inhibitors of cell division. *J. Bacteriol.* **1996**, *178*, 5080–5085. [CrossRef] [PubMed]

76. Glynn, J.M.; Froehlich, J.E.; Osteryoung, K.W. *Arabidopsis* ARC6 coordinates the division machineries of the inner and outer chloroplast membranes through interaction with PDV2 in the intermembrane space. *Plant Cell* **2008**, *20*, 2460–2470. [CrossRef] [PubMed]

77. Wang, W.; Li, J.; Sun, Q.; Yu, X.; Zhang, W.; Jia, N.; An, C.; Li, Y.; Dong, Y.; Han, F.; et al. Structural insights into the coordination of plastid division by the ARC6-PDV2 complex. *Nat. Plants* **2017**, *3*, 17011. [CrossRef] [PubMed]

78. Wang, X.; Huang, J.; Mukherjee, A.; Cao, C.; Lutkenhaus, J. Analysis of the interaction of FtsZ with itself, GTP, and FtsA. *J. Bacteriol.* **1997**, *179*, 5551–5559. [CrossRef] [PubMed]

79. Feucht, A.; Lucet, I.; Yudkin, M.D.; Errington, J. Cytological and biochemical characterization of the FtsA cell division protein of *Bacillus subtilis*. *Mol. Microbiol.* **2001**, *40*, 115–125. [CrossRef] [PubMed]

80. Krupka, M.; Rowlett, V.W.; Morado, D.; Vitrac, H.; Schoenemann, K.; Liu, J.; Margolin, W. *Escherichia coli* FtsA forms lipid-bound minirings that antagonize lateral interactions between FtsZ protofilaments. *Nat. Commun.* **2017**, *8*, 15957. [CrossRef] [PubMed]

81. Yoder, D.W.; Kadirjan-Kalbach, D.; Olson, B.J.; Miyagishima, S.Y.; Deblasio, S.L.; Hangarter, R.P.; Osteryoung, K.W. Effects of mutations in Arabidopsis *FtsZ1* on plastid division, FtsZ ring formation and positioning, and FtsZ filament morphology in vivo. *Plant Cell Physiol.* **2007**, *48*, 775–791. [CrossRef] [PubMed]

82. Nakanishi, H.; Suzuki, K.; Kabeya, Y.; Miyagishima, S.Y. Plant-specific protein MCD1 determines the site of chloroplast division in concert with bacteria-derived MinD. *Curr. Biol.* **2009**, *19*, 151–156. [CrossRef] [PubMed]

83. De Boer, P.A.; Crossley, R.E.; Rothfield, L. A division inhibitor and a topological specificity factor coded for by the minicell locus determine proper placement of the division septum in *E. coli*. *Cell* **1989**, *56*, 641–649. [CrossRef]

84. Gao, H.; Kadirjan-Kalbach, D.; Froehlich, J.E.; Osteryoung, K.W. ARC5, a cytosolic dynamin-like protein from plants, is part of the chloroplast division machinery. *Proc. Natl. Acad. Sci. USA* **2003**, *100*, 4328–4333. [CrossRef] [PubMed]

85. Miyagishima, S.Y.; Froehlich, J.E.; Osteryoung, K.W. PDV1 and PDV2 mediate recruitment of the dynamin-related protein ARC5 to the plastid division site. *Plant Cell* **2006**, *18*, 2517–2530. [CrossRef] [PubMed]

International Journal of
Molecular Sciences

MDPI

Brief Report

Two Coiled-Coil Proteins, WEB1 and PMI2, Suppress the Signaling Pathway of Chloroplast Accumulation Response that Is Mediated by Two Phototropin-Interacting Proteins, RPT2 and NCH1, in Seed Plants

Noriyuki Suetsugu [1],* and Masamitsu Wada [2]

[1] Graduate School of Biostudies, Kyoto University, Kyoto 606-8502, Japan
[2] School of Science and Engineering, Tokyo Metropolitan University, Tokyo 192-0397, Japan;
    masamitsu.wada@gmail.com
* Correspondence: suetsugu@lif.kyoto-u.ac.jp; Tel.: +81-75-753-6390; Fax: +81-75-753-6120

Received: 10 May 2017; Accepted: 4 July 2017; Published: 8 July 2017

Abstract: Chloroplast movement is induced by blue light in a broad range of plant species. Weak light induces the chloroplast accumulation response and strong light induces the chloroplast avoidance response. Both responses are essential for efficient photosynthesis and are mediated by phototropin blue-light receptors. J-DOMAIN PROTEIN REQUIRED FOR CHLOROPLAST ACCUMULATION RESPONSE 1 (JAC1) and two coiled-coil domain proteins WEAK CHLOROPLAST MOVEMENT UNDER BLUE LIGHT 1 (WEB1) and PLASTID MOVEMENT IMPAIRED 2 (PMI2) are required for phototropin-mediated chloroplast movement. Genetic analysis suggests that JAC1 is essential for the accumulation response and WEB1/PMI2 inhibit the accumulation response through the suppression of JAC1 activity under the strong light. We recently identified two phototropin-interacting proteins, ROOT PHOTOTROPISM 2 (RPT2) and NPH3/RPT2-like (NRL) PROTEIN FOR CHLOROPLAST MOVEMENT 1 (NCH1) as the signaling components involved in chloroplast accumulation response. However, the relationship between RPT2/NCH1, JAC1 and WEB1/PMI2 remained to be determined. Here, we performed genetic analysis between RPT2/NCH1, JAC1, and WEB1/PMI2 to elucidate the signal transduction pathway.

Keywords: Arabidopsis; blue light; *Marchantia*; organelle movement; phototropin

1. Introduction

Phototropins (phot) are blue-light photoreceptor kinases that mediate phototropism, leaf flattening, stomatal opening, and chloroplast movement including low light-induced chloroplast accumulation response and strong light-induced chloroplast avoidance response (herein referred to as the accumulation and avoidance response, respectively). These responses contribute to optimal photosynthetic light utilization at the organ/tissue, cellular, and organelle level [1,2]. Most land plants have two or more phototropin genes and functional differences exist between them. There are two phototropins in *Arabidopsis thaliana*, phot1 and phot2. Phototropism, leaf flattening, stomatal opening, and the accumulation response are mediated by both phot1 and phot2; however, phot1 plays a greater role in these responses, especially at lower blue-light intensities [1,2]. In contrast, the avoidance response is mediated primarily by phot2 [1,2]. Clear functional divergence of phototropins in the accumulation and avoidance response is observed in the moss *Physcomitrella patens* and the fern *Adiantum capillus-veneris* [3,4]. However, in the liverwort *Marchantia polymorpha*, which is a basal land plant, a single phototropin mediates both the accumulation and avoidance responses [5].

Thus, phototropins can intrinsically mediate all phototropin-mediated responses and functional diversification of phototropins seems to have occurred during land plant evolution concomitant with phototropin gene duplication [6].

By regulating various signaling components, such as phototropin-interacting proteins, phototropins can regulate multiple, diverse responses. For example, BLUE LIGHT SIGNALING1 kinase is a direct phototropin substrate and specifically mediates stomatal opening [7]. In addition, the two phototropin-interacting bric á brac, tramtrack and broad complex/pox virus and zinc finger (BTB/POZ) domain proteins NONPHOTOTROPIC HYPOCOTYL 3 (NPH3) and ROOT PHOTOTROPISM 2 (RPT2), which belong to the NPH3/RPT2-like (NRL) protein family, mediate phototropism and leaf flattening [8–11]. Recently, we identified a phototropin-interacting NRL protein, NRL PROTEIN FOR CHLOROPLAST MOVEMENT 1 (NCH1), and found that NCH1 specifically mediates the accumulation response [12]. NCH1 is highly similar to RPT2 than NPH3 and contains four conserved regions including a BTB/POZ domain (Figure 1a). Furthermore, functional redundancy was found between NCH1 and RPT2 for the accumulation response, but not the avoidance response [12]. These results indicate that phototropism, leaf flattening, and the accumulation response are dependent on these NRL proteins, while stomatal opening and the avoidance response are independent of these NRL proteins. In *M. polymorpha*, the RPT2/NCH1 ortholog MpNCH1 specifically mediates the accumulation response, indicating that phototropin-regulated chloroplast movement is conserved in land plants [12].

Similar to NPH3 and RPT2, NCH1 is localized on the plasma membrane and interacts with phototropins [12], but the downstream function of NCH1 as well as other NRL proteins remained to be determined. Here, we performed genetic analysis of *RPT2* and *NCH1* using triple or quadruple mutant plants between *rpt2nch1* and other mutants that were implicated in the signal transduction of chloroplast movement.

2. Results and Discussion

J-domain protein required for chloroplast accumulation response 1 (*jac1*) and *rpt2nch1* plants are both defective in the accumulation response [12,13]. JAC1 protein is a C-terminal J-domain protein similar to clathrin uncoating factor auxilin (Figure 1a) [13]. To analyze chloroplast movement in *jac1* and *rpt2nch1* in detail, we performed analysis of the light-induced changes in leaf transmittance, reflective of light-induced chloroplast movements (Figure 1b) [14]. In response to 3 μmol m^{-2} s^{-1} of blue light, which induces the accumulation response in wild type, a clear avoidance response is induced in *rpt2nch1* but not in *jac1* (Figure 1c,d) [12]. Therefore, RPT2/NCH1 could suppress the induction of the avoidance response to facilitate efficient induction of the accumulation response under low light conditions (Figure 2). Compared to wild type and *jac1*, a faster avoidance response was induced by 20 μmol m^{-2} s^{-1} of blue light in *rpt2nch1* (Figure 1c,d; one-way ANOVA followed by Tukey–Kramer multiple comparison post hoc test, $p < 0.01$ for wild type or *jac1* vs. *rpt2nch1*), although similar transmittance changes (defined as "amplitude" in Figure 1e) were observed for both *rpt2nch1* and *jac1* following 40 min of 20 μmol m^{-2} s^{-1} blue-light irradiation (Figure 1c,e; one-way ANOVA followed by Tukey–Kramer multiple comparison post hoc test, $p > 0.05$ for *jac1* vs. *rpt2nch1*). Following subsequent application of 50 μmol m^{-2} s^{-1} blue light, only a slight additional avoidance response was observed in *rpt2nch1* which contrasted with the stronger avoidance response induced in wild type (Figure 1c) [12]. This result could be attributed to prior movement of the majority of chloroplasts to the side walls in *rpt2nch1* during the former irradiation period. Interestingly, avoidance responses of similar magnitudes were induced in *jac1* in response to both 20 and 50 μmol m^{-2} s^{-1} of blue light and a decreased rate of avoidance response induction was not observed during strong light irradiation (Figure 1c). The changes observed in leaf transmittance for *rpt2nch1jac1* were intermediate between those observed for *rpt2nch1* and *jac1*.

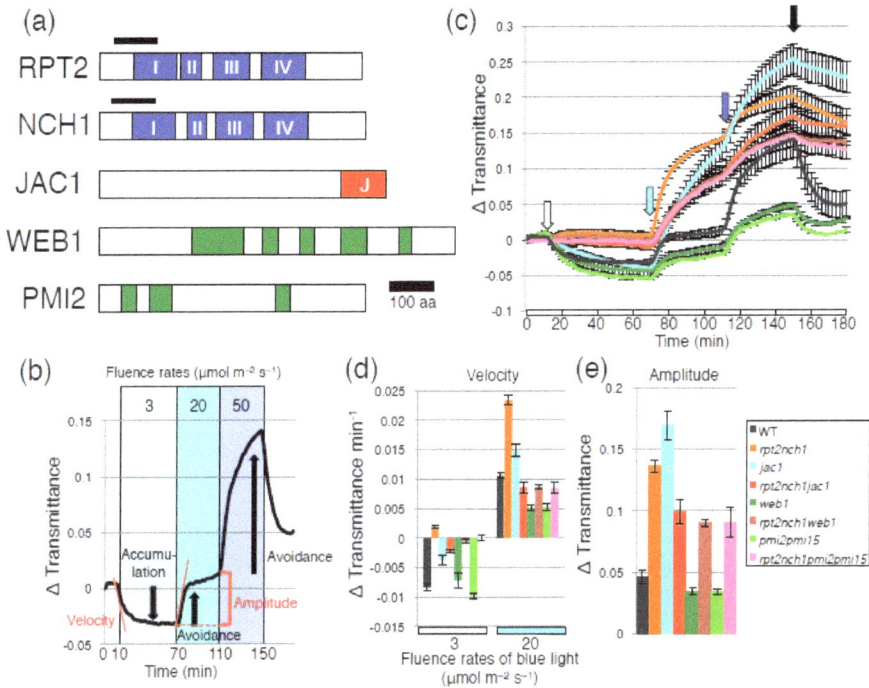

Figure 1. (a) Protein structure of ROOT PHOTOTROPISM 2 (RPT2), NRL PROTEIN FOR CHLOROPLAST MOVEMENT 1 (NCH1), J-DOMAIN PROTEIN REQUIRED FOR CHLOROPLAST ACCUMULATION RESPONSE 1 (JAC1), WEAK CHLOROPLAST MOVEMENT UNDER BLUE LIGHT 1 (WEB1), and PLASTID MOVEMENT IMPAIRED 2 (PMI2). **Blue** boxes indicate the four conserved regions of NPH3/RPT2-like (NRL) proteins. The position of the BTB/POZ domain is indicated by a **black** bar. **Red** box is a J-domain. **Green** boxes indicate the coiled-coil domains; (b) Measurement of light-induced changes in leaf transmittance as a result of chloroplast photorelocation movements. The depicted trace represents typical data collected for wild type under the various light irradiation conditions (indicated by color boxes). There is a decrease in leaf transmittance in response to 3 µmol m^{-2} s^{-1} of blue light, indicating that the accumulation response is induced (**downward arrow**). Conversely, there is an increase in leaf transmittance in response to 20 and 50 µmol m^{-2} s^{-1} of blue light, indicating that the avoidance response is induced (**upward arrows**). **Red lines** mark the initial linear fragments of leaf transmittance rate change during the first 2–6 min of the irradiation period, indicating the velocity. A **red parenthesis** marks the difference between the transmittance level observed following 60 min of 3 µmol m^{-2} s^{-1} blue-light irradiation and the transmittance level observed a following further 40 min of 20 µmol m^{-2} s^{-1} blue-light irradiation, indicating the amplitude of the avoidance response caused by 20 µmol m^{-2} s^{-1} blue-light irradiation; (c–e) Distinct chloroplast movements observed between *rpt2nch1* and *jac1*; (c) Light-induced changes in leaf transmittance of the indicated lines were measured using a custom-made plate reader system [14]. The samples were sequentially irradiated with 3, 20 and 50 µmol m^{-2} s^{-1} of continuous blue light. The beginning of each irradiation period is indicated by **white**, **cyan** and **blue** arrows, respectively. The light was extinguished after 150 min (**black arrow**); (d) The velocity of light-induced transmittance changes. (e) The amplitude of the avoidance response caused by 20 µmol m^{-2} s^{-1} blue-light irradiation. Data for wild type, *rpt2nch1*, *jac1* and *rpt2nch1jac1* from Suetsugu et al. (2016) [12] were used for comparison, because data for *web1*, *rpt2nch1web1*, *pmi2pmi15* and *rpt2nch1pmi2pmi15* were acquired in the same experiments using the same plate. Data are presented as means of three independent experiments and the error bars indicate standard errors. WT, wild type.

Int. J. Mol. Sci. **2017**, *18*, 1469

    Previously, we showed that *JAC1* mutation suppresses the defective avoidance response in *weak chloroplast movement under blue light 1 (web1)* and *plastid movement impaired 2 (pmi2)* [15]. WEB1 and PMI2 are related coiled-coil domain proteins that interact with each other (Figure 1a) [15]. Although the low light-induced accumulation response was normal in *web1* and *pmi2pmi15*, both mutant plants exhibited attenuated avoidance response under the strong light conditions (Figure 1c to e) [15]. The *jac1web1* and *jac1pmi2pmi15* showed nearly the same phenotypes as *jac1* single mutants [15]. Importantly, the weak avoidance response phenotype observed in *web1* and *pmi2pmi15* was completely suppressed in *jac1web1* and *jac1pmi2pmi15*, respectively [15]. Therefore, we hypothesized that WEB1 and PMI2 suppress JAC1 function under strong light, preventing the induction of the JAC1-dependent accumulation response and leading to efficient induction of the avoidance response (Figure 2). The weak avoidance response phenotype observed in *web1* and *pmi2pmi15* was absent in *rpt2nch1web1* and *rpt2nch1pmi2pmi15*, similar to *jac1web1* and *jac1pmi2pmi15* (Figure 1c–e). Mutation of *JAC1* suppressed *web1* and *pmi2pmi15* phenotypes, because *jac1web1* and *jac1pmi2pmi15* phenotypes are indistinguishable from *jac1* [15]. Although mutation of *RPT2* and *NCH1* largely suppressed the weak avoidance response phenotypes observed in *web1* and *pmi2pmi15*, the velocity and amplitude of the avoidance response in these mutants did not match those in *rpt2nch1* (Figure 1c–e; one-way ANOVA followed by Tukey–Kramer multiple comparison post hoc test, $p < 0.01$ for *rpt2nch1web1* or *rpt2nch1pmi2pmi15* vs. *rpt2nch1* in velocity and $p < 0.05$ for *rpt2nch1web1* or *rpt2nch1pmi2pmi15* vs. *rpt2nch1* in amplitude). The phenotypes of *rpt2nch1web1* and *rpt2nch1pmi2pmi15* were very similar to *rpt2nch1jac1* in that no detectable chloroplast movement was observed under 3 μmol m^{-2} s^{-1} of blue light and their avoidance response phenotypes were similar to *jac1* (Figure 1c,d; one-way ANOVA followed by Tukey–Kramer multiple comparison post hoc test, $p > 0.05$ for *rpt2nch1web1* or *rpt2nch1pmi2pmi15* vs. *rpt2nch1jac1*). Collectively, our results indicate that RPT2 and NCH1 are essential for the accumulation response and regulate JAC1-dependent and -independent pathways and that WEB1/PMI2 represses the signaling pathway for the accumulation response under the strong light conditions (Figure 2). However, how WEB1 and PMI2 suppress the accumulation response pathway remained to be determined. Interaction of WEB1 and/or PMI2 with JAC1 has never been detected [15]. At the least, the amounts of phototropins were normal in *web1* and *pmi2pmi15* mutant plants [15]. Further analysis of the relationship between WEB1/PMI2, RPT2/NCH1 and JAC1 is required.

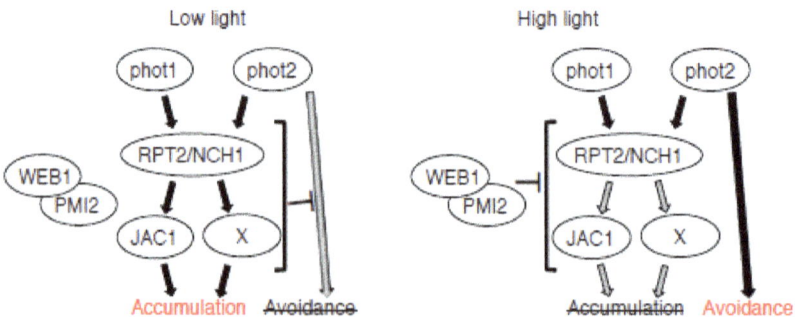

Figure 2. Working model of chloroplast photorelocation movements. The photoreceptors phot1 and phot2 mediate the accumulation response under a low light condition through RPT2 and NCH1. RPT2 and NCH1 might regulate both JAC1-dependent and -independent (X) pathways. The signaling pathway by RPT2/NCH1 and JAC1 suppresses that of the avoidance response under a low light condition. Under the high light condition, the WEB1/PMI2 complex suppresses the signaling pathway for the accumulation response that is regulated by RPT2/NCH1 and JAC1 through an unknown mechanism, resulting in the efficient induction of the avoidance response mediated by phot2. **Gray** arrows indicate the suppressed signaling pathways. **Black** arrows indicate the activated signaling pathways.

RPT2 and NCH1 are localized on the plasma membrane and interact with phototropins, indicating that RPT2 and NCH1 are the initial downstream signaling components involved in the phototropin-mediated accumulation response (Figure 2) [12]. Notably, RPT2 and NCH1 are conserved in land plants, but JAC1, WEB1 and PMI2 orthologs are found only in seed plants [16]. Thus, to maximize light utilization through chloroplast movement, land plants have evolved a sophisticated mechanism of controlling chloroplast movement by increasing the molecular components involved in blue-light signaling.

3. Materials and Methods

3.1. Arabidopsis Lines and the Growth Condition

The wild-type and mutant lines are a Columbia *gl1* background. Seeds were sown on 0.8% agar medium containing 1/3 strength Murashige & Skoog salt and 1% sucrose, and grown under white light at ca. ~100 μmol m^{-2} s^{-1} (16 h)/dark (8 h) cycle at 23 °C in an incubator. *rpt2-4nch1-1* [12], *jac1-1* [13], *web1-2* [15] and *pmi2-2pmi15-1* [15] were described previously. For *PMI2* mutant plants, *pmi2pmi15* was used, because PMI15 is closely related to PMI2 and *pmi15* exhibits a very weak defect in chloroplast movement [17]. *rpt2-4*, *nch1-1*, *pmi2-2* and *pmi15-1* are T-DNA knockout lines [12,17]. *jac1-1* carries a missense mutation [13] and *web1-2* carries a deletion of one nucleotide [15]. Western blot analysis showed that JAC1 and WEB1 proteins were not detected in *jac1-1* and *web1-2*, respectively [13,15]. Double, triple and quadruple mutants were generated by genetic crossing.

3.2. Analyses of Chloroplast Photorelocation Movements

Chloroplast photorelocation movements were analyzed by the measurement of light-induced changes in leaf transmittance as described previously [14]. Third leaves that were detached from 16-day-old seedlings were placed on 1% (*w/v*) gellan gum in a 96-well plate and then dark-adapted at least for 1 h before transmittance measurement.

3.3. Statistical Analysis

Statistical analyses were performed by one-way ANOVA followed by Tukey–Kramer multiple comparison post hoc test.

Acknowledgments: This work was supported in part by the Grant-in-Aid for Scientific Research Grants (26840097 and 15KK0254 to Noriyuki Suetsugu; 20227001, 23120523, 25120721, and 25251033 to Masamitsu Wada).

Author Contributions: Noriyuki Suetsugu conceived, designed and performed the experiments; Noriyuki Suetsugu and Masamitsu Wada analyzed the data, contributed reagents/materials/analysis tools and wrote the paper.

Conflicts of Interest: The authors declare no conflict of interest.

References

1. Christie, J.M. Phototropin blue-light receptors. *Annu. Rev. Plant Biol.* **2007**, *58*, 21–45. [CrossRef] [PubMed]
2. Suetsugu, N.; Wada, M. Evolution of three LOV blue light receptor families in green plants and photosynthetic stramenopiles: Phototropin, ZTL/FKF1/LKP2 and aureochrome. *Plant Cell Physiol.* **2013**, *54*, 8–23. [CrossRef] [PubMed]
3. Kasahara, M.; Kagawa, T.; Sato, Y.; Kiyosue, T.; Wada, M. Phototropins mediate blue and red light-induced chloroplast movements in *Physcomitrella patens*. *Plant Physiol.* **2004**, *135*, 1388–1397. [CrossRef] [PubMed]
4. Kagawa, T.; Kasahara, M.; Abe, T.; Yoshida, S.; Wada, M. Function analysis of phototropin2 using fern mutants deficient in blue light-induced chloroplast avoidance movement. *Plant Cell Physiol.* **2004**, *45*, 416–426. [CrossRef] [PubMed]
5. Komatsu, A.; Terai, M.; Ishizaki, K.; Suetsugu, N.; Tsuboi, H.; Nishihama, R.; Yamato, K.T.; Wada, M.; Kohchi, T. Phototropin encoded by a single-copy gene mediates chloroplast photorelocation movements in the liverwort *Marchantia polymorpha*. *Plant Physiol.* **2014**, *166*, 411–427. [CrossRef] [PubMed]

6.   Li, F.W.; Rothfels, C.J.; Melkonian, M.; Villarreal, J.C.; Stevenson, D.W.; Graham, S.W.; Wong, G.K.S.; Mathews, S.; Pryer, K.M. The origin and evolution of phototropins. *Front. Plant Sci.* **2015**, *6*, 637. [CrossRef] [PubMed]

7.   Takemiya, A.; Sugiyama, N.; Fujimoto, H.; Tsutsumi, T.; Yamauchi, S.; Hiyama, A.; Tada, Y.; Christie, J.M.; Shimazaki, K. Phosphorylation of BLUS1 kinase by phototropins is a primary step in stomatal opening. *Nat. Commun.* **2013**, *4*, 2094. [CrossRef] [PubMed]

8.   Motchoulski, A.; Liscum, E. Arabidopsis NPH3: A NPH1 photoreceptor-interacting protein essential for phototropism. *Science* **1999**, *286*, 961–964. [CrossRef] [PubMed]

9.   Sakai, T.; Wada, T.; Ishiguro, S.; Okada, K. RPT2: A signal transducer of the phototropic response in Arabidopsis. *Plant Cell* **2000**, *12*, 225–236. [CrossRef] [PubMed]

10.   Inoue, S.; Kinoshita, T.; Takemiya, A.; Doi, M.; Shimazaki, K. Leaf positioning of Arabidopsis in response to blue light. *Mol. Plant* **2008**, *1*, 15–26. [CrossRef] [PubMed]

11.   Harada, A.; Takemiya, A.; Inoue, S.; Sakai, T.; Shimazaki, K. Role of RPT2 in leaf positioning and flattening and a possible inhibition of phot2 signaling by phot1. *Plant Cell Physiol.* **2013**, *54*, 36–47. [CrossRef] [PubMed]

12.   Suetsugu, N.; Takemiya, A.; Kong, S.G.; Higa, T.; Komatsu, A.; Shimazaki, K.; Kohchi, T.; Wada, M. RPT2/NCH1 subfamily of NPH3-like proteins is essential for the chloroplast accumulation response in land plants. *Proc. Natl. Acad. Sci. USA* **2016**, *113*, 10424–10429. [CrossRef] [PubMed]

13.   Suetsugu, N.; Kagawa, T.; Wada, M. An auxilin-like J-domain protein, JAC1, regulates phototropin-mediated chloroplast movement in Arabidopsis. *Plant Physiol.* **2005**, *139*, 151–162. [CrossRef] [PubMed]

14.   Wada, M.; Kong, S.G. Analysis of chloroplast movement and relocation in Arabidopsis. *Methods Mol. Biol.* **2011**, *774*, 87–102. [CrossRef]

15.   Kodama, Y.; Suetsugu, N.; Kong, S.G.; Wada, M. Two interacting coiled-coil proteins, WEB1 and PMI2, maintain the chloroplast photorelocation movement velocity in Arabidopsis. *Proc. Natl. Acad. Sci. USA* **2010**, *107*, 19591–19596. [CrossRef] [PubMed]

16.   Suetsugu, N.; Wada, M. Evolution of the cp-actin-based motility system of chloroplasts in green plants. *Front. Plant Sci.* **2016**, *7*, 561. [CrossRef] [PubMed]

17.   Luesse, D.R.; DeBlasio, S.L.; Hangarter, R.P. Plastid movement impaired 2, a new gene involved in normal blue-light-induced chloroplast movements in Arabidopsis. *Plant Physiol.* **2006**, *141*, 1328–1337. [CrossRef] [PubMed]

MDPI

St. Alban-Anlage 66

4052 Basel

Switzerland

Tel. +41 61 683 77 34

Fax +41 61 302 89 18

www.mdpi.com

International Journal of Molecular Sciences Editorial Office

E-mail: ijms@mdpi.com

www.mdpi.com/journal/ijms